CALIFORNIA

Physical Science

HOLT, RINEHART AND WINSTON

A Harcourt Classroom Education Company

Austin • New York • Orlando • Atlanta • San Francisco • Boston • Dallas • Toronto • London

Staff Credits

Editorial

Robert W. Todd, Executive Editor
David F. Bowman, Managing Editor
Anne Bunce, Senior Editor
Anne Engelking, Editor
Michael Mazza, Laura Prescott, Robin Goodman (Feature Articles)

Annotated Teacher's Edition

Ken Shepardson, Bill Burnside, Kelly Graham

Ancillaries

Jennifer Childers, Senior Editor
Erin Bao, Kristen Karns, Andrew Strickler, Clay Crenshaw, Wayne Duncan, Molly Frohlich, Amy James, Monique Mayer, Traci Maxwell

Copyeditors

Steve Oelenberger, Copyediting Supervisor
Suzanne Brooks, Brooke Fugitt, Tania Hannan, Denise Nowotny

Editorial Support Staff

Christy Bear, Jeanne Graham, Rose Segrest, Tanu'e White

Editorial Permissions

Cathy Paré, Permissions Manager
Jan Harrington, Permissions Editor

Art, Design, and Photo

Book Design

Richard Metzger, Art Director
Marc Cooper, Senior Designer
Ron Bowdoin, Designer
Alicia Sullivan (ATE), Cristina Bowerman (ATE), Eric Rupprath (Ancillaries)

Image Services

Debra Schorn, Director
Elaine Tate, Art Buyer Supervisor
Sean Moynihan, Art Buyer

Photo Research

Stephanie Morris, Assistant Photo Researcher

Photo Studio

Sam Dudgeon, Senior Staff Photographer
Victoria Smith, Photo Specialist

Design New Media

Susan Michael, Art Director

Design Media

Joe Melomo, Art Director
Shawn McKinney, Designer

Production

Mimi Stockdell, Senior Production Manager
Beth Sample, Production Coordinator
Suzanne Brooks, Sara Carroll-Downs

Media Production

Kim A. Scott, Senior Production Manager
Nancy Hargis, Production Supervisor
Adriana Bardin, Production Coordinator

New Media

Jim Bruno, Senior Project Manager
Lydia Doty, Senior Project Manager
Jessica Bega, Project Manager
Armin Gutzmer, Manager Training and Technical Support
Cathy Kuhles, Nina Degollado

Design Implementation and Production

Preface, Inc.

For permission to reprint copyrighted material, grateful acknowledgment is made to the following source: *sci*LINKS is owned and provided by the National Science Teachers Association. All rights reserved.

Printed in the United States of America
ISBN 0-03-055797-6
6 7 8 048 05 04 03 02 01

Acknowledgments

Chapter Writers

Christie Borgford, Ph.D.
Professor of Chemistry
University of Alabama
Birmingham, Alabama

Andrew Champagne
Former Physics Teacher
Ashland High School
Ashland, Massachusetts

Mapi Cuevas, Ph.D.
Professor of Chemistry
Santa Fe Community College
Gainesville, Florida

Leila Dumas
Former Physics Teacher
LBJ Science Academy
Austin, Texas

Mary Kay Hemenway, Ph.D.
Research Associate and Senior Lecturer
Department of Astronomy
The University of Texas
Austin, Texas

William G. Lamb, Ph.D.
Science Teacher and Dept. Chair
Oregon Episcopal School
Portland, Oregon

Karen J. Meech, Ph.D.
Associate Astronomer
Institute for Astronomy
University of Hawaii
Honolulu, Hawaii

Sally Ann Vonderbrink, Ph.D.
Chemistry Teacher
St. Xavier High School
Cincinnati, Ohio

Lab Writers

Phillip G. Bunce
Former Physics Teacher
Bowie High School
Austin, Texas

Kenneth E. Creese
Science Teacher
White Mountain Junior High School
Rock Springs, Wyoming

William G. Lamb, Ph.D.
Science Teacher and Dept. Chair
Oregon Episcopal School
Portland, Oregon

Alyson Mike
Science Teacher
East Valley Middle School
East Helena, Montana

Joseph W. Price
Science Teacher and Dept. Chair
H. M. Browne Junior High School
Washington, D.C.

Denice Lee Sandefur
Science Teacher
Nucla High School
Nucla, Colorado

John Spadafino
Mathematics and Physics Teacher
Hackensack High School
Hackensack, New Jersey

Walter Woolbaugh
Science Teacher
Manhattan Junior High School
Manhattan, Montana

Academic Reviewers

Paul D. Asimow, Ph.D.
Postdoctoral Research Fellow
Lamont-Doherty Earth Observatory
Columbia University
Palisades, New York

G. Fritz Benedict, Ph.D.
Senior Research Scientist and Astronomer
McDonald Observatory
The University of Texas
Austin, Texas

Paul R. Berman, Ph.D.
Professor of Physics
University of Michigan
Ann Arbor, Michigan

Lawrence Bornstein, Ph.D.
Professor of Physics
New York University
New York, New York

Russell M. Brengelman, Ph.D.
Professor of Physics
Morehead State University
Morehead, Kentucky

Michael Brown, Ph.D.
Assistant Professor of Planetary Astronomy
California Institute of Technology
Pasadena, California

Wesley N. Colley, Ph.D.
Postdoctoral Fellow
Harvard-Smithsonian Center for Astrophysics
Cambridge, Massachusetts

Andrew J. Davis, Ph.D.
Manager, ACE Science Center
Department of Physics
California Institute of Technology
Pasadena, California

Peter E. Demmin, Ed.D.
Former Science Teacher and Department Chair
Amherst Central High School
Amherst, New York

Roger Falcone, Ph.D.
Professor of Physics and Department Chair
University of California
Berkeley, California

Cassandra A. Fraser, Ph.D.
Assistant Professor of Chemistry
University of Virginia
Charlottesville, Virginia

L. John Gagliardi, Ph.D.
Associate Professor of Physics and Department Chair
Rutgers University
Camden, New Jersey

Gabriele F. Giuliani, Ph.D.
Professor of Physics
Purdue University
West Lafayette, Indiana

John L. Hubisz, Ph.D.
Professor of Physics
North Carolina State University
Raleigh, North Carolina

Samuel P. Kounaves, Ph.D.
Professor of Chemistry
Tufts University
Medford, Massachusetts

Henry Krakauer, Ph.D.
Professor of Physics
The College of William and Mary
Williamsburg, Virginia

Karol Lang, Ph.D.
Associate Professor of Physics
The University of Texas
Austin, Texas

Gloria Langer, Ph.D.
Professor of Physics
University of Colorado
Boulder, Colorado

Joseph A. McClure, Ph.D.
Associate Professor of Physics
Georgetown University
Washington, D.C.

LaMoine L. Motz, Ph.D.
Coordinator of Science Education
Department of Learning Services
Oakland County Schools
Waterford, Michigan

R. Thomas Myers, Ph.D.
Professor of Chemistry Emeritus
Kent State University
Kent, Ohio

Hillary Clement Olson, Ph.D.
Research Associate
Institute for Geophysics
The University of Texas
Austin, Texas

David P. Richardson, Ph.D.
Professor of Chemistry
Thompson Chemical Laboratory
Williams College
Williamstown, Massachusetts

Gary Rottman, Ph.D.
Associate Director
Laboratory for Atmosphere and Space Physics
University of Colorado
Boulder, Colorado

Peter Sheridan, Ph.D.
Professor of Chemistry
Colgate University
Hamilton, New York

David Sprayberry, Ph.D.
Assistant Director for Observing Support
W. M. Keck Observatory
California Association for Research in Astronomy
Kamuela, Hawaii

Jack B. Swift, Ph.D.
Professor of Physics
The University of Texas
Austin, Texas

Atiq Syed, Ph.D.
Master Instructor of Mathematics and Science
Texas State Technical College
Harlingen, Texas

Acknowledgments (cont.)

Leonard Taylor, Ph.D.
Professor Emeritus
Department of Electrical Engineering
University of Maryland
College Park, Maryland

Virginia L. Trimble, Ph.D.
Professor of Physics and Astronomy
University of California
Irvine, California

Martin VanDyke, Ph.D.
Professor of Chemistry Emeritus
Front Range Community College
Westminster, Colorado

Gabriela Waschewsky, Ph.D.
Math and Science Teacher
Chicago International Charter School, South
Chicago, Illinois

Safety Reviewer

Jack A. Gerlovich, Ph.D.
Associate Professor
School of Education
Drake University
Des Moines, Iowa

Teacher Reviewers

Barry L. Bishop
Science Teacher and Dept. Chair
San Rafael Junior High School
Ferron, Utah

Paul Boyle
Science Teacher
Perry Heights Middle School
Evansville, Indiana

Daniel Bugenhagen
Science Teacher and Dept. Co-chair
Yutan Junior-Senior High School
Yutan, Nebraska

Kenneth Creese
Science Teacher
White Mountain Junior High School
Rock Springs, Wyoming

Vicky Farland
Science Teacher and Dept. Chair
Crane Junior High School
Yuma, Arizona

Rebecca Ferguson
Science Teacher
North Ridge Middle School
North Richland Hills, Texas

Laura Fleet
Science Teacher
Alice B. Landrum Middle School
Ponte Vedra Beach, Florida

Jennifer Ford
Science Teacher and Dept. Chair
North Ridge Middle School
North Richland Hills, Texas

Susan Gorman
Science Teacher
North Ridge Middle School
North Richland Hills, Texas

C. John Graves
Science Teacher
Monforton Middle School
Bozeman, Montana

Dennis Hanson
Science Teacher and Dept. Chair
Big Bear Middle School
Big Bear Lake, California

David A. Harris
Science Teacher and Dept. Chair
The Thacher School
Ojai, California

Norman E. Holcomb
Science Teacher
Marion Local Schools
Maria Stein, Ohio

Kenneth J. Horn
Science Teacher and Dept. Chair
Fallston Middle School
Fallston, Maryland

Tracy Jahn
Science Teacher
Berkshire Junior-Senior High School
Canaan, New York

Kerry A. Johnson
Science Teacher
Isbell Middle School
Santa Paula, California

David D. Jones
Science Teacher
Andrew Jackson Middle School
Cross Lanes, West Virginia

Drew E. Kirian
Science Teacher
Solon Middle School
Solon, Ohio

Harriet Knops
Science Teacher and Dept. Chair
Rolling Hills Middle School
El Dorado, California

Scott Mandel, Ph.D.
Director and Educational Consultant
Teachers Helping Teachers
Los Angeles, California

Edith C. McAlanis
Science Teacher and Dept. Chair
Socorro Middle School
El Paso, Texas

Kevin McCurdy, Ph.D.
Science Teacher
Elmwood Junior High School
Rogers, Arkansas

Kathy McKee
Science Teacher
Hoyt Middle School
Des Moines, Iowa

Alyson Mike
Science Teacher
East Valley Middle School
East Helena, Montana

Donna Norwood
Science Teacher and Dept. Chair
Monroe Middle School
Monroe, North Carolina

Gabriell DeBear Paye
Biology Teacher
West Roxbury High School
West Roxbury, Massachusetts

Geraldine Okewesa
Science Teacher
Stuart-Hobson Museum Magnet Middle School
Washington, D.C.

Joseph W. Price
Science Teacher and Dept. Chair
H. M. Browne Junior High School
Washington, D.C.

Terry J. Rakes
Science Teacher
Elmwood Junior High School
Rogers, Arkansas

Beth Richards
Science Teacher
North Middle School
Crystal Lake, Illinois

Rodney A. Sandefur
Science Teacher
Naturita Middle School
Naturita, Colorado

Helen Schiller
Science Teacher
Northwood Middle School
Taylors, South Carolina

Patricia McFarlane Soto
Science Teacher and Dept. Chair
G. W. Carver Middle School
Miami, Florida

David M. Sparks
Science Teacher
Redwater Junior High School
Redwater, Texas

Larry Tackett
Science Teacher and Dept. Chair
Andrew Jackson Middle School
Cross Lanes, West Virginia

Elsie N. Waynes
Science Teacher and Dept. Chair
R. H. Terrell Junior High School
Washington, D.C.

Walter Woolbaugh
Science Teacher
Manhattan Junior High School
Manhattan, Montana

Sharon L. Woolf
Science Teacher
Langston Hughes Middle School
Reston, Virginia

Alexis S. Wright
Middle School Science Coordinator
Rye Country Day School
Rye, New York

Lee Yassinski
Science Teacher
Sun Valley Middle School
Sun Valley, California

John Zambo
Science Teacher
Elizabeth Ustach Middle School
Modesto, California

Gordon Zibelman
Science Teacher
Drexel Hill Middle School
Drexel Hill, Pennsylvania

Contents in Brief

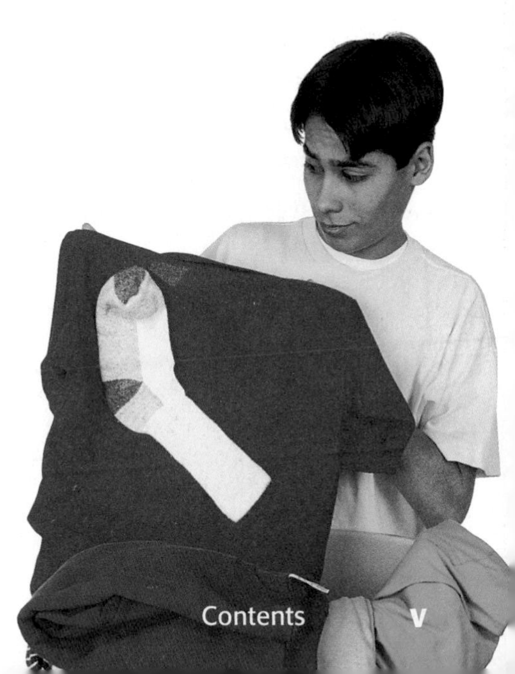

Contents

Unit 1 ... Introduction to Matter

CHAPTER 1

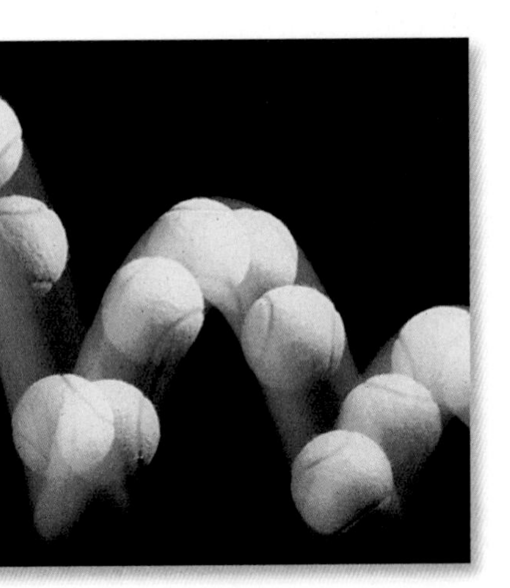

CHAPTER 2

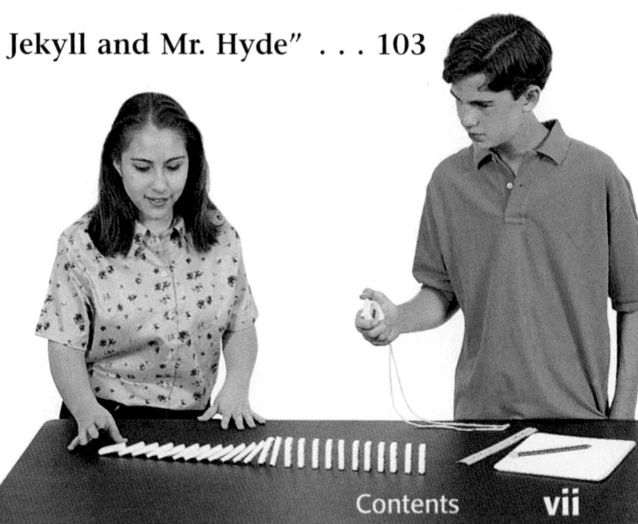

Unit 2 ··· Motion and Forces

CHAPTER 5

CHAPTER 6

CHAPTER 

Energy and Energy Resources 212

CHAPTER 

The Energy of Waves 244

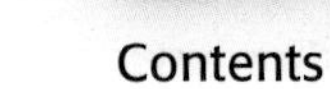

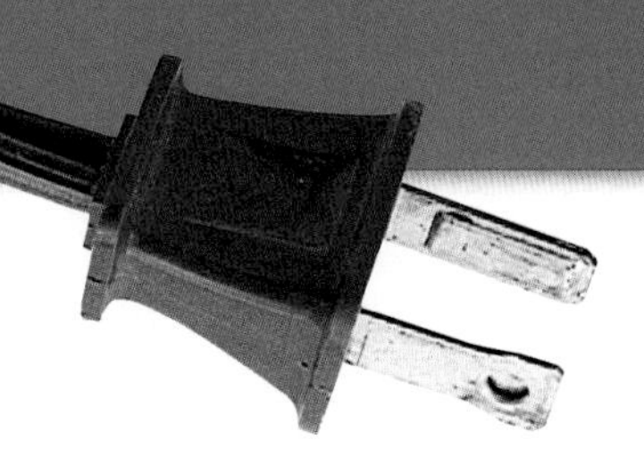

H
Hydrogen

CHAPTER 15

CHAPTER 16

H₂C=CH₂ structure: H H C=C H H

Unit 6 ... Introduction to Astronomy

The more labs, the better!

Take a minute to browse the **LabBook** located at the end of this textbook. You'll find a wide variety of exciting labs that will help you experience science firsthand. But please don't forget to be safe. Read the "Safety First!" section before starting any of the labs.

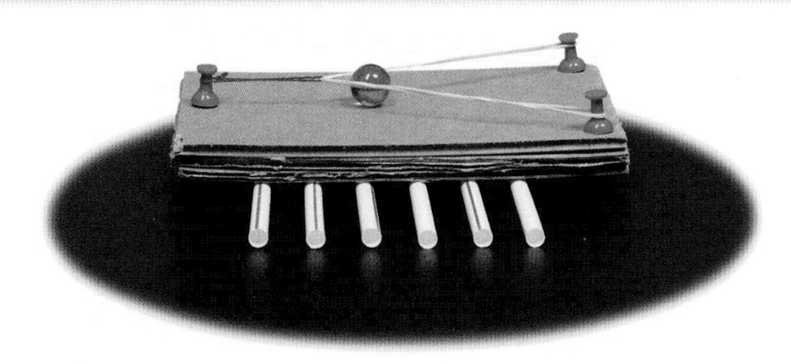

Now is the time to Investigate!

Science is a process in which investigation leads to information and understanding. The **Investigate!** at the beginning of each chapter helps you gain scientific understanding of the topic through hands-on experience.

Quick Lab

Not all laboratory investigations have to be long and involved.

The **QuickLabs** found throughout the chapters of this textbook require only a small amount of time and limited equipment. But just because they are quick, don't skimp on the safety.

÷ 5 ÷ Ω ≤ ∞ + Ω √ 9 ∞ ≤ Σ 2

MATH BREAK

Science and math go hand in hand.

The **MathBreaks** in the margins of the chapters show you many ways that math applies directly to science and vice versa.

APPLY

Science can be very useful in the real world.

It is interesting to learn how scientific information is being used in the real world. You can see for yourself in the **Apply** features. You will also be asked to apply your own knowledge. This is a good way to learn!

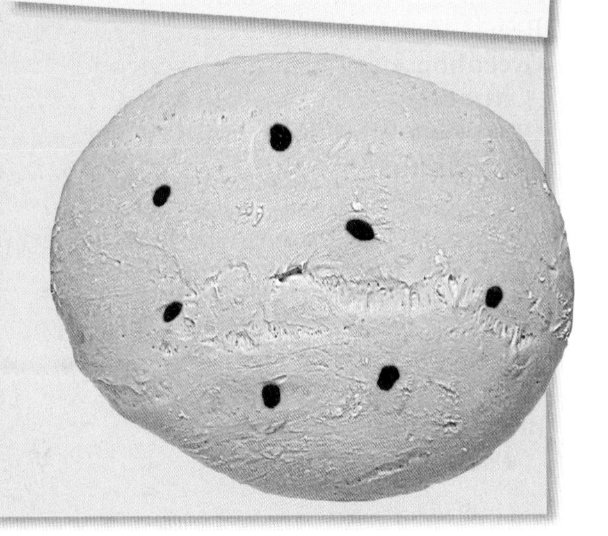

Science Connections

One science leads to another.

You may not realize it at first, but different areas of science are related to each other in many ways. Each **Connection** explores a topic from the viewpoint of another science discipline. In this way, areas of science merge to improve your understanding of the world around you.

Feature Articles

Feature articles for any appetite!

Science and technology affect us all in many ways. The following articles will give you an idea of just how interesting, strange, helpful, and action-packed science and technology are. At the end of each chapter, you will find two feature articles. Read them and you will be surprised at what you learn.

CAREERS

ACROSS THE SCIENCES

Science, Technology, and Society

EYE ON THE ENVIRONMENT

Eureka!

Health WATCH

SCIENTIFICDEBATE

UNIT

1

TIMELINE

Introduction to Matter

In this unit, you will explore a basic question that people have been pondering for centuries: What is the nature of matter? You will learn how to define the word *matter* and how to describe matter and the changes it goes through. You will also learn about the different states of matter and how to classify different arrangements of matter as elements, compounds, or mixtures. This timeline shows some of the events and discoveries that have occurred throughout history as scientists have sought to understand the nature of matter.

1661

Robert Boyle, a chemist in England, determines that elements are substances that cannot be broken down into anything simpler by chemical processes.

1712

Thomas Newcomen invents the first practical steam engine.

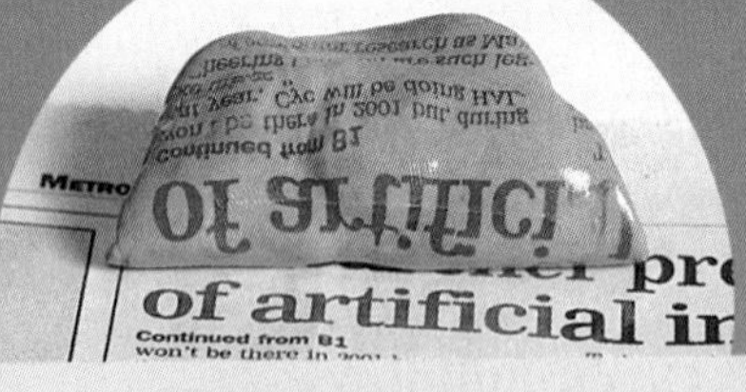

1949

Silly Putty® is sold in a toy store for the first time. The soft, gooey substance quickly becomes popular because of its strange properties, including the ability to "pick up" the print from a newspaper page.

1957

The space age begins when the Soviet Union launches *Sputnik I,* the first artificial satellite to circle the Earth.

1971

The first "pocket" calculator is introduced. It has a mass of more than 1 kg and a price of about $150—hardly the kind of pocket calculator that exists today.

1766

English chemist Henry Cavendish discovers and describes the properties of a highly flammable substance now known as hydrogen gas.

1800

Current from an electric battery is used to separate water into the elements hydrogen and oxygen for the first time.

1920

American women win the right to vote with the ratification of the 19th Amendment to the Constitution.

1928

Sir Alexander Fleming discovers that the mold *Penicillium notatum,* shown here growing on an orange, is capable of killing some types of bacteria. The antibiotic penicillin is derived from this mold.

1937

The *Hindenburg* explodes while docking in Lakehurst, New Jersey. The airship was filled with flammable hydrogen gas to make it lighter than air.

1989

An oil tanker strikes a reef in Prince William Sound, Alaska, spilling nearly 11 million gallons of oil. The floating oil injures or kills thousands of marine mammals and seabirds and damages the Alaskan coastline.

2000

The World's Fair, an international exhibition featuring exhibits and participants from around the world, is held in Hanover, Germany. The theme is "Humankind, Nature, and Technology."

CHAPTER 1 The World of Physical Science

Would You Believe . . . ?

In the fifteenth century, Leonardo da Vinci, a famous Italian artist and scientist, studied sea gulls to try to learn the secrets of flight. Although he never built a successful flying machine, he did learn a lot about the scientific principles of flight. Many other scientists have been inspired by nature—including James Czarnowski (zahr NOW SKEE) and Michael Triantafyllou (tree AHN ti FEE loo), two scientists from the Massachusetts Institute of Technology (MIT), in Cambridge.

In 1997, Czarnowski and Triantafyllou were looking for an example from nature that could inspire a new way to power boats. On a trip to the New England Aquarium, in Boston, Czarnowski was watching penguins swim through the water when he realized that the penguins could be nature's answer. So he and Triantafyllou set out to create a boat that would imitate the way a penguin swims. After much thought, many questions, and a lot of research, *Proteus* (PROH tee uhs)—the penguin boat—was born.

Less than 4 meters long, *Proteus* is powered by two car batteries. Two broad paddles, similar to a penguin's flippers, flap together as often as 200 times per minute. At its top speed, this experimental boat can "swim" through the water at 2 meters per second. If a full-sized version of *Proteus* were developed, it could cruise easily at about 45 kilometers per hour. That's faster than most cargo carriers, which make up the majority of boats on the ocean. Also, full-sized penguin boats would use less fuel, which means that they would save money and produce less pollution.

Proteus *(shown above on the Charles River) is named after the son of Poseidon, the Greek god of the sea.*

Proteus is a creative solution to a scientific problem. In this chapter, you'll learn how Czarnowski and Triantafyllou used a series of steps called the *scientific method* to develop this amazing boat.

What Do You Think?

In your ScienceLog, try to answer the following questions based on what you already know:

1. What is physical science?
2. What are some steps scientists take to answer questions?
3. What purpose do models serve?

Mission Impossible?

In this activity, you will do some creative thinking to figure out a solution to what might seem like an impossible problem.

Procedure 

1. Examine an **index card.** Take note of its size and shape. Your mission is to fit yourself through the card, as shown at right.
2. Brainstorm with a partner about possible ways to complete your mission, keeping the following guidelines in mind: You can use **scissors,** and you can fold the card, but you cannot use staples, paper clips, tape, glue, or any other form of adhesive.
3. When you and your partner have planned your strategy, write your procedure in your ScienceLog.
4. Test your strategy. Did it work? If necessary, get another index card and try again, recording your new strategy and results in your ScienceLog.
5. Share your strategies and results with other groups in your class.

Analysis

6. Why was it helpful to plan your strategy in advance?
7. How did testing your strategy help you complete your mission?
8. How did sharing your ideas with your classmates help you complete your mission? What did they do differently?

Section 1

Exploring Physical Science

NEW TERMS
physical science

OBJECTIVES
- Describe physical science as the study of energy and matter.
- Explain the role of physical science in the world around you.
- Name some careers that rely on physical science.

It's Monday morning. You're eating breakfast and trying to pull yourself out of an early morning daze. As you eat a spoonful of Crunch Blasters, your favorite cereal, you look down and notice your reflection in your spoon. Something's funny about it—it's upside down! "Why is my reflection upside down even though I'm holding the spoon right side up?" you wonder. Is your spoon playing tricks on you? Next you look at the back of the spoon. "A-ha!" you think, "Now my reflection is right side up!" However, when you look back at the inside of the spoon, your reflection is upside down again. What is it about the spoon that makes your reflection look right side up on one side and upside down on the other?

That's Science!

What would you say if someone told you that you were just doing science? You may not realize it, but that's exactly what it was. Science is all about being curious, making observations, and asking questions about those observations. For example, you noticed your reflection in your spoon and became curious about it. You observed that it was upside down, but that when you looked at the back of the spoon, your reflection was right side up. Then you asked what the two sides of the spoon had to do with your reflection. So you were definitely doing science!

Everyday Science Science is all around you, even if you're not thinking about it. Everyday actions such as putting on your sunglasses when you're outside, timing your microwave popcorn just right, and using the brakes on your bicycle all use your knowledge of science. But how do you know how to do these things? From experience—you've gained an understanding of your world by observing and discovering all your life.

Because science is all around, you might not be surprised to learn that there are different branches of science. This book is all about physical science. So just what is physical science?

Matter + Energy → Physical Science

Physical science is the study of matter and energy. Matter is the "stuff" that everything is made of—even stuff that is so small you can't see it. Your shoes, your pencil, and even the air you breathe are made of matter. And all of that matter has energy. Energy is easier to describe than to explain. For example, energy is partly responsible for rainbows in the sky, but it isn't the rainbow itself. When you throw a ball, you give the ball energy. Moving objects have energy, as you can see in **Figure 1.** Food also has energy. When you eat food, the energy in the food is transferred to you, and you can use that energy to carry out your daily activities. But energy isn't always associated with motion or food. All matter has energy, even matter that isn't moving, like that shown in **Figure 2.**

As you explore physical science, you'll learn more about the relationship between matter and energy by answering questions such as the following: Why does paper burn but gold does not? Why is it harder to throw a bowling ball than a baseball? How can water turn into steam and back to water? All of the answers have to do with matter and energy. And although it is difficult to talk about matter without talking about energy, sometimes it is useful to focus on one or the other. That's why physical science is often divided into two categories—chemistry and physics.

Figure 1 *The cheetah, the fastest land mammal, has a lot of energy when running full speed. The cheetah also uses a lot of energy to run so fast. But a successful hunt will supply the energy the cheetah needs to continue living.*

Figure 2 *All matter has energy—even this monumental stone head that is over 1.5 m tall!*

Chemistry Studying all forms of matter and how they interact is what chemistry is all about. You'll learn about the properties and structure of matter and how different substances behave under certain conditions, such as high temperature and high pressure. You'll also discover how and why matter can go through changes, such as the one shown in **Figure 3.** Check out the chart below to find out what you can learn by studying chemistry.

Figure 3 *When you wash your clothes, the detergent and the stains interact. The result? Clean clothes!*

Physics Like chemistry, physics deals with matter. But unlike chemistry, physics is mostly concerned with energy and how it affects matter. Studying different forms of energy is what physics is all about. When you study physics, you'll discover how energy can make matter do some interesting things, as shown in **Figure 4.** You'll also begin to understand aspects of your world such as motion, force, gravity, electricity, light, and heat. Check out the chart below to find out what you can learn by studying physics.

Figure 4 *When you study physics you'll learn how energy causes the motion that makes a roller coaster ride so exciting.*

Explore

List three items in your classroom that you think resulted from advances in chemistry, and list three items that you think resulted from advances in physics. Compare the two lists. What do the items on the two lists have in common? Explain why. Compare your lists with lists made by your classmates.

By studying chemistry, you can find out . . .	By studying physics, you can find out . . .
▪ why yeast makes bread dough rise before you put it in the oven.	▪ why you move to the right when the car you're in turns left.
▪ how the elements chlorine and sodium combine to form table salt, a compound.	▪ why you would weigh less on the moon than you do on Earth.
▪ why water boils at 100°C.	▪ why you see a rainbow after a rainstorm.
▪ why sugar dissolves faster in hot tea than in iced tea.	▪ how a compass works.
▪ how pollution affects our atmosphere.	▪ how your bicycle's gears help you pedal faster or slower.

Physical Science Is All Around You

Believe it or not, the things that you'll learn about matter and energy by studying physical science are important for what you'll learn in other science classes, too. Take a look below to see the role of physical science in areas that you might have thought only involved Earth science or life science.

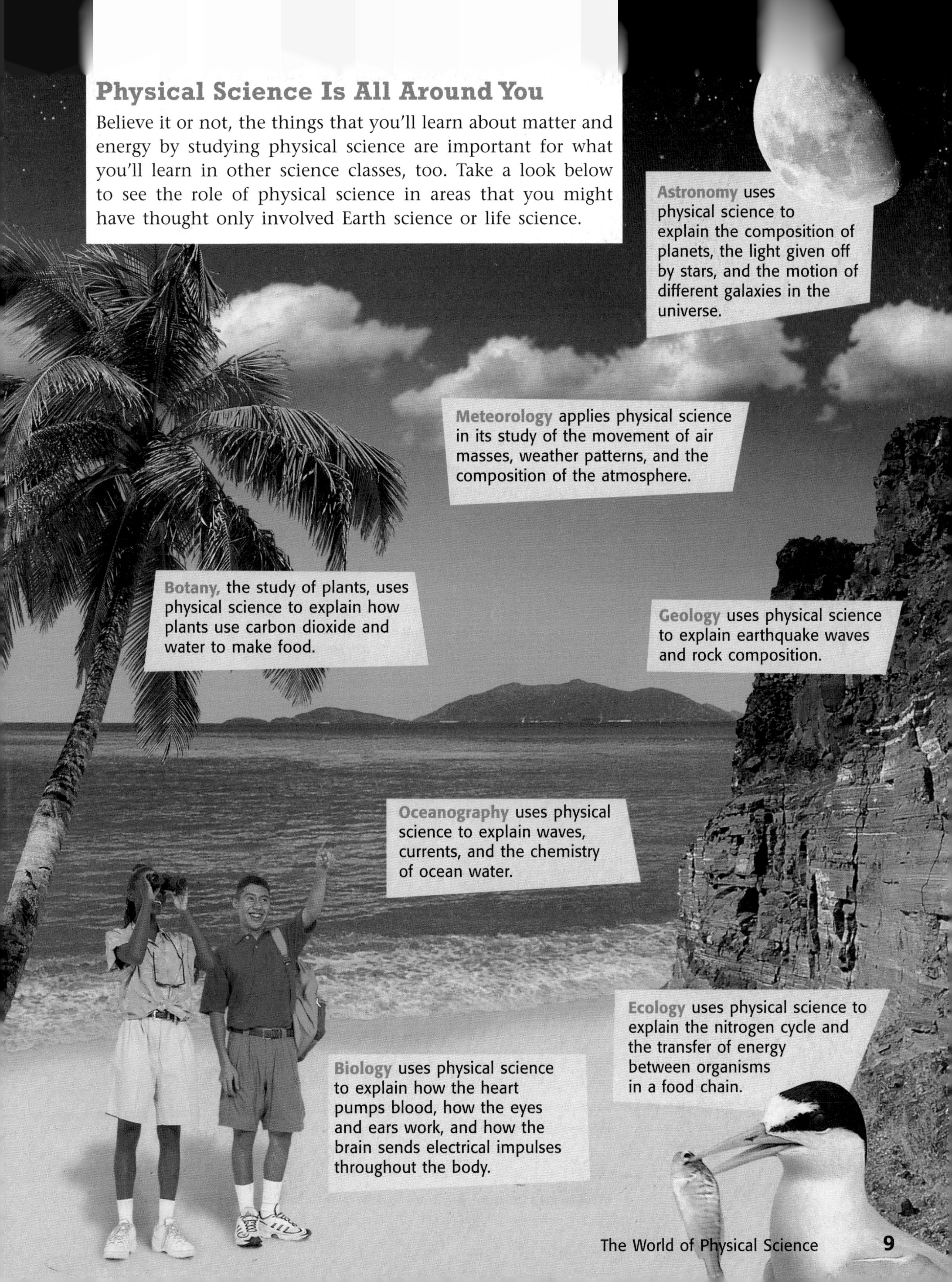

Astronomy uses physical science to explain the composition of planets, the light given off by stars, and the motion of different galaxies in the universe.

Meteorology applies physical science in its study of the movement of air masses, weather patterns, and the composition of the atmosphere.

Botany, the study of plants, uses physical science to explain how plants use carbon dioxide and water to make food.

Geology uses physical science to explain earthquake waves and rock composition.

Oceanography uses physical science to explain waves, currents, and the chemistry of ocean water.

Ecology uses physical science to explain the nitrogen cycle and the transfer of energy between organisms in a food chain.

Biology uses physical science to explain how the heart pumps blood, how the eyes and ears work, and how the brain sends electrical impulses throughout the body.

Physical Science in Action

Now that you know physical science is all around you, you may not be surprised to learn that a lot of careers rely on physical science. What's more, you don't have to be a scientist to use physical science in your job! On this page, you can see some career opportunities that involve physical science.

Gene Webb is an auto mechanic. He understands how the parts of a car engine move and how to keep cars working efficiently.

Shirley Ann Jackson has used physical science as a researcher in the semiconductor and optical physics industries. Former chair of the Nuclear Regulatory Commission, she became president of Rensselaer Polytechnic Institute in 1999.

Sung Park is a photojournalist. He knows how to use different amounts of light to ensure that his photographs capture the best story.

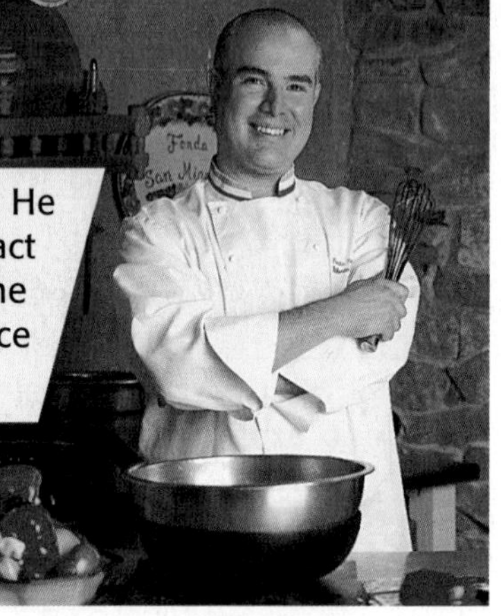

Roberto Santibanez is a chef. He knows how ingredients interact and how energy can cause the chemical changes that produce his delicious meals.

Julie Fields is a chemist who studies the structures of chemical substances found in living organisms. She investigates how these substances can be made in the laboratory and turned into products such as medicines.

Other Careers Involving Physical Science

Architect
Pharmacist
Firefighter
Engineer
Construction worker
Optician
Pilot
Electrician
Computer technician

REVIEW

1. What is physical science all about?
2. List three things you do every day that use your experience with physical science.
3. **Applying Concepts** Choose one of the careers listed in the chart at left. How do you think physical science is involved in that career?

Section 2

Using the Scientific Method

NEW TERMS

scientific method
technology
observation
hypothesis
data
theory
law

OBJECTIVES

- Identify the steps used in the scientific method.
- Give examples of technology.
- Explain how the scientific method is used to answer questions and solve problems.
- Describe how our knowledge of science changes over time.

When you hear or read about advancements in science, do you wonder how they were made? How did the scientists make their discoveries? Were they just lucky? Maybe, but chances are that it was much more than luck. The scientific method probably had a lot to do with it!

What Is the Scientific Method?

The **scientific method** is a series of steps that scientists use to answer questions and solve problems. The chart below shows the steps that are commonly used in the scientific method. Although the scientific method has several distinct steps, it is not a rigid procedure whose steps must be followed in a certain order. Scientists may use the steps in a different order, skip steps, or repeat steps. It all depends on what works best to answer the question.

The Scientific Method

- *Ask a question.*
- *Form a hypothesis.*
- *Test the hypothesis.*
- *Analyze the results.*
- *Draw conclusions.*
- *Communicate results.*

Do you remember James Czarnowski and Michael Triantafyllou, the two scientists discussed at the beginning of this chapter? What scientific problem were they trying to solve? In the next few pages, you'll learn how they used the scientific method to develop new technology—*Proteus,* the penguin boat.

Spotlight on Technology

Technology is the application of knowledge, tools, and materials to solve problems and accomplish tasks. Technology can also refer to the objects used to accomplish tasks. For example, computers, headphones, and the Internet are all examples of technology. But even things like toothbrushes, light bulbs, and pencils are examples of technology. A toothbrush helps you accomplish the task of cleaning your teeth. A light bulb solves the problem of how to read when the sun goes down. A pencil helps you accomplish the task of writing.

Science and technology are not the same thing. The goal of science is to gain knowledge about the natural world. The goal of technology is to apply scientific understanding to solve problems. Technology is sometimes called *applied science.*

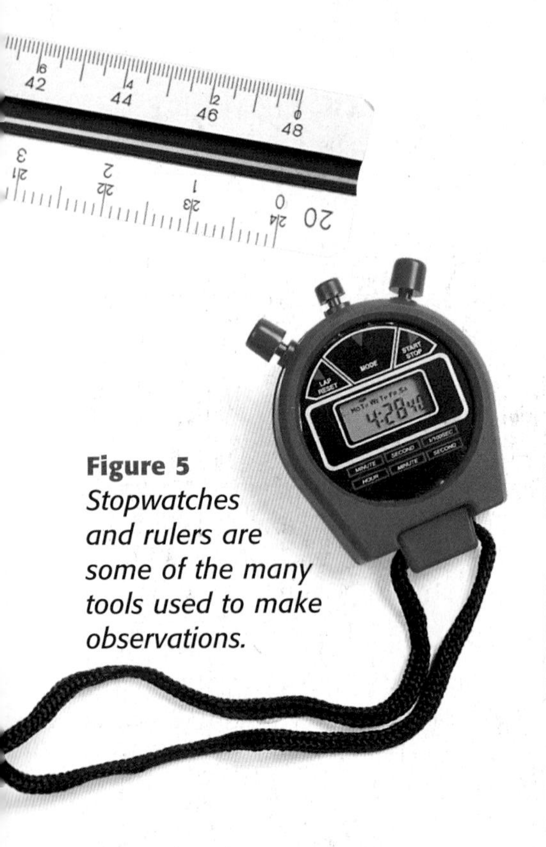

Figure 5 *Stopwatches and rulers are some of the many tools used to make observations.*

Ask a Question Asking questions helps you focus your investigation. A question identifies something you don't know but want to find out. Usually, scientists ask a question after they've made a lot of observations. An **observation** is any use of the senses to gather information. Measurements are observations that are made with instruments, such as those shown in **Figure 5.** The chart below gives you some examples of observations. Keep in mind that you can make observations at any point while using the scientific method.

Examples of Observations	
▪ The sky is blue.	▪ He is 125 centimeters tall.
▪ The ice began to melt 30 seconds after it was taken out of the freezer.	▪ Adding food coloring turned the water red. Adding bleach made the water clear again.
▪ This soda bottle has a volume of 1 liter.	▪ This brick feels heavier than this sponge.
▪ Cotton balls feel soft.	▪ Sandpaper is rough.
▪ The box was easier to move when I put it on wheels.	▪ My dog responded to the whistle, but I couldn't hear it.

So what question did the scientists who made *Proteus* ask? Czarnowski and Triantafyllou, shown in **Figure 6,** are engineers (EN juh NIRZ), scientists who put scientific knowledge to work for practical human uses. Engineers create technology. While a graduate student in the department of Ocean Engineering at the Massachusetts Institute of Technology, Czarnowski worked with Triantafyllou, his professor, on improving boat technology. Specifically, they wanted to observe boat propulsion systems and investigate how to make them work better. A propulsion system is what makes a boat move; most boats are driven by propellers.

One thing that Czarnowski and Triantafyllou were studying is the efficiency of boat propulsion systems. *Efficiency* compares energy output (the energy used to move the boat forward) with energy input (the energy supplied by the boat's engine). Czarnowski and Triantafyllou learned from their observations that boat propellers, shown in **Figure 7** on the next page, are not very efficient.

Figure 6 *James Czarnowski (left) and Michael Triantafyllou (right) made observations about how boats work in order to develop* Proteus.

Figure 7 Observations About the Efficiency of Boat Propellers

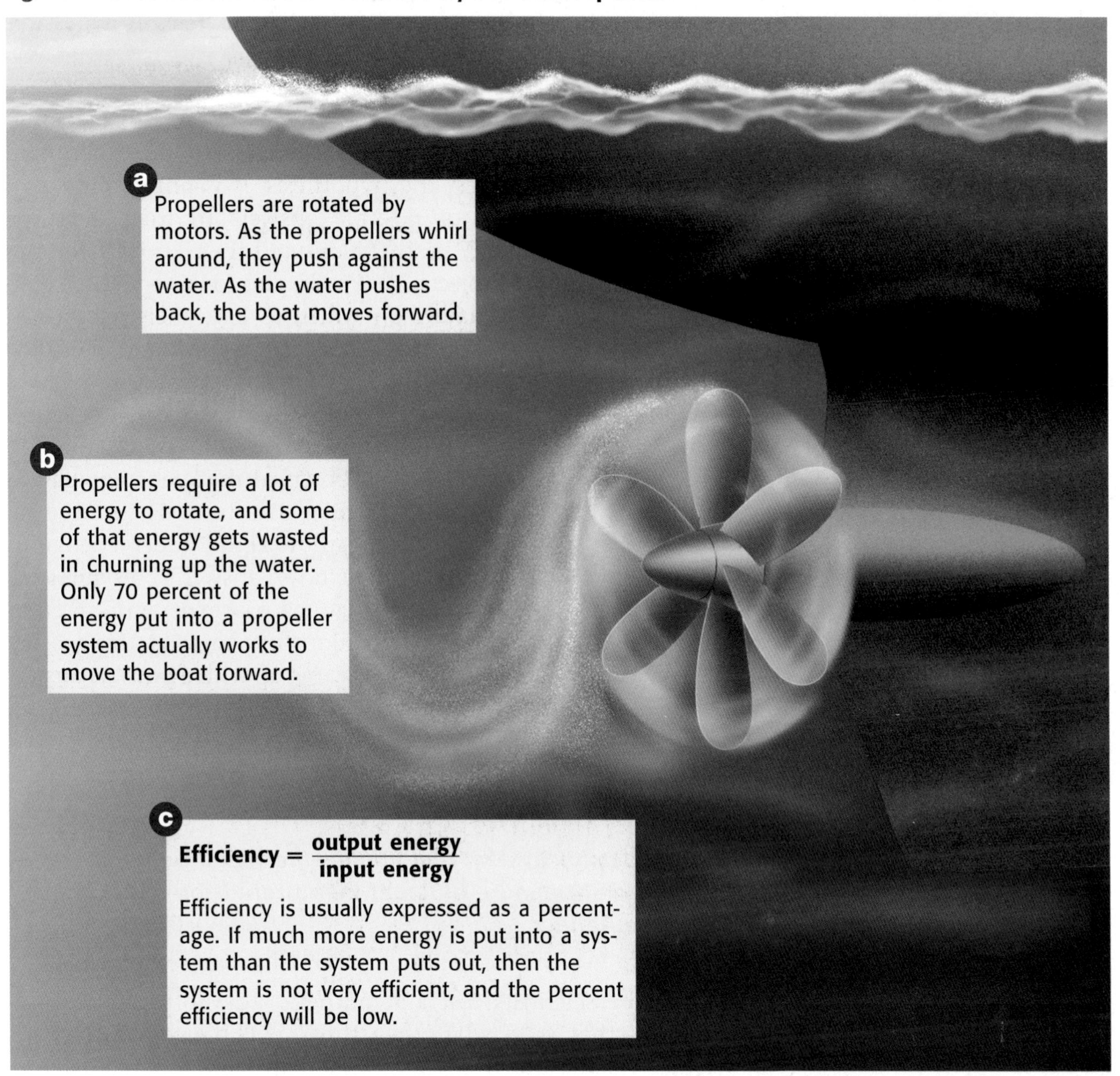

Why is boat efficiency important? Most boats are only about 70 percent efficient. Making only a small fraction of the United States' boats and ships just 10 percent more efficient would save millions of liters of fuel per year. Saving fuel means saving money, but it also means using less of the Earth's supply of fossil fuels. Based on their observations and all of this information, Czarnowski and Triantafyllou knew what they wanted to find out.

Ask some questions of your own on page 516 in the LabBook.

The Question:

How can boat propulsion systems be made more efficient?

Figure 8 *Penguins use their flippers almost like wings to "fly" underwater. As they pull their flippers toward their body, they push against the water, which propels them forward.*

Form a Hypothesis Once you've asked your question, your next step is forming a hypothesis. A **hypothesis** is a possible explanation or answer to a question. You can use what you already know and any observations that you have made to form a hypothesis. A good hypothesis is testable. If no observations or information can be gathered or if no experiment can be designed to test the hypothesis, it is untestable.

In thinking of possible answers to their question, Czarnowski and Triantafyllou used their knowledge from past boat projects. They were also looking for an example from nature on which to base their hypothesis. Czarnowski had made observations of penguins swimming at the New England Aquarium. **Figure 8** shows how penguins propel themselves. He observed how quickly and easily the penguins moved through the water. He also observed that penguins have a rigid body, similar to a boat. These observations led Czarnowski to wonder if penguins could provide an answer to his question.

Using their past experience and new observations of penguins, the two scientists came up with a possible answer to their question: a propulsion system that works the way a penguin swims!

Hypothesis:

A propulsion system that mimics the way a penguin swims will be more efficient than propulsion systems that use propellers.

life science CONNECTION

Penguins, although flightless, are better adapted to water and extreme cold than any other bird. Most of the world's 18 penguin species live and breed on islands in the subantarctic waters. Penguins can swim as fast as 40 km/h, and some can leap more than 2 m above the water.

Before scientists test a hypothesis, they often make predictions that state what they think will happen during the actual test of the hypothesis. Scientists usually state predictions in an "If . . . then . . ." format. The engineers at MIT might have made the following prediction: *If* two flippers are attached to a boat, *then* the boat will be more efficient than a boat powered by propellers.

REVIEW

1. How do scientists and engineers use the scientific method?
2. Give three examples of technology from your everyday life.
3. **Analyzing Methods** Explain how the accuracy of your observations might affect how you develop a hypothesis.

Test the Hypothesis After you form a hypothesis, you must test it to determine whether it is a reasonable answer to your question. In other words, testing helps you find out if your hypothesis is pointing you in the right direction or if it is way off the mark. Often a scientist will test a hypothesis by testing a prediction.

One way to test a hypothesis is to conduct a controlled experiment. In a controlled experiment, there is a control group and an experimental group. Both groups are the same except for one factor in the experimental group, called a *variable*. The experiment will then determine the effect that this variable has on the system.

Sometimes a controlled experiment is not possible. Stars, for example, are too far away to be used in an experiment. In such cases, you can test your hypothesis by making additional observations or by conducting research. If your scientific investigation involves creating technology to solve a problem, you test your hypothesis in a different way. You make or build what you want to test and see if it does what you expected it to do. That's just what Czarnowski and Triantafyllou did—they built *Proteus*, the penguin boat, shown in **Figure 9.**

Quick Lab

That's Swingin'!

1. Make a pendulum by tying a **piece of string** to a **ring stand** and hanging a **small mass,** such as a washer, from the end of the string.
2. Form a hypothesis about which factors (such as length of string, mass, etc.) affect the rate at which the pendulum swings.
3. In your ScienceLog, record what factors you will control and what factor will be your variable.
4. Test your hypothesis by conducting several trials, recording the number of swings made in a given time, such as 10 seconds, for each trial.
5. Was your hypothesis supported? In your ScienceLog, analyze your results.

Figure 9 Testing Penguin Propulsion

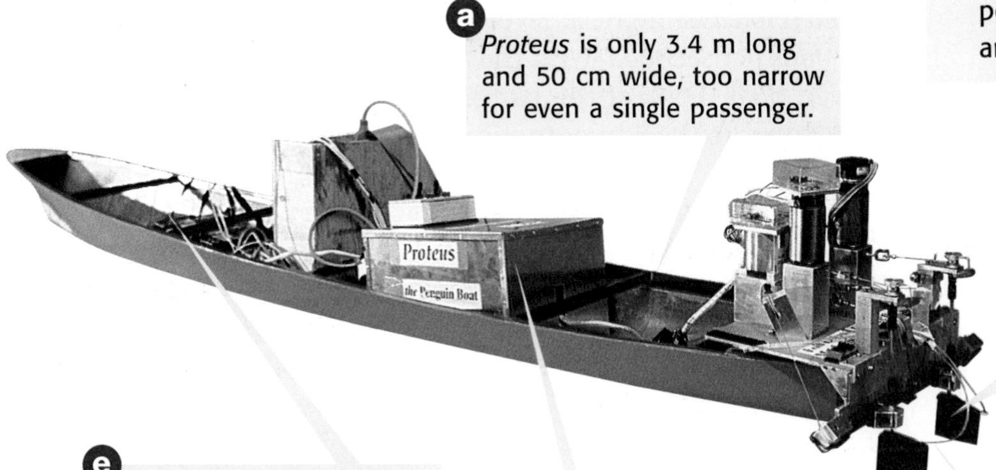

a *Proteus* is only 3.4 m long and 50 cm wide, too narrow for even a single passenger.

b *Proteus* has two flipper-like paddles, called *foils*. Both foils move out and then in, much as a penguin uses its flippers underwater.

c As the foils flap, they push water backward. The water pushes against the foils, propelling the boat forward.

d A desktop computer programs the number of times the foils flap per second.

e Each of *Proteus*'s flapping foils is driven by a motor that gets its energy from two car batteries.

Self-Check

What variable were Czarnowski and Triantafyllou testing? (See page 596 to check your answer.)

Once constructed, *Proteus* was ready to test. After a few trials in a laboratory tank, Czarnowski and Triantafyllou took *Proteus* out into the open water of the Charles River. They were ready to collect data. **Data** are any pieces of information acquired through experimentation. The engineers used an onboard computer to adjust the flapping motion and to measure how much energy the motors used. These measurements were data for the energy input. The engineers did several tests, each time changing only the flapping rate. For each test, data such as the flapping rate, energy used, and the speed achieved by the boat were carefully recorded. The output energy was determined from the speed *Proteus* achieved.

Analyze the Results After you collect and record your data and other observations, you must analyze them to determine whether the results of your test support the hypothesis. To do this, you must organize and study your data and observations. Sometimes doing calculations can help you learn more about your results. Organizing numerical data into tables and graphs makes relationships between information easier to see.

To analyze their results, Czarnowski and Triantafyllou used the data for energy input and energy output to calculate *Proteus*'s efficiency for different flapping rates. These data can be graphed as shown in the line graph in **Figure 10.** Remember that their hypothesis was that penguin propulsion would be more efficient than propeller propulsion. The scientists compared *Proteus*'s highest level of efficiency—the efficiency at 1.7 flaps per second—with the average efficiency of a propeller-driven boat. Look at the bar graph in Figure 10 to see if their data support their hypothesis.

Figure 10 Graphs of the Test Results

This line graph shows that *Proteus* was most efficient when its foils were flapping about 1.7 times per second.

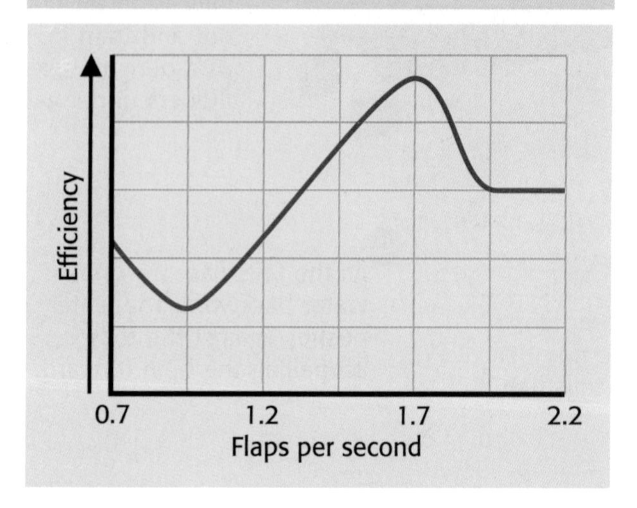

This bar graph shows that *Proteus* is 17 percent more efficient than a propeller-driven boat.

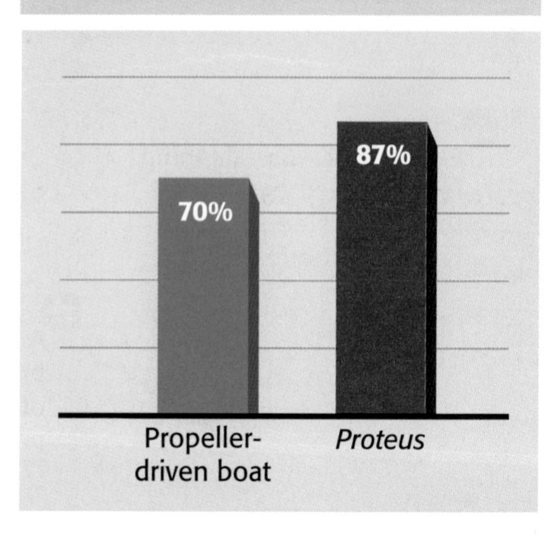

Draw Conclusions At the end of an investigation, you must draw a conclusion. You could conclude that your results supported your hypothesis, that your results did *not* support your hypothesis, or that you need more information. If your results support your hypothesis, you can ask further questions. If your results do not support your hypothesis, you should check your results or calculations for errors. You may have to modify your hypothesis or form a new one and conduct another investigation. If you find that your results neither support nor disprove your hypothesis, you may need to gather more information, test your hypothesis again, or redesign the procedure.

After Czarnowski and Triantafyllou analyzed the results of their test, they conducted many more trials. Still they found that the penguin propulsion system was more efficient than a propeller propulsion system. So they concluded that their hypothesis was supported, which led to more questions, as you can see in **Figure 11.**

Figure 11 *Could a penguin propulsion system be used on large ships, such as an oil tanker? Other scientists are conducting more research to find out!*

Communicate Results One of the most important steps in any investigation is to communicate your results. You can write a scientific paper explaining your results, make a presentation, or create a Web site. Telling others what you learned is how science keeps going. After you've completed an investigation, other scientists can conduct their own tests, modify your tests to learn something more specific, or study a new problem based on your results.

Czarnowski and Triantafyllou published their results in academic papers, but they also displayed their project and its results on the Internet. In addition, science magazines and newspapers have reported the work of these engineers. These reports allow you to conduct some research of your own about *Proteus.*

Breaking the Mold of the Scientific Method Not all scientists use the same scientific method, nor do they always follow the same steps in the same order. Why not? Sometimes you may have a clear idea about the question you want to answer. Other times, you may have to revise your hypothesis and test it again. While you should always take accurate measurements and record data correctly, you don't always have to follow the scientific method in a certain order. **Figure 12** shows you some other paths through the scientific method.

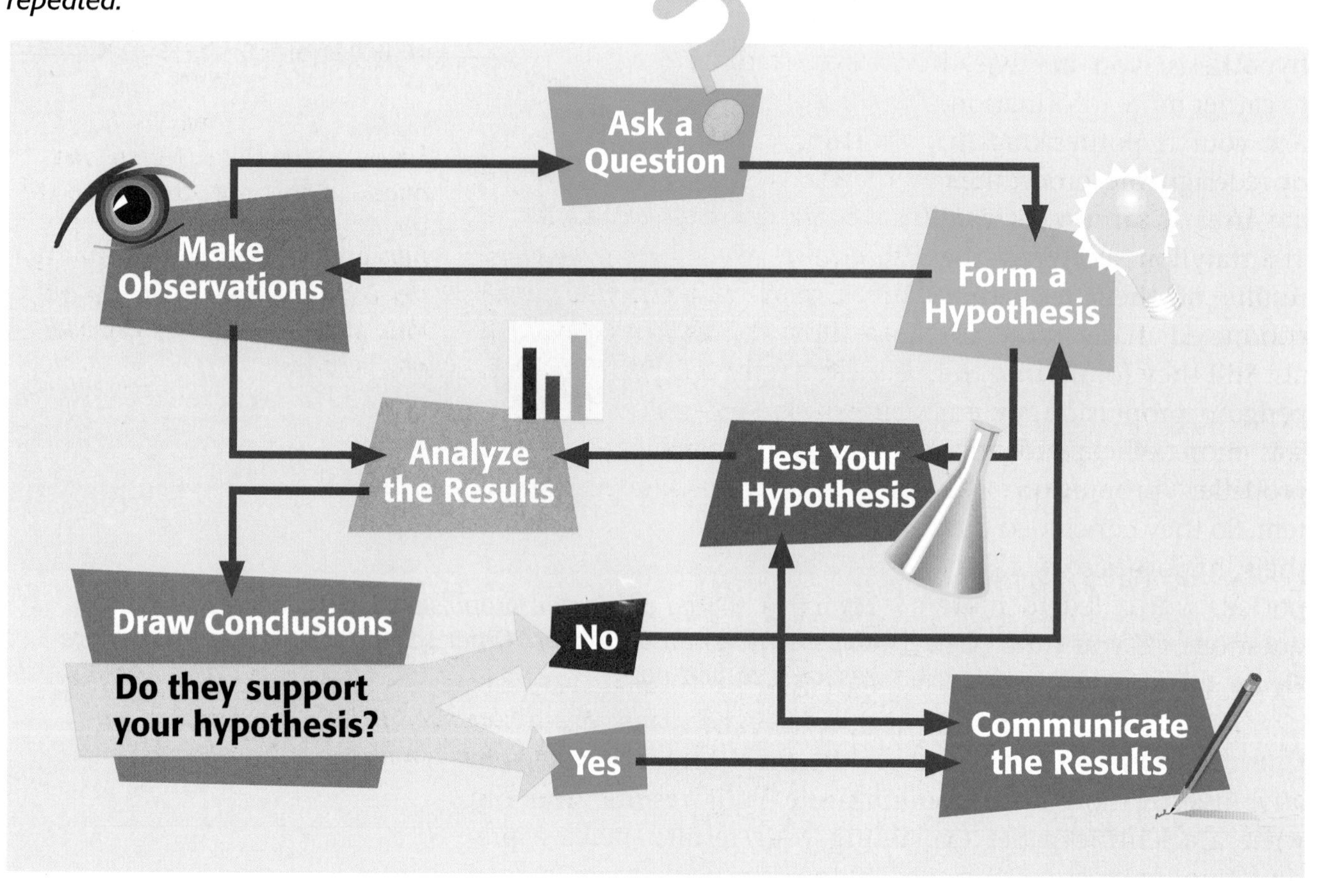

Figure 12 *Scientific investigations do not always proceed from one step of the scientific method to the next. Sometimes steps are skipped, and sometimes they are repeated.*

Building Scientific Knowledge

Using the scientific method is a way to find answers to questions and solutions to problems. But you should understand that answers are very rarely *final* answers. As our understanding of science grows, our understanding of the world around us changes. New ideas and new experiments teach us new things. Sometimes, however, an idea is supported again and again by many experiments and tests. When this happens, the idea can become a theory or even a law. As you will read on the next page, theories and laws help to build new scientific knowledge.

Turn to page 33 to discover a tale of young Einstein's encounter with some other science heavyweights.

Scientific Theories You've probably heard a detective on a TV show say, "I've got a theory about who committed the crime." Does the detective have a scientific theory? Probably not; it might be just a guess. A scientific theory is more complex than a simple guess.

In science, a **theory** is a unifying explanation for a broad range of hypotheses and observations that have been supported by testing. A theory not only can explain an observation you've made but also can predict an observation you might make in the future. Keep in mind that theories, like the one shown in **Figure 13,** can be changed or replaced as new observations are made or as new hypotheses are tested.

Figure 13 *According to the big-bang theory, the universe was once a small, hot, and dense volume of matter. About 10 to 20 billion years ago, an event called the big bang sent matter in all directions, forming the galaxies and planets.*

Scientific Laws What do you think of when you hear the word *law*? Traffic laws? Federal laws? Well, scientific laws are not like these laws. Scientific laws are determined by nature, and you can't break a scientific law!

In science, a **law** is a summary of many experimental results and observations. A law tells you how things work. Laws are not the same as theories because laws only tell you *what* happens, not *why* it happens, as shown in **Figure 14.** Although a law does not explain why something happens, the law tells you that you can expect the same thing to happen every time.

Figure 14 *Dropping a ball illustrates the law of conservation of energy. Although the ball doesn't bounce back to its original height, energy is not lost–it is transferred to the ground.*

REVIEW

1. Name the steps that can be used in the scientific method.

2. How is a theory different from a hypothesis?

3. Analyzing Ideas Describe how our knowledge of science changes over time.

Section 3

Using Models in Physical Science

NEW TERMS

model

OBJECTIVES

- Explain how models represent real objects or systems.
- Give examples of different ways models are used in science.

Think again about *Proteus*. How much like a penguin was it? Well, *Proteus* didn't have feathers and wasn't a living thing, but its "flippers" were designed to create the same kind of motion as a penguin's flippers. The MIT engineers built *Proteus* to mimic the way a penguin swims so that they could gain a greater understanding about boat propulsion. In other words, they created a *model*.

What Is a Model?

A **model** is a representation of an object or system. Models are used in science to describe or explain certain characteristics of things. Models can also be used for making predictions and explaining observations. A model is never exactly like the real object or system—if it were, it would no longer be a model. Models are particularly useful in physical science because many characteristics of matter and energy can be either hard to see or difficult to understand. You can see some examples of scientific models below.

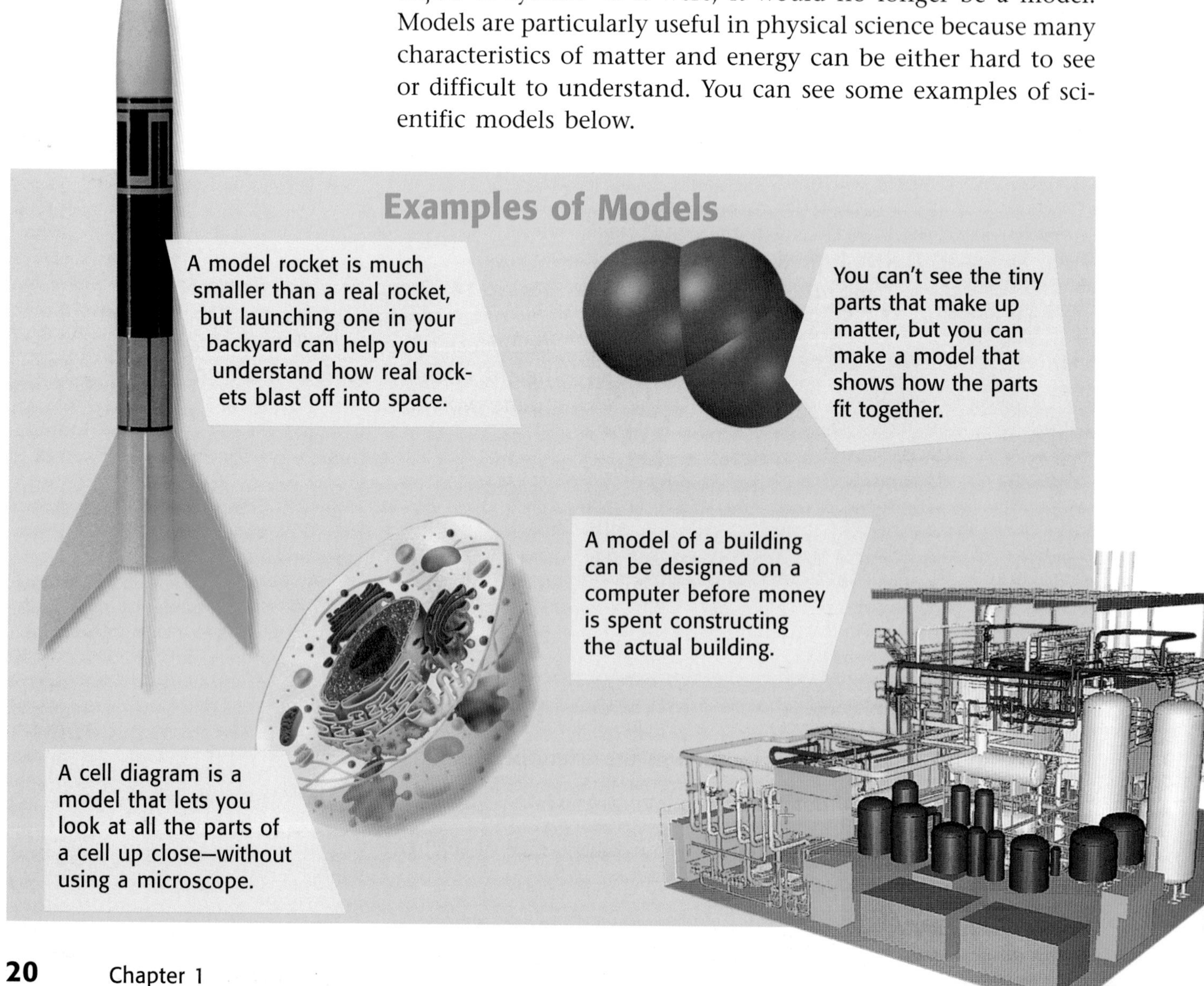

Models Help You Visualize Information

When you're trying to learn about something that you can't see or observe directly, a model can help you visualize it, or picture it in your mind. Familiar objects or ideas can help you understand something a little less familiar.

Objects as Models When you use a real object as a model for something you cannot see, the object must have characteristics similar to those of the real thing. For example, a coiled spring toy is often used as a model of sound waves. You've probably used this kind of spring toy before, so it's a familiar object. Sound waves are probably a little less familiar—after all, you can't see them. But the spring toy behaves a lot like sound waves do. So using the spring toy as a model, as shown in **Figure 15,** can make the behavior of sound waves easier to understand.

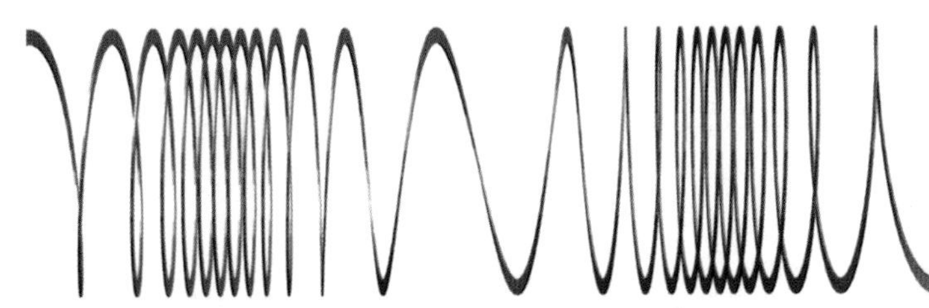

Figure 15 *A coiled spring toy can show you how air particles crowd together in parts of a sound wave.*

Ideas as Models When you're trying to understand something but don't have an object to use as a model, you can create a model from an idea. For example, when sugar dissolves in iced tea, it seems to disappear. To try to understand where the sugar went, imagine a single drop of tea magnified until it is almost as big as you are, with tiny spaces between the particles of water in the tea. Using this model, as shown in **Figure 16,** you can understand that the sugar seems to disappear because the sugar particles fit into spaces between the water particles in the tea.

Figure 16 *Just by imagining a big drop of tea, you are creating a model!*

You've probably seen a weather report on television. Think about the models that a weather reporter uses to tell you about the weather: satellite pictures, color-coded maps, and live radar images. How are these models used to represent the weather? Why do you think that sometimes weather forecasts are wrong?

Models Are Just the Right Size

How can you observe how the phases of the moon occur? That's a tough problem, because you're on Earth and you can't easily get off of the Earth to observe the moon going around it. But you can observe a model of the moon, Earth, and sun, as shown in **Figure 17.** As you can see, models can represent things that are too large to easily observe.

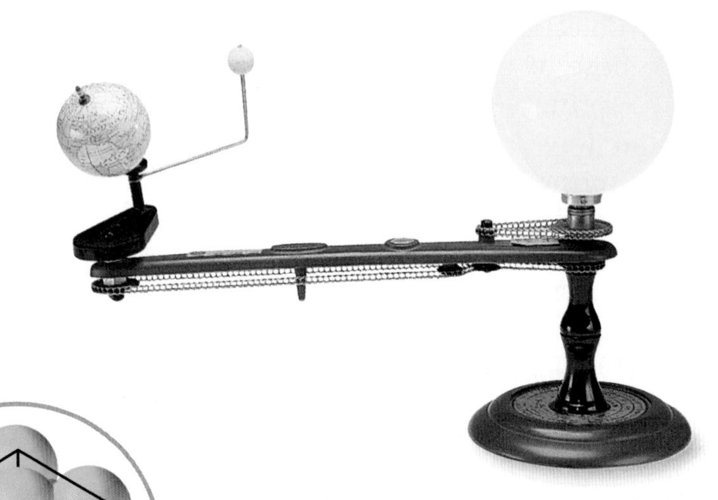

Figure 17 *Using this model, you can see how the Earth's rotation, in addition to the moon's revolution around the Earth as the Earth revolves around the sun, results in the different phases of the moon.*

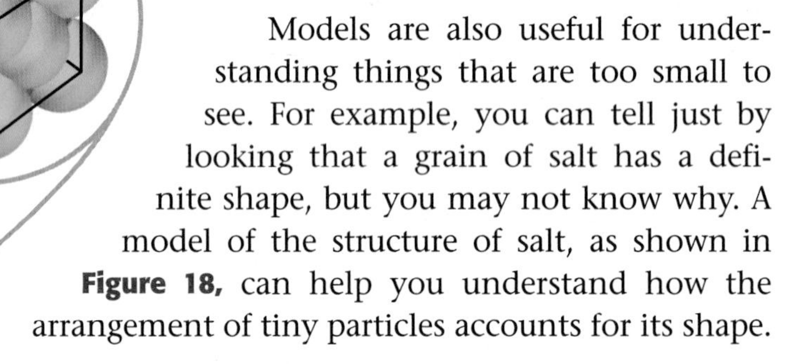

Figure 18 *The particles of matter in a grain of salt connect in a continuous pattern that forms a cube. That's why a grain of salt has a cubic shape.*

Models are also useful for understanding things that are too small to see. For example, you can tell just by looking that a grain of salt has a definite shape, but you may not know why. A model of the structure of salt, as shown in **Figure 18,** can help you understand how the arrangement of tiny particles accounts for its shape.

Models Build Scientific Knowledge

Models not only can represent scientific ideas and objects but also can be tools that you can use to conduct investigations and illustrate theories.

Testing Hypotheses The MIT engineers were trying to test their hypothesis that a boat that mimics the way a penguin swims would be more efficient than a boat powered by propellers. How did they test this hypothesis? By building a model, *Proteus*. When using the scientific method to develop new technology, testing a hypothesis often requires building a model. By conducting tests with *Proteus,* the MIT engineers tested their hypothesis and found out what factors affected the model's efficiency. Using the data they collected, they could consider building a full-sized penguin boat.

LabBook

Build a model car and test its speed on page 517 in the LabBook.

Illustrating Theories Recall that a theory explains why things happen the way they do. Sometimes, however, a theory is hard to picture. That's where models come in handy. A model is different from a theory, but a model can present a picture of what the theory explains when you cannot actually observe it. You can see an example of this in **Figure 19.**

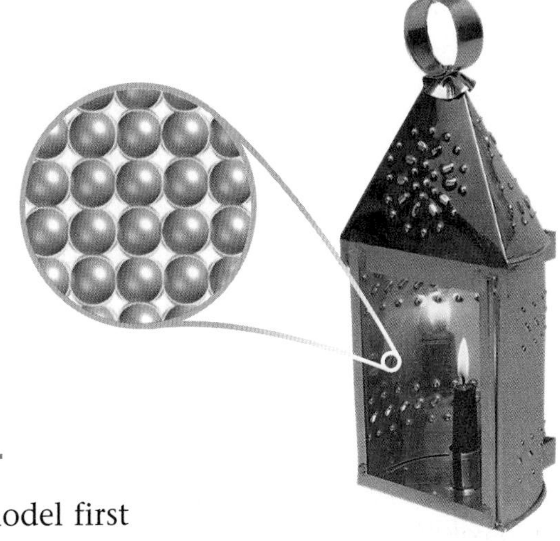

Figure 19 *This model illustrates the atomic theory, which states that all matter is made of tiny particles called atoms.*

Models Can Save Time and Money

When creating technology, scientists often create a model first so that they can test its characteristics and improve its design before building the real thing. You may recall that *Proteus* wasn't big enough to carry even a single passenger. Why didn't the MIT engineers begin by building a full-sized boat? Imagine if they had gone to all that trouble and found out that their design didn't work. What a waste! Models allow you to test ideas without having to spend the time and money necessary to make the real thing. In **Figure 20,** you can see another example of how models save time and money.

Figure 20 *Car engineers can conduct cyber-crashes, in which computer-simulated cars crash in various ways. Engineers use the results to determine which safety features to install on the car—all without damaging a single automobile.*

REVIEW

1. What is the purpose of a model?
2. Give three examples of models that you see every day.
3. **Interpreting Models** Both a globe and a flat world map model certain features of the Earth. Give an example of when you would use a globe and an example of when you would use a flat map.

Section 4

Measurement and Safety in Physical Science

NEW TERMS

meter
volume
mass
temperature
area
density

OBJECTIVES

- Explain the importance of the International System of Units.
- Determine the appropriate units to use for particular measurements.
- Describe how area and density are derived quantities.

Hundreds of years ago, different countries used different systems of measurement. In England, the standard for an inch used to be three grains of barley placed end to end. Other standardized units of the modern English system, which is used in the United States, used to be based on parts of the body, such as the foot. Such units were not very accurate because they were based on objects that varied in size.

Eventually people recognized that there was a need for a single measurement system that was simple and accurate. In the late 1700s, the French Academy of Sciences began to develop a global measurement system, now known as the International System of Units, or SI.

The International System of Units

Today most scientists in almost all countries use the International System of Units. One advantage of using SI measurements is that it helps scientists share and compare their observations and results. Another advantage of SI is that all units are based on the number 10, which makes conversions from one unit to another easy to do. The table in **Figure 21** contains the commonly used SI units for length, volume, mass, and temperature.

Figure 21 *Prefixes are used with SI units to convert them to larger or smaller units. For example* kilo *means 1,000 times, and* milli *indicates 1/1,000 times. The prefix used depends on the size of the object being measured.*

Common SI Units		
Length	**meter (m)** kilometer (km) decimeter (dm) centimeter (cm) millimeter (mm) micrometer (µm) nanometer (nm)	 1 km = 1,000 m 1 dm = 0.1 m 1 cm = 0.01 m 1 mm = 0.001 m 1 µm = 0.000001 m 1 nm = 0.000000001 m
Volume	**cubic meter (m^3)** cubic centimeter (cm^3) liter (L) milliliter (mL)	 1 cm^3 = 0.000001 m^3 1 L = 1 dm^3 = 0.001 m^3 1 mL = 0.001 L = 1 cm^3
Mass	**kilogram (kg)** gram (g) milligram (mg)	 1 g = 0.001 kg 1 mg = 0.000001 kg
Temperature	**Kelvin (K)** Celsius (°C)	0°C = 273 K 100°C = 373 K

Length How long is the construction crane shown in **Figure 22**? To describe its length, a physical scientist would use **meters** (m), the basic SI unit of length. Other SI units of length are larger or smaller than the meter by multiples of 10. For example, 1 kilometer (km) equals 1,000 meters. One meter equals 100 centimeters, or 1,000 millimeters. If you divide 1 m into 1,000 parts, each part equals 1 mm. This means that 1 mm is one-thousandth of a meter. Although that seems pretty small, some objects are so tiny that even smaller units must be used. To describe the length of a grain of salt, micrometers (μm) or nanometers (nm) are used.

Figure 22 *The length of this crane would be expressed in meters.*

Volume Imagine that you need to move some lenses to a laser laboratory. How many lenses will fit into a crate? That depends on the volume of the crate and the volume of each lens. **Volume** is the amount of space that something occupies or, as in the case of the crate, the amount of space that something contains.

Volumes of liquids are expressed in liters (L). Liters are based on the meter. A cubic meter (1 m^3) is equal to 1,000 L. So 1,000 L will fit into a box 1 m on each side. A milliliter (mL) will fit into a box 1 cm on each side. So 1 mL = 1 cm^3. Graduated cylinders are used to measure the volume of liquids.

Volumes of solid objects are expressed with cubic meters (m^3). Volumes of smaller objects can be expressed with cubic centimeters (cm^3) or cubic millimeters (mm^3). To find the volume of a crate, or any other rectangular shape, multiply the length by the width by the height. To find the volume of an irregularly shaped object, measure how much liquid that object displaces. You can see how this works in **Figure 23.**

Explore

Pick an object to use as a unit of measure. It could be a pencil, your hand, or anything else. Find how many units wide your desk is, and compare your measurement with those of your classmates. What were some of the units that your classmates used? In your ScienceLog, explain why it is important to use standard units of measure.

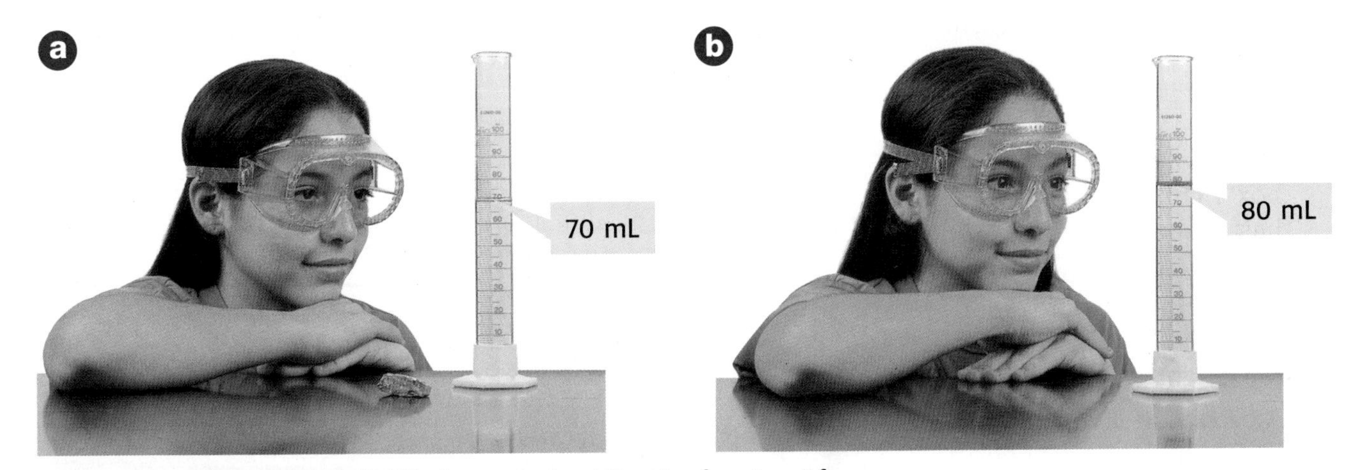

Figure 23 *This graduated cylinder contains 70 mL of water. After the rock is added, the water level moves to 80 mL. Because the rock displaces 10 mL of water, and because 1 mL = 1 cm^3, the volume of the rock is 10 cm^3.*

Mass How many cars can a bridge support? That depends on the strength of the bridge and the mass of the cars. **Mass** is the amount of matter that something is made of. The kilogram (kg) is the basic SI unit for mass and would be used to express the mass of a car. Grams (one-thousandth of a kilogram) are used to express the mass of small objects. A medium-sized apple has a mass of about 100 g. Masses of very large objects are expressed in metric tons. A metric ton equals 1,000 kg.

Figure 24 Measuring Temperature

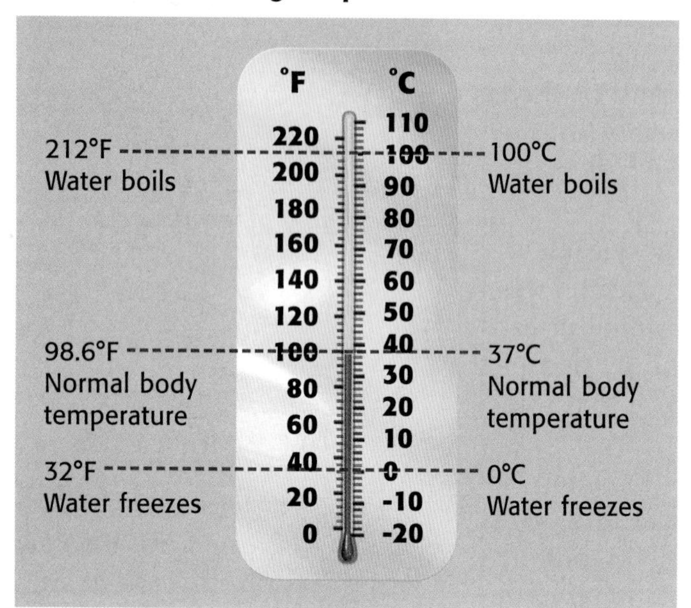

Temperature How hot is melted iron? To answer this question, a physical scientist would measure the temperature of the liquid metal. **Temperature** is a measure of how hot (or cold) something is. You are probably used to expressing temperature with degrees Fahrenheit (°F). Scientists often use degrees Celsius (°C), but the kelvin (K) is the SI unit for temperature. The thermometer in **Figure 24** compares °F with °C, the unit you will most often see in this book.

Derived Quantities

Some quantities are formed from combinations of other measurements. Such quantities are called *derived quantities.* Both area and density are derived quantities.

Area How much carpet would it take to cover the floor of your classroom? Answering this question involves finding the area of the floor. **Area** is a measure of how much surface an object has. To calculate the area of a rectangular surface, first measure the length and width, then use the following equation:

$$\text{Area} = \text{length} \times \text{width}$$

The units for area are called square units, such as m^2, cm^2, and km^2. **Figure 25** will help you understand square units.

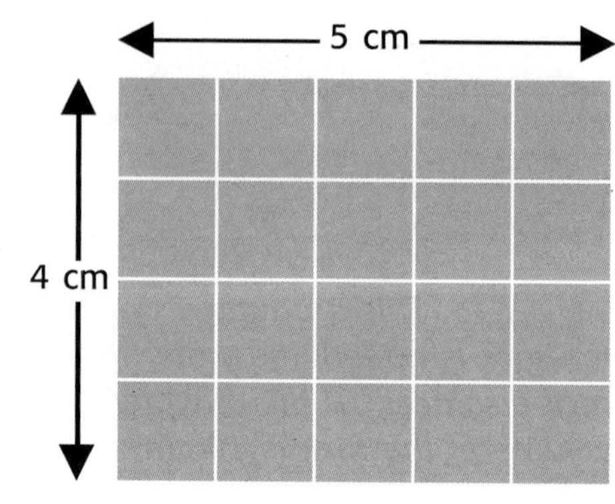

Figure 25 *The area of this rectangle is 20 cm^2. If you count the smaller squares within the rectangle, you'll count 20 squares that each measure 1 cm^2.*

MATH BREAK

Using Area to Find Volume

Area can be used to find the volume of an object according to the following equation:

$$\text{Volume} = \text{Area} \times \text{height}$$

1. What is the volume of a box 5 cm tall whose lid has an area of 9 cm^2?
2. A crate has a volume of 48 m^3. The area of its bottom side is 16 m^2. What is the height of the crate?
3. A cube with a volume of 8,000 cm^3 has a height of 20 cm. What is the area of one of its sides?

Density Another derived quantity is density. **Density** is mass per unit volume. So an object's density is the amount of matter it has in a given space. To find density (D), first measure mass (m) and volume (V). Then use the following equation:

$$D = \frac{m}{V}$$

For example, suppose you want to know the density of the gear shown at right. Its mass is 75 g and its volume is 20 cm^3. You can calculate the gear's density like this:

$$D = \frac{m}{V} = \frac{75\ \text{g}}{20\ \text{cm}^3} = 3.75\ \text{g/cm}^3$$

Safety Rules!

Physical science is exciting and fun, but it can also be dangerous. So don't take any chances! Always follow your teacher's instructions, and don't take shortcuts—even when you think there is little or no danger.

Before starting an experiment, get your teacher's permission and read the lab procedures carefully. Pay particular attention to safety information and caution statements. The chart below shows the safety symbols used in this book. Get to know these symbols and what they mean. Do this by reading the safety information starting on page 512. **This is important!** If you are still unsure about what a safety symbol means, ask your teacher.

Stay on the safe side by reading the safety information on page 512. **You must do this before doing any experiment!**

Safety Symbols		
Eye protection	Clothing protection	Hand safety
Heating safety	Electric safety	Sharp object
Chemical safety	Animal safety	Plant safety

REVIEW

1. Why is SI important?
2. Which SI unit would you use to express the height of your desk? Which SI unit would you use to express the volume of this textbook?
3. **Comparing Concepts** How is area different from volume?

Chapter Highlights

SECTION 1

Vocabulary

physical science *(p. 7)*

Section Notes

- Science is a process of making observations and asking questions about those observations.
- Physical science is the study of matter and energy and is often divided into physics and chemistry.
- Physical science is part of many other areas of science.
- Many different careers involve physical science.

SECTION 2

Vocabulary

scientific method *(p. 11)*
technology *(p. 11)*
observation *(p. 12)*
hypothesis *(p. 14)*
data *(p. 16)*
theory *(p. 19)*
law *(p. 19)*

Section Notes

- The scientific method is a series of steps that scientists use to answer questions and solve problems.
- Any information you gather through your senses is an observation. Observations often lead to questions or problems.
- A hypothesis is a possible explanation or answer to a question. A good hypothesis is testable.
- After you test a hypothesis, you should analyze your results and draw conclusions about whether your hypothesis was supported.
- Communicating your findings allows others to verify your results or continue to investigate your problem.
- A scientific theory is the result of many investigations and many hypotheses. Theories can be changed or modified by new evidence.
- A scientific law is a summary of many experimental results and hypotheses that have been supported over time.

Labs

Exploring the Unseen *(p. 516)*

☑ Skills Check

Math Concepts

AREA To calculate the area of a rectangular surface, first measure its length and width, then multiply those values. The area of a piece of notebook paper with a length of 28 cm and a width of 21.6 cm can be calculated as follows:

$$\text{Area} = \text{length} \times \text{width}$$
$$= 28 \text{ cm} \times 21.6 \text{ cm}$$
$$= 604.8 \text{ cm}^2$$

Visual Understanding

SCIENTIFIC METHOD To answer a question in science, you can use the scientific method. Review the flowchart on page 18 to see that the scientific method does not have to follow a specific order.

MODELS A model is a representation of an object or system. Look back at the examples on page 20 to learn more about different models.

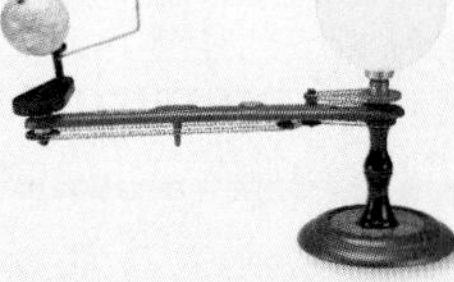

SECTION 3

Vocabulary

model *(p. 20)*

Section Notes

- Scientific models are representations of objects or systems. Models make difficult concepts easier to understand.
- Models can represent things too small to see or too large to observe directly.
- Models can be used to test hypotheses and illustrate theories.

Labs

Off to the Races! *(p. 517)*

SECTION 4

Vocabulary

meter *(p. 25)*
volume *(p. 25)*
mass *(p. 26)*
temperature *(p. 26)*
area *(p. 26)*
density *(p. 27)*

Section Notes

- The International System of Units is the standard system of measurement used around the world.
- Length, volume, mass, and temperature are quantities of measurement. Each quantity of measurement is expressed with a particular SI unit.
- Area is a measure of how much surface an object has. Density is a measure of mass per unit volume.
- Safety rules are important and must be followed at all times during scientific investigations.

Labs

Measuring Liquid Volume *(p. 518)*
Coin Operated *(p. 519)*

internetconnect

GO TO: go.hrw.com

Visit the **HRW** Web site for a variety of learning tools related to this chapter. Just type in the keyword:

KEYWORD: HSTWPS

GO TO: www.scilinks.org

Visit the **National Science Teachers Association** on-line Web site for Internet resources related to this chapter. Just type in the ***sci*LINKS** number for more information about the topic:

TOPIC: Matter and Energy — ***sci*LINKS NUMBER:** HSTP005
TOPIC: The Scientific Method — ***sci*LINKS NUMBER:** HSTP010
TOPIC: Using Models in Physical Science — ***sci*LINKS NUMBER:** HSTP015
TOPIC: SI Units — ***sci*LINKS NUMBER:** HSTP020

Chapter Review

USING VOCABULARY

For each pair of terms, explain the difference in their meanings.

1. science/technology
2. observation/hypothesis
3. theory/law
4. model/theory
5. volume/mass
6. area/density

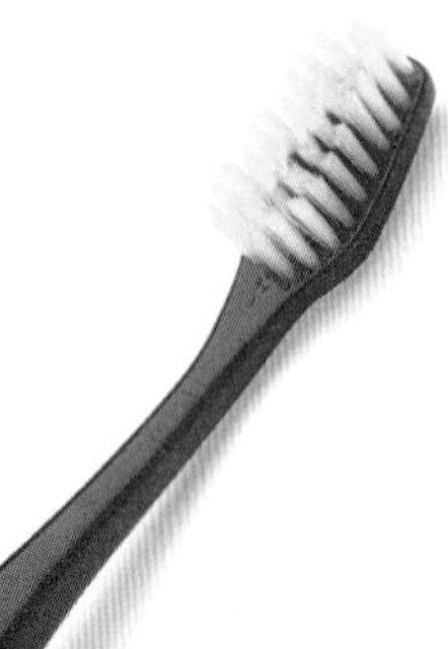

UNDERSTANDING CONCEPTS

Multiple Choice

7. Physical science is the study of
   a. matter and motion.
   b. matter and energy.
   c. energy and motion.
   d. matter and composition.

8. 10 m is equal to
   a. 100 cm.
   b. 1,000 cm.
   c. 10,000 mm.
   d. Both (b) and (c)

9. For a hypothesis to be valid, it must be
   a. testable.
   b. supported by evidence.
   c. made into a law.
   d. Both (a) and (b)

10. The statement "Sheila has a stain on her shirt" is an example of a(n)
    a. law.
    b. hypothesis.
    c. observation.
    d. prediction.

11. A hypothesis is often developed out of
    a. observations.
    b. experiments.
    c. laws.
    d. Both (a) and (b)

12. How many milliliters are in 3.5 kL?
    a. 3,500
    b. 0.0035
    c. 3,500,000
    d. 35,000

13. A map of Seattle is an example of a
    a. law.
    b. quantity.
    c. model.
    d. unit.

14. Which of the following is an example of technology?
    a. mass
    b. physical science
    c. screwdriver
    d. none of the above

Short Answer

15. Name two areas of science other than chemistry and physics, and describe how physical science has a role in those areas of science.

16. Explain why the results of one experiment are never really final results.

17. Explain why area and density are called derived quantities.

18. If a hypothesis is not testable, does that mean that it is wrong? Explain.

Concept Mapping

19. Use the following terms to create a concept map: science, scientific method, hypothesis, problems, questions, experiments, observations.

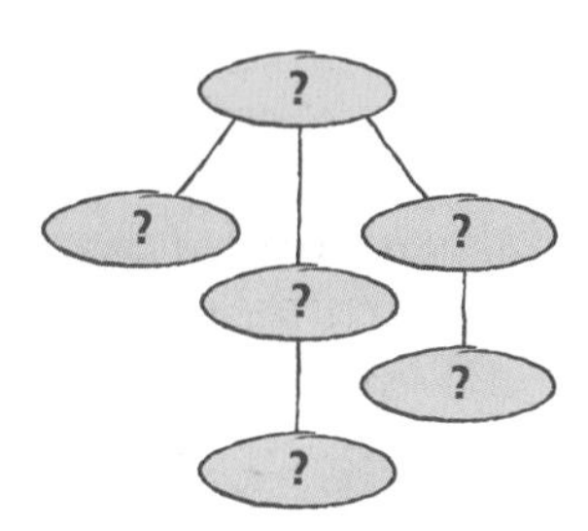

CRITICAL THINKING AND PROBLEM SOLVING

20. A tailor is someone who makes or alters items of clothing. Why might a standard system of measurement be helpful to a tailor?

21. Two classmates are having a debate about whether a spatula is an example of technology. Using what you know about science, technology, and spatulas, write a couple of sentences that will help your classmates settle their debate.

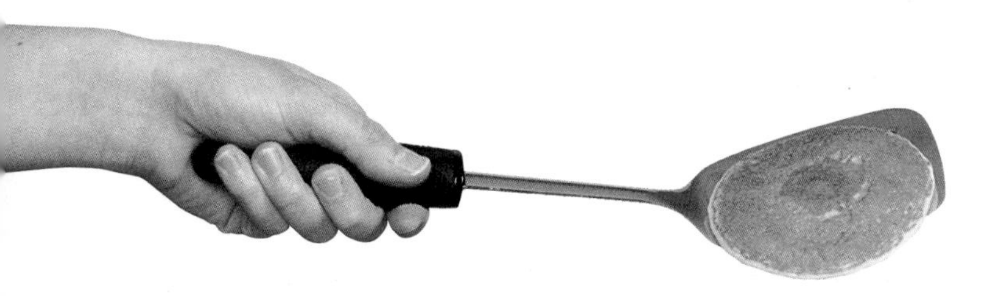

22. Imagine that you are conducting an experiment in which you are testing the effects of the height of a ramp on the speed at which a toy car goes down the ramp. What is the variable in this experiment? What factors must be controlled?

23. Suppose a classmate says, "I don't need to study physical science because I'm not going to be a scientist, and scientists are the only people who use physical science." How would you respond? (Hint: In your answer, give several examples of careers that use physical science.)

MATH IN SCIENCE

24. The cereal box at right has a mass of 340 g. Its dimensions are 27 cm × 19 cm × 6 cm.
    a. What is the volume of the box?
    b. What is its density?
    c. What is the area of the front side of the box?

INTERPRETING GRAPHICS

Examine the picture below, and answer the questions that follow:

25. How similar to the real object is this model?

26. What characteristics of the real object does this model not show?

27. Why might this model be useful?

NOW What Do You Think?

Take a minute to review your answers to the ScienceLog questions on page 5. Have your answers changed? If necessary, revise your answers based on what you have learned since you began this chapter.

CAREERS

ELECTRONICS ENGINEER

Julie Williams-Byrd uses her knowledge of physics to develop better lasers. She started working with lasers as a graduate student at Hampton University, in Virginia. Today Williams-Byrd works as an electronics engineer in the Laser Systems Branch (LSB) of NASA. She designs and builds lasers that are used to monitor phenomena in the atmosphere, such as wind and ozone.

The white light we see every day is actually composed of all the colors of the spectrum. A laser emits a very small portion of this spectrum. That is why there are blue lasers, red lasers, and so on. High-voltage sources called laser "pumps" cause laser materials to emit certain wavelengths of light, depending on the material used. A laser material, such as helium neon gas, emits radiation (light) as a result of changes in the electron energy levels of its atoms. This process gives lasers their name: **L**ight **A**mplification of the **S**timulated **E**mission of **R**adiation.

Using Scientific Models

Julie Williams-Byrd uses scientific models to predict the nature of different aspects of laser design. Different laser materials emit radiation at different wavelengths, and specific pump sources must be used to induce "lasing." "Researchers at LSB use laser models to predict output energy, wavelength, efficiency, and a host of other properties of the laser system," Williams-Byrd says.

New Technologies

Her most challenging project has been building a laser-transmitter that will be used to measure winds in the atmosphere. This system, called *Lidar,* is very much like radar except that it uses light waves instead of sound waves to bounce off objects. To measure winds, a laser beam is transmitted into the atmosphere, where it illuminates particles. A receiver looks at these particles over a period of time and determines the changes in position of the particles. Wind velocity is then determined from this information. This new technology is expected to be used in a space shuttle mission called Sparcle.

Lasers All Around Us

Although Williams-Byrd works with high-tech lasers, she points out that lasers are a part of daily life for many people. For example, lasers are used in scanners at many retail stores. Optometrists use lasers to correct nearsightedness. Some metal workers use them to cut metal. And lasers are even used to create spectacular light shows!

Going Further

▶ Can you think of any new uses for lasers? Make a list in your ScienceLog, and then do some research to find out if any of your ideas already exist.

▼ *Julie Williams-Byrd uses laser generators like this one in her work at NASA.*

Science Fiction

"Inspiration"

by Ben Bova

No matter where you are on the face of the Earth, you can pinpoint your location. Most of the time, you use map coordinates, or latitude and longitude readings. And you can give your distance above sea level. With modern technology, you can give an accurate, three-dimensional description of where you are. But what about a fourth dimension? Consider for a moment traveling in *time.* Not just getting through today and into tomorrow but actually being able to leap back and forth through time.

Novelist H. G. Wells imagined such a possibility in his novelette *The Time Machine.* When the story was published in 1895, most physicists said that the notion of traveling in time was nonsense and against all the laws of physics that govern the universe. The idea that *time* was similar to *length, width,* or *height* was foolishness. Or so they thought.

It was up to Albert Einstein, in 1905, to propose a different view of the universe. However, when Wells's story was first published, Einstein was just 16 and not a very good student. What if Einstein had been discouraged and had not pursued his interest in physics? But Einstein did look at the universe and maybe, just maybe, he had an inspiration.

Ben Bova's story "Inspiration" describes just such a possibility. Young Einstein meets Wells and the great physicist of the time, Lord Kelvin. But was the meeting just a lucky coincidence or something else entirely? Escape to the *Holt Anthology of Science Fiction,* and read "Inspiration" to find out.

CHAPTER 2

The Properties of Matter

Imagine . . .

The year is 1849. You are one of thousands of people who have come to California to prospect for gold. You left home several months ago in the hopes of striking it rich. But so far, no luck. In fact, you've decided that if you don't find gold today, you're going to pack up your things and head back home.

You swing your pickax into the granite bedrock, and a bright flash catches your eye. The flash is caused by a shiny yellow chunk sticking out of the rock. When you first started prospecting, such a sight made you catch your breath. Now you just sigh. More fool's gold, you think.

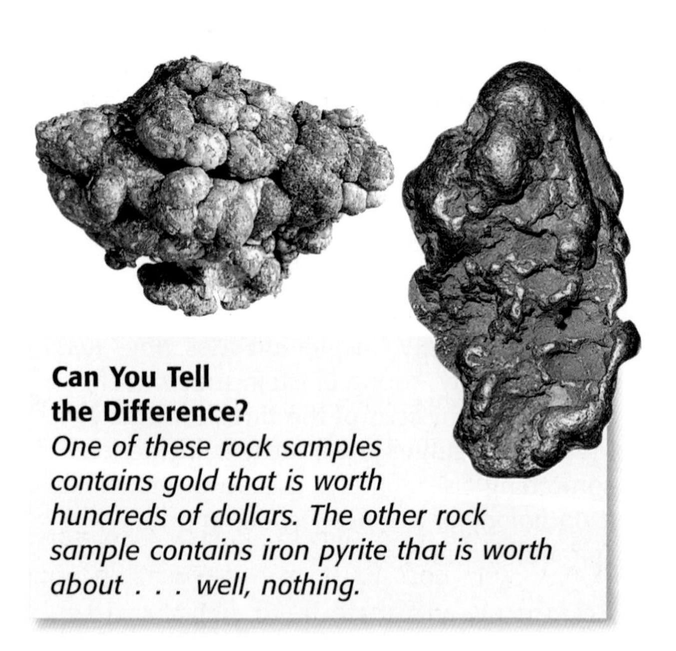

Can You Tell the Difference?
One of these rock samples contains gold that is worth hundreds of dollars. The other rock sample contains iron pyrite that is worth about . . . well, nothing.

Fool's gold is the nickname for iron pyrite (PIE RIET), a mineral that looks like gold and is found in the same areas of California where gold is found. But iron pyrite differs from gold in several ways. When hit with a hammer, iron pyrite shatters into pieces, and sparks fly everywhere. Gold just bends when it is hit, and no sparks are produced. Iron pyrite also produces foul-smelling smoke when it is heated. Gold does not.

You perform a few quick tests on your shiny find. When you hit it with a hammer, it bends but does not shatter, and no sparks are produced. When you heat it, there is no smoke or odor. You start to get excited. You'll have to perform a few more tests when you get back to town, but this time you're almost certain that you've struck gold. Congratulations! Your knowledge of the different characteristics, or properties, of fool's gold and real gold has finally paid off.

In this chapter you'll learn more about the many different properties that objects can have and why these properties are important to know.

What Do You Think?

In your ScienceLog, try to answer the following questions based on what you already know:

1. What is matter?
2. What is the difference between a physical property and a chemical property?
3. What is the difference between a physical change and a chemical change?

Sack Secrets

In this activity, you will test your skills in determining the identity of an object based on its properties.

Procedure

1. You and two or three of your classmates will receive a **sealed paper sack** with a number on it. Write the number in your ScienceLog. Inside the sack is a **mystery object.** Do not open the sack!
2. For 5 minutes, make as many observations as you can about the object. You may shake the sack, touch the object through the sack, listen to the object in the sack, smell the object through the sack, and so on. Be sure to write down your observations.

Analysis

3. At the end of 5 minutes, take a couple of minutes to discuss your findings with your partners.
4. With your partners, list the object's properties, and make a conclusion about the object's identity. Write your conclusion in your ScienceLog.
5. Share your observations, list of properties, and conclusion with the class. Now you are ready to open the sack.
6. Did you properly identify the object? If so, how? If not, why not? Write your answers in your ScienceLog, and share them with the class.

Section 1

What Is Matter?

NEW TERMS

matter
volume
meniscus
mass
gravity
weight
newton
inertia

OBJECTIVES

- Name the two properties of all matter.
- Describe how volume and mass are measured.
- Compare mass and weight.
- Explain the relationship between mass and inertia.

Here's a strange question: What do you have in common with a toaster?

Do you give up? Okay, here's another question: What do you have in common with a steaming bowl of soup or a bright neon sign?

You are probably thinking these are trick questions. After all, it is hard to imagine that a human—you—has anything in common with a kitchen appliance, some hot soup, or a glowing neon sign.

From a scientific point of view, however, you have at least one characteristic in common with these things. You, the toaster, the bowl, the soup, the steam, the glass tubing, and the glowing gas are all made of matter. In fact, everything in the universe that you can touch (even if you cannot see it) is made of matter. For example, DNA, microscopic bacteria, and even air are all made of matter. But what is matter exactly? If so many different kinds of things are made of matter, you might expect the definition of the word *matter* to be complicated. But it is really quite simple. **Matter** is anything that has volume and mass.

*Quick*Lab

Space Case

1. Crumple a **piece of paper,** and fit it tightly in the bottom of a **cup** so that it won't fall out.
2. Turn the cup upside down. Lower the cup straight down into a **large beaker or bucket** half-filled with **water** until the cup is all the way underwater.
3. Lift the cup straight out of the water. Turn the cup upright and observe the paper. Record your observations in your ScienceLog.
4. Now punch a small hole in the bottom of the cup with the point of a **pencil.** Repeat steps 2 and 3.
5. How do these results show that air has volume? Record your explanation in your ScienceLog.

Matter Has Volume

All matter takes up space. The amount of space taken up, or occupied, by an object is known as the object's **volume.** The sun, shown in **Figure 1,** has volume because it takes up space at the center of our solar system. Your fingernails have volume because they occupy space at the end of your hands. (The less you bite them, the more volume they have!) Likewise, the Statue of Liberty, the continent of Africa, and a cloud all have volume. And because these things have volume, they cannot share the same space at the same time. Even the tiniest speck of dust takes up space, and there's no way another speck of dust can fit into that space without somehow bumping the first speck out of the way. Try the QuickLab on this page to see for yourself that matter takes up space—even matter you can't see.

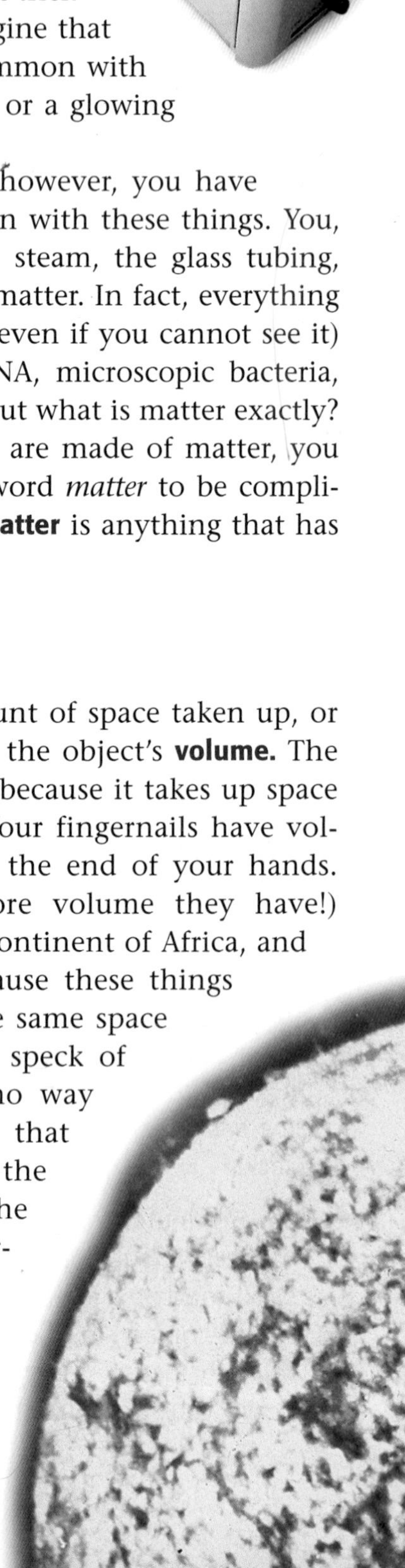

Figure 1 *The volume of the sun is about 1,000,000 (1 million) times larger than the volume of the Earth.*

Liquid Volume Locate the Great Lakes on a map of the United States. Lake Erie, the smallest of the Great Lakes, has a volume of approximately 483,000,000,000,000 (483 trillion) liters of water. Can you imagine that much liquid? Well, think of a 2 liter bottle of soda. The water in Lake Erie could fill more than 241 trillion of those bottles. That's a lot of water! On a smaller scale, a can of soda has a volume of only 355 milliliters, which is approximately one-third of a liter. The next time you see a can of soda, you can read the volume printed on the can. Or you can check its volume by pouring the soda into a large measuring cup from your kitchen, as shown in **Figure 2,** and reading the scale at the level of the liquid's surface.

Figure 2 *If the measurement is accurate, the volume measured should be the same as the volume printed on the can.*

Measuring the Volume of Liquids In your science class, you'll probably use a graduated cylinder to measure the volume of liquids. Keep in mind that the surface of a liquid in a graduated cylinder is not flat. The curve that you see at the liquid's surface has a special name—the **meniscus** (muh NIS kuhs). When you measure the volume of a liquid, you must look at the bottom of the meniscus, as shown in **Figure 3.** (A liquid in any container, including a measuring cup or a large beaker, has a meniscus. The meniscus is just too flat to see in a wider container.)

Liters (L) and milliliters (mL) are the units used most often to express the volume of liquids. The volume of any amount of liquid, from one raindrop to a can of soda to an entire ocean, can be expressed in these units.

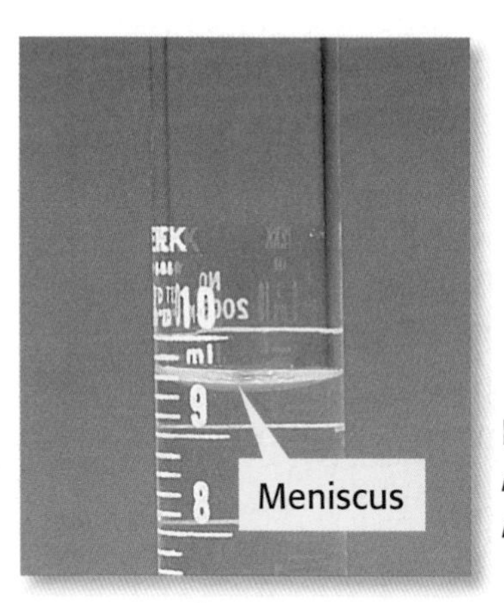

Figure 3 *To measure volume correctly, read the scale at the lowest part of the meniscus (as indicated) at eye level.*

Measuring the Volume of Solids What would you do if you wanted to measure the volume of this textbook? You cannot pour this textbook into a graduated cylinder to find the answer (sorry, no shredders allowed!). In the MathBreak activity on the next page, you will learn an easy way to find the volume of any solid object with rectangular sides.

The volume of a typical raindrop is approximately 0.09 mL, which means that it would take almost 4,000 raindrops to fill a soda can.

The volume of any solid object, from a speck of dust to the tallest skyscraper, is expressed in cubic units. The term *cubic* means "having three dimensions." (A *dimension* is simply a measurement in one direction.) The three dimensions that are used to find volume are length, width, and height, as shown in **Figure 4.**

Figure 4 *A cubic meter has a height of 1 m, a length of 1 m, and a width of 1 m, so its volume is 1 m × 1 m × 1 m = 1 m³.*

Cubic meters (m^3) and cubic centimeters (cm^3) are the units most often used to express the volume of solid items. In the unit abbreviations m^3 and cm^3, the *3* to the upper right of the unit shows that the final number is the result of multiplying three quantities of that unit.

You now know that the volumes of solids and liquids are expressed in different units. So how can you compare the volume of a solid with the volume of a liquid? For example, suppose you are interested in determining whether the volume of an ice cube is equal to the volume of water that is left when the ice cube melts. Well, lucky for you, 1 mL is equal to 1 cm^3. Therefore, you can express the volume of the water in cubic centimeters and compare it with the original volume of the solid ice cube. The volume of any liquid can be expressed in cubic units in this way. (However, keep in mind that in SI, volumes of solids are never expressed in liters or milliliters.)

Measuring the Volume of Gases How do you measure the volume of a gas? You can't hold a ruler up to a gas to measure its dimensions. You can't pour a gas into a graduated cylinder. So it's impossible, right? Wrong! A gas expands to fill its container. If you know the volume of the container that a gas is in, then you know the volume of the gas.

Matter Has Mass

Another characteristic of all matter is mass. **Mass** is the amount of matter in a given substance. For example, the Earth contains a very large amount of matter and therefore has a large mass. A peanut has a much smaller amount of matter and thus has a smaller mass. Remember, even something as small as a speck of dust is made of matter and therefore has mass.

MATH BREAK

Calculating Volume

A typical compact disc (CD) case has a length of 14.2 cm, a width of 12.4 cm, and a height of 1.0 cm. The volume of the case is the length multiplied by the width multiplied by the height:

$$14.2 \text{ cm} \times 12.4 \text{ cm} \times 1.0 \text{ cm} = 176.1 \text{ cm}^3$$

Now It's Your Turn

1. A book has a length of 25 cm, a width of 18 cm, and a height of 4 cm. What is its volume?
2. What is the volume of a suitcase with a length of 95 cm, a width of 50 cm, and a height of 20 cm?
3. For additional practice, find the volume of other objects that have square or rectangular sides. Compare your results with those of your classmates.

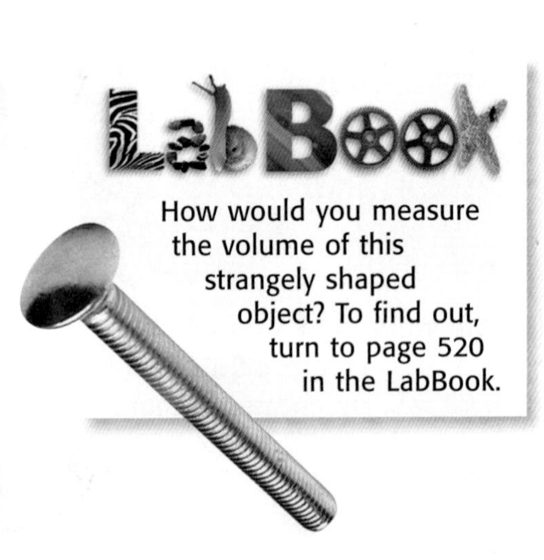

How would you measure the volume of this strangely shaped object? To find out, turn to page 520 in the LabBook.

An object's mass can be changed only by changing the amount of matter in the object. Consider the bowling ball shown in **Figure 5.** Its mass is constant because the amount of matter in the bowling ball never changes (unless you use a sledgehammer to remove a chunk of it!). Now consider the puppy. Does its mass remain constant? No, because the puppy is growing. If you measured the puppy's mass next year or even next week, you'd find that it had increased. That's because more matter—more puppy—would be present.

Figure 5 *The mass of the bowling ball does not change. The mass of the puppy increases as more matter is added—that is, as the puppy grows.*

The Difference Between Mass and Weight

Weight is different from mass. To understand this difference, you must first understand gravity. **Gravity** is a force of attraction between objects that is due to their masses. This attraction causes objects to experience a "pull" toward other objects. Because all matter has mass, all matter experiences gravity. The amount of attraction objects experience toward each other depends on two things—the masses of the objects and the distance between them, as shown in **Figure 6.**

Figure 6 How Mass and Distance Affect Gravity Between Objects

a Gravitational force (represented by the width of the arrows) is large between objects with large masses that are close together.

b Gravitational force is smaller between objects with smaller masses that are close together than between objects with large masses that are close together (as shown in **a**).

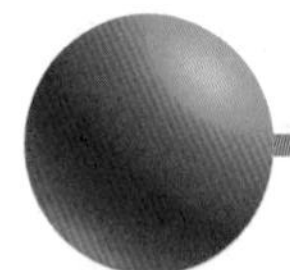

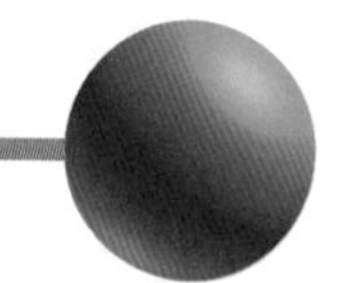

c An increase in distance reduces gravitational force between two objects. Therefore, gravitational force between objects with large masses (such as those in **a**) is less if they are far apart.

Explore

Imagine the following items resting side by side on a table: an elephant, a tennis ball, a peanut, a bowling ball, and a housefly. In your ScienceLog, list these items in order of their attraction to the Earth due to gravity, from least to greatest amount of attraction. Follow your list with an explanation of why you arranged the items in the order that you did.

The mineral calcium is stored in bones, and it accounts for about 70 percent of the mass of the human skeleton. Calcium strengthens bones, helping the skeleton to remain upright against the strong force of gravity pulling it toward the Earth.

May the Force Be with You Gravitational force is experienced by all objects in the universe all the time. But the ordinary objects you see every day have masses so small (relative to, say, planets) that their attraction toward each other is hard to detect. Therefore, the gravitational force experienced by objects with small masses is very slight. However, the Earth's mass is so large that the attraction of other objects to it is great. Therefore, gravitational force between objects and the Earth is great. In fact, the Earth is so massive that our atmosphere, satellites, the space shuttle, and even the moon experience a strong attraction toward the Earth. Gravity is what keeps you and everything else on Earth from floating into space.

So What About Weight? **Weight** is simply a measure of the gravitational force on an object. Consider the brick in **Figure 7.** The brick has mass. The Earth also has mass. Therefore, the brick and the Earth are attracted to each other. A force is exerted on the brick because of its attraction to the Earth. The weight of the brick is a measure of this gravitational force.

Now look at the sponge in Figure 7. The sponge is the same size as the brick, but its mass is much less. Therefore, the sponge's attraction toward the Earth is not as great, and the gravitational force on it is not as great. Thus, the *weight* of the sponge is less than the *weight* of the brick.

Because the attraction that objects experience decreases as the distance between them increases, the gravitational force on objects—and therefore their weight—also decreases as the distance increases. For this reason, a brick floating in space would weigh less than it does resting on Earth's surface. However, the brick's mass would be the same in space as it is on Earth.

Figure 7
This brick and sponge may be the same size, but their masses, and therefore their weights, are quite different.

Massive Confusion Back on Earth, the gravitational force exerted on an object is about the same everywhere, so an object's weight is also about the same everywhere. Because mass and weight remain constant everywhere on Earth, the terms *mass* and *weight* are often used as though they mean the same thing. But using the terms interchangeably can lead to confusion, especially if you are trying to measure these properties of an object. So remember, weight depends on mass, but weight is not the same thing as mass.

Some scientists think that there are WIMPs in space. Find out more on page 56.

Measuring Mass and Weight

The SI unit of mass is the kilogram (kg), but mass is often expressed in grams (g) and milligrams (mg) as well. These units can be used to express the mass of any object, from a single cell in your body to the entire solar system. Weight is a measure of gravitational force and must be expressed in units of force. The SI unit of force is the **newton (N).** So weight is expressed in newtons.

A newton is approximately equal to the weight of a 100 g mass on Earth. So if you know the mass of an object, you can calculate its weight on Earth. Conversely, if you know the weight of an object on Earth, you can determine its mass. **Figure 8** summarizes the differences between mass and weight.

Figure 8 Differences Between Mass and Weight

Mass is . . .

- a measure of the amount of matter in an object.
- always constant for an object no matter where the object is in the universe.
- measured with a balance (shown below).
- expressed in kilograms (kg), grams (g), and milligrams (mg).

Weight is . . .

- a measure of the gravitational force on an object.
- varied depending on where the object is in relation to the Earth (or any other large body in the universe).
- measured with a spring scale (shown above).
- expressed in newtons (N).

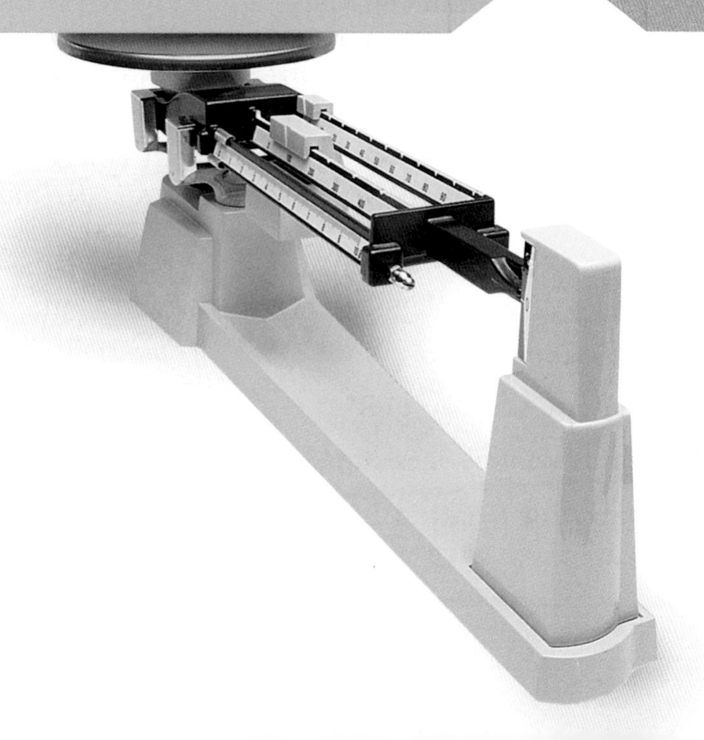

Self-Check

If all of your school books combined have a mass of 3 kg, what is their total weight in newtons? Remember that 1 kg = 1,000 g. *(See page 596 to check your answer.)*

Ordinary bathroom scales are spring scales. Many scales available today show a reading in both pounds (a common though not SI unit of weight) and kilograms. How does such a reading contribute to the confusion between mass and weight?

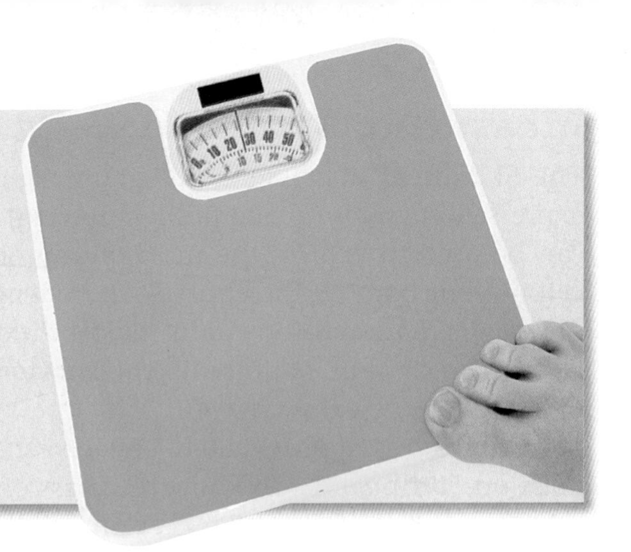

Mass Is a Measure of Inertia

Which do you think would be easier to pick up and throw, a soccer ball or a bowling ball? Well, you could probably throw the soccer ball clear across your backyard, but the bowling ball would probably not go very far. What's the difference? The difference has to do with inertia (in UHR shuh). **Inertia** is the tendency of all objects to resist any change in motion. Because of inertia, an object at rest (like the soccer ball or the bowling ball) will remain at rest until something causes it to move. Likewise, a moving object continues to move at the same speed and in the same direction unless something acts on it to change its speed or direction.

So why do we say that mass is a measure of inertia? Well, think about this: An object with a large mass is harder to start in motion and harder to stop than an object with a smaller mass. This is because the object with the large mass has greater inertia. For example, imagine that you are going to push a grocery cart that has only one potato in it. No problem, right? But suppose the grocery cart is filled with potatoes, as in **Figure 9.** Now the total mass—and the inertia—of the cart full of potatoes is much greater. It will be harder to get the cart moving and harder to stop it once it is moving. So an object with a large mass has greater inertia than an object with a smaller mass.

Figure 9 *Why is a cartload of potatoes harder to get moving than a single potato? Because of inertia, that's why!*

REVIEW

1. What are the two properties of all matter?
2. How is volume measured? How is mass measured?
3. **Analyzing Relationships** Do objects with large masses always have large weights? Explain your reasoning.

Section 2

Describing Matter

NEW TERMS

physical property
density
chemical property
physical change
chemical change

OBJECTIVES

- Give examples of matter's different properties.
- Describe how density is used to identify different substances.
- Compare physical and chemical properties.
- Explain what happens to matter during physical and chemical changes.

Have you ever heard of the game called "20 Questions"? In this game, your goal is to determine the identity of an object that another person is thinking of by asking questions about the object. The other person can respond with only a "yes" or "no." If you can identify the object after asking 20 or fewer questions, you win! If you still can't figure out the object's identity after asking 20 questions, you may not be asking the right kinds of questions.

What kinds of questions should you ask? You might find it helpful to ask questions about the properties of the object. Knowing the properties of an object can help you determine the object's identity, as shown below.

Explore

With a partner, play a game of 20 Questions. One person will think of an object, and the other person will ask yes/no questions about it. Write the questions in your ScienceLog as you go along. Put a check mark next to the questions asked about physical properties. When the object is identified or when the 20 questions are up, switch roles. Good luck!

Physical Properties

Some of the questions shown above help the asker gather information about *color* (Is it orange?), *odor* (Does it have an odor?), and *mass* and *volume* (Could I hold it in my hand?). Each of these properties is a physical property of matter. A **physical property** of matter can be observed or measured without changing the identity of the matter. For example, you don't have to change what the apple is made of to see that it is red or to hold it in your hand.

You rely on physical properties all the time. For example, physical properties help you determine whether your socks are clean (odor), whether you can fit all your books into your backpack (volume), or whether your shirt matches your pants (color). The table below lists some more physical properties that are useful in describing or identifying matter.

More Physical Properties

Physical property	Definition	Example
Thermal conductivity	The ability to transfer thermal energy from one area to another	Plastic foam is a poor conductor, so hot chocolate in a plastic-foam cup will not burn your hand.
State	The physical form in which a substance exists, such as a solid, liquid, or gas	Ice is water in its solid state.
Malleability (MAL ee uh BIL uh tee)	The ability to be pounded into thin sheets	Aluminum can be rolled or pounded into sheets to make foil.
Ductility (duhk TIL uh tee)	The ability to be drawn or pulled into a wire	Copper is often used to make wiring.
Solubility (SAHL yoo BIL uh tee)	The ability to dissolve in another substance	Sugar dissolves in water.
Density	Mass per unit volume	Lead is used to make sinkers for fishing line because lead is more dense than water.

Spotlight on Density Density is a very helpful property when you need to distinguish different substances. There are some interesting things you should know about density. Look at the definition of density in the table above—mass per unit volume. If you think back to what you learned in Section 1, you can define density in other terms: **density** is the amount of matter in a given volume, as shown in **Figure 10.**

Figure 10 *A golf ball is more dense than a table-tennis ball because the golf ball contains more matter in a similar volume.*

To find an object's density (D), first measure its mass (m) and volume (V). Then use the following equation:

$$D = \frac{m}{V}$$

Units for density are expressed using a mass unit divided by a volume unit, such as g/cm^3, g/mL, kg/m^3, and kg/L.

Using Density to Identify Substances Density is a useful property for identifying substances for two reasons. First, the density of a particular substance is always the same at a given pressure and temperature. For example, the helium in a huge airship has a density of 0.0001663 g/cm^3 at 20°C and normal atmospheric pressure. You can calculate the density of any other sample of helium at that same temperature and pressure—even the helium in a small balloon—and you will get 0.0001663 g/cm^3. Second, the density of one substance is usually different from that of another substance. Check out the table below to see how density varies among substances.

Densities of Common Substances*

Substance	Density (g/cm^3)	Substance	Density (g/cm^3)
Helium (gas)	0.0001663	Copper (solid)	8.96
Oxygen (gas)	0.001331	Silver (solid)	10.50
Water (liquid)	1.00	Lead (solid)	11.35
Iron pyrite (solid)	5.02	Mercury (liquid)	13.55
Zinc (solid)	7.13	Gold (solid)	19.32

** at 20°C and normal atmospheric pressure*

Mass = 96.6 g
Volume = 5.0 cm^3

Figure 11 *Did you find gold or fool's gold?*

Do you remember your imaginary attempt at gold prospecting? To make sure you hadn't found more fool's gold (iron pyrite), you could compare the density of a nugget from your sample, shown in **Figure 11,** with the known densities for gold and iron pyrite at the same temperature and pressure. By comparing densities, you'd know whether you'd actually struck gold or been fooled again.

MATH BREAK

Density

You can rearrange the equation for density to find mass and volume as shown below:

$$D = \frac{m}{V}$$

$$m = D \times V \qquad V = \frac{m}{D}$$

1. Find the density of a substance with a mass of 5 kg and a volume of 43 m^3.
2. Suppose you have a lead ball with a mass of 454 g. What is its volume? (Hint: Use the table at left.)
3. What is the mass of a 15 mL sample of mercury? (Hint: Use the table at left.)

BRAIN FOOD

Pennies minted before 1982 are made mostly of copper and have a density of 8.85 g/cm^3. In 1982, a penny's worth of copper began to cost more than one cent, so the U.S. Department of the Treasury began producing pennies using mostly zinc with a copper coating. Pennies minted after 1982 have a density of 7.14 g/cm^3. Check it out for yourself!

Figure 12 *The yellow liquid is the least dense, and the green liquid is the densest.*

Liquid Layers What do you think causes the liquid in **Figure 12** to look like it does? Is it magic? Is it trick photography? No, it's differences in density! There are actually four different liquids in the jar. Each liquid has a different density. Because of these differences in density, the liquids do not mix together but instead separate into layers, with the densest layer on the bottom and the least dense layer on top. The order in which the layers separate helps you determine how the densities of the liquids compare with one another.

The Density Challenge Imagine that you could put a lid on the jar in the picture and shake up the liquids. Would the different liquids mix together so that the four colors would blend into one interesting color? Maybe for a minute or two. But if the liquids are not soluble in one another, they would start to separate, and eventually you'd end up with the same four layers.

The same thing happens when you mix oil and vinegar to make salad dressing. But what do you think would happen if you added more oil? What if you added so much oil that there was several times as much oil as there was vinegar? Surely the oil would get so heavy that it would sink below the vinegar, right? Wrong! No matter how much oil you have, it will always be less dense than the vinegar, so it will always rise to the top. The same is true of the four liquids shown in Figure 12. Even if you add more yellow liquid than all of the other liquids combined, all of the yellow liquid will rise to the top. That's because density does not depend on how much of a substance you have.

Experiment for yourself with liquid layers on page 523 in the LabBook.

APPLY

The grease separator shown here is a kitchen device that cooks use to collect the best meat juices for making gravies. Based on what you know about density, describe how a grease separator works. Be sure to explain why the spout is at the bottom.

REVIEW

1. List three physical properties of water.
2. Why does a golf ball feel heavier than a table-tennis ball?
3. Describe how you can determine the relative densities of liquids.
4. **Applying Concepts** How could you determine that a coin is not pure silver?

Chemical Properties

Physical properties such as density, color, and mass are not the only properties that describe matter. **Chemical properties** describe a substance based on its ability to change into a new substance with different properties. For example, a piece of wood can be burned to create new substances (ash and smoke) with very different properties from the original piece of wood. Therefore, wood has the chemical property of *flammability*—the ability to burn. A substance that does not burn, such as gold, has the chemical property of nonflammability. Other common chemical properties include reactivity with oxygen, reactivity with acid, and reactivity with water. (The word *reactivity* just means that when two substances get together, something can happen.)

Like physical properties, chemical properties can be observed with your senses. However, chemical properties aren't as easy to observe. For example, you can observe the flammability of wood only while the wood is burning. Likewise, you can observe the nonflammability of gold only when you try to burn it and it won't burn. But a substance always has its chemical properties, even when you are not observing them. Therefore, a piece of wood is flammable even when it's not burning.

Some Chemical Properties of Car Maintenance Look at the old car shown in **Figure 13.** Its owner calls it Rust Bucket. Why has this car rusted so badly while some other cars the same age remain in great shape? Knowing about chemical properties can help answer this question.

Most car bodies are made from steel, which consists mostly of iron. Iron has many favorable physical properties, including strength, malleability, hardness, and a high melting point. Iron also has many favorable chemical properties, including nonreactivity with oil and gasoline. All in all, steel is a good material to use for car bodies. It's not perfect, however, as you can probably tell from the car shown here.

Figure 13 Rust Bucket *One unfavorable chemical property of iron is its reactivity with oxygen. When iron is exposed to oxygen, it rusts. If left unprotected, the iron will eventually rust away.*

Physical vs. Chemical Properties

Figure 14 *Substances have different physical and chemical properties.*

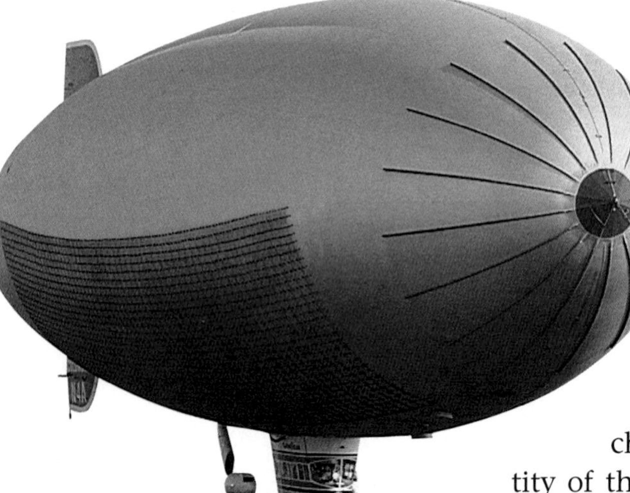

a Helium is used in airships because it is less dense than air and is nonflammable.

b If you add bleach to water that is mixed with red food coloring, the red color will disappear.

You can describe matter by both physical and chemical properties. The properties that are most useful in identifying a substance, such as density, solubility, and reactivity with acids, are its characteristic properties. The *characteristic properties* of a substance are always the same whether the sample you're observing is large or small. Scientists rely on characteristic properties to distinguish substances and to separate them from one another. **Figure 14** describes some physical and chemical properties.

It is important to remember the differences between physical and chemical properties. You can observe physical properties without changing the identity of the substance. You can observe chemical properties only in situations in which the identity of the substance could change. The table below can help you understand the distinction between physical and chemical properties.

Comparing Physical and Chemical Properties

Substance	Physical property	Chemical property
Helium	less dense than air	nonflammable
Wood	grainy texture	flammable
Baking soda	white powder	reacts with vinegar to produce bubbles
Powdered sugar	white powder	does not react with vinegar
Rubbing alcohol	clear liquid	flammable
Red food coloring	red color	reacts with bleach and loses color
Iron	malleable	reacts with oxygen
Tin	malleable	reacts with oxygen

Physical Changes Don't Form New Substances

A **physical change** is a change that affects one or more physical properties of a substance. For example, if you were to break a piece of chalk in two, you would be changing its physical properties of size and shape. But no matter how many times you break it, the chalk is still chalk. In other words, the chemical properties of the chalk remain unchanged. Each piece of chalk would still produce bubbles if you placed it in vinegar.

BRAIN FOOD

Bending a bar of tin produces a squealing sound known as a tin cry.

Melting is another example of a physical change, as you can see in **Figure 15.** Still another physical change occurs when a substance dissolves into another substance. If you dissolve sugar in water, the sugar seems to disappear into the water. But the identity of the sugar does not change. If you taste the water, you will notice that the sugar is still there. It has just undergone a physical change. See the chart below for some more examples of physical changes.

Figure 15 *It took a physical change to turn a stick of butter into the liquid butter that makes popcorn so tasty, but the identity of the butter did not change.*

More Examples of Physical Changes	
■ Freezing water for ice cubes	■ Crushing an aluminum can
■ Sanding a piece of wood	■ Bending a paper clip
■ Cutting your hair	■ Mixing oil and vinegar

Can Physical Changes Be Undone? Because physical changes do not change the identity of substances, they are often easy to undo. If you leave butter out on a warm counter, it will undergo a physical change—it will melt. Putting it back in the refrigerator will reverse this change by making it solid again. Likewise, if you create a figure, such as a dragon or a person, from a lump of clay, you drastically change the clay's shape, causing a physical change. But because the identity of the clay does not change, you can crush your creation and form the clay back into the shape it was in before.

Chemical Changes Form New Substances

A **chemical change** occurs when one or more substances are changed into entirely new substances. The new substances have a different set of properties from the original substances. Chemical changes will or will not occur as described by the chemical properties of substances. But don't confuse chemical changes with chemical properties—they are not the same thing. A chemical property describes a substance's ability to go through a chemical change; a chemical change is the actual process in which that substance changes into another substance. You can observe chemical properties only when a chemical change might occur. Try the QuickLab on this page to learn more about chemical changes.

Quick Lab

Changing Change

1. Place a folded **paper towel** in a **small pie plate.**
2. Pour **vinegar** into the pie plate until the entire paper towel is damp.
3. Place **two or three shiny pennies** on top of the paper towel.
4. Put the pie plate in a place where it won't be bothered, and wait 24 hours.
5. Describe the chemical change that took place.
6. Write your observations in your ScienceLog.

Figure 16 *Each of these ingredients has different physical and chemical properties.*

A fun (and delicious) way to see what happens during chemical changes is to bake a cake. When you bake a cake, you combine eggs, flour, sugar, butter, and other ingredients as shown in **Figure 16.** Each ingredient has its own set of properties. But if you mix them together and bake the batter in the oven, you get something completely different. The heat of the oven and the interaction of the ingredients cause a chemical change. As shown in **Figure 17,** you get a cake that has completely different properties than any of the ingredients. Some more examples of chemical changes are shown below.

Figure 17 *Chemical changes produce new substances with different properties.*

Examples of Chemical Changes

Soured milk smells bad because bacteria have formed new substances in the milk.

Effervescent tablets bubble when the citric acid and baking soda in them react in water.

The hot gas formed when hydrogen and oxygen join to make water helps blast the space shuttle into orbit.

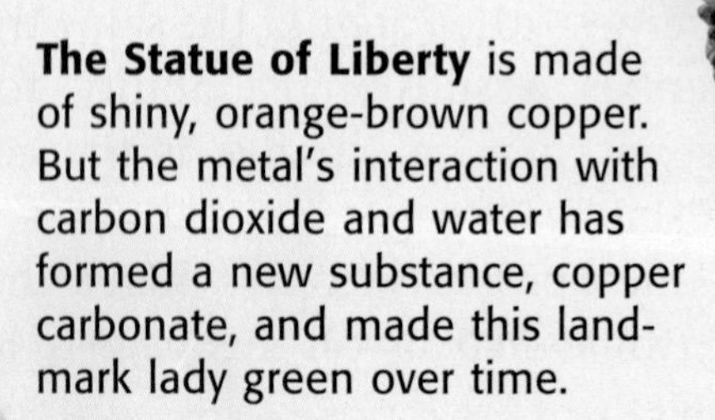

The Statue of Liberty is made of shiny, orange-brown copper. But the metal's interaction with carbon dioxide and water has formed a new substance, copper carbonate, and made this landmark lady green over time.

Clues to Chemical Changes Look back at the bottom of the previous page. In each picture, there is at least one clue that signals a chemical change. Can you find the clues? Here's a hint: chemical changes often cause color changes, fizzing or foaming, heat, or the production of sound, light, or odor.

In the cake example, you would probably smell the sweet aroma of the cake as it baked. If you looked into the oven, you would see the batter rise and turn brown. When you cut the finished cake, you would see the spongy texture created by gas bubbles that formed in the batter (if you baked it right, that is!). All of these yummy clues are signals of chemical changes. But are the clues and the chemical changes the same thing? No, the clues just result from the chemical changes.

Can Chemical Changes Be Undone? Because new substances are formed, you cannot reverse chemical changes using physical means. In other words, you can't uncrumple or iron out a chemical change. Imagine trying to un-bake the cake shown in **Figure 18** by pulling out each ingredient. No way! Most of the chemical changes in your daily life, such as a cake baking or milk turning sour, would be difficult to reverse. However, some chemical changes can be reversed under the right conditions by other chemical changes. For example, the water formed in the space shuttle's rockets could be split back into hydrogen and oxygen using an electric current.

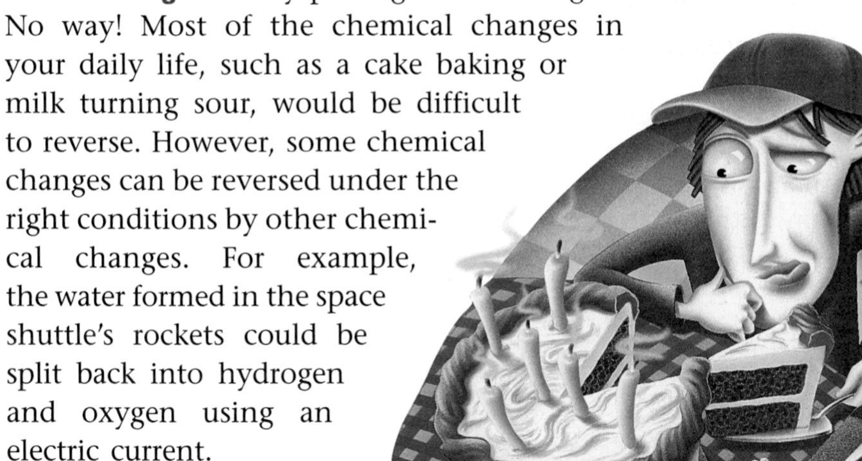

Figure 18 *Looking for the original ingredients? You won't find them—their identities have changed.*

environmental science CONNECTION

When fossil fuels are burned, a chemical change takes place involving sulfur (a substance in fossil fuels) and oxygen (from the air). This chemical change produces sulfur dioxide, a gas. When sulfur dioxide enters the atmosphere, it undergoes another chemical change by interacting with water and oxygen. This chemical change produces sulfuric acid, a contributor to acid precipitation. Acid precipitation can kill trees and make ponds and lakes unable to support life.

REVIEW

1. Classify each of the following properties as either physical or chemical: reacts with water, dissolves in acetone, is blue, does not react with hydrogen.
2. List three clues that a chemical change might be taking place.
3. **Comparing Concepts** Describe the difference between physical changes and chemical changes in terms of what happens to the matter involved in each kind of change.

Chapter Highlights

SECTION 1

Vocabulary

matter *(p. 36)*
volume *(p. 36)*
meniscus *(p. 37)*
mass *(p. 38)*
gravity *(p. 39)*
weight *(p. 40)*
newton *(p. 41)*
inertia *(p. 42)*

Section Notes

- Matter is anything that has volume and mass.
- Volume is the amount of space taken up by an object.
- The volume of liquids is expressed in liters and milliliters.
- The volume of solid objects is expressed in cubic units, such as cubic meters.
- Mass is the amount of matter in an object.
- Mass and weight are not the same thing. Weight is a measure of the gravitational force on an object, usually in relation to the Earth.
- Mass is usually expressed in milligrams, grams, and kilograms.
- The newton is the SI unit of force, so weight is expressed in newtons.
- Inertia is the tendency of all objects to resist any change in motion. Mass is a measure of inertia. The more massive an object is, the greater its inertia.

Labs

Volumania! *(p. 520)*

☑ Skills Check

Math Concepts

DENSITY To calculate an object's density, divide the mass of the object by its volume. For example, the density of an object with a mass of 45 g and a volume of 5.5 cm³ is calculated as follows:

$$D = \frac{m}{V}$$

$$D = \frac{45 \text{ g}}{5.5 \text{ cm}^3}$$

$$D = 8.2 \text{ g/cm}^3$$

Visual Understanding

MASS AND WEIGHT Mass and weight are related, but they're not the same thing. Look back at Figure 8 on page 41 to learn about the differences between mass and weight.

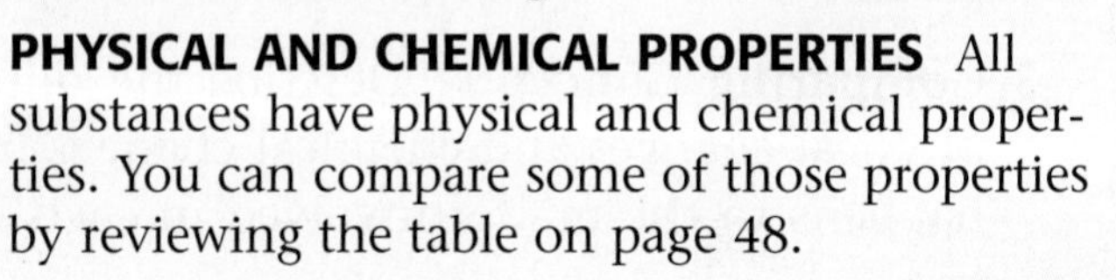

PHYSICAL AND CHEMICAL PROPERTIES All substances have physical and chemical properties. You can compare some of those properties by reviewing the table on page 48.

SECTION 2

Vocabulary

physical property *(p. 43)*
density *(p. 44)*
chemical property *(p. 47)*
physical change *(p. 48)*
chemical change *(p. 49)*

Section Notes

- Physical properties of matter can be observed without changing the identity of the matter.
- The density (mass per unit volume) of a substance is always the same at a given pressure and temperature regardless of the size of the sample of the substance.
- Chemical properties describe a substance based on its ability to change into a new substance with different properties.
- Chemical properties can be observed only when one substance might become a new substance.
- The characteristic properties of a substance are always the same whether the sample observed is large or small.
- When a substance undergoes a physical change, its identity remains the same.
- A chemical change occurs when one or more substances are changed into new substances with different properties.

Labs

Determining Density *(p. 522)*
Layering Liquids *(p. 523)*
White Before Your Eyes *(p. 524)*

internet connect

GO TO: go.hrw.com

Visit the **HRW** Web site for a variety of learning tools related to this chapter. Just type in the keyword:

KEYWORD: HSTMAT

GO TO: www.scilinks.org

Visit the **National Science Teachers Association** on-line Web site for Internet resources related to this chapter. Just type in the ***sci*LINKS** number for more information about the topic:

TOPIC: What Is Matter? ***sci*LINKS NUMBER:** HSTP030
TOPIC: Describing Matter ***sci*LINKS NUMBER:** HSTP035
TOPIC: Dark Matter ***sci*LINKS NUMBER:** HSTP040
TOPIC: Building a Better Body ***sci*LINKS NUMBER:** HSTP045

Chapter Review

USING VOCABULARY

For each pair of terms, explain the difference in their meanings.

1. mass/volume
2. mass/weight
3. inertia/mass
4. volume/density
5. physical property/chemical property
6. physical change/chemical change

UNDERSTANDING CONCEPTS

Multiple Choice

7. Which of these is *not* matter?
   a. a cloud
   b. your hair
   c. sunshine
   d. the sun

8. The mass of an elephant on the moon would be
   a. less than its mass on Mars.
   b. more than its mass on Mars.
   c. the same as its weight on the moon.
   d. None of the above

9. Which of the following is *not* a chemical property?
   a. reactivity with oxygen
   b. malleability
   c. flammability
   d. reactivity with acid

10. Your weight could be expressed in which of the following units?
    a. pounds
    b. newtons
    c. kilograms
    d. Both (a) and (b)

11. You accidentally break your pencil in half. This is an example of
    a. a physical change.
    b. a chemical change.
    c. density.
    d. volume.

12. Which of the following statements about density is true?
    a. Density depends on mass and volume.
    b. Density is weight per unit volume.
    c. Density is measured in milliliters.
    d. Density is a chemical property.

13. Which of the following pairs of objects would have the greatest attraction toward each other due to gravity?
    a. a 10 kg object and a 10 kg object, 4 m apart
    b. a 5 kg object and a 5 kg object, 4 m apart
    c. a 10 kg object and a 10 kg object, 2 m apart
    d. a 5 kg object and a 5 kg object, 2 m apart

14. Inertia increases as __?__ increases.
    a. time
    b. length
    c. mass
    d. volume

Short Answer

15. In one or two sentences, explain the different processes in measuring the volume of a liquid and measuring the volume of a solid.

16. In one or two sentences, explain the relationship between mass and inertia.

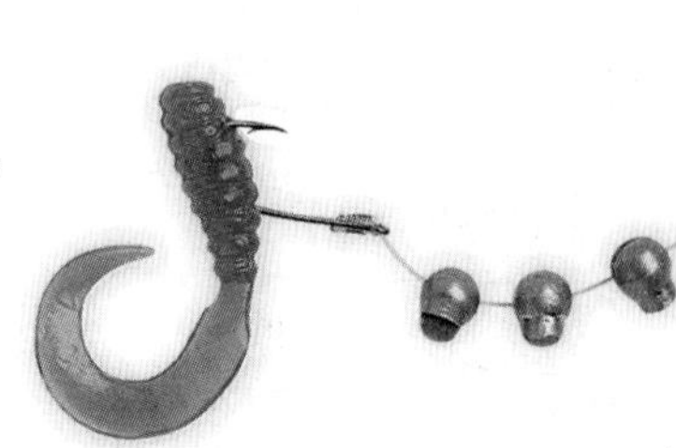

17. What is the formula for calculating density?

18. List three characteristic properties of matter.

Concept Mapping

19. Use the following terms to create a concept map: matter, mass, inertia, volume, milliliters, cubic centimeters, weight, gravity.

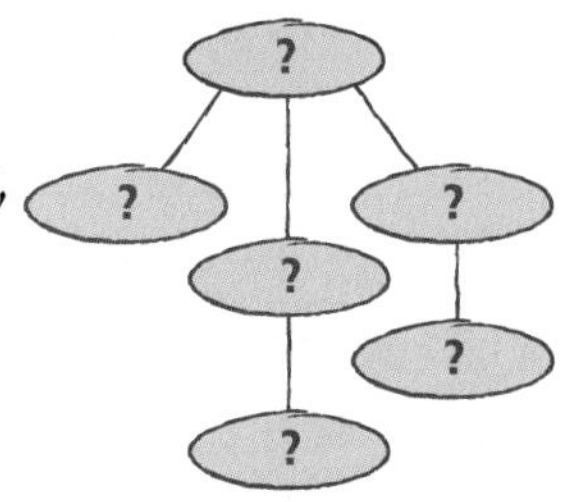

CRITICAL THINKING AND PROBLEM SOLVING

20. You are making breakfast for your picky friend, Filbert. You make him scrambled eggs. He asks, "Would you please take these eggs back to the kitchen and poach them?" What scientific reason do you give Filbert for not changing his eggs?

21. You look out your bedroom window and see your new neighbors moving in. Your neighbor bends over to pick up a small cardboard box, but he cannot lift it. What can you conclude about the item(s) in the box? Use the terms *mass* and *inertia* to explain how you came to this conclusion.

22. You may sometimes hear on the radio or on television that astronauts are "weightless" in space. Explain why this is not true.

23. People commonly use the term *volume* to describe the capacity of a container. How does this definition of volume differ from the scientific definition?

MATH IN SCIENCE

24. What is the volume of a book with the following dimensions: a width of 10 cm, a length that is two times the width, and a height that is half the width? Remember to express your answer in cubic units.

25. A jar contains 30 mL of glycerin (mass = 37.8 g) and 60 mL of corn syrup (mass = 82.8 g). Which liquid is on top? Show your work, and explain your answer.

INTERPRETING GRAPHICS

Examine the photograph below, and answer the following questions:

26. List three physical properties of this can.

27. Did a chemical change or a physical change cause the change in this can's appearance?

28. How does the density of the metal in the can compare before and after the change?

29. Can you tell what the chemical properties of the can are just by looking at the picture? Explain.

NOW What Do You Think?

Take a minute to review your answers to the ScienceLog questions on page 35. Have your answers changed? If necessary, revise your answers based on what you have learned since you began this chapter.

In the Dark About Dark Matter

What is the universe made of? Believe it or not, when astronomers try to answer this question, they still find themselves in the dark. Surprisingly, there is more to the universe than meets the eye.

A Matter of Gravity

Astronomers noticed something odd when studying the motions of galaxies in space. They expected to find a lot of mass in the galaxies. Instead, they discovered that the mass of the galaxies was not great enough to explain the large gravitational force causing the galaxies' rapid rotation. So what was causing the additional gravitational force? Some scientists think the universe contains matter that we cannot see with our eyes or our telescopes. Astronomers call this invisible matter *dark matter.*

Dark matter doesn't reveal itself by giving off any kind of electromagnetic radiation, such as visible light, radio waves, or gamma radiation. According to scientific calculations, dark matter could account for between 90 and 99 percent of the total mass of the universe! What is dark matter? Would you believe MACHOs and WIMPs?

▲ *The Large Magellanic Cloud, located 180,000 light-years from Earth*

MACHOs

Scientists recently proved the existence of *MAssive Compact Halo Objects* (MACHOs) in our Milky Way galaxy by measuring their gravitational effects. Even though scientists know MACHOs exist, they aren't sure what MACHOs are made of. Scientists suggest that MACHOs may be brown dwarfs, old white dwarfs, neutron stars, or black holes. Others suggest they are some type of strange, new object whose properties still remain unknown. Even though the number of MACHOs is apparently very great, they still do not represent enough missing mass. So scientists offer another candidate for dark matter—WIMPs.

WIMPs

Theories predict that *Weakly Interacting Massive Particles* (WIMPs) exist, but scientists have never detected them. WIMPs are thought to be massive elementary particles that do not interact strongly with matter (which is why scientists have not found them).

More Answers Needed

So far, evidence supports the existence of MACHOs, but there is little or no solid evidence of WIMPs or any other form of dark matter. Scientists who support the idea of WIMPs are conducting studies of the particles that make up matter to see if they can detect WIMPs. Other theories are that gravity acts differently around galaxies or that the universe is filled with things called "cosmic strings." Scientists admit they have a lot of work to do before they will be able to describe the universe—and all the matter in it.

On Your Own

▶ What is microlensing, and what does it have to do with MACHOs? How might the neutrino provide valuable information to scientists who are interested in proving the existence of WIMPs? Find out on your own!

Building a Better Body

Have you ever broken an arm or a leg? If so, you probably wore a cast while the bone healed. But what happens when a bone is too badly damaged to heal? In some cases, a false bone made from a metal called titanium can take the original bone's place. Could using titanium bone implants be the first step in creating bionic body parts? Think about it as you read about some of titanium's amazing properties.

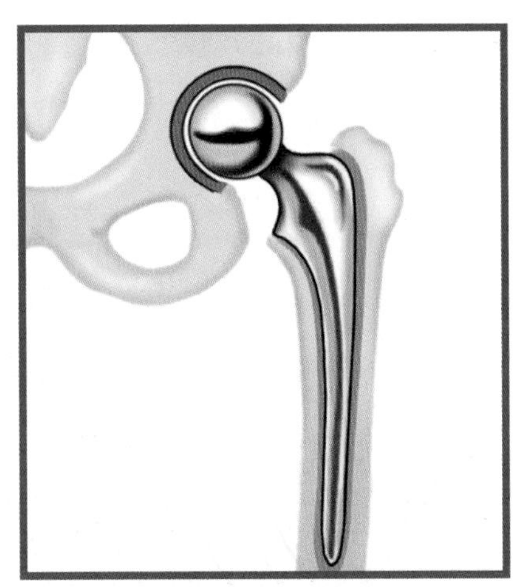

▲ *Titanium bones—even better than the real thing?*

Imitating the Original

Why would a metal like titanium be used to imitate natural bone? Well, it turns out that a titanium implant passes some key tests for bone replacement. First of all, real bones are incredibly lightweight and sturdy, and healthy bones last for many years. Therefore, a bone-replacement material has to be lightweight but also very durable. Titanium passes this test because it is well known for its strength, and it is also lightweight.

Second, the human body's immune system is always on the lookout for foreign substances. If a doctor puts a false bone in place and the patient's immune system attacks it, an infection can result. Somehow, the false bone must be able to chemically trick the body into thinking that the bone is real. Does titanium pass this test? Keep reading!

Accepting Imitation

By studying the human body's immune system, scientists found that the body accepts certain metals. The body almost always accepts one metal in particular. Yep, you guessed it—titanium! This turned out to be quite a discovery. Doctors could implant pieces of titanium into a person's body without triggering an immune reaction. A bond can even form between titanium and existing bone tissue, fusing the bone to the metal!

Titanium is shaping up to be a great bone-replacement material. It is lightweight and strong, is accepted by the body, can attach to existing bone, and resists chemical changes, such as corrosion. But scientists have encountered a slight problem. Friction can wear away titanium bones, especially those used near the hips and elbows.

Real Success

An unexpected surprise, not from the field of medicine but from the field of nuclear physics, may have solved the problem. Researchers have learned that by implanting a special form of nitrogen on the surface of a piece of metal, they can create a surface layer on the metal that is especially durable and wear resistant. When this form of nitrogen is implanted in titanium bones, the bones retain all the properties of pure titanium bones but also become very wear resistant. The new bones should last through decades of heavy use without needing to be replaced.

Think About It

▶ What will the future hold? As time goes by, doctors become more successful at implanting titanium bones. What do you think would happen if the titanium bones were to eventually become better than real bones?

CHAPTER 3 States of Matter

Imagine . . .

You are jogging along an ocean beach. An immense black storm cloud forms a short distance ahead. Suddenly there is a blinding flash of light followed by an explosion of thunder.

As the storm moves inland, you continue your jog. A short time later, you arrive at the section of beach where the storm passed. You notice an odd mark in the sand. You dig into the sand and find an object like the one shown below. You wonder what this object could be.

Lightning sometimes leaves behind a strange calling card known as a *fulgurite* (FUHL gyoo RIET). A fulgurite is a rare type of natural glass that is sometimes formed when lightning strikes silica, a mineral often found in soil or sand. Producing a temperature equal to that of the sun's surface (33,000°C), lightning melts solid silica into liquid. The silica then cools and hardens to become glass. The transformation of the silica from a solid to a liquid and back to a solid happens in the blink of an eye!

The physical changes that occur in the manufacture of glass are identical to those that occur when a fulgurite is created. The process, however, is very different.

Instead of lightning, glass makers use a large oven to heat the silica and other ingredients. Once this mixture becomes a liquid, it is removed from the oven and formed into a desired shape. The shaping process must happen quickly, before the liquid glass freezes into solid. By controlling the physical change between liquid glass and solid glass, known as a *change of state,* glass makers create the windows, light bulbs, and bottles you use every day. Read on to discover more about the states of matter.

What Do You Think?

In your ScienceLog, try to answer the following questions based on what you already know:

1. What are the four most familiar states of matter?
2. Compare the motion of the particles in a solid, a liquid, and a gas.
3. Name three ways matter changes from one state to another.

Vanishing Act

In this activity, you will use rubbing alcohol to investigate a change of state.

Procedure

1. Pour **rubbing alcohol** into a small **plastic cup** until it just covers the bottom of the cup.
2. Moisten the tip of a **cotton swab** by dipping it into the alcohol in the cup.
3. Rub the cotton swab on the palm of your hand.
4. Record your observations in your ScienceLog.
5. Wash your hands thoroughly.

Analysis

6. Explain what happened to the alcohol.
7. Did you feel a sensation of hot or cold? If so, how do you explain what you observed?
8. Record your answers in your ScienceLog.

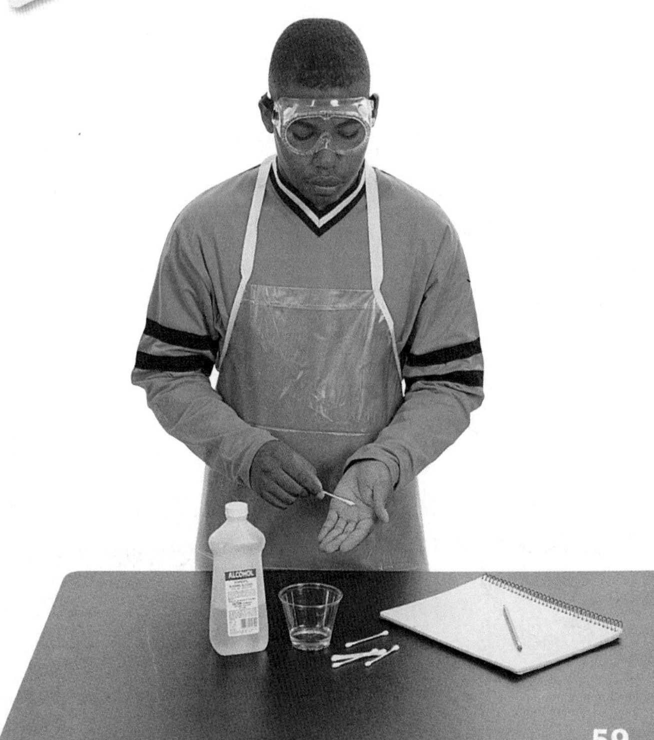

Section 1

Four States of Matter

NEW TERMS

states of matter
solid
liquid
gas
pressure
Boyle's law
Charles's law
plasma

OBJECTIVES

- Describe the properties shared by particles of all matter.
- Describe the four states of matter discussed here.
- Describe the differences between the states of matter.
- Predict how a change in pressure or temperature will affect the volume of a gas.

Figure 1 shows a model of the earliest known steam engine, invented about A.D. 60 by Hero, a scientist who lived in Alexandria, Egypt. This model also demonstrates the four most familiar states of matter: solid, liquid, gas, and plasma. The **states of matter** are the physical forms in which a substance can exist. For example, water commonly exists in three different states of matter: solid (ice), liquid (water), and gas (steam).

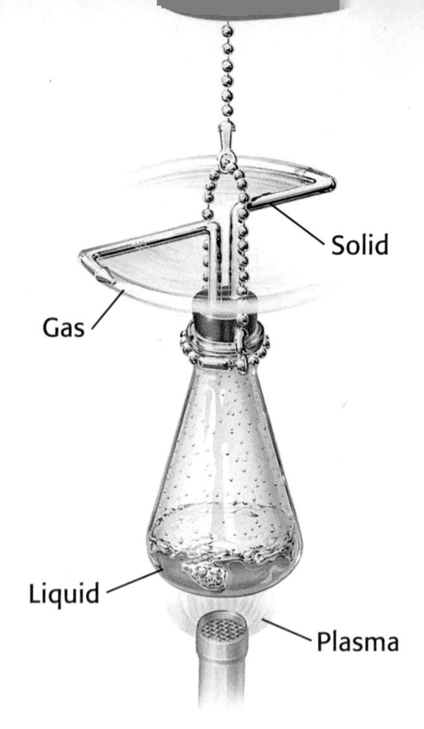

Figure 1 *This model of Hero's steam engine spins as steam escapes through the nozzles.*

Moving Particles Make Up All Matter

Matter consists of tiny particles called atoms and molecules (MAHL i KYOOLZ) that are too small to see without an amazingly powerful microscope. These atoms and molecules are always in motion and are constantly bumping into one another. The state of matter of a substance is determined by how fast the particles move and how strongly the particles are attracted to one another. **Figure 2** illustrates three of the states of matter—solid, liquid, and gas—in terms of the speed and attraction of the particles.

Figure 2 Models of a Solid, a Liquid, and a Gas

Particles of a solid do not move fast enough to overcome the strong attraction between them, so they are held tightly in place. The particles vibrate in place.

Particles of a liquid move fast enough to overcome some of the attraction between them. The particles are able to slide past one another.

Particles of a gas move fast enough to overcome nearly all of the attraction between them. The particles move independently of one another.

Solids Have Definite Shape and Volume

Look at the ship in **Figure 3.** Even though it is in a bottle, it keeps its original shape and volume. If you moved it to a larger bottle, the shape and volume of the ship would not change. Scientifically, the state in which matter has a definite shape and volume is **solid.** Because the particles of a substance in the solid state are very close together, the attraction between them is stronger than the attraction between the particles of the same substance in the liquid or gaseous state. The atoms or molecules in a solid are still moving, but they do not move fast enough to overcome the attraction between them. Each particle vibrates in place because it is locked in position by the particles around it.

Figure 3 *Because this ship is a solid, it does not take the shape of the bottle.*

Two Types of Solids Solids are often divided into two categories—*crystalline* and *amorphous* (uh MOHR fuhs). Crystalline solids have a very orderly, three-dimensional arrangement of atoms or molecules. The particles are arranged much like the seats in a movie theater. That is, the particles are in a repeating pattern of rows. Examples of crystalline solids include iron, diamond, and ice. Amorphous solids are composed of atoms or molecules that are in no particular order. The particles in an amorphous solid are arranged like people attending a concert in a park. That is, each particle is in a particular spot, but the particles are in no particular pattern. Examples of amorphous solids include rubber and wax. **Figure 4** illustrates the differences in the arrangement of particles in these two solids.

Explore

Imagine that you are a particle in a solid. Your position in the solid is your chair. In your ScienceLog, describe the different types of motion that are possible even though you cannot leave your chair.

Figure 4 *Differing arrangements of particles in crystalline solids and amorphous solids lead to different properties. Imagine trying to hit a home run with a rubber bat!*

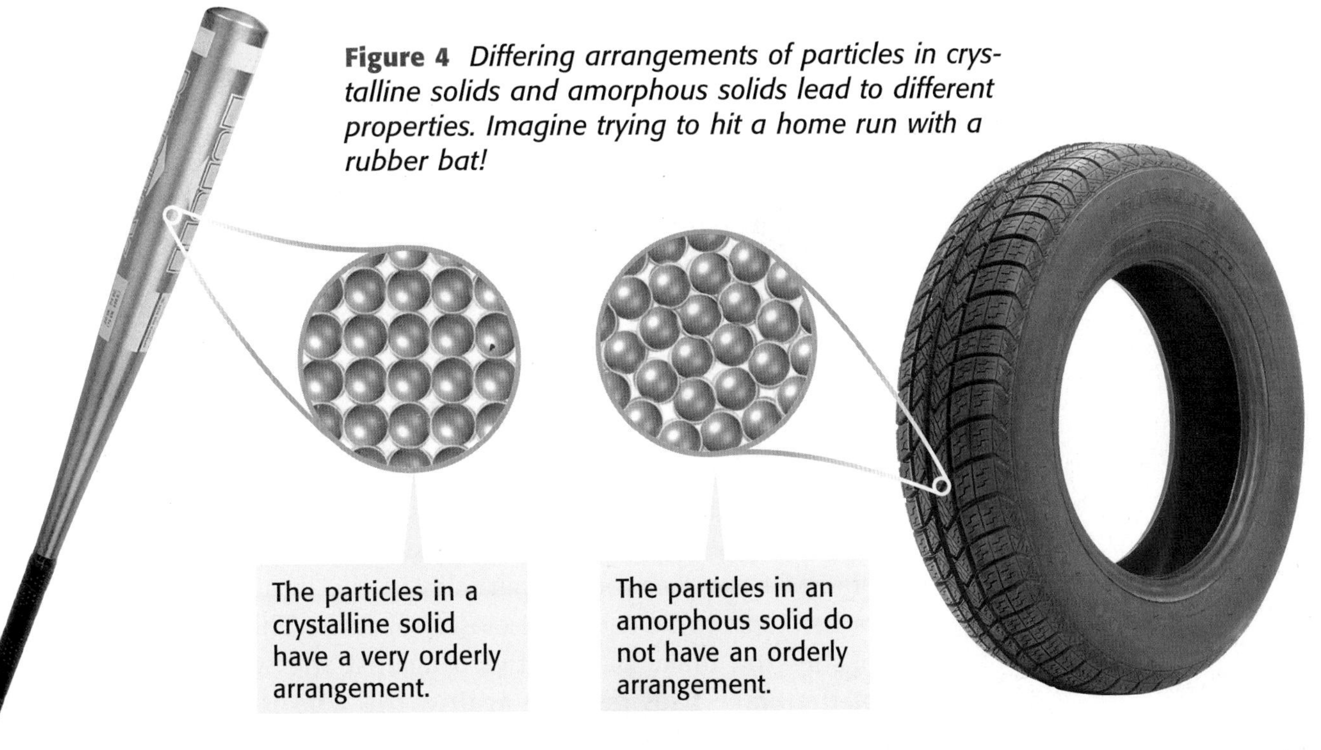

Liquids Change Shape but Not Volume

Figure 5 *Particles in a liquid slide past one another until the liquid conforms to the shape of its container.*

You already know from your own experience that a liquid will conform to the shape of whatever container it is put in. You are reminded of this every time you pour yourself a glass of juice. The state in which matter takes the shape of its container but has a definite volume is **liquid.** The atoms or molecules in liquids move fast enough to overcome some of the attractions between them. The particles slide past each other until the liquid takes the shape of its container. **Figure 5** shows how the particles in juice might look if they were large enough to see.

Even though liquids change shape, they do not readily change volume. You have also experienced this for yourself. You know that a can of soda contains a certain volume of liquid regardless of whether you pour it into a large container or a small one. **Figure 6** illustrates this point using a beaker and a graduated cylinder.

Figure 6 *Even when liquids change shape, they don't change volume.*

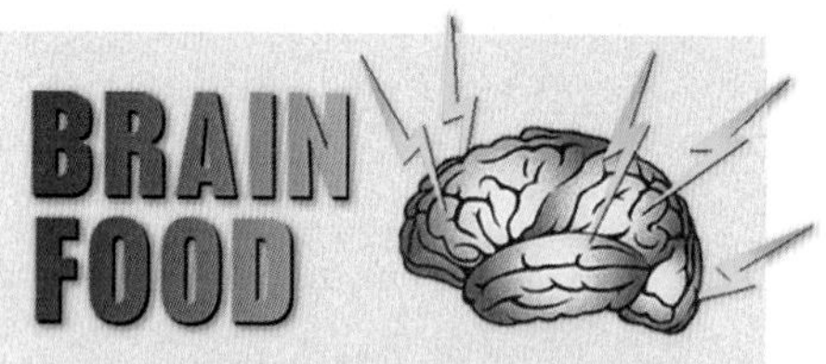

The Boeing 767 Freighter, a type of commercial airliner, has 187 km (116 mi) of hydraulic tubing.

Because the particles in liquids are close to one another, it is difficult to push them closer together. This makes liquids ideal for use in hydraulic (hie DRAW lik) systems to do work. For example, brake fluid is the liquid used in the brake systems of cars. Stepping on the brake pedal applies a force to the liquid. It is easier for the particles in the liquid to move away rather than to be squeezed closer together. Therefore, the fluid pushes the brake pads outward against the wheels. The force of the brake pads pushing against the wheels slows the car.

Two Properties of Liquids Two other important properties of liquids are *surface tension* and *viscosity* (vis KAHS uh tee). Surface tension is the force acting on the particles at the surface of a liquid that causes the liquid to form spherical drops, as shown in **Figure 7.** Different liquids have different surface tensions. For example, rubbing alcohol has a lower surface tension than water, but mercury has a higher surface tension than water.

Viscosity is a liquid's resistance to flow. In general, the stronger the attractions between a liquid's particles are, the more viscous the liquid is. Think of the difference between pouring honey and pouring water. Honey flows more slowly than water because it has a higher viscosity than water.

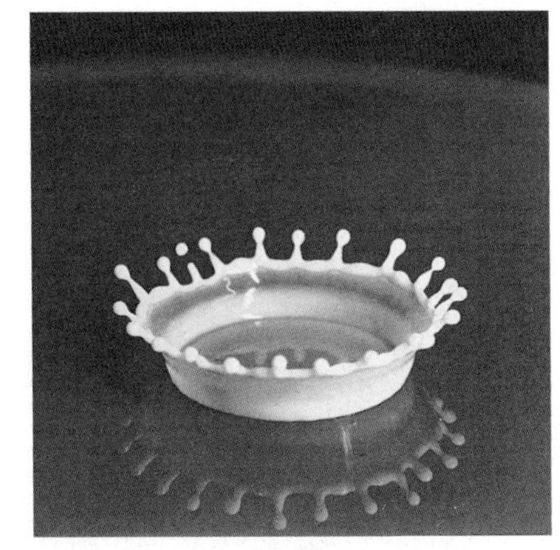

Figure 7 *Liquids form spherical drops as a result of surface tension.*

Gases Change Both Shape and Volume

The last time you saw balloons being filled with helium gas, did you wonder how many balloons could be filled from a single metal cylinder of helium? The number may surprise you. One cylinder can fill approximately 700 balloons. How is this possible? After all, the volume of the metal cylinder is equal to the volume of only about five inflated balloons.

To answer this question, you must know that the state in which matter changes in both shape and volume is **gas.** The atoms or molecules in a gas move fast enough to break away completely from one another. Therefore, the particles of a substance in the gaseous state have less attraction between them than particles of the same substance in the solid or liquid state. The particles move independently of one another, colliding frequently with one another and with the inside of the container as they spread out. So in a gas, there is empty space between particles.

The amount of empty space in a gas can change. For example, the helium in the metal cylinder consists of atoms that have been forced very close together, as shown in **Figure 8.** As the helium fills the balloon, the atoms spread out, and the amount of empty space in the gas increases. As you continue reading, you will learn how this empty space is related to pressure.

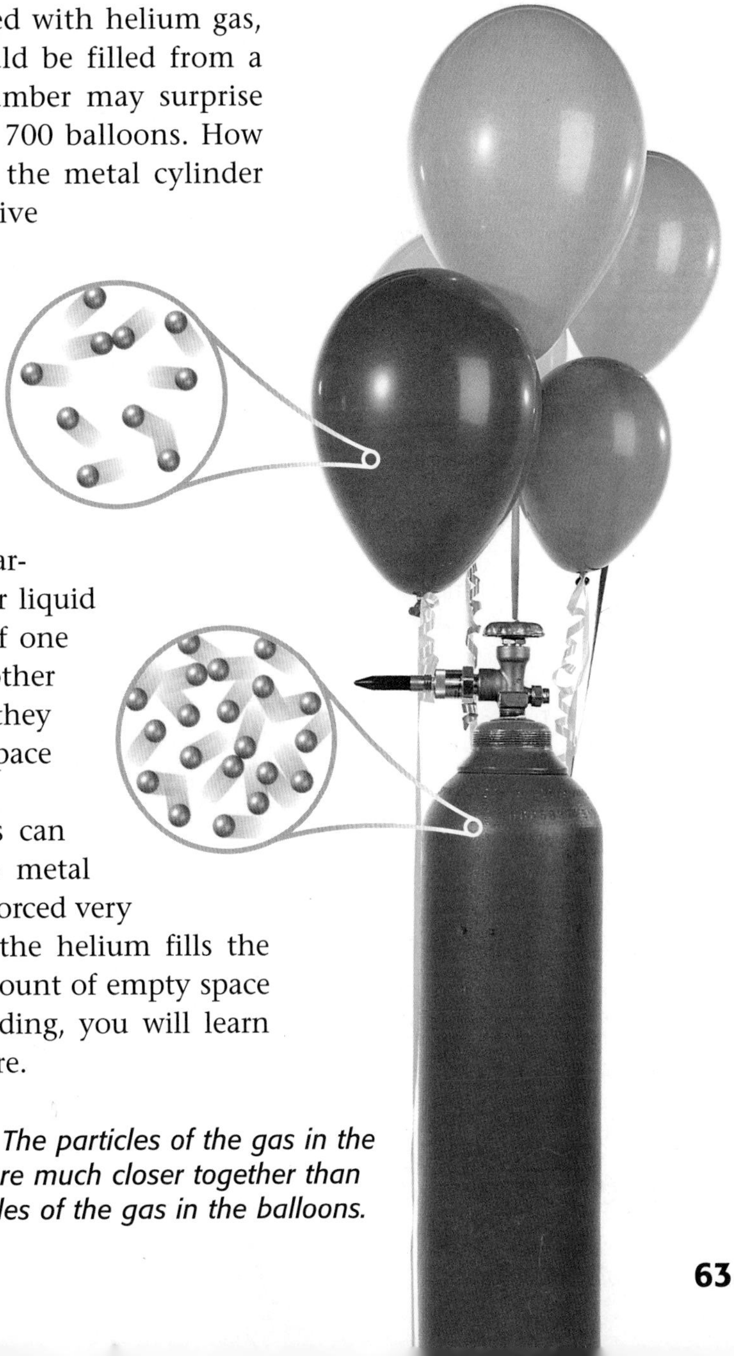

Figure 8 *The particles of the gas in the cylinder are much closer together than the particles of the gas in the balloons.*

Gas Under Pressure

Pressure is the amount of force exerted on a given area. You can think of this as the number of collisions of particles against the inside of the container. Compare the basketball with the beach ball in **Figure 9.** The balls have the same volume and contain particles of gas (air) that constantly collide with one another and with the inside surface of the balls. Notice, however, that there are more particles in the basketball than in the beach ball. As a result, more particles collide with the inside surface of the basketball than with the inside surface of the beach ball. When the number of collisions increases, the force on the inside surface of the ball increases. This increased force leads to increased pressure.

Self-Check

How would an increase in the speed of the particles affect the pressure of gas in a metal cylinder? *(See page 596 to check your answer.)*

Figure 9 *Both balls shown here are full of air, but the pressure in the basketball is higher than the pressure in the beach ball.*

The basketball has a higher pressure than the beach ball because the greater number of particles of gas are closer together. Therefore, they collide with the inside of the ball at a faster rate.

The beach ball has a lower pressure than the basketball because the lesser number of particles of gas are farther apart. Therefore, they collide with the inside of the ball at a slower rate.

REVIEW

1. List two properties that all particles of matter have in common.
2. Describe solids, liquids, and gases in terms of shape and volume.
3. Why can the volume of a gas change?
4. **Applying Concepts** Explain why you would pump up a flat basketball.

Laws Describe Gas Behavior

Earlier in this chapter, you learned about the atoms and molecules in both solids and liquids. You learned that compared with gas particles, the particles of solids and liquids are closely packed together. As a result, solids and liquids do not change volume very much. Gases, on the other hand, behave differently; their volume can change by a large amount.

It is easy to measure the volume of a solid or liquid, but how do you measure the volume of a gas? Isn't the volume of a gas the same as the volume of its container? The answer is yes, but there are other factors, such as pressure, to consider.

Boyle's Law Imagine a diver at a depth of 10 m blowing a bubble of air. As the bubble rises, its volume increases. By the time the bubble reaches the surface, its original volume will have doubled due to the decrease in pressure. The relationship between the volume and pressure of a gas is known as Boyle's law because it was first described by Robert Boyle, a seventeenth-century Irish chemist. **Boyle's law** states that for a fixed amount of gas at a constant temperature, the volume of a gas increases as its pressure decreases. Likewise, the volume of a gas decreases as its pressure increases. Boyle's law is illustrated by the model in **Figure 10.**

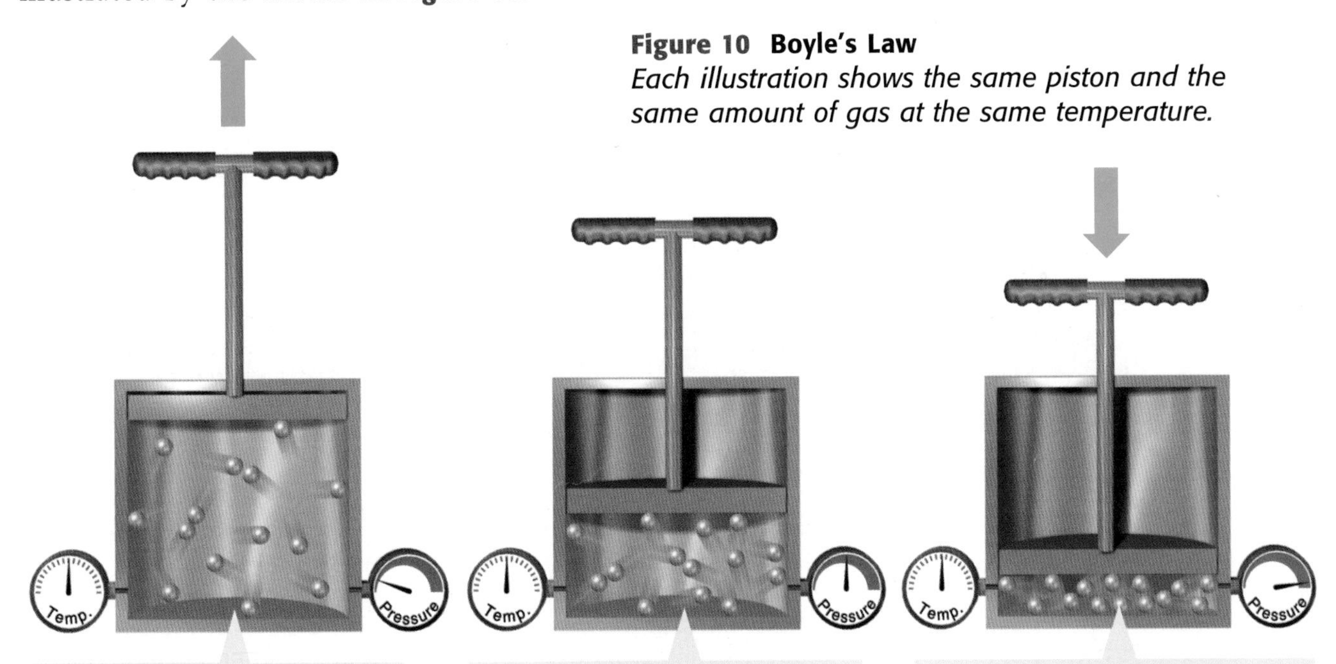

Figure 10 Boyle's Law
Each illustration shows the same piston and the same amount of gas at the same temperature.

Lifting the plunger decreases the pressure of the gas. The particles of gas collide less often with the walls of the piston as they spread farther apart. The volume of the gas increases as the pressure decreases.

Releasing the plunger allows the gas to change to an intermediate volume and pressure.

Pushing the plunger down increases the pressure of the gas. The particles of gas collide more often with the walls of the piston as they are forced closer together. The volume of the gas decreases as the pressure increases.

See Charles's law in action for yourself using a balloon on page 526 of the LabBook.

Weather balloons demonstrate a practical use of Boyle's law. A weather balloon carries equipment into the atmosphere to collect information used to predict the weather. This balloon is filled with only a small amount of gas because the pressure of the gas decreases and the volume increases as the balloon rises. If the balloon were filled with too much gas, it would pop as the volume of the gas increased.

Charles's Law An inflated balloon will also pop when it gets too hot, demonstrating another gas law—Charles's law. **Charles's law** states that for a fixed amount of gas at a constant pressure, the volume of the gas increases as its temperature increases. Likewise, the volume of the gas decreases as its temperature decreases. Charles's law is illustrated by the model in **Figure 11.** You can see Charles's law in action by putting an inflated balloon in the freezer. Wait about 10 minutes, and see what happens!

MATH BREAK

Gas Law Graphs

Each graph below illustrates a gas law. However, the variable on one axis of each graph is not labeled. Answer the following questions for each graph:

1. As the volume increases, what happens to the missing variable?
2. Which gas law is shown?
3. What label belongs on the axis?
4. Is the graph linear or nonlinear? What does this tell you?

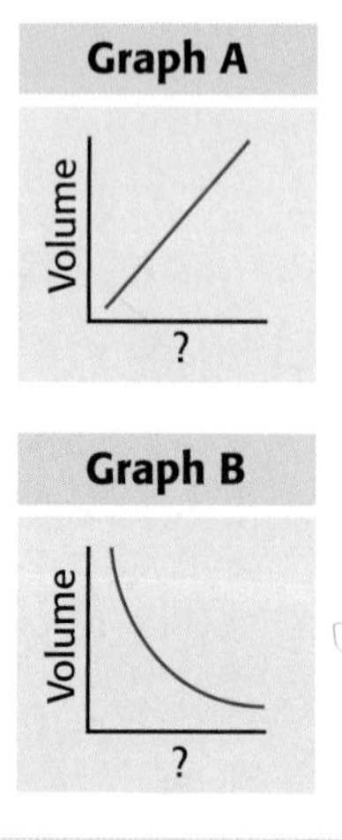

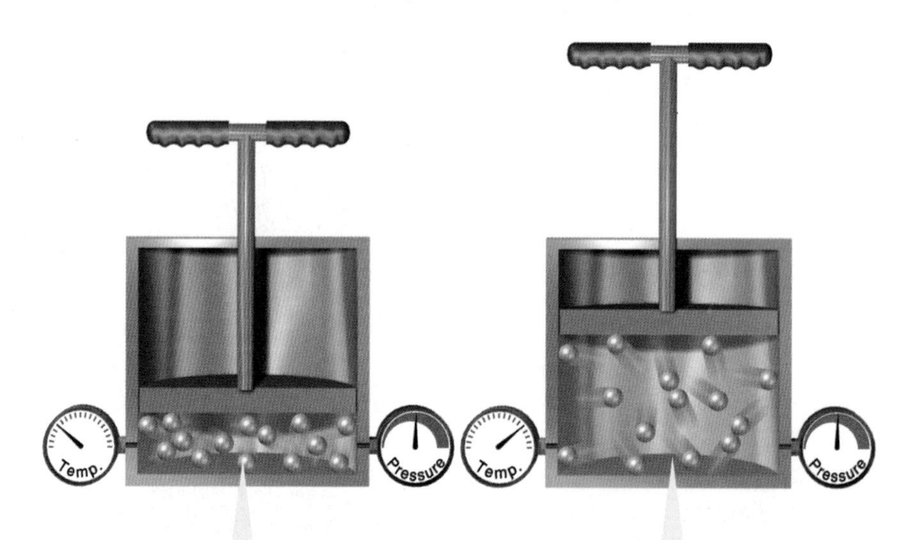

Figure 11 Charles's Law
Each illustration shows the same piston and the same amount of gas at the same pressure.

Lowering the temperature of the gas causes the particles to move more slowly. They hit the sides of the piston less often and with less force. As a result, the plunger enters the piston and the volume of the gas decreases.

Raising the temperature of the gas causes the particles to move more quickly. They hit the sides of the piston more often and with greater force. As a result, the plunger is pushed upward and the volume of the gas increases.

One of your friends overinflated the tires on her bicycle. Use Charles's law to explain why she should let out some of the air before going for a ride on a hot day.

Plasmas

The sun and other stars are made of the most common state of matter in the universe, called plasma. **Plasma** is the state of matter that does not have a definite shape or volume and whose particles have broken apart.

Plasmas have some properties that are quite different from the properties of gases. Plasmas conduct electric current, while gases do not. Electric and magnetic fields affect plasmas but do not affect gases. In fact, strong magnets are used to form a magnetic "bottle" to contain very hot plasmas that would destroy any other container.

Here on Earth, natural plasmas are found in lightning and fire. The incredible light show in **Figure 12,** called the aurora borealis (ah ROHR uh BOHR ee AL is), is a result of plasma from the sun causing gas particles in the upper atmosphere to glow. Artificial plasmas, found in fluorescent lights and plasma balls, are created by passing electric charges through gases.

Figure 12 *Auroras, like the aurora borealis seen here, form when high-energy plasma collides with gas particles in the upper atmosphere.*

REVIEW

1. When scientists record the volume of a gas, why do they also record the temperature and the pressure?
2. List two differences between gases and plasmas.
3. **Applying Concepts** What happens to the volume of a balloon left on a sunny windowsill? Explain.

Even though plasmas are rare on Earth, more than 99 percent of the known matter in the universe is in the plasma state.

Section 2

Changes of State

NEW TERMS

change of state
melting
endothermic
freezing
exothermic
vaporization
evaporation
boiling
condensation
sublimation

OBJECTIVES

- Describe how substances change from state to state.
- Explain the difference between an exothermic change and an endothermic change.
- Compare the changes of state.

A **change of state** is the conversion of a substance from one physical form to another. All changes of state are physical changes. In a physical change, the identity of a substance does not change. In **Figure 13,** the ice, liquid water, and steam are all the same substance, water. In this section, you will learn about the four changes of state illustrated in Figure 13 as well as a fifth change of state called *sublimation* (SUHB li MAY shuhn).

Figure 13 *The terms in the arrows are changes of state. Water commonly goes through the changes of state shown here.*

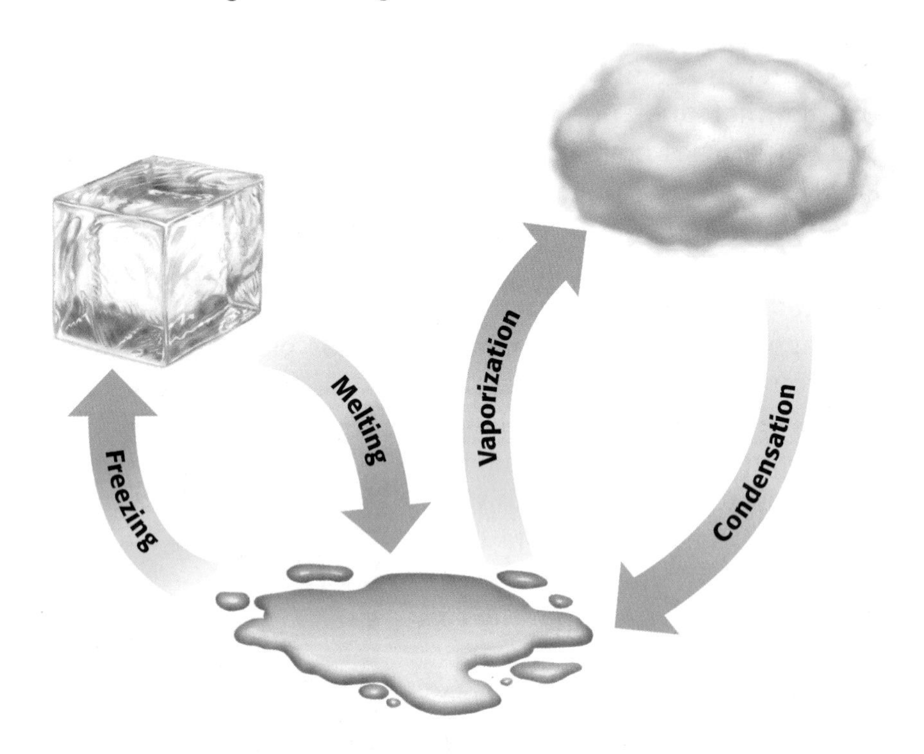

During a change of state, the energy of a substance changes. The *energy* of a substance is related to the motion of the particles of the substance. For example, the molecules in the liquid water in Figure 13 move faster than the molecules in the ice. Therefore, the liquid water has more energy than the ice.

If energy is added to a substance, the particles of the substance move faster. If energy is removed from a substance, the particles of the substance move slower. The *temperature* of a substance is a measure of the speed of the particles and therefore is a measure of the energy of a substance. For example, the steam shown above has a higher temperature than the liquid water, so the particles in the steam have more energy than the particles in the liquid water. A transfer of energy, known as *heat,* causes the temperature of a substance to change, which can lead to a change of state.

Want to learn how to get power from changes of state? Steam ahead to page 79.

Melting: Solids to Liquids

Melting is the change of state from a solid to a liquid. This is what happens when an ice cube melts. **Figure 14** shows a metal called gallium melting. What is unusual about this metal is that it melts at around 30°C. Because your normal body temperature is about 37°C, gallium would reach its melting point right in your hand!

The *melting point* of a substance is simply the temperature at which the substance changes from a solid to a liquid. The melting points of substances vary widely. As you know, the melting point of gallium is 30°C. Common table salt, however, has a melting point of 801°C.

Most substances have a unique melting point. Melting point can be used with other data to identify a substance. Because the melting point does not change with different amounts of the substance, the melting point is considered a *characteristic property* of a substance.

For a solid to melt, the particles must overcome some of their attractions to other particles. When a solid is at its melting point, any energy it absorbs increases the motion of the atoms or molecules until some of them overcome the attractions that hold them in place. Melting is an **endothermic** change because energy is absorbed, or taken in, by the substance as it changes state.

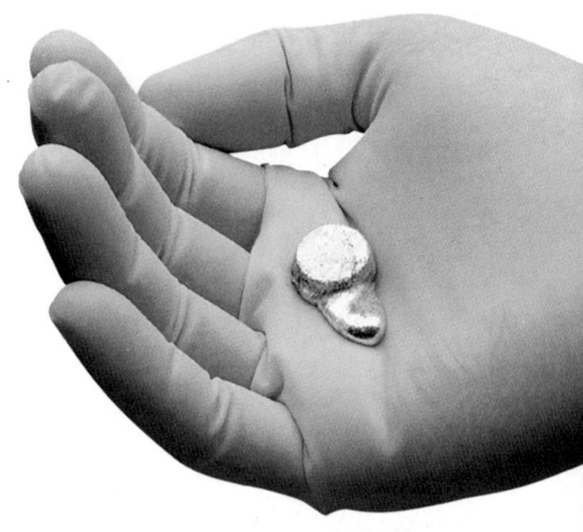

Figure 14 *Even though gallium is a metal, it would not be very useful as jewelry!*

Freezing: Liquids to Solids

Freezing is the change of state from a liquid to a solid. The temperature at which a liquid changes into a solid is its *freezing point*. Freezing is the reverse process of melting, so freezing and melting occur at the same temperature, as shown in **Figure 15.**

For a liquid to freeze, the motion of the atoms or molecules must slow to the point where attractions between them overcome the motion. The particles are pulled into a more ordered arrangement. When a liquid is at its freezing point, removing energy causes the particles to begin locking into place. Freezing is an **exothermic** change because energy is removed from, or taken out of, the substance as it changes state.

Figure 15 *Liquid water freezes at the same temperature that ice melts–0°C.*

If energy is added at 0°C, the ice will melt.

If energy is removed at 0°C, the liquid water will freeze.

Vaporization: Liquids to Gases

One way to experience vaporization (VAY puhr i ZAY shuhn) is to iron a shirt—carefully!—using a steam iron. You will notice steam coming up from the iron as the wrinkles are eliminated. This steam results from the vaporization of liquid water by the iron. **Vaporization** is simply the change of state from a liquid to a gas.

Boiling is vaporization that occurs throughout a liquid. The temperature at which a liquid boils is called its *boiling point*. Like the melting point, the boiling point is a characteristic property of a substance. The boiling point of water is 100°C, whereas the boiling point of liquid mercury is 357°C. **Figure 16** illustrates the process of boiling and a second form of vaporization, evaporation (ee VAP uh RAY shuhn).

Evaporation is vaporization that occurs at the surface of a liquid below its boiling point, as shown in Figure 16. When you perspire, your body is cooled through the process of evaporation. Perspiration is mostly water. Water absorbs energy from your skin as it evaporates. You feel cooler because your body transfers energy through heat to the water. Evaporation also explains why water in a glass placed on a table disappears after several days.

Self-Check

Is vaporization an endothermic or exothermic change? *(See page 596 to check your answer.)*

Boiling occurs in a liquid at its boiling point. As energy is added to the liquid, particles throughout the liquid move fast enough to break away from the particles around them and become a gas.

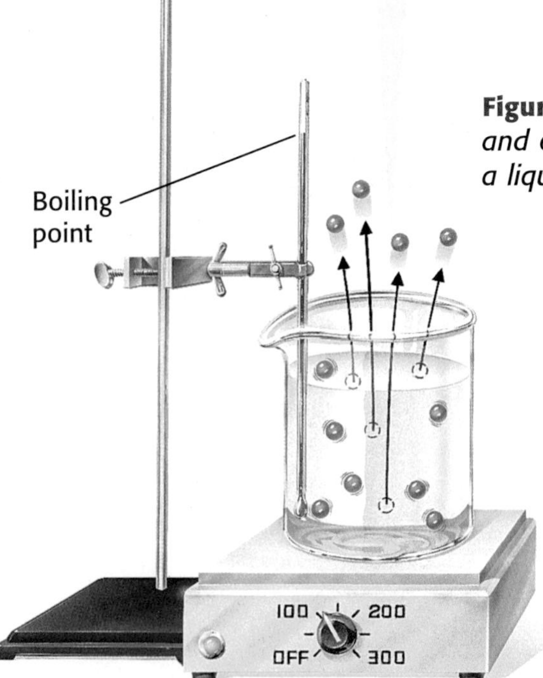

Figure 16 *Both boiling and evaporation change a liquid to a gas.*

Evaporation occurs in a liquid below its boiling point. Some particles at the surface of the liquid move fast enough to break away from the particles around them and become a gas.

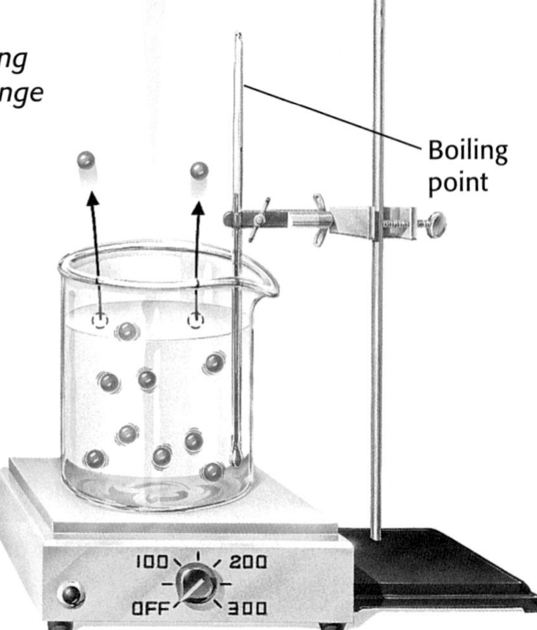

Pressure Affects Boiling Point Earlier you learned that water boils at 100°C. In fact, water only boils at 100°C at sea level, where the atmospheric pressure is 101,000 Pa. A pascal (Pa) is simply the SI unit for pressure. It is a force of 1 N exerted over an area of 1 m^2. Atmospheric pressure is caused by the weight of the gases that make up the atmosphere. Atmospheric pressure varies depending on where you are in relation to sea level because the higher you go above sea level, the less air there is above you. The atmospheric pressure is lower at higher elevations. If you were to boil water at the top of a mountain, the boiling point would be lower than 100°C. For example, Denver, Colorado, is 1.6 km (1 mi) above sea level. Water boils in Denver at about 95°C. You can make water boil at an even lower temperature by doing the QuickLab at right.

QuickLab

Boiling Water Is Cool

1. Remove the cap from a **syringe.**
2. Place the tip of the syringe in the **warm water** provided by your teacher. Pull the plunger out until you have 10 mL of water in the syringe.
3. Tightly cap the syringe.
4. Hold the syringe, and slowly pull the plunger out.
5. Observe any changes you see in the water. Record your observations in your ScienceLog.
6. Why are you not burned by the boiling water in the syringe?

Condensation: Gases to Liquids

Look at the cool glass of lemonade in **Figure 17.** Notice the beads of water on the outside of the glass. These form as a result of condensation. **Condensation** is the change of state from a gas to a liquid. The *condensation point* of a substance is the temperature at which the gas becomes a liquid and is the same temperature as the boiling point at a given pressure. Thus, at sea level, steam condenses to form water at 100°C—the same temperature at which water boils.

For a gas to become a liquid, large numbers of atoms or molecules must clump together. Particles clump together when the attraction between them overcomes their motion. For this to occur, energy must be removed from the gas to slow the particles down. Therefore, condensation is an exothermic change.

Figure 17 *Gaseous water in the air will become liquid when it contacts a cool surface.*

The amount of gaseous water that air can hold decreases as the temperature of the air decreases. As the air cools, some of the gaseous water condenses to form small drops of liquid water. These drops form clouds in the sky and fog near the ground.

Sublimation: Solids Directly to Gases

Look at the solids shown in **Figure 18.** The solid on the left is ice. Notice the drops of liquid collecting as it melts. On the right, you see carbon dioxide in the solid state, also called dry ice. It is called dry ice because instead of melting into a liquid, it goes through a change of state called sublimation. **Sublimation** is the change of state from a solid directly into a gas. Dry ice is colder than ice, and it doesn't melt into a puddle of liquid. It is often used to keep food, medicine, and other materials cold without getting them wet.

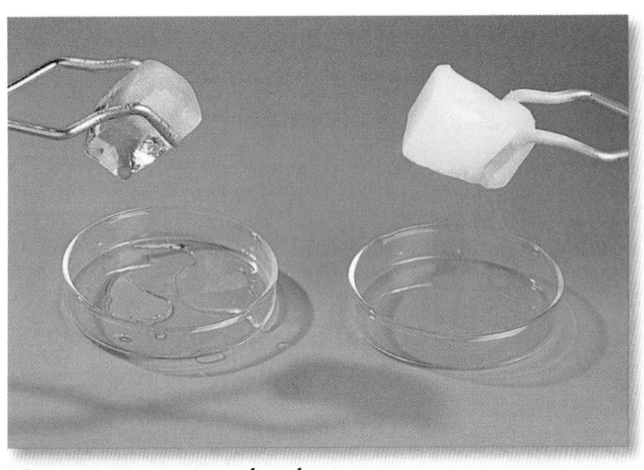

Figure 18 *Ice melts, but dry ice, on the right, turns directly into a gas.*

For a solid to change directly into a gas, the atoms or molecules must move from being very tightly packed to being very spread apart. The attractions between the particles must be completely overcome. Because this requires the addition of energy, sublimation is an endothermic change.

Comparing Changes of State

As you learned in Section 1 of this chapter, the state of a substance depends on how fast its atoms or molecules move and how strongly they are attracted to each other. A substance may undergo a physical change from one state to another by an endothermic change (if energy is added) or an exothermic change (if energy is removed). The table below shows the differences between the changes of state discussed in this section.

Summarizing the Changes of State

Change of state	Direction	Endothermic or exothermic?	Example
Melting	solid → liquid	endothermic	Ice melts into liquid water at 0°C.
Freezing	liquid → solid	exothermic	Liquid water freezes into ice at 0°C.
Vaporization	liquid → gas	endothermic	Liquid water vaporizes into steam at 100°C.
Condensation	gas → liquid	exothermic	Steam condenses into liquid water at 100°C.
Sublimation	solid → gas	endothermic	Solid dry ice sublimes into a gas at −78°C.

Temperature Change Versus Change of State

When most substances lose or absorb energy, one of two things happens to the substance: its temperature changes or its state changes. Earlier in the chapter, you learned that the temperature of a substance is a measure of the speed of the particles. This means that when the temperature of a substance changes, the speed of the particles also changes. But while a substance changes state, its temperature does not change until the change of state is complete, as shown in **Figure 19.**

Figure 19 Changing the State of Water

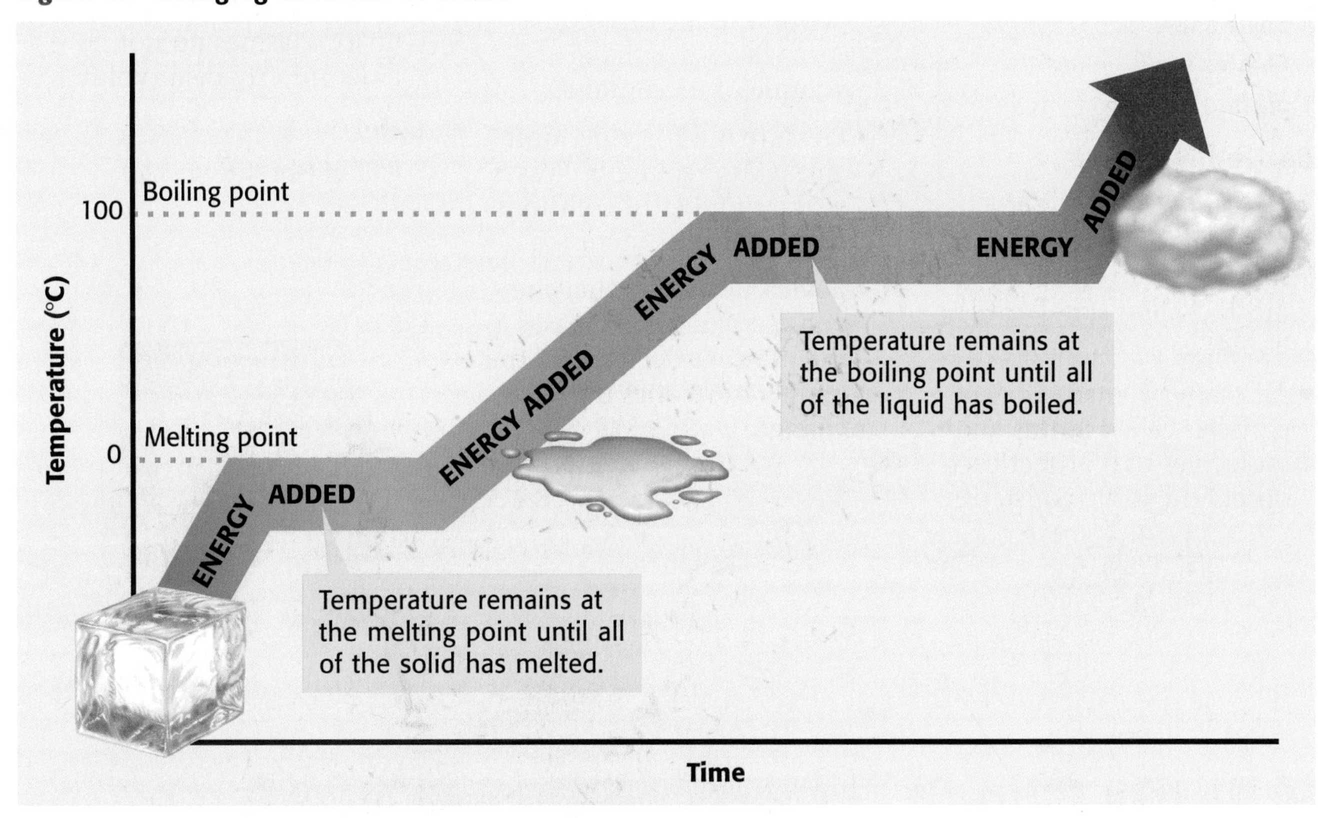

REVIEW

1. Compare endothermic and exothermic changes.
2. Classify each change of state (melting, freezing, vaporization, condensation, and sublimation) as endothermic or exothermic.
3. Describe how the motion and arrangement of particles change as a substance freezes.
4. **Comparing Concepts** How are evaporation and boiling different? How are they similar?

Chapter Highlights

SECTION 1

Vocabulary

states of matter *(p. 60)*
solid *(p. 61)*
liquid *(p. 62)*
gas *(p. 63)*
pressure *(p. 64)*
Boyle's law *(p. 65)*
Charles's law *(p. 66)*
plasma *(p. 67)*

Section Notes

- The states of matter are the physical forms in which a substance can exist. The four most familiar states are solid, liquid, gas, and plasma.
- All matter is made of tiny particles called atoms and molecules that attract each other and move constantly.
- A solid has a definite shape and volume.
- A liquid has a definite volume but not a definite shape.
- A gas does not have a definite shape or volume. A gas takes the shape and volume of its container.
- Pressure is a force per unit area. Gas pressure increases as the number of collisions of gas particles increases.
- Boyle's law states that the volume of a gas increases as the pressure decreases if the temperature does not change.
- Charles's law states that the volume of a gas increases as the temperature increases if the pressure does not change.
- Plasmas are composed of particles that have broken apart. Plasmas do not have a definite shape or volume.

Labs

Full of Hot Air! *(p. 526)*

Skills Check

Math Concepts

GRAPHING DATA The relationship between measured values can be seen by plotting the data on a graph. The top graph shows the linear relationship described by Charles's law—as the temperature of a gas increases, its volume increases. The bottom graph shows the nonlinear relationship described by Boyle's law—as the pressure of a gas increases, its volume decreases.

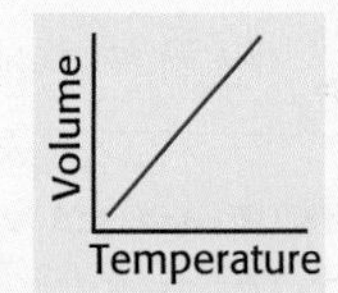

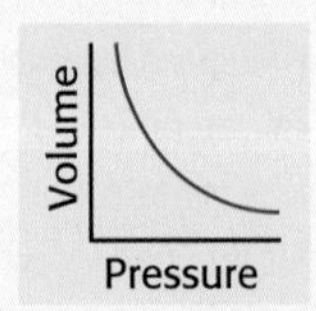

Visual Understanding

PARTICLE ARRANGEMENT Many of the properties of solids, liquids, and gases are due to the arrangement of the atoms or molecules of the substance. Review the models in Figure 2 on page 60 to study the differences in particle arrangement between the solid, liquid, and gaseous states.

SUMMARY OF THE CHANGES OF STATE Review the table on page 72 to study the direction of each change of state and whether energy is absorbed or removed during each change.

SECTION 2

Vocabulary

change of state *(p. 68)*
melting *(p. 69)*
endothermic *(p. 69)*
freezing *(p. 69)*
exothermic *(p. 69)*
vaporization *(p. 70)*
boiling *(p. 70)*
evaporation *(p. 70)*
condensation *(p. 71)*
sublimation *(p. 72)*

Section Notes

- A change of state is the conversion of a substance from one physical form to another. All changes of state are physical changes.
- Exothermic changes release energy. Endothermic changes absorb energy.
- Melting changes a solid to a liquid. Freezing changes a liquid to a solid. The freezing point and melting point of a substance are the same temperature.
- Vaporization changes a liquid to a gas. Boiling occurs throughout a liquid at the boiling point. Evaporation occurs at the surface of a liquid, at a temperature below the boiling point.
- Condensation changes a gas to a liquid.
- Sublimation changes a solid directly to a gas.
- Temperature does not change during a change of state.

Labs

Can Crusher *(p. 527)*
A Hot and Cool Lab *(p. 528)*

internetconnect

GO TO: go.hrw.com

Visit the **HRW** Web site for a variety of learning tools related to this chapter. Just type in the keyword:

KEYWORD: HSTSTA

GO TO: www.scilinks.org

Visit the **National Science Teachers Association** on-line Web site for Internet resources related to this chapter. Just type in the ***sci*LINKS** number for more information about the topic:

TOPIC: Forms and Uses of Glass — ***sci*LINKS NUMBER:** HSTP055
TOPIC: Solids, Liquids, and Gases — ***sci*LINKS NUMBER:** HSTP060
TOPIC: Natural and Artificial Plasma — ***sci*LINKS NUMBER:** HSTP065
TOPIC: Changes of State — ***sci*LINKS NUMBER:** HSTP070
TOPIC: The Steam Engine — ***sci*LINKS NUMBER:** HSTP075

Chapter Review

USING VOCABULARY

For each pair of terms, explain the difference in meaning.

1. exothermic/endothermic
2. Boyle's Law/Charles's Law
3. evaporation/boiling
4. melting/freezing

UNDERSTANDING CONCEPTS

Multiple Choice

5. Which of the following best describes the particles of a liquid?
   a. The particles are far apart and moving fast.
   b. The particles are close together but moving past each other.
   c. The particles are far apart and moving slowly.
   d. The particles are closely packed and vibrate in place.

6. Boiling points and freezing points are examples of
   a. chemical properties.
   b. physical properties.
   c. energy.
   d. matter.

7. During which change of state do atoms or molecules become more ordered?
   a. boiling
   b. condensation
   c. melting
   d. sublimation

8. Which of the following describes what happens as the temperature of a gas in a balloon increases?
   a. The speed of the particles decreases.
   b. The volume of the gas increases and the speed of the particles increases.
   c. The volume decreases.
   d. The pressure decreases.

9. Dew collects on a spider web in the early morning. This is an example of
   a. condensation.
   b. evaporation.
   c. sublimation.
   d. melting.

10. Which of the following changes of state is exothermic?
    a. evaporation
    b. sublimation
    c. freezing
    d. melting

11. What happens to the volume of a gas inside a piston if the temperature does not change but the pressure is reduced?
    a. increases
    b. stays the same
    c. decreases
    d. not enough information

12. The atoms and molecules in matter
    a. are attracted to one another.
    b. are constantly moving.
    c. move faster at higher temperatures.
    d. All of the above

13. Which of the following contains plasma?
    a. dry ice
    b. steam
    c. a fire
    d. a hot iron

Short Answer

14. Explain why liquid water takes the shape of its container but an ice cube does not.

15. Rank solids, liquids, and gases in order of decreasing particle speed.

16. Compare the density of iron in the solid, liquid, and gaseous states.

Concept Mapping

17. Use the following terms to create a concept map: states of matter, solid, liquid, gas, plasma, changes of state, freezing, vaporization, condensation, melting.

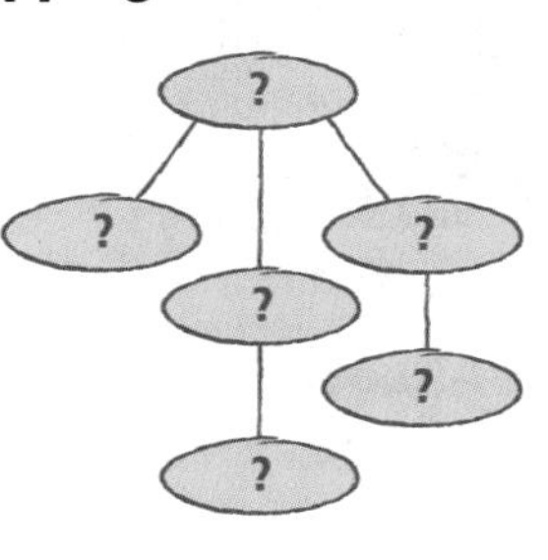

CRITICAL THINKING AND PROBLEM SOLVING

18. After taking a shower, you notice that small droplets of water cover the mirror. Explain how this happens. Be sure to describe where the water comes from and the changes it goes through.

19. In the photo below, water is being split to form two new substances, hydrogen and oxygen. Is this a change of state? Explain your answer.

20. To protect their crops during freezing temperatures, orange growers spray water onto the trees and allow it to freeze. In terms of energy lost and energy gained, explain why this practice protects the oranges from damage.

21. At sea level, water boils at 100°C, while methane boils at –161°C. Which of these substances has a stronger force of attraction between its particles? Explain your reasoning.

MATH IN SCIENCE

22. Kate placed 100 mL of water in five different pans, placed the pans on a windowsill for a week, and measured how much water evaporated. Draw a graph of her data, shown below, with surface area on the *x*-axis. Is the graph linear or nonlinear? What does this tell you?

Pan number	1	2	3	4	5
Surface area (cm^2)	44	82	20	30	65
Volume evaporated (mL)	42	79	19	29	62

23. Examine the graph below, and answer the following questions:
    a. What is the boiling point of the substance? What is the melting point?
    b. Which state is present at 30°C?
    c. How will the substance change if energy is added to the liquid at 20°C?

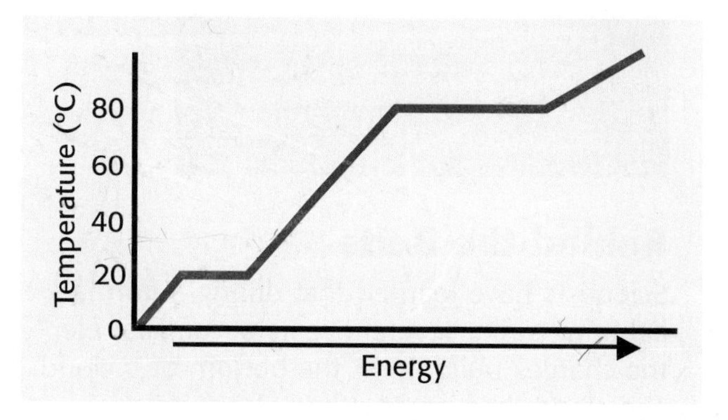

NOW What Do You Think?

Take a minute to review your answers to the ScienceLog questions on page 59. Have your answers changed? If necessary, revise your answers based on what you have learned since you began this chapter.

Science, Technology, and Society

Guiding Lightning

By the time you finish reading this sentence, lightning will have flashed more than 500 times around the world. This common phenomenon can have devastating results. Each year in the United States alone, lightning kills several hundred people and costs power companies more than $100 million. While controlling this awesome outburst of Mother Nature may seem an impossible task, scientists around the world are searching for ways to reduce destruction caused by lightning.

Behind the Bolts

Scientists have learned that during a normal lightning strike several events occur. First electric charges build up at the bottom of a cloud. The cloud then emits a line of negatively charged air particles that zigzags toward the Earth. The attraction between these negatively charged air particles and positively charged particles from objects on the ground forms a *plasma channel.* This channel is the pathway for a lightning bolt. As soon as the plasma channel is complete, BLAM!—between 3 and 20 lightning bolts separated by thousandths of a second travel along it.

A Stroke of Genius

Armed with this information, scientists have begun thinking of ways to redirect these naturally occurring plasma channels. One idea is to use laser beams. In theory, a laser beam directed into a thundercloud can charge the air particles in its path, causing a plasma channel to develop and forcing lightning to strike.

By creating the plasma channels themselves, scientists can, in a way, catch a bolt of lightning before it strikes and direct it to a safe area of the ground. So scientists simply use lasers to direct naturally occurring lightning to strike where they want it to.

A Bright Future?

Laser technology is not without its problems, however. The machines that generate laser beams are large and expensive, and they can themselves be struck by misguided lightning bolts. Also, it is not clear whether creating these plasma channels will be enough to prevent the devastating effects of lightning.

▲ ***Sometime in the future, a laser like this might be used to guide lightning away from sensitive areas.***

Find Out for Yourself

▶ Use the Internet or an electronic database to find out how rockets have been used in lightning research. Share your findings with the class.

Eureka!

Full Steam Ahead!

It was huge. It was 40 m long, about 5 m high, and it weighed 245 metric tons. It could pull a 3.28 million kilogram train at 100 km/h. It was a 4-8-8-4 locomotive, called a Big Boy, delivered in 1941 to the Union Pacific Railroad in Omaha, Nebraska. It was also one of the final steps in a 2,000-year search to harness steam power.

A Simple Observation

For thousands of years, people used wind, water, gravity, dogs, horses, and cattle to replace manual labor. But until about 300 years ago, they had limited success. Then in 1690, Denis Papin, a French mathematician and physicist, observed that steam expanding in a cylinder pushed a piston up. As the steam then cooled and contracted, the piston fell. Watching the motion of the piston, Papin had an idea: attach a water-pump handle to the piston. As the pump handle rose and fell with the piston, water was pumped.

More Uplifting Ideas

Eight years later, an English naval captain named Thomas Savery made Papin's device more efficient by using water to cool and condense the steam. Savery's improved pump was used in British coal mines. As good as Savery's pump was, the development of steam power didn't stop there!

In 1712, an English blacksmith named Thomas Newcomen improved Savery's device by adding a second piston and a horizontal beam that acted like a seesaw. One end of the beam was attached to the piston in the steam cylinder. The other end of the beam was attached to the pump piston. As the steam piston moved up and down, it created a vacuum in the pump cylinder and sucked water up from the mine. Newcomen's engine was the most widely used steam engine for more than 50 years.

Watt a Great Idea!

In 1764, James Watt, a Scottish technician, was repairing a Newcomen engine. He realized that heating the cylinder, letting it cool, then heating it again wasted an enormous amount of energy. Watt added a separate chamber where the steam could cool and condense. The two chambers were connected by a valve that let the steam escape from the boiler. This improved the engine's efficiency—the boiler could stay hot all the time!

A few years later, Watt turned the whole apparatus on its side so that the piston was moving horizontally. He added a slide valve that admitted steam first to one end of the chamber (pushing the piston in one direction) and then to the other end (pushing the piston back). This changed the steam pump into a true steam engine that could drive a locomotive the size of Big Boy!

Explore Other Inventions

▶ Watt's engine helped trigger the Industrial Revolution as many new uses for steam power were found. Find out more about the many other inventors, from tinkerers to engineers, who harnessed the power of steam.

CHAPTER 4

Elements, Compounds, and Mixtures

This Really Happened!

In the early morning hours of April 15, 1912, the *Titanic,* the largest ship ever to set sail, sank on its first voyage. The *Titanic* was considered to be unsinkable, yet more than 1,500 of its passengers and crew were killed after it hit an iceberg and sank.

How could an iceberg, which is made of ice, destroy the 2.5 cm thick steel plates that made up the *Titanic*'s hull? Analysis of a recovered piece of the *Titanic*'s hull showed that the steel contained large amounts of the element sulfur, which is a normal component of steel. However, in this case, the steel contained much more sulfur than is the standard for steel made today. This excessive amount of sulfur may have caused the steel to be brittle, much like glass. Scientists suspect that this brittle steel may have cracked on impact with the iceberg, allowing water to enter the hull.

Could something as simple as using less sulfur in the *Titanic*'s steel have prevented the ship from sinking? It is impossible to know for sure. What is known, however, is that the composition and properties of elements, compounds, and mixtures are very important in preventing future disasters. In this chapter, you will learn about elements and how they are assembled into compounds and mixtures with some very different properties.

This piece of steel hull from the Titanic *(at left) was recovered from the wreck.*

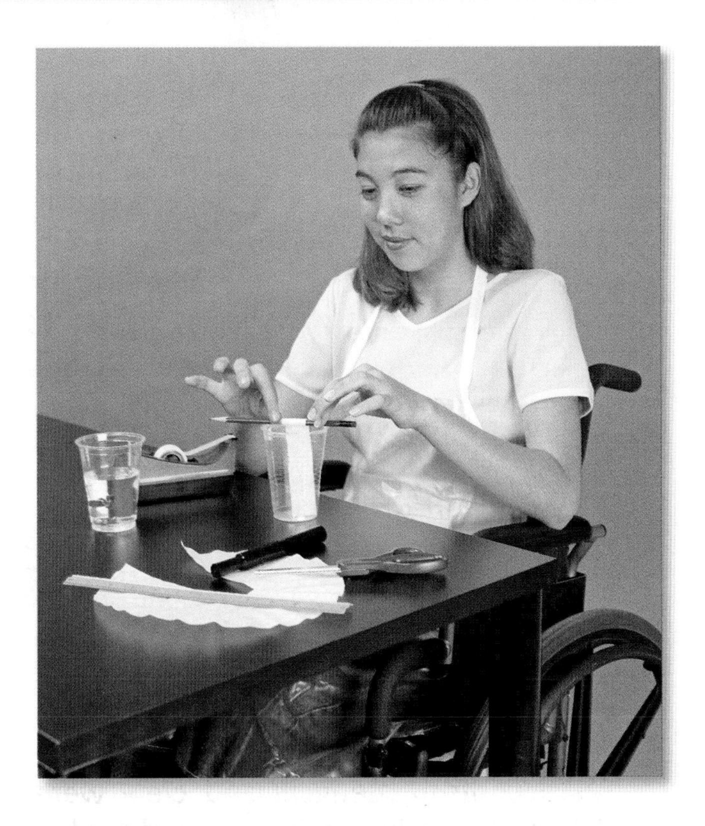

Mystery Mixture

Steel is just one of the many mixtures that you encounter every day. In fact, you might be using a mixture to write notes for this activity! Some inks used in pens and markers are a mixture of several dyes. In this lab, you will separate the parts of an ink mixture.

Procedure  

1. Cut a **3 × 15 cm strip of paper** from a **coffee filter.** Wrap one end around a **pencil** so that the other end will just touch the bottom of a **clear plastic cup** (as shown in the photo above). Secure the strip of paper to the pencil with a **piece of tape.**
2. Take the paper out of the cup. Using a **water-soluble black marker,** make a small dot in the center of the strip about 2 cm from the bottom end of the paper.
3. Pour **water** in the cup to a depth of 1 cm.
4. Carefully lower the paper into the cup so that the end is in the water but the dot you made is not underwater.
5. Watch the filter paper. Remove the paper when the water is 1 cm from the top of the paper. Record your observations in your ScienceLog.

Analysis

6. What happened as the filter paper soaked up the water? What colors were mixed to make black ink?
7. Compare your results with those of your classmates. How did the mixture in your marker compare with the mixtures in their markers?
8. Do you think the process used to create the ink involved a physical or chemical change? Explain.

Going Further

The procedure you used to separate the components of ink is a scientific technique called chromatography. Find out more about chromatography and its uses by looking in a chemistry or reference book.

What Do You Think?

In your ScienceLog, try to answer the following questions based on what you already know:

1. What is an element?
2. What is a compound? How are compounds and mixtures different?
3. What are the components of a solution called?

Section 1

Elements

NEW TERMS

element
pure substance
metals
nonmetals
metalloids

OBJECTIVES

- Describe pure substances.
- Describe the characteristics of elements, and give examples.
- Explain how elements can be identified.
- Classify elements according to their properties.

Imagine you are working as a lab technician for the Break-It-Down Corporation. Your job is to break down incoming materials into the simplest substances you can obtain. One day a material seems particularly difficult to break down. You start by crushing and grinding it. You notice that the resulting pieces are smaller, but they are still the same material. You try other physical changes, including melting, boiling, and filtering it, but the material does not change into anything simpler.

Next you try some chemical changes. For example, you pass electric current through the material. Although many substances can be broken down using electric current, this material still does not become any simpler. After recording your observations, you analyze the results of your tests. You then draw a conclusion: the substance must be an element. An **element** is a pure substance that cannot be separated into simpler substances by physical or chemical means, as shown in **Figure 1.**

Figure 1 *No matter what kind of physical or chemical change you attempt, an element cannot be changed into a simpler substance!*

An Element Has Only One Type of Particle

A **pure substance** is a substance in which there is only one type of particle. Because elements are pure substances, each element contains only one type of particle. For example, every particle (atom) in a 5 g nugget of the element gold is like every other particle of gold. The particles of a pure substance are alike no matter where that substance is found. Take a look at **Figure 2.** The element iron is a major component of many meteorites. Although a meteorite might travel more than 400 million kilometers (about 248 million miles) to reach Earth, the particles of iron in a meteorite are identical to the particles of iron in objects around your home!

Figure 2 *The atoms of the element iron are alike whether they are in a meteorite or in a common iron skillet.*

Every Element Has a Unique Set of Properties

Each element has a unique set of properties that allows you to identify it. For example, each element has its own *characteristic properties*. These properties do not depend on the amount of material present in a sample of the element. Characteristic properties include some physical properties, such as boiling point, melting point, and density, as well as chemical properties, such as reactivity with acid. The elements helium and krypton are unreactive gases. However, the density (mass per unit volume) of helium is less than the density of air. Therefore, a helium-filled balloon will float up if it is released. Krypton is more dense than air, so a krypton-filled balloon will sink to the ground if it is released.

Look at the elements cobalt, iron, and nickel, shown in **Figure 3.** Even though these three elements have some similar properties, each can be identified by its unique set of properties.

Notice that the physical properties for the three elements include melting point and density. Other physical properties, such as color, hardness, and texture, could be added to the list. Also, depending on the elements being identified, other chemical properties might be useful. For example, some elements, such as hydrogen and carbon, are flammable. Other elements, such as sodium, react immediately with oxygen. Still other elements, such as zinc, are reactive with acid.

Cobalt

Melting point is 1,495°C.
Density is 8.9 g/cm^3.
Conducts electricity and heat.
Unreactive with oxygen in the air.

Iron

Melting point is 1,535°C.
Density is 7.9 g/cm^3.
Conducts electricity and heat.
Combines slowly with oxygen in the air to form rust.

Nickel

Melting point is 1,455°C.
Density is 8.9 g/cm^3.
Conducts electricity and heat.
Unreactive with oxygen in the air.

Figure 3 *Like all other elements, cobalt, iron, and nickel can be identified by their unique combination of properties.*

Most Elements Are Combined in Nature Most elements are found combined with other elements in nature. The reason lies in their chemical properties. Most elements undergo chemical changes when combined with oxygen or water. When these elements are exposed to air or moisture, they undergo chemical changes and form more-complex substances called *compounds*. (You'll learn more about compounds in the next section.) Some elements do not react readily with water or air and can be found uncombined in nature. These elements include gold, copper, sulfur, neon, and carbon.

Elements Are Classified by Their Properties

Consider how many different breeds of dogs there are. Consider also how you tell one breed from another. Most often you can tell just by their appearance, or what might be called physical properties. **Figure 4** shows several breeds of dogs, which all happen to be terriers. Many terriers are fairly small in size and have short hair. Many also share the same basic body shape. Even the behavior of chasing small animals like rats or foxes is shared by most terriers. Although terriers are not exactly alike, they share enough common properties to be classified in the same group.

Similarly, elements are classified into groups according to their shared properties. Recall the elements iron, nickel, and cobalt. All three are shiny, and all three conduct thermal energy and electric current. Using these shared properties, scientists have grouped these three elements, along with other elements that are shiny and are good conductors, into one large group called metals. Metals are not all exactly alike, but they do have some properties in common. Likewise, you can group elements that are not shiny and are not good conductors into a different group called nonmetals.

Figure 4 *Even though these dogs are different breeds, they have enough in common to be classified as terriers.*

If You Know the Category, You Know the Properties If you have ever browsed at a music store, you know that the CDs and tapes are categorized by basic type of music. If you are interested in rock-and-roll, you would go to the rock-and-roll section. Even though you might not recognize a particular CD, you know that it must have the characteristics of rock-and-roll for it to be in this section. Otherwise, it would be kept in the classical, bluegrass, country, or some other section.

Likewise, you can predict some of the properties of an unfamiliar element by knowing the category to which it belongs. As shown in the concept map in **Figure 5,** elements are classified into three categories—metals, nonmetals, and metalloids. Cobalt, iron, and nickel are classified as metals. If you know that a particular element is a metal, you know that it shares certain properties with iron, nickel, and cobalt. The chart on the next page shows examples of elements in each category and a summary of the properties that identify the members of each category.

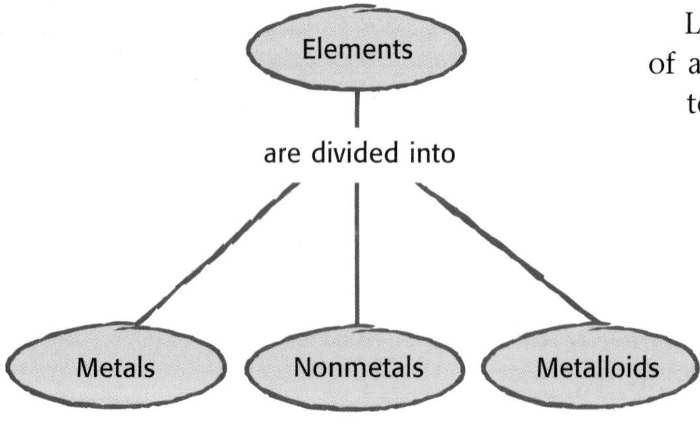

Figure 5 *Elements are divided into three categories: metals, nonmetals, and metalloids.*

The Three Major Categories of Elements

Metals

Metals are elements that are shiny and are good conductors of thermal energy and electric current. They are easily shaped into different forms because they are *malleable* (they can be hammered into thin sheets) and *ductile* (they can be drawn into thin wires). Iron has many uses in building and automobile construction. Copper is used in wires and coins.

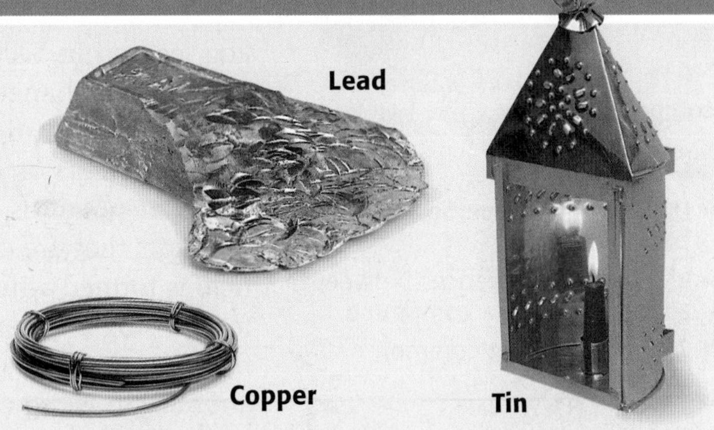

Nonmetals

Nonmetals are elements that are dull (not shiny) and that are poor conductors of thermal energy and electric current. Solid nonmetals tend to be brittle and unmalleable. Few familiar objects are made of only nonmetals. The neon used in lights is a nonmetal, as is the graphite (carbon) used in pencils.

Neon

Metalloids

Metalloids, also called semiconductors, are elements that have properties of both metals and nonmetals. Some metalloids are shiny, while others are dull. Metalloids are somewhat malleable and ductile. Some metalloids conduct thermal energy and electric current well. Other metalloids can become good conductors when they are mixed with other elements. Silicon is used to make computer chips. However, other elements must be mixed with silicon to make a working chip.

REVIEW

1. What is a pure substance?
2. List three properties that can be used to classify elements.
3. **Applying Concepts** Which category of element would be the least appropriate choice for making a container that can be dropped without shattering? Explain your reasoning.

Section 2

Compounds

NEW TERMS

compound

OBJECTIVES

- Describe the properties of compounds.
- Identify the differences between an element and a compound.
- Give examples of common compounds.

You learned in Section 1 that because most elements take part in chemical changes fairly easily, few elements are found alone in nature. Instead, most elements are found combined with other elements as compounds.

A **compound** is a pure substance composed of two or more elements that are chemically combined. In a compound, a particle is formed when particles of two or more elements join to form a single larger particle. In order for elements to combine, they must *react,* or undergo a chemical change, with one another. In **Figure 6,** you see magnesium reacting with oxygen to form a compound called magnesium oxide. The compound is a new pure substance that is different from the elements that reacted to form it.

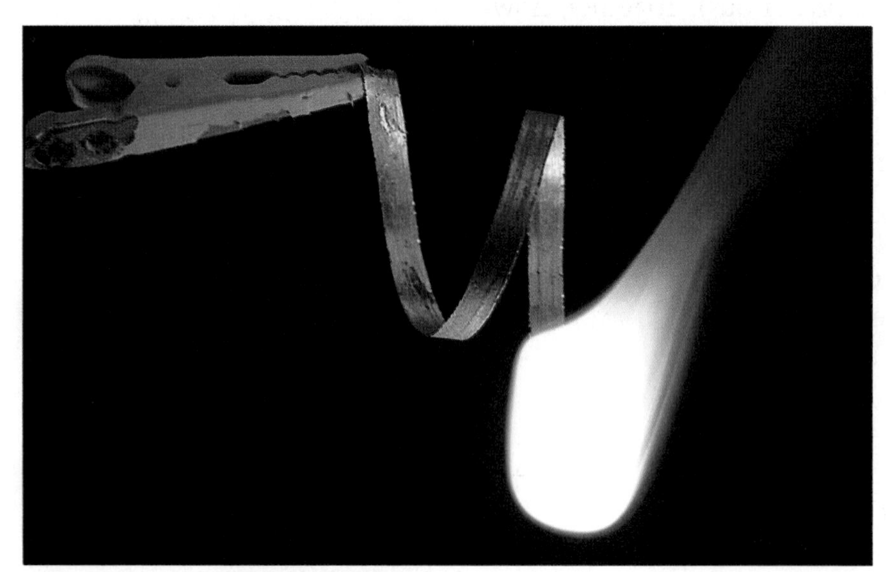

Figure 6 *As magnesium burns, it reacts with oxygen and forms the compound magnesium oxide. Magnesium is used often in the manufacture of fireworks because of the bright white light produced during this chemical change.*

Most substances you encounter every day are compounds. For example, table salt is a compound composed of the elements sodium and chlorine chemically combined. Water is a compound composed of the elements hydrogen and oxygen chemically combined. Calcium phosphate, an important compound in bones and teeth, is composed of the elements calcium, phosphorus, and oxygen chemically combined.

Elements Combine in a Definite Ratio to Form a Compound

Compounds are not random combinations of elements. When a compound forms, the elements join in a specific ratio according to their masses. For example, the ratio of the mass of hydrogen to the mass of oxygen in water is always the same—1 g of hydrogen to 8 g of oxygen. This mass ratio can be written as 1:8 or as the fraction 1/8. Every sample of water has this 1:8 mass ratio of hydrogen to oxygen. If a sample of a compound has a different mass ratio of hydrogen to oxygen, the compound cannot be water. The mass ratio of hydrogen to oxygen in the compound hydrogen peroxide is 1:16. The mass ratio of carbon to oxygen in carbon monoxide is 3:4, but in carbon dioxide the mass ratio is 3:8.

Every Compound Has a Unique Set of Properties

Each compound has a unique set of properties that allows you to distinguish it from other compounds. Like elements, each compound has its own physical properties, such as boiling point, melting point, density, and color. Compounds can also be identified by their different chemical properties. Some compounds, such as calcium carbonate found in chalk, react with acid. Others, such as hydrogen peroxide, react when exposed to light. You can see how chemical properties can be used to identify compounds in the QuickLab at right.

A compound has different properties from the elements that form it. Did you know that ordinary table salt is a compound made from two very dangerous elements? Table salt—sodium chloride—consists of sodium (which reacts violently with water) and chlorine (which is poisonous). Together, however, these elements form a harmless compound with unique properties. Take a look at **Figure 7.** Because a compound has different properties from the elements that react to form it, sodium chloride is safe to eat and dissolves (without exploding!) in water.

Another example to consider can be found in Figure 6 on the previous page. The element oxygen is a colorless gas, and the element magnesium is a silvery white metal. However, the compound magnesium oxide is a white powdery solid.

Quick Lab

Compound Confusion

1. Measure 4 g (1 tsp) of **compound A,** and place it in a **clear plastic cup.**
2. Measure 4 g (1 tsp) of **compound B,** and place it in a **second clear plastic cup.**
3. Observe the color and texture of each compound. Record your observations.
4. Add 5 mL (1 tsp) of **vinegar** to each cup. Record your observations.
5. Baking soda reacts with vinegar, while powdered sugar does not. Which of these compounds is compound A, and which is compound B?

Figure 7 *Table salt is formed when the elements sodium and chlorine join. The properties of salt are different from the properties of sodium and chlorine.*

Sodium is a soft, silvery white metal that reacts violently with water.

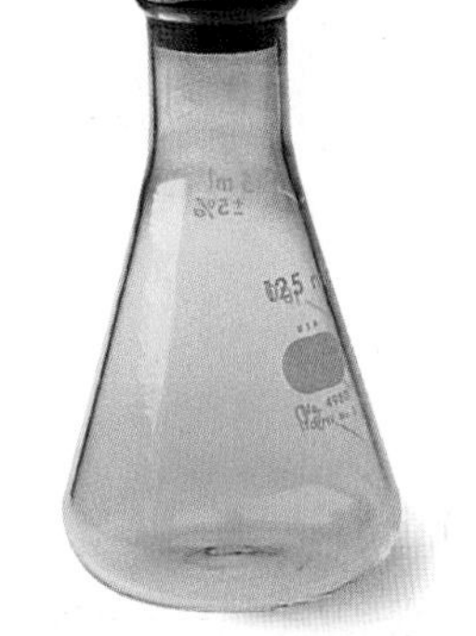

Chlorine is a poisonous, greenish yellow gas.

Sodium chloride, or table salt, is a white solid that dissolves easily in water and is safe to eat.

Self-Check

Do the properties of pure water from a glacier and from a desert oasis differ? *(See page 596 to check your answer.)*

Compounds Can Be Broken Down into Simpler Substances

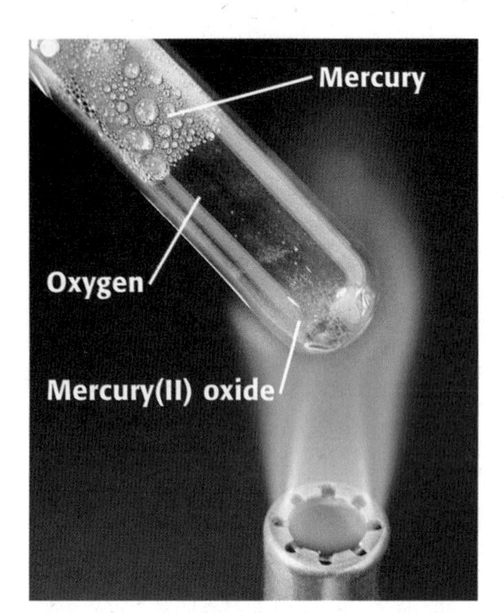

Figure 8 *Heating mercury(II) oxide causes a chemical change that separates it into the elements mercury and oxygen.*

If some elements are found only in compounds, how can pure samples of these elements be produced? Compounds can be broken down into elements through chemical changes. Look at **Figure 8.** When the compound mercury(II) oxide is heated, it breaks down into the elements mercury and oxygen. Likewise, if electricity is passed through melted table salt, the elements sodium and chlorine are produced.

Some compounds undergo chemical changes to form simpler compounds. These compounds can be broken down into elements through additional chemical changes. For example, carbonic acid is a compound that helps to give carbonated beverages their "fizz." Carbonic acid breaks down to form the compounds water and carbon dioxide gas. You can cause this chemical change by simply opening a bottle of a carbonated drink to release the pressure, as shown in **Figure 9.** The carbon dioxide and water that are formed can be further broken down into the elements carbon, oxygen, and hydrogen through additional chemical changes.

Figure 9 *Opening a carbonated drink can be messy as carbonic acid breaks down into two simpler compounds—carbon dioxide and water.*

physics CONNECTION

The process of using electric current to break compounds into simpler compounds and elements is known as electrolysis. The amount of compounds and elements produced depends on the amount of the original compound present, the amount of current, and the length of time the electric current flows. The elements aluminum and copper and the compound hydrogen peroxide are important industrial products obtained through electrolysis.

Compounds Cannot Be Broken Down by Physical Changes

The only way to break down a compound is through a chemical change. If you pour water through a filter, the water will pass through the filter unchanged. Filtration is a physical change, so it cannot be used to break down a compound. Likewise, a compound cannot be broken down by grinding it into a powder or by any other physical process.

One chemical process that can be used to break down compounds is electrolysis. In electrolysis, electric current is used to break down compounds. Electrolysis can be used to separate water into hydrogen and oxygen.

Compounds in Your World

You are always surrounded by compounds. Compounds make up the food you eat, the school supplies you use, the clothes you wear—even you! Compounds are formed and broken down every day in nature and in industry.

Figure 10 *The bumps on the roots of this pea plant are home to bacteria that form compounds from atmospheric nitrogen. The pea plant makes proteins from these compounds.*

Compounds in Nature Proteins are compounds found in all living things. The element nitrogen is needed to make proteins. Although nitrogen makes up 78 percent of the atmosphere, plants and animals cannot use nitrogen directly from the air. Some plants overcome this problem as shown in **Figure 10.** Other plants use nitrogen compounds that are in the soil. Animals get the nitrogen they need by eating plants or by eating animals that have eaten plants. As an animal digests food, the proteins in the food are broken down into smaller compounds that the animal's cells can use. That's why you need to eat your fruits and vegetables!

Another compound that plays an important role in life is carbon dioxide. You exhale carbon dioxide that was made in your body. Plants take in carbon dioxide and use it to make other compounds, including sugar.

Compounds in Industry To help plants get enough nitrogen, a nitrogen compound called ammonia is manufactured for use in fertilizers. The element nitrogen is combined with the element hydrogen to form ammonia. Plants can use ammonia as a source of nitrogen for their proteins. Other manufactured compounds are used in medicines, food preservatives, and synthetic fabrics.

The compounds found in nature are usually not the raw materials needed by industry. Often, these compounds must be broken down to form elements needed as raw material. For example, the element aluminum is used in cans, airplanes, and building materials, but it is not found alone in nature. Aluminum is produced by breaking down the compound aluminum oxide.

REVIEW

1. What is a compound?
2. What type of change is needed to break down a compound?
3. **Analyzing Ideas** A jar contains samples of the element carbon and oxygen. Does the jar contain a compound? Explain.

LabBook

Help keep the fireworks colorful on page 530 of the LabBook.

Section 3

Mixtures

NEW TERMS

mixture
solution
solute
solvent
alloys
concentration
solubility
suspension
colloid

OBJECTIVES

- Describe the properties of mixtures.
- Describe methods of separating the components of a mixture.
- Analyze a solution in terms of its solute, solvent, and concentration.
- Compare the properties of solutions, suspensions, and colloids.

Have you ever made your own pizza? You roll out the dough, add a layer of tomato sauce, then add toppings like green peppers, mushrooms, and olives—maybe even some pepperoni! Sprinkle cheese on top, and you're ready to bake. You have just created not only a pizza but also a mixture—and a delicious one at that!

Properties of Mixtures

All mixtures—even pizza—share certain properties. A **mixture** is a combination of two or more substances that are not chemically combined. Two or more materials together form a mixture if they do not react to form a compound. For example, cheese and tomato sauce do not react when they are used to make a pizza.

Substances in a Mixture Retain Their Identity Because no chemical change occurs, each substance in a mixture has the same chemical makeup it had before the mixture formed. That is, each substance in a mixture keeps its identity. In some mixtures, such as the pizza above or the piece of granite shown in **Figure 11,** you can even see the individual components. In other mixtures, such as salt water, you cannot see all the components.

Figure 11 *Colorless quartz, pink feldspar, and black mica make up the mixture granite. You can identify each component because the identity of a substance does not change in a mixture.*

Mixtures Can Be Physically Separated If you don't like mushrooms on your pizza, you can simply pick them off with your fingers. This separation is a physical change of the mixture, because the identities of the substances are not changed in the process. In contrast, compounds can be broken down only through chemical changes.

Each substance in a mixture retains most of its characteristic properties, such as density and boiling point. In fact, these properties can often be used to separate the components of a mixture. You cannot simply pick salt out of a saltwater mixture, but you can separate the salt from the water by heating the mixture. The water undergoes a physical change by vaporizing, or changing state from a liquid to a gas. The salt remains behind as a white solid. Several common techniques for separating mixtures are shown on the following page.

Common Techniques for Separating Mixtures

Distillation is a process that separates a mixture based on the boiling points of the components. Here you see pure water being distilled from a saltwater mixture. In addition to water purification, distillation is used to separate crude oil into its components, including gasoline and kerosene.

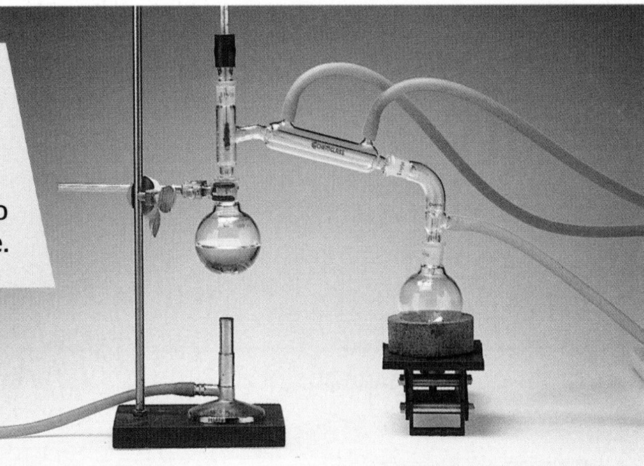

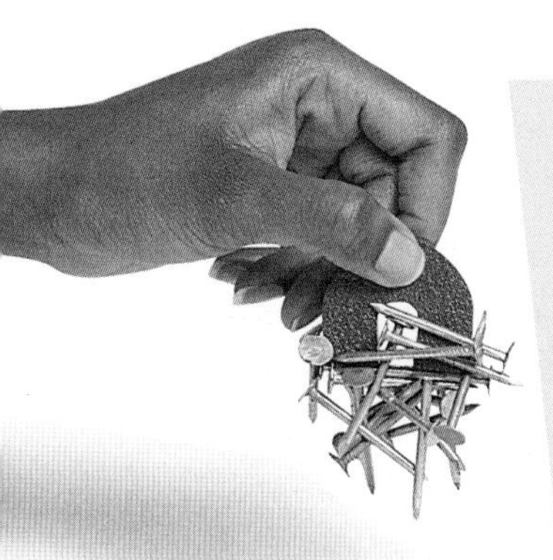

A magnet can be used to separate a mixture of the elements iron and aluminum. Iron is attracted to the magnet, but aluminum is not.

The components that make up blood are separated using a machine called a centrifuge. This machine separates mixtures according to the densities of the components.

A mixture of the compound sodium chloride (table salt) with the element sulfur requires more than one separation step.

The first step is to mix them with another compound—water. Salt dissolves in water, but sulfur does not.

In the second step, the mixture is poured through a filter. The filter traps the solid sulfur.

In the third step, the sodium chloride is separated from the water by simply evaporating the water.

Mixtures vs. Compounds	
Mixtures	**Compounds**
Components are elements, compounds, or both	Components are elements
Components keep their original properties	Components lose their original properties
Separated by physical means	Separated by chemical means
Formed using any ratio of components	Formed using a set mass ratio of components

The Components of a Mixture Do Not Have a Definite Ratio Recall that a compound has a specific mass ratio of the elements that form it. Unlike compounds, the components of a mixture do not need to be combined in a definite ratio. For example, granite that has a greater amount of feldspar than mica or quartz appears to have a pink color. Granite that has a greater amount of mica than feldspar or quartz appears black. Regardless of which ratio is present, this combination of materials is always a mixture—and it is always called granite.

Air is a mixture composed mostly of nitrogen and oxygen, with smaller amounts of other gases, such as carbon dioxide and water vapor. Some days the air has more water vapor, or is more humid, than on other days. But regardless of the ratio of the components, air is still a mixture.

The chart at left summarizes the differences between mixtures and compounds.

REVIEW

1. What is a mixture?
2. Is a mixture separated by physical or chemical changes?
3. **Applying Concepts** Suggest a procedure to separate iron filings from sawdust. Explain why this procedure works.

Solutions

A **solution** is a mixture that appears to be a single substance but is composed of particles of two or more substances that are distributed evenly amongst each other. Solutions are often described as *homogeneous mixtures* because they have the same appearance and properties throughout the mixture.

The process in which particles of substances separate and spread evenly throughout a mixture is known as *dissolving*. In solutions, the **solute** is the substance that is dissolved, and the **solvent** is the substance in which the solute is dissolved. A solute is *soluble,* or able to dissolve, in the solvent. A substance that is *insoluble,* or unable to dissolve, forms a mixture that is not homogeneous and therefore is not a solution.

Salt water is a solution. Salt is soluble in water, meaning that salt dissolves in water. Therefore, salt is the solute and water is the solvent. When two liquids or two gases form a solution, the substance with the greater volume is the solvent.

Many substances are soluble in water, including salt, sugar, alcohol, and oxygen. Water does not dissolve everything, but it dissolves so many different solutes that it is often called the universal solvent.

You may think of solutions as being liquids. And, in fact, tap water, soft drinks, gasoline, and many cleaning supplies are liquid solutions. However, solutions may also be gases, such as air, and solids, such as steel. **Alloys** are solid solutions of metals or nonmetals dissolved in metals. Brass is an alloy of the metal zinc dissolved in copper. Steel, including that used to build the *Titanic,* is an alloy made of the nonmetal carbon and other elements dissolved in iron. Look at the chart below for examples of the different states of matter used as solutes and solvents in solutions.

Self-Check

Yellow gold is an alloy made from equal parts copper and silver combined with a greater amount of gold. Identify each component of yellow gold as a solute or solvent. *(See page 596 to check your answer.)*

Examples of Different States in Solutions	
Gas in gas	Dry air (oxygen in nitrogen)
Gas in liquid	Soft drinks (carbon dioxide in water)
Liquid in liquid	Antifreeze (alcohol in water)
Solid in liquid	Salt water (salt in water)
Solid in solid	Brass (zinc in copper)

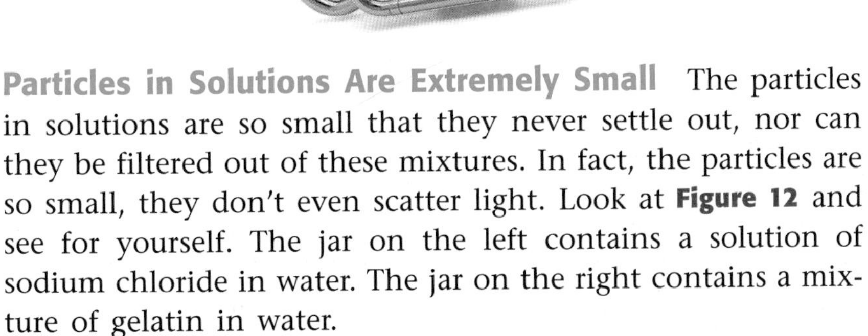

Particles in Solutions Are Extremely Small The particles in solutions are so small that they never settle out, nor can they be filtered out of these mixtures. In fact, the particles are so small, they don't even scatter light. Look at **Figure 12** and see for yourself. The jar on the left contains a solution of sodium chloride in water. The jar on the right contains a mixture of gelatin in water.

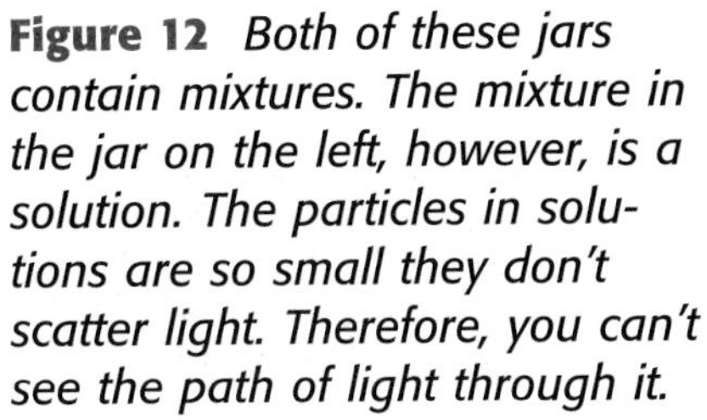

Figure 12 *Both of these jars contain mixtures. The mixture in the jar on the left, however, is a solution. The particles in solutions are so small they don't scatter light. Therefore, you can't see the path of light through it.*

÷ 5 ÷ Ω ≤ ∞ +Ω √ 9 ∞ ≤ Σ 2 +

MATHBREAK

Calculating Concentration

Many solutions are colorless. Therefore, you cannot always compare the concentrations of solutions by looking at the color—you have to compare the actual calculated concentrations. One way to calculate the concentration of a liquid solution is to divide the grams of solute by the milliliters of solvent. For example, the concentration of a solution in which 35 g of salt is dissolved in 175 mL of water is

$$\frac{35 \text{ g salt}}{175 \text{ mL water}} = 0.2 \text{ g/mL}$$

Now It's Your Turn

Calculate the concentrations of each solution below. Solution A has 55 g of sugar dissolved in 500 mL of water. Solution B has 36 g of sugar dissolved in 144 mL of water. Which solution is more dilute? Which is more concentrated?

Smelly solutions? Follow your nose and learn more on page 102.

Concentration: How Much Solute Is Dissolved? A measure of the amount of solute dissolved in a solvent is **concentration.** Concentration can be expressed in grams of solute per milliliter of solvent. Knowing the exact concentration of a solution is very important in chemistry and medicine because using the wrong concentration can be dangerous.

Solutions can be described as being *concentrated* or *dilute.* Look at **Figure 13.** Both solutions have the same amount of solvent, but the solution on the left contains less solute than the solution on the right. The solution on the left is dilute while the solution on the right is concentrated. Keep in mind that the terms *concentrated* and *dilute* do not specify the amount of solute that is actually dissolved. Try your hand at calculating concentration and describing solutions as concentrated or dilute in the MathBreak at left.

Figure 13 *The dilute solution on the left contains less solute than the concentrated solution on the right.*

A solution that contains all the solute it can hold at a given temperature is said to be *saturated.* An *unsaturated* solution contains less solute than it can hold at a given temperature. More solute can dissolve in an unsaturated solution.

Solubility: How Much Solute Can Dissolve? If you add too much sugar to a glass of lemonade, not all of the sugar can dissolve. Some of the sugar collects on the bottom of the glass. To determine the maximum amount of sugar that can dissolve, you would need to know the solubility of sugar. The **solubility** of a solute is the amount of solute needed to make a saturated solution using a given amount of solvent at a certain temperature. Solubility is usually expressed in grams of solute per 100 mL of solvent. **Figure 14** on the next page shows the solubility of several different substances in water at different temperatures.

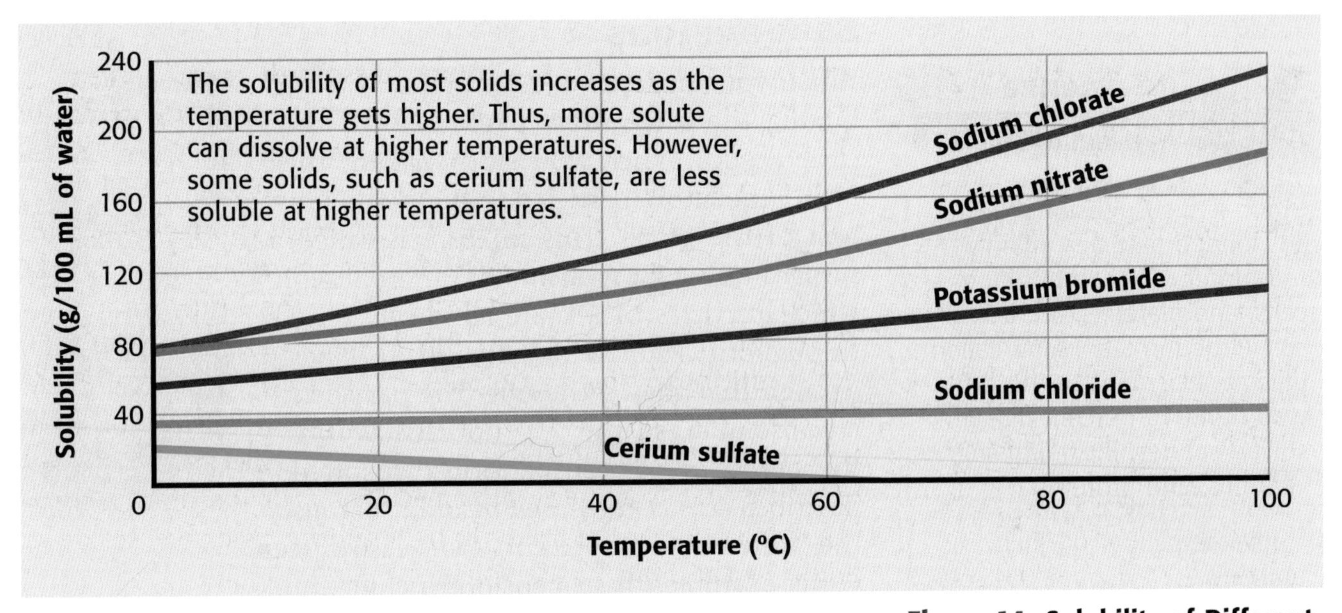

Figure 14 Solubility of Different Substances

Unlike the solubility of most solids in liquids, the solubility of gases in liquids decreases as the temperature is raised. As you heat a pot of water, bubbles of gas appear in the water long before the water begins to boil. The gases that are dissolved in the water cannot remain dissolved as the temperature increases because the solubility of the gases is lower at higher temperatures.

What Affects How Quickly Solids Dissolve in Liquids?

Many of the solutions used in your science class and in your home are formed when a solid solute is dissolved in water. Several factors affect how fast the solid will dissolve. Look at **Figure 15** to see three methods used to make a solute dissolve faster. You can see why you will enjoy a glass of lemonade sooner if you stir granulated sugar into the lemonade before adding ice!

Figure 15 *Mixing, heating, and crushing iron(III) chloride increase the speed at which it will dissolve.*

Mixing by stirring or shaking causes the solute particles to separate from one another and spread out more quickly among the solvent particles.

Heating causes particles to move more quickly. The solvent particles can separate the solute particles and spread them out more quickly.

Crushing the solute increases the amount of contact between the solute and the solvent. The particles of solute mix with the solvent more quickly.

Suspensions

When you shake up a snow globe, you are mixing the solid snow particles with the clear liquid. When you stop shaking the globe, the snow particles settle to the bottom of the globe. This mixture is called a suspension. A **suspension** is a mixture in which particles of a material are dispersed throughout a liquid or gas but are large enough that they settle out. The particles are insoluble, so they do not dissolve in the liquid or gas. Suspensions are often described as *heterogeneous mixtures* because the different components are easily seen. Other examples of suspensions include muddy water and Italian salad dressing.

life science CONNECTION

Blood is a suspension. The suspended particles, mainly red blood cells, white blood cells, and platelets, are actually suspended in a solution called plasma. Plasma is 90 percent water and 10 percent dissolved solutes, including sugar, vitamins, and proteins.

The particles in a suspension are fairly large, and they scatter or block light. This often makes a suspension difficult to see through. But the particles are too heavy to remain mixed without being stirred or shaken. If a suspension is allowed to sit undisturbed, the particles will settle out, as in a snow globe.

A suspension can be separated by passing it through a filter. The liquid or gas passes through, but the solid particles are large enough to be trapped by the filter, as shown in **Figure 16.**

Figure 16 *Dirty air is a suspension that could damage a car's engine. The air filter in a car separates dust from air to keep the dust from getting into the engine.*

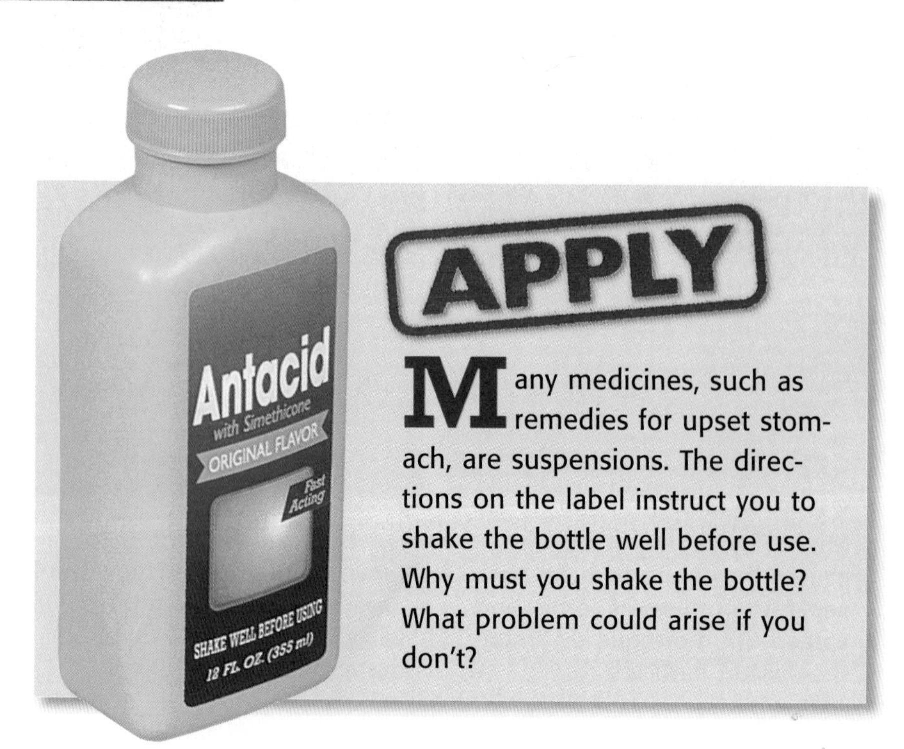

APPLY

Many medicines, such as remedies for upset stomach, are suspensions. The directions on the label instruct you to shake the bottle well before use. Why must you shake the bottle? What problem could arise if you don't?

Colloids

Some mixtures have properties of both solutions and suspensions. These mixtures are known as colloids (KAWL OYDZ). A **colloid** is a mixture in which the particles are dispersed throughout but are not heavy enough to settle out. The particles in a colloid are relatively small and are fairly well mixed. Solids, liquids, and gases can be used to make colloids. You might be surprised at the number of colloids you encounter each day. Milk, mayonnaise, stick deodorant—even the gelatin and whipped cream in **Figure 17**—are colloids. The materials that compose these products do not separate between uses because their particles do not settle out.

Figure 17 *This dessert includes two delicious examples of colloids—fruity gelatin and whipped cream.*

Although the particles in a colloid are much smaller than the particles in a suspension, they are still large enough to scatter a beam of light shined through the colloid, as shown in **Figure 18.** Finally, unlike a suspension, a colloid cannot be separated by filtration. The particles are small enough to pass through a filter.

Figure 18 *The colloid fog can create a dangerous situation for drivers. The particles in fog scatter light, making it difficult for drivers to see the road ahead.*

REVIEW

1. List two methods of making a solute dissolve faster.
2. Identify the solute and solvent in a solution made from 15 g of oxygen and 5 g of helium.
3. **Comparing Concepts** What are three differences between solutions and suspensions?

Make a colloid found in your kitchen on page 533 of the LabBook.

Chapter Highlights

SECTION 1

Vocabulary

element *(p. 82)*
pure substance *(p. 82)*
metals *(p. 85)*
nonmetals *(p. 85)*
metalloids *(p. 85)*

Section Notes

- A substance in which all the particles are alike is a pure substance.
- An element is a pure substance that cannot be broken down into anything simpler by physical or chemical means.
- Each element has a unique set of physical and chemical properties.
- Elements are classified as metals, nonmetals, and metalloids based on their properties.

SECTION 2

Vocabulary

compound *(p. 86)*

Section Notes

- A compound is a pure substance composed of two or more elements chemically combined.
- Each compound has a unique set of physical and chemical properties that are different from the properties of the elements that compose it.
- The elements that form a compound always combine in a specific ratio according to their masses.
- Compounds can be broken down into simpler substances by chemical changes.

Labs

Flame Tests *(p. 530)*

☑ Skills Check

Math Concepts

CONCENTRATION The concentration of a solution is a measure of the amount of solute dissolved in a solvent. For example, a solution is formed by dissolving 85 g of sodium nitrate in 170 mL of water. The concentration of the solution is calculated as follows:

$$\frac{85 \text{ g sodium nitrate}}{170 \text{ mL water}} = 0.5 \text{ g/mL}$$

Visual Understanding

THREE CATEGORIES OF ELEMENTS Elements are classified into metals, nonmetals, and metalloids, based on their properties. The chart on page 85 provides a summary of the properties that distinguish each category.

SEPARATING MIXTURES Mixtures can be separated through physical changes based on differences in the physical properties of their components. Review the illustrations on page 91 for some techniques for separating mixtures.

SECTION 3

Vocabulary

mixture *(p. 90)*
solution *(p. 92)*
solute *(p. 92)*
solvent *(p. 92)*
alloys *(p. 93)*
concentration *(p. 94)*
solubility *(p. 94)*
suspension *(p. 96)*
colloid *(p. 97)*

Section Notes

- A mixture is a combination of two or more substances, each of which keeps its own characteristics.
- Mixtures can be separated by physical means, such as filtration and evaporation.
- The components of a mixture can be mixed in any proportion.
- A solution is a mixture that appears to be a single substance but is composed of a solute dissolved in a solvent. Solutions do not settle, cannot be filtered, and do not scatter light.
- Concentration is a measure of the amount of solute dissolved in a solvent.

- The solubility of a solute is the amount of solute needed to make a saturated solution using a given amount of solvent at a certain temperature.
- Suspensions are heterogeneous mixtures that contain particles large enough to settle out, be filtered, and block or scatter light.
- Colloids are mixtures that contain particles too small to settle out or be filtered but large enough to scatter light.

Labs

A Sugar Cube Race! *(p. 532)*
Making Butter *(p. 533)*
Unpolluting Water *(p. 534)*

internet**connect**

GO TO: go.hrw.com

Visit the **HRW** Web site for a variety of learning tools related to this chapter. Just type in the keyword:

KEYWORD: HSTMIX

GO TO: www.scilinks.org

Visit the **National Science Teachers Association** on-line Web site for Internet resources related to this chapter. Just type in the ***sci*LINKS** number for more information about the topic:

TOPIC: The *Titanic* — ***sci*LINKS NUMBER:** HSTP080
TOPIC: Elements — ***sci*LINKS NUMBER:** HSTP085
TOPIC: Compounds — ***sci*LINKS NUMBER:** HSTP090
TOPIC: Mixtures — ***sci*LINKS NUMBER:** HSTP095

Chapter Review

USING VOCABULARY

Complete the following sentences by choosing the appropriate term from the vocabulary list to fill in each blank:

1. A __?__ has a definite ratio of components.
2. The amount of solute needed to form a saturated solution is the __?__ of the solute.
3. A __?__ can be separated by filtration.
4. A pure substance must be either a(n) __?__ or a(n) __?__.
5. Elements that are brittle and dull are __?__.
6. The substance that dissolves to form a solution is the __?__.

UNDERSTANDING CONCEPTS

Multiple Choice

7. Which of the following increases the solubility of a gas in a liquid?
   a. increasing the temperature
   b. stirring
   c. decreasing the temperature
   d. decreasing the amount of liquid

8. Which of the following best describes chicken noodle soup?
   a. element
   b. mixture
   c. compound
   d. solution

9. Which of the following does not describe elements?
   a. all the particles are alike
   b. can be broken down into simpler substances
   c. have unique sets of properties
   d. can join together to form compounds

10. A solution that contains a large amount of solute is best described as
    a. unsaturated.
    b. concentrated.
    c. dilute.
    d. weak.

11. Which of the following substances can be separated into simpler substances only by chemical means?
    a. sodium
    b. salt water
    c. water
    d. gold

12. Which of the following would not increase the rate at which a solid dissolves?
    a. decreasing the temperature
    b. crushing the solid
    c. stirring
    d. increasing the temperature

13. An element that conducts thermal energy well and is easily shaped is a
    a. metal.
    b. metalloid.
    c. nonmetal.
    d. None of the above

14. In which classification of matter are the components chemically combined?
    a. alloy
    b. colloid
    c. compound
    d. suspension

Short Answer

15. What is the difference between an element and a compound?

16. Identify the solute and solvent if nail polish is dissolved in acetone.

Concept Mapping

17. Use the following terms to create a concept map: matter, element, compound, mixture, solution, suspension, colloid.

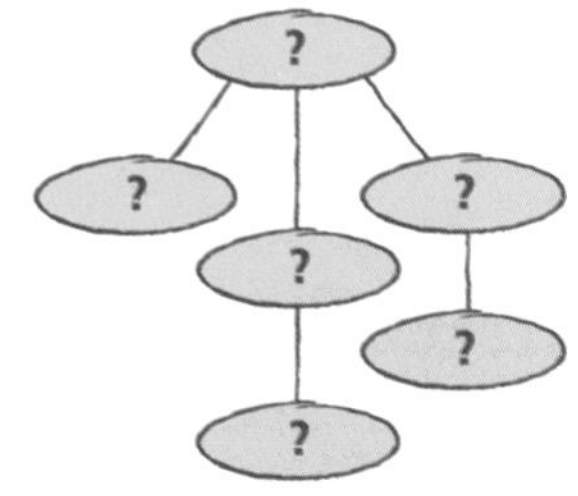

CRITICAL THINKING AND PROBLEM SOLVING

18. Describe a procedure to separate a mixture of salt, finely ground pepper, and pebbles.

19. A light green powder is heated in a test tube. A gas is given off, while the solid becomes black. In which classification of matter does the green powder belong? Explain your reasoning.

20. Why is it desirable to know the exact concentration of solutions rather than whether they are concentrated or dilute?

21. Explain the three properties of mixtures using a fruit salad as an example.

22. To keep the "fizz" in carbonated beverages after they have been opened, should you store them in a refrigerator or in a cabinet? Explain.

MATH IN SCIENCE

23. What is the concentration of a solution prepared by mixing 50 g of salt with 200 mL of water?

24. How many grams of sugar must be dissolved in 150 mL of water to make a solution with a concentration of 0.6 g/mL?

INTERPRETING GRAPHICS

25. Use Figure 14 on page 95 to answer the following questions:
    a. Can 50 g of sodium chloride dissolve in 100 mL of water at 60°C?
    b. How much cerium sulfate is needed to make a saturated solution in 100 mL of water at 30°C?
    c. Is sodium chloride or sodium nitrate more soluble in water at 20°C?

26. Dr. Sol Vent tested the solubility of a compound. The data below was collected using 100 mL of water. Graph Dr. Vent's results. To increase the solubility, would you increase or decrease the temperature? Explain.

Temperature (°C)	10	25	40	60	95
Dissolved solute (g)	150	70	34	25	15

27. What type of mixture is shown in the photo below? Explain.

NOW What Do You Think?

Take a minute to review your answers to the ScienceLog questions on page 81. Have your answers changed? If necessary, revise your answers based on what you have learned since you began this chapter.

Science, Technology, and Society

Perfume: Fragrant Solutions

Making perfumes is an ancient art. It was practiced, for example, by the ancient Egyptians, who rubbed their bodies with a substance made by soaking fragrant woods and resins in water and oil. From certain references and formulas in the Bible, we know that the ancient Israelites also practiced the art of perfume making. Other sources indicate that the art was also known to the early Chinese, Arabs, Greeks, and Romans.

▲ *Perfumes have been found in the tombs of Egyptians who lived more than 3,000 years ago.*

Only the E-scent-ials

Over time, perfume making has developed into a complicated art. A fine perfume may contain more than 100 different ingredients. The most familiar ingredients come from fragrant plants or flowers, such as sandalwood or roses. These plants get their pleasant odor from their essential oils, which are stored in tiny baglike parts called sacs. The parts of plants that are used for perfumes include the flowers, roots, and leaves. Other perfume ingredients come from animals and from man-made chemicals.

Making Scents

Perfume makers first remove essential oils from the plants using distillation or reactions with solvents. Then the essential oils are blended with other ingredients to create perfumes. Fixatives, which usually come from animals, make the other odors in the perfume last longer. Oddly enough, most natural fixatives smell awful! For example, civet musk is a foul-smelling liquid that the civet cat sprays on its enemies.

Taking Notes

When you take a whiff from a bottle of perfume, the first odor you detect is called the top note. It is a very fragrant odor that evaporates rather quickly. The middle note, or modifier, adds a different character to the odor of the top note. The base note, or end note, is the odor that lasts the longest.

▲ *Not all perfume ingredients smell good. The foul-smelling oil from the African civet cat is used as a fixative in some perfumes.*

Smell for Yourself

▶ Test a number of different perfumes and colognes to see if you can identify three different notes in each.

Science Fiction

"The Strange Case of Dr. Jekyll and Mr. Hyde"

by Robert Louis Stevenson

A violent "monkey man" murders an old gentleman. A respectable scientist, wealthy and handsome, commits suicide. Are these two tragedies connected in some way?

Dr. Henry Jekyll is an admirable member of society. He is a doctor and a scientist. Although wild as a young man, Jekyll has become cold and analytical as he has aged and pursued his scientific theories. Now he wants to understand the nature of human identity. He wants to explore the different parts of the human personality that usually fit together smoothly to make a complete person. His theory is that if he can separate his personality into "good" and "evil" parts, he can get rid of his evil side and lead a happy, useful life. So Jekyll develops a chemical mixture that will allow him to test his theory. The results are startling!

Who is the mysterious Mr. Hyde? He is not a scientist. He is a man of action and anger, who sparks fear in the hearts of those he comes in contact with. Where did he come from? What does he do? How can local residents be protected from his wrath?

Robert Louis Stevenson's story of the decent doctor Henry Jekyll and the violent Edward Hyde is a classic science-fiction story. When Jekyll mixes his "salts" and drinks his chemical mixture, he changes his life—and Edward Hyde's—completely. To find out more, read Stevenson's "The Strange Case of Dr. Jekyll and Mr. Hyde" in the *Holt Anthology of Science Fiction.*

TIMELINE

UNIT 2

Motion and Forces

It's hard to imagine a world where nothing ever moves. Without motion or forces to cause motion, life would be very dull! The relationship between force and motion is the subject of this unit. You will learn how to describe the motion of objects, how forces affect motion, and how fluids exert force. This timeline shows some events and discoveries that have occurred as scientists have worked to understand the motion of objects here on Earth and in space.

Around 250 B.C.

Archimedes, a Greek mathematician, develops the principle that bears his name. The principle relates the buoyant force on an object in a fluid to the amount of fluid the object displaces.

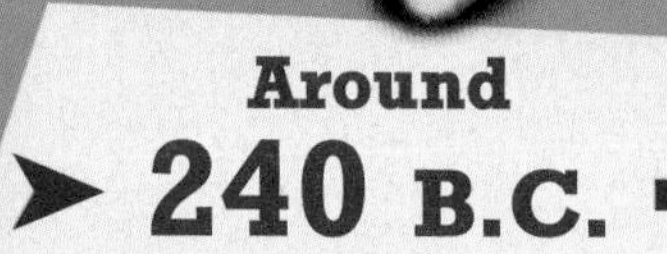

Around 240 B.C.

Chinese astronomers are the first to record a sighting of Halley's Comet.

1905

While employed as a patent clerk, German physicist Albert Einstein publishes his special theory of relativity. The theory states that the speed of light is constant, no matter what the reference point is.

1921

Bessie Coleman becomes the first African-American woman licensed to fly an airplane.

1947

While flying a Bell X-1 rocket-powered airplane, American pilot Chuck Yeager becomes the first human to travel faster than the speed of sound.

1519

Portuguese explorer Ferdinand Magellan begins the first voyage around the world.

1687

Sir Isaac Newton, a British mathematician and scientist, publishes *Principia,* a book describing his laws of motion and the law of universal gravitation.

PHILOSOPHIÆ NATURALIS PRINCIPIA MATHEMATICA.

Autore J S. NEWTON, Trin. Coll. Cantab. Soc. Matheseos Professore Lucasiano, & Societatis Regalis Sodali.

1764

In London, Wolfgang Amadeus Mozart composes his first symphony—at the age of 9.

1846

After determining that the orbit of Uranus is different from what is predicted from the law of universal gravitation, scientists discover Neptune, shown here, whose gravitational force is causing Uranus's unusual orbit.

1971

American astronaut Alan Shepard takes a break from gathering lunar data to play golf on the moon during the *Apollo 14* mission.

1990

The *Magellan* spacecraft begins orbiting Venus for a four-year mission to map the planet. The spacecraft uses the sun's gravitational force to propel it to Venus without burning much fuel.

1999

NASA launches the Mars *Polar Lander* spacecraft, one of a series sent to explore Mars.

CHAPTER 5

Matter in Motion

Would You Believe ...?

Detail of *Ball Play of the Choctaw* by George Catlin, National Museum of American Art, Smithsonian Institution, Washington D.C./ Art Resouce, N.Y.

Suppose you were told that there was once a game that could be played by as few as 5 or as many as 1,000 players. The game could be played on a small field for a few hours or on a huge tract of land for several days. The game was not just for fun—in fact, it was so important that it was often used as a substitute for war. One of the few rules was that the players couldn't touch the ball with their hands—they could only catch and throw the ball using a special stick with webbing on one end. Does this game sound unbelievable? Welcome to the history of lacrosse!

Lacrosse is a game that was originally played by Native Americans, as shown above. They called the game *baggataway* (bag AT uh way), which means "little brother of war." Although lacrosse has changed much and is now played by men and women all over the world, the game still requires special webbed sticks.

Using a lacrosse stick, a player can throw the ball at speeds well over 100 km/h. At this speed, you wouldn't want to catch the ball with your bare hands—unless you wanted to experience force firsthand (no pun intended)! You see, in order to move the ball this fast, a large force (push or pull) has to be supplied. Likewise, in order to stop a ball moving this fast, another force has to be supplied. If this force were supplied by your bare hand, it would probably hurt!

Three generations of Native American lacrosse players

Even though you may have never played lacrosse, you have certainly experienced motion and the forces that cause it or prevent it. As you read this chapter, you shouldn't have any trouble thinking of your own examples.

What Do You Think?

In your ScienceLog, try to answer the following questions based on what you already know:

1. How is motion measured?
2. What is a force?
3. How does friction affect motion?
4. How does gravity affect objects?

The Domino Derby

You are probably familiar with the term *speed*. Speed is the rate at which an object moves. In this activity, you will determine the factors that affect the speed of falling dominoes.

Procedure

1. Set up **25 dominoes** in a straight line. Try to keep equal spacing between the dominoes.
2. Using a **metric ruler,** measure the total length of your row of dominoes, and record it in your ScienceLog.
3. Using a **stopwatch,** time how long it takes for the entire row of dominoes to fall. Record the time in your ScienceLog.
4. Predict what would happen to that amount of time if you shortened or lengthened the distance between the dominoes. Write your predictions in your ScienceLog.
5. Repeat steps 2 and 3 several times, varying the distance between the dominoes each time. Be sure to try several spacings that are smaller and larger than in your original setup.

Analysis

6. Calculate the average speed for each trial by dividing the total distance (the length of the domino row) by the time taken to fall.
7. Comparing all the trials, how did the spacing between dominoes affect the average speed? Did your results confirm your predictions? If not, explain.

Going Further

Make a graph of your results. Explain why your graph has the shape that it does.

Section 1

Measuring Motion

NEW TERMS

motion
speed
velocity
acceleration

OBJECTIVES

- Identify the relationship between motion and a reference point.
- Identify the two factors that speed depends on.
- Determine the difference between speed and velocity.
- Analyze the relationship of velocity to acceleration.
- Interpret a graph showing acceleration.

Look around you—you're likely to see something in motion. Your teacher may be walking across the room, or perhaps a classmate is writing in her ScienceLog. You might even notice a bird flying outside a window. Even if you don't see anything moving, motion is still occurring all around you. Tiny air particles are whizzing around, the moon is circling the Earth, and blood is traveling through your veins and arteries!

Observing Motion

You might think that the motion of an object is easy to detect—you just have to observe the object. But there's more to it than that! You actually must observe the object in relation to another object that appears to stay in place. The object that appears to stay in place is a *reference point.* When an object changes position over time when compared with a reference point, the object is in **motion.** When an object is in motion, you can describe the direction of its motion with a reference direction. Typical reference directions are north, south, east, west, or up and down.

The Earth's surface is a common reference point for determining position and motion. Nonmoving objects on Earth's surface, such as buildings, trees, and mountains, are also useful reference points for observing motion, as shown in **Figure 1.**

A moving object can also be used as a reference point. For example, if you were on the hot-air balloon shown below, you could watch a bird flying overhead and see that it was changing position in relation to your moving balloon. Furthermore, Earth itself is a moving reference point—it is moving around the sun.

Figure 1 *During the time it took for these pictures to be taken, the hot-air balloon changed position compared with a reference point–the mountain. Therefore, the balloon was in motion.*

Speed Depends on Distance and Time

You can tell that the hot-air balloon in Figure 1 traveled a certain distance in the time interval between the two photographs. By dividing the distance traveled by the time it took to travel the distance, you can find the balloon's rate of motion. The rate at which an object moves is its **speed.** Speed depends on the distance traveled and the time taken to travel that distance. Suppose the time interval between the pictures was 10 seconds and the balloon traveled 50 m in that time. The speed (distance divided by time) of the balloon is 50 m/10 s, or 5 m/s.

The SI unit for speed is meters per second (m/s). Kilometers per hour, feet per second, and miles per hour are other units commonly used to express speed.

Determining Average Speed Most of the time, objects do not travel at constant speed. For example, you probably do not travel at a constant speed as you walk from one class to the next. Because objects do not often travel at a constant speed, it is useful to calculate an object's *average speed* using the following equation:

$$\text{Average speed} = \frac{\text{total distance}}{\text{total time}}$$

Suppose a person drives from one city to another. The blue line in the graph below shows the distance traveled every hour. Notice that the distance traveled every hour is different. This is because the speed (distance/time) is not constant—the driver changes speed often because of weather, traffic, or varying speed limits. The average speed can be calculated by adding up the total distance and dividing it by the total time:

$$\text{Average speed} = \frac{360\text{ km}}{4\text{ h}} = 90\text{ km/h}$$

The red line shows the average distance traveled each hour. The slope of this line is the average speed.

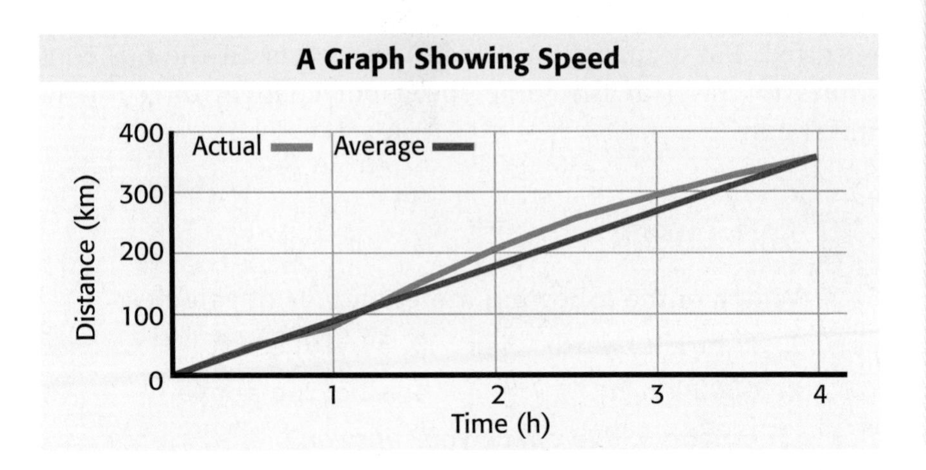

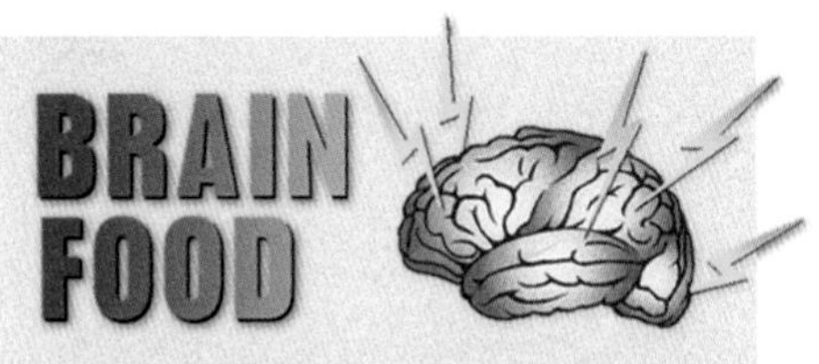

The list below shows a comparison of some interesting speeds:

Cockroach . . . . . . 1.25 m/s
Kangaroo . . . . 15 m/s
Cheetah (the fastest land animal). . . . . . . 27 m/s
Sound . . . . 330 m/s
Space shuttle . . . 10,000 m/s
Light . . . . . 300,000,000 m/s

MATH BREAK

Calculating Average Speed

Practice calculating average speed in the problems listed below:

1. If you take a walk for 1.5 hours and travel 7.5 km, what is your average speed?
2. A bird flies at a speed of 15 m/s for 10 s, 20 m/s for 10 s, and 25 m/s for 5 s. What is the bird's average speed?

Velocity: Direction Matters

Here's a riddle for you: Two birds leave the same tree at the same time. They both fly at 10 km/h for 1 hour, 15 km/h for 30 minutes, and 5 km/h for 1 hour. Why don't they end up at the same destination?

Have you figured it out? Even though the birds traveled at the same speeds for the same amounts of time, they did not end up at the same place because they went in different directions. In other words, the birds had different velocities. The speed of an object in a particular direction is the object's **velocity** (vuh LAHS uh tee).

Figure 2 *The speeds of these cars may be similar, but their velocities are different because they are going in different directions.*

Be careful not to confuse the terms *speed* and *velocity;* they do not mean the same thing. Because velocity must include direction, it would not be correct to say that an airplane's velocity is 600 km/h. However, you could say the plane's velocity is 600 km/h south. Velocity always includes a reference direction. **Figure 2** further illustrates the difference between speed and velocity.

Velocity Changes as Speed or Direction Changes You can think of velocity as the rate of change of an object's position. An object's velocity is constant only if its speed and direction don't change. Therefore, constant velocity is always along a straight line. An object's velocity will change if either its speed or direction changes. For example, if a bus traveling at 15 m/s south speeds up to 20 m/s, a change in velocity has occurred. But a change in velocity also occurs if the bus continues to travel at the same speed but changes direction to travel east.

Self-Check

Which of the following are examples of velocity?

1. 25 m/s forward
2. 1,500 km/h
3. 55 m/h south
4. all of the above

(See page 596 to check your answer.)

Combining Velocities If you're riding in the bus traveling east at 15 m/s, you and all the other passengers are also traveling at a velocity of 15 m/s east. But suppose you stand up and walk down the bus's aisle while it is moving. Are you still moving at the same velocity as the bus? No! **Figure 3** shows how you can combine velocities to determine the *resultant velocity*.

Figure 3 Determining Resultant Velocity

When you combine two velocities that are in the same direction, add them together to find the resultant velocity.

Person's resultant velocity

15 m/s east + 1 m/s east = 16 m/s east

When you combine two velocities that are in opposite directions, subtract the smaller velocity from the larger velocity to find the resultant velocity. The resultant velocity is in the direction of the larger velocity.

Person's resultant velocity

15 m/s east – 1 m/s west = 14 m/s east

REVIEW

1. What is a reference point?
2. What two things must you know to determine speed?
3. What is the difference between speed and velocity?
4. **Applying Concepts** Explain why it is important to know a tornado's velocity and not just its speed.

BRAIN FOOD

The space shuttle is always launched in the same direction that the Earth rotates, thus taking advantage of the Earth's rotational velocity (over 1,500 km/h east). This allows the shuttle to use less fuel to reach space than if it had to achieve such a great velocity on its own.

Acceleration: The Rate at Which Velocity Changes

Imagine that you are skating very quickly down the sidewalk on in-line skates. Up ahead, you see a large rock in your path. You slow down and swerve at the same time to keep from hitting the rock. A neighbor out in his yard exclaims, "That was great acceleration! I'm amazed that you could slow down and turn so quickly without falling!" You're puzzled. Doesn't *accelerate* mean to speed up—like when your parent presses the accelerator pedal in the car? But you didn't speed up—you slowed down and turned. So why did your neighbor say that you accelerated?

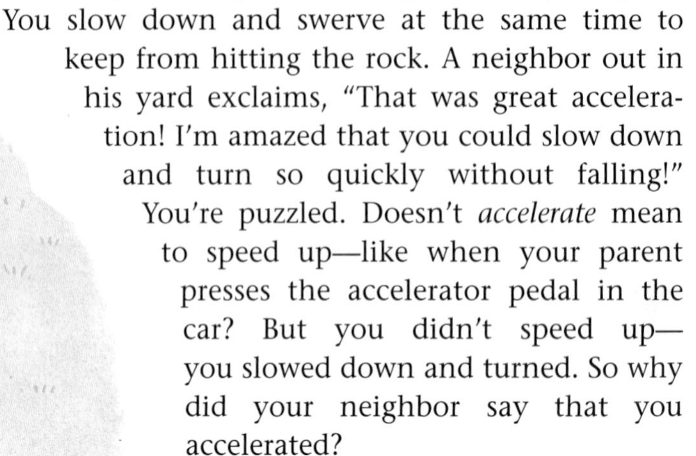

Although the word *accelerate* is commonly used in everyday language to mean "speed up," there's more to its meaning scientifically. **Acceleration** (ak SEL uhr AY shun) is the rate at which velocity changes. To *accelerate* means to change velocity. You just learned that velocity changes if speed changes, direction changes, or both. So your neighbor was right! Your speed and direction changed, so you accelerated.

Keep in mind that acceleration is not just the amount velocity changes. Acceleration tells you *how fast* velocity changes. The faster velocity changes, the greater the acceleration is.

Calculating Acceleration Acceleration is calculated using the following equation:

$$\text{Acceleration} = \frac{\text{final velocity} - \text{starting velocity}}{\text{time it takes to change velocity}}$$

The unit of measurement for velocity is meters per second (m/s), and the unit for time is seconds (s). Therefore, the unit of measurement for acceleration is meters per second per second (m/s/s).

Suppose you get on your bicycle and accelerate southward at a rate of 1 m/s/s. (Like velocity, acceleration has size and direction.) This means that every second, your southward velocity increases at a rate of 1 m/s, as shown on the next page.

Figure 4 Acceleration at 1 m/s/s South

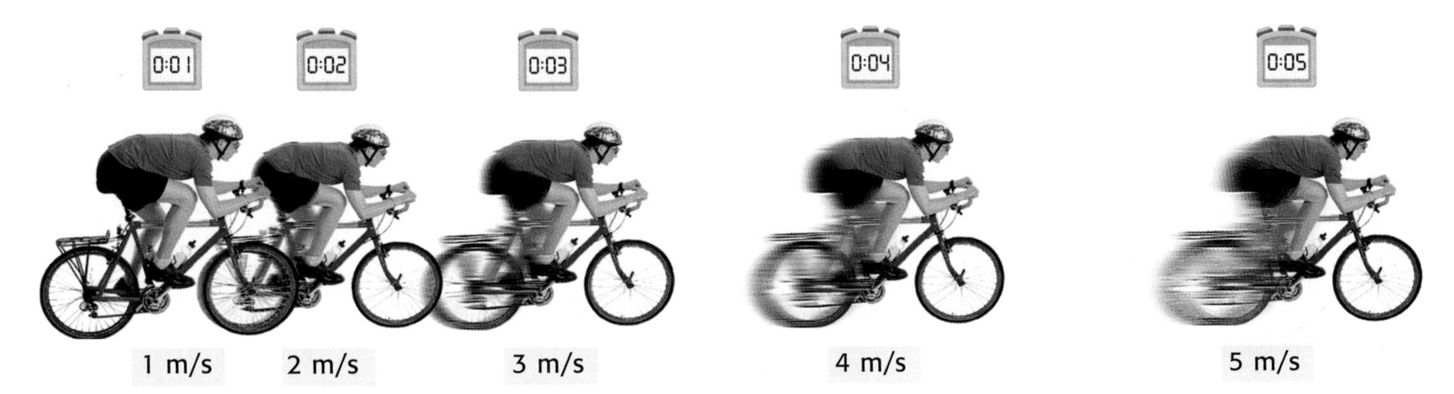

After 1 second, you have a velocity of 1 m/s south, as shown in **Figure 4.** After 2 seconds, you have a velocity of 2 m/s south. After 3 seconds, you have a velocity of 3 m/s south, and so on. If your final velocity after 5 seconds is 5 m/s south, your acceleration can be calculated as follows:

$$\text{Acceleration} = \frac{5\ \text{m/s} - 0\ \text{m/s}}{5\,\text{s}} = 1\ \text{m/s/s south}$$

You can practice calculating acceleration by doing the MathBreak shown here.

Examples of Acceleration In the example above, your velocity was originally zero and then it increased. Because your velocity changed, you accelerated. Acceleration in which velocity increases is sometimes called *positive acceleration.*

Acceleration also occurs when velocity decreases. In the skating example, you accelerated because you slowed down. Acceleration in which velocity decreases is sometimes called *negative acceleration* or *deceleration.*

Remember that velocity has direction, so velocity will change if your direction changes. Therefore, a change in direction is acceleration, even if there is no change in speed. Some more examples of acceleration are shown in the chart below.

Example of Acceleration	How Velocity Changes
A plane taking off	Increase in speed
A car stopping at a stop sign	Decrease in speed
Jogging on a winding trail	Change in direction
Wind gusting	Increase in speed
Driving around a corner	Change in direction
Standing at Earth's equator	Change in direction

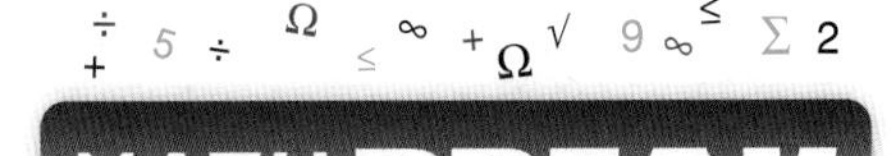

MATH BREAK

Calculating Acceleration

Use the equation shown on the previous page to do the following problems. Be sure to express your answer in m/s/s and include direction.

1. A plane passes over Point A with a velocity of 8,000 m/s north. Forty seconds later it passes over Point B at a velocity of 10,000 m/s north. What is the plane's acceleration from A to B?
2. A coconut falls from the top of a tree and reaches a velocity of 19.6 m/s when it hits the ground. It takes 2 seconds to reach the ground. What is the coconut's acceleration?

Figure 5 *The blades of this windmill are constantly changing direction as they travel in a circle. Thus, centripetal acceleration is occurring.*

Circular Motion: Continuous Acceleration Does it surprise you to find out that standing at Earth's equator is an example of acceleration? After all, you're not changing speed, and you're not changing direction . . . or are you? In fact, you are traveling in a circle as the Earth rotates. An object traveling in a circular motion is always changing its direction. Therefore, its velocity is always changing, so acceleration is occurring. The acceleration that occurs in circular motion is known as *centripetal* (sen TRIP uht uhl) *acceleration.* Another example of centripetal acceleration is shown in **Figure 5.**

Recognizing Acceleration on a Graph Suppose that you have just gotten on a roller coaster. The roller coaster moves slowly up the first hill until it stops at the top. Then you're off, racing down the hill! The graph below shows your velocity for the 10 seconds coming down the hill. You can tell from this graph that your acceleration is positive because your velocity increases as time passes. Because the graph is not a straight line, you can also tell that your acceleration is not constant for each second.

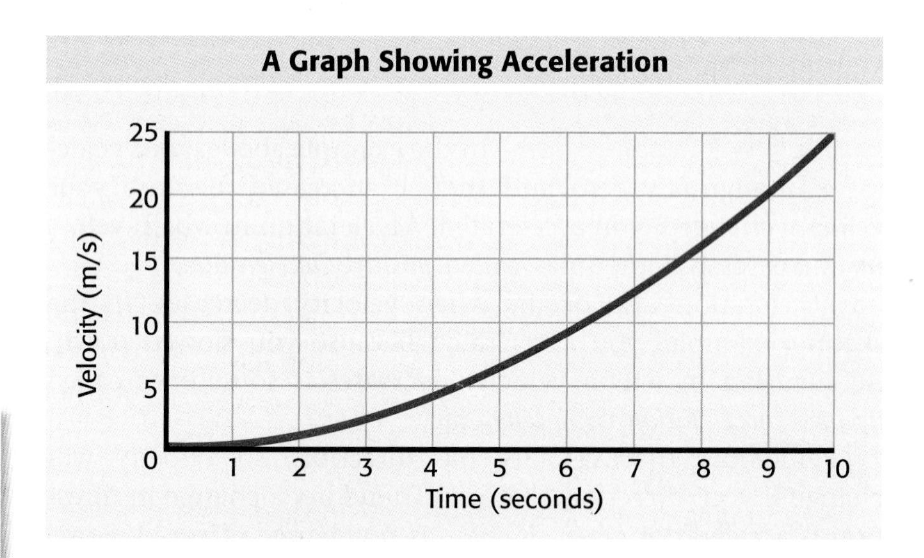

Use this simple device to "see" acceleration on page 537 of the LabBook.

REVIEW

1. What is acceleration?
2. Does a change in direction affect acceleration? Explain your answer.
3. **Interpreting Graphics** How do you think a graph of deceleration would differ from the graph shown above? Explain your reasoning.

Section 2

What Is a Force?

NEW TERMS

force
net force
newton (N)

OBJECTIVES

- Give examples of different kinds of forces.
- Determine the net force on an object.
- Compare balanced and unbalanced forces.

You often hear the word *force* in everyday conversation:

"That storm had a lot of force!"
"Our basketball team is a force to be reckoned with."
"A flat tire forced me to stop riding my bicycle."
"The inning ended with a force-out at second base."

But what exactly is a force? In science, a **force** is simply a push or a pull. All forces have both size and direction.

Forces are everywhere. In fact, any time you see something moving, you can be sure that its motion was created by a force. Scientists measure force with a unit called the **newton (N).** The more newtons, the greater the force.

Forces Act on Objects

All forces are exerted by one object on another object. For any push to occur, something has to receive the push. You can't push nothing! The same is true for any pull. When doing schoolwork, you use your fingers to pull open books or to push the buttons on a computer keyboard. In these examples, your fingers are exerting forces on the books and the keys. However, just because a force is being exerted by one object on another doesn't mean that motion will occur. For example, you are probably sitting on a chair as you read this. But the force you are exerting on the chair does not cause the chair to move. That's because the Earth is also exerting a force on the chair. In most cases, it is easy to determine where the push or pull is coming from, as shown in **Figure 6.**

Figure 6 *It is obvious that the bulldozer is exerting a force on the pile of soil. But did you know that the pile of soil also exerts a force, even when it is just sitting on the ground?*

Figure 7
Something unseen exerts a force that makes your socks cling together when they come out of the dryer. You have to exert a force to separate the socks.

However, it is not always so easy to tell what is exerting a force or what is receiving a force, as shown in **Figure 7.** You cannot see what exerts the force that pulls magnets to refrigerators, and the air you breathe is an unseen receiver of a force called *gravity.* You will learn more about gravity later in this chapter.

Forces in Combination

Often more than one force is exerted on an object at the same time. The **net force** is the force that results from combining all the forces exerted on an object. So how do you determine the net force? The examples below can help you answer this question.

Forces in the Same Direction Suppose you and a friend are asked to move a piano for the music teacher. To do this, you pull on one end of the piano, and your friend pushes on the other end. Together, your forces add up to enough force to move the piano. This is because your forces are in the same direction. **Figure 8** shows this situation. Because the forces are in the same direction, they can be added together to determine the net force. In this case, the net force is 45 N, which is plenty to move a piano—if it is on wheels, that is!

Figure 8 *When the forces are in the same direction, you add the forces together to determine the net force.*

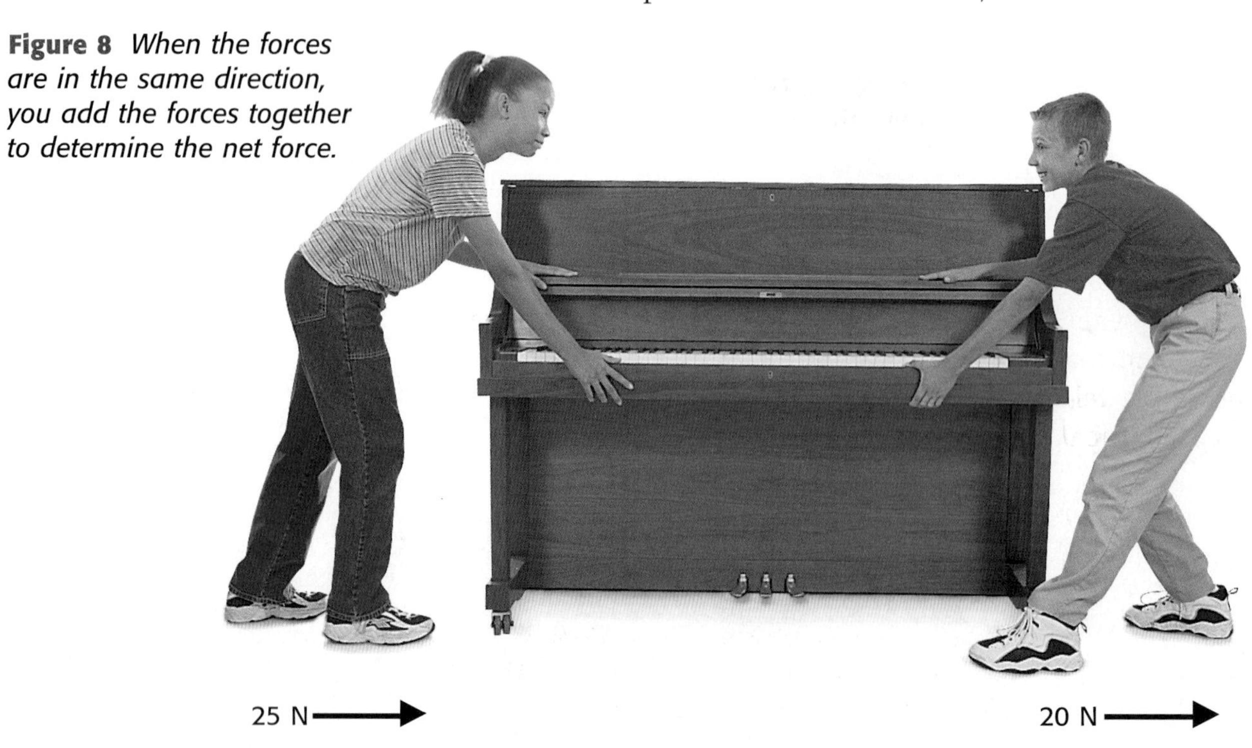

Net force
25 N + 20 N = 45 N
to the right

Forces in Different Directions Consider two dogs playing tug of war with a short piece of rope. Each is exerting a force, but in opposite directions. **Figure 9** shows this scene. Notice that the dog on the left is pulling with a force of 10 N and the dog on the right is pulling with a force of 12 N. Which dog do you think will win the tug of war?

Because the forces are in opposite directions, the net force is determined by subtracting the smaller force from the larger one. In this case, the net force is 2 N in the direction of the dog on the right. Give that dog a dog biscuit!

across the sciences CONNECTION

Every moment, forces in several directions are exerted on the Golden Gate Bridge. For example, Earth exerts a powerful downward force on the bridge while elastic forces pull and push portions of the bridge up and down. To learn how the bridge stands up to these forces, turn to page 135.

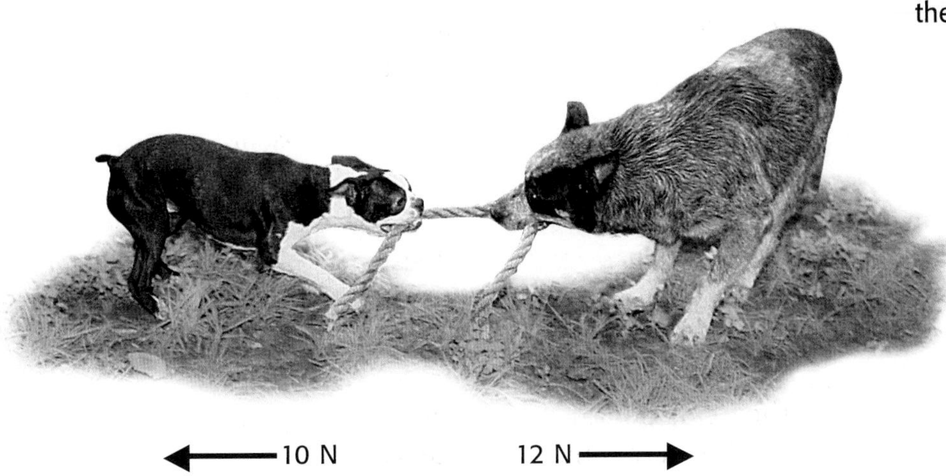

Net force
12 N − 10 N = 2 N
to the right

Figure 9 *When the forces are in different directions, you subtract the smaller force from the larger force to determine the net force.*

Unbalanced and Balanced Forces

If you know the net force on an object, you can determine the effect the force will have on the object's motion. Why? The net force tells you whether the forces on the object are balanced or unbalanced.

Unbalanced Forces Produce a Change in Motion In the examples shown in Figures 8 and 9, the net force on the object is greater than zero. When the net force on an object is not zero, the forces on the object are *unbalanced.* Unbalanced forces produce a change in motion (acceleration). In the two previous examples, the receivers of the forces—the piano and the rope—move. Unbalanced forces are necessary to cause a nonmoving object to start moving.

Self-Check

What is the net force when you combine a force of 7 N north with a force of 5 N south? *(See page 596 to check your answer.)*

Unbalanced forces are also necessary to change the motion of moving objects. For example, consider a soccer game. The soccer ball is already moving when it is passed from one player to another. When the ball reaches the second player, the player exerts an unbalanced force—a kick—on the ball. After the kick, the ball moves in a new direction and with a new speed.

Keep in mind that an object can continue to move even when the unbalanced forces are removed. A soccer ball, for example, receives an unbalanced force when it is kicked. However, the ball continues to roll along the ground long after the force of the kick has ended.

Balanced Forces Produce No Change in Motion When the forces applied to an object produce a net force of zero, the forces are *balanced.* Balanced forces do not cause a nonmoving object to start moving. Furthermore, balanced forces will not cause a change in the motion of a moving object.

Many objects around you have only balanced forces acting on them. For example, a light hanging from the ceiling does not move because the force of gravity pulling down on the light is balanced by an elastic force due to tension that pulls the light up. A bird's nest in a tree and a hat resting on your head are also examples of objects with only balanced forces acting on them. **Figure 10** shows another case where the forces on an object are balanced. Because all the forces are balanced, the house of cards does not move.

Figure 10 *The forces on this house of cards are balanced. An unbalanced force on one of the cards would cause motion—and probably a mess!*

REVIEW

1. Give four examples of a force being exerted.
2. Explain the difference between balanced and unbalanced forces and how each affects the motion of an object.
3. **Interpreting Graphics** In the picture at left, two bighorn sheep push on each other's horns. The arrow shows the direction the two sheep are moving. Describe the forces the sheep are exerting and how the forces combine to produce the sheep's motion.

Friction: A Force That Opposes Motion

NEW TERMS
friction

OBJECTIVES
- Explain why friction occurs.
- List the types of friction, and give examples of each.
- Explain how friction can be both harmful and helpful.

Picture a warm summer day. You are enjoying the day by wearing shorts and tossing a ball with your friends. By accident, one of your friends tosses the ball just out of your reach. You have to make a split-second decision to dive for it or not. You look down and notice that if you dove for it, you would most likely slide across pavement rather than the surrounding grass. What would you decide?

Unless you enjoy scraped knees, you probably would not want to slide on the pavement. The painful difference between sliding on grass and sliding on pavement has to do with friction. **Friction** is a force that opposes motion between two surfaces that are touching.

The Source of Friction

Friction occurs because the surface of any object is rough. Even surfaces that look or feel very smooth are actually covered with microscopic hills and valleys. When two surfaces are in contact, the hills and valleys of one surface stick to the hills and valleys of the other surface, as shown in **Figure 11.** This contact causes friction even when the surfaces appear smooth.

The amount of friction between two surfaces depends on many factors, including the roughness of the surfaces and the force pushing the surfaces together.

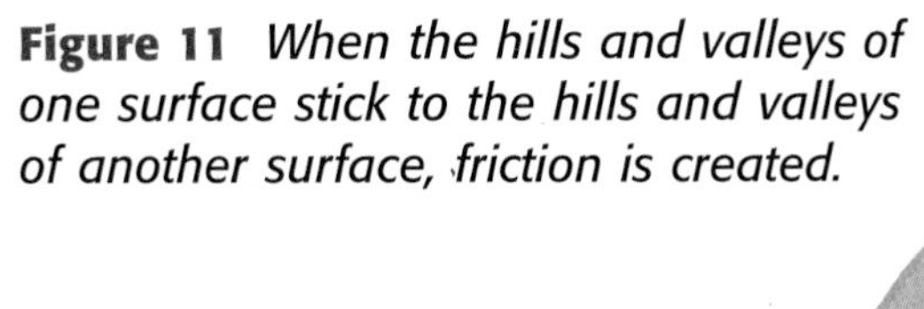

Figure 11 *When the hills and valleys of one surface stick to the hills and valleys of another surface, friction is created.*

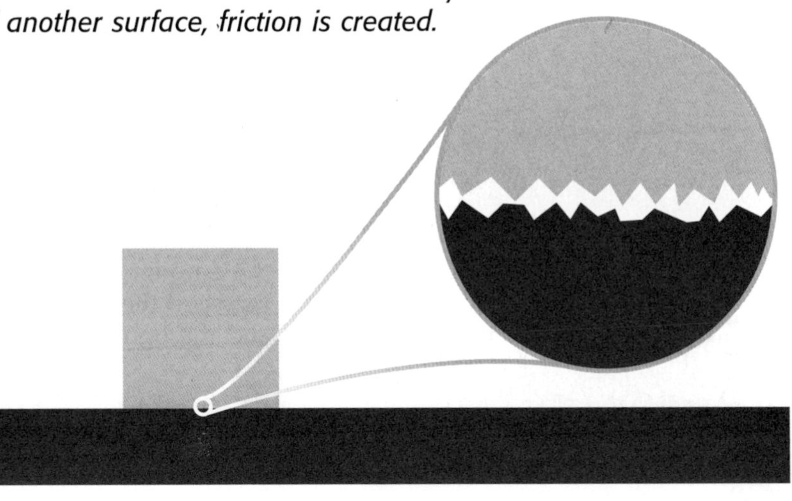

*Quick*Lab

The Friction 500

1. Make a short ramp out of a **piece of cardboard** and **one or two books** on a table.
2. Put a **toy car** at the top of the ramp and let go. If necessary, adjust the ramp height so that your car does not roll off the table.
3. Put the car at the top of the ramp again and let go. Record the distance the car travels after leaving the ramp. Do this three times, and calculate the average for your results.
4. Change the surface of the table by covering it with **sandpaper** or **cloth.** Repeat step 3. Change the surface one more time, and repeat step 3 again.
5. Which surface had the most friction? Why? What do you predict would happen if the car were heavier? Record your results and answers in your ScienceLog.

Rougher Surfaces Create More Friction Rougher surfaces have more microscopic hills and valleys. Thus, the rougher the surface, the greater the friction. Think back to the example on the previous page. Pavement is much rougher than grass. Therefore, more friction is produced when you slide on the pavement than when you slide on grass. This increased friction is more effective at stopping your sliding, but it is also more painful! On the other hand, if the surfaces are smooth, there is less friction. If you were to slide on ice instead of on grass, your landing would be even more comfortable—but also much colder!

Greater Force Creates More Friction The amount of friction also depends on the force pushing the surfaces together. If this force is increased, the hills and valleys of the surfaces can come into closer contact. This causes the friction between the surfaces to increase. Less massive objects exert less force on surfaces than more massive objects do, as illustrated in **Figure 12.** However, changing the amounts of the surfaces that touch does not change the amount of friction.

Figure 12 Force and Friction

a There is more friction between the more massive book and the table than there is between the less massive book and the table. A harder push is needed to overcome friction to move the more massive book.

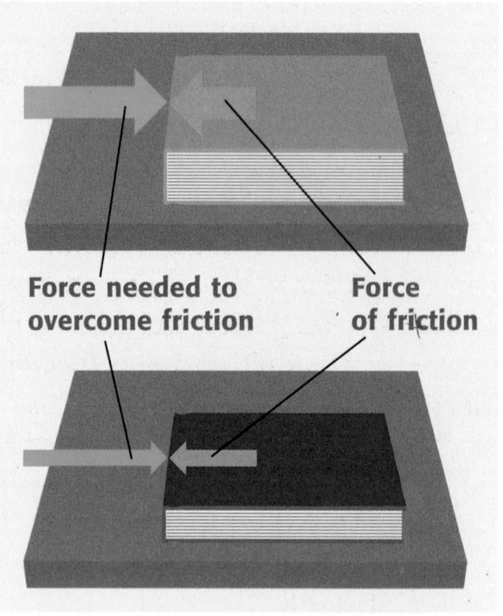

b Turning the more massive book on its edge does not change the amount of friction between the table and the book.

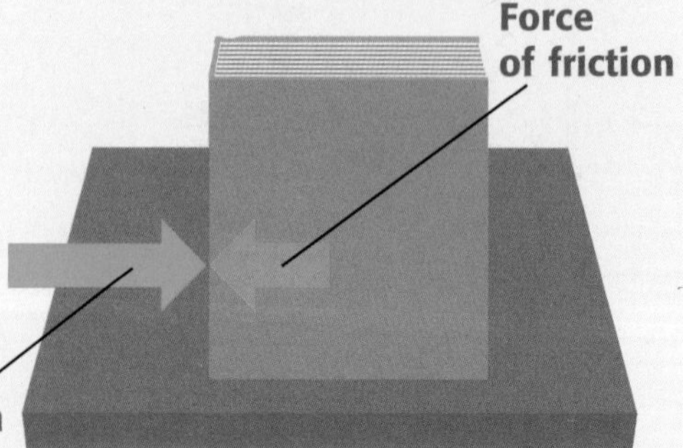

Types of Friction

The friction you observe when sliding books across a tabletop is called sliding friction. Other types of friction include rolling friction, fluid friction, and static friction. As you will learn, the name of each type of friction is a big clue as to the conditions where it can be found.

Sliding Friction If you push an eraser across your desk, the eraser will move for a short distance and then stop. This is an example of *sliding friction.* Sliding friction is very effective at opposing the movement of objects and is the force that causes the eraser to stop moving. You can feel the effect of sliding friction when you try to move a heavy dresser by pushing it along the floor. You must exert a lot of force to overcome the sliding friction, as shown in **Figure 13.**

You use sliding friction when you go sledding, when you apply the brakes on a bicycle or a car, or when you write with a piece of chalk.

Rolling Friction If the same heavy dresser were on wheels, you would have an easier time moving it. The friction between the wheels and the floor is an example of *rolling friction.* The force of rolling friction is usually less than the force of sliding friction. Therefore, it is generally easier to move objects on wheels than it is to slide them along the floor, as shown at right.

Rolling friction is an important part of almost all means of transportation. Anything with wheels—bicycles, in-line skates, cars, trains, and planes—uses rolling friction between the wheels and the ground to move forward.

Figure 13 **Comparing Sliding Friction and Rolling Friction**

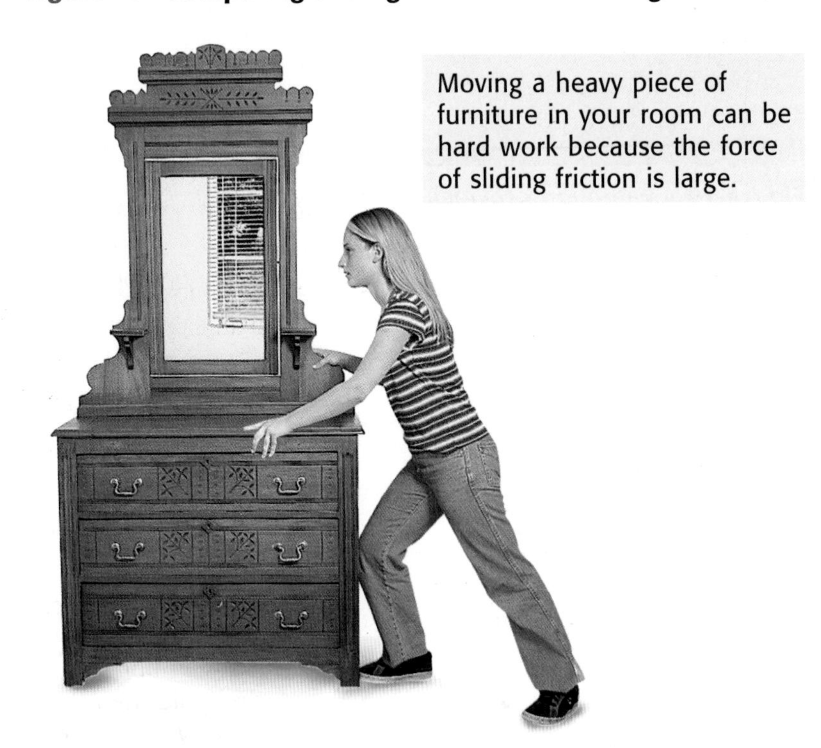

Moving a heavy piece of furniture in your room can be hard work because the force of sliding friction is large.

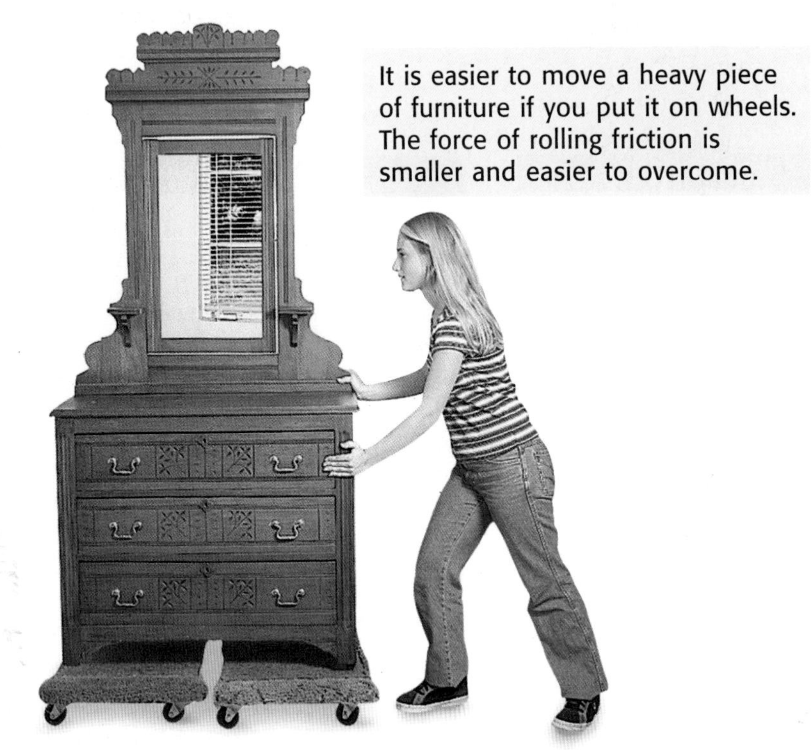

It is easier to move a heavy piece of furniture if you put it on wheels. The force of rolling friction is smaller and easier to overcome.

Figure 14 *Swimming provides a good workout because you must exert force to overcome fluid friction.*

Fluid Friction Why is it harder to walk on a freshly mopped floor than on a dry floor? The reason is that on the wet floor the sliding friction between your feet and the floor is replaced by *fluid friction* between your feet and the water. In this case, fluid friction is less than sliding friction, so the floor is slippery. You may think of *fluid* as another name for liquid, but fluids include liquids and gases; water, milk, and air are all fluids.

Fluid friction opposes the motion of objects traveling through a fluid, as illustrated in **Figure 14.** For example, fluid friction between air and a fast moving car is the largest force opposing the motion of the car. You can observe this friction by holding your hand out the window of a moving car.

Static Friction When a force is applied to an object but does not cause the object to move, *static friction* occurs. The object does not move because the force of static friction balances the force applied. Static friction disappears as soon as an object starts moving, and then another type of friction immediately occurs. Look at **Figure 15** to understand when static friction affects an object.

Figure 15 Static Friction

Force applied
Static friction

Force applied
Sliding friction

a There is no friction between the block and the table when no force is applied to the block to move it.

b If a small force—shown in blue—is exerted on the block, the block does not move because the force of static friction—shown in orange—exactly balances the force applied.

c When the force exerted on the block is greater than the force of static friction, the block starts moving. Once the block starts moving, all static friction is gone, but work must be done against sliding friction—shown in green.

Self-Check

What type of friction was involved in the imaginary situation at the beginning of this section? *(See page 596 to check your answer.)*

Friction Can Be Harmful or Helpful

Think about how friction affects a car. Without friction, the tires would not be able to push off the ground and move the car forward, the brakes would not work to stop the car, and you would not even be able to grip the door handle to get inside! Without friction, a car is useless.

However, friction can cause problems in a car too. Friction between moving engine parts increases their temperature and causes the parts to wear down. Coolant must be regularly added to the engine to keep it from overheating from friction, and engine parts need to be changed as they wear out.

Friction is both harmful and helpful to you and the world around you. Without friction, you could not do homework, play games, or even walk. Friction between your pencil and your paper is necessary for the pencil to leave a mark. Without friction, balls and other sports equipment would slip from your fingers when you tried to pick them up, and you would just slip and fall when you tried to walk. But friction is also responsible for putting holes in your socks and in the knees of your bluejeans. Friction by wind and water contributes to the erosion of the topsoil that nourishes plants. Because friction can be both harmful and helpful, it is sometimes necessary to reduce or increase friction.

Have some fun with friction! Investigate three types of friction on page 540 of the LabBook.

Some Ways to Reduce Friction One way to reduce friction is to use lubricants. *Lubricants* (LOO bri kuhnts) are substances that are applied to surfaces to reduce the friction between them. Some examples of common lubricants are motor oil, wax, and grease. **Figure 16** shows why lubricants are important to maintaining car parts.

Friction can also be reduced by switching from sliding friction to rolling friction. Ball bearings are placed between the wheels and axles of in-line skates and bicycles to make it easier for the wheels to turn by reducing friction.

BRAIN FOOD

Lubricants are usually liquids, but they can be solids or gases too. Graphite is a shiny black solid that is used in pencils. Graphite dust is very slippery and is often used as a lubricant for ball bearings in bicycle and skate wheels. An example of a gas lubricant is the air that comes out of the tiny holes of an air-hockey table.

Figure 16 *Motor oil is used as a lubricant in car engines. Without oil, engine parts would wear down quickly, as the connecting rod on the bottom has.*

Another way to reduce friction is to make surfaces that rub against each other smoother. For example, rough wood on a park bench is painful to slide across because there is a large amount of friction between your leg and the bench. Rubbing the bench with sandpaper makes it smoother and more comfortable to sit on because the friction between your leg and the bench is reduced.

Some Ways to Increase Friction One way to increase friction is to make surfaces rougher. For example, sand scattered on icy roads keeps cars from skidding. Baseball players sometimes wear textured batting gloves to increase the friction between their hands and the bat so that the bat does not fly out of their hands.

Another way to increase friction is to increase the force pushing the surfaces together. For example, you can ensure that your magazine will not blow away at the park by putting a heavy rock on it. The added mass of the rock increases the friction between the magazine and the ground. Or if you are sanding a piece of wood, you can sand the wood faster by pressing harder on the sandpaper. **Figure 17** shows another situation where friction is increased by pushing on an object.

Figure 17 *No one enjoys cleaning pans with baked-on food! To make this chore pass quickly, press down with the scrubber to increase friction.*

REVIEW

1. Explain why friction occurs.
2. Name two ways in which friction can be increased.
3. Give an example of each of the following types of friction: sliding, rolling, and fluid.
4. **Applying Concepts** Name two ways that friction is harmful and two ways that friction is helpful to you when riding a bicycle.

APPLY

The tire shown here was used for more than 80,000 km. What effect did friction have on the rubber? What kind of friction is mainly responsible for the tire's appearance? Why are car owners warned to change their car tires after using them for several thousand kilometers?

Section 4

Gravity: A Force of Attraction

NEW TERMS

gravity
mass
weight

OBJECTIVES

- Define *gravity.*
- State the law of universal gravitation.
- Describe the difference between mass and weight.

If you watch videotape of astronauts on the moon, you will notice that when the astronauts tried to walk on the lunar surface, they bounced around like beach balls instead.

Why did the astronauts—who were wearing heavy spacesuits—bounce so easily on the moon (as shown in **Figure 18**), while you must exert effort to jump a few centimeters off Earth's surface? The answer has to do with gravity. **Gravity** is a force of attraction between objects that is due to their masses. In this section, you will learn about gravity and the effects it has on objects.

Figure 18 *Because gravity is less on the moon than on Earth, walking on the moon's surface was a very bouncy experience for the Apollo astronauts.*

All Matter Is Affected by Gravity

All matter has mass. Gravity is a result of mass. Therefore, all matter experiences gravity. That is, all objects experience an attraction toward all other objects. This gravitational force "pulls" objects toward each other. Right now, because of gravity, you are being pulled toward this book, your pencil, and every other object around you.

These objects are also being pulled toward you and toward each other because of gravity. So why don't you see the effects of this attraction? In other words, why don't you notice objects moving toward each other? The reason is that the mass of most objects is too small to cause an attraction large enough to move objects toward each other. However, you are familiar with one object that is massive enough to cause a noticeable attraction—the Earth.

Scientists think seeds can "sense" gravity. The ability to sense gravity is what causes seeds to always send roots down and the green shoot up. But scientists do not understand just *how* seeds do this. Astronauts have grown seedlings during space shuttle missions to see how seeds respond to changes in gravity. So far, there are no definite answers from the results of these experiments.

Earth's Gravitational Force Is Large Compared with all the objects around you, Earth has an enormous mass. Because it is so massive, Earth's gravitational force is very large. You must constantly apply forces to overcome Earth's gravitational force. In fact, any time you lift objects or parts of your body, you are resisting this force.

Earth's gravitational force pulls everything toward the center of Earth. Because of this, the books, tables, and chairs in the room stay in place, and dropped objects fall to Earth rather than moving together or toward you.

Self-Check

What is gravity? *(See page 596 to check your answer.)*

The Law of Universal Gravitation

For thousands of years, two of the most puzzling scientific questions were "Why do objects fall toward Earth?" and "What keeps the planets in motion in the sky?" The two questions were treated as separate topics until Sir Isaac Newton (1642–1727) realized that they were two parts of the same question. Newton was a British scientist whose work on forces and light was very important to the development of science.

Legend has it that Newton made the connection between the two questions when he observed a falling apple during a summer night, as shown in **Figure 19.** He knew that unbalanced forces are necessary to move or change the motion of objects. He concluded that there had to be an unbalanced force on the apple to make it fall, just as there had to be an unbalanced force on the moon to keep it moving in a circle around Earth. He realized that the force pulling the apple to Earth and the force pulling the moon in a circular path around Earth are the same force—a force of attraction called gravity.

Newton generalized his observations on gravity in a law now known as the *law of universal gravitation.* The law of universal gravitation describes the relationships between gravitational force, mass, and distance. It is called universal because it applies to all objects in the universe, from the tiniest speck of dust to the largest star. In fact, gravitational force between gas and dust particles in space sometimes plays a part in the formation of stars and planets.

Figure 19
Newton Makes the Connection

The law of universal gravitation states the following: All objects in the universe attract each other through gravitational force. As shown in **Figure 20,** the size of the force depends on the masses of the objects and the distance between them.

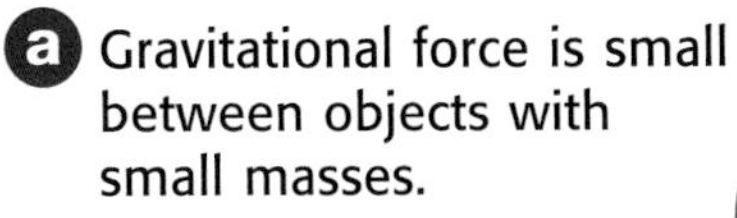

a Gravitational force is small between objects with small masses.

Figure 20 *The arrows indicate the gravitational force between the objects. The width of the arrows indicates the strength of the force.*

b Gravitational force is larger between objects with larger masses.

c If the distance between two objects is increased, the gravitational force pulling them together is reduced.

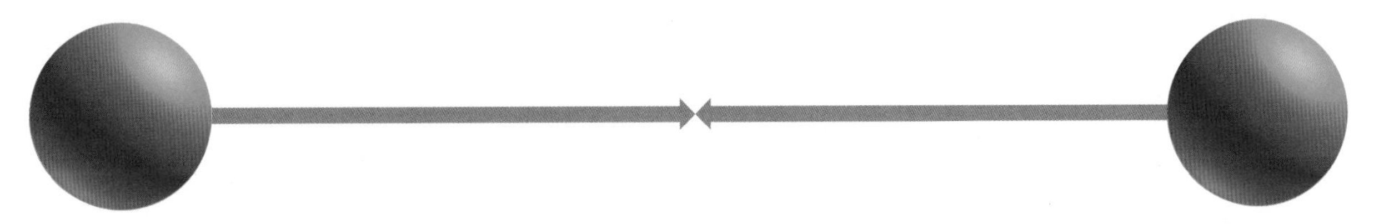

The law of universal gravitation can also be expressed mathematically. If two objects have masses m_1 and m_2 and are separated by a distance *r,* then the gravitational force can be calculated according to the following equation:

$$F = G \times \frac{m_1 \times m_2}{r^2}$$

G is a universal constant called the constant of universal gravitation. It has been measured experimentally and its value is the following:

$$G = 6.673 \times 10^{-11} \frac{\mathrm{N \bullet m^2}}{\mathrm{kg^2}}$$

Part 1: Gravitational Force Increases as Mass Increases The moon has less mass than Earth. Therefore, the moon's gravitational force is less than Earth's. Remember the astronauts on the moon? They bounced around as they walked because they were not being pulled down with as much force as they would have been on Earth.

Black holes are formed when massive stars collapse. Black holes are 10 times to 1 billion times more massive than our sun. Thus, their gravitational force is incredibly large. The gravity of a black hole is so large that an object that enters a black hole can never get out. Even light cannot escape from a black hole. Because black holes do not emit light, they cannot be seen—hence their name.

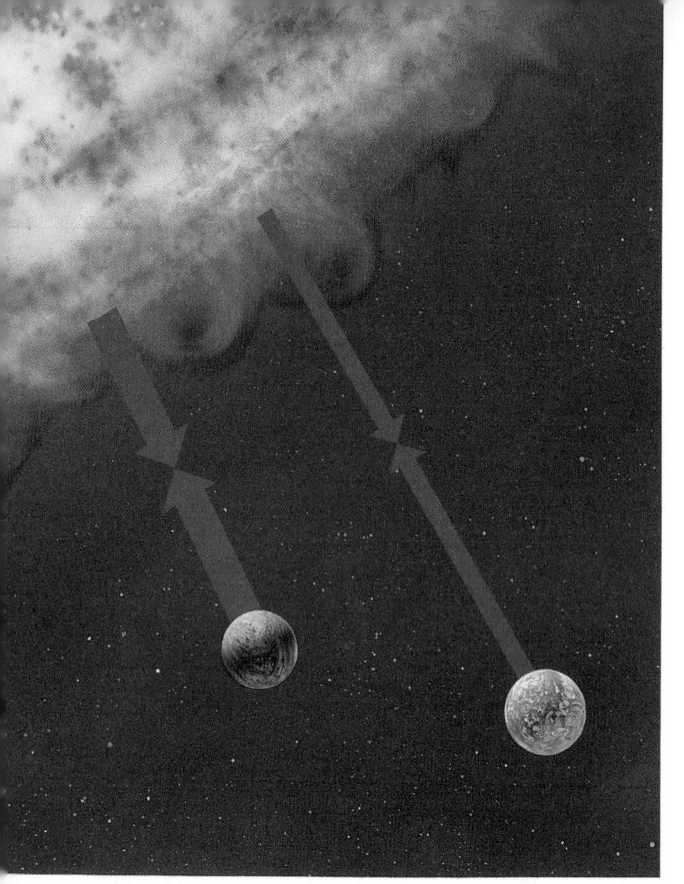

Figure 21 *Venus and Earth have approximately the same mass. However, Venus is closer to the sun. Thus, the gravity between Venus and the sun is greater than the gravity between Earth and the sun.*

Part 2: Gravitational Force Decreases as Distance Increases The gravity between you and Earth is large. Whenever you jump up, you are pulled back down by Earth's gravitational force. On the other hand, the sun is more than 300,000 times more massive than Earth. So why doesn't the sun's gravitational force affect you more than Earth's does? The reason is that the sun is so far away.

You are approximately 150 million kilometers away from the sun. At this distance, the gravity between you and the sun is very small. If there were some way you could stand on the sun (and not burn up), you would find it impossible to jump or even walk. The gravitational force acting on you would be so great that your muscles could not lift any part of your body!

Although the sun's gravitational force does not have much of an effect on your body here, it does have a big effect on Earth itself and the other planets, as shown in **Figure 21.** The gravity between the sun and the planets is large because the objects have large masses. If the sun's gravitational force did not have such an effect on the planets, the planets would not stay in orbit around the sun.

Weight Is a Measure of Gravitational Force

You have learned that gravity is a force of attraction between objects that is due to their masses. **Weight** is a measure of the gravitational force exerted on an object. When you see or hear the word *weight,* it usually refers to Earth's gravitational force on an object. But weight can also be a measure of the gravitational force exerted on objects by the moon or other planets.

You have learned that the unit of force is a newton. Because gravity is a force and weight is a measure of gravity, weight is also expressed in newtons (N). On Earth, a 100 g object, such as a medium-sized apple, weighs approximately 1 N.

Explore

Suppose you had a device that could increase or decrease the gravitational force of objects around you (including small sections of Earth). In your ScienceLog, describe what you might do with the device, what you would expect to see, and what effect the device would have on the weight of objects.

Weight and Mass Are Different Weight is related to mass, but the two are not the same. Weight changes when gravitational force changes. **Mass** is the amount of matter in an object, and its value does not change. If an object is moved to a place with a greater gravitational force—like Jupiter—its weight will increase, but its mass will remain the same. **Figure 22** shows the weight and mass of an object on Earth and a place with a smaller gravitational force.

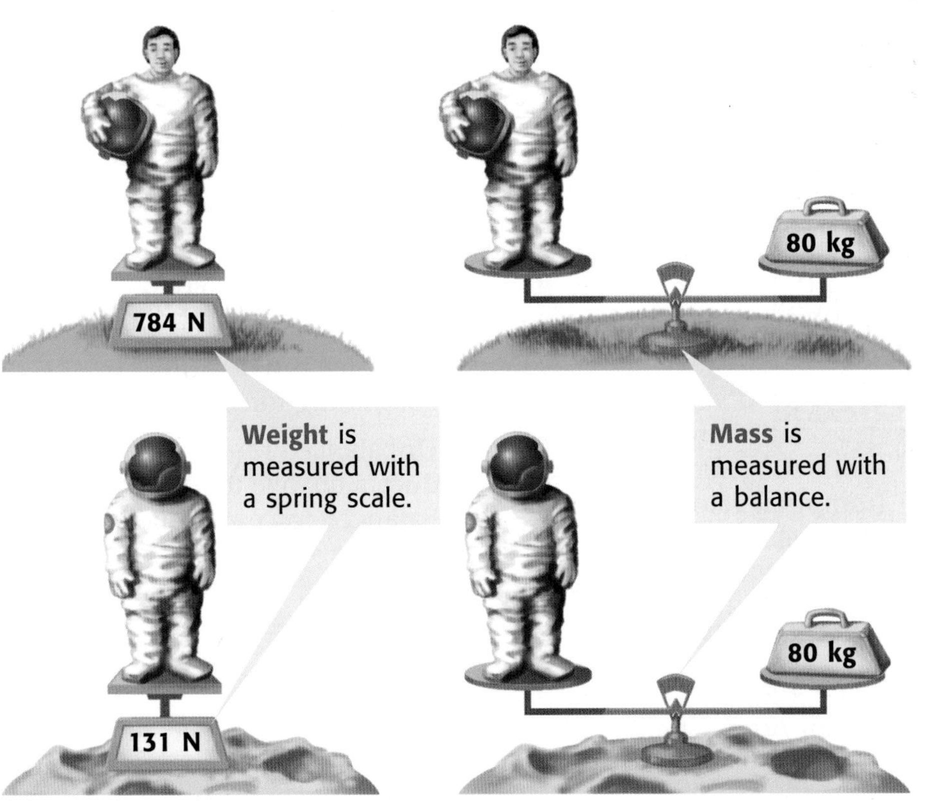

Figure 22 *Weight can change, but mass remains constant.*

The astronaut's weight and mass on Earth are shown on the spring scale and balance.

The astronaut has the same mass on the moon, but his weight is one-sixth of his weight on Earth. This is because the moon's gravitational force is one-sixth that of Earth's.

Gravitational force is about the same everywhere on Earth, so the weight of any object is about the same everywhere. Because mass and weight are constant on Earth, the terms are often used to mean the same thing. This can lead to confusion. Be sure you understand the difference!

REVIEW

1. How does the mass of an object relate to the gravitational force the object exerts on other objects?
2. How does the distance between objects affect the gravity between them?
3. **Comparing Concepts** Explain why your weight would change if you orbited Earth in the space shuttle but your mass would not.

Chapter Highlights

SECTION 1

Vocabulary

motion *(p. 108)*
speed *(p. 109)*
velocity *(p. 110)*
acceleration *(p. 112)*

Section Notes

- An object is in motion if it changes position over time when compared with a reference point.

- The speed of a moving object depends on the distance traveled by the object and the time taken to travel that distance.
- Speed and velocity are not the same thing. Velocity is speed in a given direction.
- Acceleration is the rate at which velocity changes.
- An object can accelerate by changing speed, changing direction, or both.
- Acceleration is calculated by subtracting starting velocity from final velocity, then dividing by the time required to change velocity.

Labs

Built for Speed *(p. 536)*
Detecting Acceleration *(p. 537)*

SECTION 2

Vocabulary

force *(p. 115)*
newton (N) *(p. 115)*
net force *(p. 116)*

Section Notes

- A force is a push or a pull.
- Forces are expressed in newtons.
- Force is always exerted by one object on another object.
- Net force is determined by combining forces.
- Unbalanced forces produce a change in motion. Balanced forces produce no change in motion.

☑ Skills Check

Math Concepts

ACCELERATION An object's acceleration can be determined using the following equation:

$$\text{Acceleration} = \frac{\text{final velocity} - \text{starting velocity}}{\text{time it takes to change velocity}}$$

For example, suppose a cheetah running at a velocity of 27 m/s slows down. After 15 seconds, the cheetah has stopped.

$$\text{Acceleration} = \frac{0 \text{ m/s} - 27 \text{ m/s}}{15 \text{ s}} = -1.8 \text{ m/s/s}$$

Visual Understanding

THE SOURCE OF FRICTION Even surfaces that look or feel very smooth are actually rough at the microscopic level. To understand how this roughness causes friction, review Figure 11 on page 119.

THE LAW OF UNIVERSAL GRAVITATION This law explains that the gravity between objects depends on their masses and the distance between them. Review the effects of this law by looking at Figure 20 on page 127.

SECTION 3

Vocabulary

friction *(p. 119)*

Section Notes

- Friction is a force that opposes motion.
- Friction is caused by "hills and valleys" touching on the surfaces of two objects.
- The amount of friction depends on factors such as the roughness of the surfaces and the force pushing the surfaces together.
- Four kinds of friction that affect your life are sliding friction, rolling friction, fluid friction, and static friction.
- Friction can be harmful or helpful.

Labs

Science Friction *(p. 540)*

SECTION 4

Vocabulary

gravity *(p. 125)*
weight *(p. 128)*
mass *(p. 129)*

Section Notes

- Gravity is a force of attraction between objects that is due to their masses.
- The law of universal gravitation states that all objects in the universe attract each other through gravitational force. The size of the force depends on the masses of the objects and the distance between them.
- Weight and mass are not the same. Mass is the amount of matter in an object; weight is a measure of gravitational force on an object.

Labs

Relating Mass and Weight *(p. 541)*

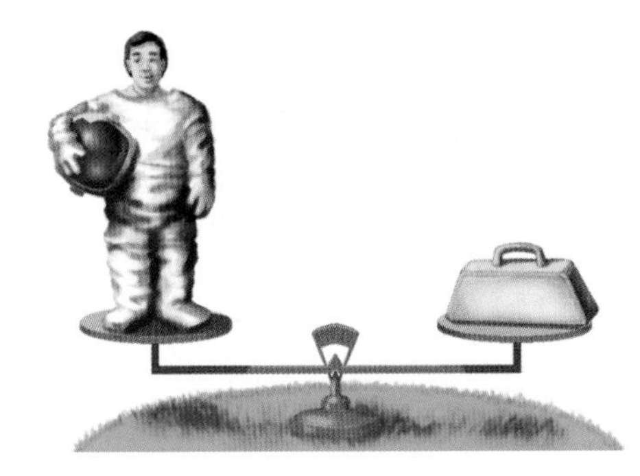

internetconnect

GO TO: go.hrw.com

Visit the **HRW** Web site for a variety of learning tools related to this chapter. Just type in the keyword:

KEYWORD: HSTMOT

GO TO: www.scilinks.org

Visit the **National Science Teachers Association** on-line Web site for Internet resources related to this chapter. Just type in the ***sci*LINKS** number for more information about the topic:

TOPIC: Measuring Motion — ***sci*LINKS NUMBER:** HSTP105
TOPIC: Force and Friction — ***sci*LINKS NUMBER:** HSTP110
TOPIC: Matter and Gravity — ***sci*LINKS NUMBER:** HSTP115
TOPIC: Virtual Reality — ***sci*LINKS NUMBER:** HSTP120
TOPIC: The Science of Bridges — ***sci*LINKS NUMBER:** HSTP125

Chapter Review

USING VOCABULARY

To complete the following sentences, choose the correct term from each pair of terms listed below:

1. ___?___ opposes motion between surfaces that are touching. (*Friction* or *Gravity*)
2. Forces are expressed in ___?___. (*newtons* or *mass*)
3. A ___?___ is determined by combining forces. (*net force* or *newton*)
4. ___?___ is the rate at which ___?___ changes. (*Velocity* or *Acceleration/velocity* or *acceleration*)

UNDERSTANDING CONCEPTS

Multiple Choice

5. A student riding her bicycle on a straight, flat road covers one block every 7 seconds. If each block is 100 m long, she is traveling at
   a. constant speed.
   b. constant velocity.
   c. 10 m/s.
   d. Both (a) and (b)
6. Friction is a force that
   a. opposes an object's motion.
   b. does not exist when surfaces are very smooth.
   c. decreases with larger mass.
   d. All of the above
7. Rolling friction
   a. is usually less than sliding friction.
   b. makes it difficult to move objects on wheels.
   c. is usually greater than sliding friction.
   d. is the same as fluid friction.
8. If Earth's mass doubled, your weight would
   a. increase because gravity increases.
   b. decrease because gravity increases.
   c. increase because gravity decreases.
   d. not change because you are still on Earth.
9. A force
   a. is expressed in newtons.
   b. can cause an object to speed up, slow down, or change direction.
   c. is a push or a pull.
   d. All of the above
10. The amount of gravity between 1 kg of lead and Earth is ______ the amount of gravity between 1 kg of marshmallows and Earth.
    a. greater than
    b. less than
    c. the same as
    d. None of the above

Short Answer

11. Describe the relationship between motion and a reference point.
12. How is it possible to be accelerating and traveling at a constant speed?
13. Explain the difference between mass and weight.

Concept Mapping

14. Use the following terms to create a concept map: speed, velocity, acceleration, force, direction, motion.

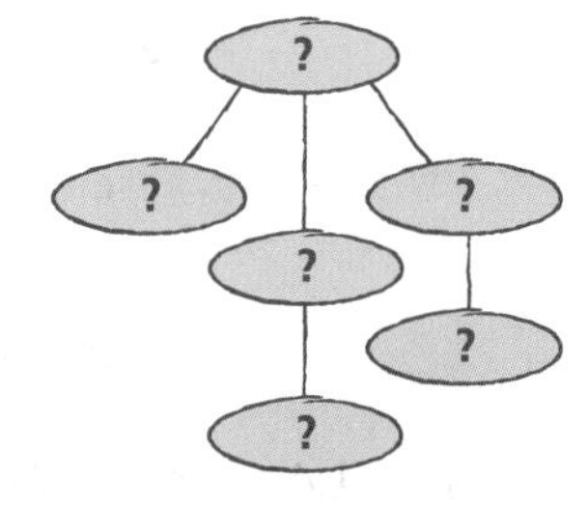

CRITICAL THINKING AND PROBLEM SOLVING

15. Your family is moving, and you are asked to help move some boxes. One box is so heavy that you must push it across the room rather than lift it. What are some ways you could reduce friction to make moving the box easier?

16. Explain how using the term *accelerator* when talking about a car's gas pedal can lead to confusion, considering the scientific meaning of the word *acceleration*.

17. Explain why it is important for airplane pilots to know wind velocity, not just wind speed, during a flight.

MATH IN SCIENCE

18. A kangaroo hops 60 m to the east in 5 seconds.
    a. What is the kangaroo's speed?
    b. What is the kangaroo's velocity?
    c. The kangaroo stops at a lake for a drink of water, then starts hopping again to the south. Every second, the kangaroo's velocity increases 2.5 m/s. What is the kangaroo's acceleration after 5 seconds?

INTERPRETING GRAPHICS

19. Is this a graph of positive or negative acceleration? How can you tell?

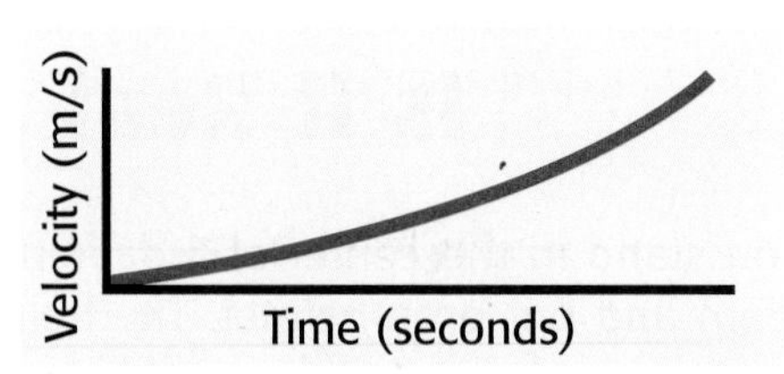

20. You know how to combine two forces that act in one or two directions. The same method you learned can be used to combine several forces acting in several directions. Examine the diagrams below, and predict with how much force and in what direction the object will move.

a

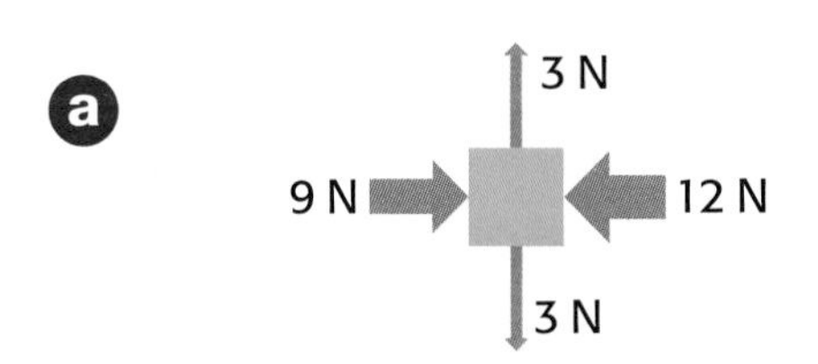

b

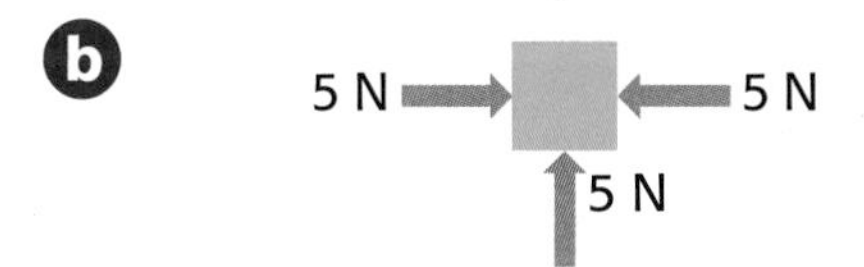

c

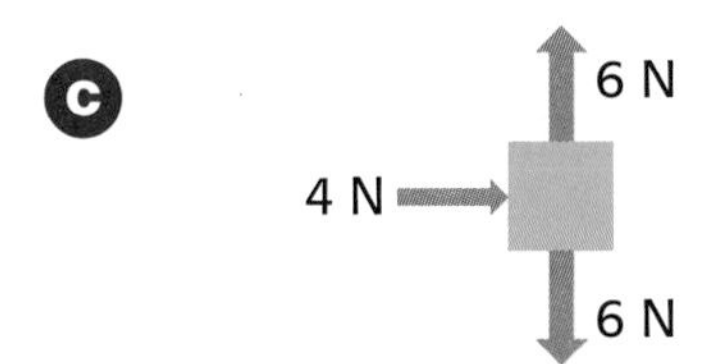

NOW What Do You Think?

Take a minute to review your answers to the ScienceLog questions on page 107. Have your answers changed? If necessary, revise your answers based on what you have learned since you began this chapter.

Science, Technology, and Society

Is It Real . . . or Is It Virtual?

You stand in the center of a darkened room and put on a helmet. The helmet covers your head and face, making it impossible for you to see or hear anything from outside. Wires run from the helmet to a series of computers, carrying information about how your head is positioned and where you are looking. Other wires carry back to you the sights and sounds the computer wants you to "see" and "hear." All of a sudden you find yourself driving a race car around a tricky course at 300 km/h. Then in another instant, you are in the middle of a rain forest staring at a live snake!

It's All an Illusion

Such simulated-reality experiences were once thought the stuff of science fiction alone. But today devices called motion simulators can stimulate the senses of sight and sound to create illusions of movement.

Virtual-reality devices, as these motion simulators are called, were first used during World War II to train pilots. Mock-ups of fighter-plane cockpits, films of simulated terrain, and a joystick that manipulated large hydraulic arms simulated the plane in "virtual flight." Today's jet pilots train with similar equipment, except the simulators use extremely sophisticated computer graphics instead of films.

Fooled You!

Virtual-reality hoods and gloves take people into a variety of "realities." Inside the hood, two small television cameras or computer-graphic images fool the wearer's sense of vision. The brain perceives the image as three-dimensional because one image is placed in front of each eye. As the images change, the computer adjusts the scene's perspective so that it appears to the viewer as though he or she is moving through the scene. When the position of the head changes, the computer adjusts the scene to account for the movement. All the while, sounds coming through the headphones trick the wearer's ears into thinking he or she is moving too.

In addition to hoods, gloves, and images, virtual-reality devices may have other types of sensors. Driving simulators, for instance, often have a steering wheel, a gas pedal, and a brake so that the participant has the sensation of driving. So whether you want spine-tingling excitement or on-the-job training, virtual reality could very well take *you* places!

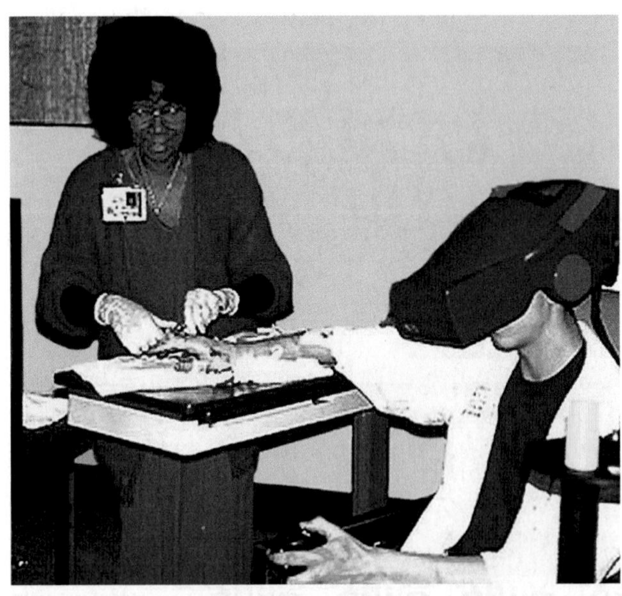

▲ *Wearing a virtual-reality helmet helps to lessen the pain this burn patient feels while his dressings are changed.*

Explore New Realities

▶ What other activities or skills could be learned or practiced with virtual reality? What are some problems with relying on this technology? Record your ideas in your ScienceLog.

PHYSICAL SCIENCE • EARTH SCIENCE

The Golden Gate Bridge

Have you ever relaxed in a hammock? If so, you may have noticed how tense the strings got when the hammock supported your weight. Now imagine a hammock 1,965 m long supporting a 20-ton roadway with more than 100,000 cars traveling along its length each day. That describes the Golden Gate Bridge! Because of the way the bridge is built, it is very much like a giant hammock.

Tug of War

The bridge's roadway is suspended from main cables 2.33 km long that sweep from one end of the bridge to the other and that are anchored at each end. Smaller cables called *hangers* connect the main cables to the roadway. Tension, the force of being pulled apart, is created as the cables are pulled down by the weight of the roadway while being pulled up by their attachment to the top of each tower.

▲ *The Golden Gate Bridge spans the San Francisco Bay.*

Towering Above

Towers 227 m tall support the cables over the long distance across San Francisco Bay, making the Golden Gate the tallest bridge in the world. The towers receive a force that is the exact opposite of tension—compression. Compression is the force of being pushed together. The main cables holding the weight of the roadway push down on the top of the towers while Earth pushes up on the bottom.

Stretching the Limits

Tension and compression are elastic forces, which means they are dependent on elasticity, the ability of an object to return to its original shape after being stretched or compressed. If an object is not very elastic, it breaks easily or becomes permanently deformed when subjected to an elastic force. The cables and towers of the Golden Gate Bridge are made of steel, a material with great elastic strength. A single steel wire 2.54 mm thick can support over half a ton without breaking!

On the Road

The roadway of the Golden Gate Bridge is subjected to multiple forces at the same time, including friction, gravity, and elastic forces. Rolling friction is caused by the wheels of each vehicle moving across the roadway's surface. Gravity pulls down on the roadway but is counteracted by the support of the towers and cables. This causes each roadway span to bend slightly and experience both tension and compression. The bottom of each span is under tension because the cables and towers pull up along the road's sides, while gravity pulls down at its center. These same forces cause compression of the top of each span. Did you ever imagine that so many forces were at work on a bridge?

Bridge the Gap

▶ Find out more about another type of bridge, such as an arch, a beam, or a cable-stayed bridge. How do forces such as friction, gravity, tension, and compression affect these types of bridges?

CHAPTER 6

Forces in Motion

Imagine . . .

You have been selected to travel on the space shuttle as NASA's first student astronaut. You're going into space! Like all astronauts, you must go through a year of training to prepare for space travel. When you are in the space shuttle, many different forces will be acting on your body that might make you dizzy or disoriented. You must get used to these forces quickly before you go into space. NASA won't be able to shorten the mission just because you don't feel well!

There are many parts to your training. For instance, there is a machine that spins you around in all directions. There is also underwater training that lets you experience what reduced gravity feels like. But the most exciting part of your training is riding on the KC-135 airplane.

The KC-135 is a modified commercial airliner that is designed to simulate what it feels like to orbit Earth in the space shuttle. The KC-135 flies upward at a steep angle, then flies downward at a 45° angle. When the airplane flies downward, the effect of reduced gravity is produced inside. As the plane "falls" out from under the passengers, the astronaut trainees inside the plane can "float," as shown above. Because the floating often makes passengers queasy, the KC-135 has earned a nickname—the Vomit Comet.

NASA scientists used their knowledge of forces, gravity, and the laws of motion to develop these training procedures. In this chapter, you will learn how gravity affects the motion of objects and how the laws of motion apply to your life.

What Do You Think?

In your ScienceLog, try to answer the following questions based on what you already know:

1. How does the force of gravity affect falling objects?
2. What is projectile motion?
3. What are Newton's laws of motion?
4. What is momentum?

Falling Water

Gravity is one of the most important forces you encounter in your daily life. Without it, objects that are thrown or dropped would never land on the ground—they would just float in space. In this activity, you will observe the effect of gravity on a falling object.

Procedure 

1. Place a **wide plastic tub** on the floor.
2. Punch a small hole in the side of a **paper cup,** near the bottom.
3. Hold your finger over the hole, and fill the cup with **water colored with food coloring.**
4. Keeping your finger over the hole, hold the cup about waist high above the tub.
5. Uncover the hole. Describe your observations in your ScienceLog.
6. Cover the hole with your finger again, and refill the cup.
7. Predict what will happen to the water if you drop the cup at the same time you uncover the hole. Write your predictions in your ScienceLog.
8. Uncover the hole, and drop the cup at the same time. Record your observations.
9. Clean up any spilled water with **paper towels.**

Analysis

10. What differences did you observe in the behavior of the water during the two trials?
11. In the second trial, how fast did the cup fall compared with the water?
12. How is the water in the cup similar to passengers in the KC-135 jet?

Section 1

Gravity and Motion

NEW TERMS

terminal velocity

free fall

projectile motion

OBJECTIVES

- Explain how gravity and air resistance affect the acceleration of falling objects.
- Explain why objects in orbit appear to be weightless.
- Describe how an orbit is formed.
- Describe projectile motion.

Suppose you drop a basketball, a baseball, and a marble at the same time from the same height. In what order do you think they would land on the ground? In ancient Greece around 400 B.C., an important philosopher named Aristotle (ER is TAWT uhl) believed that the rate at which an object falls depends on the object's mass. Imagine that you could ask Aristotle which object would land first. Based on their masses, he would predict that the basketball would land first, then the baseball, and finally the marble.

All Objects Fall with the Same Acceleration

In the late 1500s, a young Italian scientist named Galileo questioned Aristotle's idea about falling objects. Galileo proposed that all objects will land at the same time when they are dropped at the same time from the same height.

Galileo proved that the mass of an object does not affect the rate at which it falls. According to one story, Galileo did this by dropping two cannonballs of different masses from the top of the Leaning Tower of Pisa. The crowd watching from the ground was amazed to see the two cannonballs land at the same time. Whether or not this story is true, Galileo's idea changed people's understanding of gravity and falling objects.

Objects fall to the ground at the same rate because the acceleration due to gravity is the same for all objects. Does that seem odd? The force of gravity is greater between Earth and an object with a large mass than between Earth and a less massive object, so you may think that the acceleration due to gravity should be greater too. But a greater force must be applied to a large mass than to a small mass to produce the same acceleration. Thus, the difference in force is canceled by the difference in mass. **Figure 1** shows objects with different masses falling with the same acceleration.

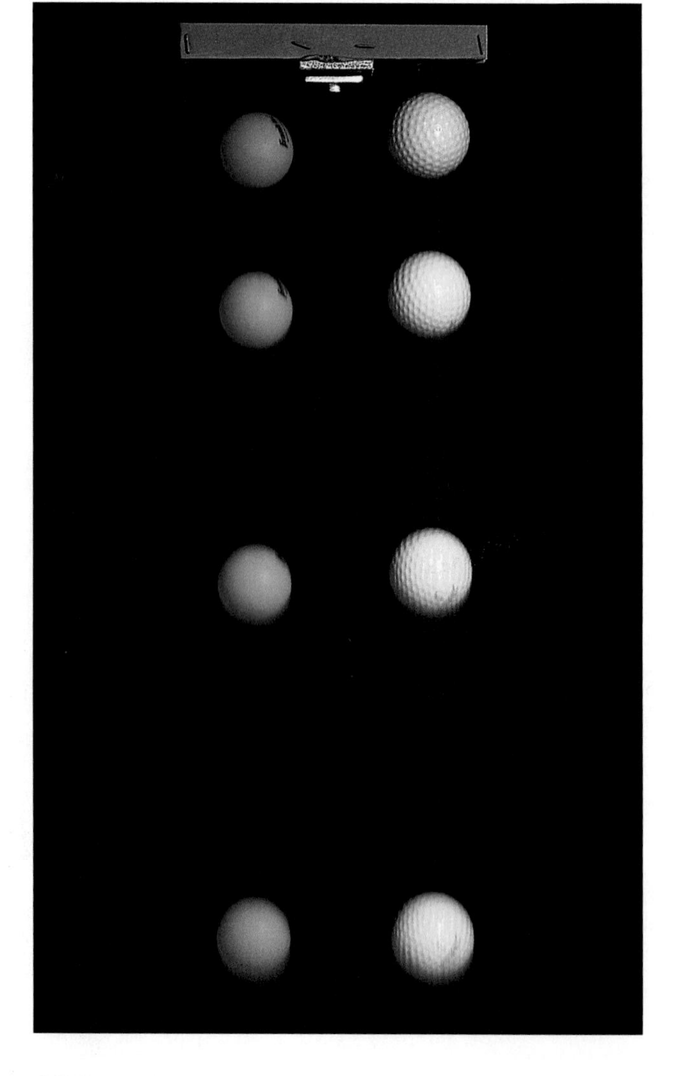

Figure 1 *A table tennis ball and a golf ball fall with the same acceleration even though they have different masses.*

All objects accelerate toward Earth at a rate of 9.8 meters per second per second, which is expressed as 9.8 m/s/s. This means that for every second that an object falls, the object's downward velocity increases by 9.8 m/s, as shown in **Figure 2.** Remember, this acceleration is the same for all objects regardless of their mass. Therefore, a basketball, baseball, and marble would land at the same time when dropped at the same time from the same height. Do the MathBreak at right to learn how to calculate the velocity of a falling object.

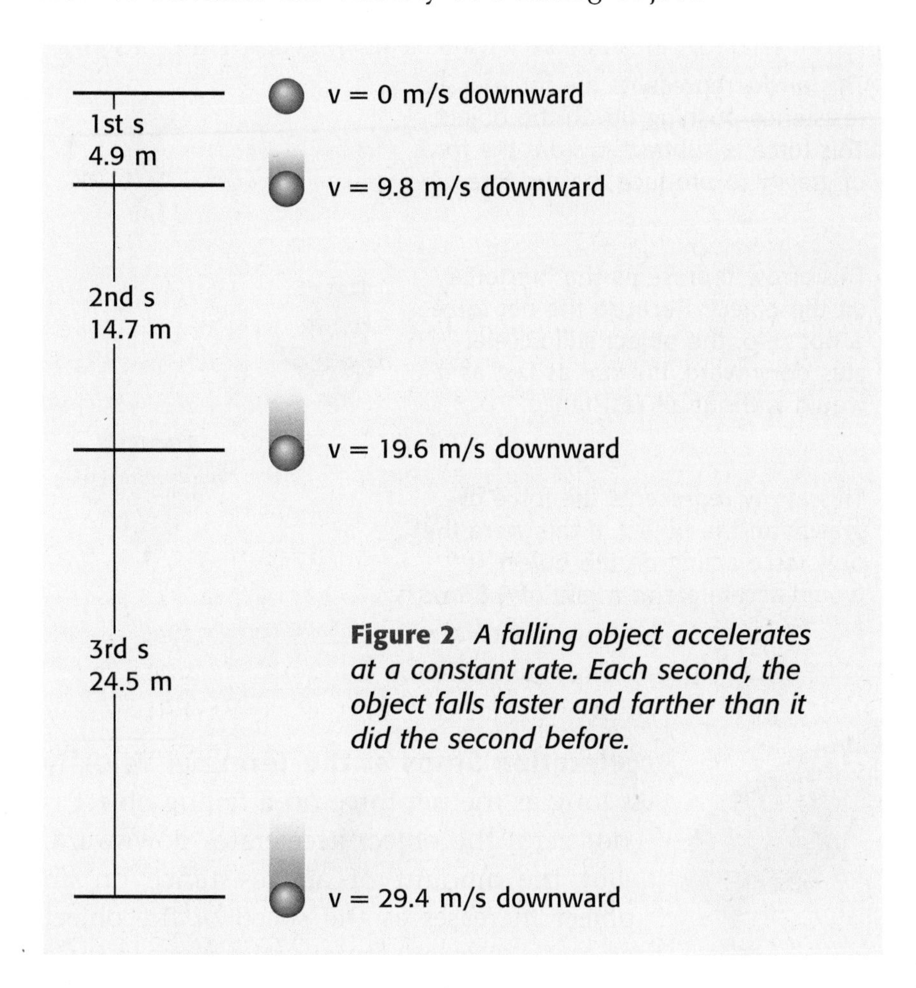

Figure 2 *A falling object accelerates at a constant rate. Each second, the object falls faster and farther than it did the second before.*

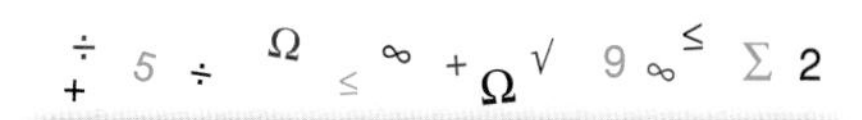

MATH BREAK

Velocity of Falling Objects

To find the change in velocity (Δv) of a falling object, multiply the acceleration due to gravity (g) by the time it takes for the object to fall in seconds (t). The equation for finding a change in velocity is as follows:

$$\Delta v = g \times t$$

For example, a stone at rest is dropped from a cliff, and it takes 3 seconds to hit the ground. Its downward velocity when it hits the ground is as follows:

$$\Delta v = 9.8 \frac{\text{m/s}}{\cancel{\text{s}}} \times 3 \cancel{\text{s}}$$
$$= 29.4 \text{ m/s}$$

Now It's Your Turn

A penny at rest is dropped from the top of a tall stairwell.

1. What is the penny's velocity after it has fallen for 2 seconds?
2. The penny hits the ground in 4.5 seconds. What is its final velocity?

Air Resistance Slows Down Acceleration

Try this simple experiment. Drop two sheets of paper—one crumpled in a tight ball and the other kept flat. Did your results contradict what you just learned about falling objects? The flat paper fell more slowly because of fluid friction that opposes the motion of objects through air. This fluid friction is also known as *air resistance.* Air resistance occurs between the surface of the falling object and the air that surrounds it.

Gravity helps make roller coasters thrilling to ride. Read about roller coaster design on page 159.

Self-Check

Which is more affected by air resistance—a leaf or an acorn? *(See page 596 to check your answer.)*

Air resistance affects some objects more than others. The amount of air resistance acting on an object depends on the size and shape of the object. Air resistance affects the flat sheet of paper more than the crumpled one, causing the flat sheet to fall more slowly than the crumpled one. Because air is all around you, any falling object you see is affected by air resistance. **Figure 3** shows the effect of air resistance on the downward acceleration of a falling object.

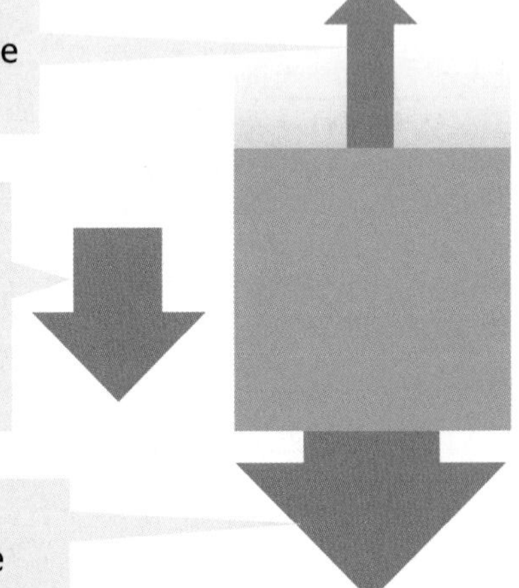

Figure 3 *The force of gravity pulls the object downward as the force of air resistance pushes it upward.*

Acceleration Stops at the Terminal Velocity

As long as the net force on a falling object is not zero, the object accelerates downward. But the amount of air resistance on an object increases as the speed of the object increases. As an object falls, the upward force of air resistance continues to increase until it exactly matches the downward force of gravity. When this happens, the net force is zero, and the object stops accelerating. The object then falls at a constant velocity, which is called the **terminal velocity.**

Sometimes the fact that falling objects have a terminal velocity is a good thing. The terminal velocity of hailstones is between 5 and 40 m/s, depending on the size of the stones. Every year cars, buildings, and vegetation are all severely damaged in hail storms. Imagine how much more destructive hail would be if there were no air resistance—hailstones would hit the Earth at velocities near 350 m/s! **Figure 4** shows another situation in which terminal velocity is helpful.

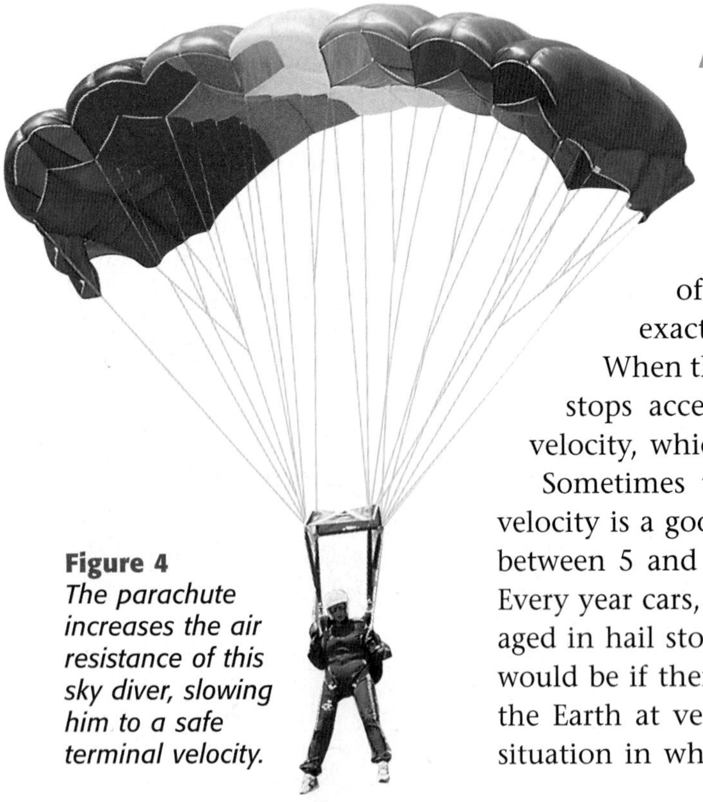

Figure 4 *The parachute increases the air resistance of this sky diver, slowing him to a safe terminal velocity.*

Free Fall Occurs When There Is No Air Resistance Sky divers are often described as being in free fall before they open their parachutes. However, that is an incorrect description, because air resistance is always acting on the sky diver.

An object is in **free fall** only if gravity is pulling it down and no other forces are acting on it. Because air resistance is a force (fluid friction), free fall can occur only where there is no air—in a vacuum (a place in which there is no matter) or in space. **Figure 5** shows objects falling in a vacuum. Because there is no air resistance, the two objects are in free fall.

Figure 5 *Air resistance normally causes a feather to fall more slowly than an apple. But in a vacuum, the feather and the apple fall with the same acceleration because both are in free fall.*

Orbiting Objects Are in Free Fall

Look at the astronaut in **Figure 6.** Why is the astronaut floating inside the space shuttle? It might be tempting to say it is because she is "weightless" in space. In fact, you may have read or heard that objects are weightless in space. However, it is impossible to be weightless anywhere in the universe.

Weight is a measure of gravitational force. The size of the force depends on the masses of objects and the distances between them. If you traveled in space far away from all the stars and planets, the gravitational force acting on you would be almost undetectable because the distance between you and other objects would be great. But you would still have mass, and so would all the other objects in the universe. Therefore, gravity would still attract you to other objects—even if just slightly—so you would still have weight.

Astronauts "float" in orbiting spaceships because of free fall. To understand this better, you need to understand what *orbiting* means and then consider the astronauts inside the ship.

Figure 6 *Astronauts appear to be weightless while floating inside the space shuttle—but they're not!*

Two Motions Combine to Cause Orbiting An object is said to be orbiting when it is traveling in a circular or nearly circular path around another object. When a spaceship orbits Earth, it is moving forward, but it is also in free fall toward Earth. **Figure 7** shows how these two motions occur together to cause orbiting.

Figure 7 How an Orbit Is Formed

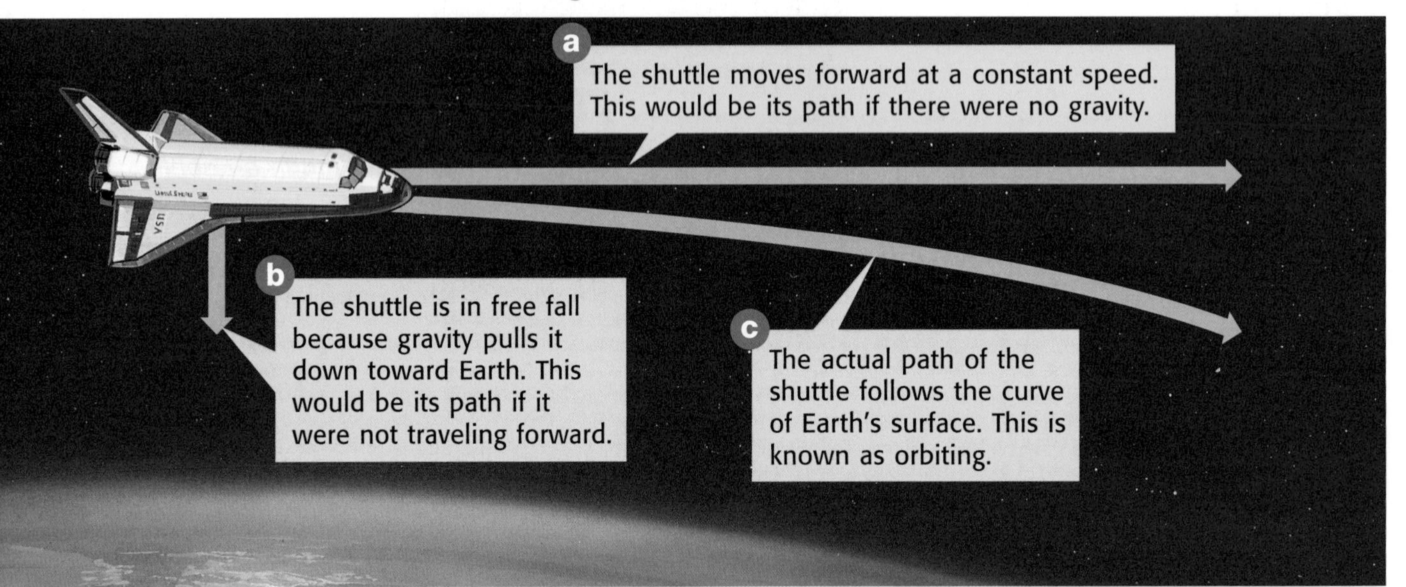

As you can see in the illustration above, the space shuttle is always falling while it is in orbit. So why don't astronauts hit their heads on the ceiling of the falling shuttle? Because they are also in free fall—they are always falling, too. Because the astronaut in Figure 6 is in free fall, she appears to be floating.

The Role of Gravity in Orbiting Besides spaceships and satellites, many other objects in the universe are in orbit. The moon orbits the Earth, Earth and the other planets orbit the sun, and many stars orbit large masses in the center of galaxies. All of these objects are traveling in a circular or nearly circular path. Remember, any object in circular motion is constantly changing direction. Because an unbalanced force is necessary to change the motion of any object, there must be an unbalanced force working on any object in circular motion.

The unbalanced force that causes objects to move in a circular path is called a *centripetal force*. Gravity provides the centripetal force that keeps objects in orbit. The word *centripetal* means "toward the center." As you can see in **Figure 8,** the centripetal force on the moon points toward the center of the circle traced by the moon's orbit.

Figure 8 *The moon stays in orbit around the Earth because Earth's gravitational force provides a centripetal force on the moon.*

Projectile Motion and Gravity

The orbit of the space shuttle around the Earth is an example of projectile (proh JEK tuhl) motion. **Projectile motion** is the curved path an object follows when thrown or propelled near the surface of the Earth. The motions of leaping frogs, thrown balls, and arrows shot from a bow are all examples of projectile motion. Projectile motion has two components—horizontal and vertical. The two components are independent; that is, they have no effect on each other. When the two motions are combined, they form a curved path, as shown in **Figure 9.**

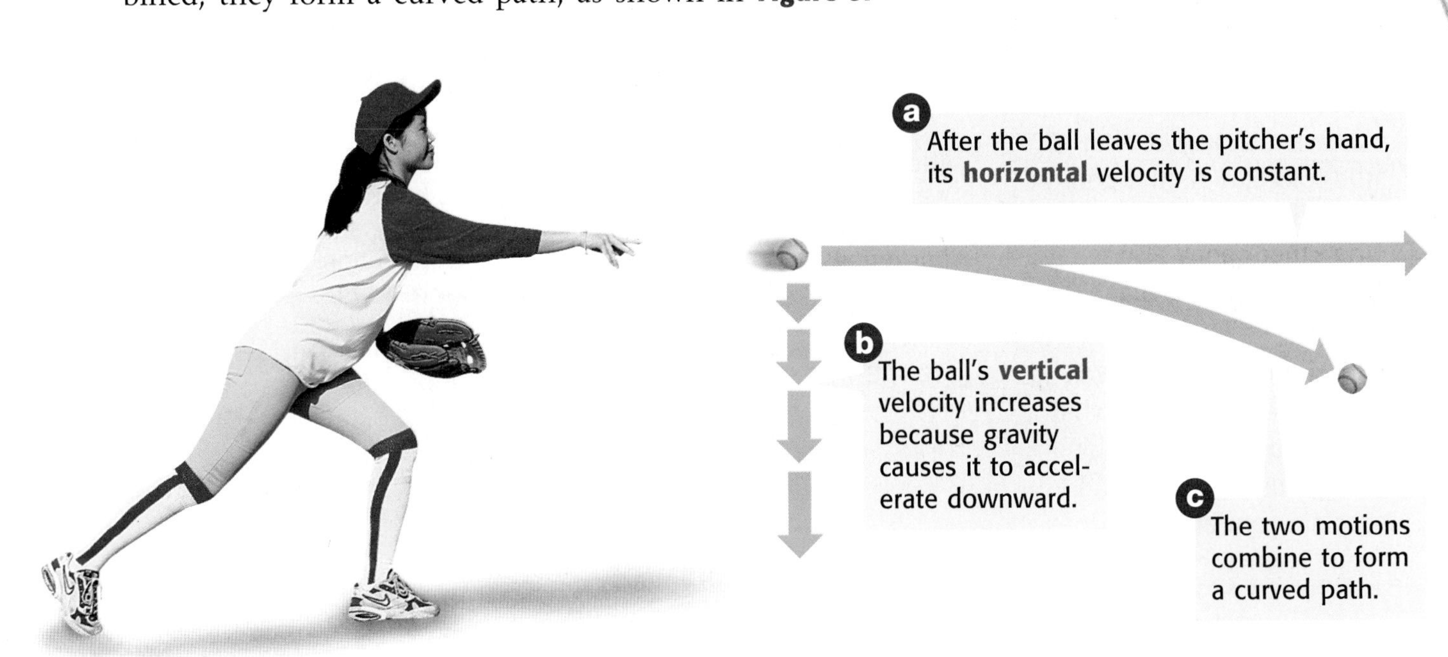

Figure 9 *Two motions combine to form projectile motion.*

Horizontal Motion When you throw a ball, your hand exerts a force on the ball that makes the ball move forward. This force gives the ball its horizontal motion. Horizontal motion is motion that is parallel to the ground.

After you let go of the ball, there are no horizontal forces acting on the ball (if you ignore air resistance). Therefore, there are no forces to change the ball's horizontal motion. Thus, the horizontal velocity of the ball is constant after the ball leaves your hand, as shown in Figure 9.

Vertical Motion When you let go of the ball, gravity pulls it downward, giving the ball vertical motion, as shown in Figure 9. Vertical motion is motion that is perpendicular (at a 90° angle) to the ground.

Examples of Objects in Projectile Motion

- A football being passed
- A leaping dancer
- Balls being juggled
- An athlete doing a high jump
- Water sprayed by a sprinkler
- A swimmer diving into water
- A hopping grasshopper

Quick Lab

Penny Projectile Motion

1. Position a **flat ruler** and **two pennies** on a desk or table as shown below:

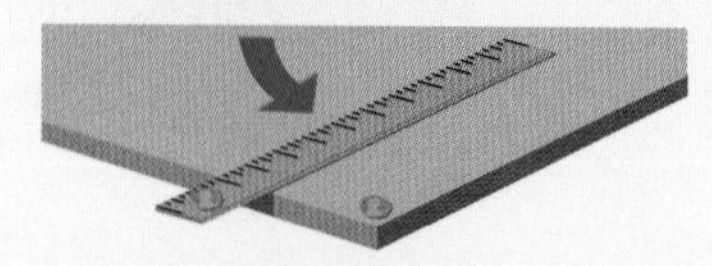

2. Hold the ruler by the end that is on the desk. Move the ruler quickly in the direction shown so that the ruler knocks the penny off the table and so that the other penny also drops. Repeat several times.
3. Which penny travels with projectile motion? In what order do the pennies hit the ground? Record and explain your answers in your ScienceLog.

Because objects in projectile motion accelerate downward, you always have to aim above a target if you want to hit it with a thrown or propelled object. This is why archers point their arrows above the bull's-eye on a target. If you aimed an arrow directly at a bull's-eye, your arrow would strike the bottom of the target rather than the middle.

Gravity pulls objects in projectile motion down with an acceleration of 9.8 m/s/s (if air resistance is ignored), just as it does all falling objects. **Figure 10** shows that the downward acceleration of a thrown object and a falling object are identical. Try the QuickLab at left to compare an object in projectile motion with a falling object.

Figure 10 Projectile Motion and Acceleration Due to Gravity

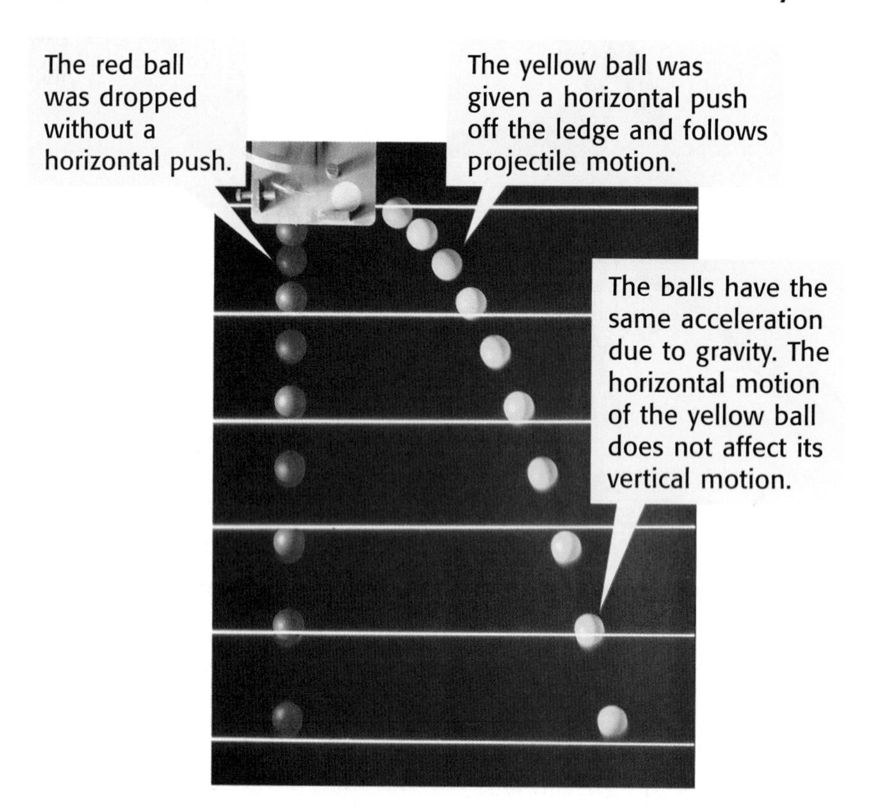

Marshmallows in projectile motion can be an uplifting experience. Turn to page 542 of the LabBook.

REVIEW

1. How does air resistance affect the acceleration of falling objects?
2. Explain why an astronaut in an orbiting spaceship floats.
3. How is an orbit formed?
4. **Applying Concepts** Think about a sport you play that involves a ball. Identify at least four different instances in which an object is in projectile motion.

Section 2

Newton's Laws of Motion

NEW TERMS

inertia
momentum

OBJECTIVES

- State and apply Newton's laws of motion.
- Compare the momentum of different objects.
- State and apply the law of conservation of momentum.

In 1686, Sir Isaac Newton published his book *Principia*. In it, he described three laws that relate forces to the motion of objects. Although he did not discover all three of the laws, he explained them in a way that helped many people understand them. Thus, the three laws are commonly known as Newton's laws of motion. In this section, you will learn about these laws and how they influence the motion of objects.

Newton's First Law of Motion

An object at rest remains at rest and an object in motion remains in motion at constant speed and in a straight line unless acted on by an unbalanced force.

Newton's first law of motion describes the motion of an object that has a net force of zero acting on it. This law may seem complicated when you first read it, but it's easy to understand when you consider its two parts separately.

Part 1: Objects at Rest What does it mean for an object to be at rest? Objects don't get tired! An object that is not moving is said to be at rest. Objects are at rest all around you. A plane parked on a runway, a chair on the floor, and a golf ball balanced on a tee are all examples of objects at rest.

Newton's first law says that objects at rest will remain at rest unless they are acted on by an unbalanced force. That means that objects will not start moving until a push or a pull is exerted on them. A plane won't soar in the air unless it is pushed by the exhaust from its jet engines, a chair won't slide across the room unless you push it, and a golf ball won't move off the tee unless struck by a golf club, as shown in **Figure 11.**

Figure 11 *A golf ball will remain at rest on a tee until it is acted on by the unbalanced force of a moving club.*

Figure 12 *Bumper cars let you have fun with Newton's first law.*

Part 2: Objects in Motion Think about riding in a bumper car at an amusement park. Your ride is pleasant as long as you are driving in an open space. But the name of the game is bumper cars, so sooner or later you are likely to run into another car, as shown in **Figure 12.**

The second part of Newton's first law explains that an object moving at a certain velocity will continue to move *forever* at the same speed and in the same direction unless some unbalanced force acts on it. Thus, your bumper car stops, but you continue to move forward until your seat belt stops you.

Friction and Newton's First Law Because an object in motion will stay in motion forever unless it is acted on by an unbalanced force, you should be able to give your desk a small push and send it sailing across the floor. If you try it, you will find that the desk quickly comes to a stop. What does this tell you?

There must be an unbalanced force that acts on the desk to stop its motion. That unbalanced force is friction. The friction between the desk and the floor works against the motion of the desk. Because of friction, it is often difficult to observe the effects of Newton's first law on the motion of everyday objects. For example, friction will cause a ball rolling on grass to slow down and stop. Friction will also make a car decelerate on a flat surface if you let up on the gas pedal. Because of friction, the motion of these objects changes.

The dummy in this crash test is wearing a seat belt, but the car does not have an air bag. Explain why Newton's first law of motion could lead to serious injuries in accidents involving cars without air bags.

Inertia Is Related to Mass Newton's first law of motion is often summed up in one sentence: Matter resists any change in motion. The tendency of all objects to resist any change in motion is called **inertia** (in UHR shuh). Due to inertia, an object at rest will remain at rest until something makes it move. Likewise, inertia is why a moving object stays in motion with the same velocity unless a force acts on it to change its speed or direction. Because Newton's first law can be explained in terms of inertia, it is sometimes called the law of inertia.

Because of inertia, you slide toward the side of a car when the driver makes a sharp turn. Inertia is also why it is impossible for a plane, car, or bicycle to stop instantaneously. Brakes must be applied well before stopping.

Mass is a measure of inertia. An object with a small mass has less inertia than an object with a large mass. Therefore, it is easier to start and to change the motion of an object with a small mass. For example, a softball has less mass and therefore less inertia than a bowling ball. Because the softball has a small amount of inertia, it is easy to pitch a softball and to change its motion by hitting it with a bat. Imagine how difficult it would be to play softball with a bowling ball! The inertia of the bowling ball would make it hard to pitch and hard to change its motion with a bat. **Figure 13** further illustrates the relationship between mass and inertia. Try the QuickLab at right to test the relationship yourself.

*Quick*Lab

First-Law Magic

1. On a table or desk, place a **large, empty plastic cup** on top of a **paper towel.**
2. Without touching the cup or tipping it over, remove the paper towel from under the cup. What did you do to accomplish this?
3. Repeat the first two steps a few times until you are comfortable with the procedure.
4. Fill the cup half full with **water,** and place the cup on the paper towel.
5. Once again, remove the paper towel from under the cup. Was it easier or harder to do this? Explain your answer in terms of mass and inertia.
6. Record your observations and explanations in your ScienceLog.

Figure 13 *Inertia makes it harder to push a car than to push a bicycle. Inertia also makes it easier to stop a moving bicycle than a moving car.*

Self-Check

When you stand while riding a bus, why do you tend to fall backward when the bus starts moving?
(See page 596 to check your answer.)

Newton's Second Law of Motion

The acceleration of an object depends on the mass of the object and the amount of force applied.

environmental science CONNECTION

Modern cars do not pollute the air as much as older cars did. One reason for this is that modern cars are lighter (less massive) than older models and have considerably smaller engines. According to Newton's second law, a less massive object requires less force to achieve the same acceleration as a more massive object. This is why a smaller car can have a smaller engine and still have acceptable acceleration. And because smaller engines use less fuel, they pollute less.

Newton's second law describes the motion of an object when an unbalanced force is acting on it. As with Newton's first law, it is easier to consider the parts of this law separately.

Part 1: Acceleration Depends on Mass Suppose you are pushing a shopping cart at the grocery store. At the beginning of your shopping trip, you have to exert only a small force on the cart to roll it quickly down the aisles. But when the cart is full, its acceleration is not as large when you exert the same amount of force as before, as shown in **Figure 14.** This example illustrates that an object's acceleration *decreases* as its mass *increases.* Conversely, an object's acceleration *increases* as its mass *decreases* when acted on by the same force.

Figure 14 *If the force applied is the same, the acceleration of the empty cart is greater than the acceleration of the full cart.*

Part 2: Acceleration Depends on Force Now suppose you give the cart a hard push, as shown in **Figure 15.** The cart will start moving faster than if you only gave it a soft push. This illustrates that an object's acceleration *increases* as the force on it *increases.* Conversely, an object's acceleration *decreases* as the force on it *decreases.*

An example of a large force causing a large acceleration is a baseball pitcher throwing a fastball. An example of a small force causing a small acceleration is the gentle tap on a golf ball that makes the ball roll slowly to the hole.

The acceleration of an object is always in the same direction as the force applied. The shopping cart, baseball, and golf ball all moved forward because the pushes were in the forward direction. To change the direction of an object, you must exert a force in the direction you want the object to go.

Figure 15 *Acceleration will increase when a larger force is exerted. The acceleration is always in the direction of the force applied.*

Expressing Newton's Second Law Mathematically The relationship of acceleration (a) to mass (m) and force (F) can be expressed mathematically with the following equation:

$$a = \frac{F}{m}$$

This equation is often rearranged to the following form:

$$F = m \times a$$

Both forms of the equation can be used to solve problems. Try the MathBreak at right to practice using the equations.

Newton's Second Law and Falling Objects In **Figure 16,** you can see how Newton's second law explains why objects fall to Earth with the same acceleration.

Figure 16 Newton's Second Law and Acceleration Due to Gravity

The **apple** has less mass, so the gravitational force on it is smaller. However, the apple also has less inertia and is easier to move.

$m = 0.102$ kg

$F = 1$ N

1 N = 1 kg · m/s/s

The **watermelon** has more mass and therefore more inertia, so it is harder to move.

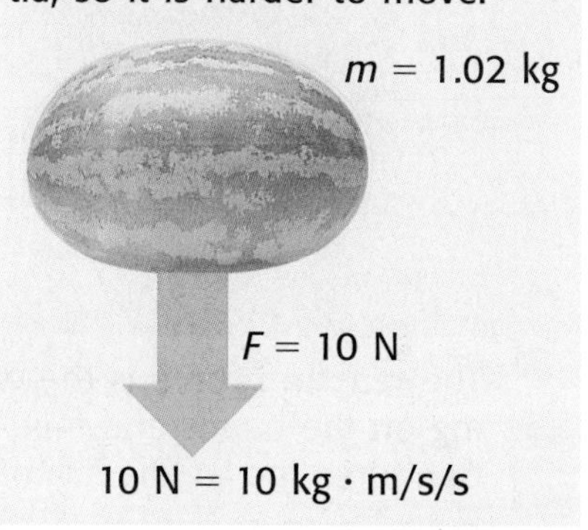

10 N = 10 kg · m/s/s

The larger weight of the watermelon is offset by its greater inertia. Thus, the accelerations of the watermelon and the apple are the same when you put the numbers into the equation $a = F/m$.

$$a = \frac{1 \cancel{kg} \cdot m/s/s}{0.102 \cancel{kg}} = 9.8 \text{ m/s/s}$$

$$a = \frac{10 \cancel{kg} \cdot m/s/s}{1.02 \cancel{kg}} = 9.8 \text{ m/s/s}$$

REVIEW

1. How is inertia related to Newton's first law of motion?
2. Name two ways to increase the acceleration of an object.
3. **Making Predictions** If the acceleration due to gravity were somehow doubled to 19.6 m/s/s, what would happen to your weight?

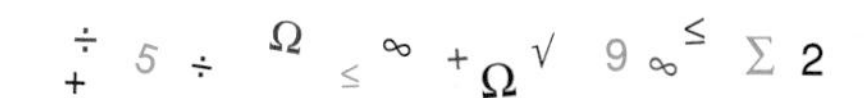

MATH BREAK

Second-Law Problems

You can rearrange the equation $F = m \times a$ to find acceleration and mass as shown below.

$$a = \frac{F}{m} \qquad m = \frac{F}{a}$$

1. What is the acceleration of a 7 kg mass if a force of 68.6 N is used to move it toward Earth?
2. What force is necessary to accelerate a 1,250 kg car at a rate of 40 m/s/s?
3. What is the mass of an object if a force of 34 N produces an acceleration of 4 m/s/s?

Newton's Third Law of Motion

Whenever one object exerts a force on a second object, the second object exerts an equal and opposite force on the first.

Newton's third law can be simply stated as follows: All forces act in pairs. If a force is exerted, another force occurs that is equal in size and opposite in direction. The law itself addresses only forces. But the way that force pairs interact affects the motion of objects.

What is meant by "forces act in pairs"? Study **Figure 17** to learn how one force pair helps propel a swimmer through water.

Figure 17 *The action force and reaction force are a pair. The two forces are equal in size but opposite in direction.*

Action and reaction force pairs occur even when there is no motion. For example, you exert a force on a chair when you sit on it. Your weight pushing down on the chair is the action force. The reaction force is the force exerted by the chair that pushes up on your body and is equal to your weight.

Force Pairs Do Not Act on the Same Object You know that a force is always exerted by one object on another object. This is true for all forces, including action and reaction forces. However, it is important to remember that action and reaction forces in a pair do not act on the same object. If they did, the net force would always be zero and nothing would ever move! To understand this better, look back at Figure 17. In this example, the action force was exerted on the water by the swimmer's hands and feet. But the reaction force was exerted on the swimmer's hands and feet by the water. The forces did not act on the same object.

Explore

Choose a sport that you enjoy playing or watching. In your ScienceLog, list five ways that Newton's laws of motion are involved in the game you selected.

The Effect of a Reaction Can Be Difficult to See Another example of a force pair is shown in **Figure 18.** Remember, gravity is a force of attraction between objects that is due to their masses. If you drop a ball off a ledge, the force of gravity pulls the ball toward Earth. This is the action force exerted by Earth on the ball. But the force of gravity also pulls Earth toward the ball. That is the reaction force exerted by the ball on Earth.

It's easy to see the effect of the action force—the ball falls to Earth. Why don't you notice the effect of the reaction force—Earth being pulled upward? To find the answer to this question, think back to Newton's second law. It states that the acceleration of an object depends on the force applied to it and on the mass of the object. The force on Earth is equal to the force on the ball, but the mass of Earth is much *larger* than the mass of the ball. Therefore, the acceleration of Earth is much *smaller* than the acceleration of the ball. The acceleration is so small that you can't even see it or feel it. Thus, it is difficult to observe the effect of Newton's third law on falling objects.

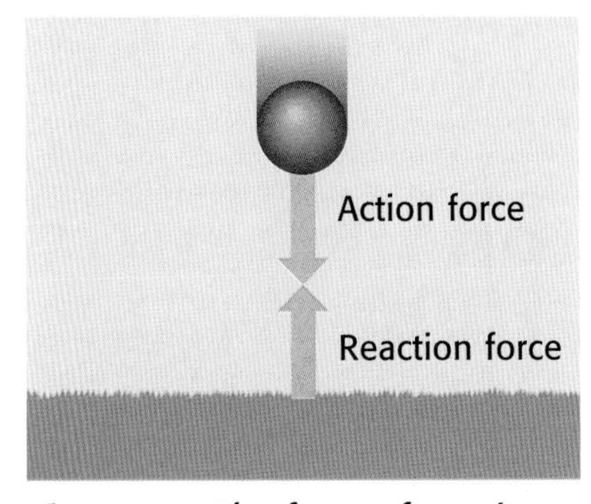

Figure 18 *The force of gravity between Earth and a falling object is a force pair.*

More Examples of Action and Reaction Force Pairs The examples below illustrate a variety of action and reaction force pairs. In each example, notice which object exerts the action and which object exerts the reaction forces.

The rabbit's legs exert a force on Earth. Earth exerts an equal force on the rabbit's legs, causing the rabbit to accelerate upward.

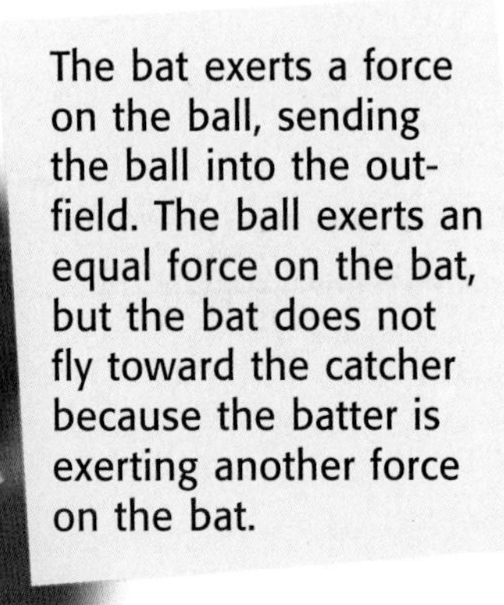

The bat exerts a force on the ball, sending the ball into the outfield. The ball exerts an equal force on the bat, but the bat does not fly toward the catcher because the batter is exerting another force on the bat.

The shuttle's thrusters push the exhaust gases downward as the gases push the shuttle upward with an equal force.

When you hit a table with your hand, your hand will hurt. This is because the table meets your hand with a force equal in size to the force you exerted.

Momentum Is a Property of Moving Objects

BRAIN FOOD

Jumping beans appear to leap into the air with no forces acting on them. However, inside each bean is a small insect larva. When the larva moves suddenly, it applies a force to the shell of the bean. The momentum of the larva is transferred to the bean, and the bean "jumps."

If a compact car and a large truck are traveling with the same velocity, it takes longer for the truck to stop than it does for the car if the same braking force is applied. Likewise, it takes longer for a fast moving car to stop than it does for a slow moving car with the same mass. The truck and the fast moving car have more momentum than the compact car and the slow moving car.

Momentum is a property of a moving object that depends on the object's mass and velocity. The more momentum an object has, the harder it is to stop the object or change its direction. Although the compact car and the truck are traveling with the same velocity, the truck has more mass and therefore more momentum, so it is harder to stop than the car. Similarly, the fast moving car has a greater velocity and thus more momentum than the slow moving car.

Momentum Is Conserved When a moving object hits another object, some or all of the momentum of the first object is transferred to the other object. If only some of the momentum is transferred, the rest of the momentum stays with the first object.

Imagine you hit a billiard ball with a cue ball so that the billiard ball starts moving and the cue ball stops, as shown in **Figure 19.** The cue ball had a certain amount of momentum before the collision. During the collision, all of the cue ball's momentum was transferred to the billiard ball. After the collision, the billiard ball moved away with the same amount of momentum the cue ball had. This example illustrates the *law of conservation of momentum.* Any time two or more objects interact, they may exchange momentum, but the total amount of momentum stays the same.

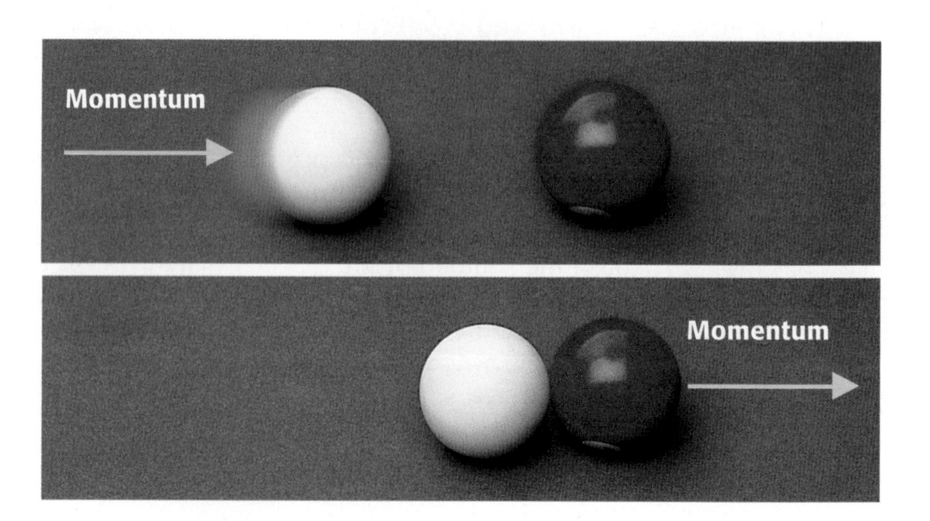

Figure 19 *The momentum before a collision is equal to the momentum after the collision.*

Bowling is another example of how conservation of momentum is used in a game. The bowling ball rolls down the lane with a certain amount of momentum. When the ball hits the pins, some of the ball's momentum is transferred to the pins and the pins move off in different directions. Furthermore, some of the pins that were hit by the ball go on to hit other pins, transferring the momentum again.

Conservation of Momentum and Newton's Third Law Conservation of momentum can be explained by Newton's third law. In the example with the billiard ball, the cue ball hit the billiard ball with a certain amount of force. This was the action force. The reaction force was the equal but opposite force exerted by the billiard ball on the cue ball. The action force made the billiard ball start moving, and the reaction force made the cue ball stop moving, as shown in **Figure 20.** Because the action and reaction forces are equal and opposite, momentum is conserved.

Figure 20 *The action force makes the billiard ball begin moving, and the reaction force stops the cue ball's motion.*

REVIEW

1. Name three action and reaction force pairs involved in doing your homework. Name what object is exerting and what object is receiving the forces.
2. Which has more momentum, a mouse running at 1 m/s north or an elephant walking at 3 m/s east? Explain your answer.
3. **Applying Concepts** When a truck pulls a trailer, the trailer and truck accelerate forward even though the action and reaction forces are the same size but in opposite directions. Why don't these forces balance each other out?

Catapult forward! Or is it backward? Find out on page 546 of the LabBook.

Chapter Highlights

SECTION 1

Vocabulary

terminal velocity *(p. 140)*
free fall *(p. 141)*
projectile motion *(p. 143)*

Section Notes

- All objects accelerate toward Earth at 9.8 m/s/s.
- Air resistance slows the acceleration of falling objects.
- An object is in free fall if gravity is the only force acting on it.
- An orbit is formed by combining forward motion and free fall.
- Objects in orbit appear to be weightless because they are in free fall.
- A centripetal force is needed to keep objects in circular motion. Gravity acts as a centripetal force to keep objects in orbit.
- Projectile motion is the curved path an object follows when thrown or propelled near the surface of Earth.
- Projectile motion has two components—horizontal and vertical. Gravity affects only the vertical motion of projectile motion.

Labs

A Marshmallow Catapult *(p. 542)*

☑ Skills Check

Math Concepts

NEWTON'S SECOND LAW The equation $a = F/m$ on page 149 summarizes Newton's second law of motion. The equation shows the relationship between the acceleration of an object, the force causing the acceleration, and the object's mass. For example, if you apply a force of 18 N to a 6 kg object, the object's acceleration is

$$a = \frac{F}{m} = \frac{18 \text{ N}}{6 \text{ kg}} = \frac{18 \cancel{\text{kg}} \cdot \text{m/s/s}}{6 \cancel{\text{kg}}} = 3 \text{ m/s/s}$$

Visual Understanding

HOW AN ORBIT IS FORMED An orbit is a combination of two motions—forward motion and free fall. Figure 7 on page 142 shows how the two motions combine to form an orbit.

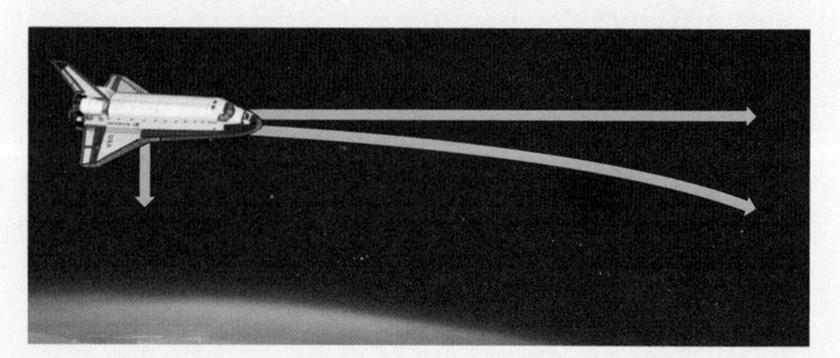

SECTION 2

Vocabulary

inertia *(p. 147)*

momentum *(p. 152)*

Section Notes

- Newton's first law of motion states that the motion of an object will not change if no unbalanced forces act on it.
- Inertia is the tendency of matter to resist a change in motion. Mass is a measure of inertia.
- Newton's second law of motion states that the acceleration of an object depends on its mass and on the force exerted on it.
- Newton's third law of motion states that whenever one object exerts a force on a second object, the second object exerts an equal and opposite force on the first.
- Momentum is the property of a moving object that depends on its mass and velocity.
- When two or more objects interact, momentum may be exchanged, but the total amount of momentum does not change. This is the law of conservation of momentum.

Labs

Blast Off! *(p. 543)*

Inertia-Rama! *(p. 544)*

Quite a Reaction *(p. 546)*

internetconnect

GO TO: go.hrw.com

Visit the **HRW** Web site for a variety of learning tools related to this chapter. Just type in the keyword:

KEYWORD: HSTFOR

GO TO: www.scilinks.org

Visit the **National Science Teachers Association** on-line Web site for Internet resources related to this chapter. Just type in the ***sci*LINKS** number for more information about the topic:

TOPIC: The Force of Gravity — ***sci*LINKS NUMBER:** HSTP130

TOPIC: Gravity and Orbiting Objects — ***sci*LINKS NUMBER:** HSTP135

TOPIC: Projectile Motion — ***sci*LINKS NUMBER:** HSTP140

TOPIC: Newton's Laws of Motion — ***sci*LINKS NUMBER:** HSTP145

Chapter Review

USING VOCABULARY

To complete the following sentences, choose the correct term from each pair of terms listed below:

1. An object in motion tends to stay in motion because it has ___?___. *(inertia* or *terminal velocity)*

2. Falling objects stop accelerating at ___?___. *(free fall* or *terminal velocity)*

3. ___?___ is the path that a thrown object follows. *(Free fall* or *Projectile motion)*

4. A property of moving objects that depends on mass and velocity is ___?___. *(inertia* or *momentum)*

5. ___?___ only occurs when there is no air resistance. *(Momentum* or *Free fall)*

UNDERSTANDING CONCEPTS

Multiple Choice

6. A feather and a rock dropped at the same time from the same height would land at the same time when dropped by
   a. Galileo in Italy.
   b. Newton in England.
   c. an astronaut on the moon.
   d. an astronaut on the space shuttle.

7. When a soccer ball is kicked, the action and reaction forces do not cancel each other out because
   a. the force of the foot on the ball is bigger than the force of the ball on the foot.
   b. the forces act on two different objects.
   c. the forces act at different times.
   d. All of the above

8. An object is in projectile motion if
   a. it is thrown with a horizontal push.
   b. it is accelerated downward by gravity.
   c. it does not accelerate horizontally.
   d. All of the above

9. Newton's first law of motion applies
   a. to moving objects.
   b. to objects that are not moving.
   c. to objects that are accelerating.
   d. Both (a) and (b)

10. Acceleration of an object
    a. decreases as the mass of the object increases.
    b. increases as the force on the object increases.
    c. is in the same direction as the force on the object.
    d. All of the above

11. A golf ball and a bowling ball are moving at the same velocity. Which has more momentum?
    a. the golf ball, because it has less mass
    b. the bowling ball, because it has more mass
    c. They both have the same momentum because they have the same velocity.
    d. There is no way to know without additional information.

Short Answer

12. Explain how an orbit is formed.

13. Describe how gravity and air resistance combine when an object reaches terminal velocity.

14. Explain why friction can make observing Newton's first law of motion difficult.

Concept Mapping

15. Use the following terms to create a concept map: gravity, free fall, terminal velocity, projectile motion, air resistance.

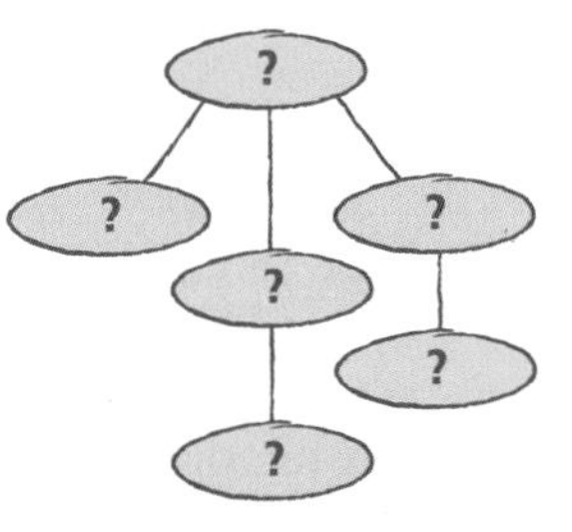

CRITICAL THINKING AND PROBLEM SOLVING

16. During a shuttle launch, about 830,000 kg of fuel is burned in 8 minutes. The fuel provides the shuttle with a constant thrust, or push off the ground. How does Newton's second law of motion explain why the shuttle's acceleration increases during takeoff?

17. When using a hammer to drive a nail into wood, you have to swing the hammer through the air with a certain velocity. Because the hammer has both mass and velocity, it has momentum. Describe what happens to the hammer's momentum after the hammer hits the nail.

18. Suppose you are standing on a skateboard or on in-line skates and you toss a backpack full of heavy books toward your friend. What do you think will happen to you and why? Explain your answer in terms of Newton's third law of motion.

MATH IN SCIENCE

19. A 12 kg rock falls from rest off a cliff and hits the ground in 1.5 seconds.
    a. Ignoring air resistance, what is the rock's velocity just before it hits the ground?
    b. What is the rock's weight after it hits the ground? (Hint: Weight is a measure of the gravitational force on an object.)

INTERPRETING GRAPHICS

20. The picture below shows a common desk toy. If you pull one ball up and release it, it hits the balls at the bottom and comes to a stop. In the same instant, the ball on the other side swings up and repeats the cycle. How does conservation of momentum explain how this toy works?

NOW What Do You Think?

Take a minute to review your answers to the ScienceLog questions on page 137. Have your answers changed? If necessary, revise your answers based on what you have learned since you began this chapter.

Eureka!

A Bat with Dimples

Wouldn't it be nice to hit a home run every time? Jeff DiTullio, a teacher at MIT, in Cambridge, Massachusetts, has found a way for you to get more bang from your bat. Would you believe *dimples?*

Building a Better Bat

If you look closely at the surface of a golf ball, you'll see dozens of tiny craterlike dimples. When air flows past these dimples, it gets stirred up. By keeping air moving near the surface of the ball, the dimples help the golf ball move faster and farther through the air.

DiTullio decided to apply this same idea to a baseball bat. His hypothesis was that dimples would allow a bat to move more easily through the air. This would help batters swing the bat faster and hit the ball harder. To test his hypothesis, DiTullio pressed hundreds of little dimples about 1 mm deep and 2 mm across into the surface of a bat.

When DiTullio tested his dimpled bat in a wind tunnel, he found that it could be swung 3 to 5 percent faster. That may not sound like much, but it could add about 5 m to a fly ball!

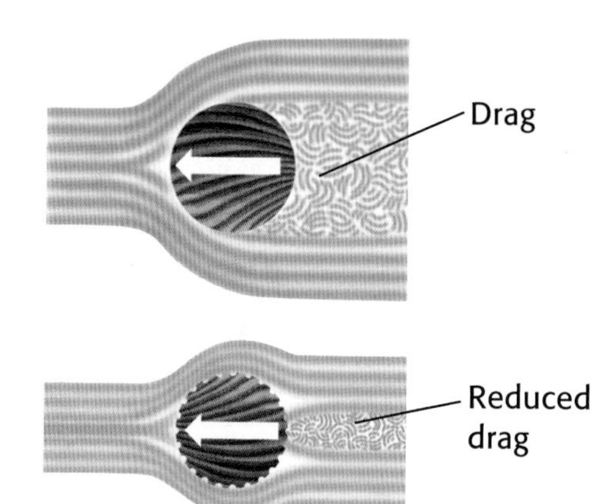

▲ *By reducing the amount of drag behind the bat, dimples help the bat move faster through the air.*

Safe . . . or Out?

As you might imagine, many baseball players would love to have a bat that could turn a long fly ball into a home run. But are dimpled baseball bats legal?

The size and shape of every piece of equipment used in Major League Baseball games are regulated. A baseball bat, for instance, must be no more than 107 cm long and no more than 7 cm across at its widest point. When DiTullio designed his dimpled bat, there was no rule stating that bats had to be smooth. But when Major League Baseball found out about the new bat, they changed the rules! Today official rules require that all bats be smooth, and they prohibit any type of "experimental" bat. Someday the rules may be revised to allow DiTullio's dimpled bat. When that happens, fans of the dimpled baseball bat will all shout, "Play ball!"

Dimple Madness

▶ Now that you know how dimples can improve baseball bats, think of other uses for dimples. How might dimples improve the way other objects move through the air? Draw a sketch of a dimpled object, and describe how the dimples improve the design.

▶ *Jeff DiTullio, pictured with his dimpled baseball bat, is an aeronautical engineer—someone who studies both the way air moves and the way things move through air.*

CAREERS

ROLLER COASTER DESIGNER

Roller coasters have fascinated **Steve Okamoto** ever since his first ride on one. "I remember going to Disneyland as a kid. My mother was always upset with me because I kept looking over the sides of the rides, trying to figure out how they worked," he laughs. To satisfy his curiosity, Okamoto became a mechanical engineer. Today he uses his scientific knowledge to design and build machines, systems, and buildings. But his specialty is roller coasters.

His West Coaster, which sits on the Santa Monica pier in Santa Monica, California, towers five stories above the Pacific Ocean. The cars on the Steel Force, at Dorney Park, in Pennsylvania, reach speeds of over 120 km/h and drop more than 60 m to disappear into a 37 m long tunnel. The Mamba, at Worlds of Fun, in Missouri, sends cars flying along as high and as fast as the Steel Force does, but it also has two giant back-to-back hills, a fast spiral, and five "camelback" humps. The camelbacks are designed to pull riders' seats out from under them, giving the riders "air time."

Coaster Motion

Roller-coaster cars really do coast along the track. A motor pulls the cars up a high hill to start the ride. After that, the cars are powered by gravity alone. As the cars roll downhill, they pick up enough speed to whiz through the rest of the curves, loops, twists, and bumps in the track.

Designing a successful coaster is no simple task. Steve Okamoto has to calculate the cars' speed and acceleration on each part of the track. "The coaster has to go fast enough to make it up the next hill," he explains. Okamoto uses his knowledge of geometry and physics to create safe but scary curves, loops, humps, and dips. Okamoto must also keep in mind that the ride's towers and structures need to be strong enough to support both the track and the speeding cars full of people. The cars themselves need special wheels to keep them locked onto the track and seat belts or bars to keep passengers safely inside. "It's like putting together a puzzle, except the pieces haven't been cut out yet," says Okamoto.

▲ *The Wild Thing, in Shakopee, Minnesota, was designed by Steve Okamoto.*

Take the Challenge

▶ Step outside for a moment. Gather some rope and a medium-sized plastic bucket half-full of water. Can you get the bucket over your head and upside down without any water escaping? How does this relate to roller coasters?

CHAPTER 7

Forces in Fluids

Imagine. . .

You're the pilot of a revolutionary new undersea vessel, *Deep Flight,* and today is the day of your first undersea voyage. Your destination: the Mariana Trench, which is the deepest spot in the ocean. The Mariana Trench is almost 11 km deep—that's deep enough to swallow Mount Everest, the tallest mountain in the world. In the history of ocean exploration, fewer than a dozen undersea vessels have ever ventured this far down. The reason? Water exerts tremendous pressure at this depth. A human subjected to this pressure would be crushed instantly. Luckily, *Deep Flight* offers you lots of protection. The hull is made of an extremely strong, hard ceramic material that can withstand the pressure.

What makes *Deep Flight* so revolutionary? *Deep Flight* actually "flies" through the water. In fact, *Deep Flight* looks a lot like an airplane with stubby wings. You can use controls to adjust the curvature of the wings, allowing you to move faster through the water.

With *Deep Flight*'s battery-powered motor and your ability to change the curvature of the wings, *Deep Flight* can reach speeds of up to 25 km/h! You can also change *Deep Flight*'s path by adjusting its wing flaps and tail fins. You can even do dives, spins, and turns. Ready to race a whale?

As futuristic as this story sounds, *Deep Flight* is a real undersea vessel that is currently being tested. Although *Deep Flight* has not yet made it to the bottom of the Mariana Trench, some scientists believe this type of undersea vessel will one day be used routinely to explore the ocean floor.

In this chapter you will explore fluids. You'll learn how pressure is exerted by water and other fluids. You'll also learn why some things sink and others float and how the curvature of a wing affects speed. Dive in!

What Do You Think?

In your ScienceLog, try to answer the following questions based on what you already know:

1. What is a fluid?
2. How is fluid pressure exerted?
3. Do moving fluids exert different forces than nonmoving fluids?

Out the Spouts

The undersea vessel *Deep Flight* was built to withstand the pressure exerted by water. In this activity you'll witness one of the effects of this pressure firsthand.

Procedure

1. With a sharp **pencil,** punch a small hole in the center of one side of an empty **cardboard milk container.**
2. Make another hole 4 cm above the center hole. Then make another hole 8 cm above the center hole.
3. With a single piece of **masking tape,** carefully cover the holes. Leave a little tape free at the bottom for easy removal.
4. Fill the container with **water,** and place it in a **large plastic tray or sink.**
5. Quickly pull the tape off the container.
6. Record your observations in your ScienceLog.

Analysis

7. Did the same thing happen at each hole after you removed the tape? If not, what do you think caused the different results? Record your answers in your ScienceLog.

Section 1

Fluids and Pressure

NEW TERMS
fluid
pressure
pascal
atmospheric pressure
density
Pascal's principle

OBJECTIVES
- Describe how fluids exert pressure.
- Analyze how fluid depth affects pressure.
- Give examples of fluids flowing from high to low pressure.
- State and apply Pascal's principle.

What does a dolphin have in common with a sea gull? What does a dog have in common with a fly? What do you have in common with all these living things? The answer is that you and all these other living things spend a lifetime moving through and even breathing fluids. A **fluid** is any material that can flow and that takes the shape of its container. Fluids include liquids (such as water and oil) and gases (such as oxygen and carbon dioxide). Fluids are able to flow because the particles in fluids, unlike the particles in solids, can move easily past each other. As you will find out, the remarkable properties of fluids allow huge ships to float, divers to explore the ocean depths, and jumbo jets to soar across the skies.

All Fluids Exert Pressure

You probably have heard the terms *air pressure, water pressure,* and *blood pressure.* Air, water, and blood are all fluids, and all fluids exert pressure. So what's pressure? Well, think about this example. When you pump up a bicycle tire, you push air into the tire. And like all matter, air is made of tiny particles that are constantly moving. Inside the tire, the air particles push against each other and against the walls of the tire, as shown in **Figure 1.** The more air you pump into the tire, the more the air particles push against the inside of your tire. Together, these pushes create a force against the tire. The amount of force exerted on a given area is **pressure.** Pressure can be calculated by dividing the force that a fluid exerts by the area over which the force is exerted:

$$\text{Pressure} = \frac{\text{Force}}{\text{Area}}$$

The SI unit for pressure is the **pascal.** One pascal (1 Pa) is the force of one newton exerted over an area of one square meter (1 N/m^2). Try the MathBreak at left to practice calculating pressure.

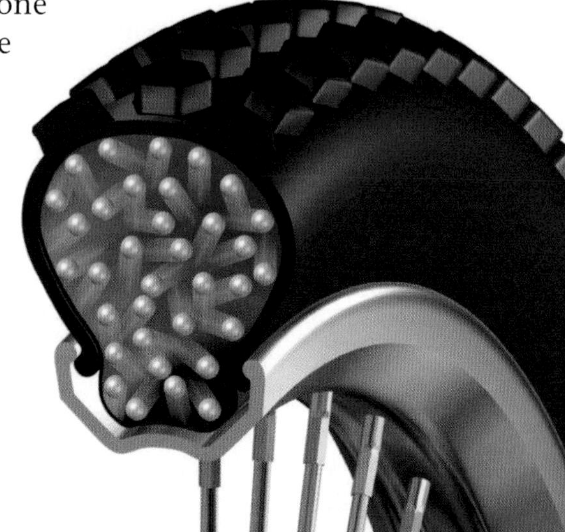

Figure 1 *The force of the air particles hitting the inner surface of the tire creates pressure, which keeps the tire inflated.*

MATH BREAK

Pressure, Force, and Area

The equation on this page can be used to find pressure or rearranged to find force or area.

$$\text{Force} = \text{Pressure} \times \text{Area}$$

$$\text{Area} = \frac{\text{Force}}{\text{Pressure}}$$

1. Find the pressure exerted by a 3,000 N crate with an area of 2 m^2.
2. Find the weight of a rock with an area of 10 m^2 that exerts a pressure of 250 Pa.

(Be sure to express your answers in the correct SI unit.)

Why Are Bubbles Round? When you blow a soap bubble, you blow in only one direction. So why doesn't the bubble get longer and longer as you blow instead of rounder and rounder? The shape of the bubble is due in part to an important property of fluids: Fluids exert pressure evenly in all directions. The air you blow into the bubble exerts pressure evenly in every direction, so the bubble expands in every direction, helping to create a sphere, as shown in **Figure 2.** This property also explains why tires inflate evenly (unless there is a weak spot in the tire).

Figure 2 *You can't blow a square bubble, because fluids exert pressure equally in every direction.*

Atmospheric Pressure

The *atmosphere* is the layer of nitrogen, oxygen, and other gases that surrounds the Earth. The atmosphere stretches about 150 km above us. If you could stack 500 Eiffel Towers on top of each other, they would come close to reaching the top of the atmosphere. However, approximately 80 percent of the gases in the atmosphere are found within 10 km of the Earth's surface. Earth's atmosphere is held in place by gravity, which pulls the gases toward Earth. The pressure caused by the weight of the atmosphere is called **atmospheric pressure.**

Atmospheric pressure is exerted on everything on Earth, including you. The atmosphere exerts a pressure of approximately 101,300 N on every square meter, or 101,300 Pa. This means that there is a weight of about 10 N (roughly the weight of a pineapple) on every square centimeter (roughly the area of the tip of your little finger) of your body. Ouch!

Why don't you feel this crushing pressure? The fluids inside your body also exert pressure, just like the air inside a balloon exerts pressure. **Figure 3** can help you understand.

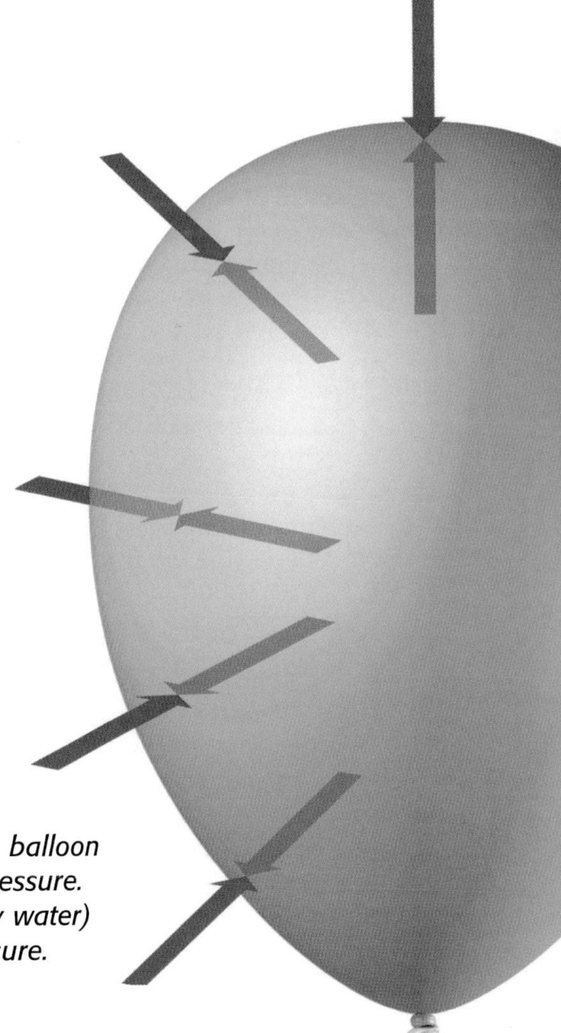

Figure 3 *The pressure exerted by the air inside a balloon keeps the balloon inflated against atmospheric pressure. Similarly, the pressure exerted by the fluid (mostly water) inside your body works against atmospheric pressure.*

Figure 4 Differences in Atmospheric Pressure

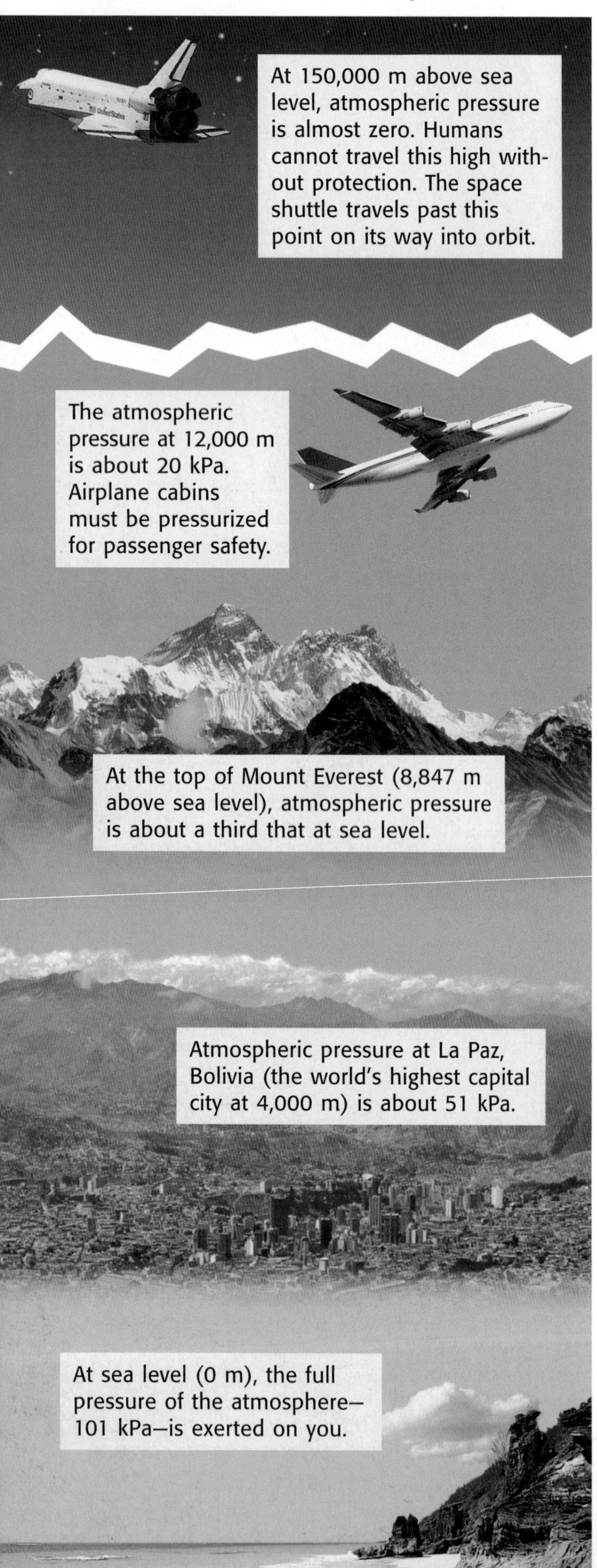

Pressure Depends on Depth Atmospheric pressure varies with different levels of the atmosphere. At the top of the atmosphere, pressure is almost nonexistent because there is no atmosphere pressing down. At the top of Mount Everest in south-central Asia (which at 8,847 m above sea level is the highest point on Earth), atmospheric pressure is about 33,000 Pa, or 33 kilopascals (kPa). People attempting to reach this height usually carry oxygen with them because most of Earth's air is below them, making it difficult to breathe. At sea level, atmospheric pressure is about 101 kPa.

As shown in **Figure 4,** pressure increases as you descend through the atmosphere. In other words, the pressure increases as the atmosphere gets "deeper." This is an important point about fluids: Pressure depends on the depth of the fluid. At lower levels of the atmosphere, there is more fluid above you being pulled by Earth's gravitational force, so there is more pressure.

If you travel to higher or lower points in the atmosphere, the fluids in your body have to adjust to maintain equal pressure. You may have experienced this if your ears have "popped" when you were in a plane taking off or a car traveling down a steep mountain road. Small pockets of air behind your eardrums contract or expand as atmospheric pressure increases or decreases. The "pop" occurs when air is released due to these pressure changes.

REVIEW

1. How do particles in a fluid exert pressure on a container?
2. Why are you not crushed by atmospheric pressure?
3. **Applying Concepts** Explain why dams on deep lakes should be thicker at the bottom than near the top.

Water Pressure

Water is a fluid; therefore, it exerts pressure, just like the atmosphere does. Water pressure also increases with depth because of gravity. Take a look at **Figure 5.** The deeper a diver goes in the water, the greater the pressure becomes because more water above the diver is being pulled by Earth's gravitational force. In addition, the atmosphere presses down on the water, so the total pressure on the diver includes water pressure as well as atmospheric pressure.

But pressure does not depend on the total amount of fluid present, only on the depth of the fluid. A swimmer would feel the same pressure swimming at 5 m below the surface of a small pond as at 5 m below the surface of an ocean, even though there is more water in the ocean.

Density Makes a Difference Water is about 1,000 times more dense than air. (Remember, **density** is the amount of matter in a certain volume, or mass per unit volume.) Because water is more dense than air, a certain volume of water has more mass—and therefore weighs more—than the same volume of air. Therefore, water exerts greater pressure than air.

For example, if you climb a 10 m tree, the decrease in atmospheric pressure is too small to notice. But if you dive 10 m underwater, the pressure on you increases to 201 kPa, which is almost twice the atmospheric pressure at the surface! Undersea vessels, such as the *Trieste* (shown here) and *Deep Flight* (which you read about earlier), must be constructed from extremely strong materials to withstand the tremendous water pressure at great depths.

Figure 5 Differences in Water Pressure

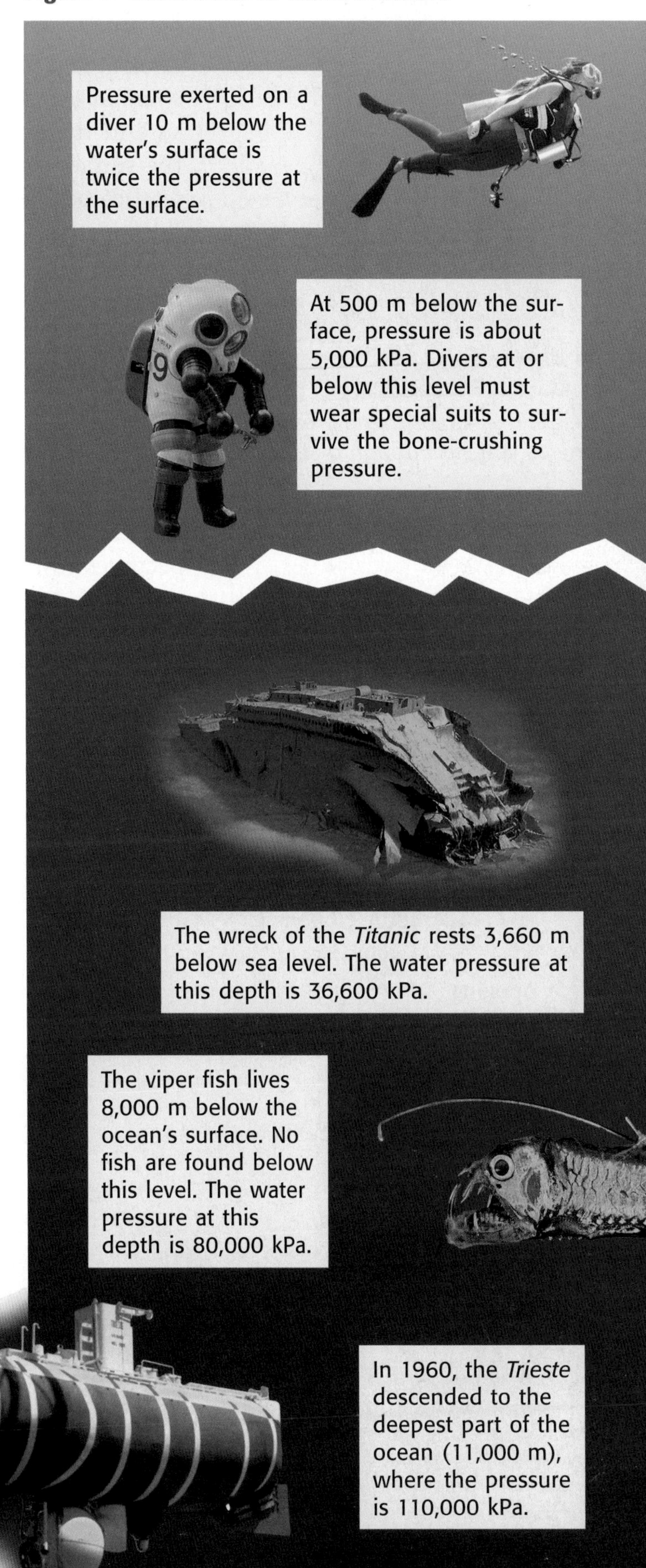

Figure 6 *Atmospheric pressure helps you sip through a straw!*

Fluids Flow from High Pressure to Low Pressure

Look at **Figure 6.** When you drink through a straw, you remove some of the air in the straw. Because there is less air, the pressure in the straw is reduced. But the atmospheric pressure on the surface of the liquid remains the same. This creates a difference between the pressure inside the straw and the pressure outside the straw. The outside pressure forces the liquid up into the straw and into your mouth. So just by sipping your drink through a straw, you can observe another important property of fluids: Fluids flow from regions of high pressure to regions of low pressure.

Go with the Flow Take a deep breath—that's fluid flowing from high to low pressure! When you inhale, a muscle increases the space in your chest, giving your lungs room to expand. This expansion lowers the pressure in your lungs so that it becomes lower than the outside air pressure. Air then flows into your lungs—from higher to lower pressure. This air carries oxygen that you need to live. **Figure 7** shows how exhaling also causes fluids to flow from higher to lower pressure. You can see this same exchange when you open a carbonated beverage or squeeze toothpaste onto your toothbrush.

QuickLab

Blown Away

1. Lay an **empty plastic soda bottle** on its side.
2. Wad **a small piece of paper** (about 4 × 4 cm) into a ball.
3. Place the paper ball just inside the bottle's opening.
4. Blow straight into the opening.
5. Record your observations in your ScienceLog.
6. Explain your results in terms of high and low fluid pressures.

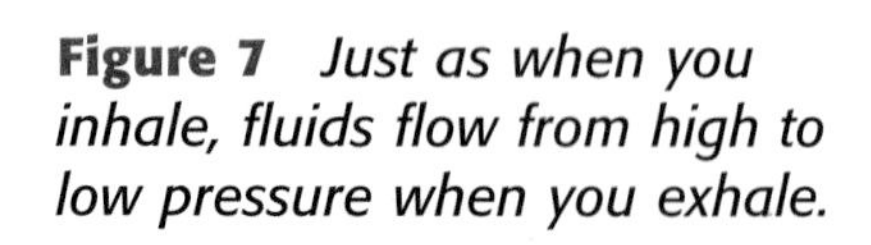

Figure 7 *Just as when you inhale, fluids flow from high to low pressure when you exhale.*

c Exhaled air carries carbon dioxide out of the lungs.

b The decrease in space causes the pressure in your lungs to increase. The air in your lungs flows from a region of higher pressure (your chest) to a region of lower pressure (outside of your body).

a When you exhale, a muscle in your chest moves upward, decreasing the space in your chest.

Pascal's Principle

Imagine that the water-pumping station in your town is modernized and can now increase the water pressure by 20 Pa. At which of the following locations will the water pressure be increased the most: the kitchen at the pumping station, a supermarket two blocks away, or a home 2 km away?

Believe it or not, the increase in water pressure will be the same—20 Pa—at each location. This is explained by Pascal's principle, named for Blaise Pascal, the seventeenth-century French scientist who discovered it. **Pascal's principle** states that a change in pressure at any point in an enclosed fluid will be transmitted equally to all parts of that fluid. Therefore, the change in pressure at the water-pumping station will be transmitted through all of the water that is pumped by the station.

Putting Pascal's Principle to Work Devices that use liquids to transmit pressure from one point to another are called *hydraulic* (hie DRAW lik) devices. Hydraulic devices use liquids because they cannot be compressed, or squeezed, into a smaller space very much. This property allows liquids to transmit pressure more efficiently than gases, which can be compressed a great deal.

Hydraulic devices can multiply forces. The brakes of a typical car are a good example. In **Figure 8,** a driver's foot exerts pressure on a cylinder of liquid. Pascal's principle tells you that this pressure is transmitted equally to all parts of the liquid-filled brake system. This liquid presses a brake pad against each wheel, and friction brings the car to a stop. The force is multiplied because the pistons that push the brake pads on each wheel are much larger than the piston that is pushed by the brake pedal.

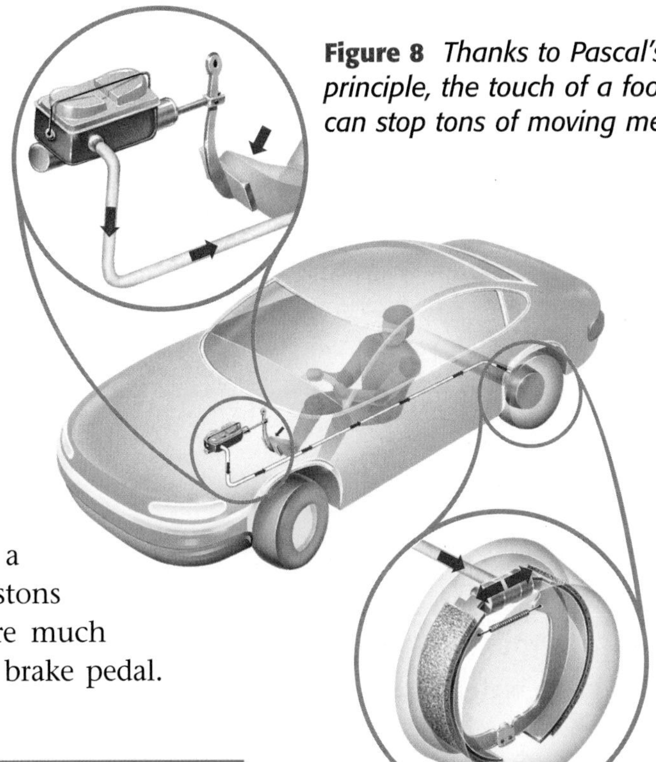

Figure 8 *Thanks to Pascal's principle, the touch of a foot can stop tons of moving metal.*

REVIEW

1. Explain how atmospheric pressure helps you drink through a straw.
2. What does Pascal's principle state?
3. **Making Predictions** When you squeeze a balloon, where is the pressure inside the balloon increased the most? Explain your answer in terms of Pascal's principle.

Section 2

Buoyant Force

NEW TERMS

buoyant force
Archimedes' principle

OBJECTIVES

- Explain the relationship between fluid pressure and buoyant force.
- Predict whether an object will float or sink in a fluid.
- Analyze the role of density in an object's ability to float.

You know that the rubber duck at right floats on water, a fluid. But *why* does the duck float? Why doesn't it sink in the water to the bottom of your bathtub? Even if you pushed the rubber duck to the bottom, it would pop back to the surface when you released it. Some force pushes the rubber duck to the top of the water. That force is **buoyant force,** the upward force that fluids exert on all matter.

Air is a fluid, so it exerts a buoyant force. But why don't you ever see rubber ducks floating in air? In this section, you'll explore the nature of buoyant force and find the answer to the question, "Why does a rubber duck float on water but sink in air?"

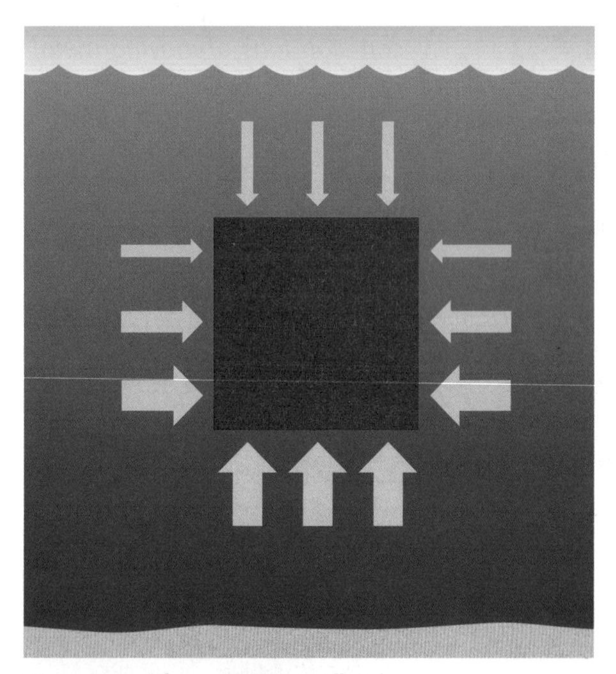

Figure 9 *There is more fluid pressure on the bottom of an object because pressure increases with depth. This results in an upward force on the object–buoyant force.*

Buoyant Force Is Caused by Differences in Fluid Pressure

Look at **Figure 9.** Water exerts fluid pressure on all sides of an object. The pressure exerted horizontally on one side of the object is equal to the pressure exerted horizontally on the opposite side. Because these equal pressures act directly opposite each other, they cancel one another. Thus, the only fluid pressures affecting the object are at the top and at the bottom. Because pressure increases with depth, the pressure on the bottom of the object is greater than the pressure at the top, as shown by the width of the arrows. Therefore, the water exerts a net upward force on the object. This upward force, caused by differences in pressure, is buoyant force.

Determining Buoyant Force How do you determine buoyant force? The answer to this question was discovered by Archimedes (ahr kuh MEE deez), a Greek mathematician who lived in the third century B.C. **Archimedes' principle** states that the buoyant force on an object in a fluid is an upward force equal to the weight of the volume of fluid that the object displaces. (*Displace* means "to take the place of.") For example, suppose the object in Figure 9 displaces 250 mL of water. The weight of that volume of displaced water is about 2.5 N. Therefore, the buoyant force on the object is 2.5 N.

Notice that the weight of the object has nothing to do with the buoyant force. Only the weight of the displaced fluid determines the buoyant force on an object. However, you do have to consider the weight of an object to understand why some objects float and others sink.

Explore

List five things that float in water and five things that sink in water. Compare the two lists. What do the objects on the "floating" list have in common? What do the objects on the "sinking" list have in common? What are the differences between the objects on the two lists?

Weight vs. Buoyant Force

An object in a fluid will sink if it has a weight greater than the weight of the fluid that is displaced. In other words, an object will sink if its weight is greater than the buoyant force acting on it. An object floats only when it displaces a volume of liquid that has a weight equal to the object's weight—that is, if the buoyant force on the object is equal to the object's weight. Since weight is simply a measure of the gravitational force on an object, you can see that buoyant force (an upward force) opposes gravity (a downward force).

A Day at the Lake The lake scene in **Figure 10** looks quite peaceful, but there are forces being exerted! The rock at the bottom of the lake weighs 75 N. It displaces 5 L of water, which has a weight of about 50 N. According to Archimedes' principle, the buoyant force is equal to the weight of the displaced water—50 N. Because the rock's weight is greater than the buoyant force, the rock sinks.

Now look at the fish. The fish weighs 12 N. It displaces a volume of water that has a weight of 12 N. Because the fish's weight is equal to the buoyant force, the fish floats in the water.

Now look at the duck. The duck weighs 9 N. The duck does not sink. What does that tell you? The buoyant force on the duck must be equal to the duck's weight. But the duck isn't even all the way underwater! Only the duck's feet, legs, and stomach have to be underwater in order to displace enough water to equal 9 N. Thus, the duck floats.

Figure 10 *Will an object sink or float? It depends on whether the buoyant force is less than or equal to the object's weight.*

MATH BREAK

How to Calculate Density

The volume of any sample of matter, no matter what state or shape, can be calculated using this equation:

$$\text{Density} = \frac{\text{Mass}}{\text{Volume}}$$

1. What is the density of a 20 cm^3 sample of liquid with a mass of 25 g?
2. A 546 g fish displaces 420 cm^3 of water. What is the density of the fish?

Suppose the duck in Figure 10 dove underwater. The duck would then displace a greater amount of water, and the buoyant force would therefore be greater. When the buoyant force on an object is greater than the object's weight, the object is *buoyed* (pushed up) out of the water until what's left underwater displaces an amount of water that equals the object's entire weight. This also explains why a rubber duck pops to the surface of the water when it is pushed to the bottom of a filled bathtub.

An Object Will Float or Sink Based on Its Density

Think again about the rock at the bottom of the lake. The rock displaces 5 L of water, which means that the volume of the rock is 5,000 cm^3. (Remember that liters are used only for fluid volumes.) But 5,000 cm^3 of rock weighs more than an equal volume of water. This is why the rock sinks. Because mass is proportional to weight on Earth, you can say that the rock has more mass per volume than water. Remember, mass per unit volume is *density.* The rock sinks because it is more dense than water. The duck floats on the surface because its mass per unit volume is less than that of water; that is, the duck is less dense than water. In Figure 10, the density of the fish is exactly equal to the density of the water. You'll see later how fish control their density to swim at any level in water.

Figure 11 *Helium in a balloon floats in air for the same reason a duck floats in water—it is less dense than the surrounding fluid.*

Most Substances Are More Dense Than Air Think back to the question about the rubber duck: "Why does it float on water but not in air?" The rubber duck floats because it is less dense than water. However, most substances are *more* dense than air—most substances contain more mass than an equal volume of air does. Therefore, there are few substances that float in air. The plastic that makes up the rubber duck is more dense than air, so the rubber duck doesn't float in air.

One substance that is less dense than air is helium, a gas. In fact, helium is more than 70 times less dense than air. A volume of helium displaces a volume of air that is much heavier than itself, so helium floats. Because of its ability to float in air, helium is used in airships and weather balloons, like the one shown in **Figure 11.**

The Mystery of Floating Steel

Steel is almost eight times more dense than water. And yet huge steel ships cruise the oceans with ease, even while carrying enormous loads. But hold on! Didn't you just learn that substances that are more dense than water will sink in water? You bet! So how does a steel ship float?

The secret is in the shape of the ship. What if a ship were just a big block of steel, as shown in **Figure 12**? If you put that steel block into water, the block would sink because it is more dense than water. For this reason, ships are built with a hollow shape, as shown below. The amount of steel in the ship is the same as in the block, but the hollow shape increases the volume of the ship. Because density is mass per volume, an increase in the ship's volume leads to a decrease in its density. Therefore, ships made of steel float because their *overall density* is less than the density of water. This is true of boats of any size, made of any material. Most ships are actually built to displace even more water than is necessary for the ship to float so that the ship won't sink when people and cargo are loaded onboard.

The *Seawise Giant* is the largest ship in the world. It is so large that crew members often use bicycles to travel around the ship.

Figure 12 A Ship's Shape Makes the Difference

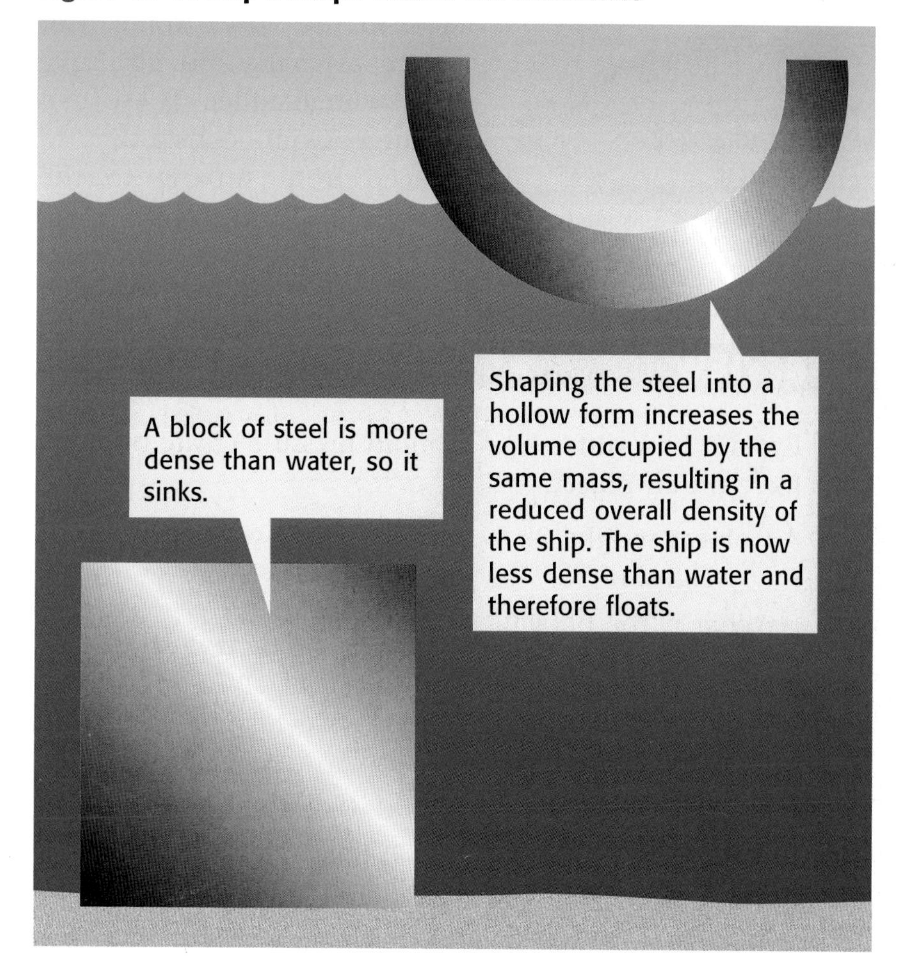

Quick Lab

Ship-Shape

1. Roll a **piece of clay** into a ball the size of a golf ball, and drop it into a **container of water**. Record your observations in your ScienceLog.
2. With your hands, flatten the ball of clay until it is a bit thinner than your little finger, and press it into the shape of a bowl or canoe.
3. Place the clay boat gently in the water. How does the change of shape affect the buoyant force on the clay? How is that change related to the average density of the clay boat? Record your answers in your ScienceLog.

The rock that makes up the Earth's continents is about 15 percent less dense than the molten (melted) mantle rock below it. Because of this difference in densities, the continents are "floating" on the mantle.

Density on the Move A submarine is a special kind of ship that can travel on the surface of the water and underwater. Submarines have special tanks that can be opened to allow sea water to flow in. This water adds mass, thus increasing the submarine's overall density so it can descend into the ocean. Crew members can control the amount of water taken in, thereby controlling the submarine's change in density and thus its depth in the ocean. Compressed air is used to blow the water out of the tanks so the submarine can rise through the water. Most submarines are built of high-strength metals that withstand water pressure. Still, most submarines can go no deeper than 400 m below the surface of the ocean.

How Is a Fish Like a Submarine? No, this is not a trick question! Like a submarine, some fish adjust their overall density in order to stay at a certain depth in the water. Most bony fish have an organ called a *swim bladder,* shown in **Figure 13.** This swim bladder is filled with gases produced in the fish's blood. The inflated swim bladder increases the fish's volume, thereby decreasing the fish's overall density and keeping it from sinking in the water. The fish's nervous system controls the amount of gas in the bladder according to the fish's depth in the water. Some fish, such as sharks, do not have a swim bladder. These fish must swim constantly to keep from sinking to the bottom of the water.

Figure 13 *Most bony fish have an organ called a swim bladder that allows the fish to adjust its overall density.*

Swim bladder

REVIEW

1. Explain how differences in fluid pressure create buoyant force on an object.
2. An object weighs 20 N. It displaces a volume of water that weighs 15 N.
   a. What is the buoyant force on the object?
   b. Will this object float or sink? Explain your answer.
3. Iron has a density of 7.9 g/cm^3. Mercury has a density of 13.6 g/cm^3. Will iron float or sink in mercury? Explain your answer.
4. **Applying Concepts** Why is it inaccurate to say that all heavy objects will sink in water?

Section 3

Bernoulli's Principle

NEW TERMS

Bernoulli's principle
lift
thrust
drag

OBJECTIVES

- Describe the relationship between pressure and fluid speed.
- Analyze the roles of lift, thrust, and drag in flight.
- Give examples of Bernoulli's principle in real-life situations.

Has this ever happened to you? You've just turned on the shower. Upon stepping into the water stream, you decide that the water pressure is not strong enough. You turn the faucet to provide more water, and all of a sudden the bottom edge of the shower curtain starts swirling around your legs. What's going on? It might surprise you that the explanation for this unusual occurrence also explains how wings help birds and planes fly and how pitchers throw curve balls.

Fluid Pressure Decreases as Speed Increases

The strange reaction of the shower curtain is caused by a property of moving fluids that was first described in the eighteenth century by Daniel Bernoulli (buhr NOO lee), a Swiss mathematician. **Bernoulli's principle** states that as the speed of a moving fluid increases, its pressure decreases. In the case of the shower curtain, the faster the water moves, the less pressure it exerts. This creates an imbalance between the pressure inside the shower curtain and the pressure outside it. Because the pressure outside is now greater than the pressure inside, the shower curtain is pushed toward the water stream.

Science in a Sink You can see Bernoulli's principle at work in **Figure 14.** A table-tennis ball is attached to a string and swung gently into a moving stream of water. Instead of being pushed back out, the ball is actually held in the moving water when the string is given a tug. Why does the ball do that? The water is moving, so it has a lower pressure than the surrounding air. The higher air pressure then pushes the ball into the area of lower pressure—the water stream. Try this at home to see for yourself!

Figure 14 *This ball is pushed by the higher pressure of the air into an area of reduced pressure—the water stream.*

Quick Lab

Breathing Bernoulli-Style

1. Hold **two pieces of paper** by their top edges, one in each hand, so that they hang next to one another about 5 cm apart.
2. Blow a steady stream of air between the two sheets of paper.
3. Record your observations in your ScienceLog. Explain the results according to Bernoulli's principle.

It's a Bird! It's a Plane! It's Bernoulli's Principle!

BRAIN FOOD

The first successful flight of an engine-driven heavier-than-air machine occurred in Kitty Hawk, North Carolina, in 1903. Orville Wright was the pilot. The plane flew only 37 m (about the length of a 737 jet) before landing, and the entire flight lasted only 12 seconds.

The most common commercial airplane in the skies today is the Boeing 737 jet. A 737 jet is almost 37 m long and has a wingspan of 30 m. Even without passengers, the plane weighs 350,000 N. That's more than 35 times heavier than an average car! How can something so big and heavy get off the ground, much less fly 10,000 m into the sky? Wing shape plays a role in helping these big planes—as well as smaller planes and even birds—achieve flight, as shown in **Figure 15.**

According to Bernoulli's principle, the faster-moving air above the wing exerts less pressure than the slower-moving air below the wing. The increased pressure that results below the wing exerts an upward force. This upward force, known as **lift,** pushes the wings (and the rest of the airplane or bird) upward against the downward pull of gravity.

Figure 15 Wing Shape Creates Differences in Air Speed

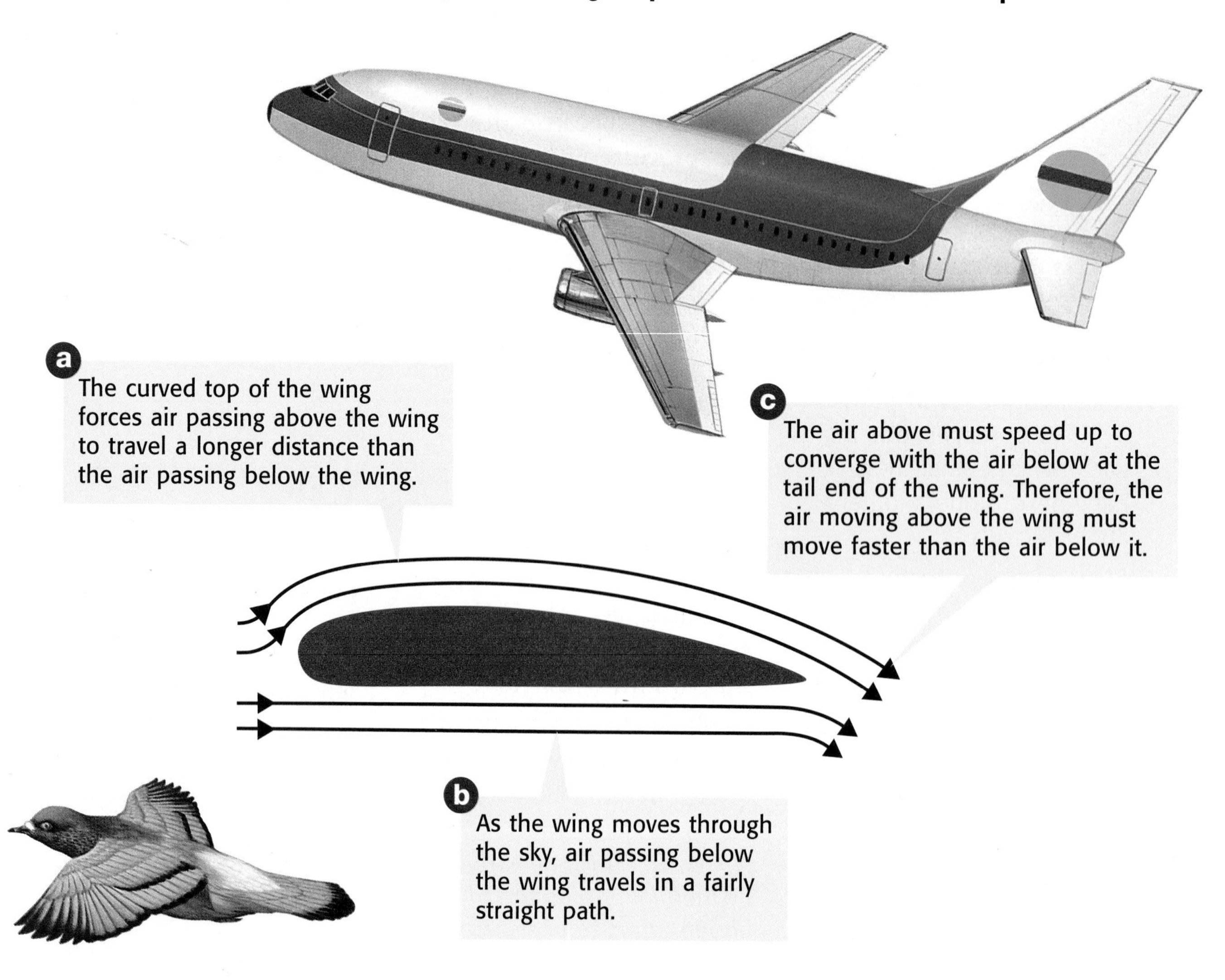

Thrust and Wing Size Determine Lift

The amount of lift created by a plane's wing is determined in part by the size of the wing and the speed at which air travels around the wing. The speed of an airplane is in large part determined by its **thrust**—the forward force produced by the plane's engine. In general, a plane with a greater amount of thrust moves faster than a plane with less thrust. This faster speed means air travels around the wing at a greater speed, which increases lift.

You can understand the relationship between wing size, thrust, and speed by thinking about a jet plane, like the one in **Figure 16.** This plane is able to fly with a relatively small wing size because its engine creates an enormous amount of thrust. This thrust pushes the plane through the sky at tremendous speeds. Therefore, the jet generates sufficient lift with small wings by moving very quickly through the air. Smaller wings keep a plane's weight low, which also contributes to speed.

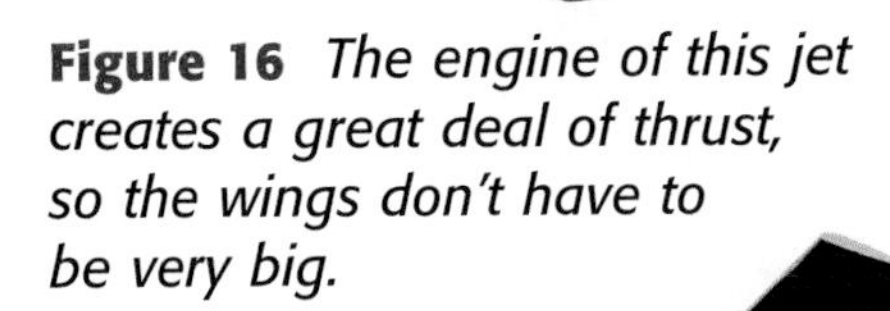

Figure 16 *The engine of this jet creates a great deal of thrust, so the wings don't have to be very big.*

Compared with the jet, a glider, like the one in **Figure 17,** has a large wing area. A glider is an engineless plane that rides rising air currents to stay in flight. Without engines, gliders produce no thrust and move more slowly than many other kinds of planes. Thus, a glider must have large wings to create the lift necessary to keep it in the air.

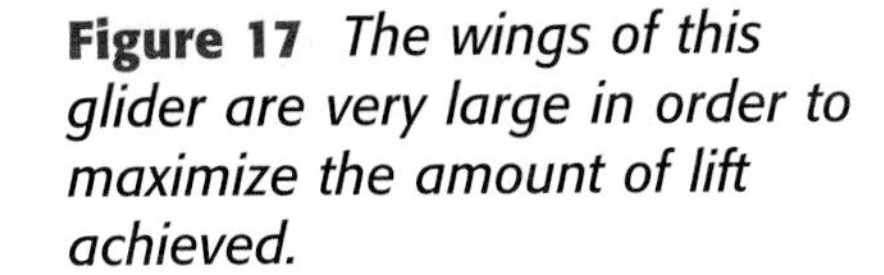

Figure 17 *The wings of this glider are very large in order to maximize the amount of lift achieved.*

Self-Check

Does air travel faster or slower over the top of a wing? *(See page 596 to check your answer.)*

Bernoulli's Principle Is for the Birds Birds don't have engines, of course, so they must flap their wings to push themselves through the air. The hawk shown at left uses its large wing size to fly with a minimum of effort. By extending its large wings to their full length and gliding on wind currents, a hawk can achieve enough lift to stay in the air while flapping only occasionally. Smaller birds must flap their wings more often to stay in the air.

Soaring science! See how wing shape affects the flight of your own airplane on page 551 of the LabBook.

Sometimes Bernoulli's principle can create a dangerous situation. At high speeds, air moving around the body of this race car could lift the car just as it lifts a plane's wing. This could cause the wheels to lose contact with the ground, sending the car out of control. To prevent this situation, an upside-down wing, or spoiler, is mounted on the rear of the car. How do spoilers help reduce the danger of accidents?

Drag Opposes Motion in Fluids

Have you ever walked into a strong wind and noticed that the wind seemed to slow you down? Fluids exert a force that opposes motion. The force that opposes or restricts motion in a fluid is called **drag.** In a strong wind, air "drags" on your clothes and body, making it difficult for you to move forward. Drag forces in flight work against the forward motion of a plane or bird and are usually caused by an irregular flow of air around the wings. An irregular or unpredictable flow of fluids is known as *turbulence.*

Lift is often reduced when turbulence causes drag. At faster speeds, drag can become a serious problem, so airplanes are equipped with ways to reduce turbulence as much as possible when in flight. For example, flaps like those shown in **Figure 18** can be used to change the shape or area of a wing, thereby reducing drag and increasing lift. Similarly, birds can adjust their wing feathers in response to turbulence to achieve greater lift.

Figure 18 *During flight, the pilot of this airplane can adjust these flaps to help increase lift.*

Wings Are Not Always Required

You don't have to look up at a bird or a plane flying through the sky to see Bernoulli's principle in your world. In fact, you've already learned how Bernoulli's principle can affect such things as shower curtains and race cars. Any time fluids are moving, Bernoulli's principle is at work. In **Figure 19,** you can see how Bernoulli's principle can mean the difference between a home run and a strike during a baseball game.

Bernoulli's principle at play—read how Frisbees® were invented on page 182.

Figure 19 *A pitcher can take advantage of Bernoulli's principle to produce a confusing curveball that is difficult for the batter to hit.*

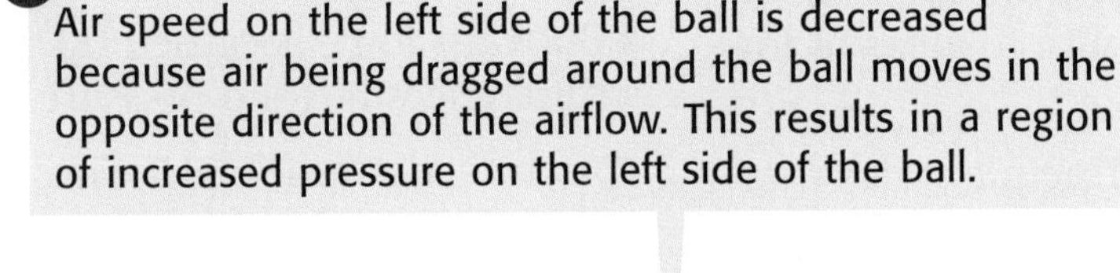

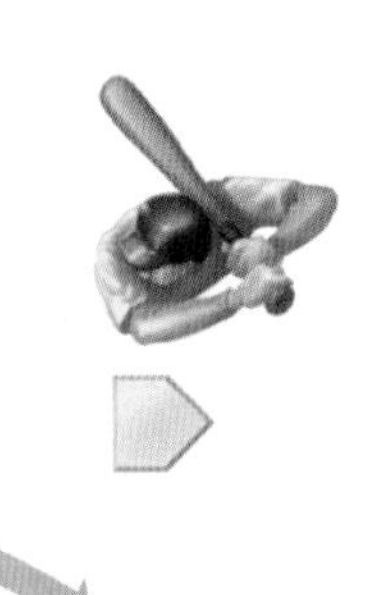

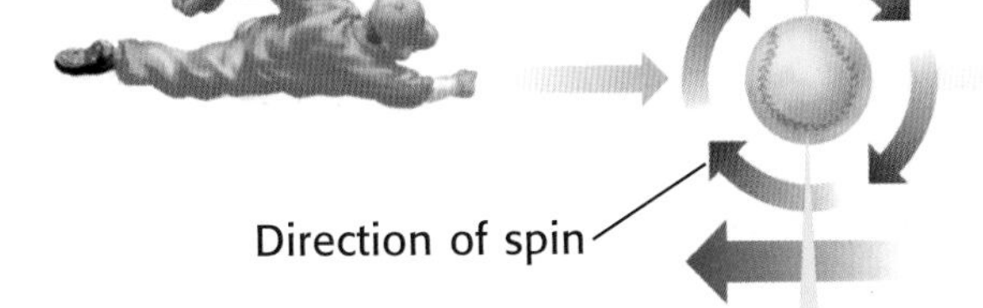

REVIEW

1. Does fluid pressure increase or decrease as fluid speed increases?
2. Explain how wing shape can contribute to lift during flight.
3. What force opposes motion through a fluid?
4. **Interpreting Graphics** When the space through which a fluid flows becomes narrow, fluid speed increases. Explain how this could lead to a collision for the two boats shown at right.

Chapter Highlights

SECTION 1

Vocabulary

fluid *(p. 162)*
pressure *(p. 162)*
pascal *(p. 162)*
atmospheric pressure *(p. 163)*
density *(p. 165)*
Pascal's principle *(p. 167)*

Section Notes

- A fluid is any material that flows and that takes the shape of its container.
- Pressure is force exerted on a given area.
- Moving particles of matter create pressure by colliding with one another and with the walls of their container.
- Fluids exert pressure equally in all directions.
- The pressure caused by the weight of Earth's atmosphere is called atmospheric pressure.
- Fluid pressure increases as depth increases.
- Fluids flow from areas of high pressure to areas of low pressure.
- Pascal's principle states that a change in pressure at any point in an enclosed fluid will be transmitted equally to all parts of the fluid.
- Hydraulic devices transmit changes of pressure through liquids.

SECTION 2

Vocabulary

buoyant force *(p. 168)*
Archimedes' principle *(p. 168)*

Section Notes

- All fluids exert an upward force called buoyant force.
- Buoyant force is caused by differences in fluid pressure.
- Archimedes' principle states that the buoyant force on an object is equal to the weight of the fluid displaced by the object.

☑ Skills Check

Math Concepts

PRESSURE If an object exerts a force of 10 N over an area of 2 m², the pressure exerted can be calculated as follows:

$$\text{Pressure} = \frac{\text{Force}}{\text{Area}} = \frac{10 \text{ N}}{2 \text{ m}^2} = \frac{5 \text{ N}}{1 \text{ m}^2}, \text{ or } 5 \text{ Pa}$$

Visual Understanding

ATMOSPHERIC PRESSURE Why aren't you crushed by atmospheric pressure? Figure 3 on page 163 can help you understand.

BUOYANT FORCE To understand how differences in fluid pressure cause buoyant force, review Figure 9 on page 168.

BERNOULLI'S PRINCIPLE AND WING SHAPE Turn to page 174 to review how a wing is often shaped to take advantage of Bernoulli's principle in creating lift.

SECTION 2

- Any object that is more dense than the surrounding fluid will sink; any object that is less dense than the surrounding fluid will float.

Labs

Fluids, Force, and Floating *(p. 548)*
Density Diver *(p. 550)*

SECTION 3

Vocabulary

Bernoulli's principle *(p. 173)*
lift *(p. 174)*
thrust *(p. 175)*
drag *(p. 176)*

Section Notes

- Bernoulli's principle states that fluid pressure decreases as the speed of a moving fluid increases.
- Wings are often shaped to allow airplanes to take advantage of decreased pressure in moving air in order to achieve flight.
- Lift is an upward force that acts against gravity.
- Lift on an airplane is determined by wing size and thrust (the forward force produced by the engine).
- Drag opposes motion through fluids.

Labs

Taking Flight *(p. 551)*

internet**connect**

GO TO: go.hrw.com

Visit the **HRW** Web site for a variety of learning tools related to this chapter. Just type in the keyword:

KEYWORD: HSTFLU

GO TO: www.scilinks.org

Visit the **National Science Teachers Association** on-line Web site for Internet resources related to this chapter. Just type in the ***sci*LINKS** number for more information about the topic.

TOPIC: Submarines and Undersea Technology — ***sci*LINKS NUMBER:** HSTP155
TOPIC: Fluids and Pressure — ***sci*LINKS NUMBER:** HSTP160
TOPIC: The Buoyant Force — ***sci*LINKS NUMBER:** HSTP165
TOPIC: Bernoulli's Principle — ***sci*LINKS NUMBER:** HSTP170

Chapter Review

USING VOCABULARY

To complete the following sentences, choose the correct term from each of the pair of terms listed below:

1. __?__ increases with the depth of a fluid. *(Pressure* or *Lift)*
2. A plane's engine produces __?__ to push the plane forward. *(thrust* or *drag)*
3. Force divided by area is known as __?__. *(density* or *pressure)*
4. The hydraulic brakes of a car transmit pressure through fluid. This is an example of __?__. *(Archimedes' principle* or *Pascal's principle)*
5. Bernoulli's principle states that the pressure exerted by a moving fluid is __?__ *(greater than* or *less than)* the pressure of fluid when it is not moving.

UNDERSTANDING CONCEPTS

Multiple Choice

6. The curve on the top of a wing
   a. causes air to travel farther in the same amount of time as the air below the wing.
   b. helps create lift.
   c. creates a low-pressure zone above the wing.
   d. All of the above

7. An object displaces a volume of fluid that
   a. is equal to its own volume.
   b. is less than its own volume.
   c. is greater than its own volume.
   d. is more dense than itself.
8. Fluid pressure is always directed
   a. up.
   b. down.
   c. sideways.
   d. in all directions.
9. If an object weighing 50 N displaces a volume of water with a weight of 10 N, what is the buoyant force on the object?
   a. 60 N
   b. 50 N
   c. 40 N
   d. 10 N
10. A helium-filled balloon will float in air because
   a. there is more air than helium.
   b. helium is less dense than air.
   c. helium is as dense as air.
   d. helium is more dense than air.
11. Materials that can flow to fit their containers include
   a. gases.
   b. liquids.
   c. both gases and liquids.
   d. neither gases nor liquids.

Short Answer

12. What two factors determine the amount of lift achieved by an airplane?
13. Where is water pressure greater, at a depth of 1 m in a large lake or at a depth of 2 m in a small pond? Explain.
14. Is there buoyant force on an object at the bottom of an ocean? Explain your reasoning.
15. Why are liquids used in hydraulic brakes instead of gases?

Concept Map

16. Use the following terms to create a concept map: fluid, pressure, depth, buoyant force, density.

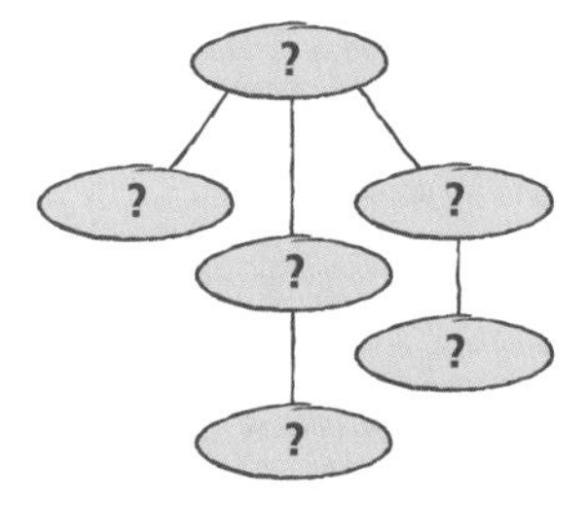

CRITICAL THINKING AND PROBLEM SOLVING

17. Compared with an empty ship, will a ship loaded with plastic-foam balls float higher or lower in the water? Explain your reasoning.

18. Inside all vacuum cleaners is a high-speed fan. Explain how this fan causes dirt to be picked up by the vacuum cleaner.

19. A 600 N clown on stilts says to two 600 N clowns sitting on the ground, "I am exerting twice as much pressure as the two of you together!" Could this statement be true? Explain your reasoning.

MATH IN SCIENCE

20. Calculate the area of a 1,500 N object that exerts a pressure of 500 Pa (N/m^2). Then calculate the pressure exerted by the same object over twice that area. Be sure to express your answers in the correct SI unit.

INTERPRETING GRAPHICS

Examine the illustration of an iceberg below, and answer the questions that follow.

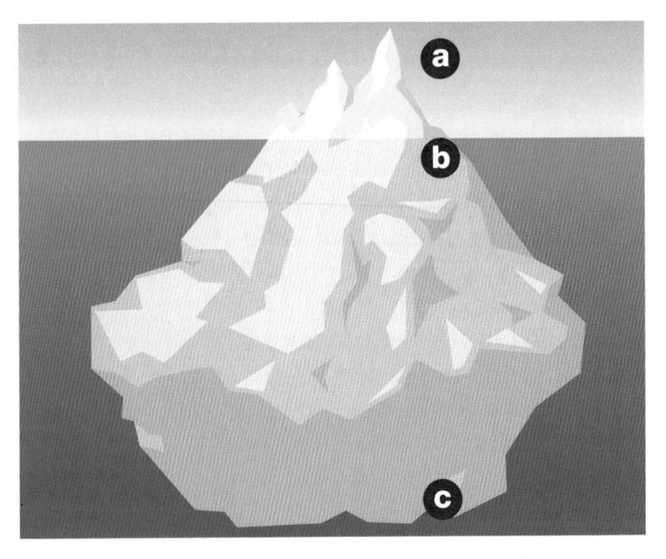

21. At what point (*a, b,* or *c*) is water pressure greatest on the iceberg?

22. How much of the iceberg has a weight equal to the buoyant force?
    a. all of it
    b. the section from *a* to *b*
    c. the section from *b* to *c*

23. How does the density of ice compare with the density of water?

24. Why do you think icebergs are so dangerous to passing ships?

NOW What Do You Think?

Take a minute to review your answers to the ScienceLog questions on page 161. Have your answers changed? If necessary, revise your answers based on what you have learned since you began this chapter.

Eureka!

Stayin' Aloft—The Story of the Frisbee®

Whoa! Nice catch! Your friend 30 m away just sent a disk spinning toward you. As you reached for it, a gust of wind floated it up over your head. With a quick jump, you snagged it. A snap of your wrist sends the disk soaring back. You are "Frisbee-ing," a game more than 100 years old. But back then, there were no plastic disks, only pie plates.

From Pie Plate...

In the late 1800s, ready-made pies baked in tin plates began to appear in stores and restaurants. A bakery near Yale University, in New Haven, Connecticut, embossed its name, Frisbie's Pies, on its pie plates. When a few fun-loving college students tossed empty pie plates, they found that the metal plates had a marvelous ability to stay in the air. Soon the students began alerting their companions of an incoming pie plate by shouting "Frisbie!" So tossing pie plates became known as Frisbie-ing. By the late 1940s, the game was played across the country.

...to Plastic

In 1947, California businessmen Fred Morrison and Warren Franscioni needed to make a little extra money. They were familiar with pie-plate tossing, and they knew the plates often cracked when they landed and developed sharp edges that caused injuries.

At the time, plastic was becoming widely available. Plastic is more durable and flexible than metal, and it isn't as likely to injure fingers. Why not make a "pie plate" out of plastic, thought Morrison and Franscioni? They did, and their idea was a huge success.

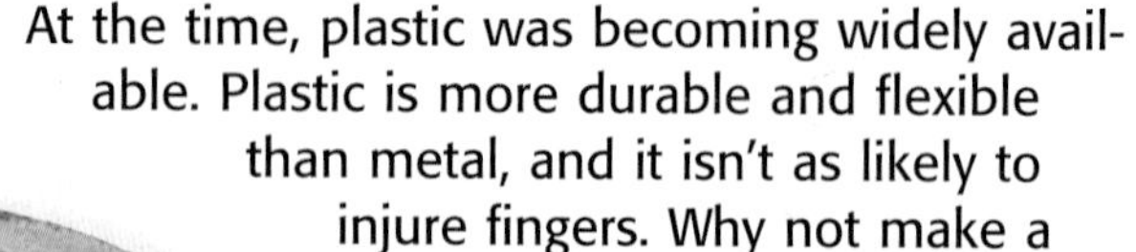

Years later, a toy company bought the rights to make the toy. One day the president of the company heard someone yelling "Frisbie!" while tossing a disk and decided to use that name, changing the spelling to "Frisbee."

Saucer Science

It looks simple, but Frisbee flight is quite complicated. It involves *thrust,* the force you give the disk to move it through the air; *angle of attack,* the slight upward tilt you give the disk when you throw it; and *lift,* the upward forces (explained by Bernoulli's principle) acting on the Frisbee to counteract gravity. But perhaps the most important aspect of Frisbee physics is *spin,* which gives the Frisbee stability as it flies. The faster a Frisbee spins, the more stable it is and the farther it can fly.

What Do You Think?

▶ From what you've learned in class, why do you think the Frisbee has a curved lip? Would a completely flat Frisbee fly as well? Why or why not? Find out more about the interesting aerodynamics of Frisbee flight. Fly a Frisbee for the class, and explain what you've learned.

Science Fiction

"Wet Behind the Ears"

by Jack C. Haldeman II

Willie Joe Thomas is a college student who lied to get into college and cheated to get a swimming scholarship. Now he is faced with a major swim meet, and his coach has told him that he has to swim or be kicked off the team. Willie Joe could lose his scholarship. What's worse, he would have to get a *job.*

"Wet Behind the Ears" is Willie Joe's story. It's the story of someone who has always taken the easy way (even if it takes more work), of someone who lies and cheats as easily as he breathes. Willie Joe could probably do things the right way, but it never even occurred to him to try it!

So when Willie Joe's roommate, Frank Emerson, announces that he has made an amazing discovery in the chemistry lab, Willie Joe doesn't much care. Frank works too hard. Frank follows the rules. Willie Joe isn't impressed.

But when he is running late for the all-important swim meet, Willie Joe remembers what Frank's new compound does. Frank said it was a "sliding compound." Willie Joe may not know chemistry, but "slippery" he understands. And Frank also said something about selling the stuff to the Navy to make its ships go faster. Hey, if it works for ships . . .

See what happens when Willie Joe tries to save his scholarship. Go to the *Holt Anthology of Science Fiction,* and read "Wet Behind the Ears," by Jack C. Haldeman II.

TIMELINE

UNIT 3

Work, Machines, and Energy

Can you imagine living in a world with no machines? In this unit, you will explore the scientific meaning of *work* and learn how machines make work easier. You will find out how energy allows you to do work and how different forms of energy can be converted into other forms of energy. You will also learn about heat and how heating and cooling systems work. This timeline shows some of the inventions and discoveries made throughout history as people have advanced their understanding of work, machines, and energy.

Around 3000 B.C.

The sail is used in Egypt. Sails use the wind rather than human power to move boats through the water.

Around 200 B.C.

Under the Han dynasty, the Chinese become one of the first civilizations to use coal as fuel.

1926

American scientist Robert Goddard launches the first rocket powered by liquid fuel. It reaches a height of 56 m and a speed of 97 km/h.

1948

Maria Telkes, a Hungarian-born physicist, designs the heating system for the first solar-heated house.

1972

The first American self-service gas station opens.

1656

Dutch scientist Christiaan Huygens invents the pendulum clock.

1776

The American colonies declare their independence from Great Britain.

1818

The first two-wheeled, rider-propelled machine is invented by German Baron Karl von Drais de Sauerbrun. Made of wood, this early machine paves the way for the invention of the bicycle.

1893

The zipper is patented.

1908

The automobile age begins with the mass production of the Ford Model T.

1988

The world's most powerful wind-powered generator begins generating electrical energy in Scotland's Orkney Islands.

2000

The 2000 Olympic Summer Games are held in Sydney, Australia.

CHAPTER 8

Work and Machines

Would You Believe . . . ?

The Great Pyramid, located in Giza (GEE zuh), Egypt, could be called the largest tombstone ever created. A monument and tomb for the pharaoh King Khufu (KOO foo), it covers an area the size of seven city blocks and rises about 40 stories high. The Great Pyramid is the largest of the three pyramids of Giza. It was built around 2600 B.C. and took less than 30 years to complete—a relatively short period of time considering that construction equipment didn't exist 4,000 years ago. So how did the Egyptians do it?

To build the Great Pyramid, the Egyptians cut and moved more than 2 million stone blocks, most averaging 2,000 kg (probably over 40 times your own mass). The blocks were cut from a stone quarry, moved near the pyramid, and then lifted into place. To finish in less than 30 years, the Egyptians would have had to cut, move, and lift about 200 blocks per day! The Egyptians did not have cranes, bulldozers, or any other heavy-duty machines. What they had were two simple machines—the inclined plane and the lever.

Archaeologists have found the remains of inclined planes, or ramps, made from mud, stone, and wood. The Egyptians pushed or pulled the blocks along these ramps to raise them to the proper height. Using ramps required less force than lifting the blocks straight up. In addition, notches in many blocks indicate that huge levers were used like giant crowbars to lift and move the heavy blocks. The workers pushed down on the lever, and the lever pushed up on a stone block, lifting it into place.

The Egyptians used simple machines to create something truly amazing. In this chapter, you'll learn about work and how machines can help make work easier.

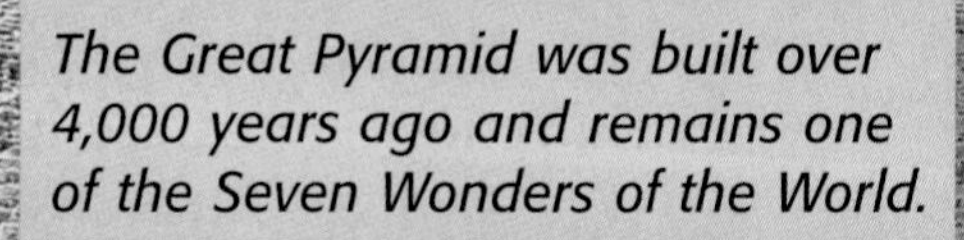

The Great Pyramid was built over 4,000 years ago and remains one of the Seven Wonders of the World.

What Do You Think?

In your ScienceLog, try to answer the following questions based on what you already know:

1. What does it mean to do work?
2. How are machines helpful when doing work?
3. What are some examples of simple machines?

Investigate!

C'mon, Lever a Little!

In this activity, you will use a simple machine, a lever, to make your task a little easier.

Procedure 

1. Gather a couple of **books,** such as dictionaries, encyclopedias, or textbooks, and stack them on a table one on top of the other. If necessary, tie some **string** around them to keep them from slipping.
2. Slide your index finger underneath the edge of the bottom book. Using the force of your finger only, try to lift one side of the books 2 or 3 cm off the table. Is it difficult? Write your observations in your ScienceLog.
3. Slide the end of a **wooden ruler** underneath the edge of the bottom book. Then slip a **large pencil eraser** under the ruler.
4. Again using the force of your index finger only, push down on the ruler and try to lift the books as shown at right.
   Caution: Push down slowly to keep the ruler and eraser from flipping.

Analysis

5. Which was easier, lifting the books with your finger or with the ruler? Explain.
6. What was different about the direction of the force your finger applied on the books compared with the force you applied on the ruler?
7. Based on your results, how would you explain the usefulness of a lever, such as a ruler?

Section 1

Work and Power

NEW TERMS

work
power
joule
watt

OBJECTIVES

- Determine when work is being done on an object.
- Calculate the amount of work done on an object.
- Explain the difference between work and power.

Suppose your science teacher has just given you a homework assignment. You have to read an entire chapter by tomorrow! Wow, that's a lot of work, isn't it? Actually, in the scientific sense, you won't be doing any work at all! How can that be?

The Scientific Meaning of *Work*

In science, **work** occurs when a force causes an object to move in the direction of the force. In the example above, you may put a lot of mental effort into doing your homework, but you won't be using a force to move an object. Therefore, in the scientific sense, you will not be doing work.

Now think about the example shown in **Figure 1.** This student is having a lot of fun, isn't she? But she is doing work, even though she is having fun. That's because she's applying a force to the bowling ball to make it move through a distance. However, it's important to understand that she is doing work on the ball only as long as she is touching it. The ball will continue to move away from her after she releases it, but she will no longer be doing work on the ball because she will no longer be applying a force to it.

Figure 1 *You might be surprised to find out that bowling is doing work!*

Working Hard or Hardly Working? You should understand that applying a force doesn't always result in work being done. Suppose your neighbor asks you to help push his stalled car. You push and push, but the car doesn't budge. Even though you may be exhausted and sweaty, you haven't done any work on the car. Why? Because the car hasn't moved. Remember, work is done on an object only when a force makes that object move. In this case, your pushing doesn't make the car move. You only do work on the car if it starts to move.

Here's another situation to think about. Suppose you're in the airport and you're late for a flight. You have to run several meters carrying a heavy suitcase. Because you're making the suitcase move, you're doing work on it, right? Wrong! For work to be done, the object must move in the same direction as the force. In this case, the motion is in a different direction than the force, as shown in **Figure 2.** So no work is done on the suitcase. However, work *is* done on the suitcase when you lift it off the ground.

You'll know that work is done on an object if two things occur: (1) the object moves as a force is applied and (2) the direction of the object's motion is the same as the direction of the force applied. The pictures and the arrows in the chart below will help you understand how to determine when work is being done on an object.

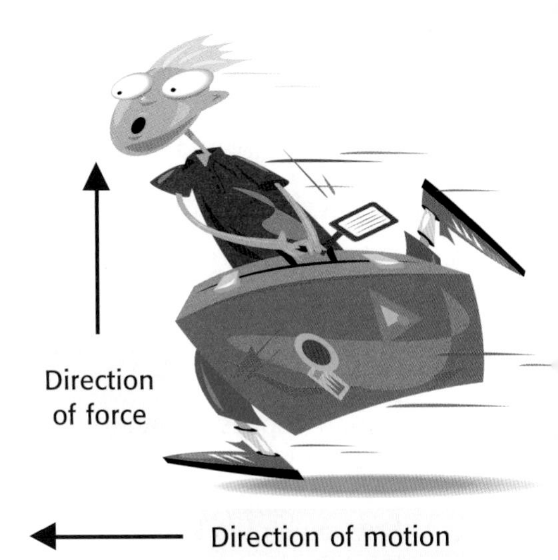

Figure 2 *You exert an upward force on the suitcase. But the motion of the suitcase is forward. Therefore, you are not doing work on the suitcase.*

Work or Not Work?

Example	Direction of force	Direction of motion	Doing work?
	→	→	Yes
	↑	→	No
	↑	↑	Yes
	↑	→	No

Self-Check

If you pulled a wheeled suitcase instead of carrying it, would you be doing work on the suitcase? Why or why not? *(See page 596 to check your answer.)*

÷ 5 ÷ Ω ≤ ∞ + Ω √ 9 ∞ ≤ Σ 2 +

MATH BREAK

Working It Out

Use the equation for work shown on this page to solve the following problems:

1. A man applies a force of 500 N to push a truck 100 m down the street. How much work does he do?
2. In which situation do you do more work?
   a. You lift a 75 N bowling ball 2 m off the floor.
   b. You lift two 50 N bowling balls 1 m off the floor.

Calculating Work

Do you do more work when you lift an 80 N barbell or a 160 N barbell? It would be tempting to say that you do more work when you lift the 160 N barbell because it weighs more. But actually, you can't answer this question with the information given. You also need to know how high each barbell is being lifted. Remember, work is a force applied through a distance. The greater the distance through which you exert a given force, the more work you do. Similarly, the greater the force you exert through a given distance, the more work you do.

The amount of work (W) done in moving an object can be calculated by multiplying the force (F) applied to the object by the distance (d) through which the force is applied, as shown in the following equation:

$$W = F \times d$$

Recall that force is expressed in newtons, and the meter is the basic SI unit for length or distance. Therefore, the unit used to express work is the newton-meter (N•m), which is more simply called the **joule (J).**

Look at **Figure 3** to learn more about calculating work. You can also practice calculating work yourself by doing the MathBreak on this page.

Figure 3 Work Depends on Force and Distance

$\boldsymbol{W = 80\text{ N} \times 1\text{ m} = 80\text{ J}}$

The force needed to lift an object is equal to the gravitational force on the object–in other words, the object's weight.

$\boldsymbol{W = 160\text{ N} \times 1\text{ m} = 160\text{ J}}$

Increasing the amount of force increases the amount of work done.

$\boldsymbol{W = 80\text{ N} \times 2\text{ m} = 160\text{ J}}$

Increasing the distance also increases the amount of work done.

Power—How Fast Work Is Done

Like *work,* the term *power* is used a lot in everyday language but has a very specific meaning in science. **Power** is the rate at which work is done. To calculate power (P), you divide the amount of work done (W) by the time (t) it takes to do that work, as shown in the following equation:

$$P = \frac{W}{t}$$

You just learned that the unit for work is the joule, and the basic unit for time is the second. Therefore, the unit used to express power is joules per second (J/s), which is more simply called the **watt (W).** So if you do 50 J of work in 5 seconds, your power is 10 J/s, or 10 W. You can calculate your own power in the QuickLab at right.

Increasing Power Power is how fast work happens. Power is increased when more work is done in a given amount of time. Power is also increased when the time it takes to do a certain amount of work is decreased, as shown in **Figure 4.**

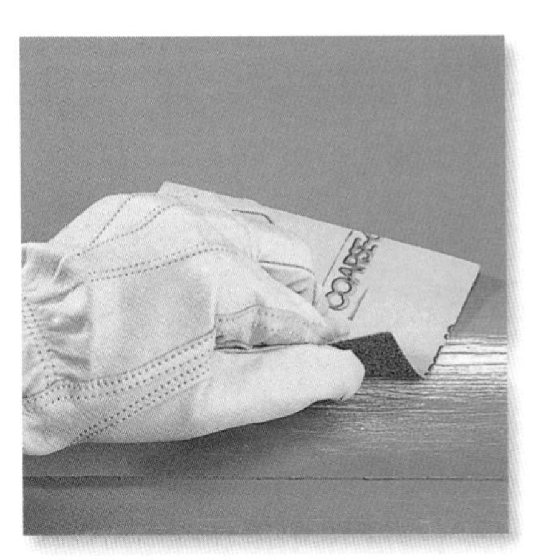

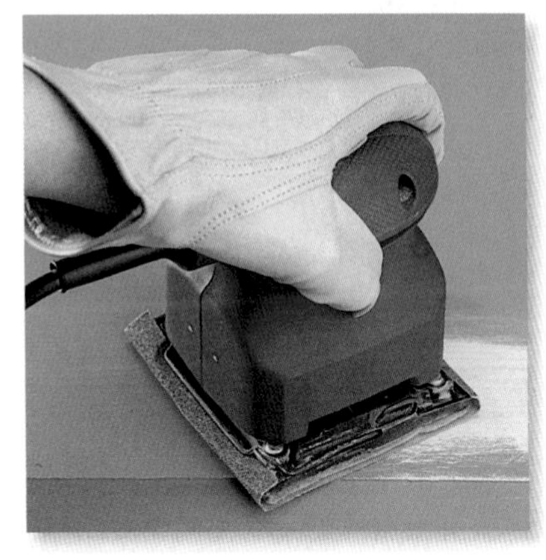

Figure 4 *No matter how fast you can sand with sandpaper, an electric sander can do the same amount of work faster. Therefore, the electric sander has more power.*

QuickLab

More Power to You

1. Use a loop of **string** to attach a **spring scale** to a **book.**
2. Slowly pull the book across a table by the spring scale. Use a **stopwatch** to determine the time this takes. In your ScienceLog, record the amount of time it took and the force used as the book reached the edge of the table.
3. With a **metric ruler,** measure the distance you pulled the book.
4. Now quickly pull the book across the same distance. Again record the time and force.
5. Calculate work and power for both trials.
6. How were the amounts of work and power affected by your pulling the book faster? Record your answers in your ScienceLog.

REVIEW

1. Work is done on a ball when a pitcher throws it. Is the pitcher still doing work on the ball as it flies through the air? Explain.
2. Explain the difference between work and power.
3. **Doing Calculations** You lift a chair that weighs 50 N to a height of 0.5 m and carry it 10 m across the room. How much work do you do on the chair?

Step up to find out more about work and power on page 552 of the LabBook.

Section 2

What Is a Machine?

NEW TERMS

machine
work input
work output
mechanical advantage
mechanical efficiency

OBJECTIVES

- Explain how a machine makes work easier.
- Describe and give examples of the force-distance trade-off that occurs when a machine is used.
- Calculate mechanical advantage.
- Explain why machines are not 100 percent efficient.

Imagine you're in the car with your mom on the way to a party when suddenly—*KABLOOM hissssss*—a tire blows out. "Now I'm going to be late!" you think as your mom pulls over to the side of the road. You watch as she opens the trunk and gets out a jack and a tire iron. Using the tire iron, she pries the hubcap off and begins to unscrew the lug nuts from the wheel. She then puts the jack under the car and turns the handle several times until the flat tire no longer touches the ground. After exchanging the flat tire with the spare, she lowers the jack and puts the lug nuts and hubcap back on the wheel. "Wow!" you think, "That wasn't as hard as I thought it would be." As your mom drops you off at the party, you think how lucky it was that she had the right equipment to change the tire.

Machines—Making Work Easier

Now imagine changing a tire without the jack and the tire iron. Would it have been so easy? No, you would have needed several people just to hold up the car! Sometimes you need a little help to do work. That's where machines come in. A **machine** is a device that helps make work easier by changing the size or direction of a force.

When you think of machines, you might think of things like cars, big construction equipment, or even computers. But not all machines are complicated or even have moving parts. In fact, the tire iron, jack, and lug nut shown above are all machines. Even the items shown in **Figure 5** are machines.

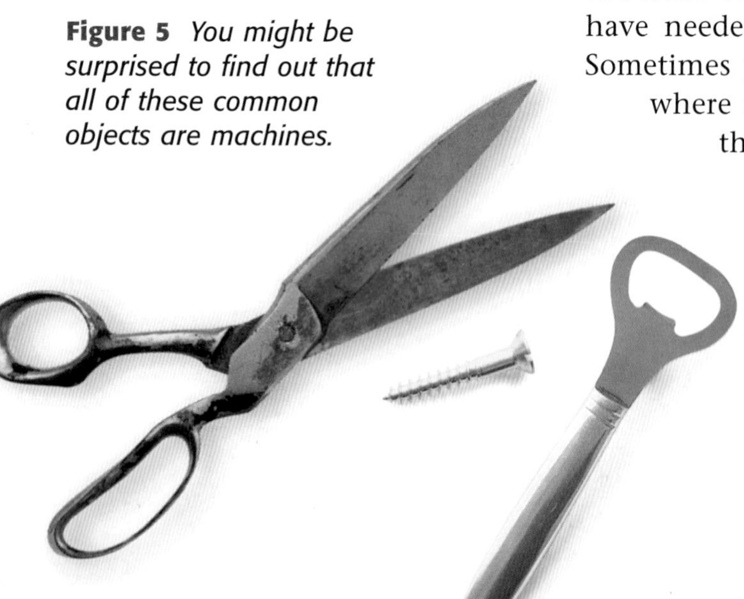

Figure 5 *You might be surprised to find out that all of these common objects are machines.*

Work In, Work Out Suppose you need to get the lid off a can of paint. What do you do? Well, one way to pry the lid off is to use the flat end of a common machine known as a screwdriver, as shown in **Figure 6.** You place the tip of the screwdriver under the edge of the lid and then push down on the handle. The other end of the screwdriver lifts the lid as you push down. In other words, you do work on the screwdriver, and the screwdriver does work on the lid. This example illustrates that two kinds of work are always involved when a machine is used—the work done on the machine and the work the machine does on another object.

Figure 6 *When you use a machine, you do work on the machine, and the machine does work on something else.*

The width of the arrows representing *input force* and *output force* indicates the relative size of the forces. The length of the arrows indicates the distance through which they are exerted.

Remember that work is a force applied through a distance. Look again at Figure 6. The work you do on a machine is called **work input.** You apply a force, called the *input force,* to the machine and move it through a distance. The work done by the machine is called **work output.** The machine applies a force, called the *output force,* through a distance. The output force opposes the forces you and the machine are working against—in this case, the weight of the lid and the friction between the can and the lid.

How Machines Help You might think that machines help you because they increase the amount of work done. But that's not true. If you multiplied the forces by the distances through which they are applied in Figure 6 (remember, $W = F \times d$), you would find that the screwdriver does *not* do more work on the lid than you do on the screwdriver. Work output can *never* be greater than work input.

Machines make work easier because they change the size or the direction of the input force. And using a screwdriver to open a paint can changes *both* the direction and size of your force. Another example of a machine making work easier is shown in **Figure 7.** As you can see, a machine does not save work. The same amount of work is involved with or without a machine.

Figure 7 *A simple plank of wood acts as a machine when it is used to help raise a load.*

Force: 450 N **Distance: 1 m**

$W = 450 \text{ N} \times 1 \text{ m} = 450 \text{ J}$

Lifting this box straight up requires an input force equal to the weight of the box.

Force: 150 N **Distance: 3 m**

$W = 150 \text{ N} \times 3 \text{ m} = 450 \text{ J}$

Using a ramp to lift the box requires an input force less than the weight of the box, but the input force must be exerted over a greater distance.

The Force-Distance Trade-off In Figure 6, the machine (the screwdriver) increases the output force relative to the input force. But the distance over which the output force is exerted is decreased. In the ramp example in Figure 7, the machine (the ramp) decreases the amount of input force necessary to do the work of lifting the box. But the distance over which the force is exerted increases. In both of these examples, the machine allows a smaller force to be applied over a longer distance.

The force-distance trade-off can be simply stated as follows: When a machine changes the size of the force, the distance through which the force is exerted must also change. Force or distance can increase, but not together. When one increases, the other must decrease. This is because the work output is never greater than the work input.

The diagram on the next page will help you better understand this force-distance trade-off. It also shows that some machines affect only the direction of the force, not the size of the force or the distance through which it is exerted.

Machines Change the Size or Direction (or Both) of a Force

Work input	Machine	Work output	Example
Small force (indicated by arrow width) applied over a long distance	Increases force	**Large force** applied over a short distance	
Large force applied over a short distance	Decreases force	**Small force** applied over a long distance	
Small force applied over a long distance	Changes direction of force	**Small force** applied over a long distance in the opposite direction	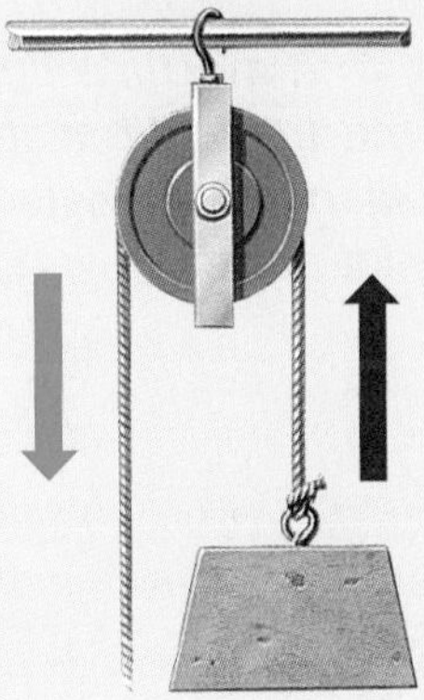
Small force applied over a long distance	Changes size and direction of force	**Large force** applied over a short distance in the opposite direction	

MATH BREAK

Finding the Advantage

1. You apply 200 N to a machine, and the machine applies 2,000 N to an object. What is the mechanical advantage?
2. You apply 10 N to a machine, and the machine applies 10 N to another object. What is the mechanical advantage? Can such a machine be useful? Why or why not?
3. Which of the following makes work easier to do?
   a. a machine with a mechanical advantage of 15
   b. a machine to which you apply 15 N and that exerts 255 N

Mechanical Advantage

Do some machines make work easier than others? Yes, because some machines can increase force more than others. A machine's **mechanical advantage** tells you how many times the machine multiplies force. In other words, it compares the input force with the output force. You can find mechanical advantage by using the following equation:

$$\text{Mechanical advantage } (MA) = \frac{\text{output force}}{\text{input force}}$$

Take a look at **Figure 8.** In this example, the output force is greater than the input force. Using the equation above, you can find the mechanical advantage of the handcart:

$$MA = \frac{500 \text{ N}}{50 \text{ N}} = 10$$

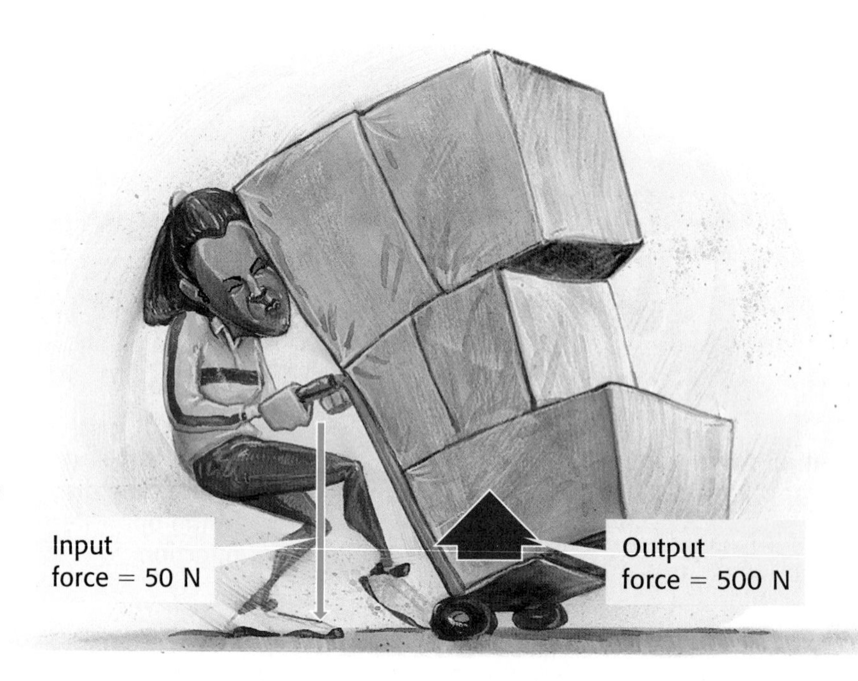

Figure 8 *A machine that has a large mechanical advantage can make lifting a heavy load a whole lot easier.*

Because the mechanical advantage of the handcart is 10, you know that the output force is 10 times bigger than the input force. The larger a machine's mechanical advantage, the easier the machine makes your work. But notice that as mechanical advantage increases (due to a greater output force), the distance that the output force moves the object decreases.

Remember that some machines don't change the size of the force—they simply change the direction of the force. In such cases, the output force is equal to the input force, and therefore the mechanical advantage is 1. Finally, some machines have a mechanical advantage that is less than 1. That means that the input force is greater than the output force. Although such a machine actually decreases your force, it does allow you to exert the force over a longer distance, as shown in **Figure 9.**

Figure 9 *With chopsticks you can pick up a big bite of food with just a little wiggle of your fingers.*

Mechanical Efficiency

As mentioned earlier, the work output of a machine can never be greater than the work input. In fact, the work output of a machine is always *less* than the work input. Why? Because some of the work done by the machine is used to overcome the friction created by the use of the machine. But keep in mind that no work is *lost.* The work output plus the work done to overcome friction equals the work input.

The less work a machine has to do to overcome friction, the more *efficient* it is. **Mechanical efficiency** (e FISH uhn see) is a comparison of a machine's work output with the work input. A machine's mechanical efficiency is calculated using the following equation:

$$\text{Mechanical efficiency} = \frac{\text{work output}}{\text{work input}} \times 100$$

The 100 in this equation means that mechanical efficiency is expressed as a percentage. Mechanical efficiency tells you what percentage of the work input gets converted into work output. No machine is 100 percent efficient, but reducing the amount of friction in a machine is a way to increase its mechanical efficiency. Inventors have tried for many years to create a machine that has no friction to overcome, but so far they have been unsuccessful. If a machine could be made that had 100 percent mechanical efficiency, it would be called an *ideal machine.*

Car manufacturers recommend regular oil changes. That's because over time, motor oil in a car's engine starts to get dark and thick and doesn't flow as well as fresh motor oil. Why do you think a car engine needs motor oil? How does getting regular oil changes improve the mechanical efficiency of a car's engine?

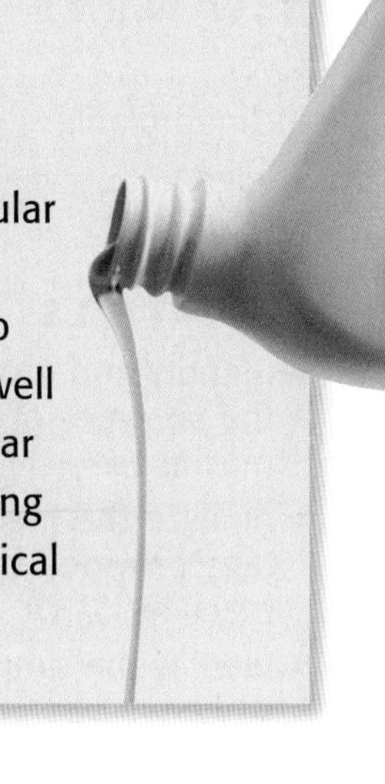

A joint is a place in the body where bones meet. For example, your elbow is a joint that connects the bone of your upper arm with the bones of your lower arm. You can bend your arm at the elbow without feeling the bones touching because of the synovial (sin OH vee uhl) fluid in the joint. This clear, thick fluid reduces friction in the joint by lubricating the ends of the separate bones, allowing for smooth, efficient movement.

REVIEW

1. Explain how using a ramp makes work easier.
2. Why is it impossible for a machine to be 100 percent efficient?
3. Suppose you exert 15 N on a machine, and the machine exerts 300 N on another object. What is the machine's mechanical advantage?
4. **Comparing Concepts** For the machine described in question 3, how does the distance through which the output force is exerted differ from the distance through which the input force is exerted?

Section 3

Types of Machines

NEW TERMS

lever
inclined plane
wedge
screw
wheel and axle
pulley
compound machine

OBJECTIVES

- Identify and give examples of the six types of simple machines.
- Analyze the mechanical advantage provided by each simple machine.
- Identify the simple machines that make up a compound machine.

All machines are constructed from these six simple machines: *lever, inclined plane, wedge, screw, wheel and axle,* and *pulley.* You've seen a couple of these machines already—a screwdriver can be used as a lever, and a ramp is an inclined plane. In the next few pages, each of the six simple machines will be discussed separately. Then you'll learn how compound machines are formed from combining simple machines.

Levers

Have you ever used the claw end of a hammer to remove a nail from a piece of wood? If so, you were using the hammer as a lever. A **lever** is a simple machine consisting of a bar that pivots at a fixed point, called a *fulcrum.* Levers are used to apply a force to a load. There are three classes of levers, based on the locations of the fulcrum, the load, and the input force.

Figure 10 A First Class Lever

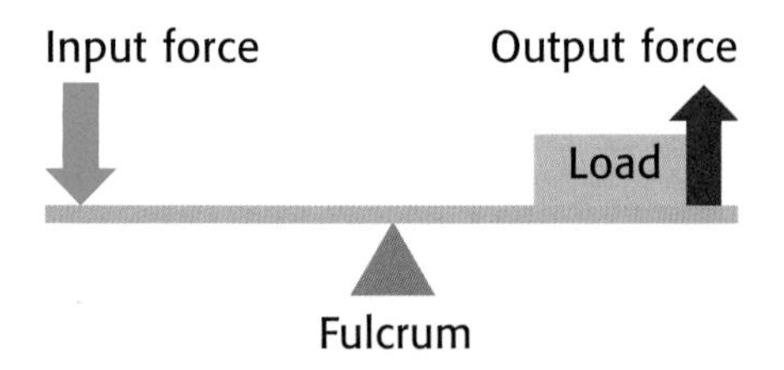

First Class Levers With a first class lever, the fulcrum is between the input force and the load, as shown in **Figure 10.** First class levers always change the direction of the input force. And depending on the location of the fulcrum, first class levers can be used to increase force or to increase distance. Some examples of first class levers are shown below.

Examples of First Class Levers

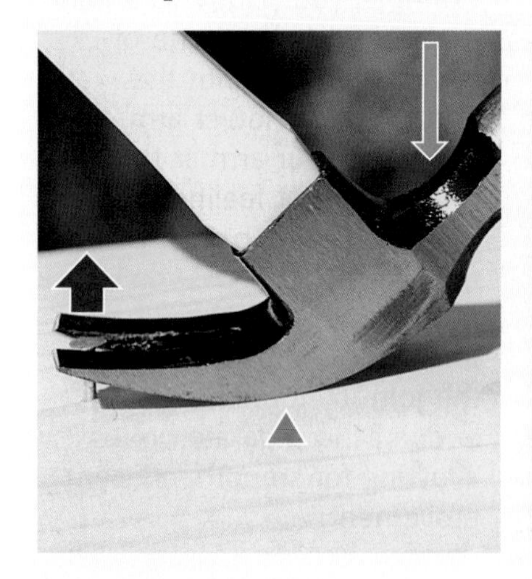

When **the fulcrum is closer to the load than to the input force,** a mechanical advantage of greater than 1 results. The output force is increased because it is exerted over a shorter distance.

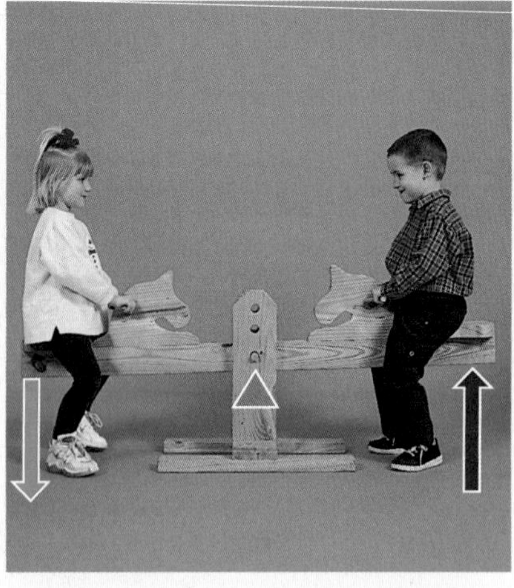

When **the fulcrum is exactly in the middle,** a mechanical advantage of 1 results. The output force is not increased because the input force's distance is not increased.

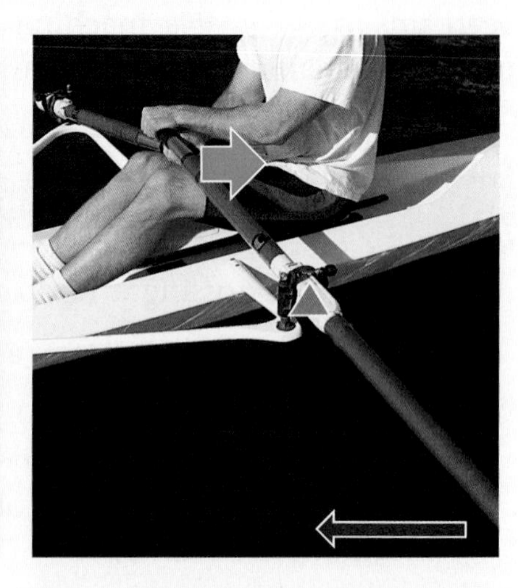

When **the fulcrum is closer to the input force than to the load,** a mechanical advantage of less than 1 results. Although the output force is less than the input force, a gain in distance occurs.

Second Class Levers With a second class lever, the load is between the fulcrum and the input force, as shown in **Figure 11.** Second class levers do not change the direction of the input force, but they allow you to apply less force than the force exerted by the load. Because the output force is greater than the input force, you must exert the input force over a greater distance. Some examples of second class levers are shown at right.

Figure 11 A Second Class Lever

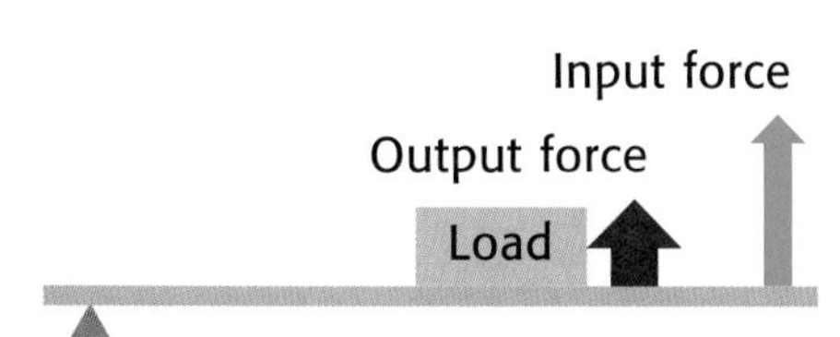

Third Class Levers With a third class lever, the input force is between the fulcrum and the load, as shown in **Figure 12.** Third class levers do not change the direction of the input force. In addition, they do *not* increase the input force. Therefore, the output force is always less than the input force. Some examples of third class levers are shown at right.

Figure 12 A Third Class Lever

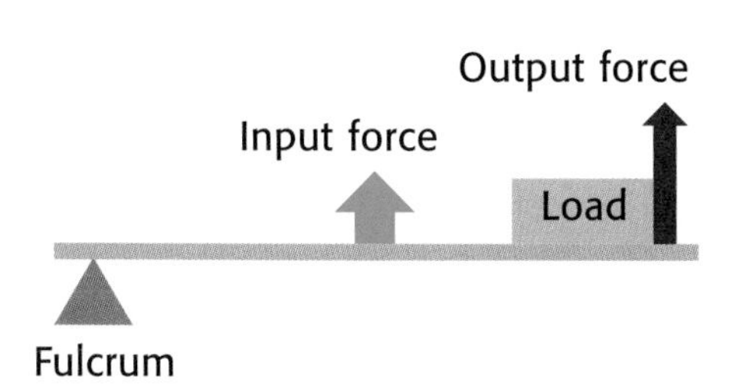

Examples of Second Class Levers

Using a second class lever results in a mechanical advantage of greater than 1. The closer the load is to the fulcrum, the more the force is increased and the greater the mechanical advantage.

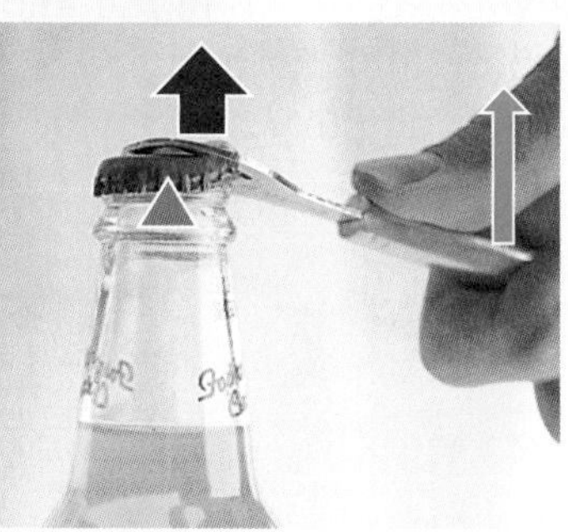

Examples of Third Class Levers

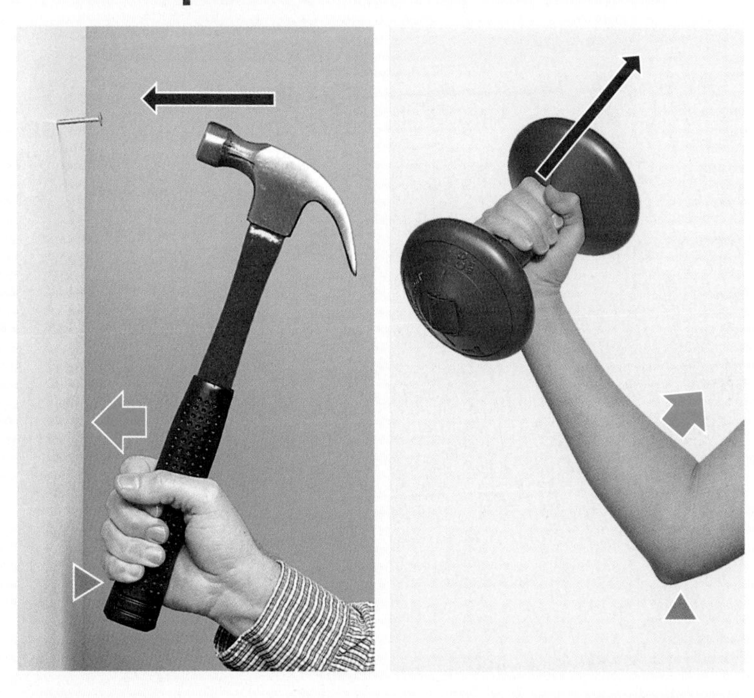

Using a third class lever results in a mechanical advantage of less than 1 because force is decreased. But third class levers are helpful because they increase the distance through which the output force is exerted.

Inclined Planes

Do you remember the story about how the Egyptians built the Great Pyramid? One of the machines they used was the *inclined plane.* An **inclined plane** is a simple machine that is a straight, slanted surface. A ramp is an example of an inclined plane.

Inclined planes can make work easier. Look at **Figure 13.** Using an inclined plane to load an upright piano into the back of a truck is easier than just lifting it into the truck. Rolling the piano into the truck along an inclined plane requires a smaller input force than is required to lift the piano into the truck. But remember that machines do not save work—therefore, the input force must be exerted over a longer distance.

BRAIN FOOD

When Napoleon Bonaparte's army invaded Egypt in 1798, one of his engineers reportedly calculated that the three pyramids of Giza contained enough stone to build a wall about 2.5 m tall and 0.3 m thick around the entire country of France.

Figure 13 *The work you do on the piano to roll it up the ramp is the same as the work you would do to lift it straight up. An inclined plane simply allows you to apply a smaller force over a greater distance.*

Mechanical Advantage of Inclined Planes The longer the inclined plane is compared with its height, the greater the mechanical advantage. The mechanical advantage (*MA*) of an inclined plane can be calculated by dividing the *length* of the inclined plane by the *height* to which the load is lifted, as shown below:

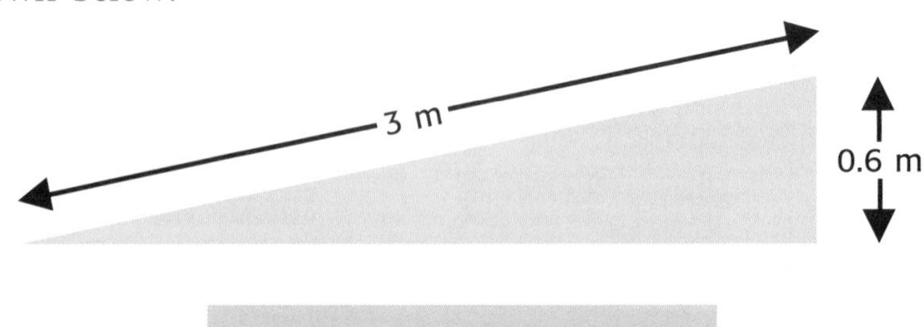

$$MA = \frac{3 \text{ m}}{0.6 \text{ m}} = 5$$

Wedges

Imagine trying to cut a watermelon in half with a spoon. It wouldn't be easy, would it? A knife is a much more useful utensil for cutting because it's a *wedge*. A **wedge** is a double inclined plane that moves. When you move a wedge through a distance, it applies a force on an object. A wedge applies an output force that is greater than your input force, but you apply the input force over a greater distance. The greater the distance you move the wedge, the greater the force it applies on the object. For example, the deeper you move a knife into a watermelon, as shown in **Figure 14,** the more force the knife applies to the two halves. Eventually, it pushes them apart. Other useful wedges include doorstops, plows, axe heads, and chisels.

Figure 14 *Wedges, which are often used to cut materials, allow you to exert your force over an increased distance.*

Mechanical Advantage of Wedges The longer and thinner the wedge is, the greater the mechanical advantage. That's why axes and knives cut better when you sharpen them—you are making the wedge thinner. Therefore, less input force is required. The mechanical advantage of a wedge can be determined by dividing the *length* of the wedge by its greatest *thickness,* as shown below.

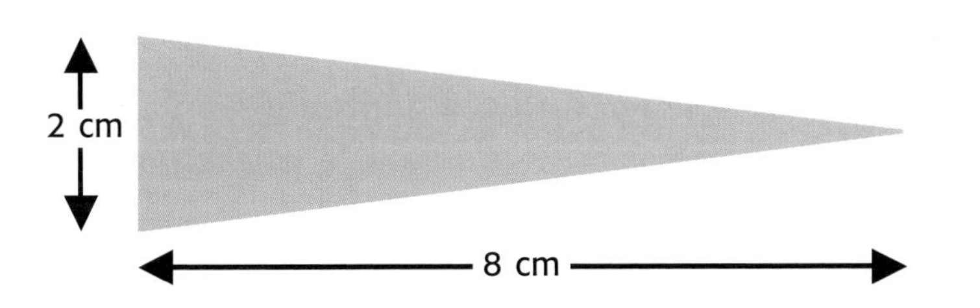

$$MA = \frac{8\text{ cm}}{2\text{ cm}} = 4$$

Screws

A **screw** is an inclined plane that is wrapped in a spiral. When a screw is rotated, a small force is applied over the long distance along the inclined plane of the screw. Meanwhile, the screw applies a large force through the short distance it is pushed. In other words, you apply a small input force over a large distance, while the screw exerts a large output force over a small distance. Screws are used most commonly as fasteners. Some examples of screws are shown in **Figure 15.**

Figure 15 Examples of Screws

When you turn a screw, you exert a small input force over a large turning distance, but the screw itself doesn't move very far. So the screw's output force is greater than your input force.

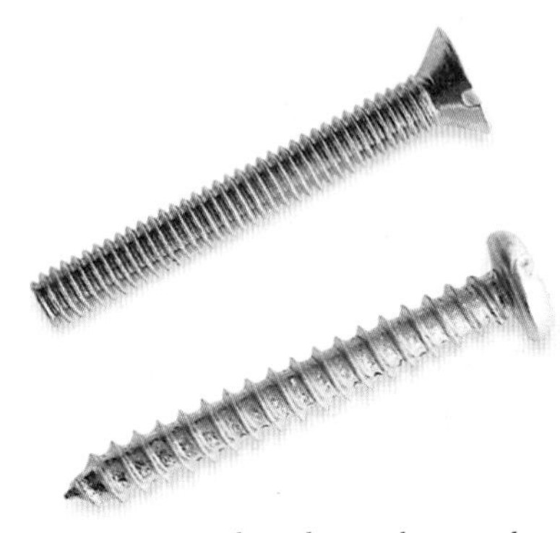

Figure 16 *The threads on the top screw are closer together and wrap more times around, so that screw has a greater mechanical advantage than the one below it.*

Mechanical Advantage of Screws If you could "unwind" the inclined plane of a screw, you would see that it is very long and has a gentle slope. Recall that the longer an inclined plane is compared with its height, the greater its mechanical advantage. Similarly, the longer the spiral on a screw is and the closer together the threads, the greater the screw's mechanical advantage, as shown in **Figure 16.**

REVIEW

1. Give an example of each of the following simple machines: first class lever, second class lever, third class lever, inclined plane, wedge, and screw.
2. A third class lever has a mechanical advantage of less than 1. Explain why it is useful for some tasks.
3. **Interpreting Graphics** Look back at Figures 6, 7, and 8 in Section 2. Identify the type of simple machine shown in each case. (If a lever is shown, identify its class.)

Wheel and Axle

Did you know that when you turn a doorknob you are using a machine? A doorknob is an example of a **wheel and axle,** a simple machine consisting of two circular objects of different sizes. A wheel can be a crank, such as the handle on a fishing reel, or it can be a knob, such as a volume knob on a radio. The axle is the smaller of the two circular objects. Doorknobs, wrenches, ferris wheels, screwdrivers, and steering wheels all use a wheel and axle. **Figure 17** shows how a wheel and axle works.

Figure 17
How a Wheel and Axle Works

a When a small input force is applied to the wheel, it rotates through a circular distance.

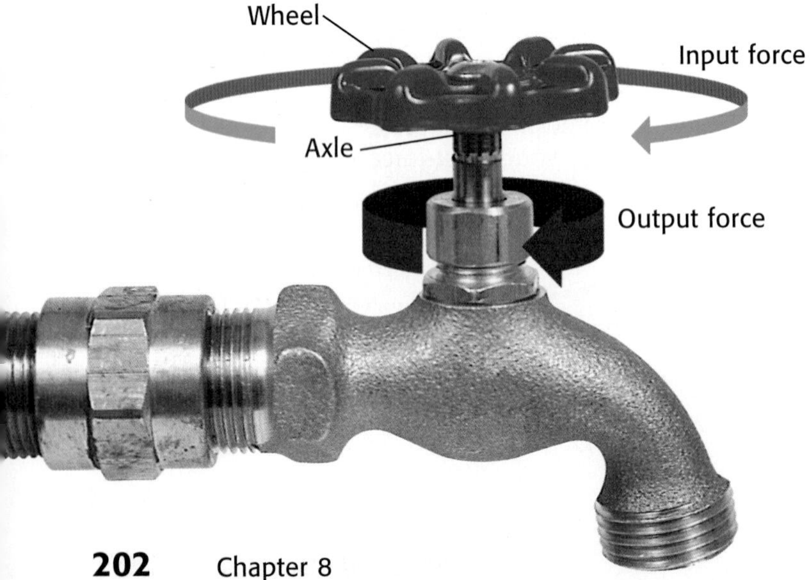

b As the wheel turns, so does the axle. But because the axle is smaller than the wheel, it rotates through a smaller distance, which makes the output force larger than the input force.

Mechanical Advantage of a Wheel and Axle The mechanical advantage of a wheel and axle can be determined by dividing the *radius* (the distance from the center to the edge) of the wheel by the radius of the axle, as shown at right. Turning the wheel results in a mechanical advantage of greater than 1 because the radius of the wheel is larger than the radius of the axle.

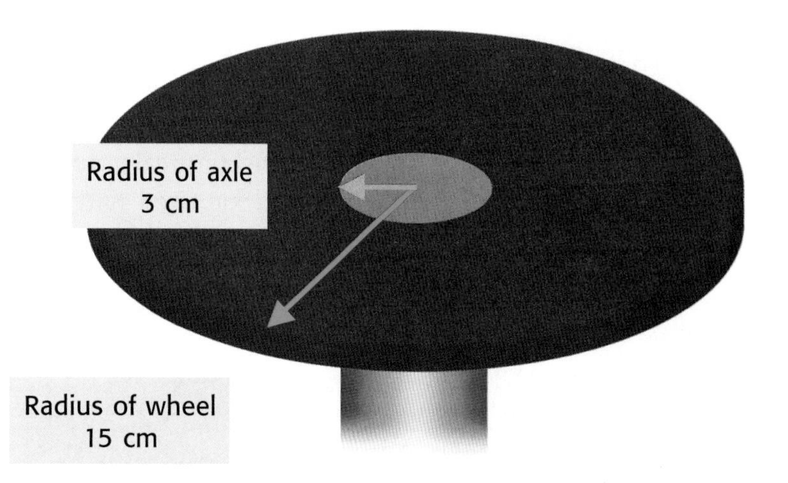

$$MA = \frac{15 \text{ cm}}{3 \text{ cm}} = 5$$

Pulleys

When you open window blinds by pulling on a cord, you're using a pulley. A **pulley** is a simple machine consisting of a grooved wheel that holds a rope or a cable. A load is attached to one end of the rope, and an input force is applied to the other end. There are two kinds of pulleys—*fixed* and *movable*. Fixed and movable pulleys can be combined to form a *block and tackle*.

Fixed Pulleys Some pulleys only change the direction of a force, as shown in **Figure 18.** This kind of pulley is called a fixed pulley. Fixed pulleys do not increase force. A fixed pulley is attached to something that does not move. By using a fixed pulley, you can pull down on the rope in order to lift the load up. This is usually easier than trying to lift the load straight up. Elevators make use of fixed pulleys.

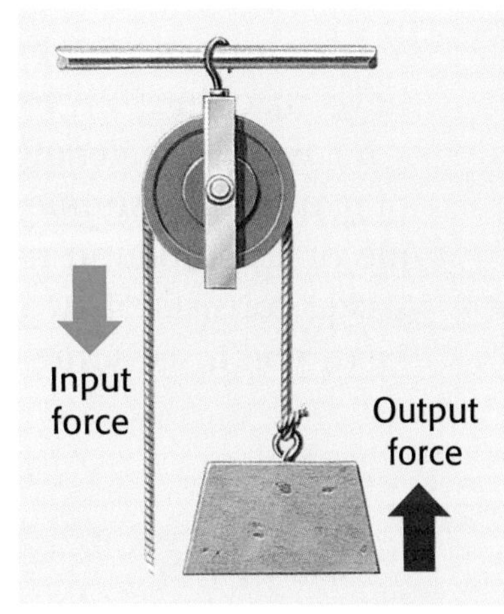

Figure 18 *A* **fixed pulley** *only spins. So the distance through which the input force and the output force are exerted—and thus the forces themselves—are the same. Therefore, a fixed pulley provides a mechanical advantage of 1.*

Movable Pulleys Unlike fixed pulleys, movable pulleys are attached to the object being moved, as shown in **Figure 19.** A movable pulley does not change a force's direction. Movable pulleys do increase force, but you must exert the input force over a greater distance than the load is moved. This is because you must make *both* sides of the rope move in order to lift the load.

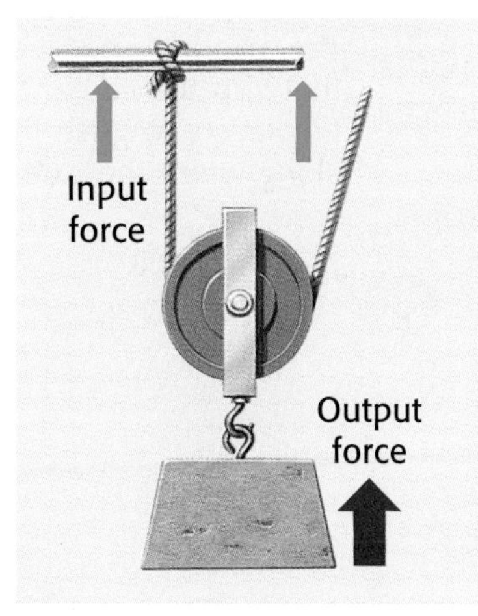

Figure 19 *A* **movable pulley** *moves up with the load as it is lifted. Force is multiplied because the combined input force is exerted over twice the distance of the output force. The mechanical advantage of a movable pulley is the number of rope segments that support the load. In this example, the mechanical advantage is 2.*

Block and Tackles When a fixed pulley and a movable pulley are used together, the pulley system is called a *block and tackle.* As shown in **Figure 20,** a block and tackle can have a large mechanical advantage if several pulleys are used. A block and tackle used within a larger pulley system is shown in **Figure 21.**

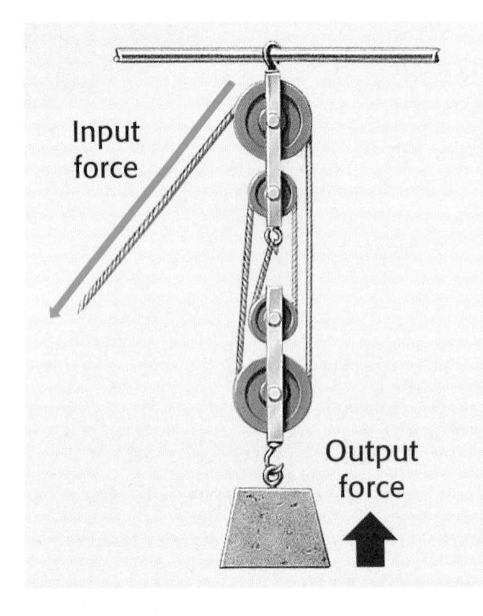

Figure 20 *The mechanical advantage of this* **block and tackle** *is 4 because there are four rope segments that support the load. This block and tackle multiplies your input force four times, but you have to pull the rope 4 m just to lift the load 1 m.*

Figure 21 *The combination of pulleys used by this crane allows it to lift heavy pieces of scrap metal.*

Compound Machines

You are surrounded by machines. As you saw earlier, you even have machines in your body! But most of the machines in your world are **compound machines,** machines that are made of two or more simple machines. You've already seen one example of a compound machine: a block and tackle. A block and tackle consists of two or more pulleys. On this page and the next, you'll see some other examples of compound machines.

Explore

List five machines that you have encountered today and indicate what type of machine each is. Try to include at least one compound machine and one machine that is part of your body.

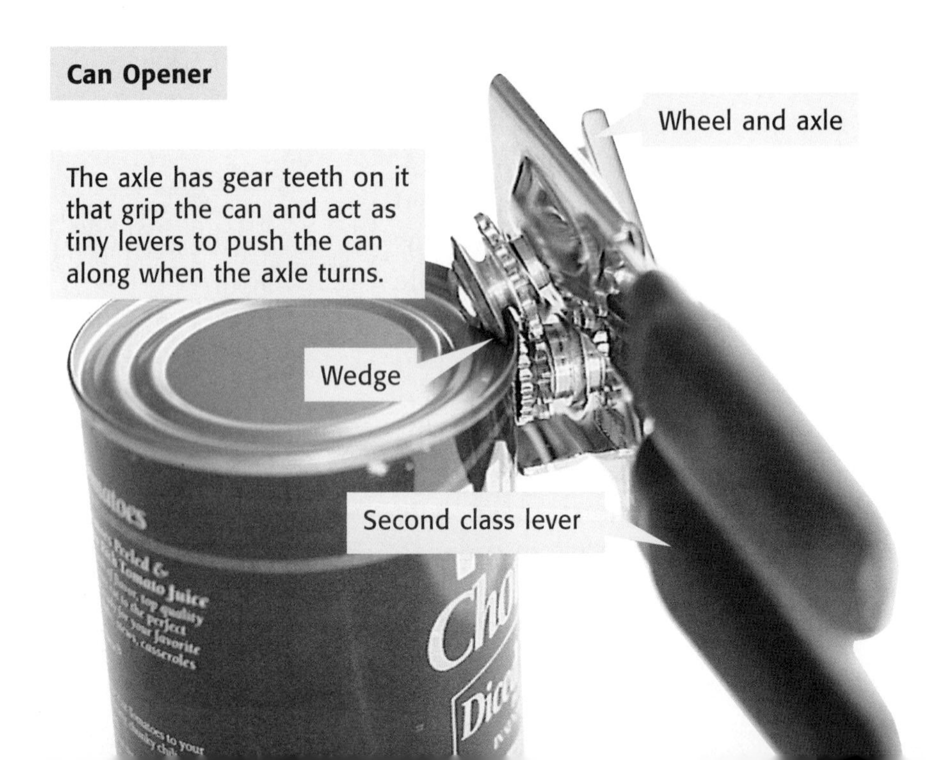

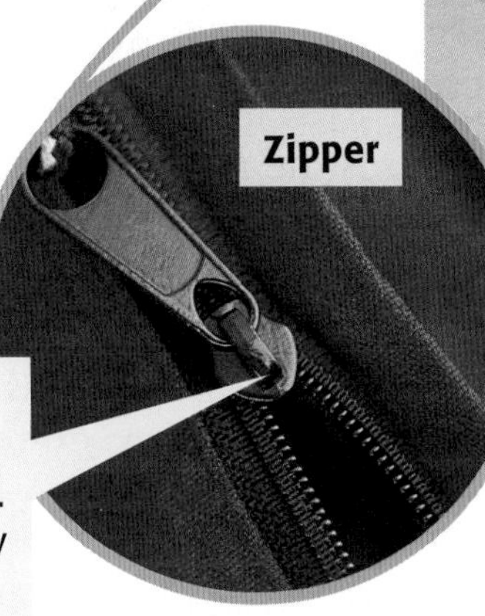

Mechanical Efficiency of Compound Machines In general, the more moving parts a machine has, the lower its mechanical efficiency. Thus the mechanical efficiency of compound machines is often quite low. For compound machines that involve many simple machines, such as automobiles and airplanes, it is very important that friction be reduced as much as possible through the use of lubrication and other techniques. Too much friction could cause heating and damage the simple machines involved, which could create safety problems and could be expensive to repair.

REVIEW

1. Give an example of a wheel and axle.
2. Identify the simple machines that make up tweezers and nail clippers.
3. **Doing Calculations** The radius of the wheel of a wheel and axle is four times greater than the radius of the axle. What is the mechanical advantage of this machine?

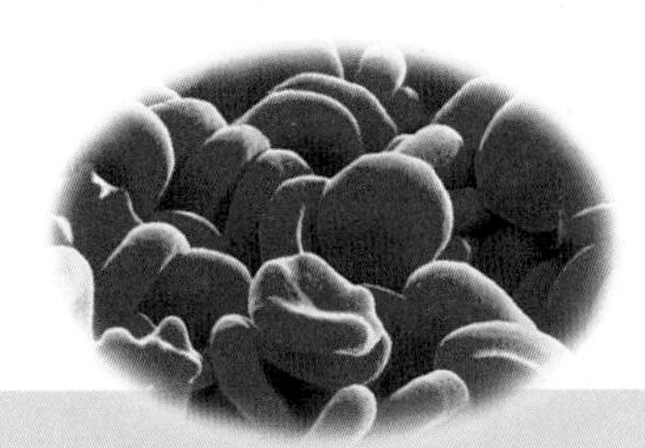

Turn to page 210 to discover how some compound machines are micromachines that are small enough to swim through your bloodstream.

Chapter Highlights

SECTION 1

Vocabulary

work *(p. 188)*
joule *(p. 190)*
power *(p. 191)*
watt *(p. 191)*

Section Notes

- Work occurs when a force causes an object to move in the direction of the force. The unit for work is the joule (J).
- Work is done on an object only when a force makes an object move and only while that force is applied.

- For work to be done on an object, the direction of the object's motion must be in the same direction as the force applied.
- Work can be calculated by multiplying force by distance.
- Power is the rate at which work is done. The unit for power is the watt (W).
- Power can be calculated by dividing the amount of work by the time taken to do that work.

Labs

A Powerful Workout *(p. 552)*

SECTION 2

Vocabulary

machine *(p. 192)*
work input *(p. 193)*
work output *(p. 193)*
mechanical advantage *(p. 196)*
mechanical efficiency *(p. 197)*

Section Notes

- A machine makes work easier by changing the size or direction (or both) of a force.
- When a machine changes the size of a force, the distance through which the force is exerted must also change. Force or distance can increase, but not together.

☑ Skills Check

Math Concepts

WORK AND POWER Suppose a woman raises a 65 N object 1.6 m in 4 s. The work done and her power can be calculated as follows:

$$W = F \times d \qquad P = \frac{W}{t}$$

$$= 65 \text{ N} \times 1.6 \text{ m} \qquad = \frac{104 \text{ J}}{4 \text{ s}}$$

$$= 104 \text{ J} \qquad = 26 \text{ W}$$

Visual Understanding

MACHINES MAKE WORK EASIER A machine can change the size or direction (or both) of a force. Review the table on page 195 to learn more about how machines make work easier.

COMPOUND MACHINES A compound machine is made of two or more simple machines. Review the examples on pages 204 and 205.

SECTION 2

- Mechanical advantage tells how many times a machine multiplies force. It can be calculated by dividing the output force by the input force.
- Mechanical efficiency is a comparison of a machine's work output with work input. Mechanical efficiency is calculated by dividing work output by work input and is expressed as a percentage.
- Machines are not 100 percent efficient because some of the work done by a machine is used to overcome friction. So work output is always less than work input.

SECTION 3

Vocabulary

lever *(p. 198)*
inclined plane *(p. 200)*
wedge *(p. 201)*
screw *(p. 201)*
wheel and axle *(p. 202)*
pulley *(p. 203)*
compound machine *(p. 204)*

Section Notes

- All machines are constructed from these six simple machines: lever, inclined plane, wedge, screw, wheel and axle, and pulley.
- Compound machines consist of two or more simple machines.
- Compound machines have low mechanical efficiencies because they have more moving parts and thus more friction to overcome.

Labs

Inclined to Move *(p. 554)*
Building Machines *(p. 555)*
Wheeling and Dealing *(p. 556)*

internet**connect**

GO TO: go.hrw.com

Visit the **HRW** Web site for a variety of learning tools related to this chapter. Just type in the keyword:

KEYWORD: HSTWRK

GO TO: www.scilinks.org

Visit the **National Science Teachers Association** on-line Web site for Internet resources related to this chapter. Just type in the ***sci*LINKS** number for more information about the topic:

TOPIC: Work and Power — ***sci*LINKS NUMBER:** HSTP180
TOPIC: Mechanical Efficiency — ***sci*LINKS NUMBER:** HSTP185
TOPIC: Simple Machines — ***sci*LINKS NUMBER:** HSTP190
TOPIC: Compound Machines — ***sci*LINKS NUMBER:** HSTP195

Chapter Review

USING VOCABULARY

For each pair of terms, explain the difference in their meanings.

1. joule/watt
2. work output/work input
3. mechanical efficiency/mechanical advantage
4. screw/inclined plane
5. simple machine/compound machine

UNDERSTANDING CONCEPTS

Multiple Choice

6. Work is being done when
   a. you apply a force to an object.
   b. an object is moving after you apply a force to it.
   c. you exert a force that moves an object in the direction of the force.
   d. you do something that is difficult.

7. The work output for a machine is always less than the work input because
   a. all machines have a mechanical advantage.
   b. some of the work done is used to overcome friction.
   c. some of the work done is used to overcome distance.
   d. power is the rate at which work is done.

8. The unit for work is the
   a. joule.
   b. joule per second.
   c. newton.
   d. watt.

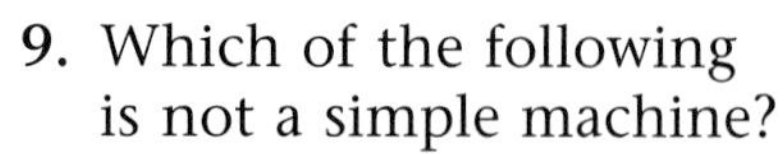

9. Which of the following is not a simple machine?
   a. a faucet handle
   b. a jar lid
   c. a can opener
   d. a seesaw

10. Power is
    a. how strong someone or something is.
    b. how much force is being used.
    c. how much work is being done.
    d. how fast work is being done.

11. The unit for power is the
    a. newton.
    b. kilogram.
    c. watt.
    d. joule.

12. A machine can increase
    a. distance at the expense of force.
    b. force at the expense of distance.
    c. neither distance nor force.
    d. Both (a) and (b)

Short Answer

13. Identify the simple machines that make up a pair of scissors.

14. In two or three sentences, explain the force-distance trade-off that occurs when a machine is used to make work easier.

15. Explain why you do work on a bag of groceries when you pick it up but not when you are carrying it.

Concept Mapping

16. Create a concept map using the following terms: work, force, distance, machine, mechanical advantage.

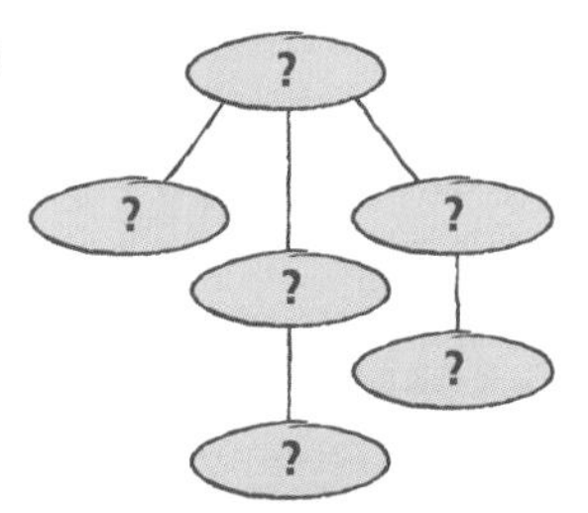

CRITICAL THINKING AND PROBLEM SOLVING

17. Why do you think levers usually have a greater mechanical efficiency than other simple machines do?

18. The winding road shown below is actually a series of inclined planes. Describe how a winding road makes it easier for vehicles to travel up a hill.

19. Why do you think you would not want to reduce the friction involved in using a winding road?

MATH IN SCIENCE

20. You and a friend together apply a force of 1,000 N to a 3,000 N automobile to make it roll 10 m in 1 minute and 40 seconds.
    a. How much work did you and your friend do together?
    b. What was your combined power?

INTERPRETING GRAPHICS

For each of the images below, identify the class of lever used and calculate the mechanical advantage.

21.

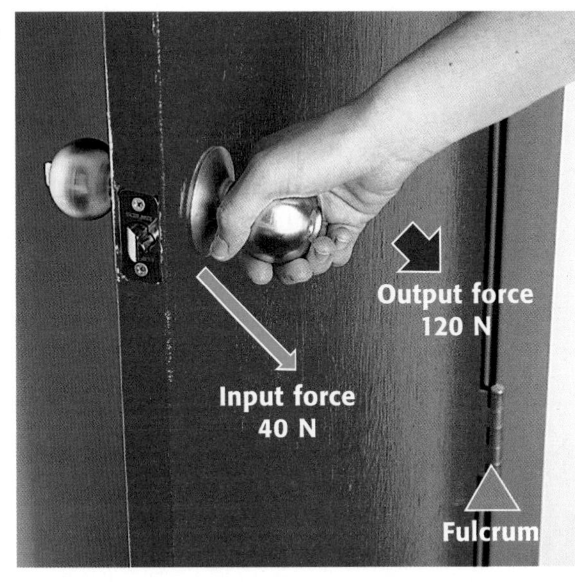

22.

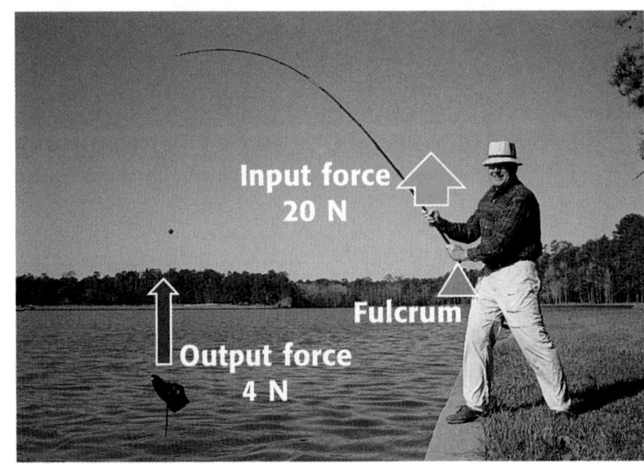

NOW What Do You Think?

Take a minute to review your answers to the ScienceLog questions on page 187. Have your answers changed? If necessary, revise your answers based on what you have learned since you began this chapter.

Science, Technology, and Society

Micromachines

The technology of making things smaller and smaller keeps growing and growing. Powerful computers can now be held in the palm of your hand. But what about motors smaller than a grain of pepper? Or gnat-sized robots that can swim through the bloodstream? These are just a couple of the possibilities for micromachines.

▲ *The earliest working micromachine had a turning central rotor.*

Microscopic Motors

Researchers have already built gears, motors, and other devices so small that you could accidentally inhale one! For example, one engineer devised a motor so small that five of the motors would fit on the period at the end of this sentence. This micromotor is powered by static electricity instead of electric current, and the motor spins at 15,000 revolutions per minute. This is about twice as fast as most automobile engines running at top speed.

Small Sensors

So far micromachines have been most useful as sensing devices. Micromechanical sensors can be used in places too small for ordinary instruments. For example, blood-pressure sensors can fit inside blood vessels and can detect minute changes in a person's blood pressure. Each sensor has a patch so thin that it bends when the pressure changes.

Cell-Sized Robots

Some scientists are investigating the possibility of creating cell-sized machines called nanobots. These tiny robots may have many uses in medicine. For instance, if nanobots could be injected into a person's bloodstream, they might be used to destroy disease-causing organisms such as viruses and bacteria. Nanobots might also be used to count blood cells or to deliver medicine.

The ultimate in micromachines would be machines created from individual atoms and molecules. Although these machines do not currently exist, scientists are already able to manipulate single atoms and molecules. For example, the "molecular man" shown below is made of individual molecules. These molecules are moved by using a scanning tunneling microscope.

A Nanobot's "Life"

► Imagine that you are a nanobot traveling through a person's body. What types of things do you think you would see? What type of work could you do? Write a story that describes what your experiences as a nanobot might be like.

► *"Molecular man" is composed of 28 carbon monoxide molecules.*

Eureka!

Wheelchair Innovators

Two recent inventions have dramatically improved the technology of wheelchairs. With these new inventions, some wheelchair riders can now control their chairs with voice commands and others can take a cruise over a sandy beach.

Voice-Command Wheelchair

At age 27, Martine Kemph invented a voice-recognition system that enables people without arms or legs to use spoken commands to operate their motorized wheelchairs. Here's how it works: The voice-recognition computer translates spoken words into digital commands, which are then directed to electric motors. These commands completely control the operating speed and direction of the motors, giving the operator total control over the chair's movement.

Kemph's system can execute spoken commands almost instantly. In addition, the system is easy to program, so each user can tailor the computer's list of commands to his or her needs.

Kemph named the computer Katalvox, using the root words *katal,* which is Greek for "to understand," and *vox,* which is Latin for "voice."

The Surf Chair

Mike Hensler was a lifeguard at Daytona Beach, Florida, when he realized that it was next to impossible for someone in a wheelchair to come onto the beach. Although he had never invented a machine before, Hensler decided to build a wheelchair that could be maneuvered across sand without getting stuck. He began spending many evenings in his driveway with a pile of lawn-chair parts, designing the chair by trial and error.

The result of Hensler's efforts looks very different from a conventional wheelchair. With huge rubber wheels and a thick frame of white PVC pipe, the Surf Chair not only moves easily over sandy terrain but also is weather resistant and easy to clean. The newest models of the Surf Chair come with optional attachments, such as a variety of umbrellas, detachable armrests and footrests, and even places to attach fishing rods.

▲ *Mike Hensler tries out his Surf Chair.*

Design One Yourself

► Can you think of any other ways to improve wheelchairs? Think about it, and put your ideas down on paper. To inspire creative thinking, consider how a wheelchair could be made lighter, faster, safer, or easier to maneuver.

CHAPTER 9 Energy and Energy Resources

Strange but True!

Vast treasure-troves of energy resources are buried at sea. These resources are more valuable than any pirate's plunder, and they've been there for millions of years. No, they're not gold doubloons—they're gas hydrates (HIE DRAYTS), energy resources that may become increasingly important in the near future.

Also known as methane hydrates, gas hydrates are icy formations of water and methane, the main component of natural gas. The methane in hydrates is produced by bacteria that help decompose organic material in the ocean. Hydrates form under the moderate pressures and low temperatures found at depths of 300–800 meters.

Gas-hydrate deposits are found under the Arctic permafrost and in marine sediments throughout the world. Large areas of hydrates have been discovered off the coasts of North Carolina and South Carolina. In just two areas that are each about the size of Rhode Island, scientists suspect there may be 37 trillion cubic meters of methane gas—70 times the amount of natural gas consumed by the United States in one year.

When brought to surface temperatures of around 15°C, the snowball-like hydrates fizz like effervescent tablets. And holding a flame near a hydrate ignites the evaporating methane, making the gas hydrate look like a burning ice cube. In both instances, a hydrate's stored energy is released. This energy could be used to drive machinery or generate electricity.

Because of the methane locked inside these icy formations, gas hydrates may become a very valuable energy resource.

One drawback to using gas hydrates is that mining them is expensive. But as more research is done, gas hydrates may begin to play a bigger role in the way energy is used every day. In this chapter, you'll learn about different forms of energy, about energy conversions, and about how you use energy resources in your daily life.

What Do You Think?

In your ScienceLog, try to answer the following questions based on what you already know:

1. What is energy?
2. How many different forms of energy are there?
3. How is energy converted from one form to another?
4. What is an energy resource?

Blast-off to Energy

In this activity, you will find out what happens when energy stored in matter is released.

Procedure

1. Use a **measuring spoon** to put a tablespoon of **baking soda** into the center of a **coffee filter.** Twist the ends of the filter tightly shut. (You can use a **twist tie** if necessary.)
2. Use a **graduated cylinder** to pour 200 mL of **water** into an empty **2 L soda bottle.** Use **another graduated cylinder** to pour 200 mL of **vinegar** into the same soda bottle. Place the bottle upright on several pieces of **newspaper** on a table or on the floor.
3. Drop the coffee filter into the soda bottle, and quickly place a **cork** in the mouth of the bottle. Take several steps back from the bottle. **Caution:** Do not lean over the bottle.
4. Observe what happens inside the bottle and what happens to the cork. Record your observations in your ScienceLog.

Analysis

5. Do you think the water, vinegar, or baking soda has energy? Explain your answer.
6. How does what happened to the cork show that the cork had energy? Where do you think the cork's energy came from?

Section 1

What Is Energy?

NEW TERMS

energy
kinetic energy
potential energy
mechanical energy

OBJECTIVES

- Explain the relationship between energy and work.
- Compare kinetic and potential energy.
- Summarize the different forms of energy.

It's match point. The crowd is dead silent. The tennis player steps up to serve. With a look of determination, she bounces the tennis ball several times. Next, in one fluid movement, she tosses the ball into the air and then slams it with her racket. The ball flies toward her opponent, who steps up and swings her racket at the ball. Suddenly, *THWOOSH!!* The ball goes into the net, and the net wiggles from the impact. Game, set, and match!!

Energy and Work—Working Together

Energy is around you all the time. So what is it exactly? In science, you can think of **energy** as the ability to do work. Work occurs when a force causes an object to move in the direction of the force. How are energy and work involved in playing tennis? In this example, the tennis player does work on her racket, the racket does work on the ball, and the ball does work on the net. Each time work is done, something is given by one object to another that allows it to do work. That "something" is energy. As you can see in **Figure 1,** work is a transfer of energy.

Because work and energy are so closely related, they are expressed in the same units—joules (J). When a given amount of work is done, the same amount of energy is involved.

Figure 1 *When one object does work on another, energy is transferred.*

Kinetic Energy Is Energy of Motion

From the tennis example on the previous page, you learned that energy is transferred from the racket to the ball. As the ball flies over the net, it has **kinetic** (ki NET ik) **energy,** the energy of motion. All moving objects have kinetic energy. Does the tennis player have kinetic energy? Definitely! She has kinetic energy when she steps up to serve and when she swings the racket. When she's standing still, she doesn't have any kinetic energy. However, the parts of her body that are moving—her eyes, her heart, and her lungs—do have some kinetic energy.

Objects with kinetic energy can do work. If you've ever gone bowling, you've done work using kinetic energy. When you throw the ball down the lane, you do work on it, transferring your kinetic energy to the ball. As a result, the bowling ball can do work on the pins. Another example of doing work with kinetic energy is shown in **Figure 2.**

Figure 2 *When you swing a hammer, you give it kinetic energy, which it uses to do work on the nail.*

Kinetic Energy Depends on Speed and Mass

An object's kinetic energy can be determined with the following equation:

$$\text{Kinetic energy} = \frac{mv^2}{2}$$

In this equation, m stands for an object's mass, and v stands for an object's speed. The faster something is moving, the more kinetic energy it has. In addition, the more massive a moving object is, the more kinetic energy it has. But which do you think has more of an effect on an object's kinetic energy, its mass or its speed? As you can see from the equation, speed is squared, so speed has a greater effect on kinetic energy than does mass. You can see an example of how kinetic energy depends on speed and mass in **Figure 3.**

Figure 3 *The red car has more kinetic energy than the green car because the red car is moving faster. But the truck has more kinetic energy than the red car because the truck is more massive.*

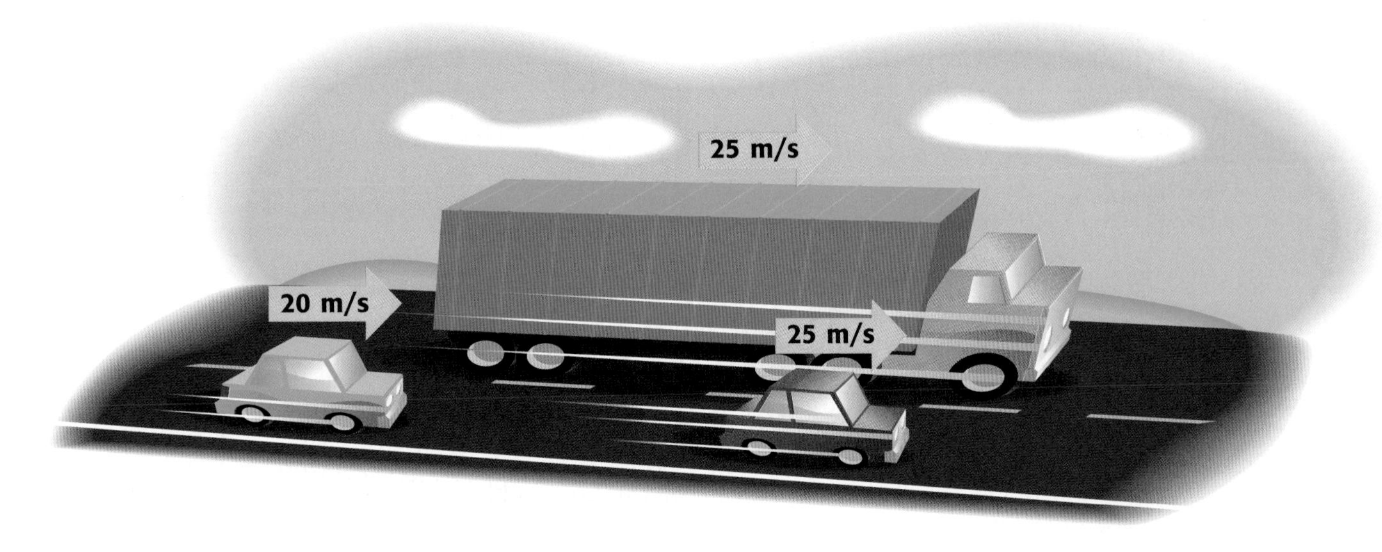

Figure 4 *The stored potential energy of the bow and string allows them to do work on the arrow when the string is released.*

Potential Energy Is Energy of Position

Not all energy involves motion. **Potential energy** is the energy an object has because of its position or shape. For example, the stretched bow shown in **Figure 4** has potential energy. The bow is not moving, but it has energy because work has been done to change its shape. A similar example of potential energy is in a stretched rubber band.

Gravitational Potential Energy Depends on Weight and Height When you lift an object, you do work on it by using a force that opposes gravitational force. As a result, you give that object *gravitational potential energy*. Books that you put on a bookshelf have gravitational potential energy, as does your backpack after you lift it onto your back. As you can see in **Figure 5,** the amount of gravitational potential energy an object has depends on its weight and its distance above Earth's surface.

Figure 5 The Effects of Weight and Height on Gravitational Potential Energy

The diver on the left weighs less and therefore has less gravitational potential energy than the diver on the right. The diver on the left did less work to climb up the platform and will do less work on the water.

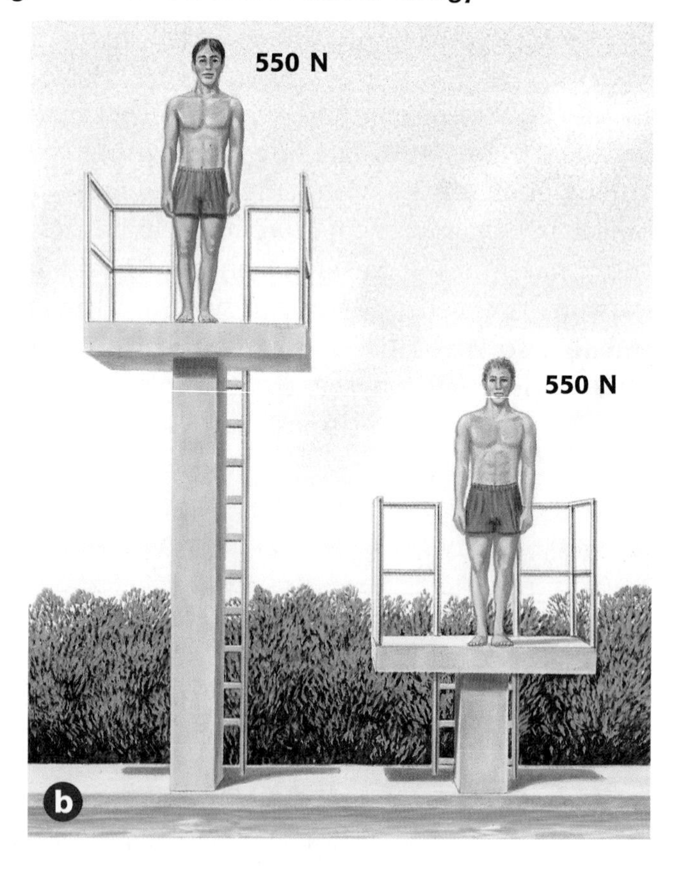

The diver on the higher platform has more gravitational potential energy than the diver on the lower platform. The diver on the higher platform did more work to climb up to the platform and will do more work on the water.

Calculating Gravitational Potential Energy You can calculate gravitational potential energy by using the following equation:

$$\text{Gravitational potential energy} = \text{weight} \times \text{height}$$

Because weight is expressed in newtons and height is expressed in meters, gravitational potential energy is expressed in newton-meters (N•m), or joules (J). So a 25 N object at a height of 3 m has 25 N × 3 m = 75 J of gravitational potential energy.

Recall that work = force × distance. Weight is the amount of force you must exert on an object in order to lift it, and height is a distance. So calculating an object's gravitational potential energy is done by calculating the amount of work done on the object to lift it to a given height. You can practice calculating gravitational potential energy as well as kinetic energy in the MathBreak at right.

÷ 5 ÷ Ω ≤ ∞ + Ω √ 9 ∞ ≤ Σ 2 +

MATH BREAK

Calculating Energy

1. What is the kinetic energy of a 4,000 kg elephant running at 3 m/s? at 4 m/s?
2. If you lift a 50 N watermelon to the top of a 2 m refrigerator, how much gravitational potential energy do you give the watermelon?

Mechanical Energy Sums It All Up

How would you describe the energy of the juggler's pins in **Figure 6**? Well, to describe their total energy, you would describe their mechanical energy. **Mechanical energy** is the total energy of motion and position of an object. Mechanical energy can be all potential energy, all kinetic energy, or some of both. The following equation defines mechanical energy as the sum of kinetic and potential energy:

$$\text{Mechanical energy} = \text{potential energy} + \text{kinetic energy}$$

When potential energy increases (or decreases), kinetic energy has to decrease (or increase) in order for mechanical energy to remain constant. So the amount of an object's kinetic or potential energy may change, but its mechanical energy remains the same. You'll learn more about these changes in the next section.

Figure 6 *As a pin is juggled, its mechanical energy is the sum of its potential energy and its kinetic energy at any point.*

REVIEW

1. How are energy and work related?
2. What is the difference between kinetic and potential energy?
3. **Applying Concepts** Explain why a high-speed collision might cause more damage to vehicles than a low-speed collision.

Forms of Energy

All energy involves either motion or position. But energy takes different forms. These forms of energy include thermal, chemical, electrical, sound, light, and nuclear energy. In the next few pages, you will learn how the different forms of energy relate to kinetic and potential energy.

Thermal Energy All matter is made of particles, atoms and molecules, that are constantly in motion. Because the particles are in motion, they have kinetic energy. *Thermal energy* is the total kinetic energy of the particles that make up an object. The more kinetic energy the particles have, the more thermal energy the object has. At higher temperatures, particles of matter move faster. The faster the particles move, the more kinetic energy they have, and the greater the object's thermal energy is. Look at **Figure 7.** Which substance do you think has the most thermal energy? The answer might surprise you. Thermal energy also depends on the number of particles in a substance.

Figure 7 *The particles in the steam have the most kinetic energy, but the ocean has the most thermal energy because it contains the most particles.*

The particles in an ice cube vibrate in fixed positions and therefore do not have a lot of kinetic energy.

The particles in ocean water move around and therefore have more kinetic energy than the particles in an ice cube.

The particles in steam move around rapidly and therefore have more kinetic energy than the particles in ocean water.

Chemical Energy What is the source of the energy in food? Food consists of chemical compounds. When compounds, such as the sugar in some foods, are formed, work is done to join, or bond, the different atoms together to form molecules. *Chemical energy* is the energy of a compound that changes as its atoms are rearranged to form new compounds. Chemical energy is a form of potential energy. Some molecules that have many atoms bonded together, such as gasoline, have a lot of chemical energy. In **Figure 8** on the next page, you can see an example of chemical energy.

Figure 8 Examples of Chemical Energy

Electrical Energy The electrical outlets in your home allow you to use electrical energy. *Electrical energy* is the energy of moving electrons. Electrons are the negatively charged particles of atoms. An atom is the smallest particle into which an element can be divided.

Suppose you plug an electrical device, such as the portable stereo shown in **Figure 9,** into an outlet and turn it on. The electrons in the wires will move back and forth, changing directions 120 times per second. As they do, energy is transferred to different parts within the stereo. The electrical energy created by moving electrons is used to do work. The work of a stereo is to produce sound.

The electrical energy available to your home is produced at power plants. Huge generators rotate magnets within coils of wire to produce electrical energy. Because the electrical energy results from the changing position of the magnet, electrical energy can be considered a form of potential energy. As soon as a device is plugged into an outlet and turned on, electrons move back and forth within the wires of the cord and within parts of the device. So electrical energy can also be considered a form of kinetic energy.

Figure 9 *The movement of electrons produces the electrical energy that a stereo uses to produce sound.*

Figure 10 *As the guitar strings vibrate, they cause particles in the air to vibrate. These vibrations transmit energy through air.*

Sound Energy You probably know that your vocal cords determine the sound of your voice. When you speak, air passes through your vocal cords, making them vibrate, or move back and forth. *Sound energy* is caused by an object's vibrations. Take a look at **Figure 10** to see how a vibrating object transmits energy through the air surrounding it.

Sound energy is a form of potential as well as kinetic energy. To make an object vibrate, work must be done to change its position. For example, when you pluck a guitar string, you stretch it and release it almost at the same time. The stretching changes the string's position. As a result, the string stores potential energy. In the release, the string uses its potential energy to move back to its original position. The moving guitar string has kinetic energy, which the string uses to do work on the air particles around it. The air particles vibrate and transmit this kinetic energy from particle to particle until the energy reaches your ears. When the vibrating air particles cause your eardrum to vibrate, you hear the sound of the guitar. Try the QuickLab on this page to learn more about sound energy.

QuickLab

Hear That Energy!

1. Make a simple drum by covering the open end of an **empty coffee can** with **wax paper.** Secure the wax paper with a **rubber band.**
2. Using the eraser end of a **pencil,** tap lightly on the wax paper. In your ScienceLog, describe how the paper responds. What do you hear?
3. Repeat step 2, but tap the paper a bit harder. In your ScienceLog, compare your results with those of step 2.
4. Cover half of the wax paper with one hand and hold it still. Now use your pencil to tap the paper. What happened? How can you describe sound energy as a form of mechanical energy?

Light Energy As shown in **Figure 11,** light is actually much more than the light that we see. *Light energy* is produced by the vibrations of electrically charged particles. Like sound vibrations, light vibrations cause energy to be transmitted. But unlike sound, the vibrations that transmit light energy don't cause other particles to vibrate. In fact, light energy can be transmitted through a vacuum (the absence of matter)!

Figure 11 Examples of Light Energy

The energy used to cook food in a microwave is a form of light energy.

Radar systems that are used for air-traffic control use a form of light energy.

Nuclear Energy What form of energy can come from a tiny amount of matter, can be used to generate electrical energy, and gives the sun its energy? It's *nuclear* (NOO klee uhr) *energy,* the energy associated with changes in the nucleus (NOO klee uhs) of an atom. Nuclear energy is produced in two ways—when two or more nuclei (NOO klee IE) join together or when the nucleus of an atom splits apart.

In the sun, shown in **Figure 12,** hydrogen nuclei join together to make a larger helium nucleus. This reaction releases a huge amount of energy, which allows the sun to light and heat the Earth. But because this reaction can take place only at temperatures upwards of 100,000,000°C, it is not yet practical to duplicate on Earth.

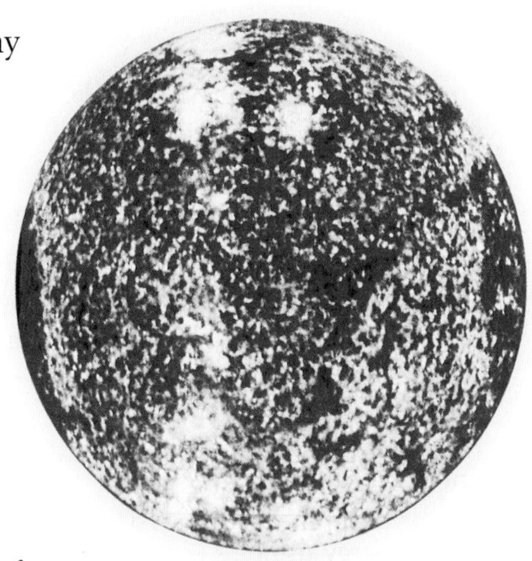

Figure 12 *Without the nuclear energy from the sun, life on Earth would not be possible.*

The nuclei of some atoms, such as uranium, store a lot of potential energy. When work is done to split these nuclei apart, that energy is released and can be used to do work. This type of nuclear energy is produced in nuclear reactors and is used to generate electrical energy at nuclear power plants, such as the one shown in **Figure 13.**

Figure 13 *In a nuclear power plant, small amounts of matter can produce large amounts of nuclear energy, which can be used to generate electrical energy.*

REVIEW

1. What determines an object's thermal energy?
2. Describe why chemical energy is a form of potential energy.
3. Explain how sound energy is produced when you beat a drum.
4. **Analyzing Relationships** When you hit a nail into a board using a hammer, the head of the nail gets warm. In terms of kinetic and thermal energy, describe why you think this happens.

Section 2

Energy Conversions

NEW TERMS

energy conversion

OBJECTIVES

- Describe an energy conversion.
- Give examples of energy conversions among the different forms of energy.
- Explain the role of machines in energy conversions.
- Explain how energy conversions make energy useful.

When you use a hammer to pound a nail into a board, you transfer your kinetic energy to the hammer, and the hammer transfers that kinetic energy to the nail. But energy is involved in other ways too. For example, sound energy is produced when you hit the nail. An energy transfer often leads to an **energy conversion,** a change from one form of energy into another. Any form of energy can be converted into any other form of energy, and often one form of energy is converted into more than one other form. In this section, you'll learn how energy conversions make your daily activities possible.

From Kinetic to Potential and Back

Take a look at **Figure 14.** Have you ever jumped on a trampoline? What types of energy are involved in this bouncing activity? Because you're moving when you jump, you have kinetic energy. And each time you jump into the air, you change your position with respect to the ground, so you also have gravitational potential energy. Another kind of potential energy is involved too—that of the trampoline stretching when you jump on it.

Figure 14 *Kinetic and potential energy are converted back and forth as you jump up and down on a trampoline.*

When you jump down, your kinetic energy is converted into the potential energy of the stretched trampoline.

The trampoline's potential energy is converted into kinetic energy, which is transferred to you, making you bounce up.

At the top of your jump, all of your kinetic energy has been converted into potential energy.

Right before you hit the trampoline, all of your potential energy has been converted back into kinetic energy.

Another example of the energy conversions between kinetic and potential energy is the motion of a pendulum (PEN dyoo luhm). Shown in **Figure 15,** a pendulum is a mass hung from a fixed point so that it can swing freely. When you lift the pendulum to one side, you do work on it, and the energy used to do that work is stored by the pendulum as potential energy. As soon as you let the pendulum go, it swings because the Earth exerts a force on it. The work the Earth does converts the pendulum's potential energy into kinetic energy.

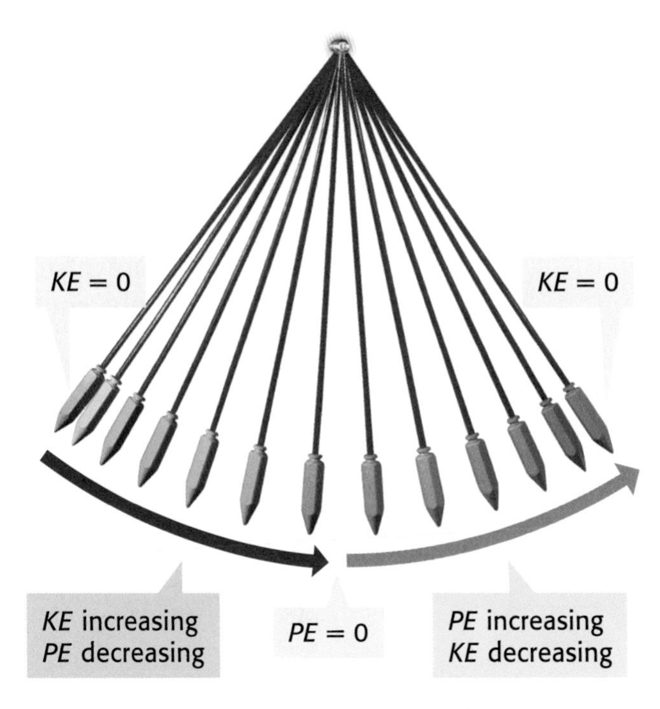

Figure 15 *A pendulum's mechanical energy is all kinetic (KE) at the bottom of its swing. At the top of its path, the pendulum's mechanical energy is all potential (PE).*

Self-Check

At what point does a roller coaster have the greatest potential energy? the greatest kinetic energy? *(See page 596 to check your answer.)*

Conversions Involving Chemical Energy

You've probably heard the expression "Breakfast is the most important meal of the day." What does this statement mean? Why does eating breakfast help you start the day? As your body digests food, it breaks down the bonds that hold the particles of food together. As a result of digestion, chemical energy is released and is available to you, as discussed in **Figure 16.**

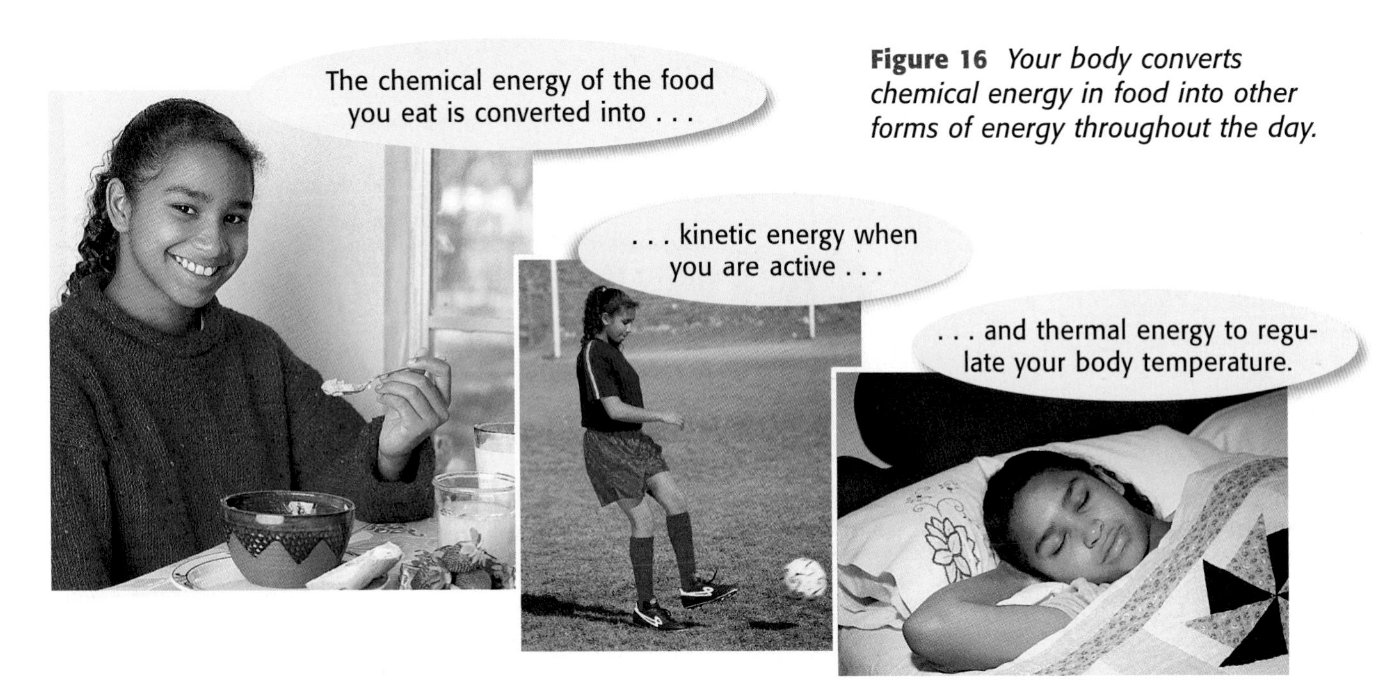

Figure 16 *Your body converts chemical energy in food into other forms of energy throughout the day.*

Would you believe that the chemical energy in the food you eat is a result of the sun's energy? It's true! When you eat fruits, vegetables, grains, or meat from animals that ate fruits, vegetables, or grains, you are taking in chemical energy that resulted from a chemical change involving the sun's energy. As shown in **Figure 17,** photosynthesis (FOH toh SIN thuh sis) uses light energy to produce new substances with chemical energy. In this way light energy is converted into chemical energy.

Figure 17 *Green plants use chlorophyll and light energy from the sun to produce the chemical energy in the food you eat.*

Photosynthesis

$$\text{carbon dioxide} + \text{water} \xrightarrow[\text{chlorophyll}]{\text{light energy}} \text{sugar} + \text{oxygen}$$

Light energy

Chlorophyll in green leaves

Carbon dioxide in the air

Sugar in food

Water in the soil

If you go camping, you probably use a stove, such as the one shown here, to prepare meals. Describe some of the energy conversions that take place when lighting the stove, cooking the food, eating the prepared meal, and then setting out on a long hike.

Conversions Involving Electrical Energy

You use electrical energy all the time—when you listen to the radio, when you make toast, and when you take a picture with a camera. Electrical energy can be easily converted into other forms of energy. **Figure 18** shows how electrical energy is converted in a hair dryer.

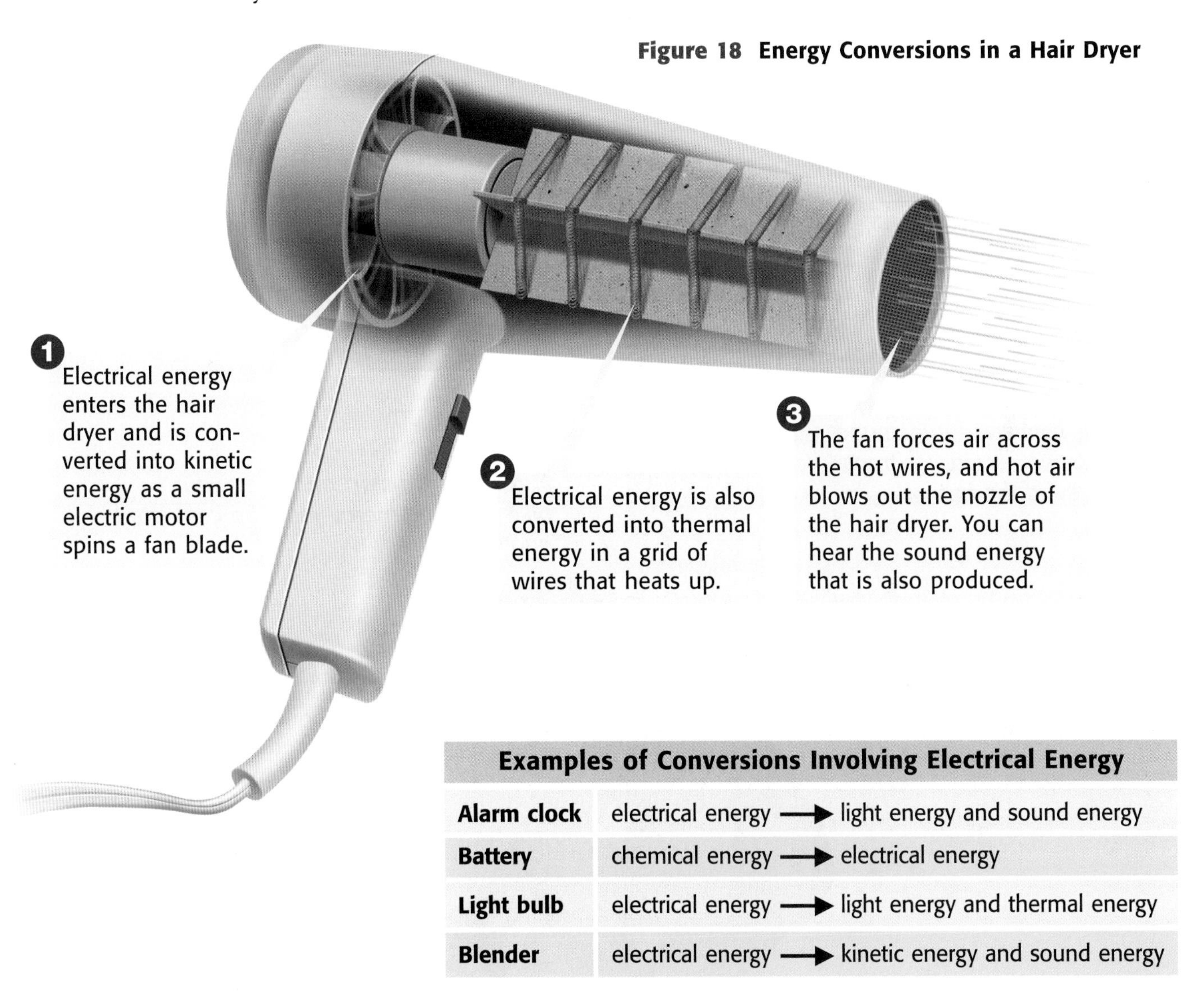

Figure 18 Energy Conversions in a Hair Dryer

Examples of Conversions Involving Electrical Energy	
Alarm clock	electrical energy → light energy and sound energy
Battery	chemical energy → electrical energy
Light bulb	electrical energy → light energy and thermal energy
Blender	electrical energy → kinetic energy and sound energy

REVIEW

1. What is an energy conversion?
2. Describe an example in which electrical energy is converted into thermal energy.
3. Describe an energy conversion involving chemical energy.
4. **Applying Concepts** Describe the kinetic-potential energy conversions that occur when you bounce a basketball.

Energy and Machines

You've been learning about energy, its different forms, and how it can undergo conversions. Another way to learn about energy is to look at how machines use energy. A machine can make work easier by changing the size or direction (or both) of the force required to do the work. Suppose you want to crack open a walnut. Using a nutcracker, like the one shown in **Figure 19,** would be much easier (and less painful) than using your fingers. You transfer your energy to the nutcracker, and it transfers energy to the nut. But the nutcracker will not transfer more energy to the nut than you transfer to the nutcracker. In addition, some of the energy you transfer to a machine can be converted by the machine into other forms of energy. Another example of how energy is used by a machine is shown in **Figure 20.**

Figure 19 *Some of the kinetic energy you transfer to a nutcracker is converted into sound energy as the nutcracker transfers energy to the nut.*

Figure 20 *To start and keep your bike moving, energy must be converted and transferred.*

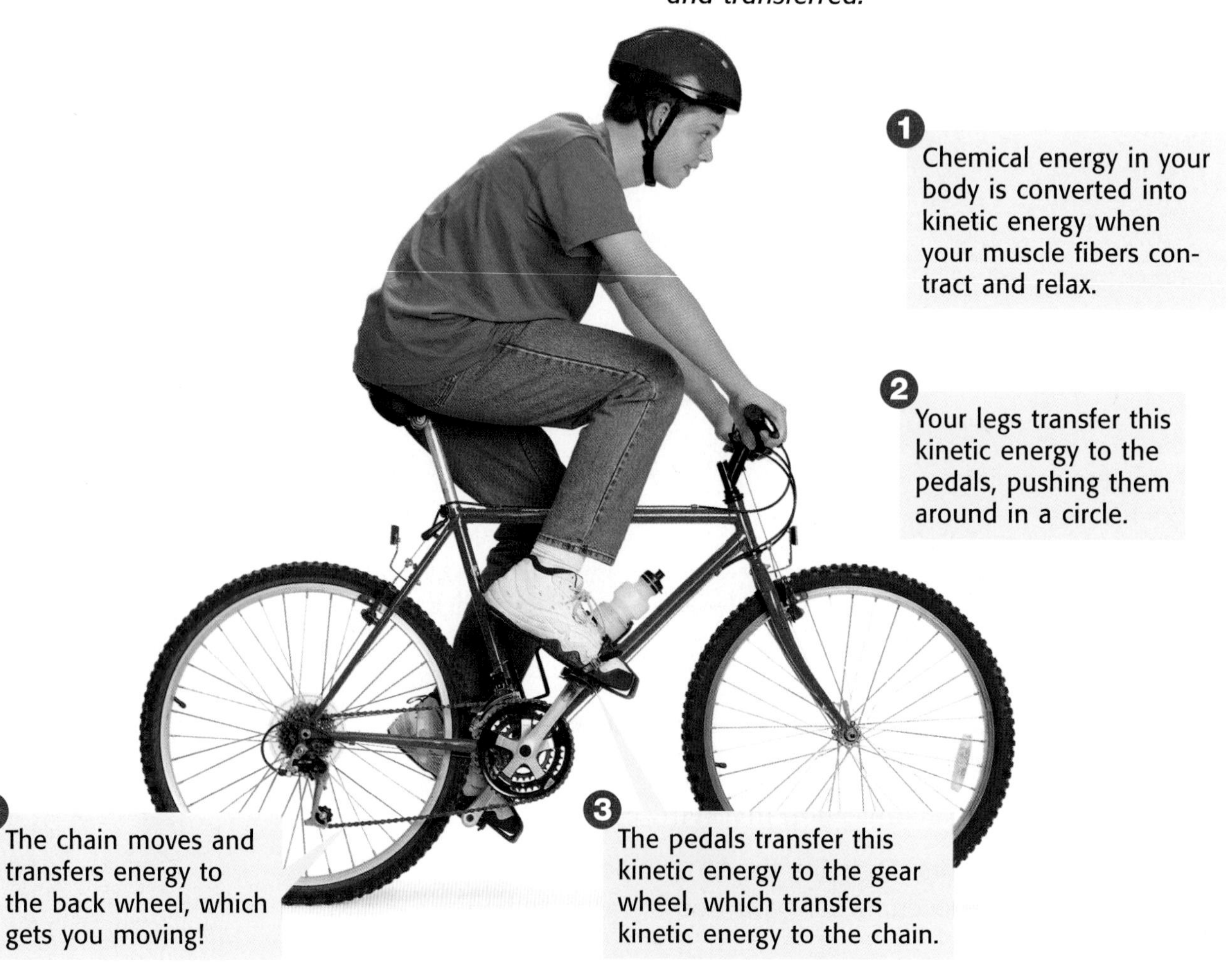

Machines Are Energy Converters As you saw in the examples on the previous page, when machines transfer energy, energy conversions can often result. For example, you can hear the sounds that your bike makes when you pedal it, change gears, or brake swiftly. That means that some of the kinetic energy being transferred gets converted into sound energy as the bike moves. Some machines are especially useful because they are energy converters. **Figure 21** shows an example of a machine specifically designed to convert energy from one form to another. In addition, the chart at right lists other machines that perform useful energy conversions.

Some Machines that Convert Energy	
▪ electric motor	▪ microphone
▪ windmill	▪ toaster
▪ doorbell	▪ dishwasher
▪ gas heater	▪ lawn mower
▪ telephone	▪ clock

Figure 21 *The continuous conversion of chemical energy into thermal energy and kinetic energy in a car's engine is necessary to make a car move.*

1 A mixture of gasoline and air enters the engine as the piston moves downward.

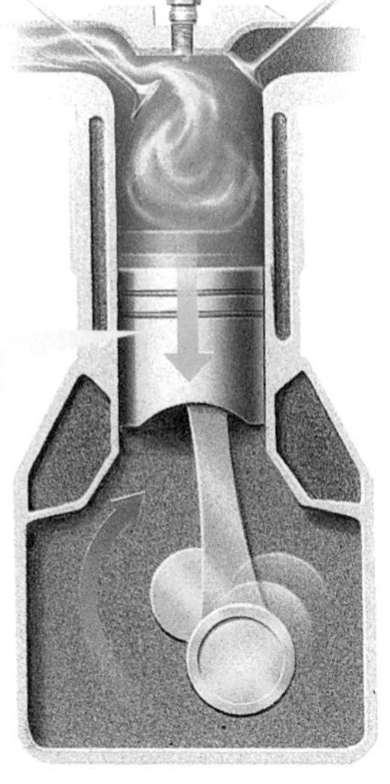

2 The kinetic energy of the crankshaft raises the piston, and the gasoline mixture is forced up toward the spark plug, which uses electrical energy to ignite the gasoline mixture.

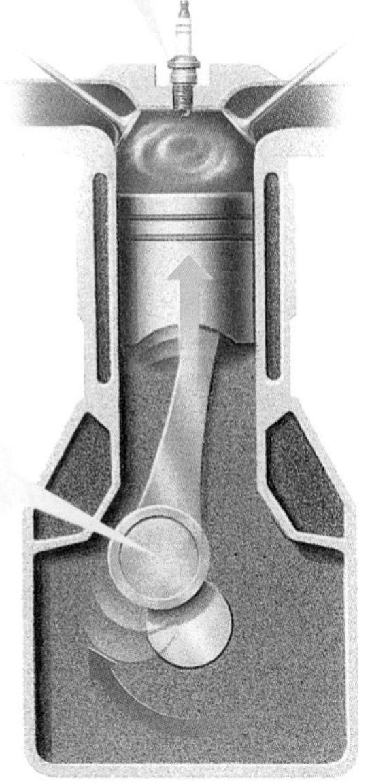

3 As the gasoline mixture burns, chemical energy is converted into thermal energy and kinetic energy, forcing the piston back down.

4 The kinetic energy of the crankshaft forces the piston up again, pushing exhaust gases out. Then the cycle repeats.

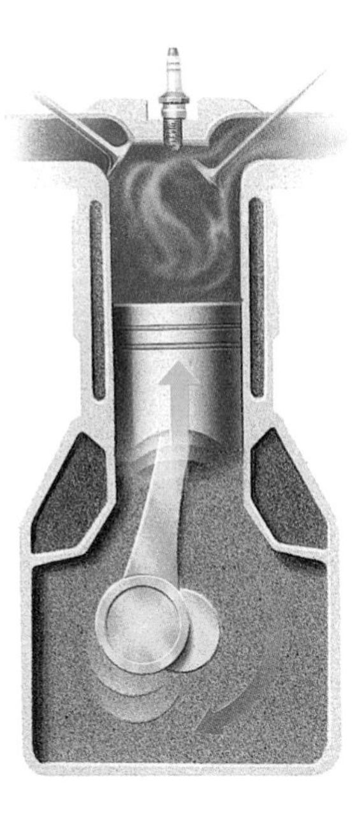

Why Energy Conversions Are Important

Everything we do is related to energy conversions. Heating our homes, getting from one place to another, reading at night, obtaining energy from a meal, growing plants, and many other activities all require energy conversions.

Figure 22 *In a wind turbine, the kinetic energy of the wind can be collected and converted into electrical energy. An electric stove can then convert that electrical energy into thermal energy that cooks your food.*

Making Energy Useful You can think of energy conversions as a way of getting energy in the form that you need. In addition, machines help harness existing energy and make that energy work for you. For example, would you ever think that the wind could help you cook a meal? Well, a wind turbine, shown in **Figure 22,** can perform an energy conversion that would allow you to do just that.

Making Conversions Efficient You may be familiar with energy efficiency (e FISH uhn see) with respect to how cars use gasoline or how your home is heated. For example, a car may be considered energy efficient if it gets good gas mileage, and your home may be energy efficient if it is well insulated.

In terms of energy conversions, *energy efficiency* is a comparison of the amount of energy before a conversion with the amount of useful energy after a conversion. For example, the energy efficiency of a light bulb would be a comparison of the electrical energy going into it with the light energy coming out of it. The less electrical energy that is converted into thermal energy instead of into light energy, the more efficient the bulb. Not all of the energy in a conversion becomes useful energy. Just as work input is always greater than work output, energy input is also always greater than energy output. But the closer the energy output is to the energy input, the more efficient the conversion is. Making energy conversions more efficient is important because greater efficiency means less waste.

Turn to page 242 to find out about buildings that are energy efficient as well as environmentally friendly.

REVIEW

1. What is the role of machines in energy conversions?
2. Give an example of a machine that is an energy converter, and explain how the machine converts one form of energy to another.
3. **Applying Concepts** A car that brakes suddenly comes to a screeching halt. Is the sound energy produced in this conversion a useful form of energy? Explain your answer.

Section 3

Conservation of Energy

NEW TERMS

friction

law of conservation of energy

OBJECTIVES

- Explain how energy is conserved within a closed system.
- Explain the law of conservation of energy.
- Give examples of how thermal energy is always a result of energy conversion.
- Explain why perpetual motion is impossible.

Many roller coasters have a mechanism that pulls the cars up to the top of the first hill, but the cars are on their own the rest of the ride. As the cars go up and down the hills on the track, their potential energy is converted into kinetic energy and back again. But the cars never return to the same height they started from. Does that mean that energy gets *lost* somewhere along the way? Nope—it just gets converted into other forms of energy.

Where Does the Energy Go?

In order to find out where a roller coaster's original potential energy goes, you have to consider more than just the hills of the roller coaster. You have to consider friction too. **Friction** is a force that opposes motion between two surfaces that are touching. For the roller coaster to move, work must be done to overcome the friction between the cars' wheels and the coaster track and between the cars and the surrounding air. The energy used to do this work comes from the original amount of potential energy that the cars have on the top of the first hill. The need to overcome friction affects the design of a roller coaster track. In **Figure 23,** you can see that the second hill will always be shorter than the first.

When energy is used to overcome friction, some of the energy is converted into thermal energy. Some of the cars' potential energy is converted into thermal energy on the way down the first hill, and then some of their kinetic energy is converted into thermal energy on the way up the second hill. So energy isn't lost at all—it just undergoes a conversion.

Figure 23 *Due to friction, not all of the cars' potential energy (PE) is converted into kinetic energy (KE) as the cars go down the first hill. In addition, not all of the cars' kinetic energy is converted into potential energy as the cars go up the second hill.*

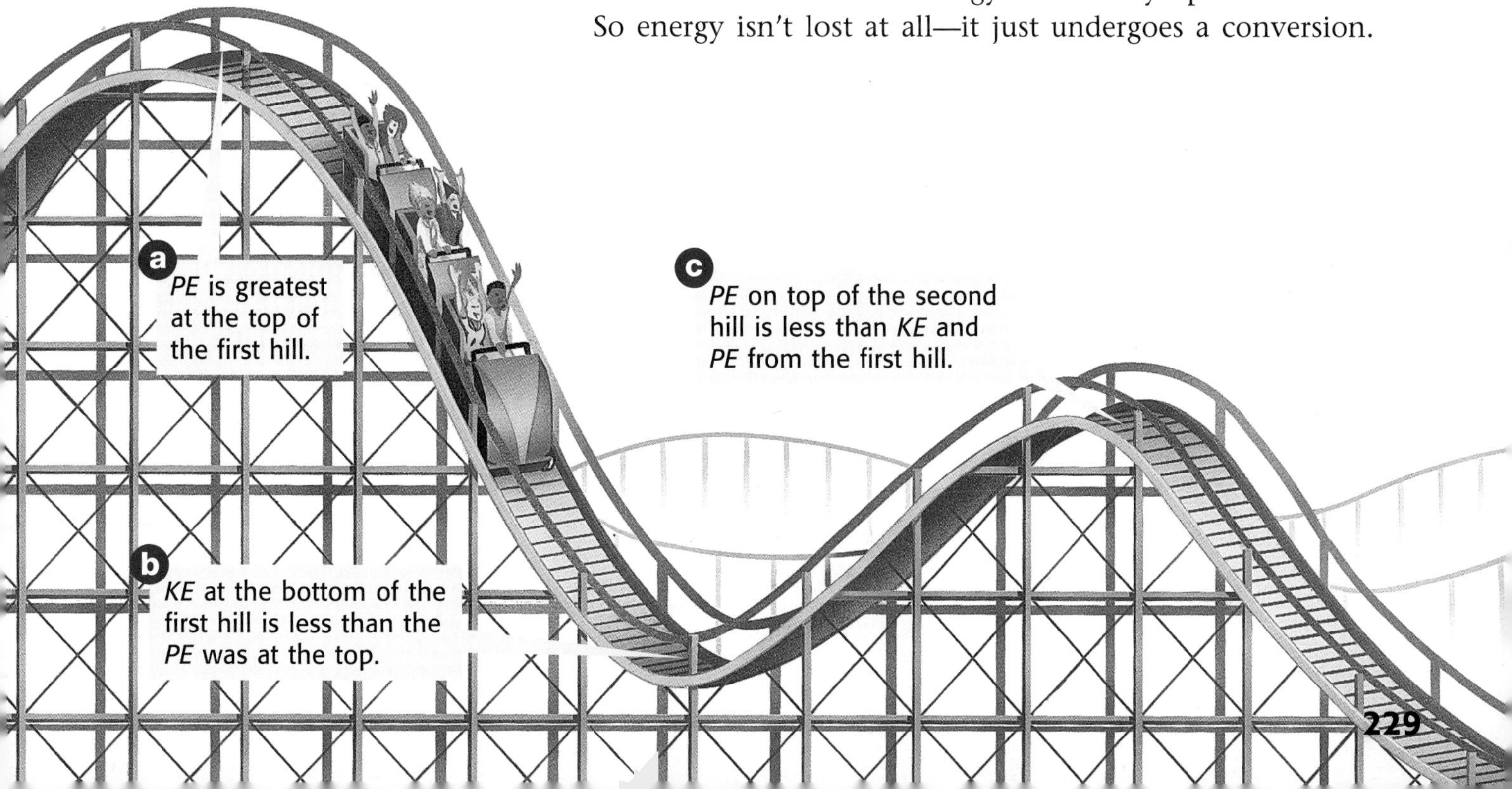

Energy Is Conserved Within a Closed System

A *closed system* is a well-defined group of objects that transfer energy between one another. For example, a closed system that involves a roller coaster consists of the track, the cars, and the surrounding air. On a roller coaster, some mechanical energy (the sum of kinetic and potential energy) is always converted into thermal energy because of friction. Sound energy is also a result of the energy conversions in a roller coaster. You can understand that energy is not lost on a roller coaster only when you consider all of the factors involved in a closed system. If you add together the cars' kinetic energy at the bottom of the first hill, the thermal energy due to overcoming friction, and the sound energy produced, you end up with the same total amount of energy as the original amount of potential energy. In other words, energy is conserved.

LabBook

Try to keep an egg from breaking while learning more about the law of conservation of energy on page 561 in the LabBook.

Law of Conservation of Energy No situation has been found where energy is not conserved. Because this phenomenon is always observed during energy conversions, it is described as a law. According to the **law of conservation of energy,** energy can be neither created nor destroyed. The total amount of energy in a closed system is always the same. Energy can be changed from one form to another, but all the different forms of energy in a system always add up to the same total amount of energy, no matter how many energy conversions occur.

Consider the energy conversions in a light bulb, shown in **Figure 24.** You can define the closed system to include the outlet, the wires, and the parts of the bulb. While not all of the original electrical energy is converted into light energy, no energy is lost. At any point during its use, the total amount of electrical energy entering the light bulb is equal to the total amount of light and thermal energy that leaves the bulb. Energy is conserved.

Figure 24 Energy Conservation in a Light Bulb

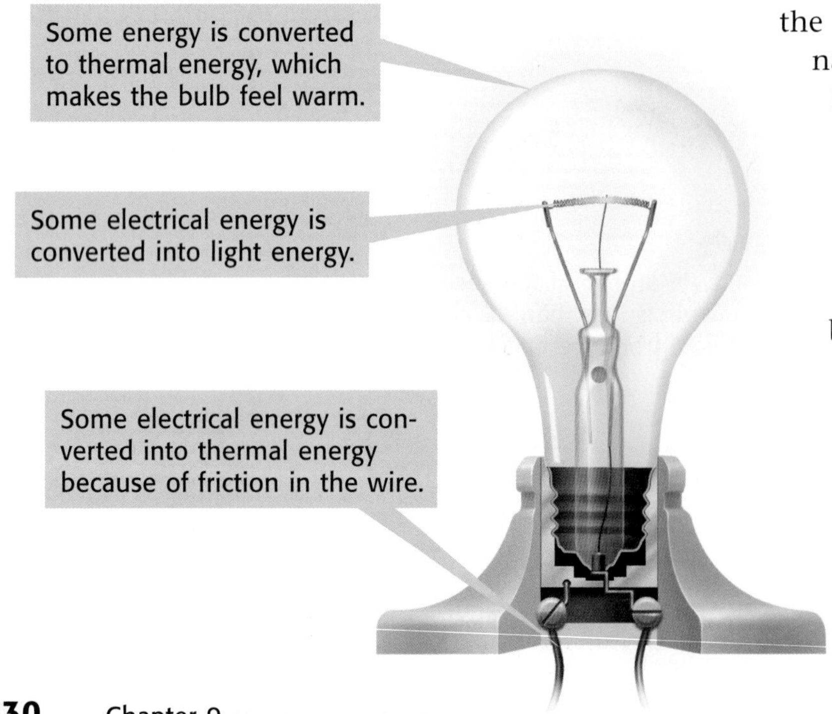

Some energy is converted to thermal energy, which makes the bulb feel warm.

Some electrical energy is converted into light energy.

Some electrical energy is converted into thermal energy because of friction in the wire.

No Conversion Without Thermal Energy

Did you notice that in the examples of closed systems mentioned so far—the roller coaster and the light bulb—some energy conversion took place that resulted in thermal energy? Any time one form of energy is converted into another form, some of the original energy always gets converted into thermal energy.

The thermal energy due to friction that results from energy conversions is not useful energy. That is, this thermal energy is not used to do work. Think about a car. You put gas into a car, but not all of the gasoline's chemical energy makes the car move. Some waste thermal energy will always result from the energy conversions. Much of this waste thermal energy exits a car engine through the radiator and the exhaust pipe.

Perpetual Motion? No Way! Throughout history, people have dreamed of constructing a machine that runs forever without any additional energy—a *perpetual* (puhr PECH oo uhl) *motion machine*. Such a machine would put out exactly as much energy as it takes in. But because some waste thermal energy always results from energy conversions, perpetual motion is impossible. Once a machine is started, its motion will continually decrease until all of its original energy is converted into thermal energy from friction and wear and tear. The only way a machine, like the device shown in **Figure 25,** can keep moving is to have a continuous supply of energy.

life science CONNECTION

Whenever you do work, you use chemical energy stored in your body that comes from food you've eaten. As you do work, some of that chemical energy is always converted into thermal energy. The more you work, the more energy you use. That's why you will feel your body heat up after performing a task, such as raking leaves, for several minutes.

a As water evaporates from the bird's wet head, fluid is drawn up from the tail, the head gets heavy, and the bird tips.

b After the bird "drinks," it flips upright, and the cycle repeats.

Figure 25 *The "drinking bird" seems to move perpetually, but thermal energy from the surrounding air is continually used to evaporate the water from its head. So it is* not *a perpetual motion machine.*

REVIEW

1. Describe the energy conversions that take place in a pendulum, and explain how energy is conserved.
2. Explain why some of the thermal energy released by a campfire is actually waste thermal energy.
3. Why is perpetual motion impossible?
4. **Analyzing Viewpoints** Imagine that you drop a ball. It bounces a few times, but then it stops. Your friend says that the ball has lost all of its energy. Using what you know about the law of conservation of energy, respond to your friend's statement.

Section 4

Energy Resources

NEW TERMS

energy resource
nonrenewable resources
fossil fuels
renewable resources

OBJECTIVES

- Name several energy resources.
- Explain how the sun is the ultimate source of energy on Earth.
- Evaluate the advantages and disadvantages of using various energy resources.

Large amounts of energy are used to light and warm our homes; to produce food, clothing, and other products; and to transport people and products from place to place. Where does all this energy come from? An **energy resource** is a natural resource that can be converted by humans into other forms of energy in order to do useful work. In this section, you will find out about several energy resources. You will also learn that the most important energy resource of all is the source responsible for most other energy resources—the sun.

Nonrenewable Resources

When an energy resource is used, it is converted into other forms of energy. Some energy resources, called **nonrenewable resources,** cannot be replaced after they are used or can be replaced only over thousands or millions of years. Fossil fuels are the most important nonrenewable resources.

Fossil Fuels Coal, petroleum, and natural gas, shown in **Figure 26,** are the most common examples of fossil fuels. **Fossil fuels** are energy resources that formed from the buried remains of plants and animals that lived millions of years ago. These plants captured and stored energy from the sun by photosynthesis. Then the animals used and stored this energy by eating the plants or by eating animals that ate plants. So fossil fuels are actually super-concentrated forms of the sun's energy.

Figure 26 Formation of Fossil Fuels

This piece of coal containing a fossil of a fern shows that coal formed from the remains of plants that lived in swamps millions of years ago.

Also called oil, petroleum was formed from the remains of organisms that lived in shallow prehistoric lakes and seas. Crushed by heavy layers of sediment and heated by the Earth, the remains were slowly changed into petroleum.

Natural gas was formed much in the same way that petroleum was formed, and it is often found along with petroleum deposits within layers of sedimentary rock.

Now, millions of years later, energy from the sun is released when fossil fuels are burned. A small amount of any fossil fuel contains a large amount of stored energy from the sun that can be converted into large amounts of other types of energy. The information below shows how important fossil fuels are to our society.

Coal

Most coal used in the United States is burned by power plants to produce steam to run electric generators.

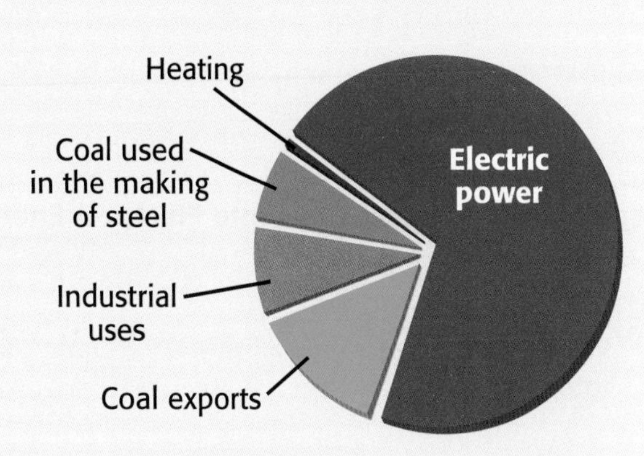

Petroleum

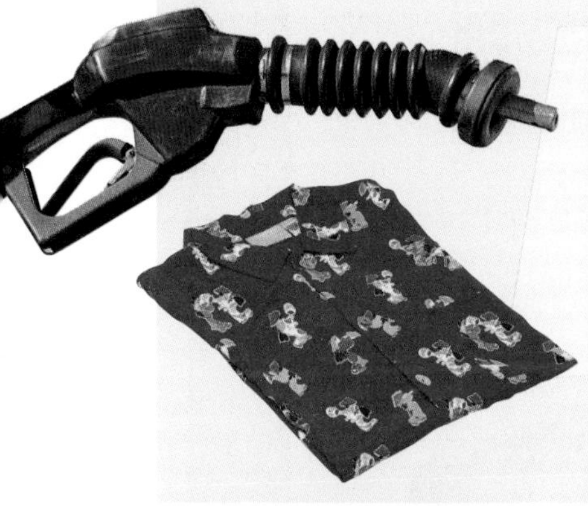

Petroleum supplies us with products such as gasoline, kerosene, and wax as well as petrochemicals, which are used to make plastics and synthetic fibers, such as rayon.

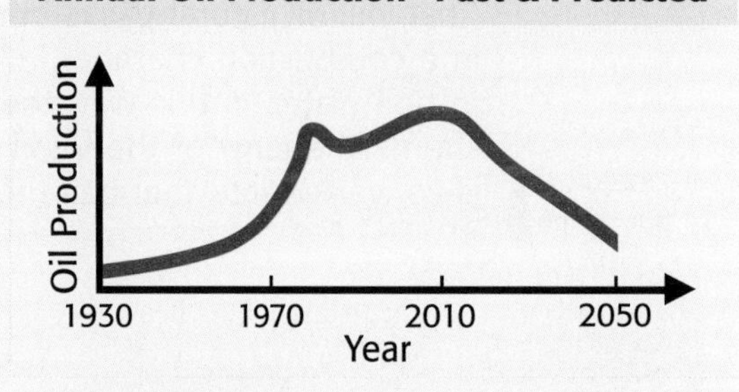

Finding alternative energy resources will become more important in years to come.

Natural Gas

Natural gas is widely used in heating systems in buildings and houses. It is also used in stoves and ovens and in vehicles as an alternative to gasoline.

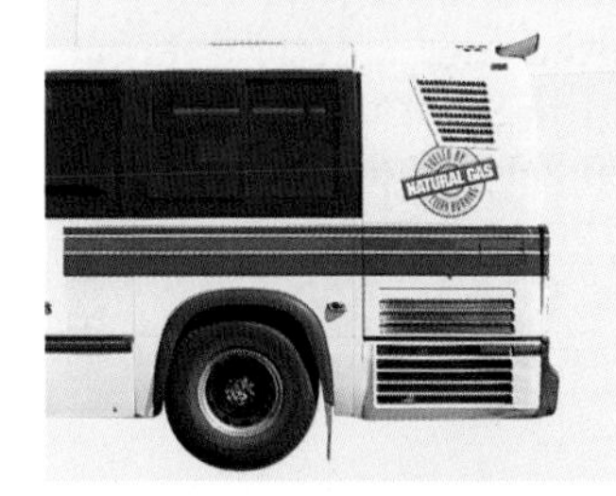

Natural gas is the cleanest burning fossil fuel.

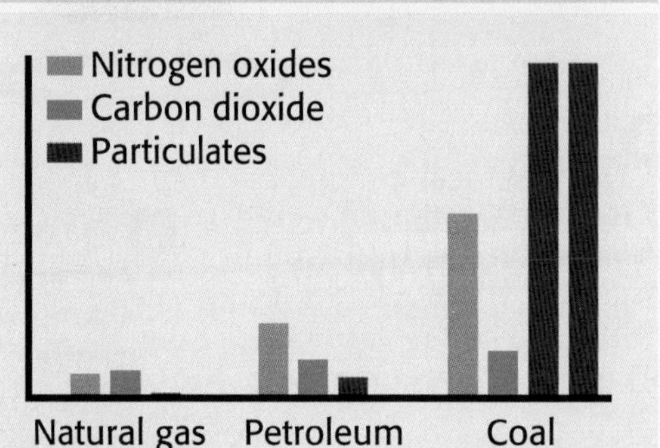

Turn to page 243 to read about a day in the life of a power-plant manager.

Electrical Energy from Fossil Fuels One way to generate electrical energy is to burn fossil fuels. In fact, fossil fuels are the primary source of electrical energy generated in the United States. Earlier in this chapter, you learned that electrical energy can result from energy conversions. Kinetic energy is converted into electrical energy by an *electric generator.* This energy conversion is part of a larger process, shown in **Figure 27,** of converting the chemical energy in fossil fuels into the electrical energy you use every day.

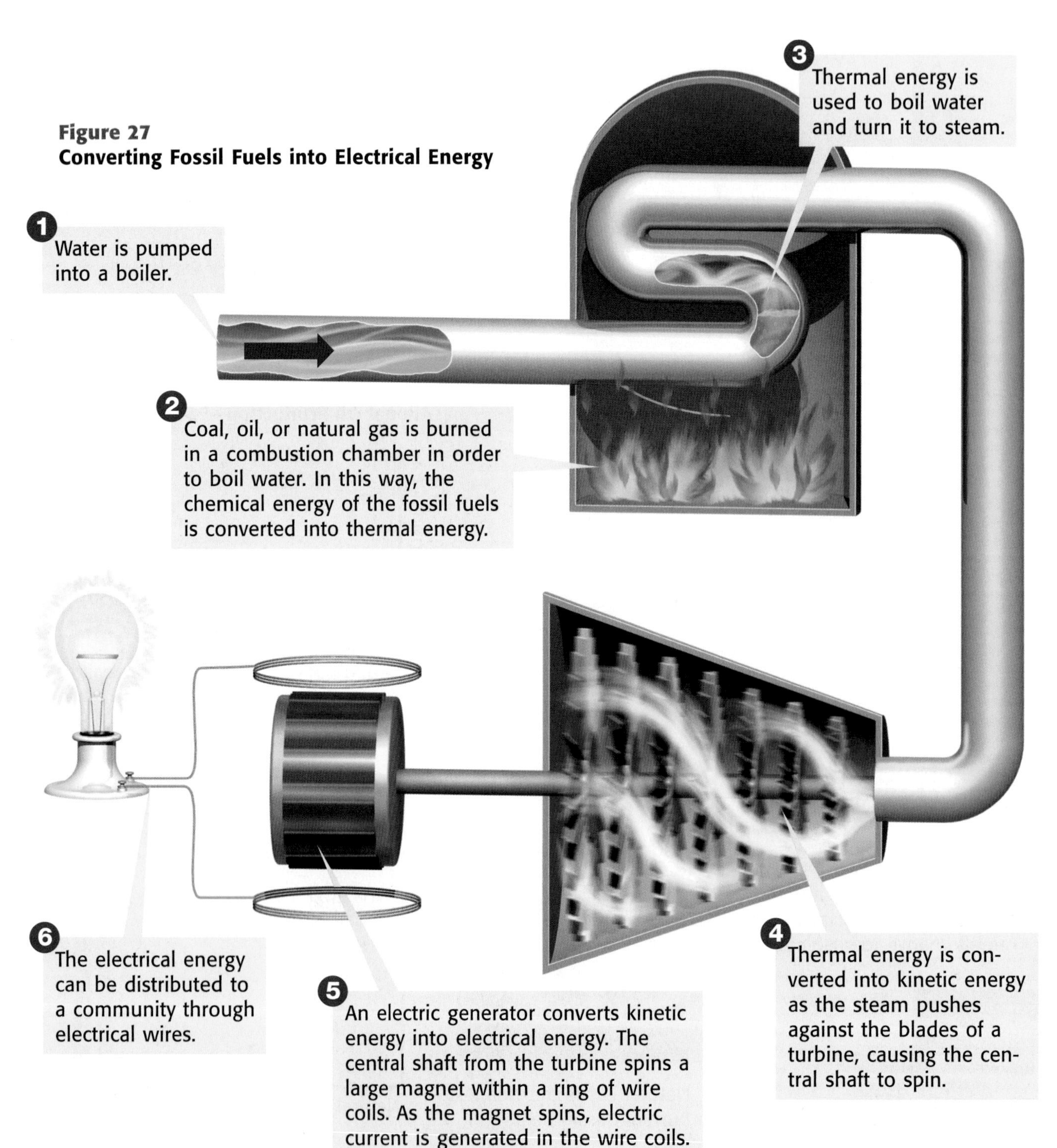

Figure 27
Converting Fossil Fuels into Electrical Energy

Nuclear Energy Another way to generate electrical energy is to use nuclear energy. Like fossil-fuel power plants, a nuclear power plant generates thermal energy that boils water to produce steam. The steam then turns a turbine, which rotates a generator that converts kinetic energy into electrical energy. However, the fuels used in nuclear power plants are different from fossil fuels. Nuclear energy is generated from radioactive elements, such as uranium, shown in **Figure 28.** In a process called *nuclear fission* (FISH uhn), the nucleus of a uranium atom is split into two smaller nuclei, releasing nuclear energy. Because the supply of these elements is limited, nuclear energy can be thought of as a nonrenewable resource.

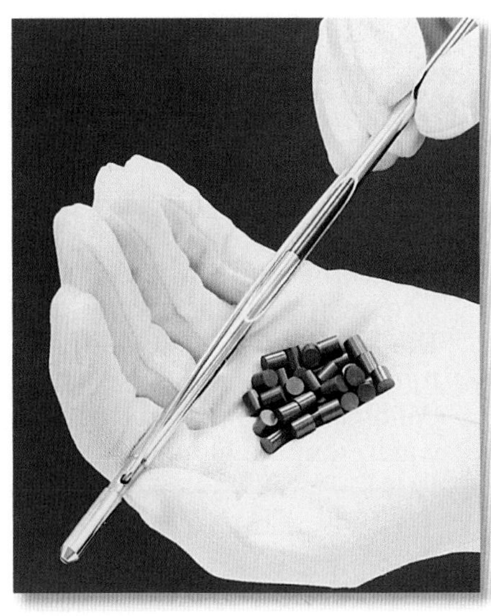

Figure 28 *Nuclear energy is an extremely concentrated energy resource. A single uranium fuel pellet contains the energy equivalent of about 1 metric ton of coal.*

Renewable Resources

Some energy resources, called **renewable resources,** can be used and replaced in nature over a relatively short period of time. Some renewable resources, such as solar energy and wind energy, are considered practically limitless.

Solar Energy

Sunlight can be converted into electrical energy through solar cells, which can be used in devices such as calculators or installed in a home to provide electrical energy.

Some houses allow sunlight into the house through large windows. The sunlight is converted into thermal energy that heats the house naturally.

Energy from Water

The sun causes water to evaporate and fall again as rain that flows through rivers. The potential energy of water in a reservoir is converted into kinetic energy as the water flows downhill through a dam.

Falling water turns a turbine in a dam, which is connected to a generator that converts kinetic energy into electrical energy. Electrical energy produced from falling water is called *hydroelectricity.*

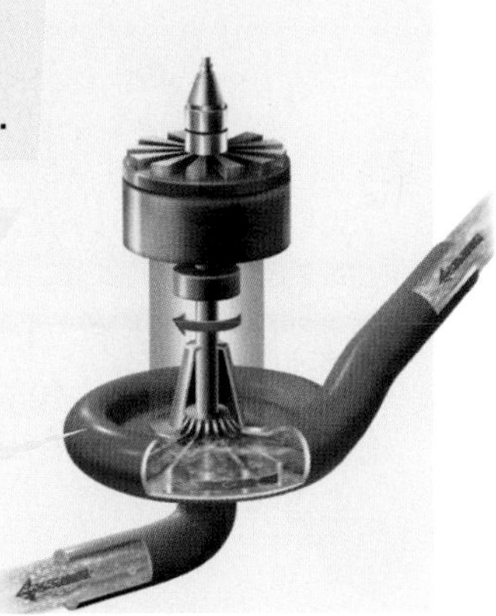

Wind Energy

Wind is caused by the sun's uneven heating of the Earth's surface, which creates currents of air. The kinetic energy of wind can turn the blades of a windmill. Windmills are often used to pump water from the ground.

A wind turbine converts kinetic energy into electrical energy by rotating a generator.

Geothermal Energy

Thermal energy resulting from the heating of Earth's crust is called *geothermal energy.* Ground water that seeps into hot spots near the surface of the Earth can form geysers.

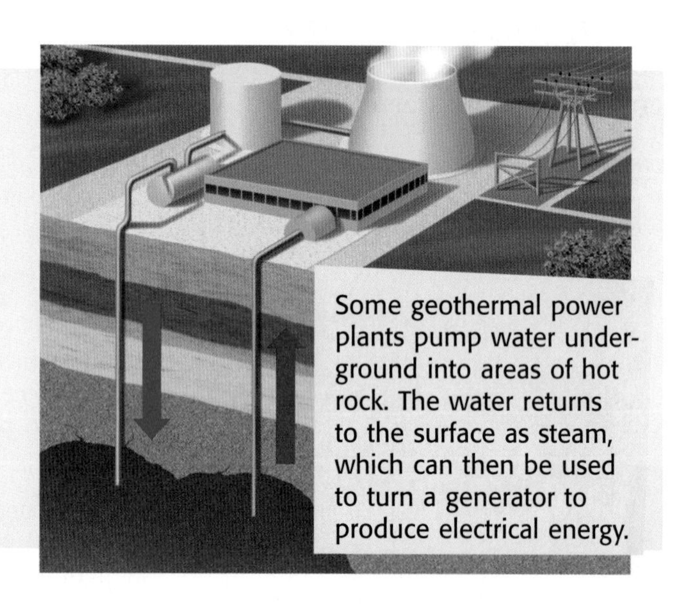

Some geothermal power plants pump water underground into areas of hot rock. The water returns to the surface as steam, which can then be used to turn a generator to produce electrical energy.

Biomass

Plants capture and store energy from the sun. Organic matter, such as plants, wood, and waste, that can be burned to release energy is called *biomass.* Nonindustrialized countries rely heavily on biomass for energy.

Certain plants can also be converted into liquid fuel. For example, corn can be used to make ethanol, which is often mixed with gasoline to make a cleaner-burning fuel for cars.

The Two Sides to Energy Resources

The table below shows how energy needs vary depending on where you live, what you need energy for, and how much you need. Sometimes one energy resource is a better choice than another.

Energy resource	Advantages	Disadvantages
Fossil fuels	▪ provide a large amount of thermal energy per unit of mass ▪ easy to get and easy to transport ▪ can be used to generate electrical energy and make products, such as plastic	▪ nonrenewable ▪ burning produces smog ▪ burning coal releases substances that can cause acid rain ▪ risk of oil spills
Nuclear	▪ very concentrated form of energy ▪ power plants do not produce smog	▪ produces radioactive waste ▪ radioactive elements are nonrenewable
Solar	▪ almost limitless source of energy ▪ does not produce pollution	▪ expensive to use for large-scale energy production ▪ only practical in sunny areas
Water	▪ renewable ▪ does not produce pollution	▪ dams disrupt a river's ecosystem ▪ available only in areas that have rivers
Wind	▪ relatively inexpensive to generate	▪ only practical in windy areas
Geothermal	▪ almost limitless source of energy ▪ power plants require little land	▪ only practical in locations near hot spots ▪ waste water can damage soil
Biomass	▪ renewable	▪ requires large areas of farmland ▪ produces smoke

REVIEW

1. How are fossil fuels and biomass similar? How are they different? Explain.
2. Why is nuclear energy a nonrenewable resource?
3. Trace electrical energy back to the sun.
4. **Interpreting Graphics** Use the pie chart at right to explain why renewable resources will become more important in years to come.

U.S. Energy Sources

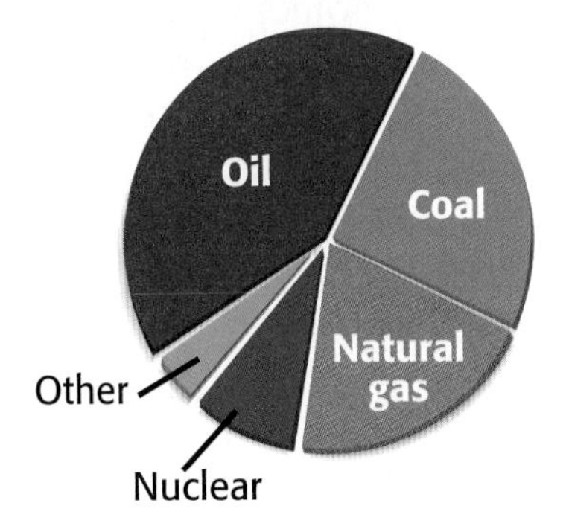

Chapter Highlights

SECTION 1

Vocabulary

energy *(p. 214)*
kinetic energy *(p. 215)*
potential energy *(p. 216)*
mechanical energy *(p. 217)*

Section Notes

- Energy is the ability to do work, and work is the transfer of energy. Both energy and work are expressed in joules.
- Kinetic energy is energy of motion and depends on speed and mass.
- Potential energy is energy of position or shape. Gravitational potential energy depends on weight and height.
- Mechanical energy is the sum of kinetic energy and potential energy.
- Thermal energy, sound energy, electrical energy, and light energy can all be forms of kinetic energy.
- Chemical energy, electrical energy, sound energy, and nuclear energy can all be forms of potential energy.

SECTION 2

Vocabulary

energy conversion *(p. 222)*

Section Notes

- An energy conversion is a change from one form of energy to another. Any form of energy can be converted into any other form of energy.
- Machines can transfer energy and convert energy into a more useful form.
- Energy conversions help to make energy useful by changing energy into the form you need.

Labs

Finding Energy *(p. 558)*
Energy of a Pendulum *(p. 560)*

☑ Skills Check

Math Concepts

GRAVITATIONAL POTENTIAL ENERGY To calculate an object's gravitational potential energy, multiply the weight of the object by its height above the Earth's surface. For example, the gravitational potential energy (*GPE*) of a box that weighs 100 N and that is sitting in a moving truck 1.5 m above the ground is calculated as follows:

$$GPE = \text{weight} \times \text{height}$$

$$GPE = 100\text{ N} \times 1.5\text{ m} = 150\text{ J}$$

Visual Understanding

POTENTIAL-KINETIC ENERGY CONVERSIONS When you jump up and down on a trampoline, potential and kinetic energy are converted back and forth. Review the picture of the pendulum on page 223 for another example of potential-kinetic energy conversions.

ENERGY RESOURCES Look back at the diagram on page 234. Converting fossil fuels into electrical energy requires several energy conversions.

SECTION 3

Vocabulary

friction *(p. 229)*

law of conservation of energy *(p. 230)*

Section Notes

- Because of friction, some energy is always converted into thermal energy during an energy conversion.
- Energy is conserved within a closed system. According to the law of conservation of energy, energy can be neither created nor destroyed.
- Perpetual motion is impossible because the energy put into a machine will eventually be converted completely into thermal energy due to friction.

Labs

Eggstremely Fragile *(p. 561)*

SECTION 4

Vocabulary

energy resource *(p. 232)*

nonrenewable resources *(p. 232)*

fossil fuels *(p. 232)*

renewable resources *(p. 235)*

Section Notes

- An energy resource is a natural resource that can be converted into other forms of energy in order to do useful work.
- Nonrenewable resources cannot be replaced after they are used or can only be replaced after long periods of time. They include fossil fuels and nuclear energy.
- Fossil fuels are nonrenewable resources formed from the remains of ancient organisms. Coal, petroleum, and natural gas are fossil fuels.
- Renewable resources can be used and replaced in nature over a relatively short period of time. They include solar energy, wind energy, energy from water, geothermal energy, and biomass.
- The sun is the ultimate source of energy on Earth.
- Depending on where you live and what you need energy for, one energy resource can be a better choice than another.

internetconnect

GO TO: go.hrw.com

Visit the **HRW** Web site for a variety of learning tools related to this chapter. Just type in the keyword:

KEYWORD: HSTENG

GO TO: www.scilinks.org

Visit the **National Science Teachers Association** on-line Web site for Internet resources related to this chapter. Just type in the ***sci*LINKS** number for more information about the topic:

TOPIC: What Is Energy? ***sci*LINKS NUMBER:** HSTP205

TOPIC: Forms of Energy ***sci*LINKS NUMBER:** HSTP210

TOPIC: Energy Conversions ***sci*LINKS NUMBER:** HSTP215

TOPIC: Nonrenewable Resources ***sci*LINKS NUMBER:** HSTP220

TOPIC: Renewable Resources ***sci*LINKS NUMBER:** HSTP225

Chapter Review

USING VOCABULARY

For each pair of terms, explain the difference in their meanings.

1. potential energy/kinetic energy
2. friction/energy conversion
3. energy conversion/law of conservation of energy
4. energy resources/fossil fuels
5. renewable resources/nonrenewable resources

UNDERSTANDING CONCEPTS

Multiple Choice

6. Kinetic energy depends on
   a. mass and volume.
   b. speed and weight.
   c. weight and height.
   d. speed and mass.

7. Gravitational potential energy depends on
   a. mass and speed.
   b. weight and height.
   c. mass and weight.
   d. height and distance.

8. Which of the following is not a renewable resource?
   a. wind energy
   b. nuclear energy
   c. solar energy
   d. geothermal energy

9. Which of the following is a conversion from chemical energy to thermal energy?
   a. Food is digested and used to regulate body temperature.
   b. Charcoal is burned in a barbecue pit.
   c. Coal is burned to boil water.
   d. all of the above

10. Machines can
    a. increase energy.
    b. transfer energy.
    c. convert energy.
    d. Both (b) and (c)

11. In every energy conversion, some energy is always converted into
    a. kinetic energy.
    b. potential energy.
    c. thermal energy.
    d. mechanical energy.

12. An object that has kinetic energy must be
    a. at rest.
    b. lifted above the Earth's surface.
    c. in motion.
    d. None of the above

13. Which of the following is *not* a fossil fuel?
    a. gasoline
    b. coal
    c. firewood
    d. natural gas

Short Answer

14. Name two forms of energy, and relate them to kinetic or potential energy.
15. Give three specific examples of energy conversions.
16. Explain how energy is conserved within a closed system.
17. How are fossil fuels formed?

Concept Mapping

18. Use the following terms to create a concept map: energy, machines, energy conversions, thermal energy, friction.

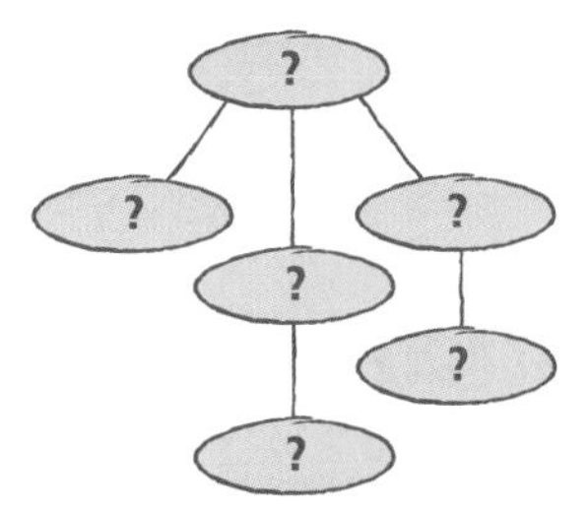

CRITICAL THINKING AND PROBLEM SOLVING

19. What happens when you blow up a balloon and release it? Describe what you would see in terms of energy.

20. After you coast down a hill on your bike, you eventually come to a complete stop unless you keep pedaling. Relate this to the reason why perpetual motion is impossible.

21. Look at the photo of the pole-vaulter below. Trace the energy conversions involved in this event, beginning with the pole-vaulter's breakfast of an orange-banana smoothie.

22. If the sun were exhausted of its nuclear energy, what would happen to our energy resources on Earth?

MATH IN SCIENCE

23. A box has 400 J of gravitational potential energy.
    a. How much work had to be done to give the box that energy?
    b. If the box weighs 100 N, how far was it lifted?

INTERPRETING GRAPHICS

24. Look at the illustration below, and answer the questions that follow.

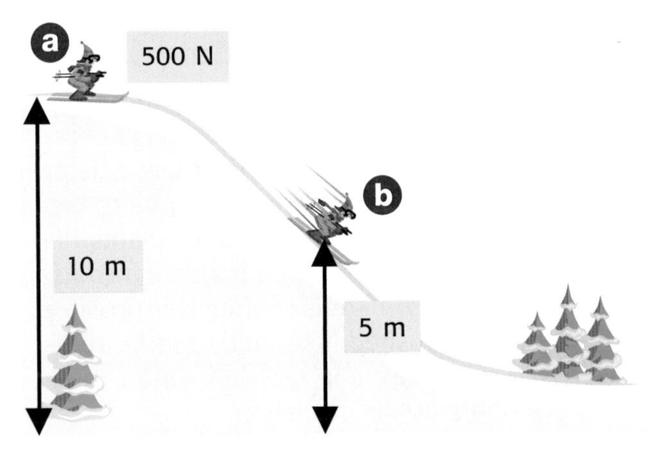

a. What is the skier's gravitational potential energy at point *A*?
b. What is the skier's gravitational potential energy at point *B*?
c. What is the skier's kinetic energy at point *B*? (Hint: mechanical energy = potential energy + kinetic energy.)

NOW What Do You Think?

Take a minute to review your answers to the ScienceLog questions on page 213. Have your answers changed? If necessary, revise your answers based on what you have learned since you began this chapter.

Green Buildings

How do you make a building green without painting it? You make sure it does as little damage to the environment as possible. *Green,* in this case, does not refer to the color of pine trees or grass. Instead, *green* means "environmentally safe." And the "green movement" is growing quickly.

Green Methods and Materials

One strategy that architects employ to turn a building green is to minimize its energy consumption. They also reduce water use wherever possible. One way to do this would be to create landscapes that use only native plants that require little watering. Green builders also use recycled building materials whenever possible. For example, crushed light bulbs can be recycled into floor tiles, and recycled cotton can replace fiberglass as insulation.

Seeing Green

Although green buildings cost more than conventional buildings to construct, they save a lot of money in the long run. For example, the Audubon Building, in Manhattan, saves $100,000 in maintenance costs every year—that is $60,000 in electricity bills alone! The building uses more than 60 percent less energy and electricity than a conventional building does. Inside, the workers enjoy natural lighting, cleaner air, and an environment that is free of unnecessary chemicals.

Some designers want to create buildings that are even more environmentally friendly than the Audubon Building. Walls can be made of straw bales or packed dirt, and landscapes can be maintained with rainwater collected from rooftops. By conserving, recycling, and reducing waste, green builders are doing a great deal to help the environment.

Design It Yourself!

▶ Design a building, a home, or even a doghouse that is made of only recycled materials. Be inventive! When you think you have the perfect design, create a scale model. Describe how your green structure saves resources.

◀ *The walls of this building are being made out of worn-out tires packed with soil. The walls will later be covered with stucco.*

CAREERS

POWER-PLANT MANAGER

As a power-plant manager, **Cheryl Mele** is responsible for almost a billion watts of electric power generation at the Decker Power Plant in Austin, Texas. More than 700 MW are produced using a steam-driven turbine system with natural gas fuel and oil as a backup fuel. Another 200 MW are generated by gas turbines. The steam-driven turbine system and gas turbines together provide enough electrical energy for many homes and businesses.

According to Cheryl Mele, her job as plant manager includes "anything that needs doing." Her training as a mechanical engineer allows her to conduct routine testing and to diagnose problems successfully. A firm believer in protecting our environment, Mele operates the plant responsibly. Mele states, "It is very important to keep the plant running properly and burning as efficiently as possible." Her previous job helping to design more-efficient gas turbines helped make her a top candidate for the job of plant manager.

The Team Approach

Mele uses the team approach to maintain the power plant. She says, "We think better as a team. We all have areas of expertise and interest, and we maximize our effectiveness." Mele observes that working together makes everyone's job easier.

Advice to Young People

Mele believes that mechanical engineering and managing a power plant are interesting careers because you get to work with many exciting new technologies. These professions are excellent choices for both men and women. In these careers you interact with creative people as you try to improve mechanical equipment to make it more efficient and reduce harm to the environment. Her advice for young people is to pursue what interests you. "Be sure to connect the math you learn to the science you are doing," she says. "This will help you to understand both."

A Challenge

► With the help of an adult, find out how much electrical energy your home uses each month. How many homes like yours could Mele's billion-watt power plant supply energy to each month?

► *Cheryl Mele manages the Decker Power Plant in Austin, Texas.*

CHAPTER 10

The Energy of Waves

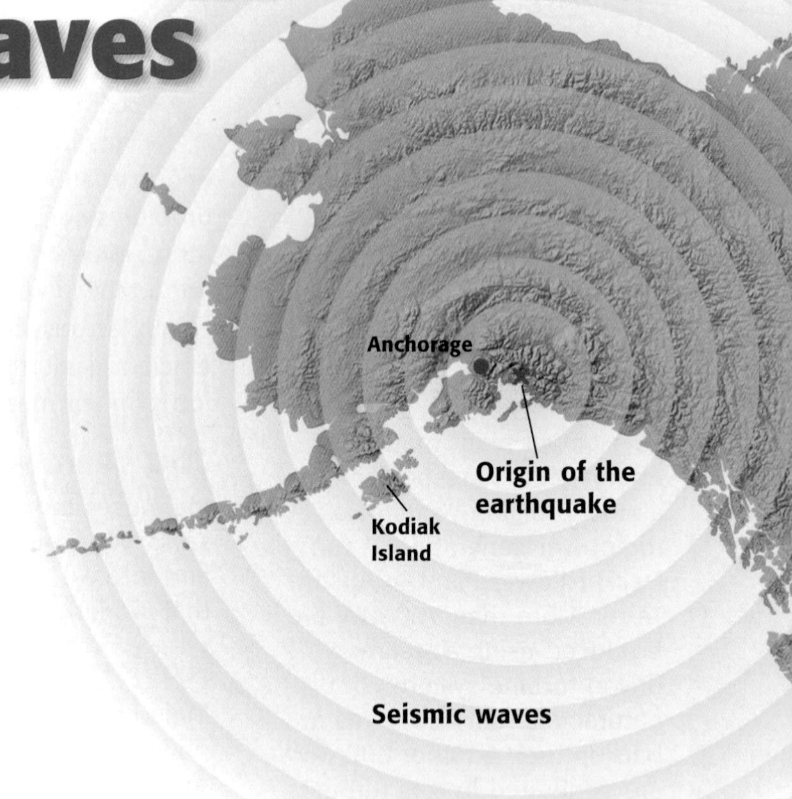

This Really Happened!

On March 27, 1964, the most powerful earthquake ever recorded on the North American continent rocked Alaska. The quake started on land near the city of Anchorage, but the seismic waves spread quickly in all directions, toppling buildings and ripping up roads.

The earthquake created a series of waves called tsunamis (tsoo NAH mees) in the Gulf of Alaska. In the deep water of the Gulf, the tsunamis were short and widely separated. But as these waves entered the shallow water surrounding Kodiak Island, off the coast of Alaska, they became taller and closer together. One of the tsunamis rose to a height of nearly 30 m! That's as tall as an eight-story building.

The powerful tsunamis pounded everything in their path. Eighty-one fishing boats were destroyed, and 86 others were damaged in the town of Kodiak. The destructive forces of the earthquake and tsunamis killed 21 people and caused $10 million in damage to Kodiak, making this the worst marine disaster in the town's 200-year history.

A tsunami is a dramatic example of the energy of waves. But waves affect your life in many common and less harmful ways. In fact, whenever you listen to music, talk with your friends, or watch a sunrise, you are experiencing the energy of waves. Read on to find out more about waves and how they affect your life every day.

Making Waves

There are many types of waves, but you are probably most familiar with water waves. All waves behave in similar ways, so you can use water waves to predict what might happen with other types of waves.

Procedure 

1. Spread **newspapers** on a table, and place a **wide pan** in the center of the table.
2. Fill the pan with **water** to a depth of 3 cm.
3. When the water is still, tap the surface of the water with the eraser end of a **pencil.** In what direction do the waves travel?
4. Tap the water again with the pencil, and watch what happens to a wave when it hits the side of the pan. Try this several times. Write your observations in your ScienceLog.
5. Place a **cork** in the center of the pan. Try to move the cork to the edge of the pan by making waves with the pencil.
6. Carefully empty the pan, and clean up any spilled water.

Analysis

7. The source of the energy of the water waves you observed was your hand. In what direction was the energy transmitted?
8. In step 4, you observed the *reflection* of water waves. How is this similar to the reflection of light off a mirror?
9. Describe the motion of the cork in step 5. Did the cork move with the wave? Explain your observations.

What Do You Think?

In your ScienceLog, try to answer the following questions based on what you already know:

1. What is a wave?
2. What properties do all waves have?
3. What can happen when waves interact?

Section 1

The Nature of Waves

NEW TERMS

wave
medium
transverse wave
longitudinal wave

OBJECTIVES

- Describe how waves transfer energy without transferring matter.
- Distinguish between waves that require a medium and waves that do not.
- Explain the difference between transverse and longitudinal waves.

Imagine that your family has just returned home from a day at the beach. You had fun, but you are hungry from playing in the ocean under a hot sun. You put some leftover pizza in the microwave for dinner, and you turn on the radio. Just then, the phone rings. It's your best friend calling to find out if you've done your math homework yet.

In the events described above, how many different waves were present? Believe it or not, at least five can be identified! Can you name them? Here's a hint: A **wave** is any disturbance that transmits energy through matter or space. Okay, here are the answers: water waves in the ocean; microwaves inside the microwave oven; light waves from the sun; radio waves transmitted to the radio; and sound waves from the radio, telephone, and voices. Don't worry if you didn't get very many. You will be able to name them all after you read this section.

Waves Carry Energy

Energy can be carried away from its source by a wave. However, the material through which the wave travels does not move with the energy. For example, sound waves often travel through air, but the air does not travel with the sound. If air were to travel with sound, you would feel a rush of air every time you heard the phone ring! **Figure 1** illustrates how waves carry energy but not matter.

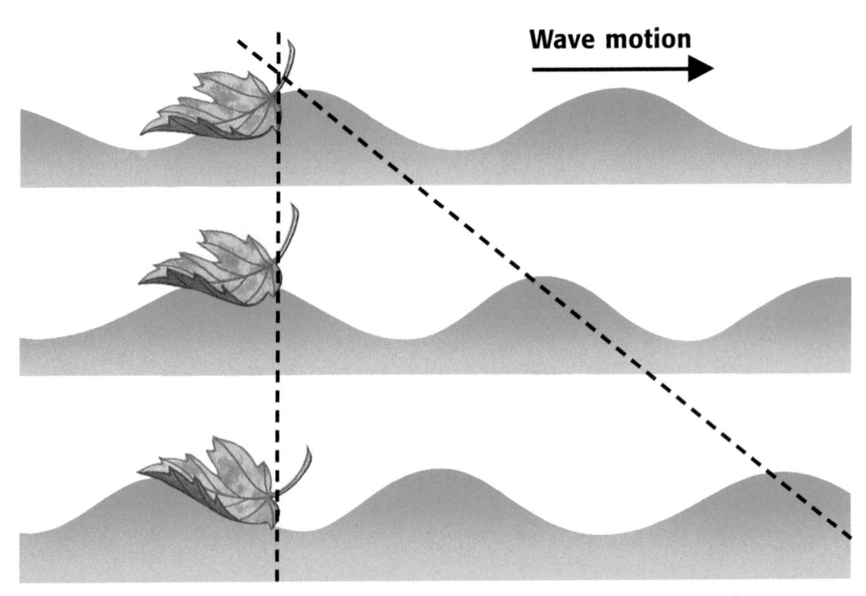

Figure 1 *Waves on a pond move toward the shore, but the water and the leaf floating on the surface do not move with the wave.*

QuickLab

Energetic Waves

1. Tie one end of a piece of **rope or string** to the back of a **chair.**
2. Hold the other end in one hand, and stand away from the chair so that the rope is almost straight but is not pulled tight.
3. Move the rope up and down quickly to create a single wave. Try this several times.
4. Which way does the wave move? How does the rope move compared with the movement of the wave?
5. Where does the energy of the wave come from?
6. Record your observations and answers in your ScienceLog.

As a wave travels, it uses its energy to do work on everything in its path. For example, the waves in a pond do work on the water to make it move up and down. The waves also do work on anything floating on the water's surface—for example, boats and ducks bob up and down with waves. But they don't move in the same direction as the waves.

Energy Transfer Through a Medium Some waves transfer energy through the vibration of the particles (atoms or molecules) in a medium. A **medium** is a substance through which a wave can travel. A medium can be a solid, a liquid, or a gas. The plural of *medium* is *media*.

When a particle vibrates (moves back and forth, as in **Figure 2**), it has energy. It can pass its energy to a particle next to it. After the energy is transferred to the second particle, the second particle will vibrate in a way similar to the first particle. In this way, energy is transmitted through a medium.

Figure 2 *A vibration is one complete back-and-forth motion of an object. The pendulum on this clock is vibrating.*

Sound waves require a medium. Air is the medium through which sound waves usually travel, but sound can also travel through liquids, solids, and other gases. Sound energy travels by the vibration of particles in liquids, solids, and gases. If there are no particles to vibrate, no sound is possible. For example, if you put an alarm clock inside a jar and remove all the air from the jar to create a vacuum, you will not be able to hear the clock ring.

Other waves that require a medium include ocean waves, which travel through water, and waves on guitar and cello strings. **Figure 3** shows the effect of another wave that requires a medium. Waves that require a medium are called *mechanical waves*.

Figure 3 *Seismic waves travel through the ground. This field has been rippled by the seismic waves of the 1964 earthquake in Alaska.*

Energy Transfer Without a Medium Some waves can transfer energy without traveling through a medium. Visible light is an example of a wave that doesn't require a medium. Other examples include microwaves produced by microwave ovens, TV and radio signals, and X rays used by dentists and doctors. Waves that do not require a medium are called *electromagnetic waves.*

Although electromagnetic waves do not require a medium, they can travel through substances such as air, water, and glass. However, they travel fastest through empty space. Light from the sun is a type of electromagnetic wave. **Figure 4** shows that light waves from the sun can travel through both space and matter to support life on Earth.

BRAIN FOOD

In 1946, an American engineer named Percy Spencer was working on a radar defense weapon when he accidentally discovered that microwaves could be used to cook food. He stepped in front of a microwave-generating tube and noticed that the chocolate bar in his pocket melted suddenly. He then placed some popcorn kernels in front of the microwave beam and watched as the popcorn popped all over his lab. The first countertop microwave oven was made in 1967. Today microwave ovens are found in more than 90 percent of American homes.

Figure 4 *Light waves from the sun travel more than 100 million kilometers through nearly empty space, then more than 300 km through the atmosphere, and then another 10 m through water to support life in and around a coral reef.*

Self-Check

How do mechanical waves differ from electromagnetic waves? *(See page 596 to check your answer.)*

Types of Waves

Waves can be divided into categories based on the direction in which the particles of the medium vibrate compared with the direction in which the waves travel. This may sound complex now, but it will become clear as you read on. The two main types of waves are transverse waves and longitudinal (LAHN juh TOOD nuhl) waves. In certain conditions, a transverse wave and a longitudinal wave can combine to form another type of wave, called a surface wave.

Transverse Waves Waves in which the particles vibrate with an up-and-down motion are called **transverse waves.** *Transverse* means "moving across." The particles in a transverse wave move across, or perpendicular to, the direction that the wave is traveling. To be *perpendicular* means to be "at right angles." Try the MathBreak at right to practice identifying perpendicular lines.

A wave moving on a rope is an example of a transverse wave. In **Figure 5,** you can see that the points along the rope vibrate perpendicular to the direction the wave is traveling. The highest point of a transverse wave is called a *crest,* and the lowest point between each crest is called a *trough.*

Although electromagnetic waves do not travel by vibrating particles in a medium, all electromagnetic waves are classified as transverse waves.

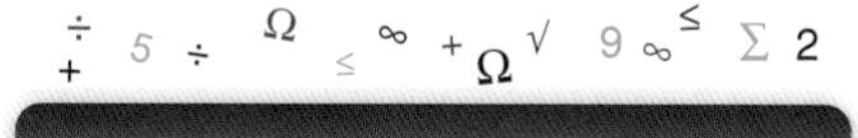

MATH BREAK

Perpendicular Lines

If the angle between two lines is 90°, the lines are said to be perpendicular. The figure below shows a set of perpendicular lines.

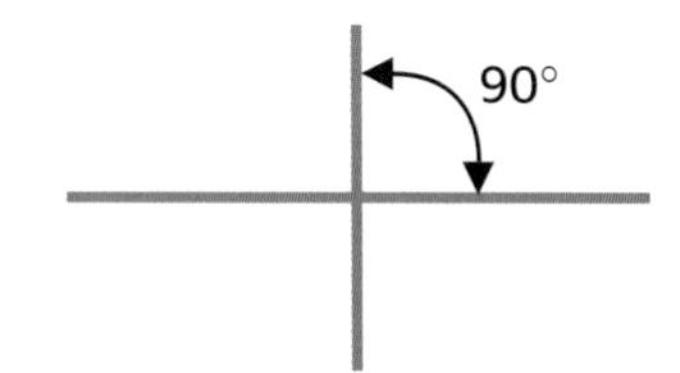

Look at the objects around you. Identify five objects with perpendicular lines or edges and five objects that do not have perpendicular lines. Sketch these objects in your ScienceLog.

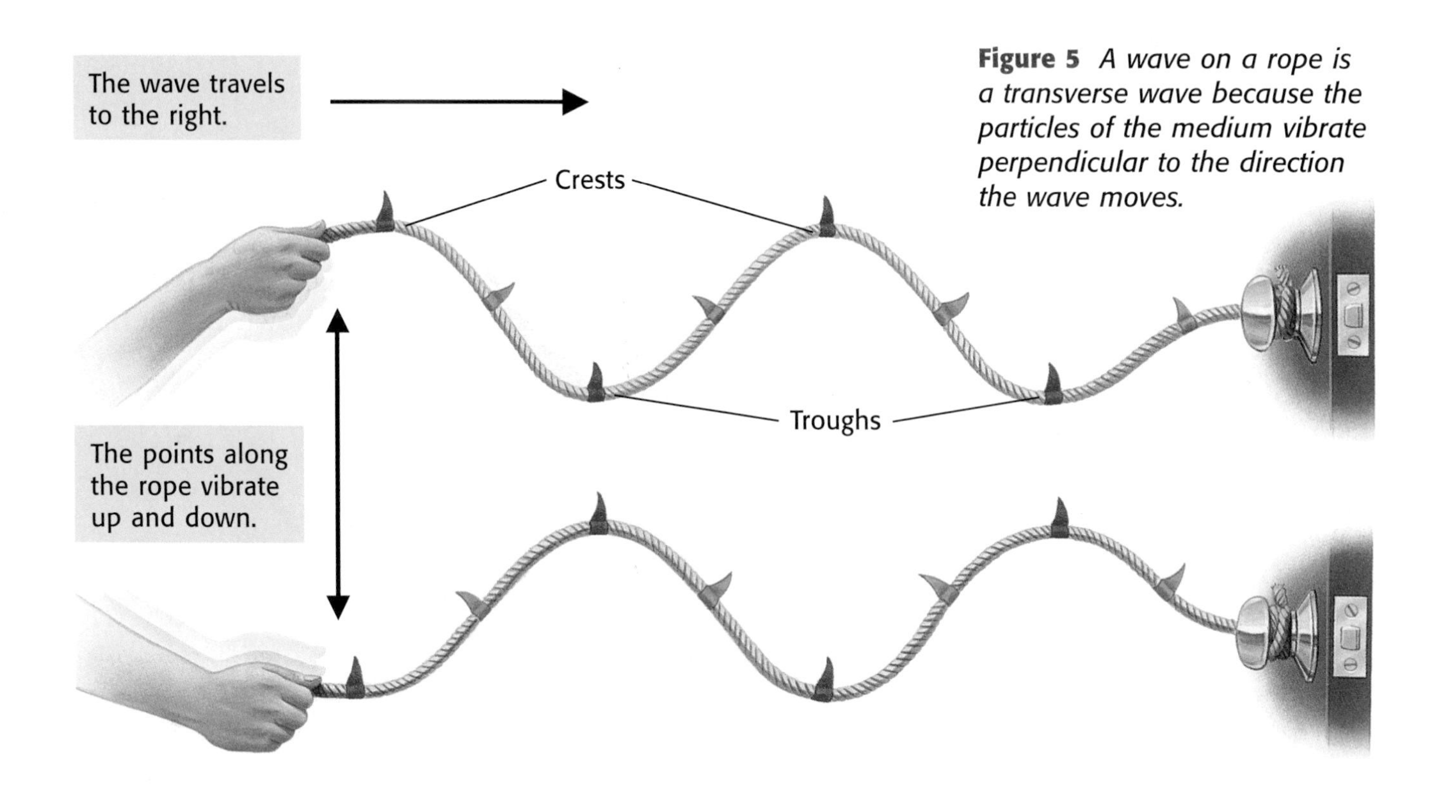

Figure 5 *A wave on a rope is a transverse wave because the particles of the medium vibrate perpendicular to the direction the wave moves.*

Longitudinal Waves In a **longitudinal wave,** the particles of the medium vibrate back and forth along the path that the wave travels. You can create a longitudinal wave on a spring, as shown in **Figure 6.**

When you push on the end of the spring, the coils of the spring are crowded together. A section of a longitudinal wave where the particles are crowded together is called a *compression.* When you pull back on the end of the spring, the coils are less crowded than normal. A section where the particles are less crowded than normal is called a *rarefaction* (RER uh FAK shuhn).

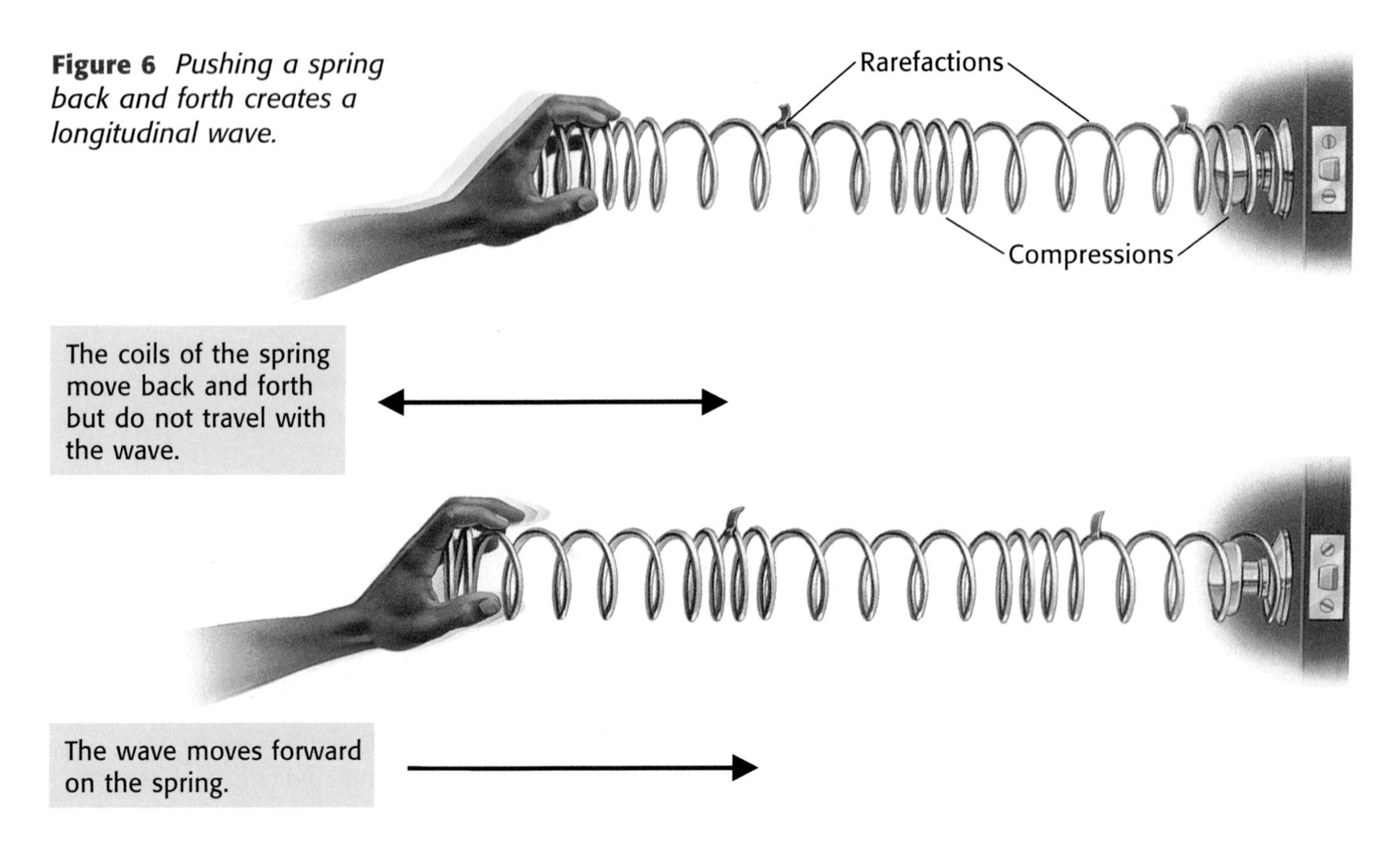

Figure 6 *Pushing a spring back and forth creates a longitudinal wave.*

Compressions and rarefactions travel along a longitudinal wave much in the way the crests and troughs of a transverse wave move from one end to the other, as shown in **Figure 7.**

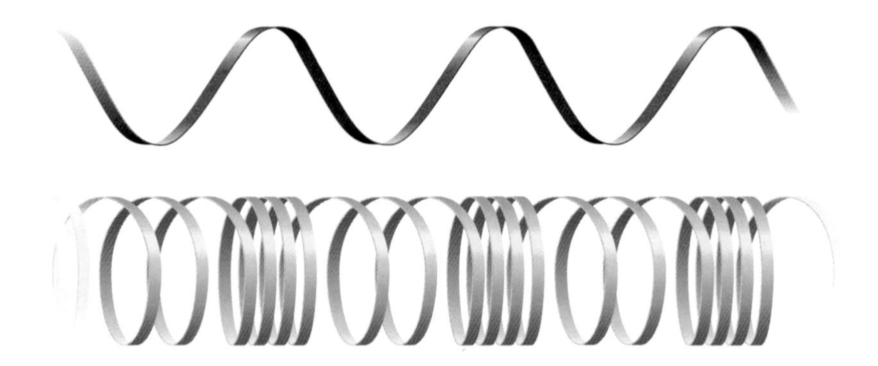

Figure 7 *The compressions of a longitudinal wave are like the crests of a transverse wave, and the rarefactions are like troughs.*

A sound wave is an example of a longitudinal wave. Sound waves travel by compressions and rarefactions of air particles. **Figure 8** shows how a vibrating drumhead creates these compressions and rarefactions.

Figure 8 *Sound energy is carried away from a drum in a longitudinal wave. Compressions and rarefactions follow each other through the air.*

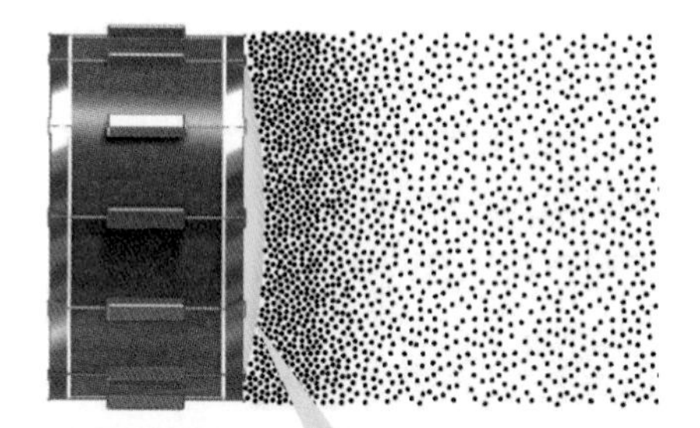

When the drumhead moves out after being hit, a compression is created in the air particles.

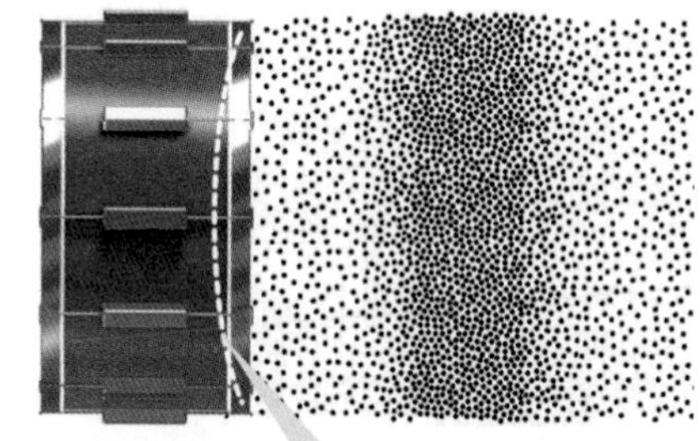

When the drumhead moves back in, a rarefaction is created.

earth science CONNECTION

Earthquakes are caused by seismic waves, which are generated by the sudden movement of the ground along a fault line. The longitudinal seismic waves are fast-moving and cause the first tremors. Slower-moving transverse seismic waves cause more damage than longitudinal waves. Surface seismic waves are the slowest but are the most destructive because they cause the ground to roll like water in oceans.

Combinations of Waves When waves occur at or near the boundary between two media, a transverse wave and a longitudinal wave can combine to form a *surface wave.* An example is shown in **Figure 9.** Surface waves look like transverse waves, but the particles of the medium in a surface wave move in circles rather than up and down. The particles move forward at the crest of each wave and move backward at the trough. The arrows in Figure 9 show the movement of particles in a surface wave.

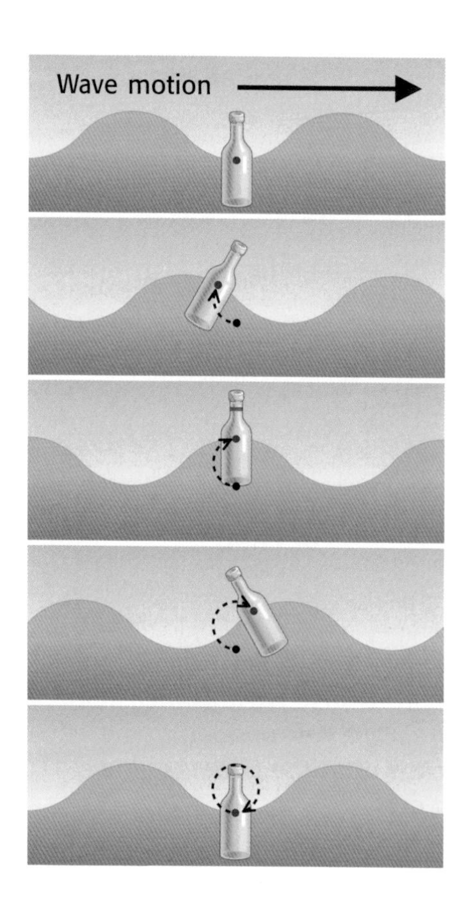

Figure 9 *Ocean waves are surface waves because they travel at the water's surface, where the water meets the air. A floating bottle shows the motion of particles in a surface wave.*

REVIEW

1. Describe how transverse waves differ from longitudinal waves.
2. Why can't you cause a floating leaf to move to the edge of a pond by throwing stones behind it?
3. Explain why supernova explosions in space can be seen but not heard on Earth.
4. **Applying Concepts** Sometimes people at a sports event do "the wave." Do you think this is a real example of a wave? Why or why not?

Section 2

Properties of Waves

NEW TERMS
amplitude
frequency
wavelength
wave speed

OBJECTIVES
- Identify and describe four wave properties.
- Explain how amplitude and frequency are related to the energy of a wave.

Imagine that you are canoeing on a lake. You decide to stop paddling for a while and relax in the sunshine. You notice that the breeze makes small waves on the water. These waves are short and close together, and they have little effect on the canoe. Suddenly, the calm is broken by a speedboat that roars past you. The speedboat creates tall, widely spaced waves that cause your canoe to rock wildly. So much for relaxation!

Just like matter, waves have properties that are useful for description and comparison. In this case, you could compare properties such as the height of the waves and the distance between the waves. In this section, you will learn about the properties of waves and how to measure them.

Amplitude

If you tie one end of a rope to the back of a chair, you can create waves by moving the other end up and down. If you move the rope a small distance, you will make a short wave. If you move the rope a greater distance, you will make a tall wave.

The property of waves that is related to the height of a wave is known as amplitude. The **amplitude** of a wave is the maximum distance the wave vibrates from its rest position. The rest position is where the particles of a medium stay when there are no disturbances.

The larger the amplitude is, the taller the wave is. Therefore, the wave you made by moving the rope a greater distance had a larger amplitude than the wave you made by moving the rope a small distance. **Figure 10** shows how the amplitude of a transverse wave is measured.

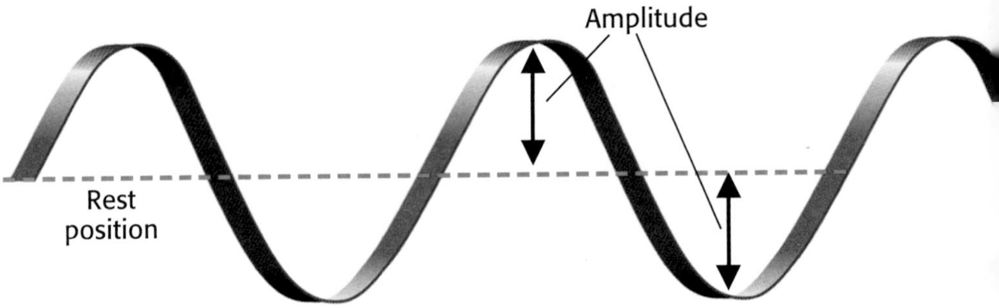

Figure 10 *The amplitude of a transverse wave is measured from the rest position to the crest or to the trough of the wave.*

Larger Amplitude Means More Energy When using a rope to make waves, you have to work harder to create a wave with a large amplitude than to create one with a small amplitude. This is because it takes more energy to move the rope farther from its rest position. Therefore, a wave with a large amplitude carries more energy than a wave with a small amplitude, as shown in **Figure 11.**

A tsunami can have a very large amplitude and thus carry a large amount of energy. This Japanese woodcut shows how the energy of a tsunami can easily toss boats on the ocean.

Figure 11 *The amplitude of a wave depends on the amount of energy.*

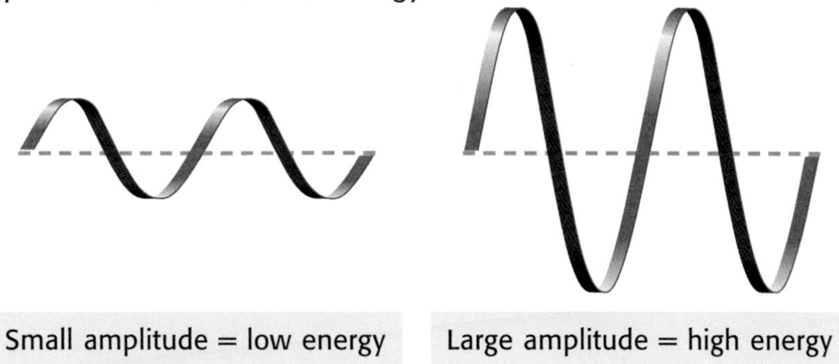

Wavelength

Another property of waves is wavelength. A **wavelength** is the distance between any two adjacent crests or compressions in a series of waves. The distance between two adjacent troughs or rarefactions is also a wavelength. In fact, the wavelength can be measured from any point on one wave to the corresponding point on the next wave. All of the measurements will be equal, as shown in **Figure 12.**

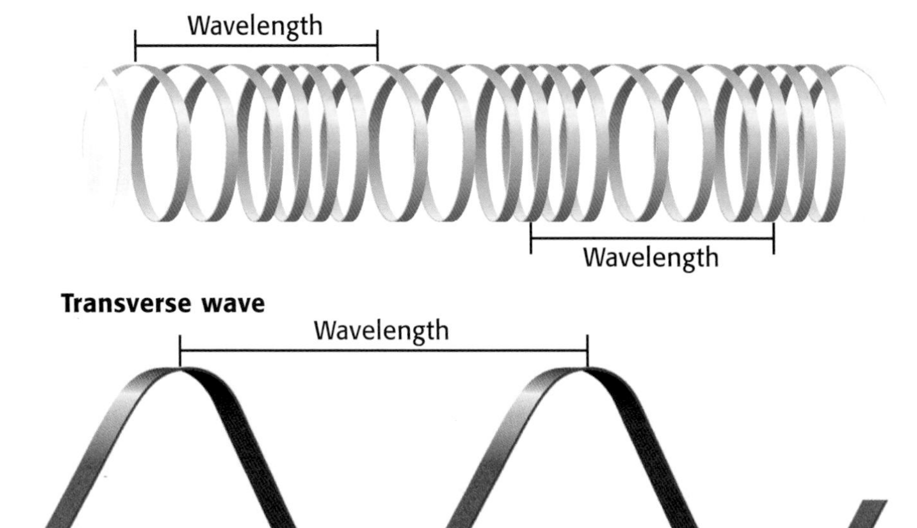

Figure 12 *Wavelengths measured from any two corresponding points are the same for a given wave.*

QuickLab

Springy Waves

1. Hold a **coiled spring toy** on the floor between you and a classmate so that the spring is straight. This is the rest position.
2. Move one end of the spring from side to side at a constant rate. The number of times you move it in a complete cycle (back and forth) each second is the frequency.
3. Keeping the frequency the same, increase the amplitude. What did you have to do? How did the change in amplitude affect the wavelength?
4. Now shake the spring back and forth about twice as fast (to double the frequency). What happens to the wavelength?
5. Record your observations and answers in your ScienceLog.

Frequency

Think about making rope waves again. The number of waves that you can make in 1 second depends on how quickly you move the rope. If you move the rope slowly, you make only a small number of waves each second. If you move it quickly, you make a large number of waves.

The number of waves produced in a given amount of time is the **frequency** of the wave. Frequency can be measured by counting either the number of crests or the number of troughs that pass a point in a certain amount of time. If you were measuring the frequency of a longitudinal wave, you would count the number of compressions or rarefactions. Frequency is usually expressed in *hertz* (Hz). For waves, one hertz equals one wave per second (1 Hz = 1/s). The frequency of a wave is related to its wavelength, as shown in **Figure 13.** Try the QuickLab at left to see the relationships between frequency, wavelength, and amplitude.

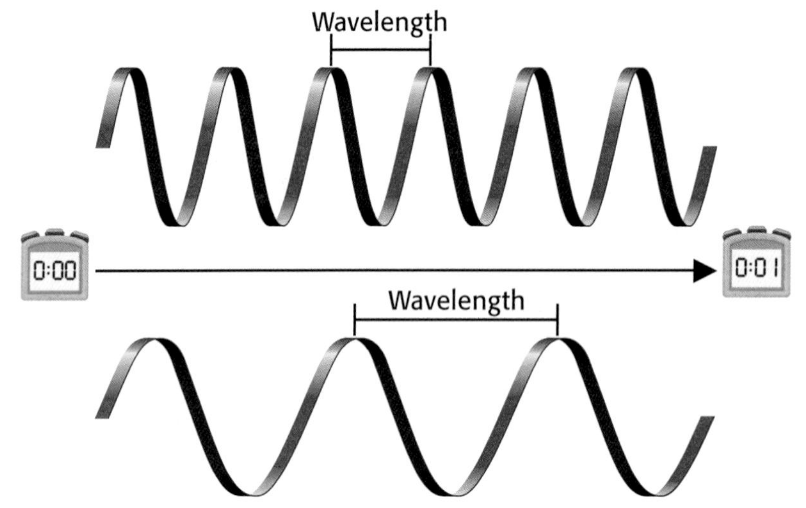

Figure 13 *At a given speed, the frequency increases as the wavelength decreases. Conversely, the frequency decreases as the wavelength increases.*

Some sound waves used by elephants for communication have frequencies that are too low for humans to hear. Turn to page 267 to learn more.

Higher Frequency Means More Energy When you make waves on a rope, you need to work harder to create waves with high frequencies than to create waves with low frequencies. This is because it takes more energy to vibrate the rope quickly than to vibrate the rope slowly. If the amplitudes are equal, high-frequency waves carry more energy than low-frequency waves. In Figure 13, the top wave carries more energy than the bottom wave.

Since frequency and wavelength are so closely related, you can also relate the amount of energy carried by a wave to the wavelength. In general, a wave with a short wavelength carries more energy than a wave with a long wavelength.

Wave Speed

Another property of waves is **wave speed**—the speed at which a wave travels. Speed is the distance traveled over time, so wave speed can be found by measuring the distance a single crest or compression travels in a given amount of time.

The speed of a wave depends on the medium in which the wave is traveling. For example, the wave speed of sound in air is about 340 m/s, but the wave speed of sound in steel is about 5,200 m/s. If the medium in which the wave is traveling is kept at a constant temperature and pressure, the speed of the wave is constant.

Wave speed can be calculated using wavelength and frequency. The relationship between wave speed (v), wavelength (λ, the Greek letter lambda), and frequency (f) is expressed in the following equation:

$$v = \lambda \times f$$

You can see in **Figure 14** how this equation can be used to determine wave speed. Try the MathBreak to practice using this equation.

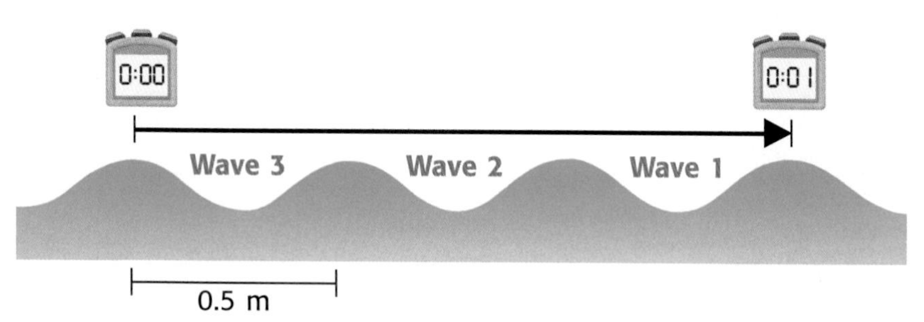

Figure 14 *To calculate wave speed, multiply the wavelength by the number of waves that pass in 1 second (frequency).*

$\lambda = 0.5$ m $\quad f = 3$ Hz (3/s)
$v = 0.5 \text{ m} \times 3 \text{ Hz} = 1.5$ m/s

REVIEW

1. Draw a transverse wave, and identify its amplitude and wavelength.
2. What is the speed (v) of a wave that has a wavelength (λ) of 2 m and a frequency (f) of 6 Hz?
3. **Inferring Conclusions** Compare the amplitudes and frequencies of the two types of waves discussed at the beginning of this section, and infer which type of wave carried the most energy. Explain your answer.

Spring into action! Find the speed of waves on a spring toy. Turn to page 564 of the LabBook.

MATH BREAK

Wave Calculations

The equation for wave speed can be rearranged to determine wavelength (λ) or frequency (f).

$\lambda = \frac{v}{f} \qquad f = \frac{v}{\lambda}$

You can determine the wavelength of a wave with a speed of 20 m/s and a frequency of 4 Hz like this:

$$\lambda = \frac{v}{f}$$
$$\lambda = 20 \text{ m/s} \div 4 \text{ Hz}$$
$$\lambda = \frac{20 \text{ m}}{\cancel{s}} \times \frac{1 \cancel{s}}{4}$$
$$\lambda = 5 \text{ m}$$

Now It's Your Turn

1. What is the frequency of a wave if it has a speed of 12 cm/s and a wavelength of 3 cm?
2. A wave has a frequency of 5 Hz and a wave speed of 18 m/s. What is its wavelength?

Section 3

Wave Interactions

NEW TERMS

reflection
refraction
diffraction
interference
standing wave
resonance

OBJECTIVES

- Describe reflection, refraction, diffraction, and interference.
- Compare destructive interference with constructive interference.
- Describe resonance, and give examples.

Imagine you are stargazing on a clear summer night. A full moon is high in the sky, and the stars are twinkling brilliantly, as shown in **Figure 15.** The sky is so clear you can find constellations (groupings of stars), such as the Big Dipper and Cassiopeia, and planets, such as Venus and Mars.

All stars, including the sun, produce light. But planets and the moon don't produce light. So why do they shine so brightly? Light from the sun *reflects* off the planets and the moon. Reflection is one of the wave interactions that you will learn about in this section.

Figure 15 *A wave interaction is responsible for this beautiful evening scene.*

Waves Bounce Back During Reflection

Reflection occurs when a wave bounces back after striking a barrier. All waves—including water, sound, and light waves—can be reflected. The reflection of water waves is shown in **Figure 16.** Reflected sound waves are called *echoes,* and light waves reflecting off an object allow you to see that object. For example, light waves from the sun are reflected when they strike the surface of the moon. These reflected waves allow us to enjoy moonlit nights like the one shown above.

Figure 16 *These water waves are reflecting off the side of the container.*

Changing Speed Bends Waves During Refraction

Try this simple experiment: place a pencil in a half-filled glass of water. Now look at the pencil from the side. The pencil appears to be broken into two pieces! But when you take the pencil out of the water, it is perfectly fine.

What you observed in this experiment was the result of the refraction of light waves. **Refraction** is the bending of a wave as it passes at an angle from one medium to another.

Remember that the speed of a wave varies depending on the medium in which the wave is traveling. So when a wave moves from one medium to another, the wave's speed changes. When a wave enters a new medium at an angle, the part of the wave that enters first begins traveling at a different speed from the rest of the wave. This causes the wave to bend, as shown in **Figure 17.**

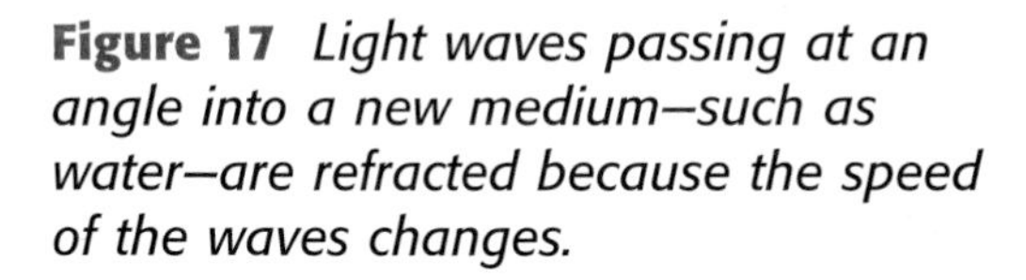

Figure 17 *Light waves passing at an angle into a new medium—such as water—are refracted because the speed of the waves changes.*

Self-Check

Will a light wave refract if it enters a new medium perpendicular to the surface? Explain. ***(See page 596 to check your answer.)***

Waves Bend Around Barriers or Through Openings During Diffraction

Suppose you are walking down a city street and you hear music. The sound seems to be coming from around the corner, but you cannot see who is playing the music because the building on the corner blocks your view. Why is it that sound waves travel around a corner better than light waves do?

Most of the time, waves travel in straight lines. For example, a beam of light from a flashlight is fairly straight. But in some circumstances, waves curve or bend when they reach the edge of an object. The bending of waves around a barrier or through an opening is known as **diffraction.**

Explore

Light waves diffract around corners of buildings much less than sound waves. Imagine what would happen if light waves diffracted around corners much more than sound waves. Write a paragraph describing how this would change what you see and hear as you walk around town.

Figure 18 Diffraction of Waves

When the barrier or opening is the same size as or is smaller than the wavelength of an approaching wave, the amount of diffraction is large.

If the barrier or opening is larger than the wavelength of the wave, there is only a small amount of diffraction.

The amount of diffraction a wave experiences depends on its wavelength and the size of the barrier or opening the wave encounters, as shown in **Figure 18.** You can hear music around the corner of a building because sound waves have long wavelengths and are able to diffract around corners. However, you cannot see who is playing the music because the wavelengths of light waves are much smaller than the building, so light is not diffracted very much.

Overlapping Waves Cause Interference

You know that all matter has volume. Therefore, objects cannot occupy the same space at the same time. But because waves are energy and not matter, more than one wave can exist in the same place at the same time. In fact, two waves can meet, share the same space, and pass through each other! When two or more waves share the same space, they overlap. The result of two or more waves overlapping is called **interference. Figure 19** shows one situation where waves occupy the same space.

Figure 19 *When sound waves from several instruments combine through interference, the result is a wave with a larger amplitude, which means a louder sound.*

Constructive Interference Increases Amplitude *Constructive interference* occurs when the crests of one wave overlap the crests of another wave or waves. The troughs of the waves also overlap. An example of constructive interference is shown in **Figure 20.** When waves combine in this way, the result is a new wave with higher crests and deeper troughs than the original waves. In other words, the resulting wave has a larger amplitude than the original waves had.

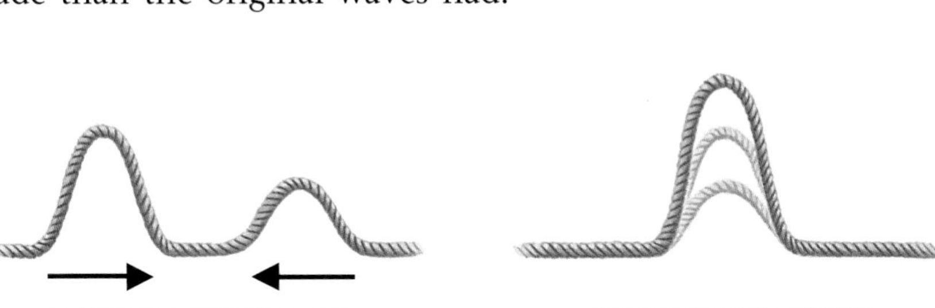

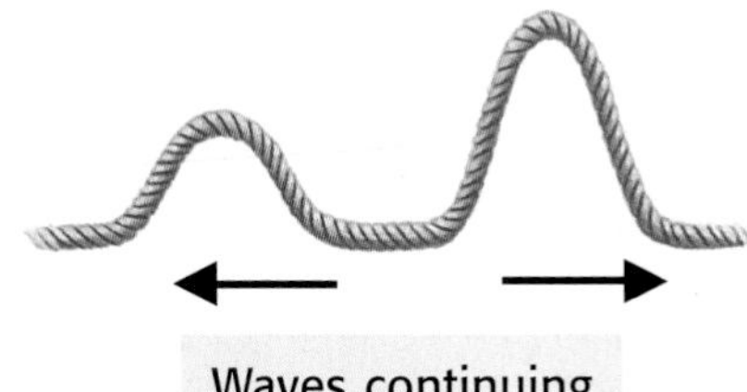

Figure 20 *When waves combine by constructive interference, the resulting wave has an amplitude that is larger than those of the original waves. After the waves interfere, they continue traveling in their original directions.*

Destructive Interference Decreases Amplitude *Destructive interference* occurs when the crests of one wave and the troughs of another wave overlap. The resulting wave has a smaller amplitude than the original waves had. What do you think happens when the waves involved in destructive interference have the same amplitude? Find out in **Figure 21.**

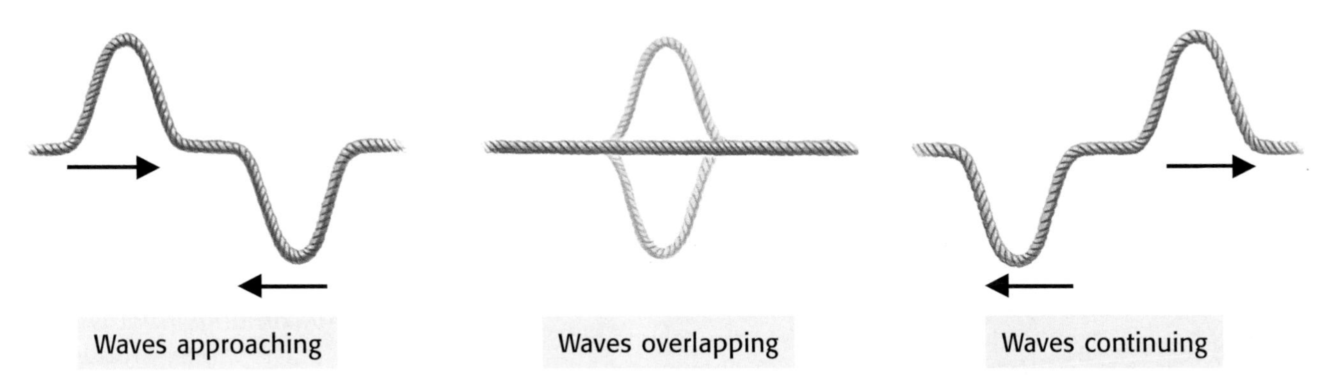

Figure 21 *When two waves with the same amplitude combine by destructive interference, they cancel each other out. This is called total destructive interference.*

Movie theaters use large screens and several speakers to make your moviegoing experience exciting. Theater designers know that increasing the amplitude of sound waves increases the volume of the sound. In terms of interference, how do you think the positioning of the speakers adds to the excitement?

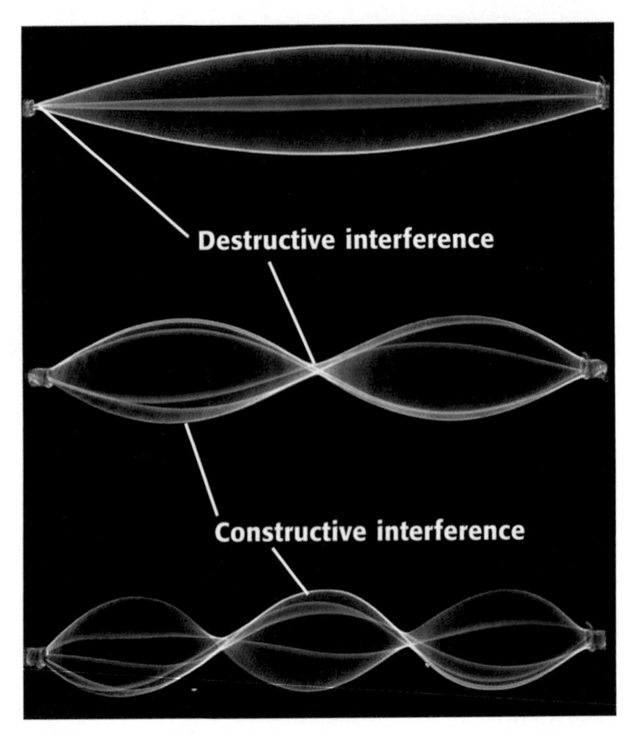

Figure 22 *When you move a rope at certain frequencies, you can create different standing waves.*

Interference Can Create Standing Waves

If you tie one end of a rope to the back of a chair and move the other end up and down, the waves you create travel down the rope and are reflected back. If you move the rope at certain frequencies, the rope appears to vibrate in loops, as shown in **Figure 22.** The loops result from the interference between the wave you created and the reflected wave. The resulting wave is called a standing wave. A **standing wave** is a wave that forms a stationary pattern in which portions of the wave are at the rest position due to total destructive interference and other portions have a large amplitude due to constructive interference. However, it only *looks* as if the wave is standing still. In reality, waves are traveling in both directions. Standing waves can be formed with transverse waves, as shown here, as well as with longitudinal waves.

One Object Causes Another to Vibrate During Resonance

As shown above, standing waves can occur at more than one frequency. The frequencies at which standing waves are produced are called *resonant frequencies*. When an object vibrating at or near the resonant frequency of a second object causes the second object to vibrate, **resonance** occurs. A resonating object absorbs energy from the vibrating object and therefore vibrates, too. An example of resonance is shown in **Figure 23.**

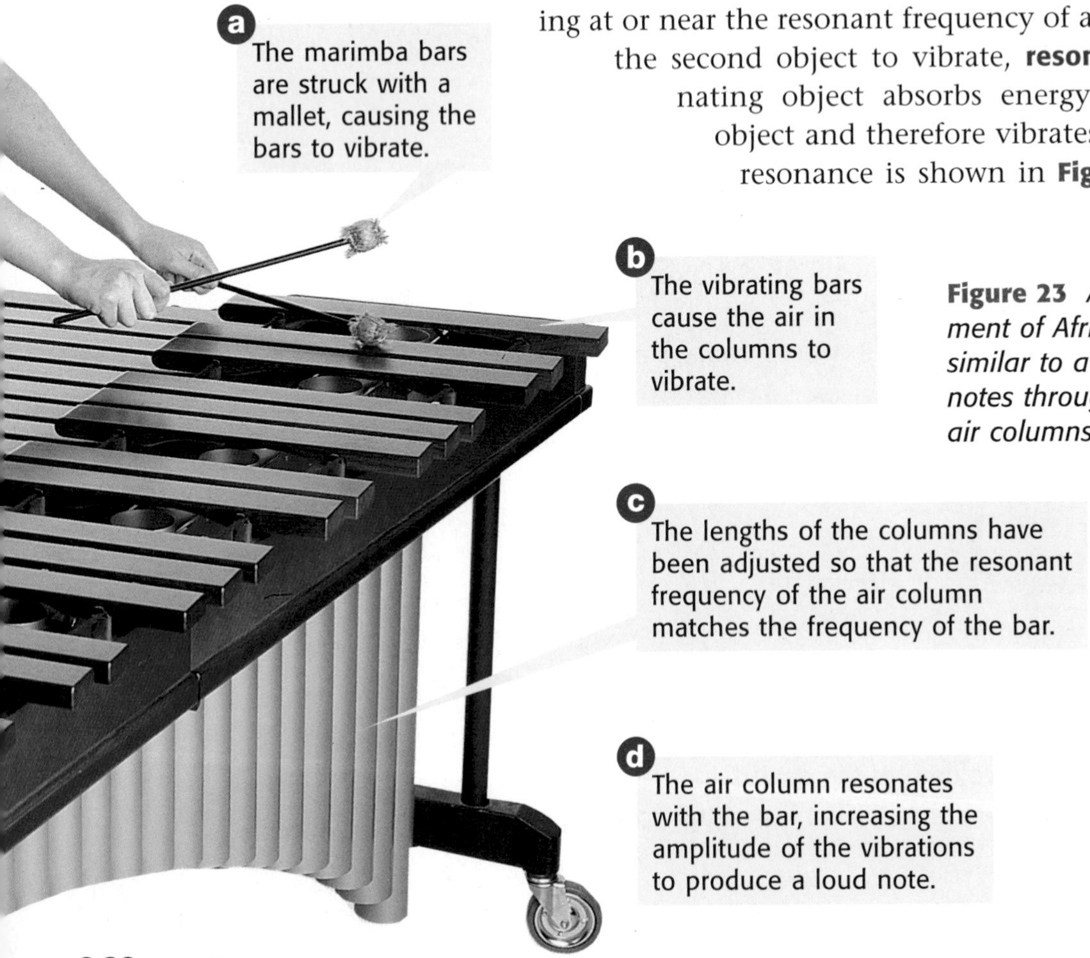

a The marimba bars are struck with a mallet, causing the bars to vibrate.

b The vibrating bars cause the air in the columns to vibrate.

c The lengths of the columns have been adjusted so that the resonant frequency of the air column matches the frequency of the bar.

d The air column resonates with the bar, increasing the amplitude of the vibrations to produce a loud note.

Figure 23 *A marimba, an instrument of African origin that is similar to a xylophone, produces notes through the resonance of air columns.*

Although resonance can be helpful for producing music, it can also lead to some destructive events. Resonance was partially responsible for the destruction of the Tacoma Narrows Bridge, in Washington. The bridge opened in July 1940 and soon earned the nickname Galloping Gertie because of its wavelike motions, shown in **Figure 24.** These motions were created by wind that blew across the bridge. The wind caused vibrations that were close to a resonant frequency of the bridge. Because the bridge was in resonance, the bridge absorbed a large amount of energy from the wind, which caused it to vibrate with a large amplitude.

Figure 24 *People who drove across the Tacoma Narrows Bridge said it was like riding a big roller coaster.*

On November 7, 1940, a supporting cable slipped, and the bridge began to twist. The twisting of the bridge, combined with high winds, further increased the amplitude of the bridge's motion. Within hours, the amplitude became so great that the bridge collapsed, as shown in **Figure 25.** Luckily, all the people on the bridge that day were able to escape before it crashed into the river below.

Figure 25 *The twisting motion that led to the destruction of the bridge was partially caused by resonance.*

REVIEW

1. Name two wave interactions that can occur when a wave encounters a barrier.
2. Describe what happens when a wave is refracted.
3. **Inferring Relationships** Sometimes when music is played loudly, you can feel your body shake. Explain what is happening in terms of resonance.

BRAIN FOOD

Resonance was responsible for the collapse of a bridge near Manchester, England, in 1831. Cavalry troops marched across the bridge in rhythm with its resonant frequency. This caused the bridge to vibrate with a large amplitude and eventually to fall. Since that time, all troops are ordered to "break step" when they cross a bridge.

Chapter Highlights

SECTION 1

Vocabulary

wave *(p. 246)*
medium *(p. 247)*
transverse wave *(p. 249)*
longitudinal wave *(p. 250)*

Section Notes

- A wave is a disturbance that transmits energy.
- A medium is a substance through which a wave can travel. The particles of a medium do not travel with the wave.
- Waves that require a medium are called mechanical waves. Waves that do not require a medium are called electromagnetic waves.
- Particles in a transverse wave vibrate perpendicular to the direction the wave travels.
- Particles in a longitudinal wave vibrate back and forth in the same direction that the wave travels.
- Transverse and longitudinal waves can combine to form surface waves.

SECTION 2

Vocabulary

amplitude *(p. 252)*
wavelength *(p. 253)*
frequency *(p. 254)*
wave speed *(p. 255)*

Section Notes

- Amplitude is the maximum distance the particles in a wave vibrate from their rest position. Large-amplitude waves carry more energy than small-amplitude waves.
- Wavelength is the distance between two adjacent crests (or compressions) of a wave.
- Frequency is the number of waves that pass a given point in a given amount of time. High-frequency waves carry more energy than low-frequency waves.

☑ Skills Check

Math Concepts

WAVE-SPEED CALCULATIONS The relationship between wave speed (v), wavelength (λ), and frequency (f) is expressed by the equation:

$$v = \lambda \times f$$

For example, if a wave has a wavelength of 1 m and a frequency of 6 Hz (6/s), the wave speed is calculated as follows:

$$v = 1 \text{ m} \times 6 \text{ Hz} = 1 \text{ m} \times 6/\text{s}$$
$$v = 6 \text{ m/s}$$

Visual Understanding

TRANSVERSE AND LONGITUDINAL WAVES Two common types of waves are transverse waves (shown below) and longitudinal waves. Study Figure 5 on page 249 and Figure 6 on page 250 to review the differences between these two types of waves.

SECTION 2

- Wave speed is the speed at which a wave travels. Wave speed can be calculated by multiplying the wavelength by the wave's frequency.

Labs

Wave Energy and Speed *(p. 562)*

Wave Speed, Frequency, and Wavelength *(p. 564)*

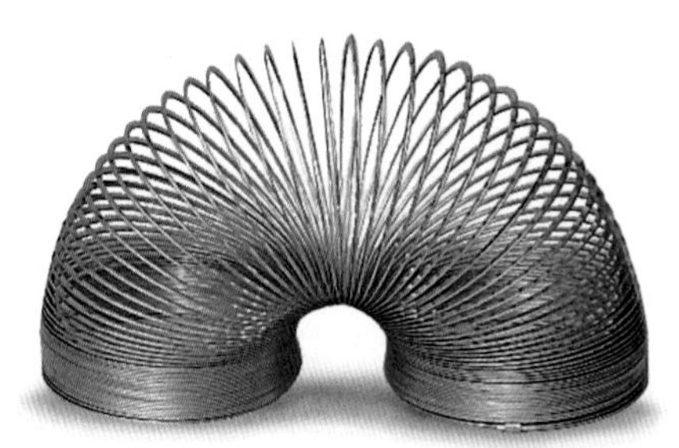

SECTION 3

Vocabulary

reflection *(p. 256)*
refraction *(p. 257)*
diffraction *(p. 257)*
interference *(p. 258)*
standing wave *(p. 260)*
resonance *(p. 260)*

Section Notes

- Waves bounce back after striking a barrier during reflection.
- Refraction is the bending of a wave when it passes from one medium to another at an angle.
- Waves bend around barriers or through openings during diffraction. The amount of diffraction depends on the wavelength of the waves and the size of the barrier or opening.
- The result of two or more waves overlapping is called interference.
- Amplitude increases during constructive interference and decreases during destructive interference.
- Standing waves are waves in which portions of the wave do not move and other portions move with a large amplitude.
- Resonance occurs when a vibrating object causes another object to vibrate at one of its resonant frequencies.

Chapter Review

USING VOCABULARY

For each pair of terms, explain the difference in their meaning.

1. longitudinal wave/transverse wave
2. frequency/wave speed
3. wavelength/amplitude
4. reflection/refraction
5. constructive interference/destructive interference

UNDERSTANDING CONCEPTS

Multiple Choice

6. As the wavelength increases, the frequency
   a. decreases.
   b. increases.
   c. remains the same.
   d. increases, then decreases.

7. Which wave interaction explains why sound waves can be heard around corners?
   a. reflection
   b. refraction
   c. diffraction
   d. interference

8. Refraction occurs when a wave enters a new medium at an angle because
   a. the frequency changes.
   b. the amplitude changes.
   c. the wave speed changes.
   d. None of the above

9. The speed of a wave with a frequency of 2 Hz (2/s), an amplitude of 3 m, and a wavelength of 10 m is
   a. 0.2 m/s.
   b. 5 m/s.
   c. 12 m/s.
   d. 20 m/s.

10. Waves transfer
    a. matter.
    b. energy.
    c. particles.
    d. water.

11. A wave that is a combination of longitudinal and transverse waves is a
    a. sound wave.
    b. light wave.
    c. rope wave.
    d. surface wave.

12. The wave property that is related to the height of a wave is the
    a. wavelength.
    b. amplitude.
    c. frequency.
    d. wave speed.

13. During constructive interference,
    a. the amplitude increases.
    b. the frequency decreases.
    c. the wave speed increases.
    d. All of the above

14. Waves that don't require a medium are
    a. longitudinal waves.
    b. electromagnetic waves.
    c. surface waves.
    d. mechanical waves.

Short Answer

15. Draw a transverse and a longitudinal wave. Label a crest, a trough, a compression, a rarefaction, and wavelengths. Also label the amplitude on the transverse wave.

16. What is the relationship between frequency, wave speed, and wavelength?

17. Explain how two waves can cancel each other out.

Concept Mapping

18. Use the following terms to create a concept map: wave, refraction, transverse wave, longitudinal wave, wavelength, wave speed, diffraction.

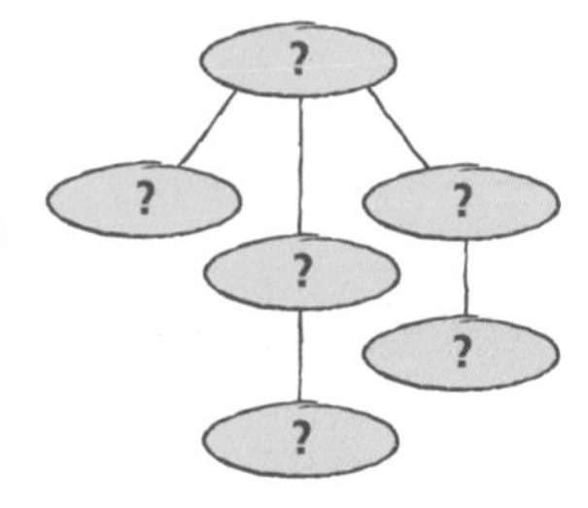

CRITICAL THINKING AND PROBLEM SOLVING

19. After you set up stereo speakers in your school's music room, you notice that in certain areas of the room the sound from the speakers is very loud and in other areas the sound is very soft. Explain how interference causes this.

20. You have lost the paddles for the canoe you rented, and the canoe has drifted to the center of the pond. You need to get the canoe back to shore, but you do not want to get wet by swimming in the pond. Your friend on the shore wants to throw rocks behind the canoe to create waves that will push the canoe toward shore. Will this solution work? Why or why not?

21. Some opera singers have voices so powerful they can break crystal glasses! To do this, they sing one note very loudly and hold it for a long time. The walls of the glass move back and forth until the glass shatters. Explain how this happens in terms of resonance.

MATH IN SCIENCE

22. A fisherman in a rowboat notices that one wave crest passes his fishing line every 5 seconds. He estimates the distance between the crests to be 2 m and estimates the crests of the waves to be 0.4 m above the troughs. Using these data, determine the amplitude and wave speed of the waves. Remember that wave speed is calculated with the formula $v = \lambda \times f$.

INTERPRETING GRAPHICS

23. Rank the waves below from highest energy to lowest energy, and explain your reasoning.

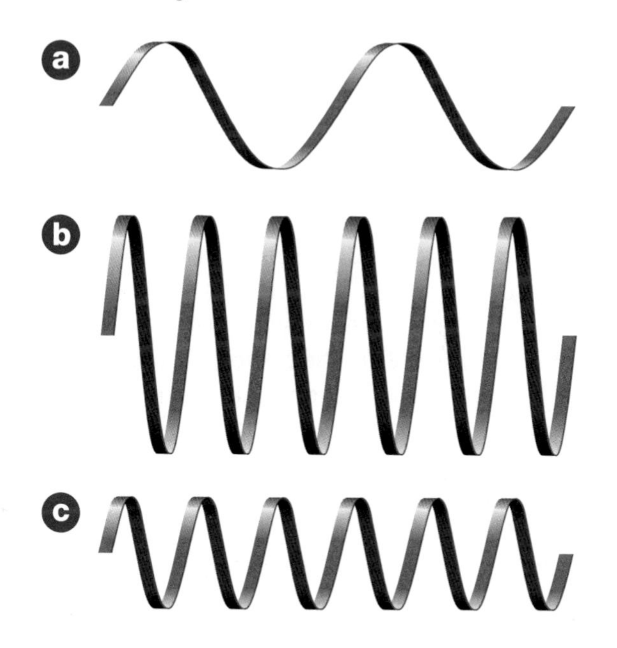

NOW What Do You Think?

Take a minute to review your answers to the ScienceLog questions on page 245. Have your answers changed? If necessary, revise your answers based on what you have learned since you began this chapter.

Science, Technology, and Society

The Ultimate Telescope

The largest telescopes in the world don't depend on visible light, lenses, or mirrors. Instead, they collect radio waves from the far reaches of outer space. One radio telescope, called the Very Large Array (VLA), is located in a remote desert in New Mexico.

▲ *Only a few of the 27 radio telescopes of the VLA, near Datil, New Mexico, can be seen in this photograph.*

From Radio Waves to Computer Images

Objects in space give off radio waves that radio telescopes collect. A bowl-shaped dish called a reflector focuses the radio waves onto a small radio antenna hung over the center of the dish. The antenna converts the waves into electric signals. The signals are relayed to a radio receiver, where they are amplified and recorded on tape that can be read by a computer. The computer combines the signals to create an image of the source of the radio waves.

A Marvel at "Seeing"

Radio telescopes have some distinct advantages over optical telescopes. They can "see" objects that are as far as 13 billion light-years away. They can even detect objects that don't release any light at all. Radio telescopes can be used in any kind of weather, can receive signals through atmospheric pollution, and can even penetrate the cosmic dust and gas clouds that occupy vast areas of space. However, radio telescopes must be large in order to be accurate.

Telescope Teamwork

The VLA is an array of 27 separate radio telescopes mounted on railroad tracks and electronically linked by computers. Each of the 27 reflectors is 25 m in diameter. When they operate together, they work like a single telescope with a diameter of 47 km! Using the VLA, astronomers have been able to explore distant galaxies, pulsars, quasars, and possible black holes.

A system of telescopes even larger than the VLA has been used. In the Very Long Baseline Array (VLBA), radio telescopes in different parts of the world all work together. The result is a telescope that is almost as large as the Earth itself!

What Do They See?

▶ Find out about some of the objects "seen" by the VLA, such as pulsars, quasars, and possible black holes. Prepare a report or create a model of one of the objects, and make a presentation to your class. Use diagrams and photographs to make your presentation more interesting.

Sounds of Silence

It's morning on the African savanna. Suddenly, without a sound, a family of elephants stops eating and begins to move off. At the same moment, about 6 km away, other members of the same family move off in a direction that will reunite them with the first group. How did the groups know when it was time to go?

Do You Hear What I Hear?

Elephants do much of their communicating by infrasound. This is sound energy with a frequency too low to be heard by humans. These infrasonic conversations take place through deep, soft rumblings produced by the animals. Though humans can't hear the sounds, elephants as far as 10 km away respond quickly to the messages.

Because scientists couldn't hear the elephant "conversations," they couldn't understand how the animals coordinated their activities. Of course, the elephants, which have superb low-frequency hearing, heard the messages clearly. It turns out that much elephant behavior is affected by infrasonic messages. For instance, one kind of rumble from a mother to her calf tells the calf it is all right to nurse. Another rumble, from the group's leader, is the "time to move on" message. Still another infrasonic message may be sent to other elephant groups in the area, warning them of danger.

Radio Collars

Once scientists learned about elephants' infrasonic abilities, they devised ways to study the sounds. Researchers developed radio collars for individual animals to wear. The collars are connected to a computer that helps researchers identify which elephant sent the message. The collars also record the messages. This information helps scientists understand both the messages and the social organization of the group.

Let's Talk

Elephants have developed several ways to "talk" to each other. For example, they greet each other by touching trunks and tusks. And elephants have as many as 25 vocal calls, including the familiar bellowing trumpet call (a sign of great excitement). In other situations, they use chemical signals.

▲ *Two elephants greeting each other*

Recently, researchers recording elephant communications found that when elephants vocalize their low-frequency sounds, they create seismic waves. Elephant messages sent by these underground energy waves may be felt more than 8 km away. Clearly, there is a lot more to elephant conversations than meets the ear!

On Your Own

► Elephants are very intelligent and highly sociable. Find out more about the complex social structure of elephant groups. Why is it important for scientists to understand how elephants communicate with each other? How can that understanding help elephants?

CHAPTER 11

Introduction to Electricity

Strange but True!

The most shocking of all fish tales concerns the electric eel, a freshwater fish of Central America and South America that can produce powerful jolts of electrical energy. Electric discharges from this 2.5 m long creature are strong enough to stun and kill smaller fish and frogs in the water. The eel can then swallow its motionless prey whole. Early travelers to the Amazon River basin wrote that, in shallow pools, the eels' electric discharges could knock horses and humans over.

How does the electric eel perform its shocking feat? Within this fish's long body are a series of electroplates, modified muscle tissues that generate low voltages. The electricity produced by one wafer-thin electroplate is small. But eels have 5,000 to 6,000 electroplates connected together and can therefore produce a high voltage. In laboratory experiments, the bursts of voltage from a fully grown eel have been measured at around 600 volts. That's five times the voltage of an electrical outlet—all from the cells of a single fish! The eel's thick, leathery skin prevents the eel from electrocuting itself while zapping prey.

Now that you know what one amazing fish can do with electricity, read on to learn what people have accomplished with this versatile form of energy.

Charge over Matter

Because you don't have electroplates in your body, you cannot produce high voltages like an electric eel. However, you can make electrically charged objects and use them to pick up other objects.

Procedure

1. Cut **6–8 small squares of tissue paper.** Each square should be about 2 × 2 cm. Place the squares on your desk.
2. Hold a **plastic comb** close to the paper squares. Describe what, if anything, happens.
3. Now rub the comb with a piece of **silk cloth** for about 30 seconds.
4. Hold the comb close to the tissue-paper squares, but don't touch them. Describe what happens. If nothing happens, rub the comb for a little while longer and try again.

What Do You Think?

In your ScienceLog, try to answer the following questions based on what you already know:

1. What is static electricity, and how is it formed?
2. How is electrical energy produced?
3. What is a circuit, and what parts make up a circuit?

5. Now hold a **metal rod** close to the tissue-paper squares, and observe what happens.
6. Rub the rod with the silk cloth, and then hold the rod close to the tissue-paper squares. Describe what, if anything, happens.

Analysis

7. When you rub the comb with the cloth, you give the comb a negative electric charge. Why do you think this allowed you to pick up tissue-paper squares?
8. How were your results for steps 2 and 4 different? Why do you think they were different?
9. What other objects do you think you can use to pick up tissue-paper squares?

Section 1

Electric Charge and Static Electricity

NEW TERMS

law of electric charges
electric force
conduction
induction
conductor
insulator
static electricity
electric discharge

OBJECTIVES

- State and give examples of the law of electric charges.
- Describe three ways an object can become charged.
- Compare conductors with insulators.
- Give examples of static electricity and electric discharge.

Have you ever reached out to open a door and received a shock from the knob? You may have been surprised, and your finger or hand probably felt tingly afterward. On dry days, you can easily produce shocks by shuffling your feet on a carpet and then lightly touching a metal object. These shocks are a result of a buildup of static electricity. But what is static electricity, and how is it formed? To answer these questions, you need to learn about charge.

Atoms and Charge

To investigate charge, you must know a little about the nature of matter. All matter is composed of very small particles called atoms. Atoms are made of even smaller particles called protons, neutrons, and electrons, as shown in **Figure 1.** One important difference between protons, neutrons, and electrons is that protons and electrons are charged particles and neutrons are not.

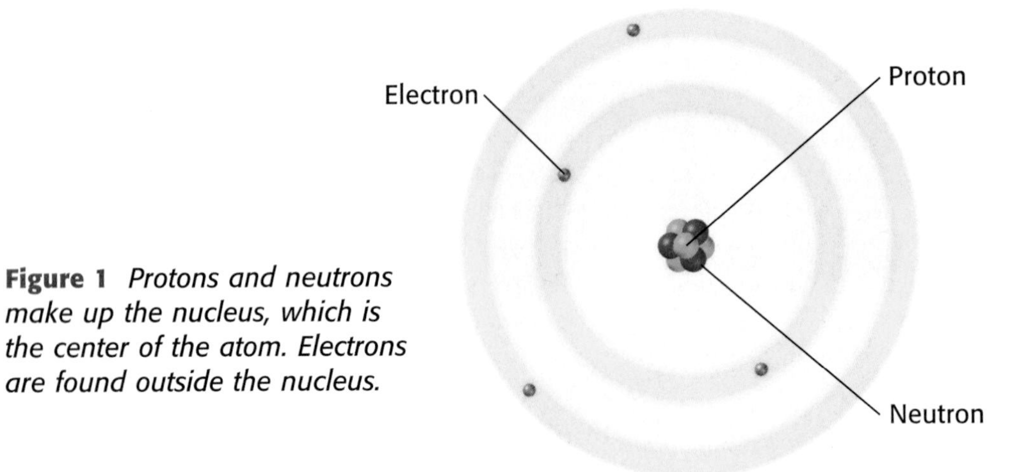

Figure 1 *Protons and neutrons make up the nucleus, which is the center of the atom. Electrons are found outside the nucleus.*

Charges Can Exert Forces Charge is a physical property that is best understood by describing how charged objects interact with each other. A charged object exerts a force—a push or a pull—on other charged objects. There are two types of charge—positive and negative. The force between two charged objects varies depending on whether the objects have the same type of charge or opposite charges, as shown in **Figure 2.** The charged balls in Figure 2 illustrate the **law of electric charges,** which states that like charges repel and opposite charges attract.

Protons are positively charged, and electrons are negatively charged. Because protons and electrons are oppositely charged, protons and electrons are attracted to each other. If this attraction didn't exist, electrons would fly away from the nucleus of the atom.

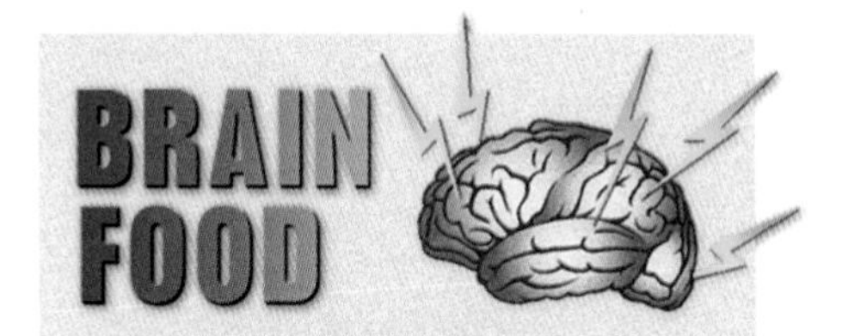

Car manufacturers take advantage of the law of electric charges when painting cars. The car bodies are given a positive charge. Then the paint droplets are given a negative charge as they exit the spray gun. The negatively charged paint droplets are attracted to the positively charged car body, so most of the paint droplets hit the car body and less paint is wasted.

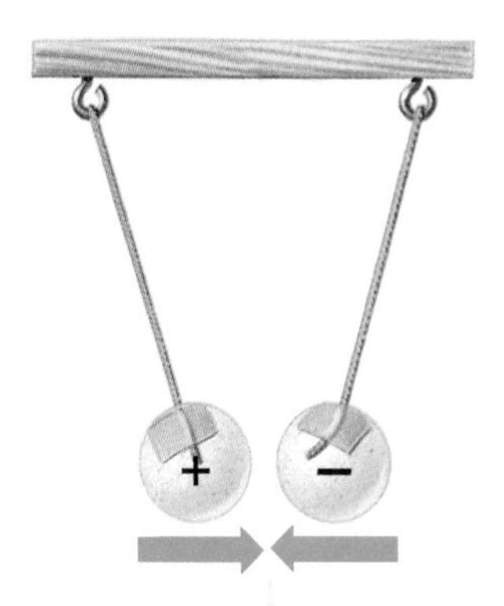

Figure 2 *The law of electric charges states that like charges repel and opposite charges attract.*

Objects that have opposite charges are attracted to each other, and the force between the objects pulls them together.

Objects that have the same charge are repelled, and the force between the objects pushes them apart.

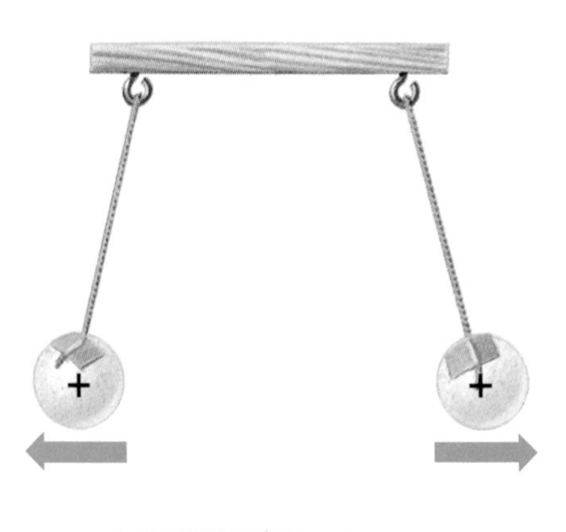

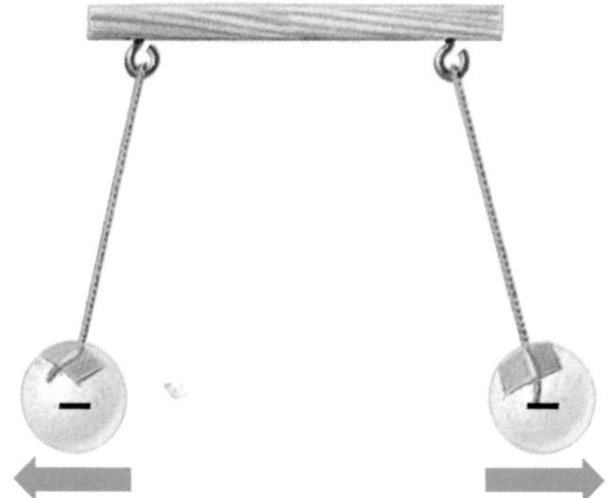

The Electric Force and the Electric Field The force between charged objects is an **electric force.** The strength of the electric force is determined by two factors. One factor is the size of the charges. The greater the charges are, the greater the electric force is. The other factor that determines the strength of the electric force is the distance between the charges. The closer together the charges are, the greater the electric force is.

The electric force exists because charged particles have electric fields around them. An *electric field* is a region around a charged particle that can exert a force on another charged particle. If a charged particle is in the electric field of another charged particle, it is attracted or repelled by the electric force exerted on it.

Charge It!

Although an atom contains charged particles, the atom itself does not have a charge. Atoms contain an equal number of protons and electrons. Therefore, the positive and negative charges cancel each other out, and the atom has no overall charge. If the atoms of an object have no charge, how can the object become charged? Objects become charged because the atoms in the objects can gain or lose electrons. If the atoms of an object lose electrons, the object becomes positively charged. If the atoms gain electrons, the object becomes negatively charged. There are three common ways for an object to become charged—friction, conduction, and induction. When an object is charged by any method, no charges are created or destroyed. The charge on any object can be detected by a device called an electroscope.

Friction Rubbing two objects together can cause electrons to be "wiped" from one object and transferred to the other. If you rub a plastic ruler with a cloth, electrons are transferred from the cloth to the ruler. Because the ruler gains electrons, the ruler becomes negatively charged. Conversely, because the cloth loses electrons, the cloth becomes positively charged. **Figure 3** shows a fun example of objects becoming charged by friction.

Figure 3 *When you rub a balloon against your hair, electrons from your hair are transferred to the balloon.*

After the electrons are transferred, the balloon is negatively charged and your hair is positively charged.

Your hair and the balloon are attracted to each other because they are oppositely charged.

Conduction Charging by **conduction** occurs when electrons are transferred from one object to another by direct contact. For example, if you touch an uncharged piece of metal with a positively charged glass rod, electrons from the metal will move to the glass rod. Because the metal loses electrons, it becomes positively charged. **Figure 4** shows what happens when you touch a negatively charged object to an uncharged object.

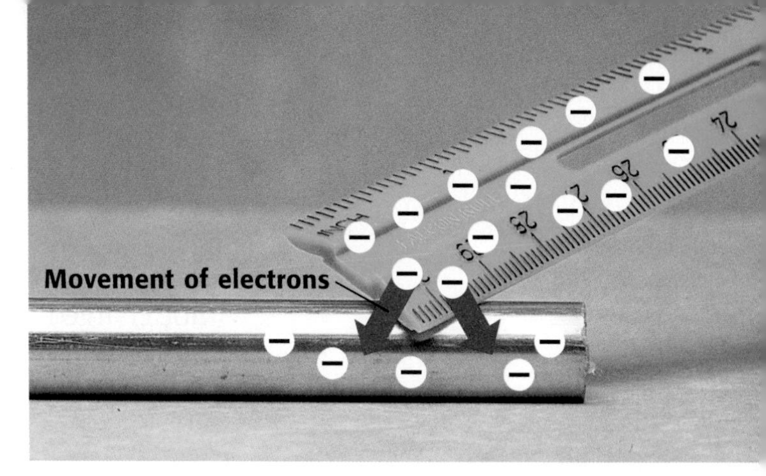

Figure 4 *Touching a negatively charged plastic ruler to an uncharged metal rod causes the electrons in the ruler to travel to the rod. The rod becomes negatively charged by conduction.*

Induction Charging by **induction** occurs when charges in an uncharged object are rearranged without direct contact with a charged object. For example, when a positively charged object is near a neutral object, the electrons in the neutral object are attracted to the positively charged object and move toward it. This movement produces a region of negative charge on the neutral object. **Figure 5** shows what happens when you hold a negatively charged balloon close to a neutral wall.

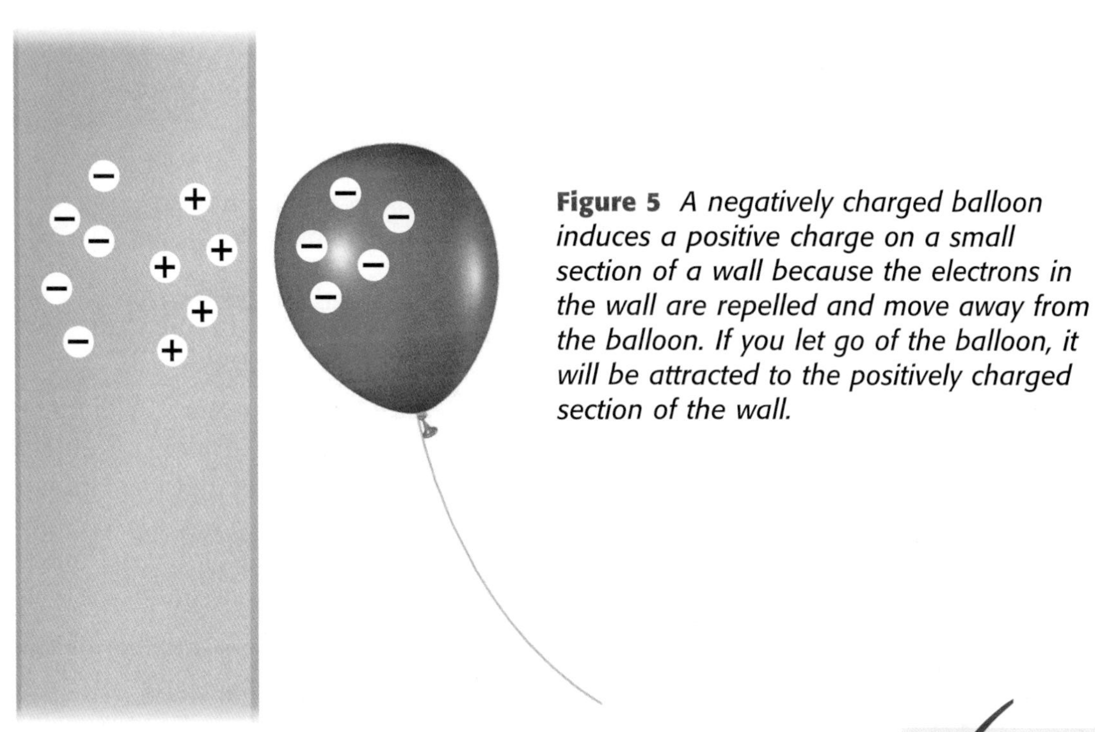

Figure 5 *A negatively charged balloon induces a positive charge on a small section of a wall because the electrons in the wall are repelled and move away from the balloon. If you let go of the balloon, it will be attracted to the positively charged section of the wall.*

Conservation of Charge When you charge objects by any method, no charges are created or destroyed. Electrons simply move from one atom to another, producing objects or regions with different charges. If you could count all the protons and all the electrons of all the atoms before and after charging an object, you would find that the numbers of protons and electrons do not change. Because charges are not created or destroyed, charge is said to be conserved.

Self-Check

Plastic wrap clings to food containers because the wrap has a charge. Explain how plastic wrap becomes charged. *(See page 596 to check your answer.)*

Detecting Charge To determine if an object has a charge, you can use a device called an *electroscope*. An electroscope is a glass flask that contains a metal rod inserted through a rubber stopper. There are two metal leaves at the bottom of the rod. The leaves hang straight down when the electroscope is not charged but spread apart when it is charged, as shown in **Figure 6.**

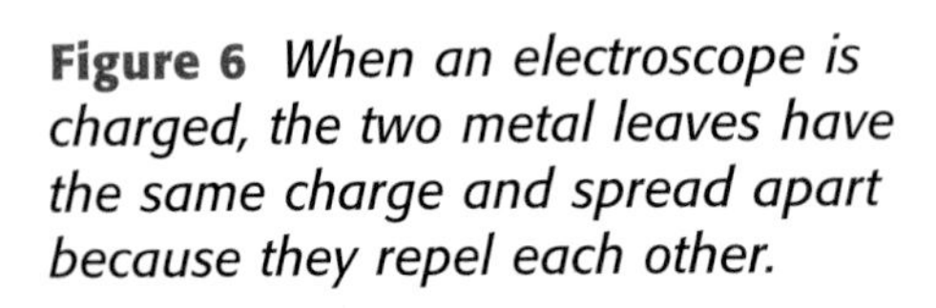

Figure 6 *When an electroscope is charged, the two metal leaves have the same charge and spread apart because they repel each other.*

Electrons from a negatively charged plastic ruler move to the electroscope and travel down the rod. The metal leaves become negatively charged and spread apart.

A positively charged glass rod attracts the electrons in the metal rod, causing the electrons to travel up the rod. The metal leaves become positively charged and spread apart.

REVIEW

1. Describe how an object is charged by friction.
2. Compare charging by conduction and induction.
3. **Inferring Conclusions** Suppose you are conducting experiments using an electroscope. You touch an object to the top of the electroscope, the metal leaves spread apart, and you determine that the object has a charge. However, you cannot determine the type of charge (positive or negative) the object has. Explain why not.

Moving Charges

Have you ever noticed that the cords that connect electrical devices to outlets are always covered in plastic, while the prongs that fit into the socket are always metal? Both plastic and metal are used to make electrical cords because they differ in their ability to transmit charges. In fact, most materials can be divided into two groups based on how easily charges travel through them. The two groups are conductors and insulators.

Conductors A **conductor** is a material in which charges can move easily. Most metals are good conductors because some of the electrons in metals are free to move about. Copper, silver, aluminum, and mercury are good conductors.

Conductors are used to make wires and other objects that transmit charges. For example, the prongs on a lamp's cord are made of metal so that charges can move in the cord and transfer energy to light the lamp.

Not all conductors are metal. Household, or "tap," water conducts charges very well. Because tap water is a conductor, you can receive an electric shock from charges traveling in it. Therefore, you should avoid using electrical devices (such as the one in **Figure 7**) near water unless they are specially designed to be waterproof.

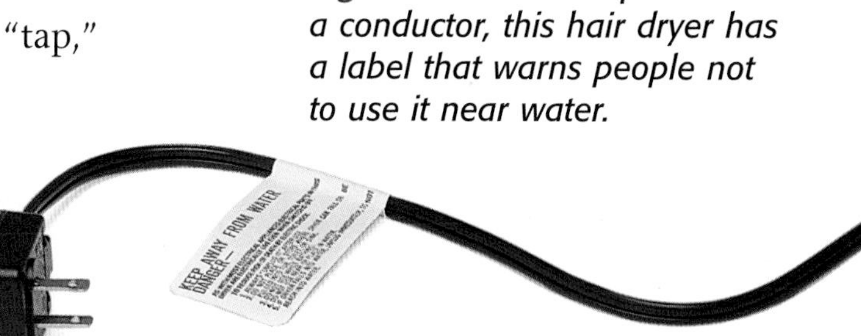

Figure 7 *Because tap water is a conductor, this hair dryer has a label that warns people not to use it near water.*

Insulators An **insulator** is a material in which charges cannot easily move. Insulators do not conduct charges very well because electrons are tightly bound to the atoms of the insulator and cannot flow freely. Plastic, rubber, glass, wood, and air are all good insulators.

Wires used to conduct electric charges are usually covered with an insulating material. The insulator prevents charges from leaving the wire and protects you from electric shock.

Static Electricity

After taking your clothes out of the dryer, you sometimes find clothing stuck together. When this happens, you might say that the clothes stick together because of static electricity. **Static electricity** is the buildup of electric charges on an object.

When something is *static,* it is not moving. The charges that create static electricity do not move away from the object they are stuck to. Therefore, the object remains charged. For example, your clothes are charged by friction as they rub against each other inside a dryer. Positive charges build up on some clothes, and negative charges build up on other clothes. Because clothing is an insulator, the charges stay on each piece of clothing, creating static electricity. You can see the result of static electricity in **Figure 8.**

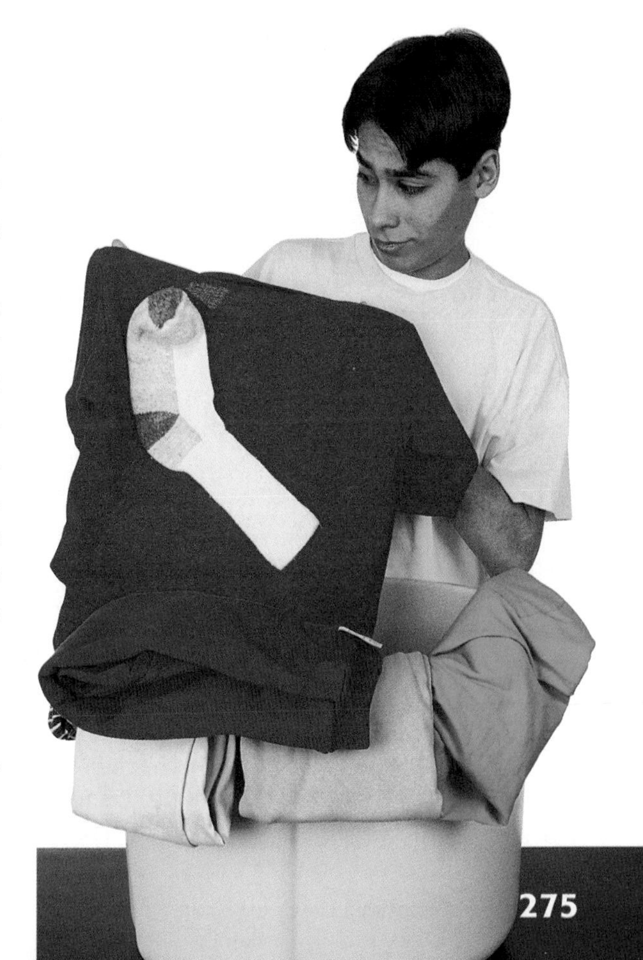

Figure 8 *Oppositely charged pieces of clothing are caused by static electricity. The clothes stick together because their charges attract each other.*

Electric Discharge Charges that build up as static electricity on an object eventually leave the object. The loss of static electricity as charges move off an object is called **electric discharge.** Sometimes electric discharge occurs slowly as charges are transferred from an object to water molecules in the air. For example, clothes stuck together by static electricity will eventually separate on their own because their electric charges are transferred to water molecules in the air over time.

Sometimes electric discharge occurs quickly and may be accompanied by a flash of light, a shock, or a cracking noise. For example, when you walk on a carpet with rubber-soled shoes, friction causes electrons to transfer from the carpet to you. Therefore, negative charges build up in your body. When you touch a metal doorknob, the negative charges in your body move quickly to the doorknob. Because the electric discharge happens quickly, you feel a shock.

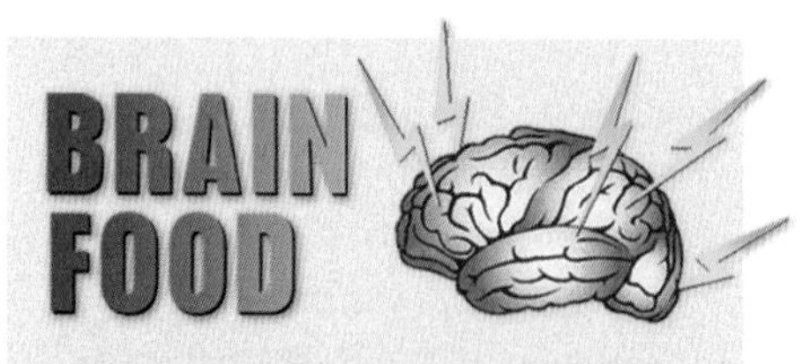

Although 70–80 percent of people struck by lightning survive, many suffer from long-term side effects such as memory loss, dizziness, and sleep disorders.

Lightning One of the most dramatic examples of electric discharge is lightning. Benjamin Franklin was the first to discover that lightning is a form of electricity. During a thunderstorm, Franklin flew a kite connected to a wire and successfully stored charge from a bolt of lightning. How does lightning form from a buildup of static electricity? **Figure 9** shows the answer.

Figure 9 How Lightning Forms

a During a thunderstorm, movement of air and water droplets within a thundercloud cause a negative charge to build up at the bottom of the cloud. A positive charge builds up at the top of the cloud.

b The negative charge at the bottom of the cloud induces a positive charge on the ground. Because of the large charge difference between the cloud and the ground, a rapid electric discharge—called lightning—occurs.

c Because different parts of clouds have different charges, lightning can also occur within and between clouds.

In addition to discovering that lightning is a form of electricity, Benjamin Franklin invented the lightning rod. A *lightning rod* is a pointed rod connected to the ground by a wire. Lightning usually strikes the highest point in a charged area because it is the easiest path for the charges to reach the ground. Therefore, lightning rods are always mounted so that they "stick out" and are the tallest point on a building, as shown in **Figure 10.**

Sprites and elves aren't just creatures in fairy tales! Read about how they are related to lightning on page 299.

A lightning rod works on the principle of grounding. Grounding provides a path for electric charges to travel to the Earth. The Earth can conduct charges and is very large. Because the Earth is so large, it can give up or absorb electric charges without being damaged. Objects that are in contact with the Earth are grounded. When lightning strikes a lightning rod, the electric charges are carried safely to the Earth through the wire. Therefore, lightning rods prevent lightning from damaging buildings by directing the lightning's charge to the Earth.

Unfortunately, anything that sticks out in an area can provide an easy path for lightning to follow. Trees and people in open areas are at risk of being struck by lightning. This is why it is particularly dangerous to be at the beach or on a golf course during a lightning storm. It is also dangerous to stand under a tree during a storm because the charge from lightning striking a tree can also travel through your body.

Figure 10 *Lightning strikes the lightning rod rather than the building because the lightning rod is the tallest point on the building.*

REVIEW

1. What is static electricity? Give an example of static electricity.
2. How is the shock you receive from a metal doorknob similar to a bolt of lightning?
3. **Applying Concepts** When you use an electroscope, you touch a charged object to a metal rod that is held in place with a rubber stopper. Why is it important to touch the object to the metal rod and not to the rubber stopper?

Section 2

Electrical Energy

NEW TERMS

cell
battery
potential difference
photocell
thermocouple

OBJECTIVES

- Explain how a cell produces an electric current.
- Describe how the potential difference is related to electric current.
- Describe how photocells and thermocouples produce electrical energy.

Imagine living without electrical energy. You could not watch television or listen to a portable radio, and you could not even turn on a light bulb to help you see in the dark! *Electrical energy*—the energy of electric charges—provides people with many comforts and conveniences. A flow of charges is called an *electric current.* Electric currents can be produced in many ways. One common way to produce electric current is through chemical reactions in a battery.

Batteries Are Included

In science, energy is defined as the ability to do work. Energy cannot be created or destroyed, it can only be converted into other types of energy. A **cell** is a device that produces an electric current by converting chemical energy into electrical energy. A **battery** also converts chemical energy into electrical energy and is made of several cells.

Parts of a Cell Every cell contains a mixture of chemicals that conducts a current; the mixture is called an *electrolyte* (ee LEK troh LIET). Chemical reactions in the electrolyte convert chemical energy into electrical energy. Every cell also contains a pair of electrodes made from two different conducting materials that are in contact with the electrolyte. An *electrode* (ee LEK TROHD) is part of a cell through which charges enter or exit. **Figure 11** shows how a cell produces an electric current.

Figure 11 *This cell has a zinc electrode and a copper electrode dipped in a liquid electrolyte.*

a A chemical reaction leaves extra electrons on the zinc electrode. Therefore, the zinc electrode has a negative charge.

b A different chemical reaction causes electrons to be pulled off the copper electrode, making the copper electrode positively charged.

c If the electrodes are connected by a wire, charges will flow from the negative zinc electrode through the wire to the positive copper electrode, producing an electric current.

Types of Cells Cells are divided into two groups—wet cells and dry cells. Wet cells, such as the cell shown in Figure 11, contain liquid electrolytes. A car battery is made of several wet cells that use sulfuric acid as the electrolyte.

Dry cells work in a similar way, but dry cells contain electrolytes that are solid or pastelike. The cells used in portable radios and flashlights are examples of dry cells.

LabBook

Potatoes aren't just for eating anymore! Learn how to use a potato to produce an electric current on page 567 of the LabBook.

You can make your own cell by inserting strips of zinc and copper into a lemon. The electric current produced when the metal strips are connected is strong enough to power a small clock, as shown in **Figure 12.**

Figure 12 *This cell uses the juice of a lemon as an electrolyte and uses strips of zinc and copper as electrodes.*

Bring on the Potential

So far you have learned that cells and batteries can produce electric currents. But why does the electric current exist between the two electrodes? The electric current exists because a chemical reaction causes a difference in charge between the two electrodes. The difference in charge means that an electric current can be produced by the cell to provide energy. The energy per unit charge is called the **potential difference** and is expressed in volts (V).

As long as there is a potential difference between the electrodes of a cell and there is a wire connecting them, charges will flow through the cell and the wire, creating an electric current. The current depends on the potential difference. The greater the potential difference is, the greater the current is. **Figure 13** shows batteries and cells with different potential differences.

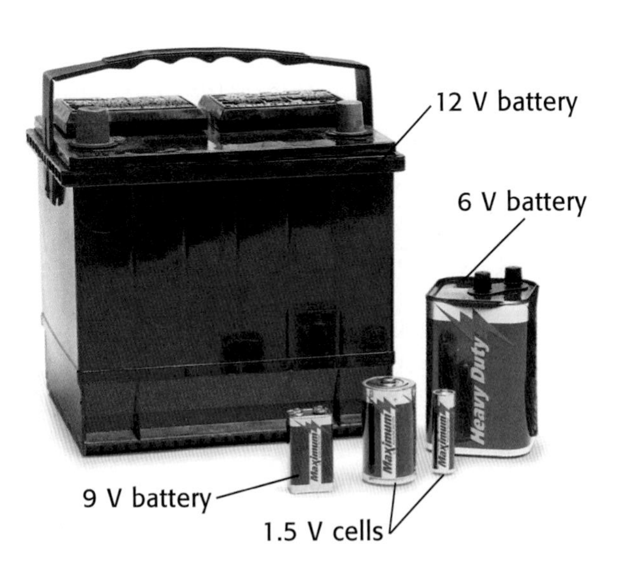

Figure 13 *Batteries are made with different potential differences. The potential difference of a battery depends on the number of cells it contains.*

Other Ways of Producing Electrical Energy

The conversion of chemical energy to electrical energy in batteries is not the only way electrical energy can be generated. Several technological devices have been developed to convert different types of energy into electrical energy for use every day. For example, generators convert kinetic energy into electrical energy. Two other devices that produce electrical energy are solar panels and thermocouples.

Solar Panels Have you ever wondered how a solar-powered calculator works? If you look above the display of the calculator, you will see a dark strip called a solar panel. This panel is made of several photocells. A **photocell** is the part of a solar panel that converts light into electrical energy.

Photocells contain silicon atoms. When light strikes the photocell, electrons are ejected from the silicon atoms. If light continues to shine on the photocell, electrons will be steadily emitted. The ejected electrons are gathered into a wire to create an electric current.

Thermocouples Thermal energy can be converted to electrical energy by a **thermocouple.** A simple thermocouple is made by joining wires made of two different metals into a loop, as shown in **Figure 14.** The loop is a path for charges to flow through. Thermocouples are used to monitor the temperature of car engines, furnaces, and ovens.

Figure 14 A Simple Thermocouple

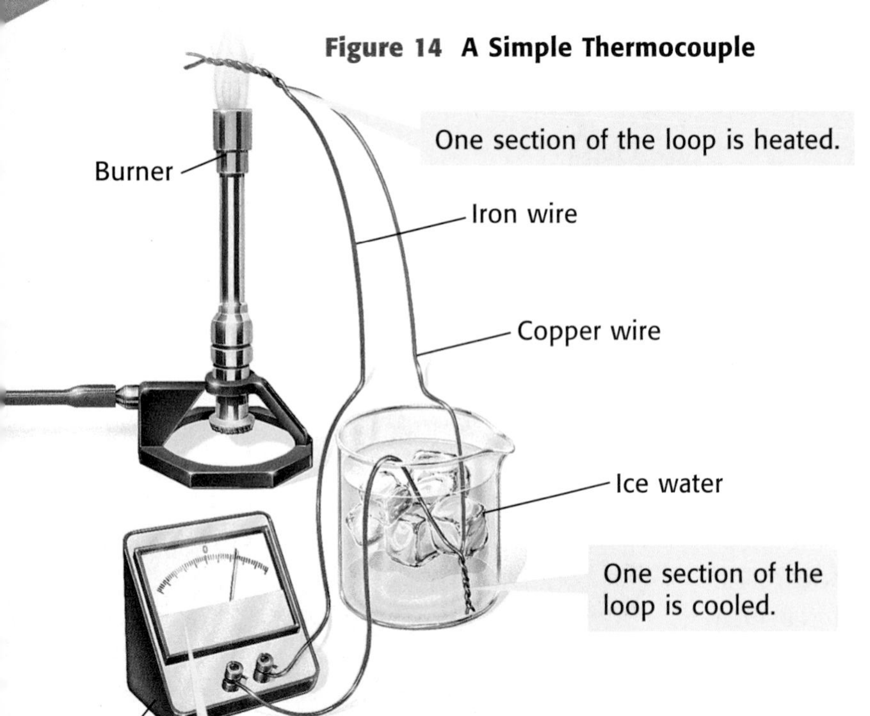

The temperature difference causes charges to flow through the loop. The greater the temperature difference is, the greater the current.

REVIEW

1. Name the parts of a cell, and explain how they work together to produce an electric current.
2. How do the currents produced by a 1.5 V flashlight cell and a 12 V car battery compare?
3. **Inferring Conclusions** Why do you think some solar calculators contain batteries?

Section 3

Electric Current

NEW TERMS

current
resistance
voltage
electric power

OBJECTIVES

- Describe electric current.
- Identify the four factors that determine the resistance of an object.
- Explain how current, voltage, and resistance are related by Ohm's law.
- Describe how electric power is related to electrical energy.

So far you have read how electrical energy can be generated by a variety of methods. A battery produces electrical energy very effectively, but even batteries have limitations. All the lamps, radios, and appliances in your home require electrical energy to function, and you can't plug them all into a battery! Electric power plants provide most of the electrical energy used every day. In this section, you will learn more about electric current and about the electrical energy you use at home.

Current Revisited

In the previous section, you learned that electric current is a continuous flow of charge. **Current** is more precisely defined as the rate at which charge passes a given point. The higher the current is, the more charge passes the point each second. The unit for current is the *ampere* (A), which is sometimes called amp for short. In equations, the symbol for current is the letter *I*.

Charge Ahead! When you flip a light switch, the light comes on instantly. Many people think it's because electrons travel through the wire at the speed of light. In fact, it's because an electric field is created at close to the speed of light.

Flipping the light switch sets up an electric field in the wire that connects to the light bulb. The electric field causes the free electrons in the wire to move, as illustrated in **Figure 15.** Because the electric field is created so quickly, the electrons start moving through the wire at practically the same instant. You can think of the electric field as a kind of command to the electrons to "Charge ahead!" The light comes on instantly because the electrons simultaneously obey this command. So the current that causes the bulb to light up is established very quickly, even though individual electrons move quite slowly. In fact, it may take a single electron over an hour to travel 1 m through a wire.

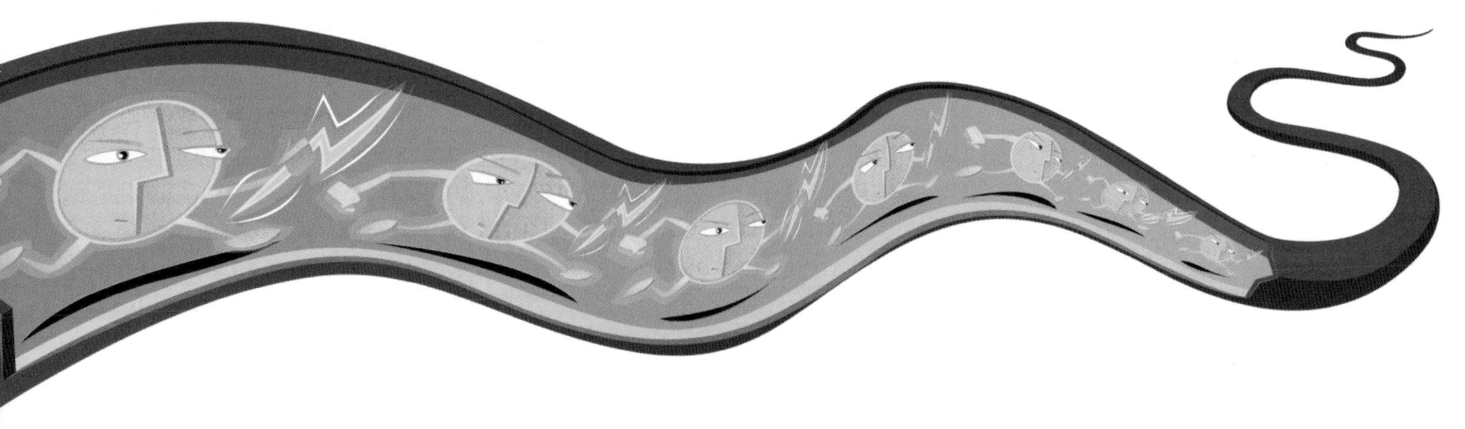

Figure 15 *Electrons moving in a wire make up current, a continuous flow of charge.*

Let's See, AC/DC . . . There are two different types of electric current—direct current (DC) and alternating current (AC). In *direct current* the charges always flow in the same direction. In *alternating current* the charges continually switch from flowing in one direction to flowing in the reverse direction. **Figure 16** illustrates the difference between DC and AC.

The electric current produced by batteries and cells is DC, but the electric current from outlets in your home is AC. Both types of electric current can be used to provide electrical energy. For example, if you connect a flashlight bulb to a battery, the light bulb will light. You can light a household light bulb by attaching it to a lamp and turning the lamp switch on.

Alternating current is used in homes because it is more practical for transferring electrical energy. In the United States, the alternating current provided to households changes directions 120 times each second.

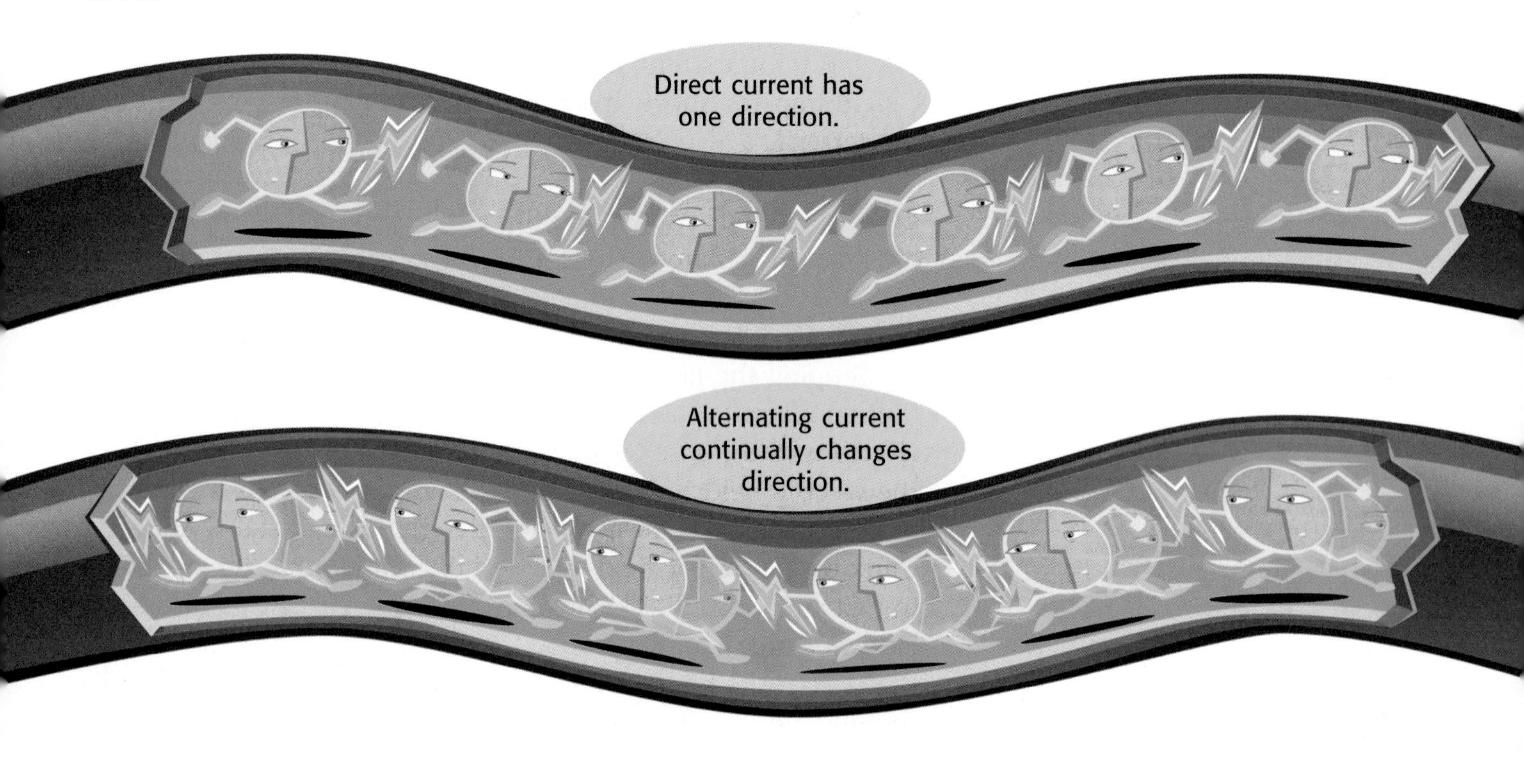

Figure 16 *Unlike DC, charges continually change direction in AC.*

Voltage

The current in a wire is determined by voltage. **Voltage** is the difference in energy per unit charge as a charge moves between two points in the path of a current. Voltage is another word for potential difference. Because voltage is the same as potential difference, voltage is expressed in volts. The symbol for voltage is the letter *V*. You can think of voltage as the amount of energy released as a charge moves between two points in the path of a current. The higher the voltage is, the more energy is released per charge. The current depends on the voltage. The greater the voltage is, the greater the current is.

In the United States, electrical outlets usually supply a voltage of 120 V. Therefore, most electrical devices that are plugged into outlets are designed to run on 120 V. Such devices include televisions, toasters, lamps, and alarm clocks. Devices that run on batteries or cells usually need a lower voltage. For example, a portable radio needs only 3 V to work. Compare this with the voltage created by the eel in **Figure 17.**

Pacemaker cells in the heart produce low electric currents at regular intervals to make the heart beat. During a heart attack, pacemaker cells do not work together and the heart beats irregularly. To correct this, doctors sometimes "jump start" the heart by creating a high voltage across a person's chest, which forces the pacemaker cells to act together, restoring a regular heartbeat.

Figure 17 *An electric eel can create a voltage of more than 600 V!*

Resistance

In addition to voltage, resistance also determines the current in a wire. **Resistance** is the opposition to the flow of electric charge. Resistance is expressed in ohms (Ω, the Greek letter *omega*). In equations, the symbol for resistance is the letter R.

You can think of resistance as "electrical friction." The higher the resistance of a material is, the lower the current in it. Therefore, as resistance increases, current decreases if the voltage is kept the same. An object's resistance varies depending on the object's material, thickness, length, and temperature.

Material Good conductors, such as copper, have low resistance. Poorer conductors, such as iron, have higher resistance. The resistance of insulators is so high that electric charges cannot flow in them.

Materials with low resistance are used to make wires and other objects that are used to transfer electrical energy from place to place. For example, most of the electrical cords in your house contain copper wires. However, sometimes it is useful to pass a current through a material with high resistance, as shown in **Figure 18.**

Figure 18 *The filaments of light bulbs are made from tungsten. Tungsten's high resistance causes electrical energy to be converted to light and thermal energy when charges flow through it.*

Thickness and Length To understand how the thickness and length of a wire affect the wire's resistance, consider the model in **Figure 19.** The pipe filled with gravel represents a wire, and the water flowing through the pipe represents electric charges. This analogy illustrates that thick wires have less resistance than thin wires and that long wires have more resistance than short wires.

Figure 19 *Gravel in a pipe is like resistance in a wire. Just as gravel makes it more difficult for water to flow through the pipe, resistance makes it more difficult for electric charges to flow in a wire.*

A thick pipe has less resistance than a thin pipe because there are more spaces between pieces of gravel for water to flow through.

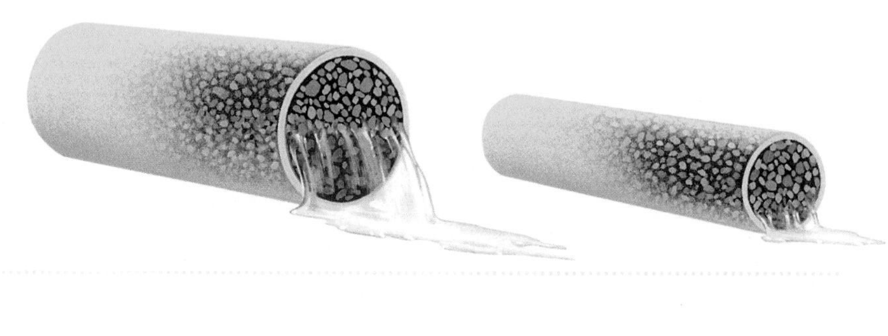

A short pipe has less resistance than a long pipe because the water does not have to work its way around as many pieces of gravel.

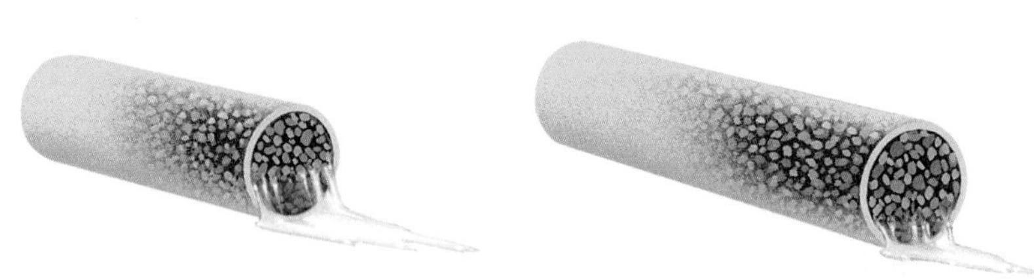

Temperature Resistance also depends somewhat on temperature. In general, the resistance of metals increases as temperature increases. This happens because atoms move faster at higher temperatures and get in the way of the flowing electric charges.

If you cool certain materials to an extremely low temperature, resistance will drop to nearly 0 Ω. Materials in this state are called *superconductors.* A small superconductor is shown in **Figure 20.** Superconductors can be useful because very little energy is wasted when electric charges travel in them. However, so much energy is necessary to cool them that superconductors are not practical for everyday use.

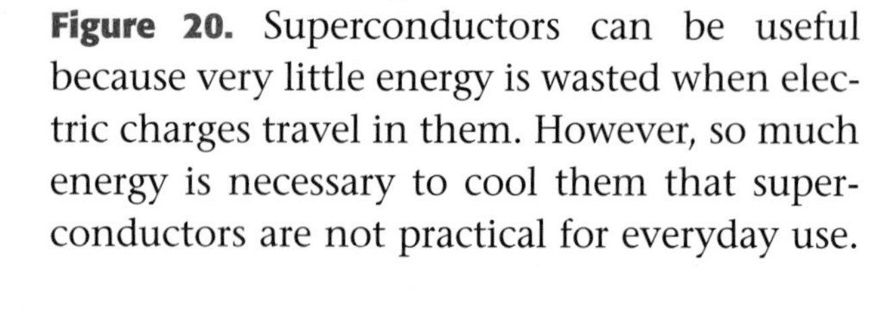

Figure 20 *One interesting property of superconductors is that they repel magnets. The superconductor in this photo is repelling the magnet so strongly that the magnet is floating.*

Ohm's Law: Putting It All Together

So far, you have learned about current, voltage, and resistance. You also learned that voltage and resistance affect current. The relationship between current, voltage, and resistance is given by the following equation:

$$\text{current} = \frac{\text{voltage}}{\text{resistance}}, \quad \text{or} \quad I = \frac{V}{R}$$

This equation was first determined by the German professor Georg Ohm, and it is therefore called *Ohm's law.* Ohm's law also shows the relationship between the units of current, voltage, and resistance.

$$\text{amperes (A)} = \frac{\text{volts (V)}}{\text{ohms }(\Omega)}$$

You can use Ohm's law to find the current produced in a wire if you know the voltage applied and the resistance of the wire. For example, if 30 V is applied to a wire with a resistance of 60 Ω, the current is calculated as follows:

$$I = \frac{V}{R} = \frac{30 \text{ V}}{60 \ \Omega} = 0.5 \text{ A}$$

Ohm's law can be rearranged and used to find voltage or resistance.

$$V = I \times R \text{ and } R = \frac{V}{I}$$

You can practice using all three arrangements of Ohm's law by doing the MathBreak at right.

MATH BREAK

Using Ohm's Law

If you know two values, you can calculate the third using Ohm's law. For example, the voltage needed to produce a 2 A current in a resistance of 12 Ω is calculated as follows:

$$V = I \times R$$
$$V = 2 \text{ A} \times 12 \ \Omega$$
$$V = 24 \text{ V}$$

Now It's Your Turn

1. Find the resistance of a wire if a voltage of 10 V produces a current of 0.5 A.
2. Find the current produced if a voltage of 36 V is applied to a resistance of 4 Ω.

Electric Power

You probably hear the word *power* used in different ways. Power can be used to mean force, strength, or energy. In science, power is the rate at which work is done. **Electric power** is the rate at which electrical energy is used to do work. The unit for power is the watt (W), and the symbol for power is the letter *P.* Electric power is calculated with the following equation:

$$\text{power} = \text{voltage} \times \text{current}, \quad \text{or} \quad P = V \times I$$

For the units:

$$\text{watts (W)} = \text{volts (V)} \times \text{amperes (A)}$$

If you have ever changed a light bulb, you are probably familiar with watts. Light bulbs, such as those shown at left, have labels such as "60 W," "75 W," or "120 W." As electrical energy is supplied to a light bulb, the light bulb glows. As power increases, the bulb burns brighter and brighter because more electrical energy is converted to light energy. That is why a 120 W bulb burns brighter than a 60 W bulb.

Another unit of power that is commonly used is the kilowatt (kW). One kilowatt is equal to 1,000 W. Kilowatts are used to express high values of power, such as the power needed to heat a house. The table shows the power ratings of some appliances you use every day.

Power Ratings of Household Appliances

Appliance	Power (W)
Clothes dryer	4,000
Toaster	1,100
Hair dryer	1,000
Refrigerator/freezer	600
Color television	200
Radio	100
Clock	3

Measuring Electrical Energy

Electric power companies sell electrical energy to homes and businesses. Such companies determine how much a household or business has to pay based on power and time. For example, the amount of electrical energy used by a household depends on the power of the electrical devices in the house and how long those devices were on. The equation for electrical energy is as follows:

$$\text{electrical energy} = \text{power} \times \text{time}, \quad \text{or} \quad E = P \times t$$

Self-Check

How much electrical energy is used by a color television that stays on for 2 hours? *(See page 596 to check your answer.)*

Because households use a large amount of electrical energy during a day, electric companies usually calculate electric energy by multiplying the power in kilowatts by the time in hours. Therefore, the unit of electrical energy is usually kilowatt-hours (kWh). If a household used 2,000 W (2 kW) of power in 3 hours, that means that 6 kWh of energy was used.

Electric power companies use electric meters such as the one shown at right to determine the number of kilowatt-hours of energy used by a household. Meters are often located outside houses and apartment buildings so someone from the power company can read them.

The amount of electrical energy used by an appliance depends on the power rating of the appliance and how long it is on. For example, a clock has a power rating of 3 W, and it is on 24 hours a day. Therefore, the clock uses 72 Wh (3 W × 24 hours), or 0.072 kWh, of energy a day. Using the information in the table on the previous page and an estimate of how long each appliance is on during a day, determine which appliances use the most energy and which use the least. Based on your findings, describe what you can do to use less energy.

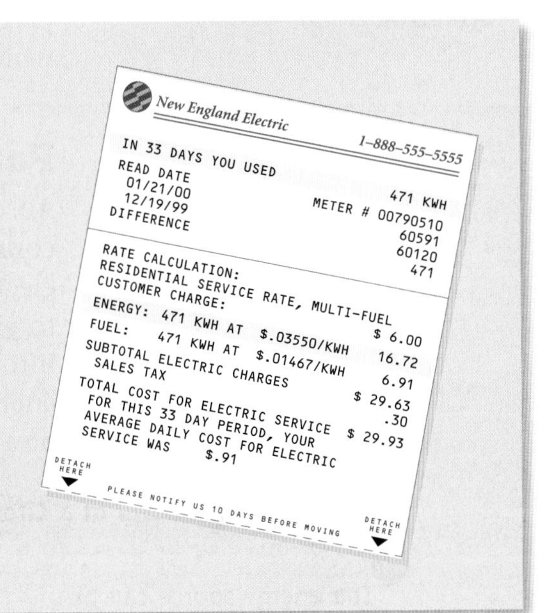

New England Electric 1-888-555-5555

IN 33 DAYS YOU USED 471 KWH

READ DATE METER # 00790510

01/21/00 60591

12/19/99 60120

DIFFERENCE 471

RATE CALCULATION:

RESIDENTIAL SERVICE RATE, MULTI-FUEL

CUSTOMER CHARGE: $ 6.00

ENERGY: 471 KWH AT $.03550/KWH 16.72

FUEL: 471 KWH AT $.01467/KWH 6.91

SUBTOTAL ELECTRIC CHARGES $ 29.63

SALES TAX .30

TOTAL COST FOR ELECTRIC SERVICE $ 29.93

FOR THIS 33 DAY PERIOD, YOUR AVERAGE DAILY COST FOR ELECTRIC SERVICE WAS $.91

DETACH HERE

PLEASE NOTIFY US 10 DAYS BEFORE MOVING

DETACH HERE

REVIEW

1. What is electric current?
2. How does increasing the voltage affect the current?
3. How does an electric power company calculate electrical energy from electric power?
4. **Making Predictions** Which wire would have the lowest resistance: a long, thin iron wire at a high temperature or a short, thick copper wire at a low temperature?
5. **Doing Calculations** Use Ohm's law to find the voltage needed to produce a current of 3 A in a device with a resistance of 9 Ω.

Section 4

Electric Circuits

NEW TERMS

circuit
series circuit
load
parallel circuit

OBJECTIVES

- Name the three essential parts of a circuit.
- Compare series circuits with parallel circuits.
- Explain how fuses and circuit breakers protect your home against short circuits and circuit overloads.

Imagine that you are lost in a forest. You need to find your way back to camp, where your friends are waiting for you. Unfortunately, there are no trails to follow, so you don't know which way to go. Just as you need a trail to follow in order to return to camp, electric charges need a path to follow in order to travel from an outlet or a battery to the device it provides energy to. A path that charges follow is called a circuit.

A circuit, however, is not exactly the same as a trail in a forest. A trail may begin in one place and end in another. But a circuit always begins and ends in the same place, forming a loop. Because a circuit forms a loop, it is said to be a closed path. So an electric **circuit** is a complete, closed path through which electric charges flow.

Parts of a Circuit

All circuits consist of an energy source, a load, and wires to connect the other parts together. A **load** is a device that uses electrical energy to do work. All loads offer some resistance to electric currents and cause the electrical energy to change into other forms of energy such as light energy or kinetic energy. **Figure 21** shows some examples of the different parts of a circuit.

Figure 21 Parts of a Circuit

a The energy source can be a battery, a photocell, a thermocouple, or an electric generator at a power plant.

b Wires connect the other parts of a circuit together. Wires are usually made of conducting materials with low resistance, such as copper.

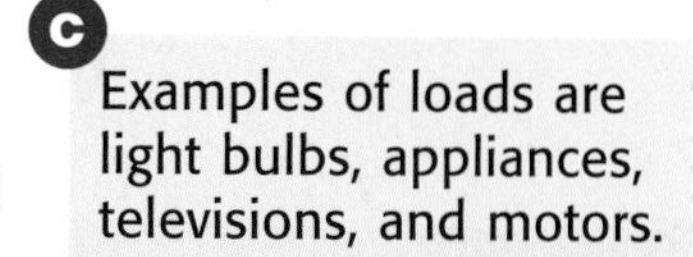

c Examples of loads are light bulbs, appliances, televisions, and motors.

Sometimes a circuit also contains a switch. A switch is used to open and close a circuit. Usually a switch is made of two pieces of conducting material, one of which can be moved, as shown in **Figure 22.** For charges to flow through a circuit, the switch must be closed, or "turned on." If a switch is open, or "off," the loop of the circuit is broken, and no charges can flow through the circuit. Light switches, power buttons on radios, and even the keys on calculators and computers work this way.

Believe it or not, your body is controlled by a large electric circuit. Electrical impulses from your brain control all the muscles and organs in your body. The food you eat is the energy source for your body's circuit, your nerves are the wires, and your muscles and organs are the loads.

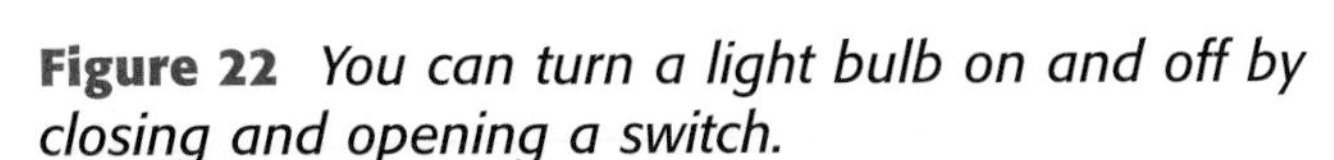

Figure 22 *You can turn a light bulb on and off by closing and opening a switch.*

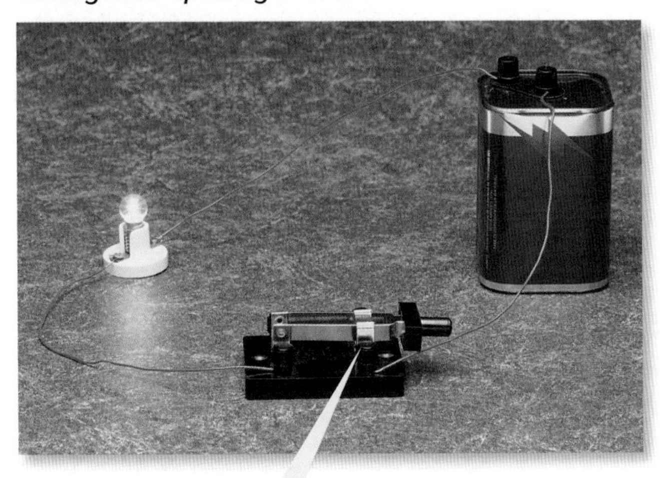

When the switch is closed, the two pieces of conducting material touch, allowing the electric charges to flow through the circuit.

When the switch is open, the gap between the two pieces of conducting material prevents the electric charges from traveling through the circuit.

Self-Check

Is a microwave oven an example of a load? Why or why not? *(See page 596 to check your answer.)*

Types of Circuits

Look around the room for a moment, and count the number of objects that use electrical energy. You probably found several objects, such as lights, a clock, and maybe a computer. All of the objects you counted are loads in a large circuit that may include several rooms in the building. In fact, most circuits contain more than one load. The loads in a circuit can be connected in two different ways—in series or in parallel.

Series Circuits A **series circuit** is a circuit in which all parts are connected in a single loop. The charges traveling through a series circuit must flow through each part and can only follow one path. **Figure 23** shows an example of a series circuit.

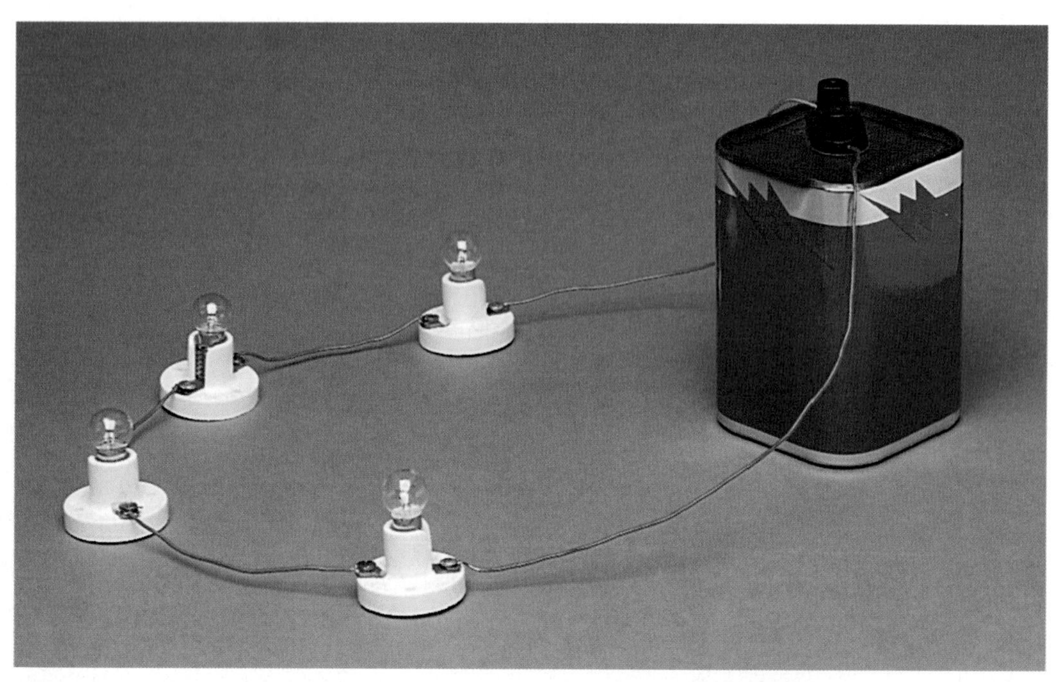

Figure 23 *The charges flow from the battery through each light bulb (load) and finally back to the battery.*

All the loads in a series circuit share the same current. Because the current in all the light bulbs in Figure 23 is the same, the light bulbs glow with the same brightness. However, if you added more light bulbs, the resistance of the entire circuit would increase and the current would decrease. Therefore, all the bulbs would be dimmer.

One disadvantage of series circuits is that all the loads in the circuit must be turned on and in working condition in order for charge to flow. For example, some older strings of holiday lights are wired as a series circuit. If all the light bulbs are good, all the light bulbs will shine. But if one of the light bulbs burns out, the circuit is broken, and all the lights go out. Imagine trying to figure out which bulb is burned out!

Although modern strings of holiday lights are also wired as a series circuit, the bulbs in the strings are designed to prevent this problem. Inside each bulb is a jumper—a piece of insulated wire that connects the wires leading to the filament. Normally, no charges pass through the jumper. But when a bulb burns out, the insulation burns off the jumper, and charges flow through the jumper. In this way, the jumper repairs the break in the circuit so the other bulbs stay lit.

Quick Lab

A Series of Circuits

1. Connect a **6 V battery** and **two flashlight bulbs** in a series circuit. Draw a picture of your circuit in your ScienceLog.
2. Add another **flashlight bulb** in series with the other two bulbs. How does the brightness of the light bulbs change?
3. Replace one of the light bulbs with a **burned-out light bulb.** What happens to the other lights in the circuit?

Parallel Circuits Think about what would happen if all the lights in your home were connected in series. If you needed a light on in your room, all the other lights in the house would have to be turned on too! Luckily, circuits in buildings are wired in parallel rather than in series. A **parallel circuit** is a circuit in which different loads are on separate branches. Because there are separate branches, the charges travel through more than one path. **Figure 24** shows a parallel circuit.

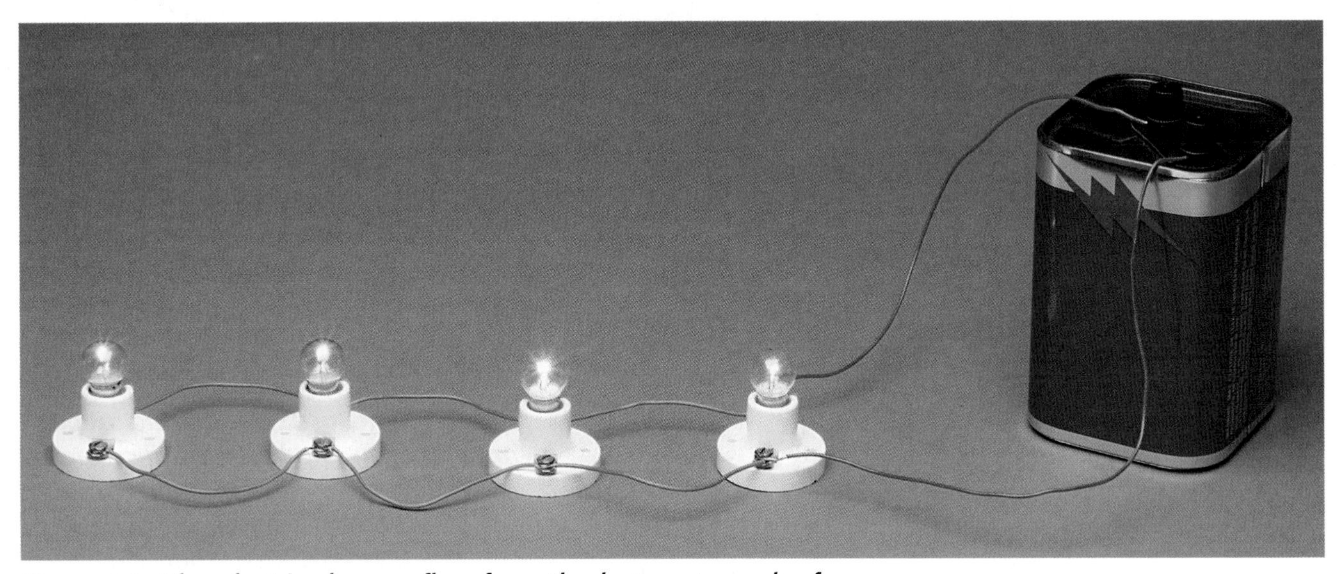

Figure 24 *The electric charges flow from the battery to each of the bulbs separately and then flow back to the battery.*

Unlike a series circuit, the loads in a parallel circuit do not have the same current in them. Instead, each load in a parallel circuit uses the same voltage. For example, the full voltage of the battery is applied to each bulb in Figure 24. As a result, each light bulb glows at full brightness, no matter how many bulbs are connected in parallel. You can connect loads that require different currents to the same parallel circuit. For example, you can connect a hair dryer, which requires a high current to operate, to the same circuit as a lamp, which requires less current.

The advantage that parallel circuits have over series circuits is that each loop of a parallel circuit can function by itself. If a break occurs in one of the circuit loops, none of the other loops will be affected. Therefore, charges will still run through the other loops, and the loads on the other loops will still work. Because of this, switches can be placed in the different branches of a parallel circuit in order to turn on each load individually. This makes it possible to use one light or appliance at a time.

QuickLab

A Parallel Lab

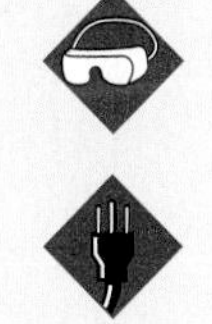

1. Connect a **6 V battery** and **two flashlight bulbs** in a parallel circuit. Draw a picture of your circuit in your ScienceLog.
2. Add another **flashlight bulb** in parallel with the other two bulbs. How does the brightness of the light bulbs change?
3. Replace one of the light bulbs with a **burned-out light bulb.** What happens to the other lights in the circuit?

Household Circuits

In every home, several circuits connect lights, major appliances, and outlets throughout the building. Most household circuits are parallel circuits that can have several loads attached to them. The circuits branch out from a breaker box or a fuse box that acts as the "electrical headquarters" for the building. Each branch receives a standard voltage, which is 120 V in the United States.

Mayday! Circuit Failure! Broken wires or water can cause electrical appliances to short-circuit. A short circuit occurs when charges bypass the loads in the circuit. When the loads are bypassed, the resistance of the circuit drops, and the current in the circuit increases. If the current increases too much, it can produce enough heat to start a fire. **Figure 25** shows how a short circuit might occur.

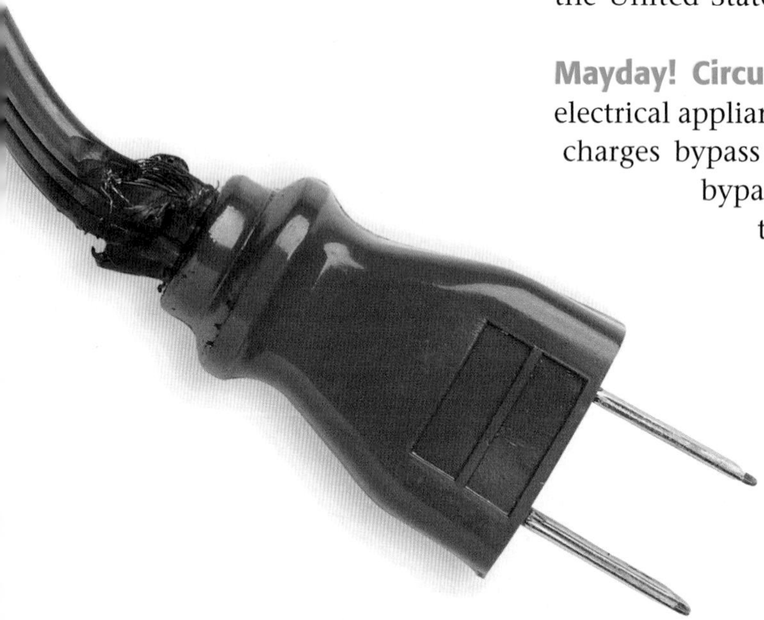

Figure 25 *If the insulating plastic around a cord is broken, the two wires inside can touch. The charges can then travel from one wire to another without traveling through the load. This is a short circuit!*

Circuits also may fail if they are overloaded. A circuit is overloaded when too many loads, or electrical devices, are attached to it. Each time you add a load to a parallel circuit, the entire circuit draws more current. If too many loads are attached to one circuit, the current increases to an unsafe level that can cause the temperature of the wires to increase and cause a fire. **Figure 26** shows a situation that can cause a circuit overload.

Figure 26 *Plugging too many devices into one outlet can cause a circuit to overload.*

Circuit Safety Because short circuits and circuit overloads can be so dangerous, safety features are built into the circuits in your home. The two most commonly used safety devices are fuses and circuit breakers. In houses, all the fuses or circuit breakers are located in a fuse box or a breaker box.

A fuse contains a thin strip of metal through which the charges for a circuit flow. If the current in the circuit is too high, the metal in the fuse warms up and melts, as shown in **Figure 27.** A break or gap in the circuit is produced, and the charges stop flowing. This is referred to as blowing a fuse. After a fuse is blown, you must replace it with a new fuse in order for the charges to flow through the circuit again.

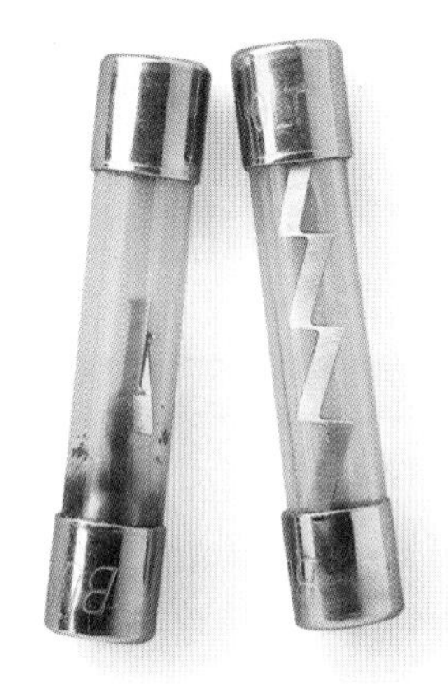

Figure 27 *The blown fuse on the left must be replaced with a new fuse, such as the one on the right.*

Circuit breakers are used more commonly than fuses because they are more convenient. A circuit breaker is a switch that automatically opens if the current in the circuit is too high. Inside a circuit breaker is a strip of metal that bends when it is heated. If the current in a circuit is too high, the metal warms up and bends away from the wires in the circuit, producing a break in the circuit. Open circuit breakers can be closed easily by flipping a switch inside the breaker box once the problem has been corrected.

A device that acts like a miniature circuit breaker is a ground fault circuit interrupter (GFCI). A GFCI, like the one shown in **Figure 28,** provides protection by comparing the current in one side of an outlet with the current in the other side. If there is even a small difference, the GFCI opens the circuit. To close the circuit, you must push the RESET button.

Figure 28 *GFCI devices are usually found on outlets in bathrooms and kitchens to protect you from electric shock.*

REVIEW

1. Name and describe the three essential parts of a circuit.
2. Why are switches useful in a circuit?
3. What is the difference between series circuits and parallel circuits?
4. How do fuses and circuit breakers protect your home against electrical fires?
5. **Developing Hypotheses** Whenever you turn on the portable heater in your room, the circuit breaker for the circuit in your room opens and all the lights go out. Propose two possible reasons for why this occurs.

Chapter Highlights

SECTION 1

Vocabulary

law of electric charges *(p. 271)*
electric force *(p. 271)*
conduction *(p. 273)*
induction *(p. 273)*
conductor *(p. 275)*
insulator *(p. 275)*
static electricity *(p. 275)*
electric discharge *(p. 276)*

Section Notes

- The law of electric charges states that like charges repel and opposite charges attract.
- The electric force varies depending on the size of the charges exerting the force and the distance between them.
- Objects become charged when they gain or lose electrons. Objects may become charged by friction, conduction, or induction.
- Charges are not created or destroyed and are said to be conserved.
- An electroscope can be used to detect charges.
- Charges move easily in conductors but do not move easily in insulators.
- Static electricity is the buildup of electric charges on an object. Static electricity is lost through electric discharge. Lightning is a form of electric discharge.
- Lightning rods work by grounding the electric charge carried by lightning.

Labs

Stop the Static Electricity! *(p. 566)*

SECTION 2

Vocabulary

cell *(p. 278)*
battery *(p. 278)*
potential difference *(p. 279)*
photocell *(p. 280)*
thermocouple *(p. 280)*

Section Notes

- Batteries are made of cells that convert chemical energy to electrical energy.
- Electric currents can be produced when there is a potential difference.
- Photocells and thermocouples are devices used to produce electrical energy.

Labs

Potato Power *(p. 567)*

Skills Check

Math Concepts

OHM'S LAW Ohm's law, shown on page 285, describes the relationship between current, voltage, and resistance. If you know two of the values, you can always calculate the third. For example, the current in a wire with a resistance of 4 Ω produced by a voltage of 12 V is calculated as follows:

$$I = \frac{V}{R} = \frac{12\text{ V}}{4\ \Omega} = 3\text{ A}$$

Visual Understanding

SERIES AND PARALLEL CIRCUITS There are two types of circuits—series and parallel. The charges in a series circuit follow only one path, but the charges in a parallel circuit follow more than one path. Look at Figures 23 and 24 on pages 290–291 to review series and parallel circuits.

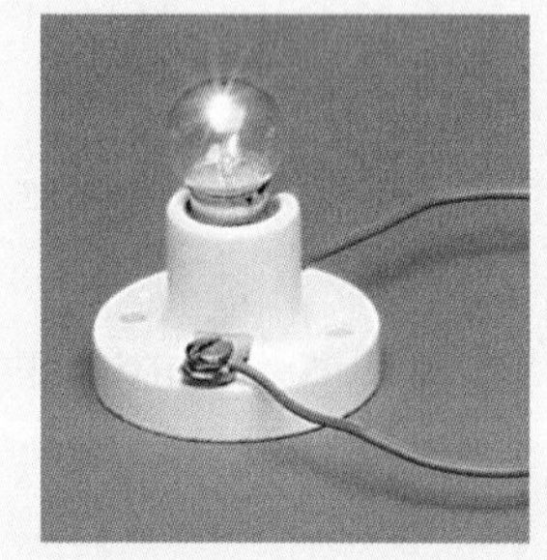

SECTION 3

Vocabulary

current *(p. 281)*
voltage *(p. 282)*
resistance *(p. 283)*
electric power *(p. 285)*

Section Notes

- Electric current is a continuous flow of charge caused by the motion of electrons.
- Voltage is the same as potential difference. As voltage increases, current increases.
- An object's resistance varies depending on the object's material, thickness, length, and temperature. As resistance increases, current decreases.
- Ohm's law describes the relationship between current, resistance, and voltage.
- Electric power is the rate at which electrical energy does work and is expressed in watts or kilowatts.
- Electrical energy is electric power multiplied by time. It is usually expressed in kilowatt-hours.

SECTION 4

Vocabulary

circuit *(p. 288)*
load *(p. 288)*
series circuit *(p. 290)*
parallel circuit *(p. 291)*

Section Notes

- Circuits consist of an energy source, a load, wires, and sometimes a switch.
- All parts of a series circuit are connected in a single loop.
- The loads in a parallel circuit are on separate branches.
- Circuits can fail because of a short circuit or circuit overload.
- Fuses or circuit breakers protect your home against circuit failure.

Labs

Circuitry 101 *(p. 568)*

internet**connect**

GO TO: go.hrw.com

Visit the **HRW** Web site for a variety of learning tools related to this chapter. Just type in the keyword:

KEYWORD: HSTELE

GO TO: www.scilinks.org

Visit the **National Science Teachers Association** on-line Web site for Internet resources related to this chapter. Just type in the ***sci*LINKS** number for more information about the topic:

TOPIC: Static Electricity — ***sci*LINKS NUMBER:** HSTP405
TOPIC: Electrical Energy — ***sci*LINKS NUMBER:** HSTP410
TOPIC: Electric Current — ***sci*LINKS NUMBER:** HSTP415
TOPIC: Electric Circuits — ***sci*LINKS NUMBER:** HSTP420

Chapter Review

USING VOCABULARY

To complete the following sentences, choose the correct term from each pair of terms listed below:

1. A __?__ converts chemical energy into electrical energy. *(battery* or *photocell)*
2. Charges flow easily in a(n) __?__. *(insulator* or *conductor)*
3. __?__ is the opposition to the flow of electric charge. *(Resistance* or *Electric power)*
4. A __?__ is a complete, closed path through which charges flow. *(load* or *circuit)*
5. Lightning is a form of __?__. *(static electricity* or *electric discharge)*

UNDERSTANDING CONCEPTS

Multiple Choice

6. If two charges repel each other, the two charges must be
   a. positive and positive.
   b. positive and negative.
   c. negative and negative.
   d. Either (a) or (c)
7. A device that can convert chemical energy to electrical energy is a
   a. lightning rod.
   b. cell.
   c. light bulb.
   d. All of the above

8. Which of the following wires has the lowest resistance?
   a. a short, thick copper wire at 25°C
   b. a long, thick copper wire at 35°C
   c. a long, thin copper wire at 35°C
   d. a short, thick iron wire at 25°C
9. An object becomes charged when the atoms in the object gain or lose
   a. protons.
   b. neutrons.
   c. electrons.
   d. All of the above
10. A device used to protect buildings from electrical fires is a(n)
    a. electric meter.
    b. circuit breaker.
    c. fuse.
    d. Both (b) and (c)
11. In order to produce a current from a cell, the electrodes of the cell must
    a. have a potential difference.
    b. be in a liquid.
    c. be exposed to light.
    d. be at two different temperatures.
12. What type of current comes from the outlets in your home?
    a. direct current
    b. alternating current
    c. electric discharge
    d. static electricity

Short Answer

13. List and describe the three essential parts of a circuit.
14. Name the two factors that affect the strength of electric force, and explain how they affect electric force.
15. Describe how direct current differs from alternating current.

Concept Mapping

16. Use the following terms to create a concept map: electric current, battery, charges, photocell, thermocouple, circuit, parallel circuit, series circuit.

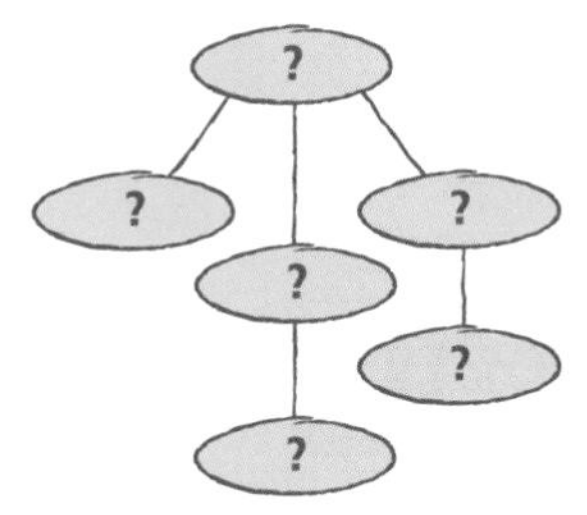

CRITICAL THINKING AND PROBLEM SOLVING

17. Your science classroom was rewired over the weekend. On Monday, you notice that the electrician may have made a mistake. In order for the fish-tank bubbler to work, the lights in the room must be on. And if you want to use the computer, you must turn on the overhead projector. Describe what mistake the electrician made with the circuits in your classroom.

18. You can make a cell using an apple, a strip of copper, and a strip of silver. Explain how you would construct the cell, and identify the parts of the cell. What type of cell is formed? Explain your answer.

19. Your friend shows you a magic trick. She rubs a plastic comb with a piece of silk and holds it close to a stream of water. When the comb is close to the water, the water bends toward the comb. Explain how this trick works. (Hint: Think about how objects become charged.)

MATH IN SCIENCE

Use Ohm's law to solve the following problems:

20. What voltage is needed to produce a 6 A current through a resistance of 3 Ω?

21. Find the current produced when a voltage of 60 V is applied to a resistance of 15 Ω.

22. What is the resistance of an object if a voltage of 40 V produces a current of 5 A?

INTERPRETING GRAPHICS

23. Classify the objects in the photograph below as conductors or insulators.

NOW What Do You Think?

Take a minute to review your answers to the ScienceLog questions on page 269. Have your answers changed? If necessary, revise your answers based on what you have learned since you began this chapter.

Science, Technology, and Society

Riding the Electric Rails

For more than 100 years, the trolley, or streetcar, was a popular way to travel around a city. Then, beginning in the 1950s, most cities ripped up their trolley tracks to make way for automobiles. Today, trolleys are making a comeback around the world.

From Horse Power to Electric Power

In 1832, the first trolleys, called *horsecars,* were pulled by horses through the streets of New York. Soon horsecars were used in most large cities in the United States. However, using horses for power presented several problems. Among other things, the horses were slow and required special attention and constant care. So inventors began looking for other sources of power.

In 1888, Frank J. Sprague developed a way to operate trolleys with electrical energy. These electric trolleys ran on a metal track and were connected by a pole to an overhead power line. Electric charges flowed down the pole to motors in the trolley. A wheel at the top of the pole, called a *shoe,* rolled along the power line, allowing the trolley to move along its track without losing contact with its power source. The charges passed through the motor and then returned to a power generator by way of the metal track.

▲ *The horsecar was a popular mode of travel in many cities during the early 1900s.*

Taking It to the Streets

By World War I, more than 40,000 km of electric-trolley tracks were in use in the United States. The trolley's popularity helped shape American cities because businesses were built along the trolley lines. But competition from cars and buses grew over the next decade, and many trolley lines were abandoned.

By the 1980s, nearly all of the trolley lines had been shut down. But by then, people were looking for new ways to cut down on the pollution, noise, and traffic problems caused by automobiles and buses. Trolleys provided one possible solution. Because they run on electrical energy, they create little pollution, and because many people can ride on a single trolley, they cut down on traffic.

Today, a new form of trolley is being used in a number of major cities. These light-rail transit vehicles are quieter, faster, and more economical than the older trolleys. They usually run on rails alongside the road and contain new systems, such as automated brakes and speed controls.

Think About It!

► Because trolleys operate on electrical energy, does this mean that they don't create any pollution? Explain your answer.

▲ *Many cities across the country now use light-rail systems for public transportation.*

Sprites and Elves

Imagine you are a pilot flying a plane on a moonless night. About 80 km away, you notice a powerful thunderstorm and see the lightning move *between* the clouds and Earth. This makes sense because you know that all weather activity takes place in the lowest layer of Earth's atmosphere, which is called the troposphere. But all of a sudden, a ghostly red glow stretches many kilometers *above* the storm clouds and *into* the stratosphere!

▲ *Sprites (left) and elves (right) are strange electric discharges in the atmosphere.*

Capturing Sprites

In 1989, scientists at the University of Minnesota followed the trail of many such reports. They captured the first image of this strange, red-glowing lightning using a video camera. Since then, photographs from space shuttles, airplanes, telescopes, and observers on the ground have identified several types of wispy electrical glows. Two of these types were named sprites and elves because, like the mythical creatures, they last only a few thousandths of a second and disappear just as the eye begins to see them.

Photographs show that sprites and elves occur only when ordinary lightning is discharged from a cloud. Sprites are very large, extending from the cloud tops at about 15 km altitude to as high as 95 km. They are up to 50 km wide. Elves are expanding disks of red light, probably caused by an electromagnetic pulse from lightning or sprites. Elves can be 200 km across and appear at altitudes above 90 km.

What Took So Long?

It is likely that sprites and elves have been occurring for thousands of years but went unrecorded. This is because they are produced with only about 1 percent of lightning flashes. They also last for a short period of time and are very faint. Since they occur above thunderclouds, where few people can see, observers are more often distracted by the brighter lightning below.

Still, scientists are not surprised to learn that electric discharges extend up from clouds. There is a large potential difference between thunderclouds and the ionosphere, a higher atmospheric level. The ionosphere is an electrically conductive layer of the atmosphere that provides a path for these electric discharges.

Search and Find

▶ Would you like to find sprites on your own? (Elves disappear too quickly.) Go with an adult, avoid being out in a thunderstorm, and remember:

- It must be completely dark, and your eyes must adjust to the total darkness.
- Viewing is best when a large thunderstorm is 48 to 97 km away, with no clouds in between.
- Block out the lightning below the clouds with dark paper so that you can still see above the clouds.
- Be patient.

Report sightings to a university geophysical department. Scientists need more information to fully understand how these discharges affect the chemical and electrical workings of our atmosphere.

UNIT 4

Thousands of years ago, people began asking the question, "What is matter made of?" This unit follows the discoveries and ideas that have led to our current theories about what makes up matter. You will learn about the atom—the building block of all matter—and its structure. You will also learn how the periodic table is used to classify and organize elements according to patterns in atomic structure and other properties. This timeline illustrates some of the events that have brought us to our current understanding of atoms and of the periodic table in which they are organized.

TIMELINE

The Atom

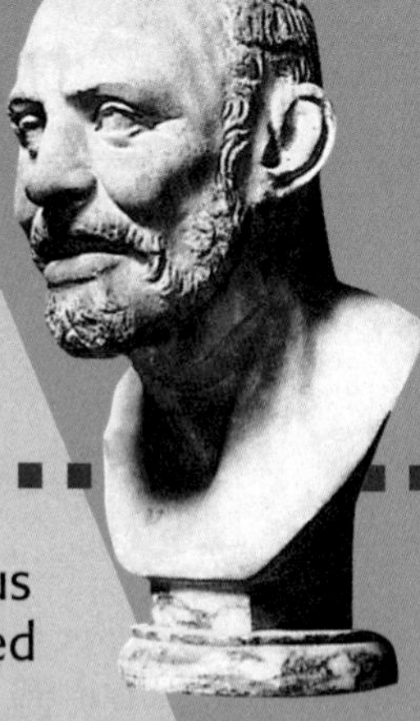

Around 400 B.C.

The Greek philosopher Democritus proposes that small particles called atoms make up all matter.

1911

Ernest Rutherford, a physicist from New Zealand, discovers the positively charged nucleus of the atom.

1932

The neutron, one of the particles in the nucleus of an atom, is discovered by British physicist James Chadwick.

1945

The United Nations is formed. Its purpose is to maintain world peace and develop friendly relations between countries.

1964

Scientists propose the idea that smaller particles make up protons and neutrons. The particles are named quarks after a word used by James Joyce in his book *Finnegans Wake*.

1803

British scientist and school teacher John Dalton reintroduces the concept of atoms with evidence to support his ideas.

1848

James Marshall finds gold while building Sutter's Mill, starting the California gold rush.

1869

Russian chemist Dmitri Mendeleev develops a periodic table that organizes the elements known at the time.

1897

British scientist J. J. Thomson identifies electrons as particles that are present in every atom.

1898

British scientists Sir William Ramsay and Morris W. Travers discover three elements—krypton, neon, and xenon—in three months. The periodic table developed by Mendeleev helps guide their research.

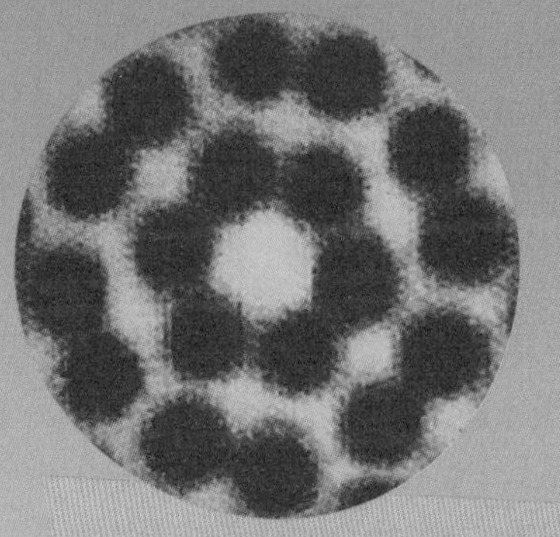

1981

Scientists in Switzerland develop a scanning tunneling microscope, which is used to see atoms for the first time.

1989

Germans celebrate when the Berlin Wall ceases to function as a barrier between East and West Germany.

1996

Another element is added to the periodic table after a team of German scientists synthesize an atom containing 112 protons in its nucleus.

CHAPTER 12

Introduction to Atoms

Would You Believe . . . ?

Tiny atoms have something in common with huge dinosaurs. In both cases, scientists have had to try to understand something they could not observe firsthand!

No one has ever seen a living dinosaur. So how did the special-effects crew for the movie *Jurassic Park* know what the *Tyrannosaurus rex* model, shown above, should look like? Scientists have determined the appearance of *T. rex* by studying fossilized skeletons. Based on fossil evidence, scientists theorize that these now-extinct creatures had big hind legs, small front legs, a long whip-like tail, and an enormous mouth full of dagger-shaped teeth.

However, theories of how *T. rex* walked have been harder to develop because there is no way to see a dinosaur in motion. For many years, most scientists thought that *T. rex* (and all dinosaurs) plodded slowly like big, lazy lizards.

However, after studying well-preserved dinosaur tracks, like those shown below, and noticing skeletal similarities between certain dinosaur fossils and living creatures such as the ostrich, many scientists now theorize that *T. rex* and its dinosaur cousins could turn on the speed. Some scientists estimate that *T. rex* had bursts of speed of 32 km/h (20 mi/h)!

Theories about *T. rex* and other dinosaurs have changed gradually over many years based on indirect evidence, such as dinosaur tracks. Likewise, our theory of the atom has changed and grown over thousands of years as scientists have uncovered more evidence about the atom, even though they were unable to see an atom directly. In this chapter, you'll learn about the development of the atomic theory and our current understanding of atomic structure.

Where Is It?

Theories about the internal structure of atoms were developed by aiming moving particles at atoms. In this activity you will develop an idea about the location and size of a hidden object by rolling marbles at the object.

What Do You Think?

In your ScienceLog, try to answer the following questions based on what you already know:

1. What are some ways that scientists have described the atom?
2. What are the parts of the atom, and how are they arranged?
3. How are atoms of all elements alike?

Procedure

1. Place a rectangular piece of **cardboard** on **four books or blocks** so that each corner of the cardboard rests on a book or block.
2. Ask your teacher to place the **unknown object** under the cardboard. Be sure that you do not see it.
3. Place a **large piece of paper** on top of the cardboard.
4. Gently roll a **marble** under the cardboard, and record on the paper the position where the marble enters and exits and the direction it travels.
5. Continue rolling the marble from different directions to determine the shape and location of the object.
6. Write down all your observations in your ScienceLog.

Analysis

7. Form a conclusion about the object's shape, size, and location. Record your conclusion in your ScienceLog.

Section 1

Development of the Atomic Theory

NEW TERMS

atom
theory
electrons
model
nucleus
electron clouds

OBJECTIVES

- Describe some of the experiments that led to the current atomic theory.
- Compare the different models of the atom.
- Explain how the atomic theory has changed as scientists have discovered new information about the atom.

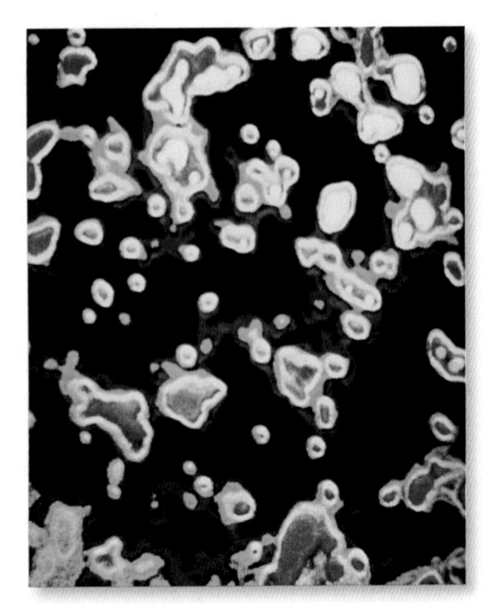

The photo at right shows uranium atoms magnified 3.5 million times by a scanning tunneling microscope. An **atom** is the smallest particle into which an element can be divided and still be the same substance. Atoms make up elements; elements combine to form compounds. Because all matter is made of elements or compounds, atoms are often called the building blocks of matter.

Before the scanning tunneling microscope was invented, in 1981, no one had ever seen an atom. But the existence of atoms is not a new idea. As you will find out, our understanding of atoms has been developing for more than 2,000 years. How is this possible? The answer has to do with theories. A **theory** is a unifying explanation for a broad range of hypotheses and observations that have been supported by testing. In this section, you will take a short trip through history to see for yourself how our understanding of atoms developed even before we could observe them directly. Your first stop—ancient Greece.

Democritus Proposes the Atom

Look at the silver coin shown in **Figure 1.** Imagine that you cut the coin in half, then cut those halves in half, and so on. Could you keep cutting the pieces in half forever, or would you eventually end up with a particle that you could not cut?

Around 440 B.C., a Greek philosopher named Democritus (di MAHK ruh tuhs) proposed that in such a situation, you would end up with an "uncuttable" particle. He called this particle an *atom* (from the Greek word *atomos,* meaning "indivisible"). Democritus proposed that all atoms are small, hard particles made of a single material formed into different shapes and sizes. He also claimed that atoms are always moving and that they form different materials by joining together.

Figure 1 *Democritus thought the smallest particle in an object like this silver coin was an atom. This coin was in use during Democritus's time.*

Aristotle (ER is TAHT uhl), a Greek philosopher who lived from 384 to 322 B.C., disagreed with Democritus's ideas. He believed that you could keep cutting an object in half over and over and never end up with an indivisible particle. Although Aristotle's ideas were eventually proved incorrect, he had such a strong influence on popular belief that Democritus's ideas were largely ignored for centuries.

In 342 or 343 B.C., King Phillip II of Macedon appointed Aristotle to be a tutor for his son, Alexander. Alexander later conquered Greece and the Persian Empire (in what is now Iran) and became known as Alexander the Great.

Dalton Creates an Atomic Theory Based on Experiments

By the late 1700s, scientists had learned that elements combine in specific proportions to form compounds. These proportions are based on the mass of the elements in the compounds. For example, hydrogen and oxygen always combine in the same proportion to form water. John Dalton, a British chemist and school teacher, wanted to know why. He performed experiments with different substances. His results demonstrated that elements combine in specific proportions because they are made of individual atoms. After many experiments and observations, Dalton, shown in **Figure 2,** published his own atomic theory in 1803. His theory stated the following:

- **All substances are made of atoms. Atoms are small particles that cannot be created, divided, or destroyed.**
- **Atoms of the same element are exactly alike, and atoms of different elements are different.**
- **Atoms join with other atoms to make new substances.**

It took many years for scientists to accept Dalton's atomic theory, but toward the end of the nineteenth century scientists agreed that his theory explained many of their observations. However, as new information was discovered that could not be explained by Dalton's ideas, the atomic theory changed. The theory was revised to more correctly explain the atom. As you read on, you will learn how Dalton's theory has changed, step by step, into the current atomic theory.

Figure 2 *John Dalton developed his atomic theory from observations gathered from many experiments.*

Thomson Finds Electrons in the Atom

In 1897, a British scientist named J. J. Thomson made a discovery that identified an error in Dalton's theory. Using relatively simple equipment (compared with modern scientific equipment), Thomson discovered that there are small particles *inside* the atom. Therefore, atoms can be divided into even smaller parts. Atoms are not indivisible, as proposed by Dalton.

Thomson experimented with a cathode-ray tube, as shown in **Figure 3.** He discovered that the direction of the beam was affected by electrically charged plates. Notice in the illustration that the plate marked with a positive sign, which represents a positive charge, attracts the beam. Because the beam was pulled toward a positive charge, Thomson concluded that the beam was made of particles with a negative electric charge.

Figure 3 Thomson's Cathode-Ray Tube Experiment

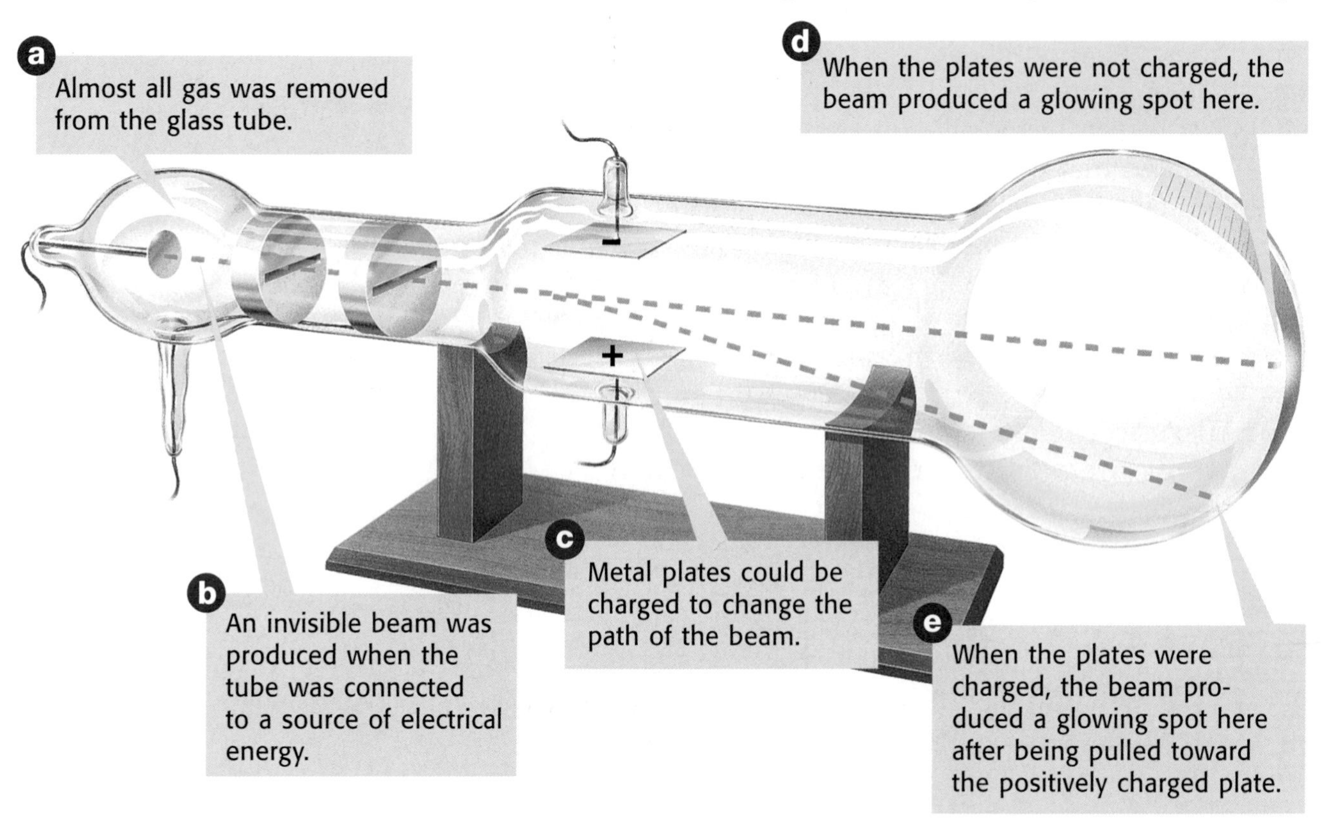

Just What Is Electric Charge?

Have you ever rubbed a balloon on your hair? The properties of your hair and the balloon seem to change, making them attract one another. To describe these observations, scientists say that the balloon and your hair become "charged." There are two types of charges, positive and negative. Objects with opposite charges attract each other, while objects with the same charge push each other away. When Thomson observed that the beam was pulled toward a positively charged plate, he concluded that the particles in the beam must be negatively charged.

Thomson repeated his experiment several times and found that the particle beam behaved in exactly the same way each time. He called the particles in the beam corpuscles (KOR PUHS uhls). His results led him to conclude that corpuscles are present in every type of atom and that all corpuscles are identical. The negatively charged particles found in all atoms are now called **electrons.**

BRAIN FOOD

The word *electron* comes from a Greek word meaning "amber." The ancient Greeks knew that a piece of amber (the solidified sap from ancient trees) could attract small bits of straw and paper after being rubbed with cloth.

Like Plums in a Pudding Thomson knew that electrons were a part of atoms and that Dalton's belief that atoms could not be divided was therefore incorrect. Thomson revised the atomic theory to account for the presence of electrons, but he still did not know how electrons are arranged inside atoms. In addition, chemists knew that atoms have no overall charge, so Thomson realized that positive charges must be present to balance the negative charges of the electrons. But just as Thomson didn't know the location of the electrons, he didn't know the location of the positive charges. He proposed a model to describe a possible structure of the atom. A **model** is a representation of an object or system. A model is different from a theory in that a model presents a picture of what the theory explains.

In Thomson's model, illustrated in **Figure 4,** the atom is a positively charged blob of material with electrons scattered throughout. This model came to be known as the plum-pudding model, named for an English dessert that was popular at the time. The electrons could be compared to the plums that were found throughout the pudding. Today you might call Thomson's model the chocolate-chip-ice-cream model; electrons in the atom could be compared to the chocolate chips found throughout the ice cream!

Thomson proposed that the atom is mostly positively charged material.

In Thomson's model, electrons are small, negatively charged particles located throughout the positive material.

Figure 4 *Thomson's plum-pudding model of the atom is shown above. A modern version of Thomson's model might be chocolate-chip ice cream.*

REVIEW

1. What discovery demonstrated that atoms are not the smallest particles?
2. What did Dalton do in developing his theory that Democritus did not do?
3. **Analyzing Methods** Why was it important for Thomson to repeat his experiment?

Rutherford Opens an Atomic "Shooting Gallery"

Find out about Melissa Franklin, a modern atom explorer, on page 323.

In 1909, a former student of Thomson's named Ernest Rutherford decided to test Thomson's theory. He designed an experiment to investigate the structure of the atom. He aimed a beam of small, positively charged particles at a thin sheet of gold foil. These particles were larger than *protons,* even smaller positive particles identified in 1902. Even though the gold foil was thinner than the foil used to wrap a stick of chewing gum, it was still about 10,000 atoms thick! **Figure 5** shows a diagram of Rutherford's experiment. To find out where the particles went after being "shot" at the gold foil, Rutherford surrounded the foil with a screen coated with zinc sulfide, a substance that glowed when struck by the particles.

Figure 5 Rutherford's Gold Foil Experiment

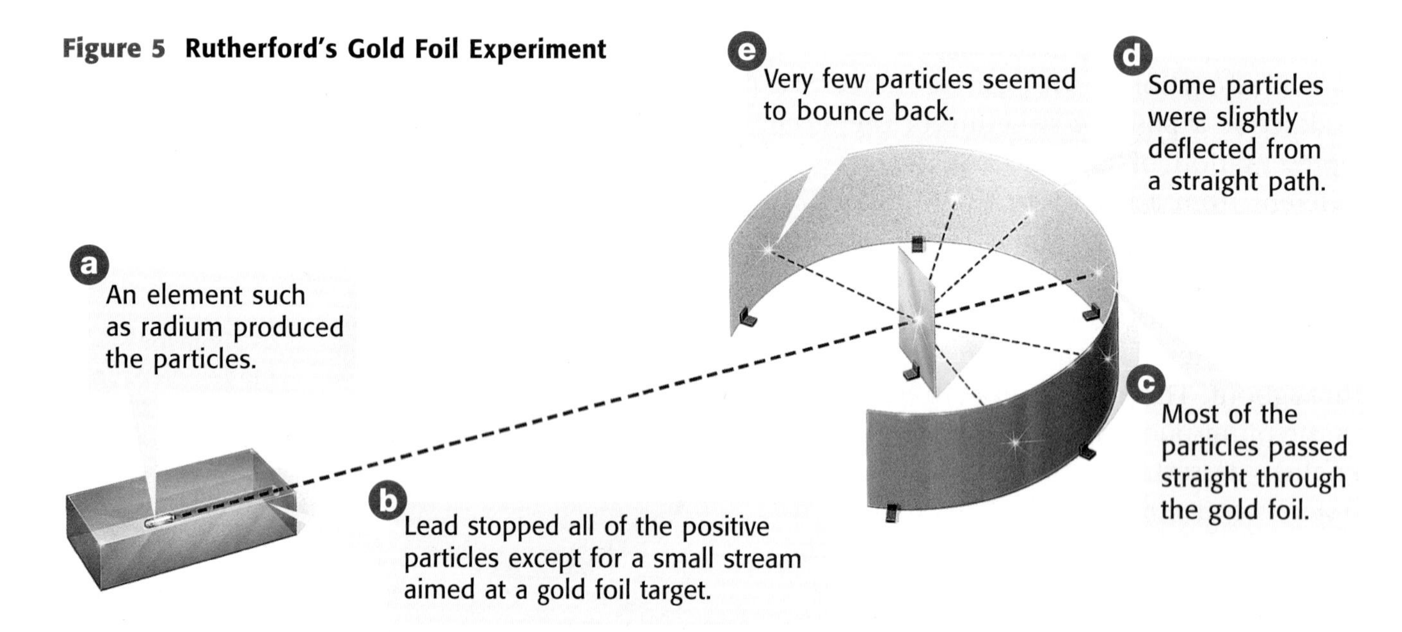

Rutherford thought that if atoms were soft "blobs" of material, as suggested by Thomson, then the particles would pass through the gold and continue in a straight line. Most of the particles did just that. But to Rutherford's great surprise, some of the particles were deflected (turned to one side) a little, some were deflected a great deal, and occasionally a particle seemed to bounce back. When describing his amazement, Rutherford reportedly said,

> *"It was quite the most incredible event that has ever happened to me in my life. It was almost as if you fired a fifteen-inch shell into a piece of tissue paper and it came back and hit you."*

Rutherford Presents a New Atomic Model It was obvious to Rutherford that the plum-pudding model of the atom did not explain his results. In 1911, he revised the atomic theory. Rutherford concluded that because almost all of the particles had passed through the gold foil, atoms are mostly empty space. He proposed that the lightweight, negative electrons move in the empty space.

To explain the deflection of the other particles, Rutherford proposed that in the center of the atom is a tiny, extremely dense, positively charged region called the **nucleus** (NOO klee uhs). The Rutherford model is illustrated in **Figure 6.** From the results of his experiment, Rutherford reasoned that positively charged particles that passed close by the nucleus were pushed away from their straight-line path by the positive charges in the nucleus. (Remember, opposite charges attract, and like charges repel.) Occasionally, a particle would head straight for a nucleus and be pushed almost straight back in the direction from which it came.

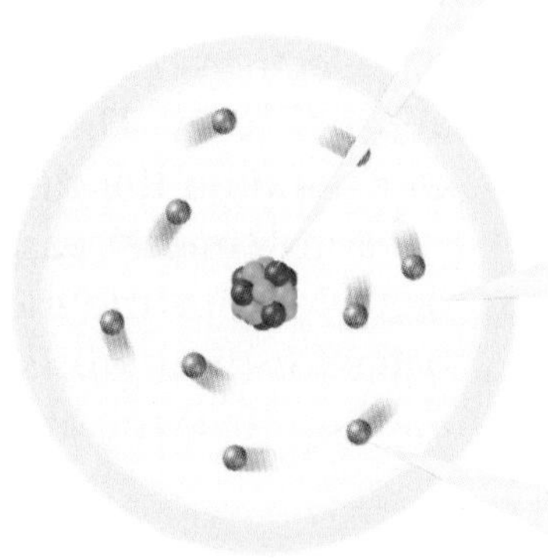

Because a few particles were deflected by the foil, Rutherford proposed that the atom has a small, dense, positively charged nucleus. Most of the atom's mass is concentrated here.

Rutherford proposed that because most particles passed straight through the gold foil, the atom is mostly empty space through which electrons travel.

Rutherford suspected that electrons travel around the nucleus like planets around the sun, but he could not explain the exact arrangement of the electrons.

Figure 6 *The results of Rutherford's experiment led to a new model of the atom.*

From the results of Rutherford's experiment, he calculated that the diameter of the nucleus was 100,000 times smaller than the diameter of the gold atom. To imagine how small this is, look at **Figure 7.**

Figure 7 *The diameter of this pinhead is 100,000 times smaller than the diameter of the stadium. Likewise, the diameter of a nucleus is 100,000 times smaller than the diameter of an atom.*

Self-Check

Why did Thomson believe the atom contains positive charges? *(See page 596 to check your answer.)*

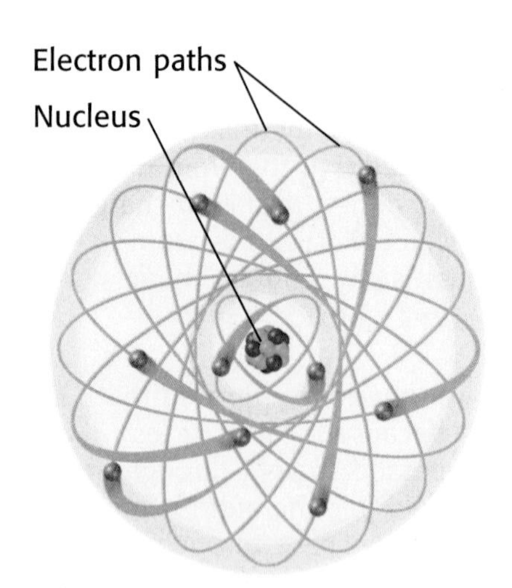

Figure 8 *Bohr proposed that electrons move in paths at certain distances around the nucleus.*

Bohr States That Electrons Can Jump Between Levels

The next step in understanding the atom came just 2 years later, from a Danish scientist who worked with Rutherford. In 1913, Niels Bohr suggested that electrons travel around the nucleus in definite paths. These paths are located in levels at certain distances from the nucleus, as illustrated in **Figure 8.** Bohr proposed that no paths are located between the levels, but electrons can jump from a path in one level to a path in another level. Think of the levels as rungs on a ladder. You can stand *on* the rungs of a ladder but not *between* the rungs.

Bohr's model was a valuable tool in predicting some atomic behavior. But the model was too simple to explain all of the behavior of atoms, so scientists continued to study the atom and improve the atomic theory.

The Modern Theory: Electron Clouds Surround the Nucleus

Many twentieth-century scientists have contributed to our current understanding of the atom. An Austrian physicist named Erwin Schrödinger (1887–1961) and a German physicist named Werner Heisenberg (1901–1976) made particularly important contributions. Their work further explained the nature of electrons in the atom. For example, electrons do not travel in definite paths as Bohr suggested. In fact, the exact path of a moving electron cannot be predicted. According to the current theory, there are regions inside the atom where electrons are *likely* to be found—these regions are called **electron clouds.** Electron clouds are related to the paths described in Bohr's model. The electron-cloud model of the atom is illustrated in **Figure 9.**

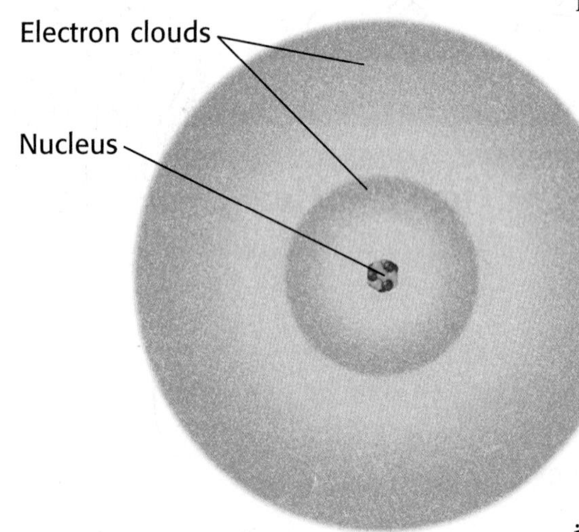

Figure 9 *In the current model of the atom, regions of the atom called electron clouds are the most likely places to find electrons.*

REVIEW

1. In what part of an atom is most of its mass located?
2. What are two differences between the atomic theory described by Thomson and that described by Rutherford?
3. **Comparing Concepts** Identify the difference in how Bohr's theory and the modern theory describe the location of electrons.

Section 2

The Atom

NEW TERMS

protons
atomic mass unit (amu)
neutrons
atomic number
isotopes
mass number
atomic mass

OBJECTIVES

- Compare the charge, location, and relative mass of protons, neutrons, and electrons.
- Calculate the number of particles in an atom using the atomic number, mass number, and overall charge.
- Calculate the atomic mass of elements.

In the last section, you learned how the atomic theory developed through centuries of observation and experimentation. Now it's time to learn about the atom itself. In this section, you'll learn about the particles inside the atom, and you'll learn about the forces that act on those particles. But first you'll find out just how small an atom really is.

How Small Is an Atom?

The photograph below shows the pattern that forms when a beam of electrons is directed at a sample of aluminum. By analyzing this pattern, scientists can determine the size of an atom. Analysis of similar patterns for many elements has shown that aluminum atoms, which are average-sized atoms, have a diameter of about 0.00000003 cm. That's three hundred-millionths of a centimeter. That is so small that it would take a stack of 50,000 aluminum atoms to equal the thickness of a sheet of aluminum foil from your kitchen!

As another example, consider an ordinary penny. Believe it or not, a penny contains about 2×10^{22} atoms, which can be written as 20,000,000,000,000,000,000,000 atoms, of copper and zinc. That's twenty thousand billion billion atoms—4,000,000,000,000 times more atoms than there are people on Earth! So if there are that many atoms in a penny, each atom must be very small. You can get a better idea of just how small an atom is in **Figure 10.**

Figure 10 *If you could enlarge a penny until it was as wide as the continental United States, each of its atoms would be only about 3 cm in diameter—about the size of this table-tennis ball.*

What's Inside an Atom?

As tiny as an atom is, it consists of even smaller particles—protons, neutrons, and electrons—as shown in the model in **Figure 11.** (The particles represented in the figures are not shown in their correct proportions because the electrons would be too small to see.) Protons and neutrons make up the nucleus, which is the center of the atom. Electrons are found outside the nucleus.

Electrons are negatively charged particles found in electron clouds outside the nucleus. The size of the electron clouds determines the size of the atom.

Protons are positively charged particles in the nucleus of an atom.

The **nucleus** is the small, dense, positively charged center of the atom. It contains most of the atom's mass.

Neutrons are particles in the nucleus of an atom that have no charge.

The diameter of the nucleus is 1/100,000 the diameter of the atom.

Figure 11 *An atom consists of three different types of particles–protons, neutrons, and electrons.*

Particle Profile	
Name:	proton
Charge:	positive
Mass:	1 amu
Location:	nucleus

Particle Profile	
Name:	neutron
Charge:	none
Mass:	1 amu
Location:	nucleus

The Nucleus **Protons** are the positively charged particles of the nucleus. It was these particles that repelled Rutherford's "bullets." All protons are identical, and each proton has a positive charge. The mass of a proton is approximately 1.7×10^{-24}g, which can also be written as 0.0000000000000000000000017g. Because the masses of particles in atoms are so small, scientists developed a unit of measurement for them. The SI unit used to measure the masses of particles in atoms is the **atomic mass unit (amu).** Scientists assigned each proton a mass of 1 amu.

Neutrons are the particles of the nucleus that have no charge. All neutrons are identical. Neutrons are slightly more massive than protons, but the difference in mass is so small that neutrons are also given a mass of 1 amu.

Protons and neutrons are the most massive particles in an atom, yet the nucleus they form has a very small volume. In other words, the nucleus is very dense. In fact, if it were possible for a nucleus to have a volume of 1 cm^3—the volume of an average grape—that nucleus would have a mass greater than 9 million metric tons!

Outside of the Nucleus *Electrons* are the negatively charged particles in atoms. The current atomic theory states that electrons are found moving around the nucleus within electron clouds. The charges of protons and electrons are opposite but equal in size. Therefore, whenever there are equal numbers of protons and electrons, their charges cancel out. An atom has no overall charge and is described as being neutral. If the number of electrons is different from the number of protons, the atom becomes a charged particle called an *ion* (IE ahn). Ions are positively charged if the protons outnumber the electrons, and they are negatively charged if the electrons outnumber the protons.

Electrons are very small in mass compared with protons and neutrons. It takes more than 1,800 electrons to equal the mass of 1 proton. In fact, the mass of an electron is so small that it is usually considered to be zero.

Particle Profile

Name: electron
Charge: negative
Mass: almost zero
Location: electron clouds

REVIEW

1. What particles form the nucleus?
2. Explain why atoms are neutral.
3. **Summarizing Data** Why do scientists say that most of the mass of an atom is located in the nucleus?

LabBook

Help wanted! Elements-4-U needs qualified nucleus builders. Report to page 570 of the LabBook.

How Do Atoms of Different Elements Differ?

There are 112 different elements, each of which is made of different atoms. What makes atoms different from each other? To find out, imagine that it's possible to "build" an atom by putting together protons, neutrons, and electrons.

It's easiest to start with the simplest atom. Protons and electrons are found in all atoms, and the simplest atom consists of just one of each. It's so simple it doesn't even have a neutron. Put just one proton in the center of the atom for the nucleus. Then put one electron in the electron cloud, as shown in the model in **Figure 12.** Because positive and negative charges in this atom cancel each other out, your atom is neutral. Congratulations! You have just made the simplest atom—a hydrogen atom.

Proton
Electron

Figure 12 *The simplest atom has one proton and one electron.*

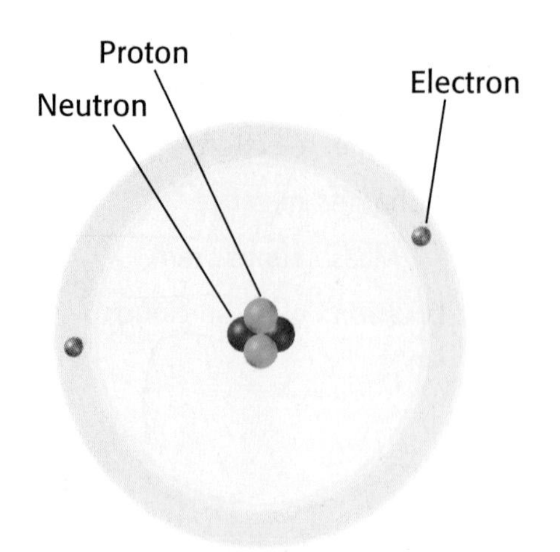

Figure 13 *A helium nucleus must have neutrons in it to keep the protons from moving apart.*

Now for Some Neutrons Now build an atom containing two protons. This time you will find that you must have some neutrons around to hold the protons together. Both of the protons are positively charged, so they repel one another. You cannot cram them together to form a nucleus unless you put some neutrons there to counteract the repulsion. For this atom, two neutrons will do. Your new atom will have two protons and two neutrons making up the nucleus and two electrons zipping around outside the nucleus, as shown in the model in **Figure 13.** This is an atom of the element helium.

You could continue combining particles, building all of the 112 known elements. You could build a carbon atom using 6 protons, 6 neutrons, and 6 electrons; you could build an oxygen atom using 8 protons, 9 neutrons, and 8 electrons; or you could build an iron atom using 26 protons, 30 neutrons, and 26 electrons. You could even build a gold atom with 79 protons, 118 neutrons, and 79 electrons! As you can see, an atom does not have to have equal numbers of protons and neutrons.

Hydrogen is the most abundant element in the universe. It is the fuel for the sun and other stars. It is currently believed that there are roughly 2,000 times more hydrogen atoms than oxygen atoms and 10,000 times more hydrogen atoms than carbon atoms.

The Number of Protons Determines the Element How can you tell which elements these atoms represent? The key is the number of protons. The number of protons in the nucleus of an atom is the **atomic number** of that atom. Each element is composed of atoms that all have the same atomic number. Every hydrogen atom has only one proton in its nucleus, so hydrogen has an atomic number of 1. Every carbon atom has six protons in its nucleus, so carbon has an atomic number of 6.

Are All Atoms of an Element the Same?

Imagine you're back in the atom-building workshop. This time you'll make an atom that has one proton, one electron, and one neutron, as shown in **Figure 14.** This new atom has one proton—what does that tell you? Its atomic number is 1, so it is hydrogen. This atom is neutral because there are equal numbers of protons and electrons. However, this hydrogen atom's nucleus has two particles; therefore, this atom has a greater mass than the first hydrogen atom you made. What you have is another isotope (IE suh TOHP) of hydrogen.

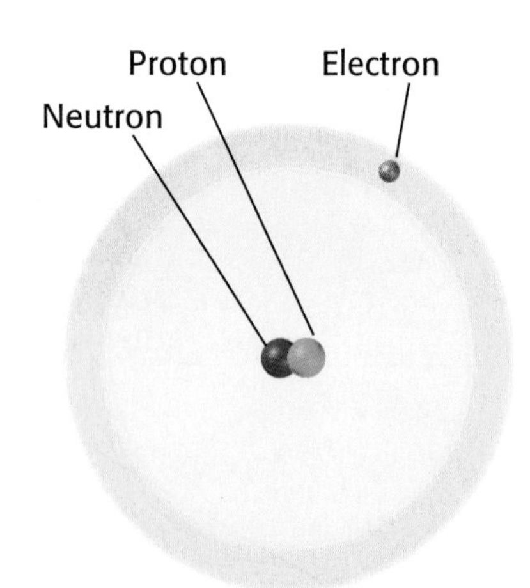

Figure 14 *The atom in this model and the one in Figure 12 are isotopes because each has one proton but a different number of neutrons.*

Isotopes are atoms that have the same number of protons but have different numbers of neutrons. Each element has a limited number of isotopes that occur naturally. Atoms that are isotopes of each other are always the same element because the number of protons in each atom is the same.

Some isotopes of an element have unique properties because they are unstable. An unstable atom is an atom whose nucleus can change its composition. This type of isotope is *radioactive.* However, isotopes of an element share most of the same chemical and physical properties. For example, the most common oxygen isotope has 8 neutrons in the nucleus, but other isotopes have 9 or 10 neutrons. All three isotopes are colorless, odorless gases at room temperature. Each isotope has the chemical property of combining with a substance as it burns and even behaves the same in chemical changes in your body.

APPLY

Oxygen reacts, or undergoes a chemical change, with the hot filament in a light bulb, causing the bulb to burn out quickly. The element argon does not react with the filament, so a light bulb filled with argon does not burn out as quickly as one that contains oxygen. Three isotopes of argon occur in nature. Do you think all three isotopes have the same effect on the filament when used in light bulbs? Explain your reasoning.

How Can You Tell One Isotope from Another? You can identify each isotope of an element by its mass number. The **mass number** is the sum of the protons and neutrons in an atom. Electrons are not included in an atom's mass number because their mass is so small that they have very little effect on the atom's total mass. Look at the boron isotope models shown in **Figure 15** to see how to calculate an atom's mass number.

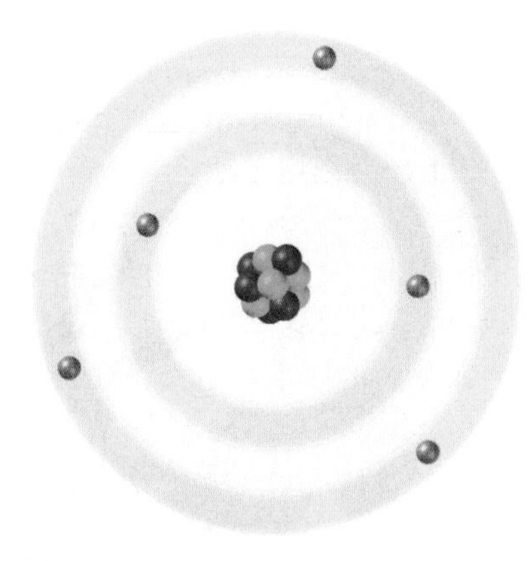

Protons: 5
Neutrons: 5
Electrons: 5
Mass number =
protons + neutrons = 10

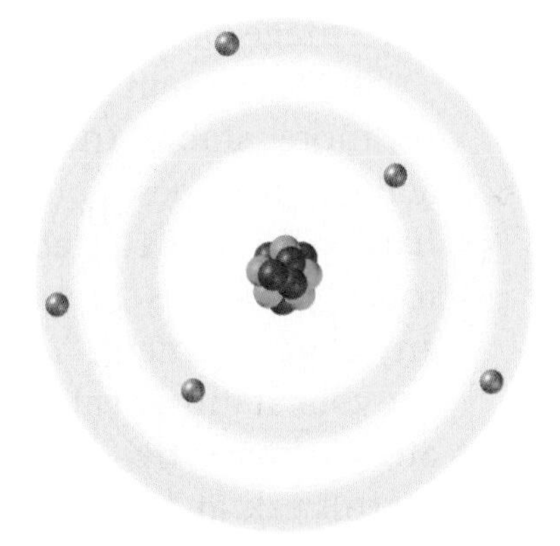

Protons: 5
Neutrons: 6
Electrons: 5
Mass number =
protons + neutrons = 11

Figure 15 *Each of these boron isotopes has five protons. But because each has a different number of neutrons, each has a different mass number.*

Explore

Draw diagrams of hydrogen-2, helium-3, and carbon-14. Show the correct number and location of each type of particle. For the electrons, simply write the total number of electrons in the electron cloud. Use colored pencils or markers to represent the protons, neutrons, and electrons.

To identify a specific isotope of an element, write the name of the element followed by a hyphen and the mass number of the isotope. A hydrogen atom with one proton and no neutrons has a mass number of 1. Its name is hydrogen-1. Hydrogen-2 has one proton and one neutron in the nucleus. The carbon isotope with a mass number of 12 is called carbon-12. If you know that the atomic number for carbon is 6, you can calculate the number of neutrons in carbon-12 by subtracting the atomic number from the mass number. For carbon-12, the number of neutrons is 12 − 6, or 6.

12	**Mass number**
−6	**Number of protons (atomic number)**
6	**Number of neutrons**

How Do You Calculate the Mass of an Element?

Most elements found in nature contain a mixture of different isotopes. For example, all copper is composed of copper-63 atoms and copper-65 atoms. The term *atomic mass* describes the mass of a mixture of isotopes. **Atomic mass** is the weighted average of the masses of all the naturally occurring isotopes of an element. A weighted average accounts for the percentages of each isotope that are present. Copper, including the copper in the Statue of Liberty (shown in **Figure 16**), is 69 percent copper-63 and 31 percent copper-65. The atomic mass of copper is 63.6 amu. Because the atomic mass is closer to 63 than to 65, you can tell the percentage of copper-63 is greater than the percentage of copper-65.

Most elements have two or more stable (nonradioactive) isotopes found in nature. Tin has 10 stable isotopes, which is more than any other element. You can try your hand at calculating atomic mass by doing the MathBreak at left.

Figure 16 *The copper used to make the Statue of Liberty includes both copper-63 and copper-65. Copper's atomic mass is 63.6 amu.*

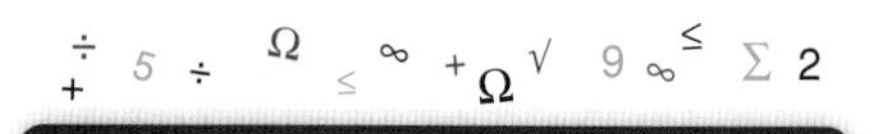

MATH BREAK

Atomic Mass

To calculate the atomic mass of an element, multiply the mass number of each isotope by its percentage abundance in decimal form. Then add these amounts together to find the atomic mass. For example, chlorine-35 makes up 76 percent (its percentage abundance) of all the chlorine in nature, and chlorine-37 makes up the other 24 percent. The atomic mass of chlorine is calculated as follows:

$$(35 \times 0.76) = 26.6$$
$$(37 \times 0.24) = +8.9$$
$$35.5 \text{ amu}$$

Now It's Your Turn

Calculate the atomic mass of boron, which occurs naturally as 20 percent boron-10 and 80 percent boron-11.

What Forces Are at Work in Atoms?

You have seen how atoms are composed of protons, neutrons, and electrons. But what are the *forces* (the pushes or pulls between two objects) acting between these particles? Four basic forces are at work everywhere, including within the atom—gravity, the electromagnetic force, the strong force, and the weak force. These forces are discussed below.

Forces in the Atom

Gravity Probably the most familiar of the four forces is *gravity.* Gravity acts between all objects all the time. The amount of gravity between objects depends on their masses and the distance between them. Gravity pulls objects, such as the sun, Earth, cars, and books, toward one another. However, because the masses of particles in atoms are so small, the force of gravity within atoms is very small.

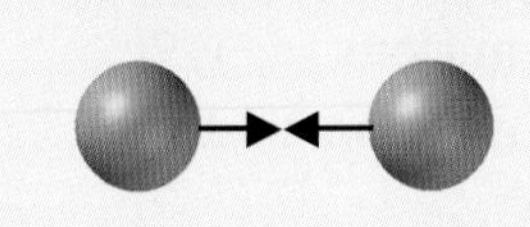

Electromagnetic Force As mentioned earlier, objects that have the same charge repel each other, while objects with opposite charge attract each other. This is due to the *electromagnetic force*. Protons and electrons are attracted to each other because they have opposite charges. The electromagnetic force holds the electrons around the nucleus.

Particles with the same charges repel each other.

Particles with opposite charges attract each other.

Strong Force Protons push away from one another because of the electromagnetic force. A nucleus containing two or more protons would fly apart if it were not for the *strong force*. At the close distances between protons in the nucleus, the strong force is greater than the electromagnetic force, so the nucleus stays together.

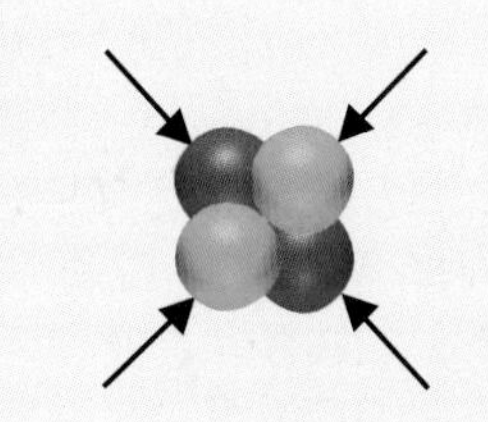

Weak Force The *weak force* is an important force in radioactive atoms. In certain unstable atoms, a neutron can change into a proton and an electron. The weak force plays a key role in this change.

REVIEW

1. List the charge, location, and mass of a proton, a neutron, and an electron.
2. Determine the number of protons, neutrons, and electrons in an atom of aluminum-27.
3. **Doing Calculations** The metal thallium occurs naturally as 30 percent thallium-203 and 70 percent thallium-205. Calculate the atomic mass of thallium.

Chapter Highlights

SECTION 1

Vocabulary

atom *(p. 304)*
theory *(p. 304)*
electrons *(p. 307)*
model *(p. 307)*
nucleus *(p. 309)*
electron clouds *(p. 310)*

Section Notes

- Atoms are the smallest particles of an element that retain the properties of the element.
- In ancient Greece, Democritus argued that atoms were the smallest particles in all matter.
- Dalton proposed an atomic theory that stated the following: Atoms are small particles that make up all matter; atoms cannot be created, divided, or destroyed; atoms of an element are exactly alike; atoms of different elements are different; and atoms join together to make new substances.

- Thomson discovered electrons. His plum-pudding model described the atom as a lump of positively charged material with negative electrons scattered throughout.
- Rutherford discovered that atoms contain a small, dense, positively charged center called the nucleus.
- Bohr suggested that electrons move around the nucleus at only certain distances.
- According to the current atomic theory, electron clouds are where electrons are most likely to be in the space around the nucleus.

Skills Check

Math Concepts

ATOMIC MASS The atomic mass of an element takes into account the mass of each isotope and the percentage of the element that exists as that isotope. For example, magnesium occurs naturally as 79 percent magnesium-24, 10 percent magnesium-25, and 11 percent magnesium-26. The atomic mass is calculated as follows:

$$\begin{aligned}(24 \times 0.79) &= 19.0\\ (25 \times 0.10) &= 2.5\\ (26 \times 0.11) &= +\underline{2.8}\\ & 24.3 \text{ amu}\end{aligned}$$

Visual Understanding

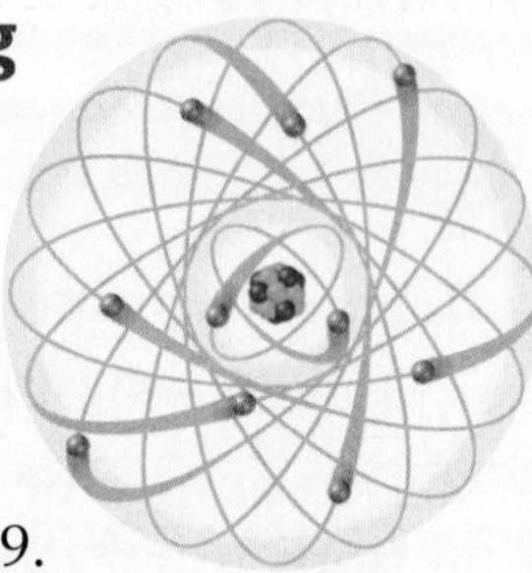

ATOMIC MODELS The atomic theory has changed over the past several hundred years. To understand the different models of the atom, look over Figures 2, 4, 6, 8, and 9.

PARTS OF THE ATOM Atoms are composed of protons, neutrons, and electrons. To review the particles and their placement in the atom, study Figure 11 on page 312.

SECTION 2

Vocabulary

protons *(p. 312)*
atomic mass unit (amu) *(p. 312)*
neutrons *(p. 312)*
atomic number *(p. 314)*
isotopes *(p. 315)*
mass number *(p. 315)*
atomic mass *(p. 316)*

Section Notes

- A proton is a positively charged particle with a mass of 1 amu.
- A neutron is a particle with no charge that has a mass of 1 amu.
- An electron is a negatively charged particle with an extremely small mass.
- Protons and neutrons make up the nucleus. Electrons are found in electron clouds outside the nucleus.
- The number of protons in the nucleus of an atom is the atomic number. The atomic number identifies the atoms of a particular element.
- Isotopes of an atom have the same number of protons but have different numbers of neutrons. Isotopes share most of the same chemical and physical properties.
- The mass number of an atom is the sum of the atom's neutrons and protons.
- The atomic mass is an average of the masses of all naturally occurring isotopes of an element.
- The four forces at work in an atom are gravity, the electromagnetic force, the strong force, and the weak force.

Labs

Made to Order *(p. 570)*

internetconnect

GO TO: go.hrw.com

Visit the **HRW** Web site for a variety of learning tools related to this chapter. Just type in the keyword:

KEYWORD: HSTATS

GO TO: www.scilinks.org

Visit the **National Science Teachers Association** on-line Web site for Internet resources related to this chapter. Just type in the ***sci*LINKS** number for more information about the topic:

TOPIC: Development of the Atomic Theory — ***sci*LINKS NUMBER:** HSTP255
TOPIC: Modern Atomic Theory — ***sci*LINKS NUMBER:** HSTP260
TOPIC: Inside the Atom — ***sci*LINKS NUMBER:** HSTP265
TOPIC: Isotopes — ***sci*LINKS NUMBER:** HSTP270

Chapter Review

USING VOCABULARY

The statements below are false. For each statement, replace the underlined word to make a true statement.

1. Electrons are found in the nucleus of an atom.
2. All atoms of the same element contain the same number of neutrons.
3. Protons have no electrical charge.
4. The atomic number of an element is the number of protons and neutrons in the nucleus.
5. The mass number is an average of the masses of all naturally occurring isotopes of an element.

UNDERSTANDING CONCEPTS

Multiple Choice

6. The discovery of which particle proved that the atom is not indivisible?
   a. proton
   b. neutron
   c. electron
   d. nucleus

7. In his gold foil experiment, Rutherford concluded that the atom is mostly empty space with a small, massive, positively charged center because
   a. most of the particles passed straight through the foil.
   b. some particles were slightly deflected.
   c. a few particles bounced back.
   d. All of the above

8. How many protons does an atom with an atomic number of 23 and a mass number of 51 have?
   a. 23
   b. 28
   c. 51
   d. 74

9. An atom has no overall charge if it contains equal numbers of
   a. electrons and protons.
   b. neutrons and protons.
   c. neutrons and electrons.
   d. None of the above

10. Which statement about protons is true?
    a. Protons have a mass of 1/1840 amu.
    b. Protons have no charge.
    c. Protons are part of the nucleus of an atom.
    d. Protons circle the nucleus of an atom.

11. Which statement about neutrons is true?
    a. Neutrons have a mass of 1 amu.
    b. Neutrons circle the nucleus of an atom.
    c. Neutrons are the only particles that make up the nucleus.
    d. Neutrons have a negative charge.

12. Which of the following determines the identity of an element?
    a. atomic number
    b. mass number
    c. atomic mass
    d. overall charge

13. Isotopes exist because atoms of the same element can have different numbers of
    a. protons.
    b. neutrons.
    c. electrons.
    d. None of the above

Short Answer

14. Why do scientific theories change?
15. What force holds electrons in atoms?
16. In two or three sentences, describe the plum-pudding model of the atom.

Concept Mapping

17. Use the following terms to create a concept map: atom, nucleus, protons, neutrons, electrons, isotopes, atomic number, mass number.

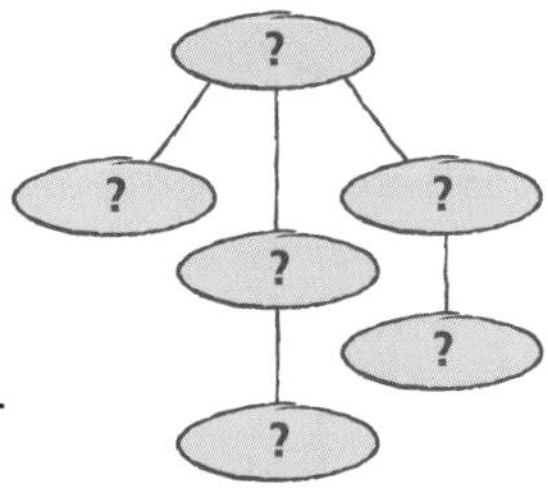

CRITICAL THINKING AND PROBLEM SOLVING

18. Particle accelerators, like the one shown below, are devices that speed up charged particles in order to smash them together. Sometimes the result of the collision is a new nucleus. How can scientists determine whether the nucleus formed is that of a new element or that of a new isotope of a known element?

19. John Dalton made a number of statements about atoms that are now known to be incorrect. Why do you think his atomic theory is still found in science textbooks?

MATH IN SCIENCE

20. Calculate the atomic mass of gallium consisting of 60 percent gallium-69 and 40 percent gallium-71.

21. Calculate the number of protons, neutrons, and electrons in an atom of zirconium-90, which has an atomic number of 40.

INTERPRETING GRAPHICS

22. Study the models below, and answer the questions that follow:

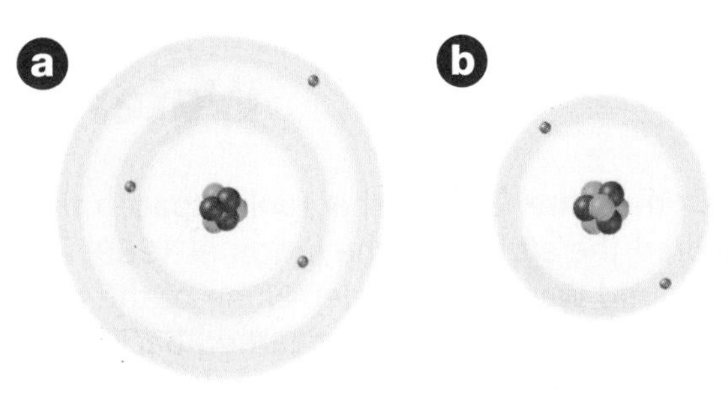

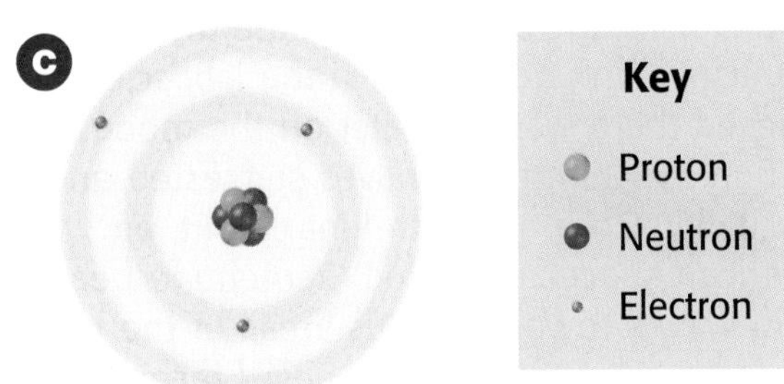

a. Which models represent isotopes of the same element?
b. What is the atomic number for (a)?
c. What is the mass number for (b)?

23. Predict how the direction of the moving particle in the figure below will change, and explain what causes the change to occur.

NOW What Do You Think?

Take a minute to review your answers to the ScienceLog questions on page 303. Have your answers changed? If necessary, revise your answers based on what you have learned since you began this chapter.

Water on the Moon?

When the astronauts of the Apollo space mission explored the surface of the moon in 1969, all they found was rock powder. None of the many samples of moon rocks they carried back to Earth contained any hint of water. Because the astronauts didn't see water on the moon and scientists didn't detect any in the lab, scientists believed there was no water on the moon.

Then in 1994, radio waves suggested another possibility. On a 4-month lunar jaunt, an American spacecraft called *Clementine* beamed radio waves toward various areas of the moon, including a few craters that never receive sunlight. Mostly, the radio waves were reflected by what appeared to be ground-up rock. However, in part of one huge, dark crater, the radio waves were reflected as if by . . . *ice.*

▲ *The* Lunar Prospector *spacecraft may have found water on the moon.*

Hunting for Hydrogen Atoms

Scientists were intrigued by *Clementine's* evidence. Two years later, another spacecraft, *Lunar Prospector,* traveled to the moon. Instead of trying to detect water with radio waves, *Prospector* scanned the moon's surface with a device called a *neutron spectrometer* (NS). A neutron spectrometer counts the number of slow neutrons bouncing off a surface. When a neutron hits something about the same mass as itself, it slows down. As it turns out, the only thing close to the mass of a neutron is an *atom* of the lightest of all elements, hydrogen. So when the NS located high concentrations of slow-moving neutrons on the moon, it indicated to scientists that the neutrons were crashing into hydrogen atoms.

As you know, water consists of two atoms of hydrogen and one atom of oxygen. The presence of hydrogen atoms on the moon is more evidence that water may exist there.

How Did It Get There?

Some scientists speculate that the water molecules came from comets (which are 90 percent water) that hit the moon more than 4 billion years ago. Water from comets may have landed in the frigid, shadowed craters of the moon, where it mixed with the soil and froze. The Aitken Basin, at the south pole of the moon, where much of the ice was detected, is more than 12 km deep in places. Sunlight never touches most of the crater. And it is very cold—temperatures there may fall to −229°C. The conditions seem right to lock water into place for a very long time.

Think About Lunar Life

▶ Do some research on conditions on the moon. What conditions would humans have to overcome before we could establish a colony there?

CAREERS

EXPERIMENTAL PHYSICIST

In the course of a single day, you could find **Melissa Franklin** operating a huge drill, giving a tour of her lab to a 10-year-old, putting together a gigantic piece of electronic equipment, or even telling a joke. Then you'd see her really get down to business—studying the smallest particles of matter in the universe.

Melissa Franklin is an experimental physicist. "I am trying to understand the forces that describe how everything in the world moves—especially the smallest things," she explains. "I want to find the things that make up all matter in the universe and then try to understand the forces between them."

Other scientists rely on her to test some of the most important hypotheses in physics. For instance, Franklin and her team recently contributed to the discovery of a particle called the top quark. (Quarks are the tiny particles that make up protons and neutrons.)

Physicists had theorized that the top quark might exist but had no evidence. Franklin and more than 450 other scientists worked together to prove the existence of the top quark. Finding it required the use of a massive machine called a particle accelerator. Basically, a particle accelerator smashes particles together, and then scientists look for the remains of the collision. The physicists had to build some very complicated machines to detect the top quark, but the discovery was worth the effort. Franklin and the other researchers have earned the praise of scientists all over the world.

Getting Her Start

"I didn't always want to be a scientist, but what happens is that when you get hooked, you really get hooked. The next thing you know, you're driving forklifts and using overhead cranes while at the same time working on really tiny, incredibly complicated electronics. What I do is a combination of exciting things. It's better than watching TV."

It isn't just the best students who grow up to be scientists. "You can understand the ideas without having to be a math genius," Franklin says. Anyone can have good ideas, she says, absolutely anyone.

Don't Be Shy!

▶ Franklin also has some good advice for young people interested in physics. "Go and bug people at the local university. Just call up a physics person and say, 'Can I come visit you for a couple of hours?' Kids do that with me, and it's really fun." Why don't you give it a try? Prepare for the visit by making a list of questions you would like answered.

▲ *This particle accelerator was used in the discovery of the top quark.*

CHAPTER 13

The Periodic Table

Would You Believe . . . ?

Suppose someone told you that the small animal shown above—a yellow-spotted rock hyrax—is genetically related to an elephant. Impossible, you say? But it's true! Even though this animal looks more like a rabbit or a rodent, scientists have determined through DNA studies that the closest relatives of the hyrax are aardvarks, sea cows, and elephants. Biologists have uncovered similar genetic links between other seemingly different species.

Scientists have also discovered that many different-looking elements, like those shown at right, actually have common properties. For almost 150 years, scientists have organized elements by observing the similarities (both obvious and not so obvious) between them. One scientist in particular—a Russian named Dmitri Mendeleev (MEN duh LAY uhf)—organized the known elements in such a way that a repeating pattern emerged. Mendeleev actually used this pattern to predict the properties of elements that had not even been discovered! His method of organization became known as the periodic table.

The modern periodic table is arranged somewhat differently than Mendeleev's, but it is still a useful tool for organizing the known elements and predicting the properties of elements still unknown. Read on to learn about the development of this remarkable table and the patterns it reveals.

Although solid iodine and liquid bromine have very different appearances, they have similar chemical properties.

Placement Pattern

Just as with animals, scientists have found patterns among the elements. You too can find patterns—right in your classroom! By gathering and analyzing information about your classmates, you can determine the pattern behind a new seating chart your teacher has created.

Procedure

1. In your ScienceLog, draw a seating chart for the classroom arrangement designated by your teacher. Write the name of each of your classmates in the correct place on the chart.
2. Write information about yourself, such as your name, date of birth, hair color, and height, in the space that represents you on the chart.
3. Starting with the people around you, ask questions to gather the same type of information about them. Write information about each person in the corresponding spaces on the seating chart.

Analysis

4. In your ScienceLog, identify a pattern within the information you gathered that could be used to explain the order of people in the seating chart. If you cannot find a pattern, collect more information, and look again.
5. Test your pattern by gathering information from a person you did not talk to before.
6. If the new information does not support your pattern, reanalyze your data, and collect more information as needed to determine another pattern.

What Do You Think?

In your ScienceLog, try to answer the following questions based on what you already know:

1. How are elements organized in the periodic table?
2. Why is the table of the elements called "periodic"?
3. What one property is shared by elements in a group?

Going Further

The science of classifying organisms is called *taxonomy.* Find out more about the way the Swedish scientist Carolus Linnaeus classified organisms.

Section 1

Arranging the Elements

NEW TERMS

periodic
periodic law
period
group

OBJECTIVES

- Describe how elements are arranged in the periodic table.
- Compare metals, nonmetals, and metalloids based on their properties and on their location in the periodic table.
- Describe the difference between a period and a group.

Imagine you go to a new grocery store in your neighborhood to buy a box of cereal. You are surprised by what you find when you walk into the store. None of the aisles are labeled, and there is no pattern to the products on the shelves! In frustration, you think it might take you days to find your cereal.

Some scientists probably felt a similar frustration before 1869. By that time, more than 60 different elements had been discovered and described. However, the elements were not organized in any special way. But in 1869, the elements were organized into a table in much the same way products are arranged (usually!) by shelf and aisle in a grocery store.

Discovering a Pattern

In the 1860s, a Russian chemist named Dmitri Mendeleev began looking for patterns among the properties of the known elements. He wrote the names and properties of these elements on small pieces of paper. He included information such as density, appearance, atomic mass, melting point, and any information he had about the compounds formed from the element. He then arranged and rearranged the pieces of paper, as shown in **Figure 1.** After much thought and work, he determined that there was a repeating pattern to the properties of the elements when the elements were arranged in order of increasing atomic mass.

Figure 1 *By playing "chemical solitaire" on long train rides, Mendeleev organized the elements according to their properties.*

The Properties of Elements Are Periodic Mendeleev saw that the properties of the elements were **periodic,** meaning they had a regular, repeating pattern. Many things that are familiar to you are periodic. For example, the days of the week are periodic because they repeat in the same order every 7 days.

When the elements were arranged in order of increasing atomic mass, similar chemical and physical properties were observed in every eighth element. Mendeleev's arrangement of the elements came to be known as a periodic table because the properties of the elements change in a periodic way.

Explore

In your ScienceLog, make a list of five things that are periodic. Explain which repeating property causes each one to be periodic.

Predicting Properties of Missing Elements Look at Mendeleev's periodic table in **Figure 2.** Notice the question marks. Mendeleev recognized that there were elements missing. Instead of questioning his arrangement, he boldly predicted that elements yet to be discovered would fill the gaps. He also predicted the properties of the missing elements by using the pattern of properties in the periodic table. When one of the missing elements, gallium, was discovered a few years later, its properties matched Mendeleev's predictions very well, and other scientists became interested in his work. Since that time, all of the missing elements on Mendeleev's periodic table have been discovered. In the chart below, you can see Mendeleev's predictions for another missing element—germanium—and the actual properties of that element.

ставляемаго мною начала ко всей совокупности элементовъ, пай которыхъ извѣстенъ съ достовѣрностiю. На этотъ разъ я и желалъ преимущественно найдти общую систему элементовъ. Вотъ этотъ опытъ:

			Ti=50	Zr=90	?=180.
			V=51	Nb=94	Ta=182.
			Cr=52	Mo=96	W=186.
			Mn=55	Rh=104,4	Pt=197,4
			Fe=56	Ru=104,4	Ir=198.
			Ni=Co=59	Pl=106,6,	Os=199.
H=1			Cu=63,4	Ag=108	Hg=200.
	Be=9,4	Mg=24	Zn=65,2	Cd=112	
	B=11	Al=27,4	?=68	Ur=116	Au=197?
	C=12	Si=28	?=70	Su=118	
	N=14	P=31	As=75	Sb=122	Bi=210
	O=16	S=32	Se=79,4	Te=128?	
	F=19	Cl=35,5	Br=80	I=127	
Li=7	Na=23	K=39	Rb=85,4	Cs=133	Tl=204
		Ca=40	Sr=87,6	Ba=137	Pb=207.
		?=45	Ce=92		
		?Er=56	La=94		
		?Yt=60	Di=95		
		?In=75,6	Th=118?		

а потому приходится въ разныхъ рядахъ имѣть различное измѣненіе разностей, чего нѣтъ въ главныхъ числахъ предлагаемой таблицы. Или же придется предполагать при составленіи системы очень много недостающихъ членовъ. То и другое мало выгодно. Мнѣ кажется притомъ, наиболѣе естественнымъ составить

Figure 2 *Mendeleev used question marks to indicate some elements that he believed would later be identified.*

Properties of Germanium		
	Mendeleev's predictions	**Actual properties**
Atomic mass	72	72.6
Density	5.5 g/cm^3	5.3 g/cm^3
Appearance	dark gray metal	gray metal
Melting point	high melting point	937°C

Changing the Arrangement

Mendeleev noticed that a few elements in the table were not in the correct place according to their properties. He thought that the calculated atomic masses were incorrect and that more accurate atomic masses would eventually be determined. However, new measurements of the atomic masses showed that the masses were in fact correct.

The mystery was solved in 1914 by a British scientist named Henry Moseley (MOHZ lee). From the results of his experiments, Moseley was able to determine the number of protons—the atomic number—in an atom. When he rearranged the elements by atomic number, every element fell into its proper place in an improved periodic table.

Since Moseley's rearrangement of the elements, more elements have been discovered. The discovery of each new element has supported the periodic law, considered to be the basis of the periodic table. The **periodic law** states that the chemical and physical properties of elements are periodic functions of their atomic numbers. The modern version of the periodic table is shown on the following pages.

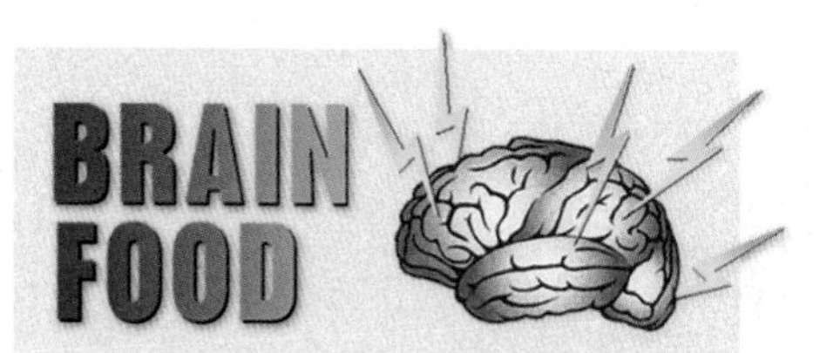

Moseley was 26 when he made his discovery. His work allowed him to predict that only three elements were yet to be found between aluminum and gold. The following year, as he fought for the British in World War I, he was killed in action at Gallipoli, Turkey. The British government no longer assigns scientists to combat duty.

Periodic Table of the Elements

Each square on the table includes an element's name, chemical symbol, atomic number, and atomic mass.

Atomic number — 6
Chemical symbol — C
Element name — Carbon
Atomic mass — 12.0

The background color indicates the type of element. Carbon is a nonmetal.

The color of the chemical symbol indicates the physical state at room temperature. Carbon is a solid.

Background
- Metals
- Metalloids
- Nonmetals

Chemical symbol
- Solid
- Liquid
- Gas

	Group 1	Group 2	Group 3	Group 4	Group 5	Group 6	Group 7	Group 8	Group 9
Period 1	1 H Hydrogen 1.0								
Period 2	3 Li Lithium 6.9	4 Be Beryllium 9.0							
Period 3	11 Na Sodium 23.0	12 Mg Magnesium 24.3							
Period 4	19 K Potassium 39.1	20 Ca Calcium 40.1	21 Sc Scandium 45.0	22 Ti Titanium 47.9	23 V Vanadium 50.9	24 Cr Chromium 52.0	25 Mn Manganese 54.9	26 Fe Iron 55.8	27 Co Cobalt 58.9
Period 5	37 Rb Rubidium 85.5	38 Sr Strontium 87.6	39 Y Yttrium 88.9	40 Zr Zirconium 91.2	41 Nb Niobium 92.9	42 Mo Molybdenum 95.9	43 Tc Technetium (97.9)	44 Ru Ruthenium 101.1	45 Rh Rhodium 102.9
Period 6	55 Cs Cesium 132.9	56 Ba Barium 137.3	57 La Lanthanum 138.9	72 Hf Hafnium 178.5	73 Ta Tantalum 180.9	74 W Tungsten 183.8	75 Re Rhenium 186.2	76 Os Osmium 190.2	77 Ir Iridium 192.2
Period 7	87 Fr Francium (223.0)	88 Ra Radium (226.0)	89 Ac Actinium (227.0)	104 Rf Rutherfordium (261.1)	105 Db Dubnium (262.1)	106 Sg Seaborgium (263.1)	107 Bh Bohrium (262.1)	108 Hs Hassium (265)	109 Mt Meitnerium (266)

A row of elements is called a period.

A column of elements is called a group or family.

Lanthanides	58 Ce Cerium 140.1	59 Pr Praseodymium 140.9	60 Nd Neodymium 144.2	61 Pm Promethium (144.9)	62 Sm Samarium 150.4
Actinides	90 Th Thorium 232.0	91 Pa Protactinium 231.0	92 U Uranium 238.0	93 Np Neptunium (237.0)	94 Pu Plutonium 244.1

These elements are placed below the table to allow the table to be narrower.

This zigzag line reminds you where the metals, nonmetals, and metalloids are.

Group 10	Group 11	Group 12	Group 13	Group 14	Group 15	Group 16	Group 17	Group 18
								2 **He** Helium 4.0
			5 **B** Boron 10.8	6 **C** Carbon 12.0	7 **N** Nitrogen 14.0	8 **O** Oxygen 16.0	9 **F** Fluorine 19.0	10 **Ne** Neon 20.2
			13 **Al** Aluminum 27.0	14 **Si** Silicon 28.1	15 **P** Phosphorus 31.0	16 **S** Sulfur 32.1	17 **Cl** Chlorine 35.5	18 **Ar** Argon 39.9
28 **Ni** Nickel 58.7	29 **Cu** Copper 63.5	30 **Zn** Zinc 65.4	31 **Ga** Gallium 69.7	32 **Ge** Germanium 72.6	33 **As** Arsenic 74.9	34 **Se** Selenium 79.0	35 **Br** Bromine 79.9	36 **Kr** Krypton 83.8
46 **Pd** Palladium 106.4	47 **Ag** Silver 107.9	48 **Cd** Cadmium 112.4	49 **In** Indium 114.8	50 **Sn** Tin 118.7	51 **Sb** Antimony 121.8	52 **Te** Tellurium 127.6	53 **I** Iodine 126.9	54 **Xe** Xenon 131.3
78 **Pt** Platinum 195.1	79 **Au** Gold 197.0	80 **Hg** Mercury 200.6	81 **Tl** Thallium 204.4	82 **Pb** Lead 207.2	83 **Bi** Bismuth 209.0	84 **Po** Polonium (209.0)	85 **At** Astatine (210.0)	86 **Rn** Radon (222.0)
110 **Uun** Ununnilium (271)	111 **Uuu** Unununium (272)	112 **Uub** Ununbium (277)						

The names and symbols of elements 110–112 are temporary. They are based on the atomic number of the element. The official name and symbol will be approved by an international committee of scientists.

63 **Eu** Europium 152.0	64 **Gd** Gadolinium 157.3	65 **Tb** Terbium 158.9	66 **Dy** Dysprosium 162.5	67 **Ho** Holmium 164.9	68 **Er** Erbium 167.3	69 **Tm** Thulium 168.9	70 **Yb** Ytterbium 173.0	71 **Lu** Lutetium 175.0
95 **Am** Americium (243.1)	96 **Cm** Curium (247.1)	97 **Bk** Berkelium (247.1)	98 **Cf** Californium (251.1)	99 **Es** Einsteinium (252.1)	100 **Fm** Fermium (257.1)	101 **Md** Mendelevium (258.1)	102 **No** Nobelium (259.1)	103 **Lr** Lawrencium (262.1)

A number in parentheses is the mass number of the most stable isotope of that element.

Finding Your Way Around the Periodic Table

At first glance, you might think studying the periodic table is like trying to explore a thick jungle without a guide—it would be easy to get lost! However, the table itself contains a lot of information that will help you along the way.

Classes of Elements Elements are classified as metals, nonmetals, and metalloids, according to their properties. The number of electrons in the outer energy level of an atom also helps determine which category an element belongs in. The zigzag line on the periodic table can help you recognize which elements are metals, which are nonmetals, and which are metalloids.

Metals

Most of the elements in the periodic table are metals. Metals are found to the left of the zigzag line on the periodic table. Atoms of most metals have few electrons in their outer energy level, as shown at right.

Most metals are solid at room temperature. Mercury, however, is a liquid. Some additional information on properties shared by most metals is shown below.

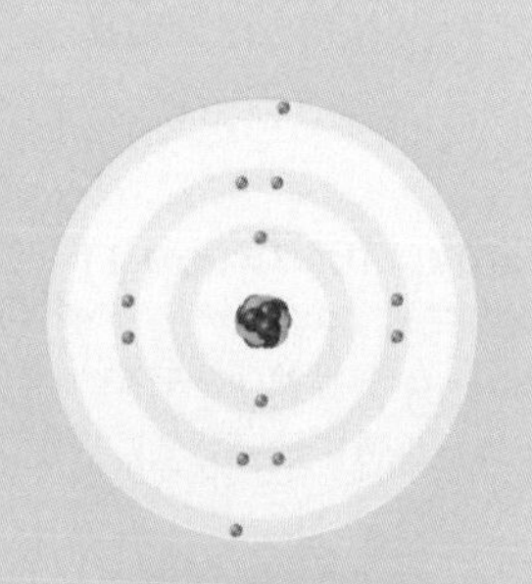

A model of a magnesium atom

Most metals are good conductors of thermal energy. This iron griddle conducts thermal energy from a stovetop to cook your favorite foods.

Most metals are malleable, meaning that they can be flattened with a hammer without shattering. Aluminum is flattened into sheets to make cans and foil.

Most metals are ductile, which means that they can be drawn into thin wires. All metals are good conductors of electric current. The wires in the electrical devices in your home are made from the metal copper.

Metals tend to be shiny. You can see a reflection in a mirror because light reflects off the shiny surface of a thin layer of silver behind the glass.

Nonmetals

Nonmetals are found to the right of the zigzag line on the periodic table. Atoms of most nonmetals have an almost complete set of electrons in their outer level, as shown at right. (Atoms of one group of nonmetals, the noble gases, have a complete set of electrons, with most having eight electrons in their outer energy level.)

More than half of the nonmetals are gases at room temperature. The properties of nonmetals are the opposite of the properties of metals, as shown below.

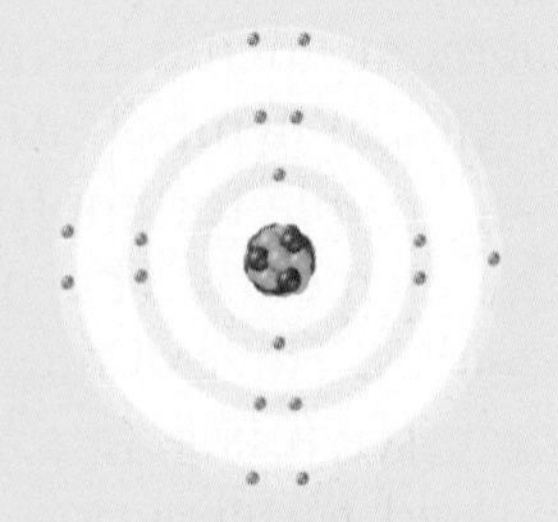

A model of a chlorine atom

Quick Lab

Conduction Connection

1. Fill a **clear plastic cup** with **hot water.**
2. Stand a piece of **copper wire** and a **graphite lead** from a mechanical pencil in the water.
3. After 1 minute, touch the top of each object. Record your observations.
4. Which material conducted thermal energy the best? Why?

Sulfur, like most nonmetals, is not shiny.

Nonmetals are not malleable or ductile. In fact, solid nonmetals, like carbon (shown here in the graphite of the pencil lead), are brittle and will break or shatter when hit with a hammer.

Nonmetals are poor conductors of thermal energy and electric current. If the gap in a spark plug is too wide, the nonmetals nitrogen and oxygen in the air will stop the spark, and a car's engine will not run.

Metalloids

Metalloids, also called semiconductors, are the elements that border the zigzag line on the periodic table. Atoms of metalloids have about a half-complete set of electrons in their outer energy level, as shown at right.

Metalloids have some properties of metals and some properties of nonmetals, as shown below.

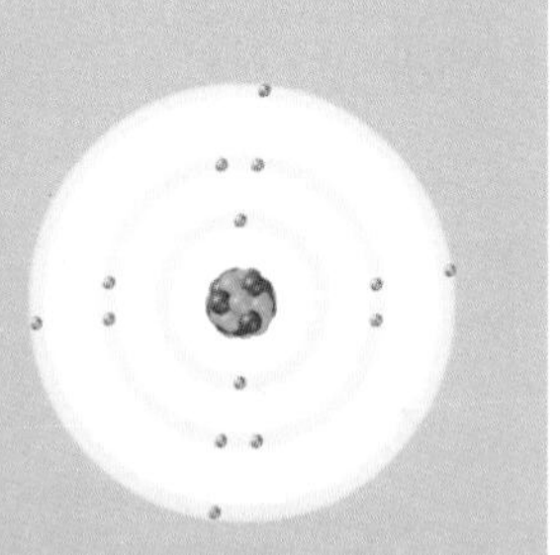

A model of a silicon atom

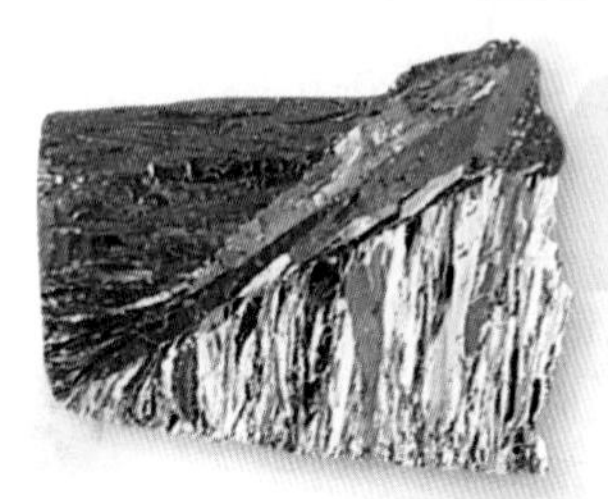

Tellurium is shiny, but it is also brittle and is easily smashed into a powder.

Boron is almost as hard as diamond, but it is also very brittle. At high temperatures, boron is a good conductor of electric current.

Explore

Draw a line down a sheet of paper to divide it into two columns. Look at the elements with atomic numbers 1 through 10 on the periodic table. Write all the chemical symbols and names that follow one pattern in one column on your paper and all chemical symbols and names that follow a second pattern in the second column. Write a sentence describing each pattern you found.

Each Element Is Identified by a Chemical Symbol Each square on the periodic table contains information about an element, including its atomic number, atomic mass, name, and chemical symbol. An international committee of scientists is responsible for approving the names and chemical symbols of the elements. The names of the elements come from many sources. For example, some elements are named after important scientists (mendelevium, einsteinium), and others are named for geographical regions (germanium, californium).

The chemical symbol for each element usually consists of one or two letters. The first letter in the symbol is always capitalized, and the second letter, if there is one, is always written in lowercase. The chart below lists the patterns that the chemical symbols follow, and the Explore activity will help you investigate two of those patterns further.

Writing the Chemical Symbols

Pattern of chemical symbols	Examples
first letter of the name	S—sulfur
first two letters of the name	Ca—calcium
first letter and third or later letter of the name	Mg—magnesium
letter(s) of a word other than the English name	Pb—lead (from the Latin *plumbum,* meaning "lead")
first letter of root words that stand for the atomic number (used for elements whose official names have not yet been chosen)	Uun—ununnilium (uhn uhn NIL ee uhm) (for atomic number 110)

You can create your own well-rounded periodic table using coins, washers, and buttons on page 572 of the LabBook.

Look at the periodic table shown here. How is it the same as the periodic table you saw earlier? How is it different? Explain why it is important for scientific communication that the chemical symbols used are the same around the world.

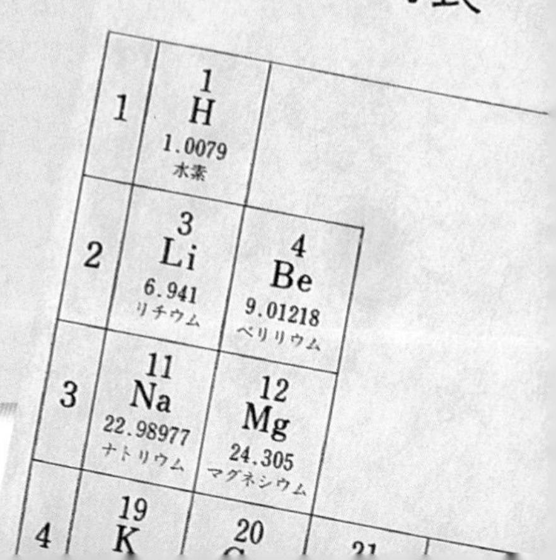

Rows Are Called Periods Each horizontal row of elements (from left to right) on the periodic table is called a **period.** For example, the row from lithium (Li) to neon (Ne) is Period 2. A row is called a period because the properties of elements in a row follow a repeating, or periodic, pattern as you move across each period. The physical and chemical properties of elements, such as conductivity and the number of electrons in the outer level of atoms, change gradually from those of a metal to those of a nonmetal in each period. Therefore, elements at opposite ends of a period have very different properties from one another, as shown in **Figure 3.**

BRAIN FOOD

To remember that a period goes from left to right across the periodic table, just think of reading a sentence. You read from left to right across the page until you come to a period.

Figure 3 *The elements in a row become less metallic from left to right.*

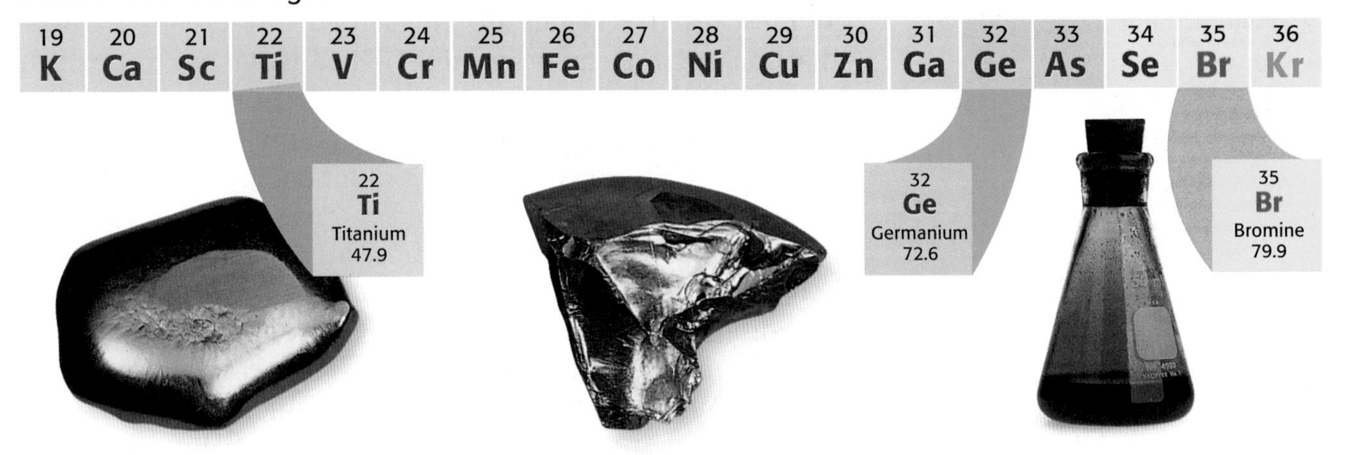

Elements at the left end of a period, such as titanium, are very metallic in their properties.

Elements farther to the right, like germanium, are less metallic in their properties.

Elements at the far right end of a period, such as bromine, are nonmetallic in their properties.

Columns Are Called Groups Each column of elements (from top to bottom) on the periodic table is called a **group.** Elements in the same group often have similar chemical and physical properties. For this reason, sometimes a group is also called a family. You will learn more about each group in the next section.

REVIEW

1. Compare a period and a group on the periodic table.
2. How are the elements arranged in the modern periodic table?
3. **Comparing Concepts** Compare metals, nonmetals, and metalloids in terms of their electrical conductivity.

Section 2

Grouping the Elements

NEW TERMS

alkali metals
alkaline-earth metals
halogens
noble gases

OBJECTIVES

- Explain why elements in a group often have similar properties.
- Describe the properties of the elements in the groups of the periodic table.

You probably know a family with several members that look a lot alike. Or you may have a friend whose little brother or sister acts just like your friend. Members of a family often—but not always—have a similar appearance or behavior. Likewise, the elements in a family or group in the periodic table often—but not always—share similar properties. The properties are similar because the atoms of the elements have the same number of electrons in their outer energy level.

Groups 1 and 2: Very Reactive Metals

The most reactive metals are the elements in Groups 1 and 2. What makes an element reactive? The answer has to do with electrons in the outer energy level of atoms. Atoms will often take, give, or share electrons with other atoms in order to have a complete set of electrons in their outer energy level. Elements whose atoms undergo such processes are *reactive* and combine to form compounds. Elements whose atoms need to take, give, or share only one or two electrons to have a filled outer level tend to be very reactive.

The elements in Groups 1 and 2 are so reactive that they are only found combined with other elements in nature. To study the elements separately, the naturally occurring compounds must first be broken apart through chemical changes.

Group 1: Alkali Metals

3 Li Lithium
11 Na Sodium
19 K Potassium
37 Rb Rubidium
55 Cs Cesium
87 Fr Francium

Group contains: Metals
Electrons in the outer level: 1
Reactivity: Very reactive
Other shared properties: Soft; silver-colored; shiny; low density

Alkali (AL kuh LIE) **metals** are soft enough to be cut with a knife, as shown in **Figure 4.** The densities of the alkali metals are so low that lithium, sodium, and potassium are actually less dense than water.

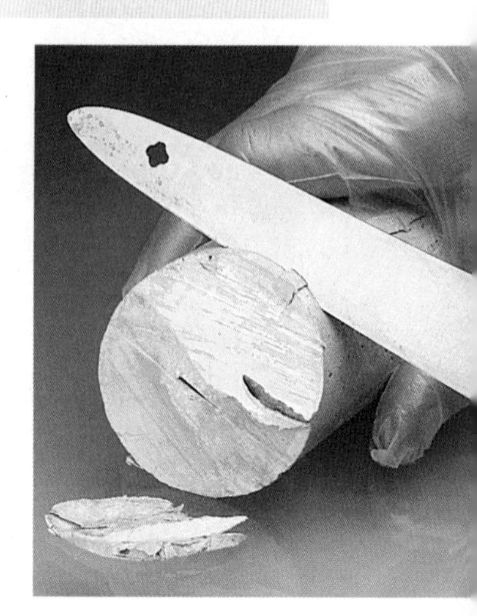

Figure 4 *Metals so soft that they can be cut with a knife? Welcome to the alkali metals.*

Although the element hydrogen appears above the alkali metals on the periodic table, it is not considered a member of Group 1. It will be described separately at the end of this section.

Alkali metals are the most reactive of the metals. This is because their atoms can easily give away the single electron in their outer level. For example, alkali metals react violently with water, as shown in **Figure 5.** Alkali metals are usually stored in oil to prevent them from reacting with water and oxygen in the atmosphere.

The compounds formed from alkali metals have many uses. Sodium chloride (table salt) can be used to add flavor to your food. Sodium hydroxide can be used to unclog your drains. Potassium bromide is one of several potassium compounds used in photography.

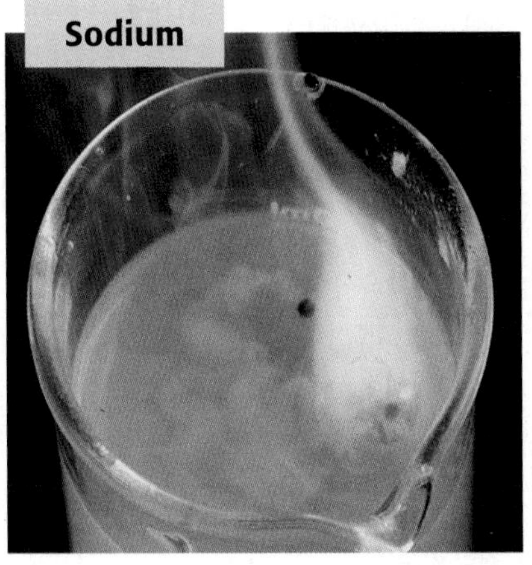

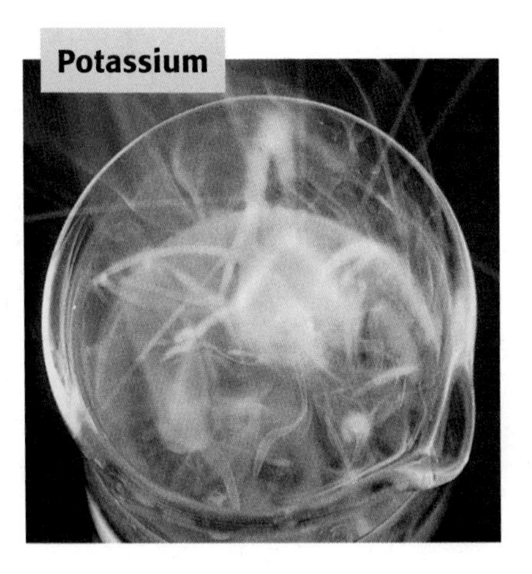

Figure 5 *As alkali metals react with water, they form hydrogen gas.*

Group 2: Alkaline-earth Metals

4 **Be** Beryllium
12 **Mg** Magnesium
20 **Ca** Calcium
38 **Sr** Strontium
56 **Ba** Barium
88 **Ra** Radium

Group contains: Metals
Electrons in the outer level: 2
Reactivity: Very reactive, but less reactive than alkali metals
Other shared properties: Silver-colored; more dense than alkali metals

Alkaline-earth metals are not as reactive as alkali metals because it is more difficult for atoms to give away two electrons than to give away only one when joining with other atoms.

The alkaline-earth metal magnesium is often mixed with other metals to make low-density materials used in airplanes. Compounds of alkaline-earth metals also have many uses. For example, compounds of calcium are found in cement, plaster, chalk, and even you, as shown in **Figure 6.**

Figure 6 *Smile! Calcium, an alkaline-earth metal, is an important component of a compound that makes your bones and teeth healthy.*

Groups 3–12: Transition Metals

Groups 3–12 do not have individual names. Instead, these groups are described together under the name *transition metals.*

Group contains: Metals
Electrons in the outer level: 1 or 2
Reactivity: Less reactive than alkaline-earth metals
Other shared properties: Shiny; good conductors of thermal energy and electric current; higher densities and melting points (except for mercury) than elements in Groups 1 and 2

21 **Sc** Scandium	22 **Ti** Titanium	23 **V** Vanadium	24 **Cr** Chromium	25 **Mn** Manganese	26 **Fe** Iron	27 **Co** Cobalt	28 **Ni** Nickel	29 **Cu** Copper	30 **Zn** Zinc
39 **Y** Yttrium	40 **Zr** Zirconium	41 **Nb** Niobium	42 **Mo** Molybdenum	43 **Tc** Technetium	44 **Ru** Ruthenium	45 **Rh** Rhodium	46 **Pd** Palladium	47 **Ag** Silver	48 **Cd** Cadmium
57 **La** Lanthanum	72 **Hf** Hafnium	73 **Ta** Tantalum	74 **W** Tungsten	75 **Re** Rhenium	76 **Os** Osmium	77 **Ir** Iridium	78 **Pt** Platinum	79 **Au** Gold	80 **Hg** Mercury
89 **Ac** Actinium	104 **Rf** Rutherfordium	105 **Db** Dubnium	106 **Sg** Seaborgium	107 **Bh** Bohrium	108 **Hs** Hassium	109 **Mt** Meitnerium	110 **Uun** Ununnilium	111 **Uuu** Unununium	112 **Uub** Ununbium

The atoms of transition metals do not give away their electrons as easily as atoms of the Group 1 and Group 2 metals do, making transition metals less reactive than the alkali metals and the alkaline-earth metals. The properties of the transition metals vary widely, as shown in **Figure 7.**

Figure 7 *Transition metals have a wide range of physical and chemical properties.*

Mercury is used in thermometers because, unlike the other transition metals, it is in the liquid state at room temperature.

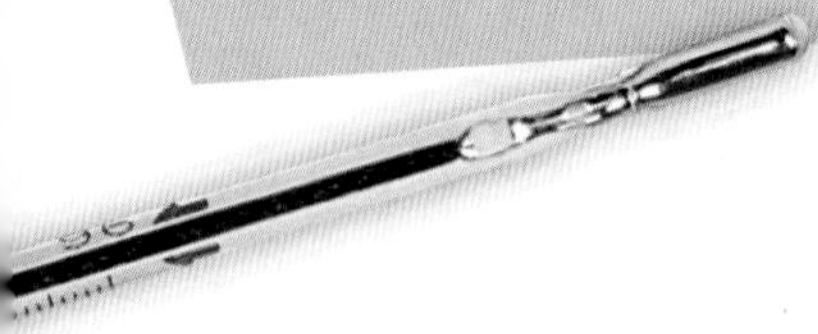

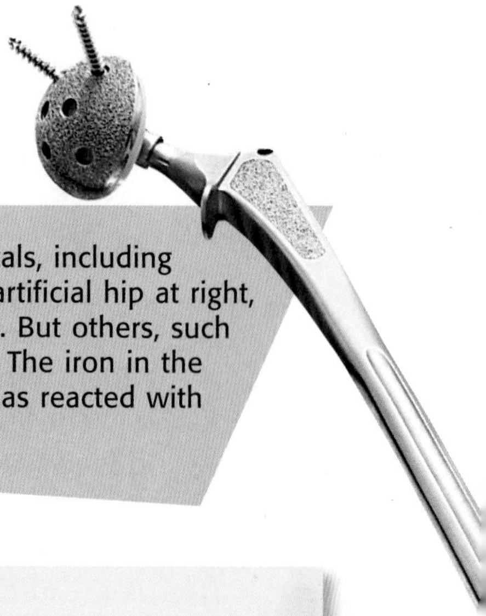

Some transition metals, including the titanium in the artificial hip at right, are not very reactive. But others, such as iron, are reactive. The iron in the steel trowel above has reacted with oxygen to form rust.

Many transition metals are silver-colored—but not all! This gold ring proves it!

Self-Check

Why are alkali metals more reactive than alkaline-earth metals? *(See page 596 to check your answer.)*

57 La Lanthanum 138.9
89 Ac Actinium (227.0)

Lanthanides and Actinides Some transition metals from Periods 6 and 7 are placed at the bottom of the periodic table to keep the table from being too wide. The properties of the elements in each row tend to be very similar.

Lanthanides	58 Ce	59 Pr	60 Nd	61 Pm	62 Sm	63 Eu	64 Gd	65 Tb	66 Dy	67 Ho	68 Er	69 Tm	70 Yb	71 Lu
Actinides	90 Th	91 Pa	92 U	93 Np	94 Pu	95 Am	96 Cm	97 Bk	98 Cf	99 Es	100 Fm	101 Md	102 No	103 Lr

Elements in the first row are called *lanthanides* because they follow the transition metal lanthanum. The lanthanides are shiny, reactive metals. Some of these elements are used to make different types of steel. An important use of a compound of one lanthanide element is shown in **Figure 8.**

Elements in the second row are called *actinides* because they follow the transition metal actinium. All atoms of actinides are radioactive, which means they are unstable. The atoms of a radioactive element can change into atoms of a different element. Elements listed after plutonium, element 94, do not occur in nature but are instead produced in laboratories. You might have one of these elements in your home. Very small amounts of americium (AM uhr ISH ee uhm), element 95, are used in some smoke detectors.

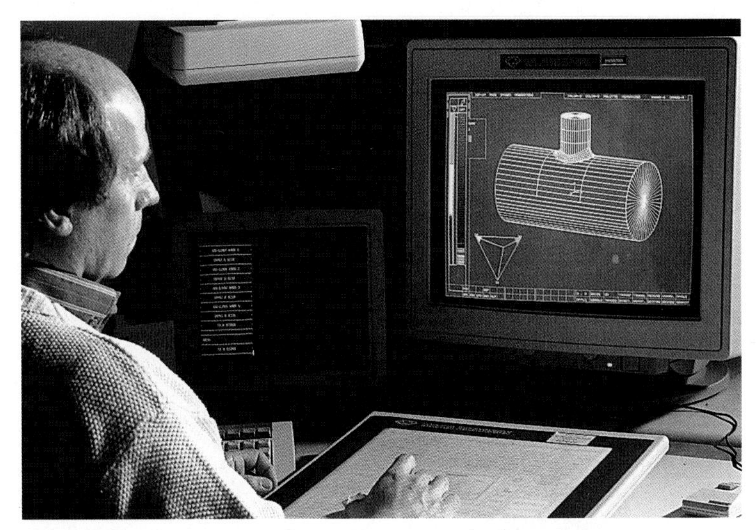

Figure 8 *Seeing red? The color red appears on a computer monitor because of a compound formed from europium that coats the back of the screen.*

REVIEW

1. What are two properties of the alkali metals?
2. What causes the properties of elements in a group to be similar?
3. **Applying Concepts** Why are neither the alkali metals nor the alkaline-earth metals found uncombined in nature?

Groups 13–16: Groups That Include Metalloids

Moving from Group 13 across to Group 16, the elements shift from metals to nonmetals. Along the way, you find the metalloids. These elements have some properties of metals and some properties of nonmetals.

Group 13: Boron Group

5 **B** Boron
13 **Al** Aluminum
31 **Ga** Gallium
49 **In** Indium
81 **Tl** Thallium

Group contains: One metalloid and four metals
Electrons in the outer level: 3
Reactivity: Reactive
Other shared properties: Solid at room temperature

The most common element from Group 13 is aluminum. In fact, aluminum is the most abundant metal in Earth's crust. Until the 1880s, it was considered a precious metal because the process used to produce pure aluminum was very expensive. In fact, aluminum was even more valuable than gold, as shown in **Figure 9.**

Today, the process is not as difficult or expensive. Aluminum is now an important metal used in making lightweight automobile parts and aircraft, as well as foil, cans, and wires.

Figure 9 *During the 1850s and 1860s, Emperor Napoleon III of France, nephew of Napoleon Bonaparte, used aluminum dinnerware because aluminum was more valuable than gold!*

Recycling aluminum uses less energy than obtaining aluminum in the first place. Aluminum must be separated from bauxite, a mixture containing naturally occurring compounds of aluminum. Twenty times more electrical energy is required to separate aluminum from bauxite than to recycle used aluminum.

Group 14: Carbon Group

6 **C** Carbon
14 **Si** Silicon
32 **Ge** Germanium
50 **Sn** Tin
82 **Pb** Lead

Group contains: One nonmetal, two metalloids, and two metals
Electrons in the outer level: 4
Reactivity: Varies among the elements
Other shared properties: Solid at room temperature

The metalloids silicon and germanium are used to make computer chips. The metal tin is useful because it is not very reactive. A tin can is really made of steel coated with tin. The tin is less reactive than the steel, and it keeps the steel from rusting.

The nonmetal carbon can be found uncombined in nature, as shown in **Figure 10.** Carbon forms a wide variety of compounds. Some of these compounds, including proteins, fats, and carbohydrates, are essential to life on Earth.

A particle of carbon shaped like a soccer ball? You'll get a kick out of reading about buckyballs on page 347.

Figure 10 *Diamonds and soot have very different properties, yet both are natural forms of carbon.*

Diamond is the hardest material known. It is used as a jewel and on cutting tools such as saws, drills, and files.

Soot—formed from burning oil, coal, and wood—is used as a pigment in paints and crayons.

Group 15: Nitrogen Group

7 **N** Nitrogen	
15 **P** Phosphorus	
33 **As** Arsenic	
51 **Sb** Antimony	
83 **Bi** Bismuth	

Group contains: Two nonmetals, two metalloids, and one metal
Electrons in the outer level: 5
Reactivity: Varies among the elements
Other shared properties: All but nitrogen are solid at room temperature.

Nitrogen, which is a gas at room temperature, makes up about 80 percent of the air you breathe. Nitrogen removed from air is reacted with hydrogen to make ammonia for fertilizers.

Although nitrogen is unreactive, phosphorus is extremely reactive, as shown in **Figure 11.** In fact, phosphorus is only found combined with other elements in nature.

Figure 11
Simply striking a match on the side of this box causes chemicals on the match to react with phosphorus on the box and begin to burn.

Group 16: Oxygen Group

8 **O** Oxygen	
16 **S** Sulfur	
34 **Se** Selenium	
52 **Te** Tellurium	
84 **Po** Polonium	

Group contains: Three nonmetals, one metalloid, and one metal
Electrons in the outer level: 6
Reactivity: Reactive
Other shared properties: All but oxygen are solid at room temperature.

Oxygen makes up about 20 percent of air. Oxygen is necessary for substances to burn, such as the chemicals on the match in Figure 11. Sulfur, another common member of Group 16, can be found as a yellow solid in nature. The principal use of sulfur is to make sulfuric acid, the most widely used compound in the chemical industry.

Figure 12 *Physical properties of some halogens at room temperature are shown here.*

Groups 17 and 18: Nonmetals Only

The elements in Groups 17 and 18 are nonmetals. The elements in Group 17 are the most reactive nonmetals, but the elements in Group 18 are the least reactive nonmetals. In fact, the elements in Group 18 normally won't react at all with other elements.

Group 17: Halogens

9 F Fluorine
17 Cl Chlorine
35 Br Bromine
53 I Iodine
85 At Astatine

Group contains: Nonmetals
Electrons in the outer level: 7
Reactivity: Very reactive
Other shared properties: Poor conductors of electric current; react violently with alkali metals to form salts; never found uncombined in nature

Halogens are very reactive nonmetals because their atoms need to gain only one electron to have a complete outer level. The atoms of halogens combine readily with other atoms, especially metals, to gain that missing electron.

Although the chemical properties of the halogens are similar, the physical properties are quite different, as shown in **Figure 12.**

Both chlorine and iodine are used as disinfectants. Chlorine is used to treat water, while iodine mixed with alcohol is used in hospitals.

Group 18: Noble Gases

2 He Helium
10 Ne Neon
18 Ar Argon
36 Kr Krypton
54 Xe Xenon
86 Rn Radon

Group contains: Nonmetals
Electrons in the outer level: 8 (2 for helium)
Reactivity: Unreactive
Other shared properties: Colorless, odorless gases at room temperature

Noble gases are unreactive nonmetals. Because the atoms of the elements in this group have a complete set of electrons in their outer level, they do not need to lose or gain any electrons. Therefore, they do not react with other elements under normal conditions.

All of the noble gases are found in Earth's atmosphere in small amounts. Argon, the most abundant noble gas in the atmosphere, makes up almost 1 percent of the atmosphere.

BRAIN FOOD

The term *noble gases* describes the nonreactivity of these elements. Just as nobles, such as kings and queens, did not often mix with common people, the noble gases do not normally react with other elements.

The nonreactivity of the noble gases makes them useful. Ordinary light bulbs last longer when filled with argon than they would if filled with a reactive gas. Because argon is unreactive, it does not react with the metal filament in the light bulb even when the filament gets hot. The low density of helium causes blimps and weather balloons to float, and its nonreactivity makes helium safer to use than hydrogen. One popular use of noble gases that does *not* rely on their nonreactivity is shown in **Figure 13.**

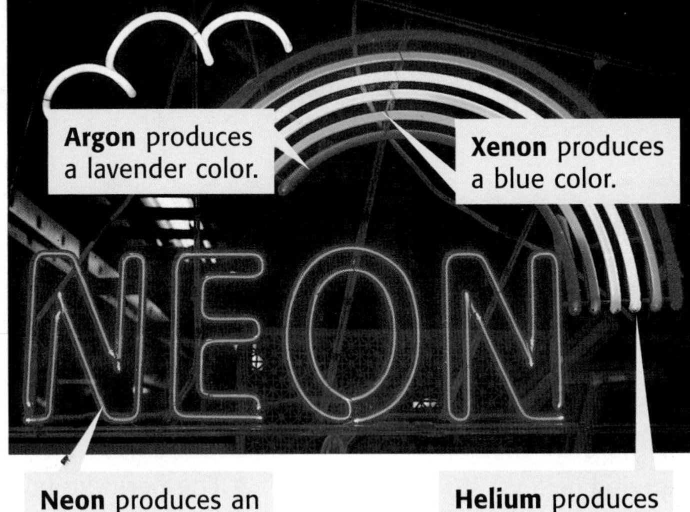

Figure 13 *Besides neon, other noble gases are often used in "neon" lights.*

Hydrogen Stands Apart

1 H Hydrogen	**Electrons in the outer level:** 1 **Reactivity:** Reactive **Other properties:** Colorless, odorless gas at room temperature; low density; reacts explosively with oxygen

The properties of hydrogen do not match the properties of any single group, so hydrogen is set apart from the other elements in the table.

Hydrogen is placed above Group 1 in the periodic table because atoms of the alkali metals also have only one electron in their outer level. Atoms of hydrogen, like atoms of alkali metals, can give away one electron when joining with other atoms. However, hydrogen's physical properties are more like the properties of nonmetals than of metals. As you can see, hydrogen really is in a group of its own.

Hydrogen is the most abundant element in the universe. Hydrogen's reactive nature makes it useful as a fuel in rockets, as shown in **Figure 14.**

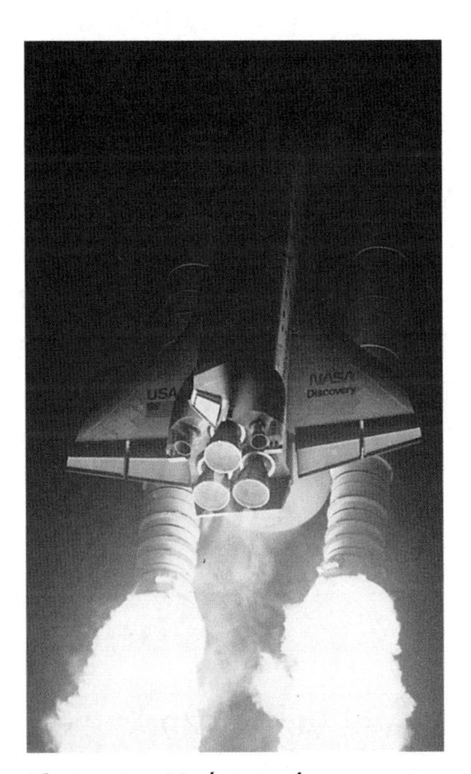

Figure 14 *Hydrogen is a reactive nonmetal. As hydrogen burns, it joins with oxygen, and the hot water vapor that forms pushes the rocket up.*

REVIEW

1. In which group are the unreactive nonmetals found?
2. What are two properties of the halogens?
3. **Making Predictions** In the future, a new noble gas may be synthesized. Predict its atomic number and properties.
4. **Comparing Concepts** Compare the element hydrogen with the alkali metal sodium.

Chapter Highlights

SECTION 1

Vocabulary

periodic *(p. 326)*
periodic law *(p. 327)*
period *(p. 333)*
group *(p. 333)*

Section Notes

- Mendeleev developed the first periodic table. He arranged elements in order of increasing atomic mass. The properties of elements repeated in an orderly pattern, allowing Mendeleev to predict properties for elements that had not yet been discovered.
- Moseley rearranged the elements in order of increasing atomic number.
- The periodic law states that the chemical and physical properties of elements are periodic functions of their atomic numbers.
- Elements in the periodic table are divided into metals, metalloids, and nonmetals.
- Each element has a chemical symbol that is recognized around the world.
- A horizontal row of elements is called a period. The elements gradually change from metallic to nonmetallic from left to right across each period.
- A vertical column of elements is called a group or family. Elements in a group usually have similar properties.

Labs

Create a Periodic Table *(p. 572)*

☑ Skills Check

Visual Understanding

PERIODIC TABLE OF THE ELEMENTS Scientists rely on the periodic table as a resource for a large amount of information. Review the periodic table on pages 328–329. Pay close attention to the labels and the key; they will help you understand the information presented in the table.

CLASSES OF ELEMENTS Identifying an element as a metal, nonmetal, or metalloid gives you a better idea of the properties of that element. Review the figures on pages 330–331 to understand how to use the zigzag line on the periodic table to identify the classes of elements and to review the properties of elements in each category.

SECTION 2

Vocabulary

alkali metals *(p. 334)*
alkaline-earth metals *(p. 335)*
halogens *(p. 340)*
noble gases *(p. 340)*

Section Notes

- The alkali metals (Group 1) are the most reactive metals. Atoms of the alkali metals have one electron in their outer level.
- The alkaline-earth metals (Group 2) are less reactive than the alkali metals. Atoms of the alkaline-earth metals have two electrons in their outer level.
- The transition metals (Groups 3–12) include most of the well-known metals as well as the lanthanides and actinides located below the periodic table.
- Groups 13–16 contain the metalloids along with some metals and nonmetals. The atoms of the elements in each of these groups have the same number of electrons in their outer level.
- The halogens (Group 17) are very reactive nonmetals. Atoms of the halogens have seven electrons in their outer level.
- The noble gases (Group 18) are unreactive nonmetals. Atoms of the noble gases have a complete set of electrons in their outer level.
- Hydrogen is set off by itself because its properties do not match the properties of any one group.

internet**connect**

GO TO: go.hrw.com

Visit the **HRW** Web site for a variety of learning tools related to this chapter. Just type in the keyword:

KEYWORD: HSTPRT

GO TO: www.scilinks.org

Visit the **National Science Teachers Association** on-line Web site for Internet resources related to this chapter. Just type in the ***sci*LINKS** number for more information about the topic:

TOPIC: The Periodic Table — ***sci*LINKS NUMBER:** HSTP280
TOPIC: Metals — ***sci*LINKS NUMBER:** HSTP285
TOPIC: Metalloids — ***sci*LINKS NUMBER:** HSTP290
TOPIC: Nonmetals — ***sci*LINKS NUMBER:** HSTP295
TOPIC: Buckminster Fuller and the Buckyball — ***sci*LINKS NUMBER:** HSTP300

Chapter Review

pg 84

USING VOCABULARY

Complete the following sentences by choosing the appropriate term from each pair of terms listed below.

1. Elements in the same vertical column in the periodic table belong to the same __?__. (*group* or *period*)
2. Elements in the same horizontal row in the periodic table belong to the same __?__. (*group* or *period*)
3. The most reactive metals are __?__. (*alkali metals* or *alkaline-earth metals*)
4. Elements that are unreactive are called __?__. (*noble gases* or *halogens*)

UNDERSTANDING CONCEPTS

Multiple Choice

5. An element that is a very reactive gas is most likely a member of the
   a. noble gases.
   b. alkali metals.
   c. halogens.
   d. actinides.
6. Which statement is true?
   a. Alkali metals are generally found in their uncombined form.
   b. Alkali metals are Group 1 elements.
   c. Alkali metals should be stored under water.
   d. Alkali metals are unreactive.

7. Which statement about the periodic table is false?
   a. There are more metals than nonmetals.
   b. The metalloids are located in Groups 13 through 16.
   c. The elements at the far left of the table are nonmetals.
   d. Elements are arranged by increasing atomic number.
8. One property of most nonmetals is that they are
   a. shiny.
   b. poor conductors of electric current.
   c. flattened when hit with a hammer.
   d. solids at room temperature.
9. Which is a true statement about elements?
   a. Every element occurs naturally.
   b. All elements are found in their uncombined form in nature.
   c. Each element has a unique atomic number.
   d. All of the elements exist in approximately equal quantities.
10. Which is NOT found on the periodic table?
    a. The atomic number of each element
    b. The symbol of each element
    c. The density of each element
    d. The atomic mass of each element

Short Answer

11. Why was Mendeleev's periodic table useful?
12. How is Moseley's basis for arranging the elements different from Mendeleev's?
13. How is the periodic table like a calendar?
14. Describe the location of metals, metalloids, and nonmetals on the periodic table.

Concept Mapping

15. Use the following terms to create a concept map: periodic table, elements, groups, periods, metals, nonmetals, metalloids.

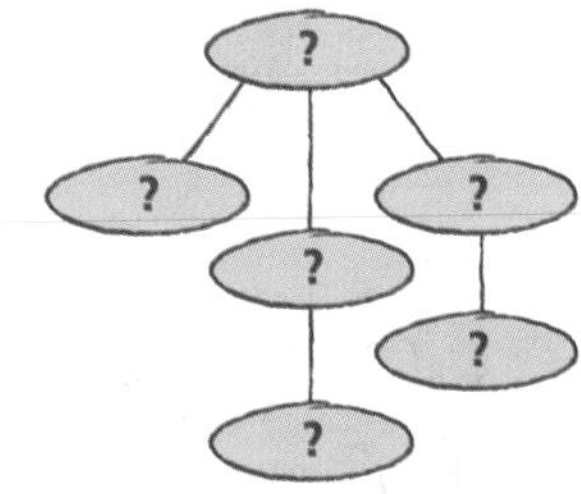

CRITICAL THINKING AND PROBLEM SOLVING

16. When an element with 115 protons in its nucleus is synthesized, will it be a metal, a nonmetal, or a metalloid? Explain.

17. Look at Mendeleev's periodic table in Figure 2. Why was Mendeleev not able to make any predictions about the noble gas elements?

18. Your classmate offers to give you a piece of sodium he found while hiking. What is your response? Explain.

19. Determine the identity of each element described below:
    a. This metal is very reactive, has properties similar to magnesium, and is in the same period as bromine.
    b. This nonmetal is in the same group as lead.
    c. This metal is the most reactive metal in its period and cannot be found uncombined in nature. Each atom of the element contains 19 protons.

MATH IN SCIENCE

20. The chart below shows the percentages of elements in the Earth's crust.

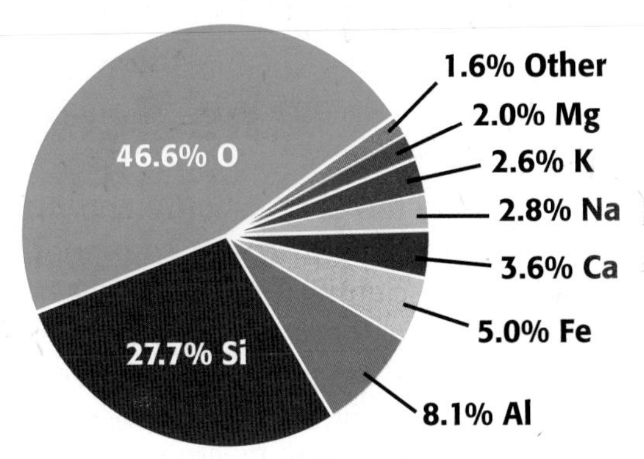

Excluding the "Other" category, what percentage of the Earth's crust is
a. alkali metals?
b. alkaline-earth metals?

INTERPRETING GRAPHICS

21. Study the diagram below to determine the pattern of the images. Predict the missing image, and draw it. Identify which properties are periodic and which properties are shared within a group.

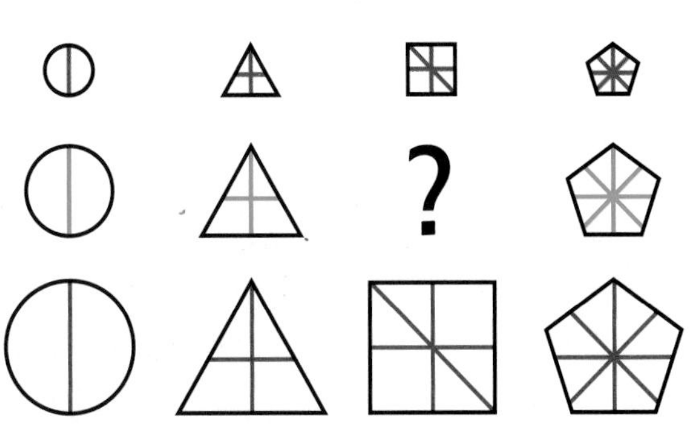

NOW What Do You Think?

Take a minute to review your answers to the ScienceLog questions on page 325. Have your answers changed? If necessary, revise your answers based on what you have learned since you began this chapter.

Science, Technology, and Society

The Science of Fireworks

What do the space shuttle and the Fourth of July have in common? The same scientific principles that help scientists launch a space shuttle also help pyrotechnicians create spectacular fireworks shows. The word *pyrotechnics* comes from the Greek words for "fire art." Explosive and dazzling, a fireworks display is both a science and an art.

An Ancient History

More than 1,000 years ago, Chinese civilizations made black powder, the original gunpowder used in pyrotechnics. They used the powder to set off firecrackers and primitive missiles. Black powder is still used today to launch fireworks into the air and to give fireworks an explosive charge. Even the ingredients—saltpeter (potassium nitrate), charcoal, and sulfur—haven't changed since ancient times.

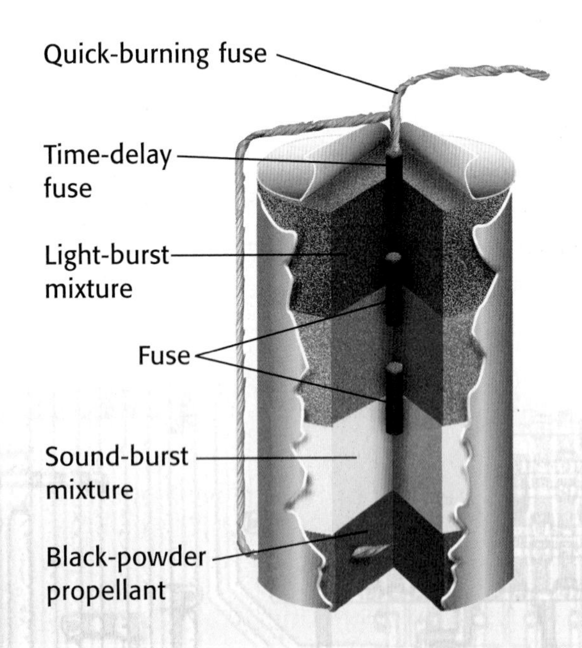

▲ *Cutaway view of a typical firework. Each shell creates a different type of display.*

Snap, Crackle, Pop!

The shells of fireworks contain the ingredients that create the explosions. Inside the shells, black powder and other chemicals are packed in layers. When ignited, one layer may cause a bright burst of light while a second layer produces a loud booming sound. The shell's shape affects the shape of the explosion. Cylindrical shells produce a trail of lights that looks like an umbrella. Round shells produce a star-burst pattern of lights.

The color and sound of fireworks depend on the chemicals used. To create colors, chemicals like strontium (for red), magnesium (for white), and copper (for blue) can be mixed with the gunpowder.

Explosion in the Sky

Fireworks are launched from metal, plastic, or cardboard tubes. Black powder at the bottom of the shell explodes and shoots the shell into the sky. A fuse begins to burn when the shell is launched. Seconds later, when the explosive chemicals are high in the air, the burning fuse lights another charge of black powder. This ignites the rest of the ingredients in the shell, causing an explosion that lights up the sky!

Bang for Your Buck

▶ The fireworks used during New Year's Eve and Fourth of July celebrations can cost anywhere from $200 to $2,000 apiece. Count the number of explosions at the next fireworks show you see. If each of the fireworks cost just $200 to produce, how much would the fireworks for the entire show cost?

WEIRD SCIENCE

BUCKYBALLS

Researchers are scrambling for the ball—the buckyball, that is. This special form of carbon has 60 carbon atoms linked together in a shape much like a soccer ball. Scientists are having a field day trying to find new uses for this unusual molecule.

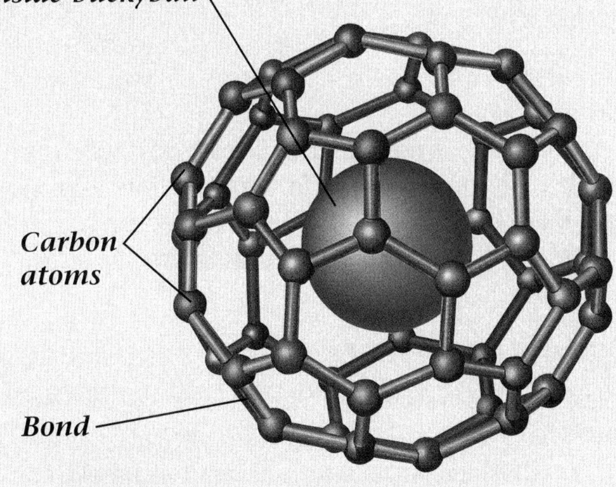

▲ *The buckyball, short for buckminsterfullerene, was named after architect Buckminster Fuller.*

The Starting Lineup

Named for architect Buckminster Fuller, buckyballs resemble the geodesic domes that are characteristic of the architect's work. Excitement over buckyballs began in 1985 when scientists projected light from a laser onto a piece of graphite. In the soot that remained, researchers found a completely new kind of molecule! Buckyballs are also found in the soot from a candle flame. Some scientists claim to have detected buckyballs in outer space. In fact, one hypothesis suggests that buckyballs might be at the center of the condensing clouds of gas, dust, and debris that form galaxies.

The Game Plan

Ever since buckyballs were discovered, chemists have been busy trying to identify the molecules' properties. One interesting property is that substances can be trapped inside a buckyball. A buckyball can act like a cage that surrounds smaller substances, such as individual atoms. Buckyballs also appear to be both slippery and strong. They can be opened to insert materials, and they can even link together in tubes.

How can buckyballs be used? They may have a variety of uses, from carrying messages through atom-sized wires in computer chips to delivering medicines right where the body needs them. Making tough plastics and cutting tools are uses that are also under investigation. With so many possibilities, scientists expect to get a kick out of buckyballs for some time!

The Kickoff

▶ A soccer ball is a great model for a buckyball. On the model, the places where three seams meet correspond to the carbon atoms on a buckyball. What represents the bonds between carbon atoms? Does your soccer-ball model have space for all 60 carbon atoms? You'll have to count and see for yourself.

UNIT 5

TIMELINE

Interactions of Matter

In this unit you will study the interactions through which matter can change its identity. You will learn how atoms bond with one another to form compounds and how different types of bonds account for differences in compounds. You will also learn how atoms join in different combinations to form new substances through chemical reactions. Finally, you will learn about the properties of several categories of compounds. This timeline includes some of the events leading to the current understanding of these interactions of matter.

1828

Urea, a compound found in urine, is produced in a laboratory. Until this time, chemists had believed that compounds created by living organisms could not be produced in the laboratory.

1858

German chemist Friedrich August Kekulé suggests that carbon forms four chemical bonds and can form long chains of carbon bonded to itself.

1964

Dr. Martin Luther King, Jr., American civil rights leader, is awarded the Nobel Peace Prize.

1969

The *Nimbus III* weather satellite is launched by the United States, representing the first civilian use of nuclear batteries.

1979

Public fear about nuclear power grows after an accident occurs at the Three Mile Island nuclear power station, in Pennsylvania.

1867

Swedish chemist Alfred Nobel develops dynamite. Dynamite's explosive power is a result of the decomposition reaction of nitroglycerin.

1898

The United States defeats Spain in the Spanish-American War.

1942

The first nuclear chain reaction is carried out in a squash court under the football stadium at the University of Chicago.

1903

Marie Curie, Pierre Curie, and Henri Becquerel are awarded the Nobel Prize in physics for the discovery of radioactivity.

1996

Evidence of organic compounds in a meteorite leads scientists to speculate that life may have existed on Mars more than 3.6 billion years ago.

2001

The first total solar eclipse of the millenium occurs on June 21.

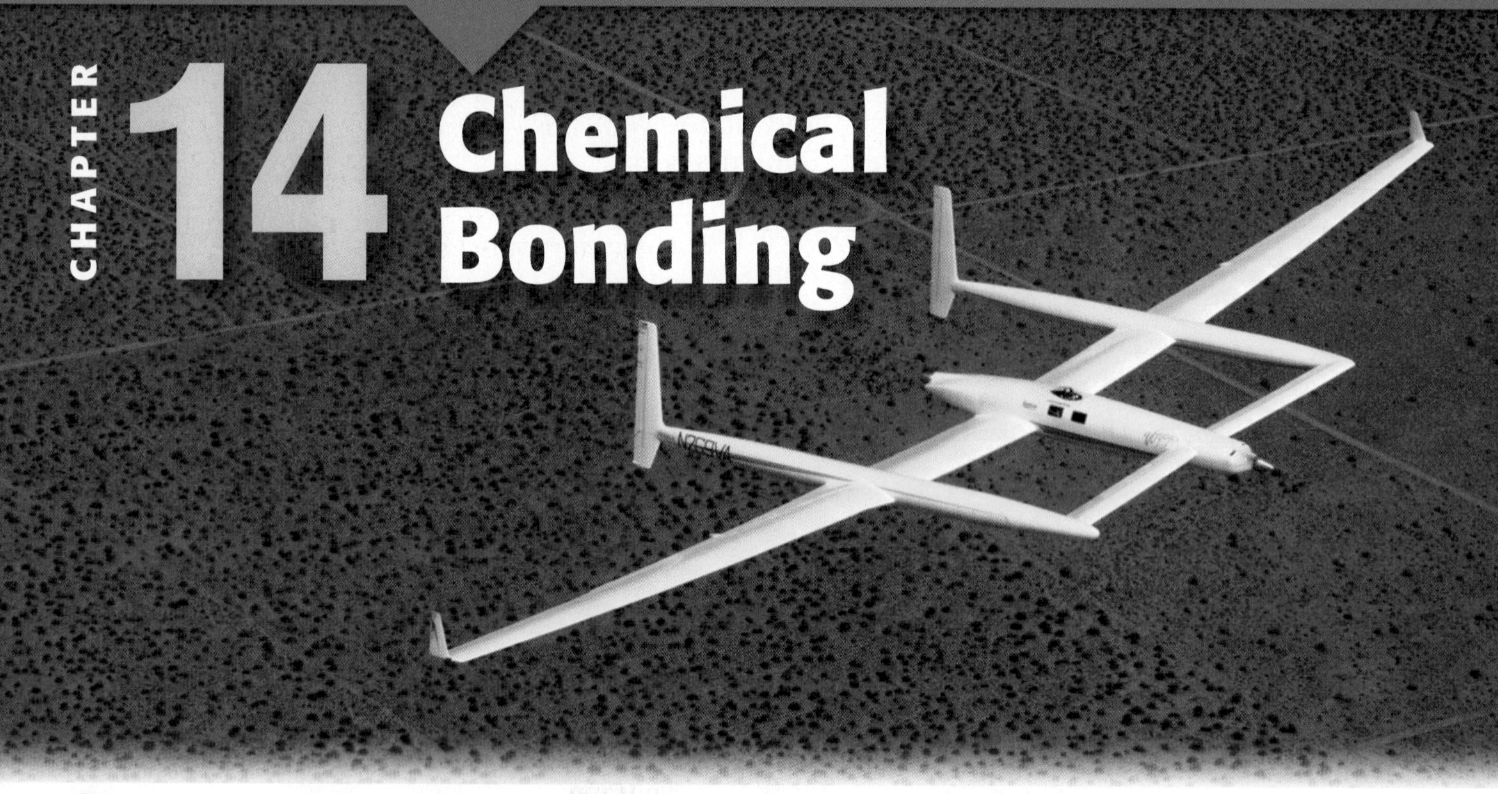

CHAPTER 14 Chemical Bonding

Strange but True!

In 1987, pilots Richard Rutan and Jeana Yeager performed a record-breaking feat. They flew the *Voyager* aircraft, shown above, around the world without refueling. The trip lasted just over 9 days. In order to carry enough fuel for the trip, the plane had to be as lightweight as possible. The designers knew that using fewer bolts than usual to attach parts would make the airplane lighter. But without the bolts, what would hold the parts together? The designers decided to replace the bolts with glue!

Not just any glue would do. They used superglue. When superglue is applied, it combines with water from the air to form *chemical bonds.* A chemical bond is a force of attraction that holds atoms together. The particles of superglue squeeze into the materials being glued. The materials stick together as if they were one material. Superglue is so strong that the weight of a two-ton elephant cannot separate two metal plates glued together with just a few drops!

Along with hundreds of household uses, superglue also has many uses in industry and medicine. For example, to make shoes stronger and lighter, manufacturers can replace some of the stitching with superglue. To repair a cracked tooth, dentists can apply superglue to hold the tooth together.

Superglue was discovered by a scientist in the early 1950s who was trying to develop a new plastic for the cockpit bubble of a jet plane.

Chemical bonding is responsible for the ways all materials behave—the properties of materials. In this chapter, you will learn about the different types of bonds that hold atoms together and how those bonds affect the properties of the materials.

What Do You Think?

In your ScienceLog, try to answer the following questions based on what you already know:

1. What is a chemical bond?
2. How are ionic bonds different from covalent bonds?
3. How are the properties of metals related to the type of bonds in them?

From Glue to Goop

Particles of glue can bond to other particles and hold objects together. Different types of bonds create differences in the properties of substances. In this activity, you will see how the formation of bonds causes an interesting change in the properties of a very common material—white glue.

Procedure

1. Fill a **small paper cup** 1/4 full of **white glue.** Observe the properties of the glue, and record your observations in your ScienceLog.
2. Fill a second **small paper cup** 1/4 full of **borax solution.**
3. Pour the borax solution into the cup containing the white glue, and stir well using a **plastic spoon.**
4. When it becomes too thick to stir, remove the material from the cup and knead it with your fingers. Observe the properties of the material, and record your observations in your ScienceLog.

Analysis

5. Compare the properties of the glue with those of the new material.
6. The properties of the new material resulted from the bonds between the borax and the particles in the glue. If too little borax were used, in what way would the properties of the material have been different?

Section 1

Electrons and Chemical Bonding

NEW TERMS
chemical bonding
chemical bond
theory
valence electrons

OBJECTIVES
- Describe chemical bonding.
- Identify the number of valence electrons in an atom.
- Predict whether an atom is likely to form bonds.

Have you ever stopped to consider that by using just the 26 letters of the alphabet, you make all of the words you use every day? Even though the number of letters is limited, their ability to be combined in different ways allows you to make an enormous number of words.

Now look around the room. Everything around you—desks, chalk, paper, even your friends—is made of atoms of elements. How can so many substances be formed from about 100 elements? In the same way that words can be formed by combining letters, different substances can be formed by combining atoms.

Atoms Combine Through Chemical Bonding

The atoms of just three elements—carbon, hydrogen, and oxygen—combine in different patterns to form the substances sugar, alcohol, and citric acid. **Chemical bonding** is the joining of atoms to form new substances. The properties of these new substances are different from those of the original elements. A force of attraction that holds two atoms together is called a **chemical bond.** As you will see, chemical bonds involve the electrons in the atoms.

Atoms and the chemical bonds that connect them cannot be observed with your eyes. During the past 150 years, scientists have performed many experiments that have led to the development of a theory of chemical bonding. Remember that a **theory** is a unifying explanation for a broad range of hypotheses and observations that have been supported by testing. The use of models helps people to discuss the theory of how and why atoms form chemical bonds.

Electron Number and Organization

To understand how atoms form chemical bonds, you first need to know how many electrons are in a particular atom and how the electrons in an atom are organized.

Explore

In your ScienceLog, write down the term *chemical bonding.* Then write down as many different words as you can that are formed from the letters in these two words.

Why are the amino acids that are chemically bonded together to form your proteins all left-handed? Read about one cosmic explanation on page 370.

The number of electrons in an atom can be determined from the atomic number of the element. Remember that the atomic number is the number of protons in an atom. Because atoms have no charge, the atomic number also represents the number of electrons in the atom. Equal numbers of positively charged protons and negatively charged electrons are needed to make the overall charge of the atom zero.

The electrons are organized in levels in an atom. These levels are usually called energy levels. The levels farther from the nucleus contain electrons that have more energy than levels closer to the nucleus. The arrangement of electrons within the energy levels of a chlorine atom is shown in **Figure 1.**

Figure 1 Electron Arrangement in an Atom

a The first energy level is closest to the nucleus and can hold up to 2 electrons.

b Electrons will enter the second energy level only after the first level is full. The second energy level can hold up to 8 electrons.

c The third energy level in this model of a chlorine atom contains only 7 electrons, for a total of 17 electrons in the atom. This outer level of the atom is not full.

Outer-Level Electrons Are the Key to Bonding As you just saw in Figure 1, a chlorine atom has a total of 17 electrons. When a chlorine atom bonds to another atom, not all of these electrons are used to create the bond. Most atoms form bonds using only the electrons in their outermost energy level. The electrons in the outermost energy level of an atom are called **valence** (VAY luhns) **electrons.** Thus, a chlorine atom has 7 valence electrons. You can see the valence electrons for atoms of some other elements in **Figure 2.**

Figure 2 *Valence electrons are the electrons in the outermost energy level of an atom.*

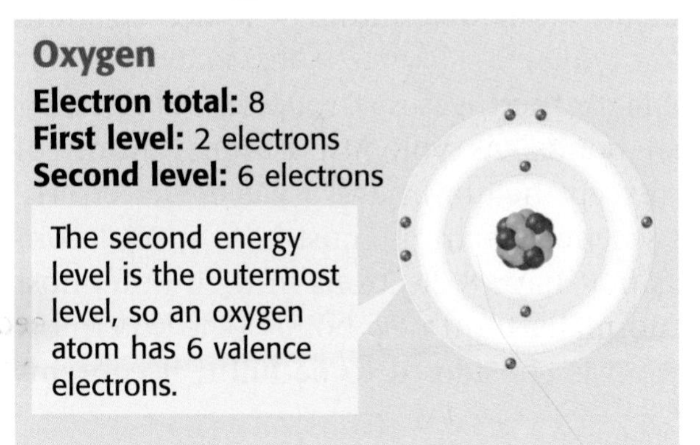

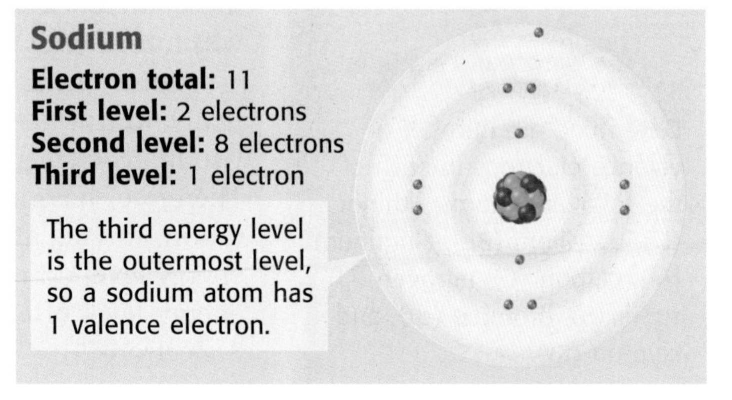

Valence Electrons and the Periodic Table You can determine the number of valence electrons in Figure 2 because you have a model to look at. But what if you didn't have a model? You have a tool that helps you determine the number of valence electrons for some elements—the periodic table!

Remember that elements in a group often have similar properties, including the number of electrons in the outermost energy level of their atoms. The number of valence electrons for many elements is related to the group number, as shown in **Figure 3.**

Figure 3 Determining the Number of Valence Electrons

Atoms of elements in **Groups 1 and 2** have the same number of valence electrons as their group number.

Atoms of elements in **Groups 13–18** have 10 fewer valence electrons than their group number. However, helium atoms have only 2 valence electrons.

Atoms of elements in **Groups 3–12** do not have a general rule relating their valence electrons to their group number.

1	2	3	4	5	6	7	8	9	10	11	12	13	14	15	16	17	18
H																	He
Li	Be											B	C	N	O	F	Ne
Na	Mg											Al	Si	P	S	Cl	Ar
K	Ca	Sc	Ti	V	Cr	Mn	Fe	Co	Ni	Cu	Zn	Ga	Ge	As	Se	Br	Kr
Rb	Sr	Y	Zr	Nb	Mo	Tc	Ru	Rh	Pd	Ag	Cd	In	Sn	Sb	Te	I	Xe
Cs	Ba	La	Hf	Ta	W	Re	Os	Ir	Pt	Au	Hg	Tl	Pb	Bi	Po	At	Rn
Fr	Ra	Ac	Rf	Db	Sg	Bh	Hs	Mt	Uun	Uuu	Uub						

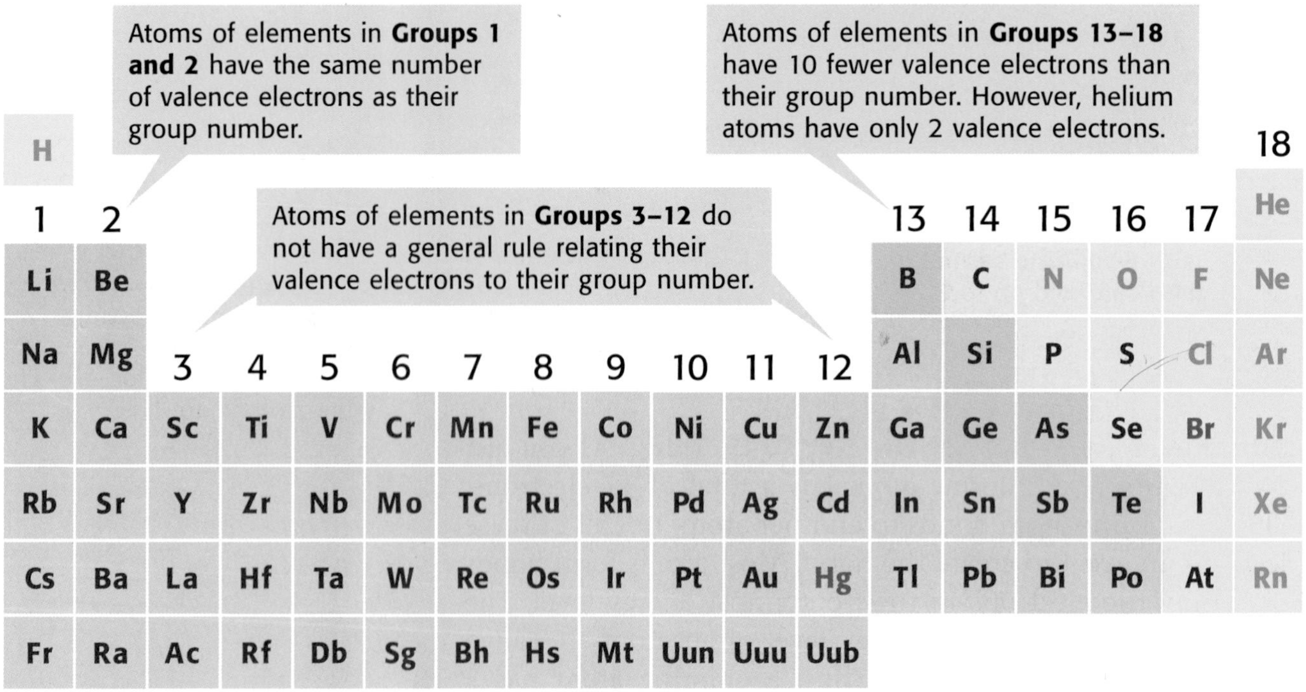

To Bond or Not to Bond

Atoms do not all bond in the same manner. In fact, some atoms rarely bond at all! The number of electrons in the outermost energy level of an atom determines whether an atom will form bonds.

Atoms of the noble, or inert, gases (Group 18) do not normally form chemical bonds. As you just learned, atoms of Group 18 elements (except helium) have 8 valence electrons. Therefore, having 8 valence electrons must be a special condition. In fact, atoms that have 8 electrons in their outermost energy level do not normally form new bonds. The outermost energy level of an atom is considered to be full if it contains 8 electrons.

Explore

Determine the number of valence electrons in each of the following atoms: lithium (Li), beryllium (Be), aluminum (Al), carbon (C), nitrogen (N), sulfur (S), bromine (Br), and krypton (Kr).

Atoms Bond to Have a Filled Outermost Level An atom that has fewer than 8 valence electrons is more reactive, or more likely to form bonds, than an atom with 8 valence electrons. Atoms bond by gaining, losing, or sharing electrons in order to have a filled outermost energy level with 8 valence electrons. **Figure 4** describes the ways in which atoms can achieve a filled outermost energy level.

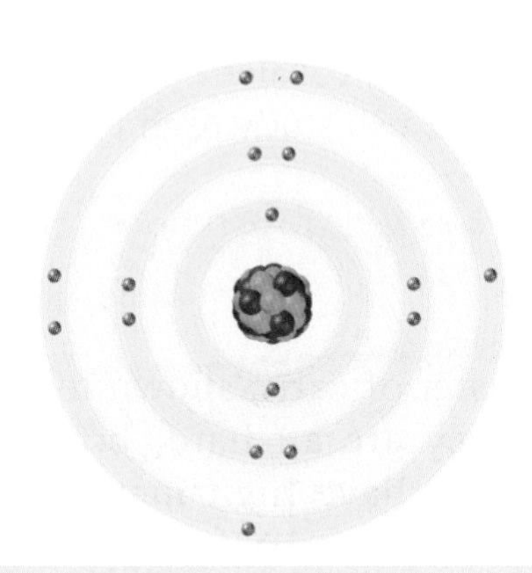

Figure 4 *These atoms achieve a full set of valence electrons in different ways.*

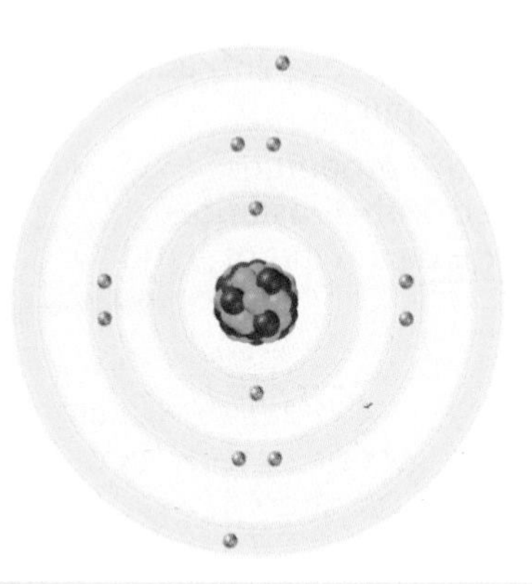

Sulfur
An atom of sulfur has 6 valence electrons. It can have 8 valence electrons by sharing 2 electrons with or gaining 2 electrons from other atoms to fill its outermost energy level.

Magnesium
An atom of magnesium has 2 valence electrons. It can have a full outer level by losing 2 electrons. The second energy level becomes the outermost energy level and contains a full set of 8 electrons.

A Full Set—with Two? Not all atoms need 8 valence electrons for a filled outermost energy level. Helium atoms need only two valence electrons. With only two electrons in the entire atom, the first energy level (which is also the outermost energy level) is full. Atoms of the elements hydrogen and lithium form bonds with other atoms in order to have the same number of electrons as helium atoms have.

REVIEW

1. What is a chemical bond?
2. What are valence electrons?
3. How many valence electrons does a silicon atom have?
4. Predict how atoms with 5 valence electrons will achieve a full set of valence electrons.
5. **Interpreting Graphics** At right is a diagram of a fluorine atom. Will fluorine form bonds? Explain.

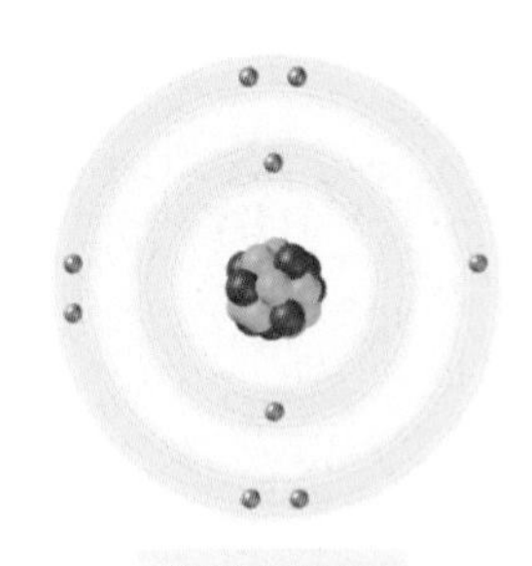

Fluorine

Section 2

Types of Chemical Bonds

NEW TERMS

ionic bond
ions
crystal lattice
covalent bond
molecule
metallic bond

OBJECTIVES

- Describe ionic, covalent, and metallic bonding.
- Describe the properties associated with substances containing each type of bond.

As you have learned, atoms bond by gaining, losing, or sharing electrons. Once bonded, most atoms have a filled outermost energy level containing eight valence electrons. Atoms are less reactive when they have a filled outermost energy level. The way in which atoms interact through their valence electrons determines the type of bond that forms. The three types of bonds are ionic (ie AHN ik), covalent (KOH VAY luhnt), and metallic. In this section, you will study each type of bond, starting with ionic bonds.

Ionic Bonds

Seashells, table salt, and plaster of Paris, shown in **Figure 5,** have much in common. They are all hard, brittle solids at room temperature, they all have high melting points, and they all contain ionic bonds. An **ionic bond** is the force of attraction between oppositely charged ions. **Ions** are charged particles that form during chemical changes when one or more valence electrons transfer from one atom to another.

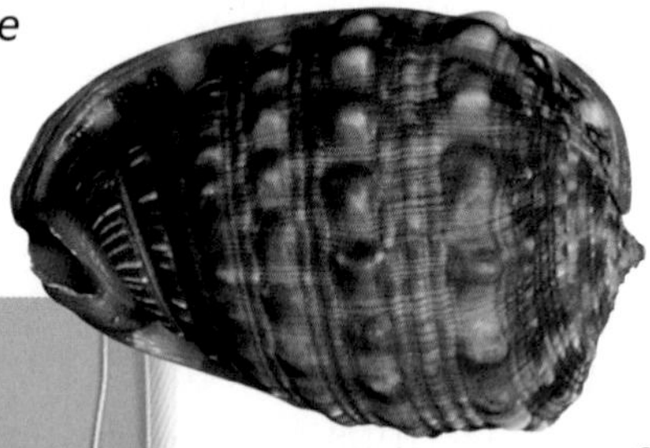

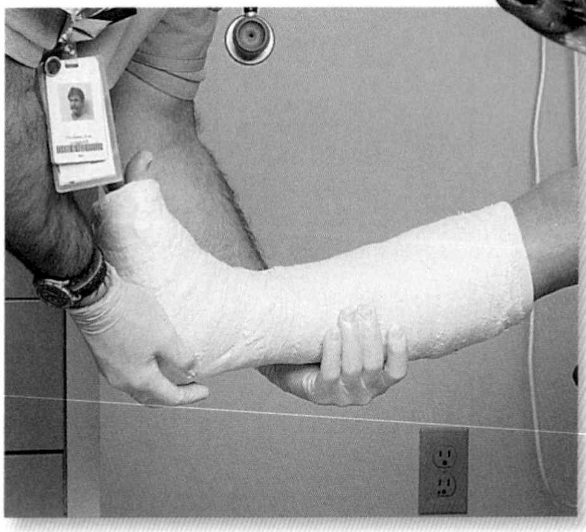

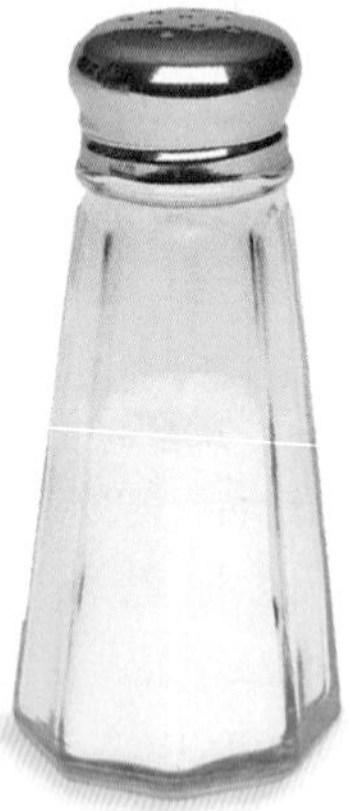

Figure 5 *Calcium carbonate in seashells, sodium chloride in table salt, and calcium sulfate used to make plaster of Paris casts all contain ionic bonds.*

Remember that in an atom, the number of electrons equals the number of protons, so the negative charges and positive charges cancel each other. Therefore, atoms are neutral. A transfer of electrons between atoms changes the number of electrons in each atom, while the number of protons stays the same. The negative charges and positive charges no longer cancel out, and the atoms become ions. Although an atom cannot gain (or lose) electrons without another atom nearby to lose (or gain) electrons, it is often easier to study the formation of ions one at a time.

Atoms That Lose Electrons Form Positive Ions Ionic bonds form during chemical reactions when atoms that have a stronger attraction for electrons pull electrons away from other atoms. The atoms that lose electrons form ions that have fewer electrons than protons. The positive charges outnumber the negative charges in the ions. Thus, the ions that are formed when atoms lose electrons have an overall positive charge.

Atoms of most metals have few electrons in their outer energy level. When metal atoms bond with other atoms, the metal atoms tend to lose these valence electrons and form positive ions. For example, look at the model in **Figure 6.** An atom of sodium has one valence electron. When a sodium atom loses this electron to another atom, it becomes a sodium ion. A sodium ion has a charge of 1+ because it contains 1 more proton than electrons. To show the difference between a sodium atom and a sodium ion, the chemical symbol for the ion is written as Na^+. Notice that the charge is written to the upper right of the chemical symbol. Figure 6 also shows a model for the formation of an aluminum ion.

Figure 6 Forming Positive Ions

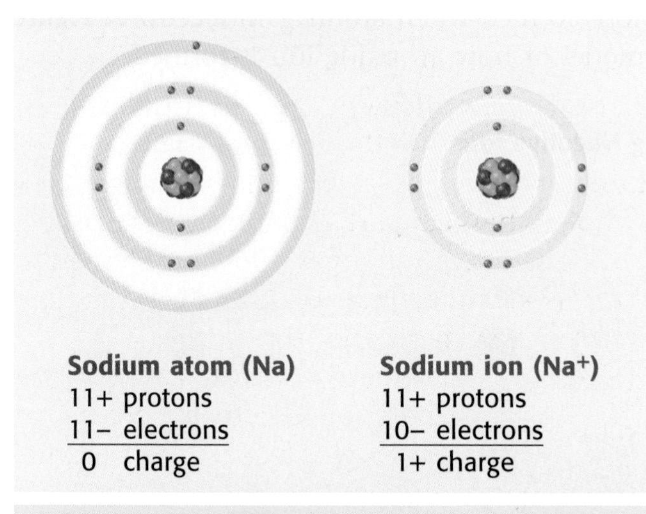

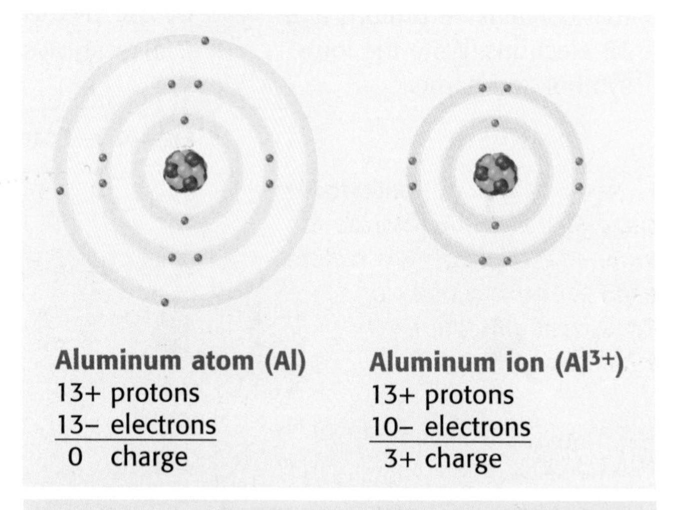

Here's How It Works: During chemical reactions, a sodium atom can lose its 1 electron in the third energy level to another atom. The filled second level becomes the outermost level, so the resulting sodium ion has 8 valence electrons.

Here's How It Works: During chemical reactions, an aluminum atom can lose its 3 electrons in the third energy level to another atom. The filled second level becomes the outermost level, so the resulting aluminum ion has 8 valence electrons.

The Energy of Losing Electrons When an atom loses electrons, energy is needed to overcome the attraction between the electrons and the protons in the atom's nucleus. Removing electrons from atoms of metals requires only a small amount of energy, so metal atoms lose electrons easily. In fact, the energy needed to remove electrons from atoms of elements in Groups 1 and 2 is so low that these elements react very easily and can be found only as ions in nature. On the other hand, removing electrons from atoms of nonmetals requires a large amount of energy. Rather than give up electrons, these atoms gain electrons when they form ionic bonds.

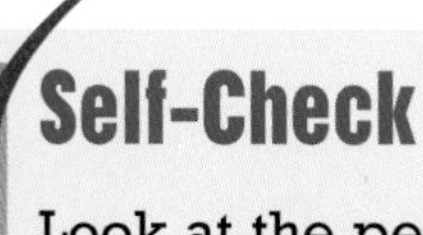

Self-Check

Look at the periodic table, and determine which noble gas has the same electron arrangement as a sodium ion. *(See page 596 to check your answer.)*

÷ 5 ÷ Ω ≤ ∞ + Ω √ 9 ∞ ≤ Σ 2

MATH BREAK

Charge!

Calculating the charge of an ion is the same as adding integers (positive or negative whole numbers or zero) with opposite signs. You write the number of protons as a positive integer and the number of electrons as a negative integer and then add the integers. Calculate the charge of an ion that contains 16 protons and 18 electrons. Write the ion's symbol and name.

Atoms That Gain Electrons Form Negative Ions Atoms that gain electrons from other atoms during chemical reactions form ions that have more electrons than protons. The negative charges outnumber the positive charges, giving each of these ions an overall negative charge.

The outermost energy level of nonmetal atoms is almost full. Only a few electrons are needed to fill the outer level, so atoms of nonmetals tend to gain electrons from other atoms. For example, look at the model in **Figure 7.** An atom of chlorine has 7 valence electrons. When a chlorine atom gains 1 electron to complete its outer level, it becomes an ion with a 1– charge called a chloride ion. The symbol for the chloride ion is Cl^-. Notice that the name of the negative ion formed from chlorine has the ending *-ide*. This ending is used for the names of the negative ions formed when atoms gain electrons. Figure 7 also shows a model of how an oxide ion is formed.

Figure 7 Forming Negative Ions

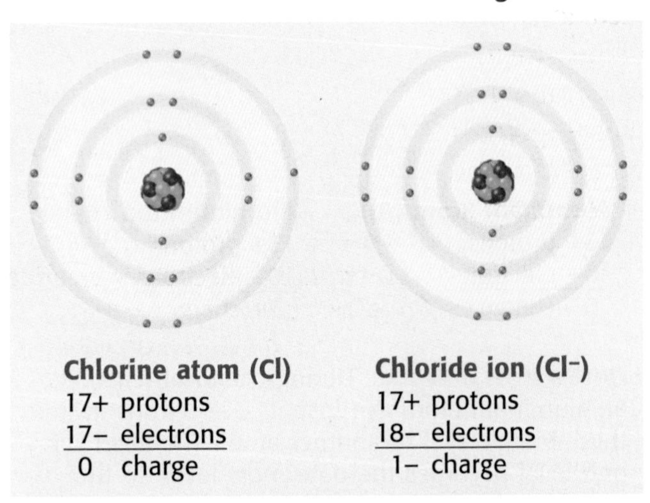

Here's How It Works: During chemical reactions, a chlorine atom gains 1 electron in the third energy level from another atom. A chloride ion is formed with 8 valence electrons. Thus, its outermost energy level is filled.

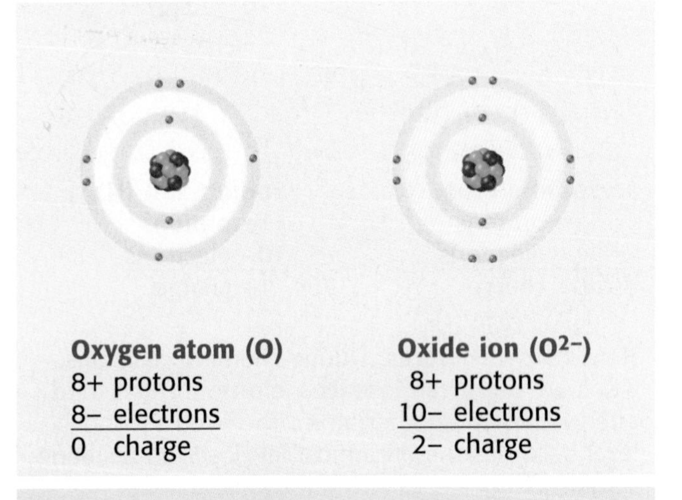

Here's How It Works: During chemical reactions, an oxygen atom gains 2 electrons in the second energy level from another atom. An oxide ion is formed with 8 valence electrons. Thus, its outermost energy level is filled.

The Energy of Gaining Electrons Atoms of most nonmetals fill their outermost energy level by gaining electrons. Energy is given off by most nonmetal atoms during this process. The more easily an atom gains an electron, the more energy an atom gives off. Atoms of the Group 17 elements (the halogens) give off the most energy when they gain an electron. The halogens, such as fluorine and chlorine, are extremely reactive nonmetals because they release a large amount of energy.

Ions Bond to Form a Crystal Lattice When a metal reacts with a nonmetal, the same number of electrons is lost by the metal atoms as is gained by the nonmetal atoms. Even though the ions that bond are charged, the compound formed is neutral. The charges of the positive ions and the negative ions cancel each other through ionic bonding. An ionic bond is an example of a special kind of attraction, called electrostatic attraction, that causes opposite electrical charges to stick together. Another example is static cling, as shown in **Figure 8.**

Figure 8 *Like ionic bonds, static cling is the result of the attraction between opposite charges.*

The ions that make up an ionic compound are bonded in a repeating three-dimensional pattern called a **crystal lattice** (KRI stuhl LAT is). In ionic compounds, such as table salt, the ions in the crystal lattice are arranged as alternating positive and negative ions, forming a solid. Each ion is bordered on every side by an ion with the opposite charge. The model in **Figure 9** shows a small part of a crystal lattice. The arrangement of bonded ions in a crystal lattice determines the shape of the crystals of an ionic compound.

The strong force of attraction between bonded ions in a crystal lattice gives ionic compounds certain properties, including a high melting point and boiling point. Ionic compounds tend to be brittle solids at room temperature. Ionic compounds usually break apart when hit with a hammer because as the ions move, ions with like charges line up and repel one another.

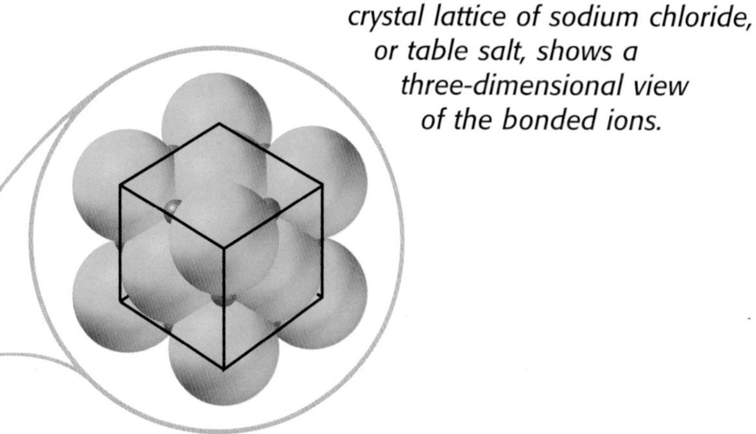

Figure 9 *This model of the crystal lattice of sodium chloride, or table salt, shows a three-dimensional view of the bonded ions.*

REVIEW

1. How does an atom become a negative ion?
2. What are two properties of ionic compounds?
3. **Applying Concepts** Which group of elements lose 2 valence electrons when their atoms form ionic bonds? What charge would the ions formed have?

Covalent Bonds

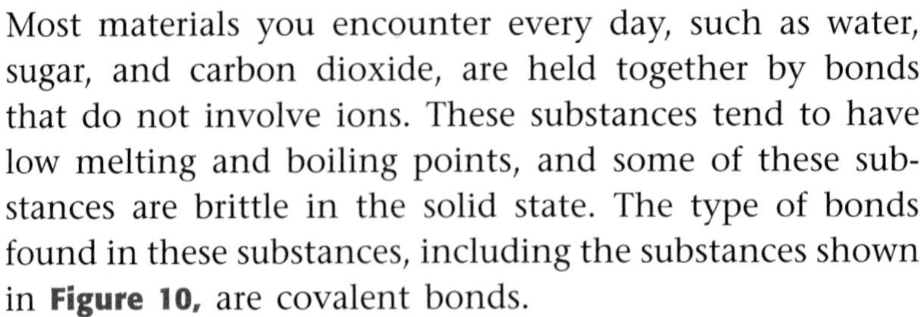

Most materials you encounter every day, such as water, sugar, and carbon dioxide, are held together by bonds that do not involve ions. These substances tend to have low melting and boiling points, and some of these substances are brittle in the solid state. The type of bonds found in these substances, including the substances shown in **Figure 10,** are covalent bonds.

Figure 10 *Covalent bonds join the atoms that make up this plastic ball, the rubber covering on the paddle, the cotton fibers in clothes, and even many of the substances that make up the human body!*

A **covalent bond** is the force of attraction between the nuclei of atoms and the electrons shared by the atoms. Based on experiments and observations, the current theory about covalent bonding is that it occurs between atoms that require a large amount of energy to remove an electron. When two atoms of nonmetals bond, too much energy is required for either atom to lose an electron, so no ions are formed. Rather than transferring electrons to complete their outermost energy levels, two nonmetal atoms bond by sharing electrons with one another, as shown in the model in **Figure 11.**

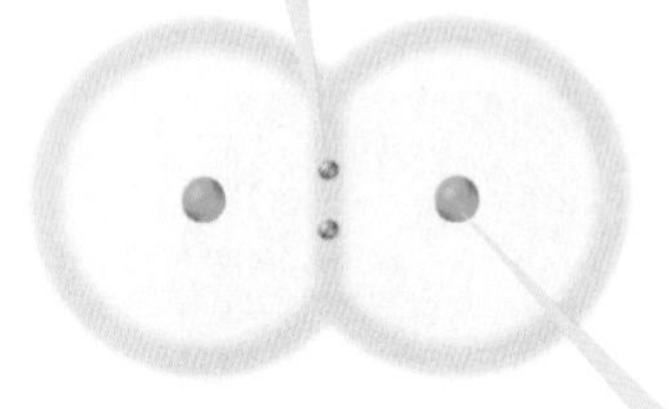

Figure 11 *By sharing electrons in a covalent bond, each hydrogen atom (the smallest atom known) has a full outermost energy level containing two electrons.*

Covalently Bonded Atoms Make Up Molecules The particles of substances containing covalent bonds differ from those containing ionic bonds. Ionic compounds consist of ions organized in a crystal. Covalent compounds consist of individual particles called molecules (MAHL i KYOOLZ). A **molecule** is a neutral group of atoms held together by covalent bonds. Each molecule is separate from other molecules of the substance. In Figure 11, you saw a model of a hydrogen molecule, which is composed of two hydrogen atoms covalently bonded. However, most molecules are composed of atoms of two or more elements. The models in **Figure 12** show two ways to represent the covalent bonds in a molecule.

Make models of molecules out of marshmallows on page 574 of the LabBook.

Figure 12 Covalent Bonds in a Water Molecule

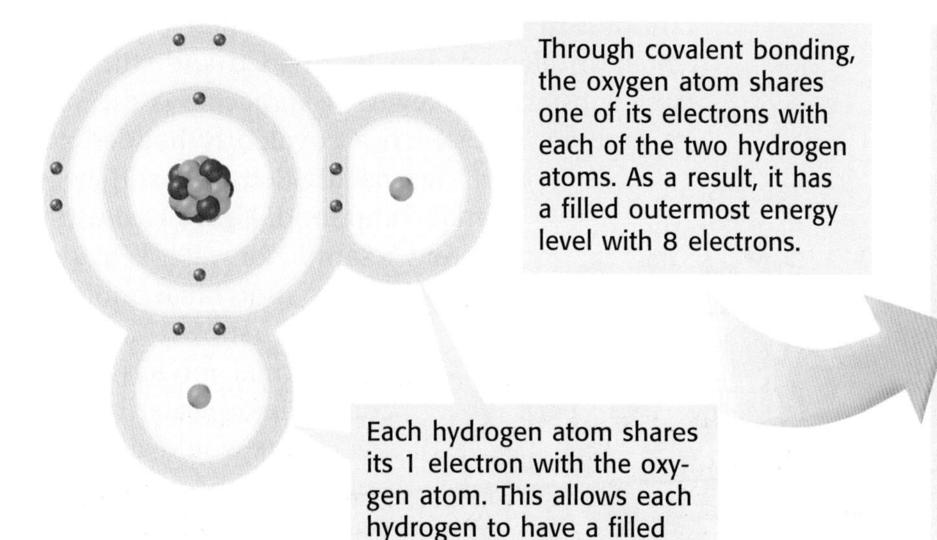

Through covalent bonding, the oxygen atom shares one of its electrons with each of the two hydrogen atoms. As a result, it has a filled outermost energy level with 8 electrons.

Each hydrogen atom shares its 1 electron with the oxygen atom. This allows each hydrogen to have a filled outer level with 2 electrons.

:Ö:H
H

Another way to show covalent bonds is to draw an electron-dot diagram. An electron-dot diagram shows only the outermost level of electrons for each atom. But you can still see how electrons are shared between the atoms.

Making Electron-Dot Diagrams

An electron-dot diagram is a model that shows only the valence electrons in an atom. Electron-dot diagrams are helpful when predicting how atoms might bond. You draw an electron-dot diagram by writing the symbol of the element and placing the correct number of dots around it. This type of model can help you to better understand bonding by showing the number of valence electrons and how atoms share electrons to fill their outermost energy levels, as shown below.

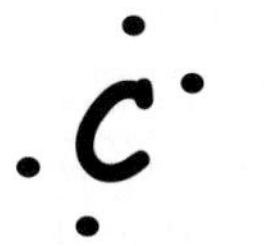

Carbon atoms have 4 valence electrons, so 4 dots are placed around the symbol. A carbon atom needs 4 more electrons for a filled outermost energy level.

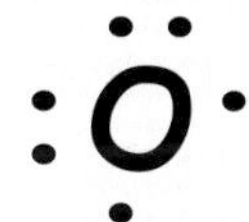

Oxygen atoms have 6 valence electrons, so 6 dots are placed around the symbol. An oxygen atom needs only 2 more electrons for a filled outermost energy level.

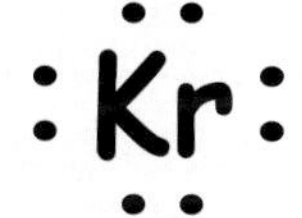

The noble gas krypton has a full set of 8 valence electrons in its atoms. Thus, krypton is nonreactive because its atoms do not need any more electrons.

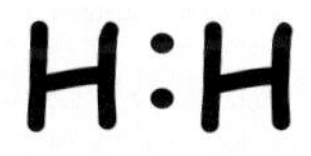

This electron-dot diagram represents hydrogen gas, the same substance shown in the model in Figure 11.

Self-Check

1. How many dots does the electron-dot diagram of a sulfur atom have?
2. How is a covalent bond different from an ionic bond?

(See page 596 to check your answers.)

Figure 13 *The water in this fishbowl is made up of many tiny water molecules. Each molecule is the smallest particle that still has the chemical properties of water.*

A Molecule Is the Smallest Particle of a Covalent Compound An atom is the smallest particle into which an element can be divided and still be the same substance. Likewise, a molecule is the smallest particle into which a covalently bonded compound can be divided and still be the same compound. **Figure 13** illustrates how a sample of water is made up of many individual molecules of water (shown as three-dimensional models). If you could divide water over and over, you would eventually end up with a single molecule of water. However, if you separated the hydrogen and oxygen atoms that make up a water molecule, you would no longer have water.

Explore

Try your hand at drawing electron-dot diagrams for a molecule of chlorine (a diatomic molecule) and a molecule of ammonia (one nitrogen atom bonded with three hydrogen atoms).

The Simplest Molecules All molecules are composed of at least two covalently bonded atoms. The simplest molecules, known as *diatomic molecules,* consist of two atoms bonded together. Some elements are called diatomic elements because they are found in nature as diatomic molecules composed of two atoms of the element. Hydrogen is a diatomic element, as you saw in Figure 11. Oxygen, nitrogen, and the halogens fluorine, chlorine, bromine, and iodine are also diatomic. By sharing electrons, both atoms of a diatomic molecule can fill their outer energy level, as shown in **Figure 14.**

Figure 14 Models of a Diatomic Fluorine Molecule

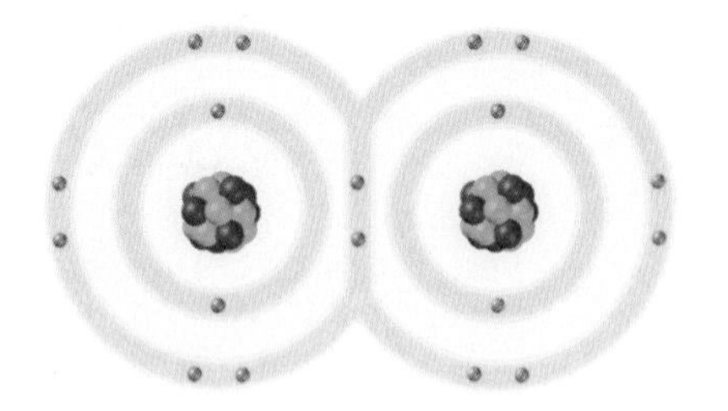

Two covalently bonded fluorine atoms have filled outermost energy levels. The pair of electrons shared by the atoms are counted as valence electrons for each atom.

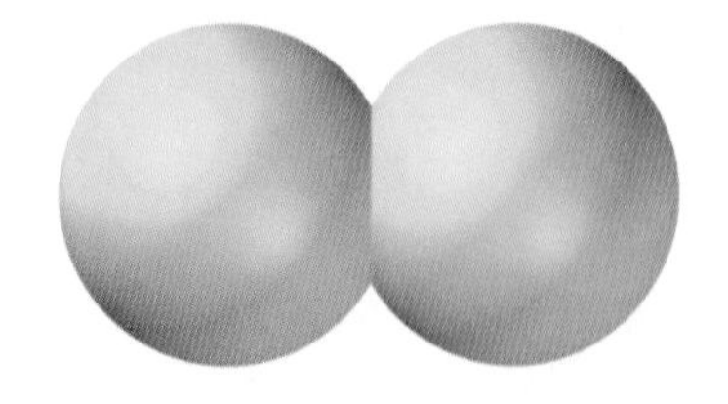

This is a three-dimensional model of a fluorine molecule.

More-Complex Molecules Diatomic molecules are the simplest—and some of the most important—of all molecules. You could not live without diatomic oxygen molecules. But other important molecules are much more complex. Gasoline, plastic, and even proteins in the cells of your body are examples of complex molecules. Carbon atoms are the basis of many of these complex molecules. Each carbon atom needs to make 4 covalent bonds to have 8 valence electrons. These bonds can be with atoms of other elements or with other carbon atoms, as shown in the model in **Figure 15.**

Proteins perform many functions throughout your body, such as digesting your food, building components of your cells, and transporting nutrients to each cell. A single protein can have a chain of 7,000 atoms of carbon and nitrogen with atoms of other elements covalently bonded to it.

Figure 15 *A granola bar contains the covalent compound sucrose, or table sugar. A molecule of sucrose is composed of carbon atoms (green spheres), hydrogen atoms (blue spheres), and oxygen atoms (red spheres) joined by covalent bonds.*

Metallic Bonds

Look at the unusual metal sculpture shown in **Figure 16.** Notice that some metal pieces have been flattened, while other metal pieces have been shaped into wires. How could the artist change the shape of the metal into all of these different forms without breaking the metal into pieces? A metal can be shaped because of the presence of a special type of bond called a metallic bond. A **metallic bond** is the force of attraction between a positively charged metal ion and the electrons in a metal. (Remember that metal atoms tend to lose electrons and form positively charged ions.)

Figure 16 *The different shapes of metal in this sculpture are possible because of the bonds that hold the metal together.*

*Quick*Lab

Bending with Bonds

1. Straighten out a **wire paper clip.** Record the result in your ScienceLog.
2. Bend a **piece of chalk.** Record the result in your ScienceLog.
3. Chalk is composed of calcium carbonate, a compound containing ionic bonds. What type of bonds are present in the paper clip?
4. In your ScienceLog, explain why you could change the shape of the paper clip but could not bend the chalk without breaking it.

Electrons Move Throughout a Metal Our scientific understanding of the bonding in metals is that the metal atoms get so close to one another that their outermost energy levels overlap. This allows their valence electrons to move throughout the metal from the energy level of one atom to the energy levels of the atoms nearby. The atoms form a crystal much like the ions associated with ionic bonding. However, the negative charges (electrons) in the metal are free to move about. You can think of a metal as being made up of positive metal ions with enough valence electrons "swimming" about to keep the ions together and to cancel the charge of the ions, as shown in **Figure 17.** The ions are held together because metallic bonds extend throughout the metal in all directions.

Figure 17 *The moving electrons are attracted to the metal ions, forming metallic bonds.*

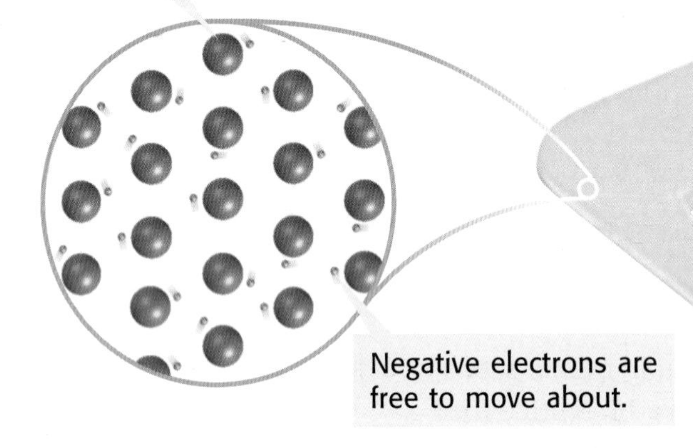

Explaining Metallic Properties You encounter metallic properties every day, such as when you turn on a lamp or wrap leftovers in aluminum foil. The abilities to conduct electrical energy and to be flattened and shaped without breaking are two properties of metals that result from metallic bonding.

When you turn on a lamp, electrons move within the wire. These moving electrons are the valence electrons in the metal. Because these electrons are not connected to any one atom in the wire, they can move freely within the wire.

Metals have a fairly high density because the metal atoms are closely packed. But because the atoms can be rearranged, metals can be shaped into useful forms. The properties of *ductility* (the ability to be drawn into wires) and *malleability* (the ability to be hammered into sheets) describe a metal's ability to be reshaped. For example, copper is made into wires for use in electrical cords. Aluminum can be pounded into thin sheets and made into aluminum foil and cans.

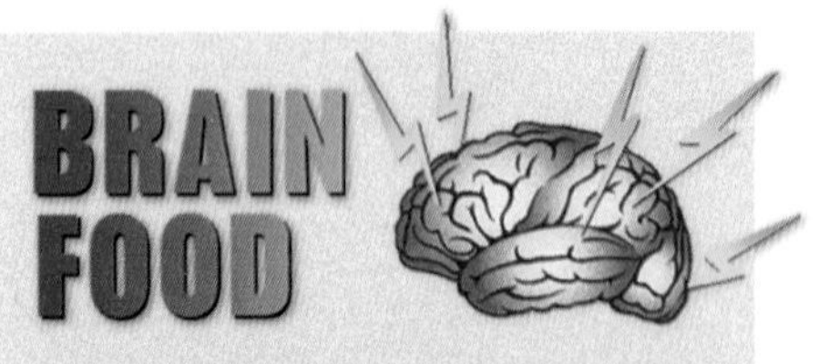

Gold can be pounded out to make a foil only a few atoms thick. A piece of gold the size of the head of a pin can be beaten into a thin "leaf" that would cover this page!

When the shape of a piece of metal is changed, the metal ions shift position in the crystal. You might expect the metal to break apart as the ions push away from one another. However, even in their new positions, the positive ions are surrounded by and attracted to the electrons, as shown in **Figure 18.** On the other hand, ionic compounds do break apart when hit with a hammer because neither the positive ions nor the negative ions are free to move.

Figure 18 *The shape of a metal can be changed without breaking because metallic bonds occur in many directions.*

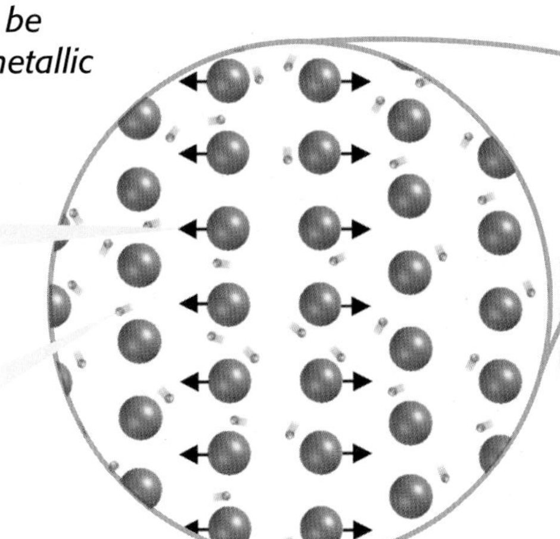

The repulsion between the positively charged metal ions increases as the ions are pushed closer to one another.

The moving electrons maintain the metallic bonds no matter how the shape of the metal changes.

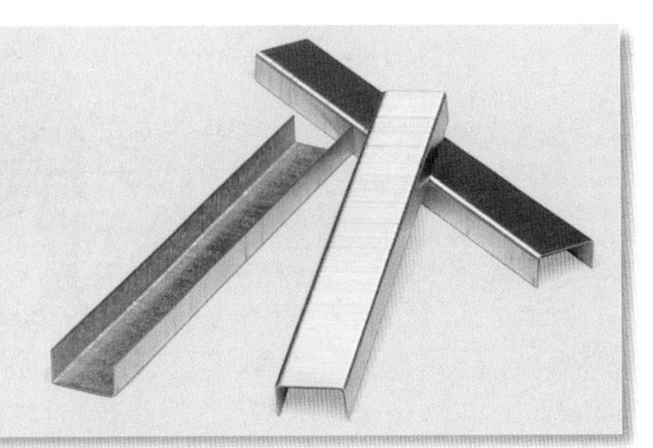

Although they are not very glamorous, metal staples are very useful in holding things such as sheets of paper together. Explain how the metallic bonds in a staple allow it to change shape so that it can function properly.

REVIEW

1. What happens to electrons in covalent bonding?
2. What type of element is most likely to form covalent bonds?
3. What is a metallic bond?
4. What is one difference between a metallic bond and a covalent bond?
5. **Interpreting Graphics** The electron-dot diagram at right is not yet complete. Which atom needs to form another covalent bond? How do you know?

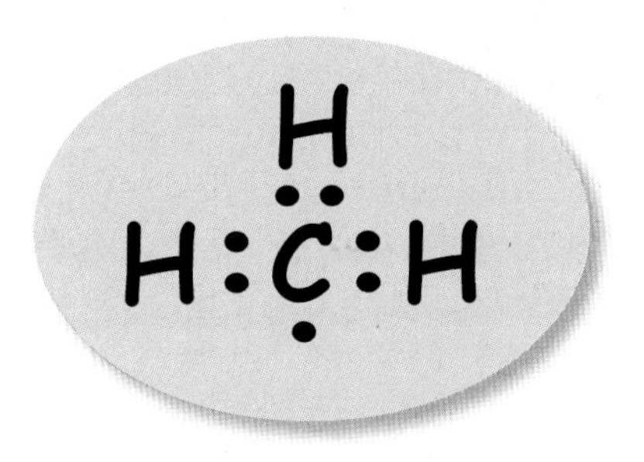

Chapter Highlights

SECTION 1

Vocabulary

chemical bonding *(p. 352)*
chemical bond *(p. 352)*
theory *(p. 352)*
valence electrons *(p. 353)*

Section Notes

- Chemical bonding is the joining of atoms to form new substances. A chemical bond is a force of attraction that holds two atoms together.
- Valence electrons are the electrons in the outermost energy level of an atom. These electrons are used to form bonds.
- Most atoms form bonds by gaining, losing, or sharing electrons until they have 8 valence electrons. Atoms of hydrogen, lithium, and helium need only 2 electrons to fill their outermost level.

SECTION 2

Vocabulary

ionic bond *(p. 356)*
ions *(p. 356)*
crystal lattice *(p. 359)*
covalent bond *(p. 360)*
molecule *(p. 360)*
metallic bond *(p. 363)*

Section Notes

- In ionic bonding, electrons are transferred between two atoms. The atom that loses electrons becomes a positive ion. The atom that gains electrons becomes a negative ion. The force of attraction between these oppositely charged ions is an ionic bond.
- Ionic bonding usually occurs between atoms of metals and atoms of nonmetals.

☑ Skills Check

Math Concepts

CALCULATING CHARGE To calculate the charge of an ion, you must add integers with opposite signs. The total positive charge of the ion (the number of protons) is written as a positive integer. The total negative charge of the ion (the number of electrons) is written as a negative integer. For example, the charge of an ion containing 11 protons and 10 electrons would be calculated as follows:

$$(11+) + (10-) = 1+$$

Visual Understanding

DETERMINING VALENCE ELECTRONS Knowing the number of valence electrons in an atom is important in predicting how it will bond with other atoms. Review Figure 3 on page 354 to learn how an element's location on the periodic table helps you determine the number of valence electrons in an atom.

FORMING IONS Turn back to Figures 6 and 7 on pages 357–358 to review how ions are formed when atoms lose or gain electrons.

SECTION 2

- Energy is needed to remove electrons from metal atoms to form positive ions. Energy is released when most nonmetal atoms gain electrons to form negative ions.
- In covalent bonding, electrons are shared by two atoms. The force of attraction between the nuclei of the atoms and the shared electrons is a covalent bond.
- Covalent bonding usually occurs between atoms of nonmetals.
- Electron-dot diagrams are a simple way to represent the valence electrons in an atom.
- Covalently bonded atoms form a particle called a molecule. A molecule is the smallest particle of a compound with the chemical properties of the compound.
- Diatomic elements are the only elements found in nature as diatomic molecules consisting of two atoms of the same element covalently bonded together.
- In metallic bonding, the outermost energy levels of metal atoms overlap, allowing the valence electrons to move throughout the metal. The force of attraction between a positive metal ion and the electrons in the metal is a metallic bond.
- Many properties of metals, such as conductivity, ductility, and malleability, result from the freely moving electrons in the metal.

Labs

Covalent Marshmallows *(p. 574)*

internet**connect**

GO TO: go.hrw.com

Visit the **HRW** Web site for a variety of learning tools related to this chapter. Just type in the keyword:

KEYWORD: HSTBND

GO TO: www.scilinks.org

Visit the **National Science Teachers Association** on-line Web site for Internet resources related to this chapter. Just type in the ***sci*LINKS** number for more information about the topic:

TOPIC: The Electron ***sci*LINKS NUMBER:** HSTP305
TOPIC: The Periodic Table ***sci*LINKS NUMBER:** HSTP310
TOPIC: Types of Chemical Bonds ***sci*LINKS NUMBER:** HSTP315
TOPIC: Properties of Metals ***sci*LINKS NUMBER:** HSTP320

Chapter Review

USING VOCABULARY

To complete the following sentences, choose the correct term from each pair of terms listed below.

1. The force of attraction that holds two atoms together is a ____. (*crystal lattice* or *chemical bond*)
2. Charged particles that form when atoms transfer electrons are ____. (*molecules* or *ions*)
3. The force of attraction between the nuclei of atoms and shared electrons is a(n) ____. (*ionic bond* or *covalent bond*)
4. Electrons free to move throughout a material are associated with a(n) ____. (*ionic bond* or *metallic bond*)
5. Shared electrons are associated with a ____. (*covalent bond* or *metallic bond*)

UNDERSTANDING CONCEPTS

Multiple Choice

6. Which element has a full outermost energy level containing only two electrons?
   a. oxygen (O)
   b. hydrogen (H)
   c. fluorine (F)
   d. helium (He)
7. Which of the following describes what happens when an atom becomes an ion with a 2– charge?
   a. The atom gains 2 protons.
   b. The atom loses 2 protons.
   c. The atom gains 2 electrons.
   d. The atom loses 2 electrons.

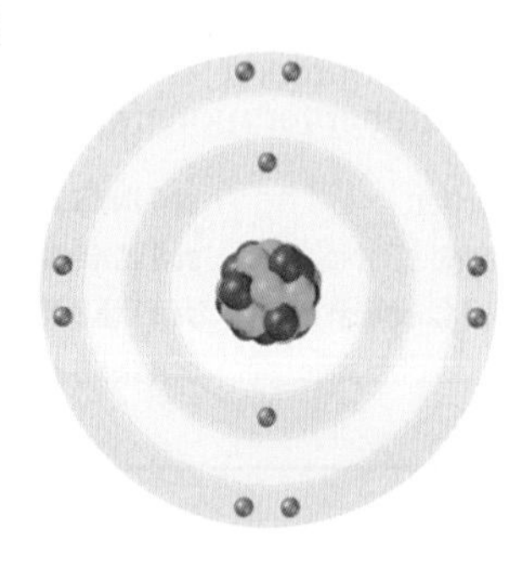

8. The properties of ductility and malleability are associated with which type of bonds?
   a. ionic
   b. covalent
   c. metallic
   d. none of the above
9. In which area of the periodic table do you find elements whose atoms easily gain electrons?
   a. across the top two rows
   b. across the bottom row
   c. on the right side
   d. on the left side
10. What type of element tends to lose electrons when it forms bonds?
    a. metal
    b. metalloid
    c. nonmetal
    d. noble gas
11. Which pair of atoms can form an ionic bond?
    a. sodium (Na) and potassium (K)
    b. potassium (K) and fluorine (F)
    c. fluorine (F) and chlorine (Cl)
    d. sodium (Na) and neon (Ne)

Short Answer

12. List two properties of covalent compounds.
13. Explain why an iron ion is attracted to a sulfide ion but not to a zinc ion.
14. Using your knowledge of valence electrons, explain the main reason so many different molecules are made from carbon atoms.
15. Compare the three types of bonds based on what happens to the valence electrons of the atoms.

Concept Mapping

16. Use the following terms to create a concept map: chemical bonds, ionic bonds, covalent bonds, metallic bonds, molecule, ions.

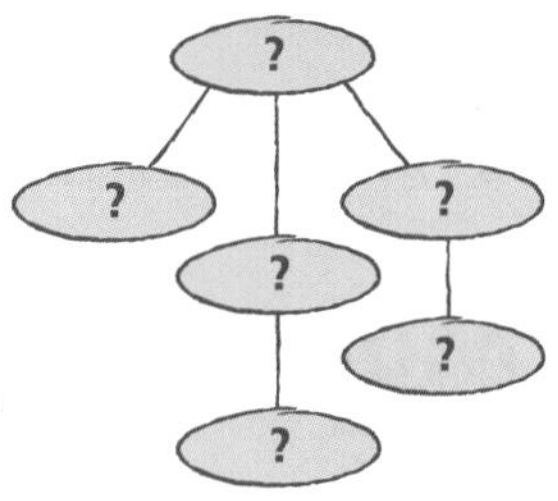

CRITICAL THINKING AND PROBLEM SOLVING

17. Predict the type of bond each of the following pairs of atoms would form:
    a. zinc (Zn) and zinc (Zn)
    b. oxygen (O) and nitrogen (N)
    c. phosphorus (P) and oxygen (O)
    d. magnesium (Mg) and chlorine (Cl)

18. Draw electron-dot diagrams for each of the following atoms, and state how many bonds it will have to make to fill its outer energy level.
    a. sulfur (S)
    b. nitrogen (N)
    c. neon (Ne)
    d. iodine (I)
    e. silicon (Si)

19. Does the substance being hit in the photo below contain ionic or metallic bonds? Explain.

MATH IN SCIENCE

20. For each atom below, write the number of electrons it must gain or lose to have 8 valence electrons. Then calculate the charge of the ion that would form.
    a. calcium (Ca)
    b. phosphorus (P)
    c. bromine (Br)
    d. sulfur (S)

INTERPRETING GRAPHICS

Look at the picture of the wooden pencil below, and answer the following questions.

21. In which part of the pencil are metallic bonds found?

22. List three materials composed of molecules with covalent bonds.

23. Identify two differences between the properties of the metallically bonded material and one of the covalently bonded materials.

NOW What Do You Think?

Take a minute to review your answers to the ScienceLog questions on page 351. Have your answers changed? If necessary, revise your answers based on what you have learned since you began this chapter.

Left-Handed Molecules

Some researchers think that light from a newly forming star 1,500 light-years away (1 light-year is equal to about 9.6 trillion kilometers) may hold the answer to an Earthly riddle that has been puzzling scientists for over 100 years!

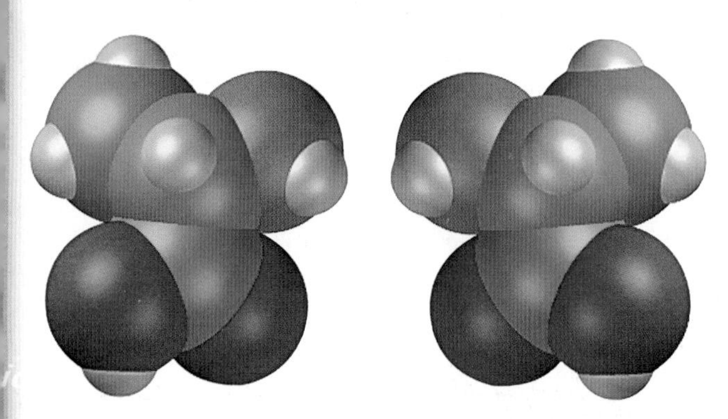

▲ *Molecules, such as the carbon molecules shown above, often come in two mirror-image forms, just as hands do.*

We Are All Lefties!

In 1848, Louis Pasteur discovered that carbon-containing molecules come in left-handed and right-handed forms. Each of the molecules is an exact mirror image of the other, just as each of your hands is a mirror image of the other. These molecules are made of the same elements, but they differ in the elements' arrangement in space.

Shortly after Pasteur's discovery, researchers stumbled across an interesting but unexplained phenomenon—all organisms, including humans, are made almost entirely of left-handed molecules! Chemists were puzzled by this observation because when they made amino acids in the laboratory, the amino acids came out in equal numbers of right- and left-handed forms. Scientists also found that organisms cannot even use the right-handed form of the amino acids to make proteins! For years, scientists have tried to explain this. Why are biological molecules usually left-handed and not right-handed?

Cosmic Explanation

Astronomers recently discovered that a newly forming star in the constellation Orion emits a unique type of infrared light. Infrared light has a wavelength longer than the wavelenth of visible light. The wave particles of this light spiral through space like a corkscrew. This light spirals in only one direction. Researchers suspect that this light might give clues to why all organisms are lefties.

Laboratory experiments show that depending on the direction of the ultraviolet light spirals, either left-handed or right-handed molecules are destroyed. Scientists wonder if a similar type of light may have been present when life was beginning on Earth. Such light may have destroyed most right-handed molecules, which explains why life's molecules are left-handed.

Skeptics argue that the infrared light has less energy than the ultraviolet light used in the laboratory experiments and thus is not a valid comparison. Some researchers, however, hypothesize that both infrared and ultraviolet light may be emitted from the newly forming star that is 1,500 light-years away.

Find Out More

▶ The French chemist Pasteur discovered left-handed and right-handed molecules in tartaric acid. Do some research to find out more about Pasteur and his discovery. Share your findings with the class.

Eureka!

Here's Looking At Ya'!

To some people, just the thought of putting small pieces of plastic in their eyes is uncomfortable. But for millions of others, those little pieces of plastic, known as contact lenses, are a part of daily life. So what would you think about putting a piece of glass in your eye instead? Strangely enough, the humble beginning of the contact lens began with doing just that—inserting a glass lens right in the eye! Ouch!

Molded Glass

Early developers of contact lenses had only glass to use until plastics were discovered. In 1929, a Hungarian physician named Joseph Dallos came up with a way to make a mold of the human eye. This was a critical step in the development of contact lenses. He used these molds to make a glass lens that followed the shape of the eye rather than laying flat against it. In combination with the eye's natural lens, light was refocused to improve a person's eyesight. As you can probably guess, glass lenses weren't very comfortable.

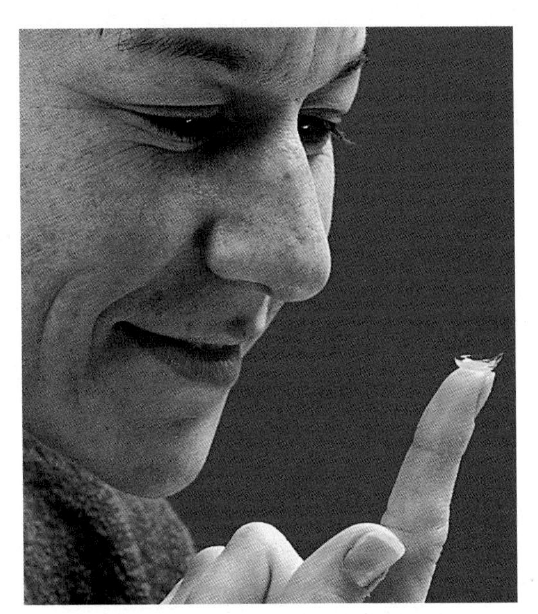

▲ *Does the thought of putting something in your eye make you squirm?*

Still Too Hard

Seven years later, an American optometrist, William Feinbloom, introduced contact lenses made of hard plastic. Plastic was a newly developed material made from long, stable chains of carbon, hydrogen, and oxygen molecules called polymers. But polymers required a lot of work to make. Chemists heated short chains, forcing them to chemically bond to form a longer, more-stable polymer. The whole process was also expensive. To make matters worse, the hard-plastic lenses made from polymers weren't much more comfortable than the glass lenses.

How About Spinning Plastic Gel?

In an effort to solve the comfort problem, Czech chemists Otto Wichterle and Drahoslav Lim invented a water-absorbing plastic gel. The lenses made from this gel were soft and pliable, and they allowed air to pass through the lens to the eye. These characteristics made the lenses much more comfortable to wear than the glass lenses.

Wichterle solved the cost problem by developing a simple and inexpensive process to make the plastic-gel lenses. In this process, called spin casting, liquid plastic is added to a spinning mold of an eye. When the plastic forms the correct shape, it is treated with ultraviolet and infrared light, which hardens the plastic. Both plastic gel and spin casting were patented in 1963, becoming the foundation for the contact lenses people wear today.

Look Toward the Future

► What do you think contact lenses might be like in 20 years? Let your imagination run wild. Sometimes the strangest ideas are the best seeds of new inventions!

CHAPTER

15 Chemical Reactions

Imagine . . .

A car slams into a wall at 97 km/h (60 mph). Although both occupants are wearing seat belts, one suffers a crushing blow to the head as he strikes the dashboard. The other occupant suffers only minor bruises thanks to the presence of an air bag. Fortunately, no one was really injured because this was just a crash test using dummies. The results of this test could lead to the design of better air bags.

The key to an air bag's success during a crash is the speed at which it inflates. Inside the bag is a gas generator that contains the compounds sodium azide, potassium nitrate, and silicon dioxide. At the moment of a crash, an electronic sensor in the vehicle detects the sudden decrease in speed. The sensor sends a small electric current to the gas generator, providing the *activation energy,* or the energy needed to start the reaction, to the chemicals in the gas generator.

The rate, or speed, at which the reaction occurs is very fast. In 1/25 of a second—less than the blink of an eye—the gas formed in the reaction inflates the bag. The air bag fills upward and outward. By filling the space between a person and the car's dashboard, the air bag protects him or her from injury.

Designers of air bags must understand a lot about chemical reactions. In this chapter, you will learn about the different types of chemical reactions. You will learn the clues that will help you identify when a chemical reaction is taking place. You will also learn about the factors that affect the rate of a reaction.

The reaction between vinegar and baking soda produces a gas. However, the reaction is too slow for use in an air bag.

Reaction Ready

The reactions that occur in an air bag produce the gas that fills the bag. In fact, the production of a gas is often a sign that a chemical reaction is taking place. In this activity, you will observe a reaction and identify signs that indicate that a reaction is taking place.

Procedure   

1. In a **large, sealable plastic bag,** place one plastic spoonful of **baking soda** and two spoonfuls of **calcium chloride.**
2. Fill a **plastic film canister** two-thirds full with **water.**
3. Carefully place the canister in the bag without spilling the water. Squeeze the air out of the bag, and seal it tightly.
4. Tip the canister over, and mix the contents of the bag.

Analysis

5. Observe the contents of the bag. Record your observations in your ScienceLog.
6. What evidence did you see that a chemical reaction was taking place?

What Do You Think?

In your ScienceLog, try to answer the following questions based on what you already know:

1. What clues can help you identify a chemical reaction?
2. What are some types of chemical reactions?
3. How can you change the rate of a chemical reaction?

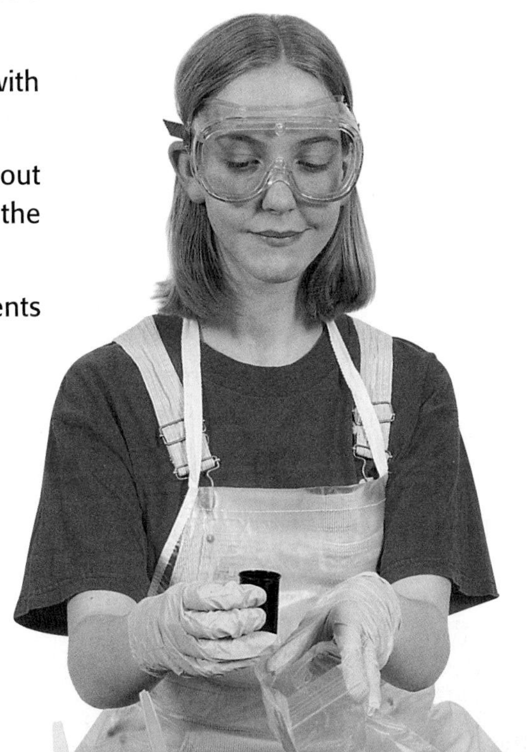

Section 1

Forming New Substances

NEW TERMS

chemical reaction
chemical formula
subscript
chemical equation
reactants
products
coefficient
law of conservation of mass

OBJECTIVES

- Identify the clues that indicate a chemical reaction might be taking place.
- Interpret and write simple chemical formulas.
- Interpret and write simple balanced chemical equations.
- Explain how a balanced equation illustrates the law of conservation of mass.

Each fall, an amazing transformation takes place. Leaves change color, as shown in **Figure 1.** Vibrant reds, oranges, and yellows that had been hidden by green all year are seen as the temperatures get cooler and the hours of sunlight become fewer. What is happening to cause this change? Leaves have a green color as a result of a compound called chlorophyll (KLOR uh FIL). Each fall, the chlorophyll undergoes a chemical change and forms simpler substances that have no color. The green color of the chlorophyll no longer hides them, so the red, orange, and yellow colors in the leaves can be seen.

Figure 1 *The change of color in the fall is a result of chemical changes in the leaves.*

Chemical Reactions

The chemical change that occurs as chlorophyll breaks down into simpler substances is one example of a chemical reaction. A **chemical reaction** is the process by which one or more substances undergo change to produce one or more different substances. These new substances have different chemical and physical properties from the original substances. Many of the changes you are familiar with are chemical reactions, including the ones shown in **Figure 2.**

Figure 2 Examples of Chemical Reactions

The substances that make up baking powder undergo a chemical reaction when mixed with water. One new substance that forms is carbon dioxide gas, which causes the bubbles in this muffin.

Once ignited, gasoline reacts with oxygen gas in the air. The new substances that form, carbon dioxide and water, push against the pistons in the engine to keep the car moving.

Clues to Chemical Reactions How can you tell when a chemical reaction is taking place? There are several clues that indicate when a reaction might be occurring. The more of these clues you observe, the more likely it is that the change is a chemical reaction. Several of these clues are described below.

Some Clues to Chemical Reactions

Gas Formation
The formation of gas bubbles is a clue that a chemical reaction might be taking place. For example, bubbles of carbon dioxide are produced when hydrochloric acid is placed on a piece of limestone. Hydrogen gas is produced when a metal reacts with an acid.

Solid Formation
Sometimes a solid forms when two solutions react. A solid formed in a solution as a result of a chemical reaction is called a *precipitate* (pruh SIP uh TAYT). Here you see potassium chromate solution being added to a silver nitrate solution. The dark red solid is a precipitate of silver chromate.

Color Change
Chlorine bleach is great for removing the color from stains on white clothes. But don't spill it on your jeans. The bleach reacts with the blue dye on the fabric, causing the color of the material to change.

Energy Change
Energy is released during some chemical reactions. A fire heats a room and provides light. Electrical energy is released when chemicals in a battery react. During some other chemical reactions, energy is absorbed. Chemicals on photographic film react when they absorb energy from light shining on the film.

Breaking and Making Bonds New substances are formed in a chemical reaction because chemical bonds in the starting substances break, atoms rearrange, and new bonds form to make the new substances. Look at the model in **Figure 3** to understand how this process occurs.

Figure 3
Reaction of Hydrogen and Chlorine

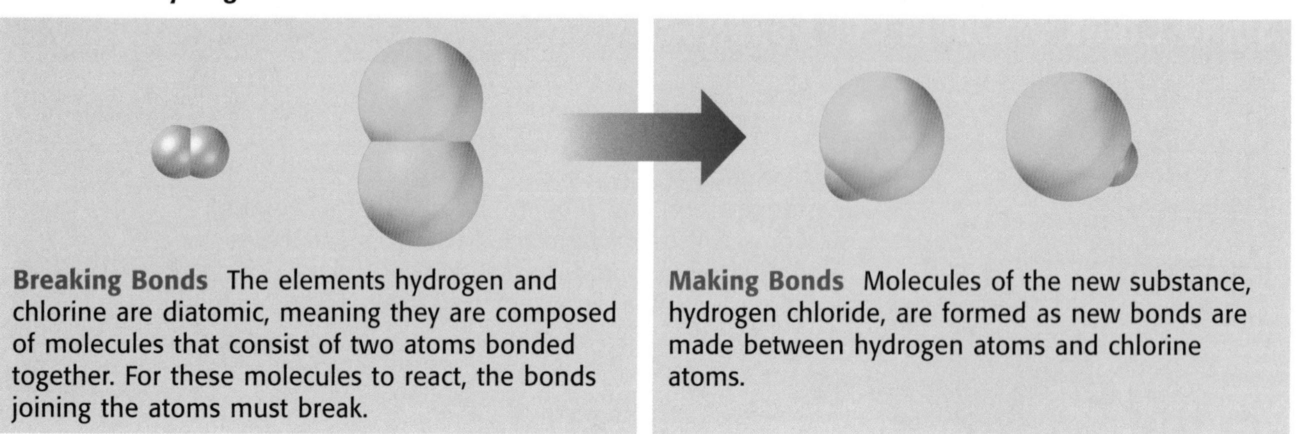

Breaking Bonds The elements hydrogen and chlorine are diatomic, meaning they are composed of molecules that consist of two atoms bonded together. For these molecules to react, the bonds joining the atoms must break.

Making Bonds Molecules of the new substance, hydrogen chloride, are formed as new bonds are made between hydrogen atoms and chlorine atoms.

Chemical Formulas

Remember that a chemical symbol is a shorthand method of identifying an element. A **chemical formula** is a shorthand notation for a compound or a diatomic element using chemical symbols and numbers. A chemical formula indicates the chemical makeup by showing how many of each kind of atom is present in a molecule.

The chemical formula for water, H_2O, tells you that a water molecule is composed of two atoms of hydrogen and one atom of oxygen. The small number *2* in the formula is a subscript. A **subscript** is a number written below and to the right of a chemical symbol in a formula. When no subscript is written after a symbol, as with the oxygen in water's formula, only one atom of that element is present. **Figure 4** shows two more chemical formulas and what they mean.

Figure 4 *A chemical formula shows the number of atoms of each element present.*

O_2

Oxygen is a diatomic element. Each molecule of oxygen gas is composed of two atoms of oxygen bonded together.

$C_6H_{12}O_6$

Every molecule of glucose (the sugar formed by plants during photosynthesis) is composed of six atoms of carbon, twelve atoms of hydrogen, and six atoms of oxygen.

MATH BREAK

Counting Atoms

Some chemical formulas contain two or more chemical symbols enclosed by parentheses. When counting atoms in these formulas, multiply everything inside the parentheses by the subscript as though they were part of a mathematical equation. For example, $Ca(NO_3)_2$ contains:

- 1 calcium atom
- 2 nitrogen atoms (2×1)
- 6 oxygen atoms (2×3)

Now It's Your Turn

Determine the number of atoms of each element in the formulas $Mg(OH)_2$ and $Al_2(SO_4)_3$.

Writing Formulas for Covalent Compounds You can often write a chemical formula if you know the name of the substance. Remember that covalent compounds are usually composed of two nonmetals. The names of covalent compounds use prefixes to tell you how many atoms of each element are in the formula. A *prefix* is a syllable or syllables joined to the beginning of a word. Each prefix used in a chemical name represents a number, as shown in the table at right. **Figure 5** demonstrates how to write a chemical formula from the name of a covalent compound.

Prefixes Used in Chemical Names

mono-	1	hexa-	6
di-	2	hepta-	7
tri-	3	octa-	8
tetra-	4	nona-	9
penta-	5	deca-	10

Figure 5 *The formulas of these covalent compounds can be written using the prefixes in their names.*

Carbon dioxide

The lack of a prefix indicates 1 carbon atom.

The prefix *di-* indicates 2 oxygen atoms.

CO_2

Dinitrogen monoxide

The prefix *di-* indicates 2 nitrogen atoms.

The prefix *mono-* indicates 1 oxygen atom.

N_2O

Self-Check

How many atoms of each element make up Na_2SO_4? *(See page 596 to check your answer.)*

Writing Formulas for Ionic Compounds If the name of a compound contains the name of a metal and a nonmetal, the compound is probably ionic. To write the formula for an ionic compound, you must make sure the compound's overall charge is zero. In other words, the formula must have subscripts that cause the charges of the ions to cancel out. (Remember that the charge of many ions can be determined by looking at the periodic table.) **Figure 6** demonstrates how to write a chemical formula from the name of an ionic compound.

Figure 6 *The formula of an ionic compound is written by using enough of each ion to make the overall charge zero.*

Sodium chloride

A sodium ion has a 1+ charge.

A chloride ion has a 1− charge.

$NaCl$

One sodium ion and one chloride ion have an overall charge of $(1+) + (1-) = 0$

Magnesium chloride

A magnesium ion has a 2+ charge.

A chloride ion has a 1− charge.

$MgCl_2$

One magnesium ion and two chloride ions have an overall charge of $(2+) + 2(1-) = 0$

Explore

Determine whether each of the following compounds is covalent or ionic, and write the chemical formula for each: sulfur trioxide, calcium fluoride, phosphorus pentachloride, dinitrogen trioxide, and lithium oxide.

Chemical Equations

Figure 7 *The symbols on this music are understood around the world—just like chemical symbols!*

A composer writing a piece of music, like the one in **Figure 7,** must communicate to the musician what notes to play, how long to play each note, and in what style each note should be played. The composer does not use words to describe what must happen. Instead, he or she uses musical symbols to communicate in a way that can be easily understood by anyone in the world who can read music.

Similarly, people who work with chemical reactions need to communicate information about reactions clearly to other people throughout the world. Describing reactions using long descriptive sentences would require translations into other languages. Chemists have developed a method of describing reactions that is short and easily understood by anyone in the world who understands chemical formulas. A **chemical equation** is a shorthand description of a chemical reaction using chemical formulas and symbols. Because each element's chemical symbol is understood around the world, a chemical equation needs no translation.

Figure 8 *Charcoal is used to cook food on a barbecue. When carbon in charcoal reacts with oxygen in the air, the primary product is carbon dioxide, as shown in the chemical equation in Figure 9.*

Reactants Yield Products Consider the example of carbon reacting with oxygen to yield carbon dioxide, as shown in **Figure 8.** The starting materials in a chemical reaction are **reactants** (ree AKT uhnts). The substances formed from a reaction are **products.** In this example, carbon and oxygen are reactants, and carbon dioxide is the product formed. The parts of the chemical equation for this reaction are described in **Figure 9.**

Figure 9 The Parts of a Chemical Equation

The formulas of the reactants are written before the arrow.

The formulas of the products are written after the arrow.

A plus sign separates the formulas of two or more reactants or products from one another.

The arrow, also called the yields sign, separates the formulas of the reactants from the formulas of the products.

The symbol or formula for each substance in the reaction must be written correctly. For a compound, determine if it is a covalent compound or an ionic compound, and write the appropriate formula. For an element, use the proper chemical symbol, and be sure to use a subscript of 2 for the diatomic elements. (The seven diatomic elements are hydrogen, nitrogen, oxygen, fluorine, chlorine, bromine, and iodine.) An equation with an incorrect chemical symbol or formula will not accurately describe the reaction. In fact, even a simple mistake in a symbol or formula can make a huge difference, as shown in **Figure 10.**

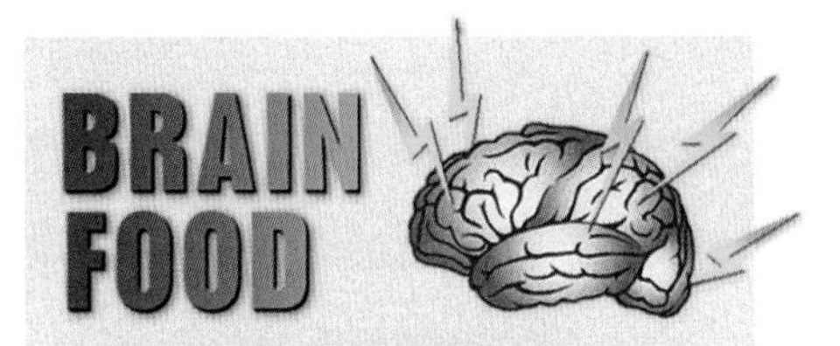

Hydrogen gas, H_2, is an important fuel that may help reduce air pollution. Because water is the only product formed as hydrogen burns, there is little air pollution from vehicles that use hydrogen as fuel.

Figure 10 *The symbols and formulas shown here are similar, but confusing them while writing an equation would cause you to indicate the wrong substance.*

The chemical formula for the compound carbon dioxide is CO_2. Carbon dioxide is a colorless, odorless gas that you exhale.

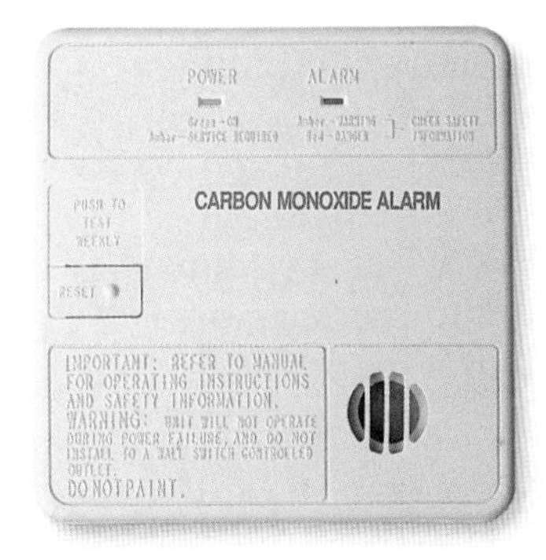

The chemical formula for the compound carbon monoxide is CO. Carbon monoxide is a colorless, odorless, poisonous gas.

The chemical symbol for the element cobalt is Co. Cobalt is a hard, bluish gray metal.

Self-Check

When calcium bromide reacts with chlorine, bromine and calcium chloride are produced. Write an equation to describe this reaction. Identify each substance as either a reactant or a product. *(See page 596 to check your answers.)*

An Equation Must Be Balanced In a chemical reaction, every atom in the reactants becomes part of the products. Atoms are never lost or gained in a chemical reaction. When writing a chemical equation, you must show that the number of atoms of each element in the reactants equals the number of atoms of those elements in the products by writing a balanced equation.

÷ 5 ÷ Ω ≤ ∞ + Ω √ 9 ∞ ≤ Σ 2 +

MATH BREAK

Balancing Act

When balancing a chemical equation, you must place coefficients in front of an entire chemical formula, never in the middle of a formula. Notice where the coefficients are in the balanced equation below:

$F_2 + 2KCl \rightarrow 2KF + Cl_2$

Now It's Your Turn

Write balanced equations for the following:

$HCl + Na_2S \rightarrow H_2S + NaCl$
$Al + Cl_2 \rightarrow AlCl_3$

Writing a balanced equation requires the use of coefficients (KOH uh FISH uhnts). A **coefficient** is a number placed in front of a chemical symbol or formula. When counting atoms, you multiply a coefficient by the subscript of each of the elements in the formula that follows it. Thus, $2CO_2$ represents 2 carbon dioxide molecules containing a total of 2 carbon atoms and 4 oxygen atoms. Coefficients are used when balancing equations because the subscripts in the formulas cannot be changed. Changing a subscript changes the formula so that it no longer represents the correct substance. **Figure 11** shows how to use coefficients to balance an equation. After you learn how to use coefficients, you can practice balancing chemical equations by doing the MathBreak at left.

Figure 11 *Follow these steps to write a balanced equation for $H_2 + O_2 \rightarrow H_2O$.*

1 Count the atoms of each element in the reactants and in the products. You can see that there are fewer oxygen atoms in the products than in the reactants.

$H_2 + O_2 \longrightarrow H_2O$

Reactants **Products**

H = 2 O = 2 H = 2 O = 1

2 To balance the oxygen atoms, place the coefficient *2* in front of water's formula. This gives you 2 oxygen atoms in both the reactants and the products. But now there are too few hydrogen atoms in the reactants.

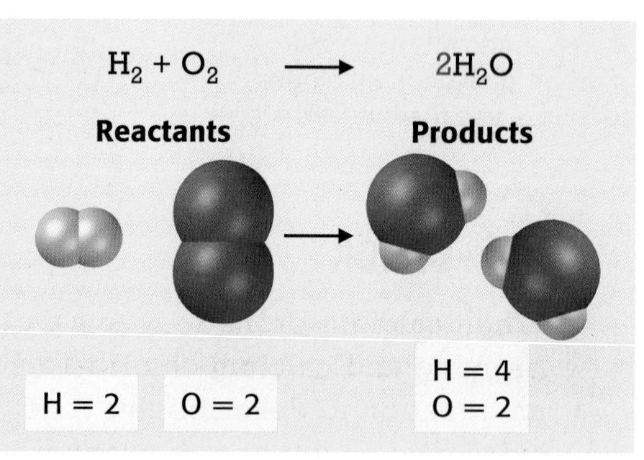

3 To balance the hydrogen atoms, place the coefficient *2* in front of hydrogen's formula. But just to be sure your answer is correct, always double-check your work!

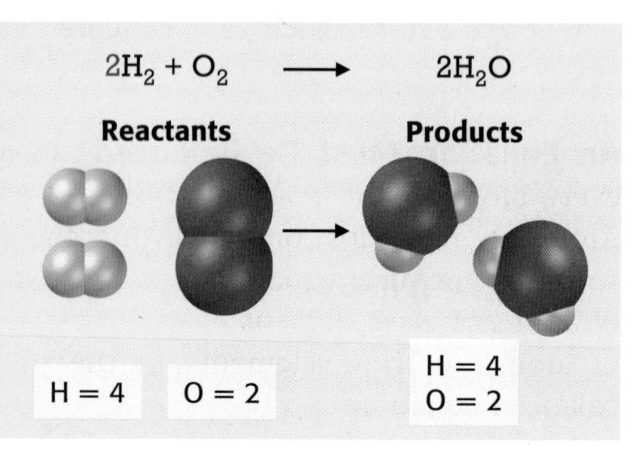

LabBook

Become a better balancer of chemical equations on page 576 of the LabBook.

Mass Is Conserved—It's a Law! The practice of balancing equations is a result of the work of a French chemist, Antoine Lavoisier (luh vwa ZYAY). In the 1700s, Lavoisier performed experiments in which he carefully measured and compared the masses of the substances involved in chemical reactions. He determined that the total mass of the reactants equaled the total mass of the products. Lavoisier's work led to the **law of conservation of mass,** which states that mass is neither created nor destroyed in ordinary chemical and physical changes. Thus, a chemical equation must show the same number and kind of atom on both sides of the arrow. The law of conservation of mass is demonstrated in **Figure 12.** You can explore this law for yourself in the QuickLab at right.

Figure 12 *In this demonstration, magnesium in the flashbulb of a camera reacts with oxygen. Notice that the mass is the same before and after the reaction takes place.*

REVIEW

1. List four clues that a chemical reaction is occurring.
2. How many atoms of each element make up $2Na_3PO_4$?
3. Write the chemical formulas for carbon tetrachloride and calcium bromide.
4. Explain how a balanced chemical equation illustrates that mass is never lost or gained in a chemical reaction.
5. **Applying Concepts** Write the balanced chemical equation for methane, CH_4, reacting with oxygen gas to produce water and carbon dioxide.

*Quick*Lab

Mass Conservation

1. Place about 5 g (1 tsp) of **baking soda** into a **sealable plastic bag.**
2. Place about 5 mL (1 tsp) of **vinegar** into a **plastic film canister,** and close the lid.
3. Use a **balance** to determine the masses of the bag with baking soda and the canister with vinegar, and record both values in your ScienceLog.
4. Place the canister into the plastic bag. Squeeze the air out of the bag, and tightly seal the bag.
5. Carefully open the lid of the canister while it is in the bag. Pour the vinegar onto the baking soda, and mix them.
6. When the reaction has stopped, use the same balance used in step 3 to determine the total mass of the bag and its contents.
7. Compare the mass of the materials before and after the reaction.

Section 2

Types of Chemical Reactions

NEW TERMS
synthesis reaction
decomposition reaction
single-replacement reaction
double-replacement reaction

OBJECTIVES

- Describe four types of chemical reactions.
- Classify a chemical equation as one of the four types of chemical reactions described here.

Imagine having to learn 50 chemical reactions. Sound tough? Well, there are thousands of known chemical reactions. It would be impossible to remember them all. But there is help! Remember that the elements are divided into categories based on their properties. In a similar way, reactions can be classified according to their similarities.

Many reactions can be grouped into one of four categories: synthesis (SIN thuh sis), decomposition, single replacement, and double replacement. By dividing reactions into these categories, you can better understand the patterns of how reactants become products. As you learn about each type of reaction, study the models provided to help you recognize each type of reaction.

Synthesis Reactions

A **synthesis reaction** is a reaction in which two or more substances combine to form a single compound. For example, the synthesis reaction in which the compound magnesium oxide is produced is seen in **Figure 13.** (This is the same reaction that occurs in the flashbulb in Figure 12.) One way to remember what happens in each type of reaction is to imagine people at a dance. A synthesis reaction would be modeled by two people joining to form a dancing couple, as shown in **Figure 14.**

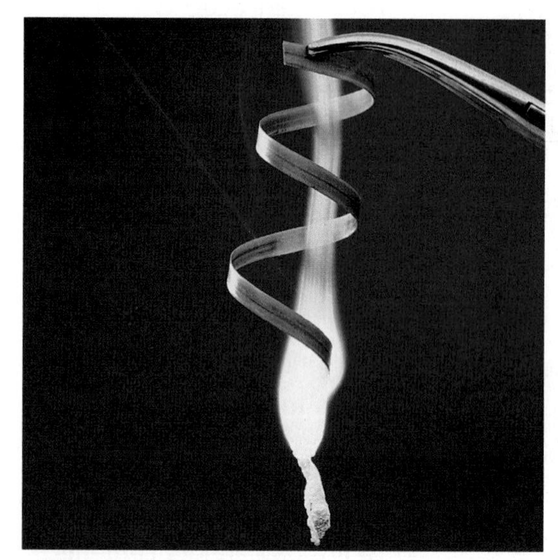

Figure 13 *The synthesis reaction that occurs when magnesium reacts with oxygen in the air forms the compound magnesium oxide.*

Figure 14 *A model for the synthesis reaction of sodium reacting with chlorine to form sodium chloride is shown below.*

$$2Na + Cl_2 \longrightarrow 2NaCl$$

Decomposition Reactions

A **decomposition reaction** is a reaction in which a single compound breaks down to form two or more simpler substances. The decomposition of water is shown in **Figure 15.** Decomposition is the reverse of synthesis. The dance model would represent a decomposition reaction as a dancing couple splitting up, as shown in **Figure 16.**

Figure 15 *Water can be decomposed into the elements hydrogen and oxygen through electrolysis.*

Figure 16 *A model for the decomposition reaction of carbonic acid to form water and carbon dioxide is shown below.*

$$H_2CO_3 \longrightarrow H_2O + CO_2$$

Single-Replacement Reactions

A **single-replacement reaction** is a reaction in which an element takes the place of another element that is part of a compound. The products of single-replacement reactions are a new compound and a different element. The dance model for single-replacement reactions is a person who cuts in on a couple dancing. A new couple is formed and a different person is left alone, as shown in **Figure 17.**

Figure 17 *A model for a single-replacement reaction of zinc reacting with hydrochloric acid to form zinc chloride and hydrogen is shown below.*

$$Zn + 2HCl \longrightarrow ZnCl_2 + H_2$$

Remember that some elements are more reactive than others. In a single-replacement reaction, a more-reactive element can replace a less-reactive one from a compound. However, the opposite reaction does not occur, as shown in **Figure 18.**

Figure 18 *More-reactive elements replace less-reactive elements in single-replacement reactions.*

$Cu + 2AgNO_3 \rightarrow 2Ag + Cu(NO_3)_2$
Copper is more reactive than silver and replaces it.

$Ag + Cu(NO_3)_2 \rightarrow$ No reaction
Silver is less reactive than copper and cannot replace it.

Double-Replacement Reactions

A **double-replacement reaction** is a reaction in which ions in two compounds switch places. One of the products of this reaction is often a gas or a precipitate. A double-replacement reaction in the dance model would be two couples dancing and switching partners, as shown in **Figure 19.**

Figure 19 *A model for the double-replacement reaction of sodium chloride reacting with silver nitrate to form sodium nitrate and the precipitate silver chloride is shown below.*

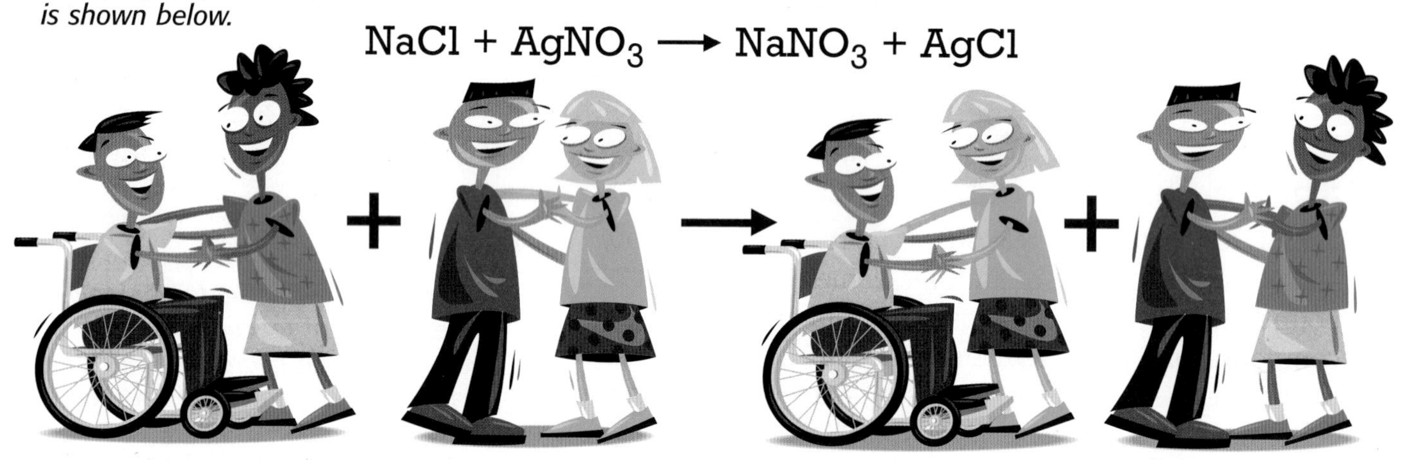

REVIEW

1. What type of reaction does each of the following equations represent?
   a. $FeS + 2HCl \longrightarrow FeCl_2 + H_2S$
   b. $NH_4OH \longrightarrow NH_3 + H_2O$
2. Which type of reaction always has an element and a compound as reactants?
3. **Comparing Concepts** Compare synthesis and decomposition reactions.

Section 3

Energy and Rates of Chemical Reactions

NEW TERMS

exothermic
endothermic
law of conservation of energy
activation energy
catalyst
inhibitor

OBJECTIVES

- Compare exothermic and endothermic reactions.
- Explain activation energy.
- Interpret an energy diagram.
- Describe the factors that affect the rate of a reaction.

You just learned one method of classifying chemical reactions. In this section, you will learn how to classify reactions in terms of the energy associated with the reaction and learn how to change the rate at which the reaction occurs.

Every Reaction Involves Energy

All chemical reactions involve chemical energy. Remember that during a reaction, chemical bonds in the reactants break as they absorb energy. As new bonds form in the products, energy is released. Energy is released or absorbed in the overall reaction depending on how the chemical energy of the reactants compares with the chemical energy of the products.

Energy Is Released in Exothermic Reactions If the chemical energy of the reactants is greater than the chemical energy of the products, the difference in energy is released during the reaction. A chemical reaction in which energy is released or removed is called **exothermic.** *Exo* means "go out" or "exit," and *thermic* means "heat" or "energy." The energy can be released in several different forms, as shown in **Figure 20.** The energy released in an exothermic reaction is often written as a product in a chemical equation, as in this equation:

$$2Na + Cl_2 \longrightarrow 2NaCl + energy$$

Figure 20 Types of Energy Released in Reactions

Energy in the form of light is released in the exothermic reaction taking place in these necklaces and light sticks.

Electrical energy is released in the exothermic reaction taking place in the dry cells in this flashlight.

Energy that keeps you warm and lights your way is released in the exothermic reaction taking place in a campfire.

Photosynthesis is an endothermic process in which light energy from the sun is used to produce glucose, a simple sugar. The equation that describes photosynthesis is as follows:

$6CO_2 + 6H_2O + \text{energy} \longrightarrow C_6H_{12}O_6 + 6O_2$

The cells in your body use glucose to get the energy they need through cellular respiration, an exothermic process described by the reverse of the above reaction:

$C_6H_{12}O_6 + 6O_2 \longrightarrow 6CO_2 + 6H_2O + \text{energy}$

Energy Is Absorbed in Endothermic Reactions If the chemical energy of the reactants is less than the chemical energy of the products, the difference in energy is absorbed during the reaction. A chemical reaction in which energy is absorbed is called **endothermic.** *Endo* means "go in," and *thermic* means "heat" or "energy." The energy absorbed in an endothermic reaction is often written as a reactant in a chemical equation, as in this equation:

$$2H_2O + \text{energy} \longrightarrow 2H_2 + O_2$$

Energy Is Conserved—It's a Law! You learned that mass is never created or destroyed in chemical reactions. The same holds true for energy. The **law of conservation of energy** states that energy can be neither created nor destroyed. The energy released in exothermic reactions was originally stored in the reactants. And the energy absorbed in endothermic reactions does not just vanish. It is stored in the products that form. If you could carefully measure all the energy in a reaction, you would find that the total amount of energy (of all types) is the same before and after the reaction.

Activation Energy Gets a Reaction Started A match can be used to light a campfire—but only if the match is lit! A strike-anywhere match, like the one shown in **Figure 21,** has all the reactants it needs to be able to burn. And though the chemicals on a match are intended to react and burn, they will not ignite by themselves. Energy is needed to start the reaction. The minimum amount of energy needed for substances to react is called **activation energy.**

Figure 21 *Rubbing the tip of this strike-anywhere match on a rough surface provides the energy needed to get the chemicals to react.*

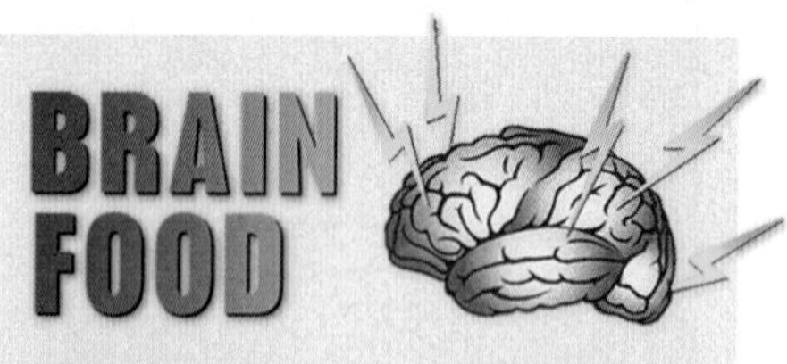

Matches rubbing together in a box could provide the activation energy to light a strike-anywhere match. Safety matches, which must be struck on a strike plate on the box, were developed to prevent such accidents.

The friction of striking a match heats the substances on the match, breaking bonds in the reactants and allowing the new bonds in the products to form. Chemical reactions require some energy to get started. An electric spark in a car's engine provides activation energy to begin the burning of gasoline. Light can also provide the activation energy for a reaction. You can better understand activation energy and the differences between exothermic reactions and endothermic reactions by studying the diagrams in **Figure 22.**

Figure 22 Energy Diagrams

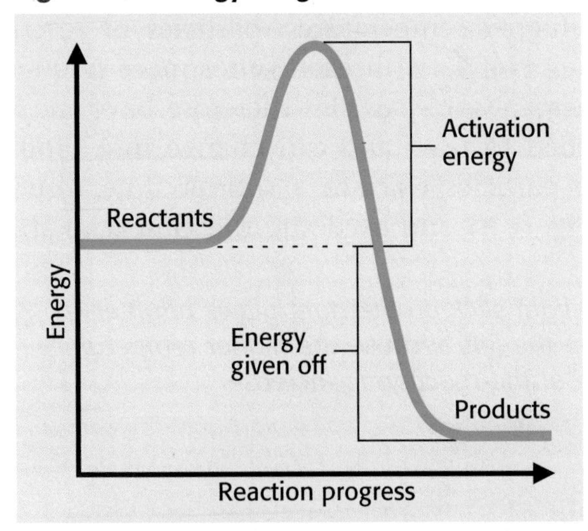

Exothermic Reaction Once begun, an exothermic reaction can continue to occur, as in a fire. The energy released as the product forms continues to supply the activation energy needed for the substances to react.

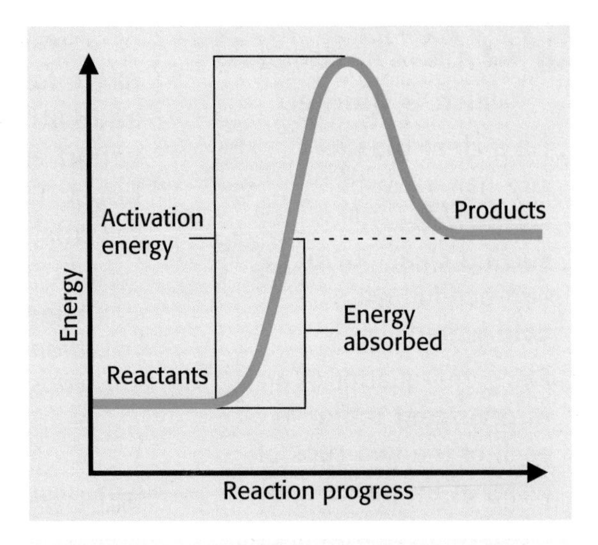

Endothermic Reaction An endothermic reaction requires a continuous supply of energy. Energy must be absorbed to provide the activation energy needed for the substances to react.

Hydrogen peroxide is used as a disinfectant for minor scrapes and cuts because it decomposes to produce oxygen gas and water, which help cleanse the wound. The decomposition of hydrogen peroxide is an exothermic reaction. Explain why hydrogen peroxide must be stored in a dark bottle to maintain its freshness. (HINT: What type of energy would be blocked by this type of container?)

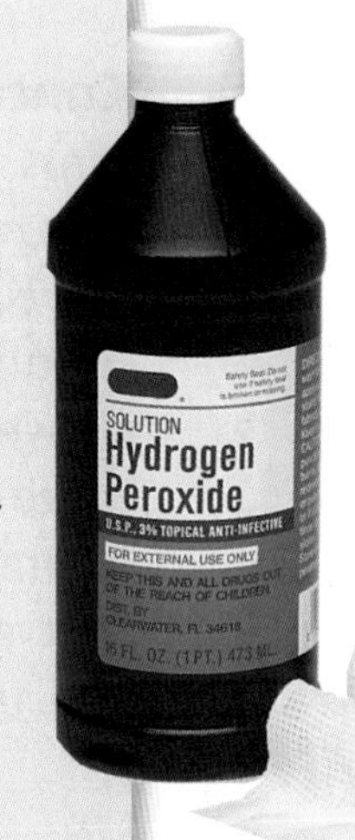

Factors Affecting Rates of Reactions

You can think of a reaction as occurring only if the particles of reactants collide when they have enough energy to break the appropriate bonds. The rate of a reaction is a measure of how rapidly the reaction takes place. Four factors that affect the rate of a reaction are temperature, concentration, surface area, and the presence of a catalyst or inhibitor.

Fighting fires with slime? Read more about it on page 394.

*Quick*Lab

Which Is Quicker?

1. Fill a **clear plastic cup** half-full with **warm water.** Fill a **second clear plastic cup** half-full with **cold water.**
2. Place one-quarter of an **effervescent tablet** in each of the two cups of water at the same time.
3. Observe the reaction, and record your observations in your ScienceLog.
4. In which cup did the reaction occur at a greater rate? What evidence supports your answer?

Temperature An increase in temperature increases the rate of a reaction. At higher temperatures, particles of reactants move faster, so they collide with each other more frequently and with more energy. More particles therefore have the activation energy needed to react and can change into products faster. Thus, more particles react in a shorter time. You can see this effect in **Figure 23** and by doing the QuickLab at left.

Figure 23 *The light stick on the right glows brighter than the one on the left because the higher temperature causes the rate of the reaction to increase.*

Concentration Generally, increasing the concentration of reactants increases the rate of a reaction, as shown in **Figure 24.** *Concentration* is a measure of the amount of one substance dissolved in another. Increasing the concentration increases the number of reactant particles present and decreases the distance between them. The reactant particles collide more often, so more particles react each second. Increasing the concentration is similar to having more people in a room. The more people that are in the room, the more frequently they will collide and interact.

Do you feel as though you are not up to speed on controlling the rate of a reaction? Then hurry over to page 580 of the LabBook.

Figure 24 *The reaction on the right produces bubbles of hydrogen gas at a faster rate because the concentration of hydrochloric acid used is higher.*

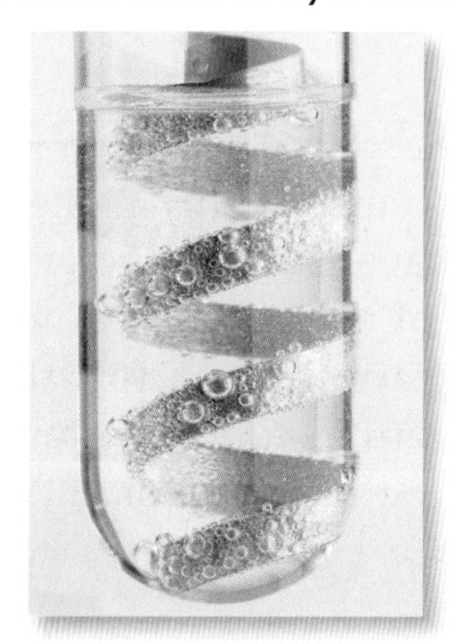

Surface Area Increasing the surface area, or the amount of exposed surface, of solid reactants increases the rate of a reaction. Grinding a solid into a powder exposes more particles of the reactant to other reactant particles. The number of collisions between reactant particles increases, increasing the rate of the reaction. You can see the effect of increasing the surface area in the QuickLab at right.

Catalysts and Inhibitors Some reactions would be too slow to be useful without a catalyst (KAT uh LIST). A **catalyst** is a substance that speeds up a reaction without being permanently changed. A catalyst lowers the activation energy of a reaction. The lower energy needed to start the reaction allows the reaction to occur more rapidly. Most reactions in your body are sped up using catalysts called enzymes. Catalysts are even found in cars, as seen in **Figure 25.**

An **inhibitor** is a substance that slows down or stops a chemical reaction. Preservatives added to foods are inhibitors that slow down reactions in the bacteria or fungus that can spoil food. Many poisons are also inhibitors.

Figure 25 *This catalytic converter contains platinum and palladium—two catalysts used to treat automobile exhaust. They increase the rate of reactions that make the car's exhaust less polluting.*

Quick Lab

I'm Crushed!

1. Fill **two clear plastic cups** half-full with **room-temperature water.**
2. Fold a **sheet of paper** around one-quarter of an **effervescent tablet.** Carefully crush the tablet.
3. Get another one-quarter of an effervescent tablet. Carefully pour the crushed tablet into one cup, and place the uncrushed tablet in the second cup.
4. Observe the reaction, and record your observations in your ScienceLog.
5. In which cup did the reaction occur at a greater rate? What evidence supports your answer?
6. Explain why the water in each cup must have the same temperature.

REVIEW

1. How does the rate of a reaction change when the temperature is decreased?
2. What is activation energy?
3. List four ways to increase the rate of a reaction.
4. **Comparing Concepts** Compare exothermic and endothermic reactions.
5. **Interpreting Graphics** Does the energy diagram at right show an exothermic or an endothermic reaction? How can you tell?

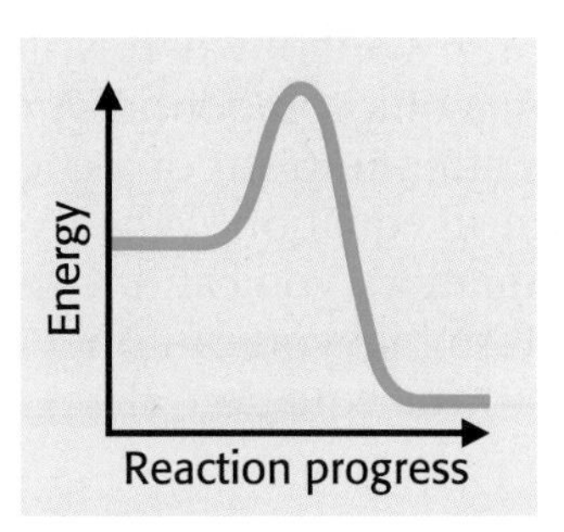

Chapter Highlights

SECTION 1

Vocabulary

chemical reaction *(p. 374)*
chemical formula *(p. 376)*
subscript *(p. 376)*
chemical equation *(p. 378)*
reactants *(p. 378)*
products *(p. 378)*
coefficient *(p. 380)*
law of conservation of mass *(p. 381)*

Section Notes

- Chemical reactions form new substances with different properties than the starting substances.
- Clues that a chemical reaction is taking place include formation of a gas or solid, a color change, and an energy change.
- A chemical formula tells the composition of a compound using chemical symbols and subscripts. Subscripts are small numbers written below and to the right of a symbol in a formula.
- Chemical formulas can sometimes be written from the names of covalent compounds and ionic compounds.
- A chemical equation describes a reaction using formulas, symbols, and coefficients.
- A balanced equation uses coefficients to illustrate the law of conservation of mass, that mass is neither created nor destroyed during a chemical reaction.

Labs

Finding a Balance *(p. 576)*

SECTION 2

Vocabulary

synthesis reaction *(p. 382)*
decomposition reaction *(p. 383)*
single-replacement reaction *(p. 383)*
double-replacement reaction *(p. 384)*

Section Notes

- Many chemical reactions can be classified as one of four types by comparing reactants with products.
- In synthesis reactions, the reactants form a single product.
- In decomposition reactions, a single reactant breaks apart into two or more simpler products.

Skills Check

Math Concepts

SUBSCRIPTS AND COEFFICIENTS A subscript is a number written below and to the right of a chemical symbol when writing the chemical formula of a compound. A coefficient is a number written in front of a chemical formula in a chemical equation. When you balance a chemical equation, you cannot change the subscripts in a formula; you can only add coefficients, as seen in the equation $2H_2 + O_2 \longrightarrow 2H_2O$.

Visual Understanding

REACTION TYPES It can be challenging to identify which type of reaction a particular chemical equation represents. Review four reaction types by studying Figures 14, 16, 17, and 19.

SECTION 2

- In single-replacement reactions, a more-reactive element takes the place of a less-reactive element in a compound. No reaction will occur if a less-reactive element is placed with a compound containing a more-reactive element.
- In double-replacement reactions, ions in two compounds switch places. A gas or precipitate is often formed.

Labs

Putting Elements Together *(p. 578)*

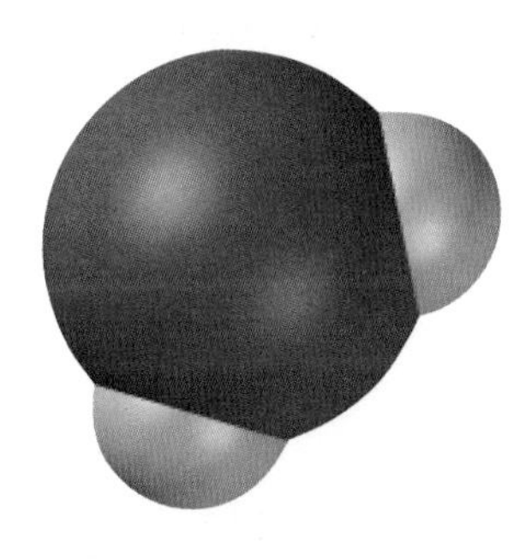

SECTION 3

Vocabulary

exothermic *(p. 385)*
endothermic *(p. 386)*
law of conservation of energy *(p. 386)*
activation energy *(p. 386)*
catalyst *(p. 389)*
inhibitor *(p. 389)*

Section Notes

- Energy is released in exothermic reactions. The energy released can be written as a product in a chemical equation.
- Energy is absorbed in endothermic reactions. The energy absorbed can be written as a reactant in a chemical equation.
- The law of conservation of energy states that energy is neither created nor destroyed.
- Activation energy is the energy needed to start a chemical reaction.
- Energy diagrams indicate whether a reaction is exothermic or endothermic by showing whether energy is given off or absorbed during the reaction.
- The rate of a chemical reaction is affected by temperature, concentration, surface area, and the presence of a catalyst or inhibitor.
- Raising the temperature, increasing the concentration, increasing the surface area, and adding a catalyst can increase the rate of a reaction.

Labs

Cata-what? Catalyst! *(p. 577)*
Speed Control *(p. 580)*

internet**connect**

GO TO: go.hrw.com

Visit the **HRW** Web site for a variety of learning tools related to this chapter. Just type in the keyword:

KEYWORD: HSTREA

GO TO: www.scilinks.org

Visit the **National Science Teachers Association** on-line Web site for Internet resources related to this chapter. Just type in the ***sci*LINKS** number for more information about the topic:

TOPIC: Chemical Reactions	***sci*LINKS NUMBER:** HSTP330
TOPIC: Chemical Formulas	***sci*LINKS NUMBER:** HSTP335
TOPIC: Chemical Equations	***sci*LINKS NUMBER:** HSTP340
TOPIC: Exothermic and Endothermic Reactions	***sci*LINKS NUMBER:** HSTP345

Chapter Review

USING VOCABULARY

To complete the following sentences, choose the correct term from each pair of terms listed below.

1. Adding a(n) ____ will slow down a chemical reaction. (*catalyst* or *inhibitor*)
2. A chemical reaction that gives off light is called ____. (*exothermic* or *endothermic*)
3. A chemical reaction that forms one compound from two or more substances is called a ____. (*synthesis reaction* or *decomposition reaction*)
4. The *2* in the formula Ag_2S is a ____. (*subscript* or *coefficient*)
5. The starting materials in a chemical reaction are ____. (*reactants* or *products*)

UNDERSTANDING CONCEPTS

Multiple Choice

6. Balancing a chemical equation so that the same number of atoms of each element is found in both the reactants and the products is an illustration of
   a. activation energy.
   b. the law of conservation of energy.
   c. the law of conservation of mass.
   d. a double-replacement reaction.
7. What is the correct chemical formula for calcium chloride?
   a. $CaCl$
   b. $CaCl_2$
   c. Ca_2Cl
   d. Ca_2Cl_2
8. In which type of reaction do ions in two compounds switch places?
   a. synthesis
   b. decomposition
   c. single-replacement
   d. double-replacement
9. Which is an example of the use of activation energy?
   a. plugging in an iron
   b. playing basketball
   c. holding a lit match to paper
   d. eating
10. Enzymes in your body act as catalysts. Thus, the role of enzymes is to
    a. increase the rate of chemical reactions.
    b. decrease the rate of chemical reactions.
    c. help you breathe.
    d. inhibit chemical reactions.

Short Answer

11. Classify each of the following reactions:
    a. $Fe + O_2 \longrightarrow Fe_2O_3$
    b. $Al + CuSO_4 \longrightarrow Al_2(SO_4)_3 + Cu$
    c. $Ba(CN)_2 + H_2SO_4 \longrightarrow BaSO_4 + HCN$
12. Name two ways that you could increase the rate of a chemical reaction.

13. Acetic acid, a compound found in vinegar, reacts with baking soda to produce carbon dioxide, water, and sodium acetate. Without writing an equation, identify the reactants and the products of this reaction.

Concept Mapping

14. Use the following terms to create a concept map: chemical reaction, chemical equation, chemical formulas, reactants, products, coefficients, subscripts.

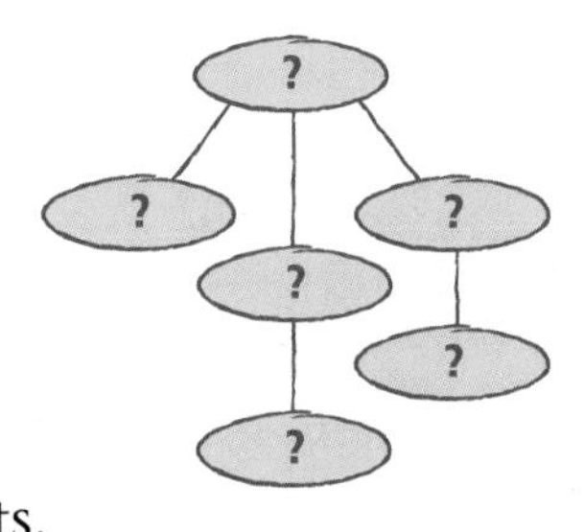

CRITICAL THINKING AND PROBLEM SOLVING

15. Your friend is very worried by rumors he has heard about a substance called dihydrogen monoxide. What could you say to your friend to calm his fears? (Be sure to write the formula of the substance.)

16. As long as proper safety precautions have been taken, why can explosives be transported long distances without exploding?

MATH IN SCIENCE

17. Calculate the number of atoms of each element shown in each of the following:
    a. $CaSO_4$
    b. $4NaOCl$
    c. $Fe(NO_3)_2$
    d. $2Al_2(CO_3)_3$

18. Write balanced equations for the following:
    a. $Fe + O_2 \longrightarrow Fe_2O_3$
    b. $Al + CuSO_4 \longrightarrow Al_2(SO_4)_3 + Cu$
    c. $Ba(CN)_2 + H_2SO_4 \longrightarrow BaSO_4 + HCN$

19. Write and balance chemical equations from each of the following descriptions:
    a. Bromine reacts with sodium iodide to form iodine and sodium bromide.
    b. Phosphorus reacts with oxygen gas to form diphosphorus pentoxide.
    c. Lithium oxide decomposes to form lithium and oxygen.

INTERPRETING GRAPHICS

20. What evidence in the photo supports the claim that a chemical reaction is taking place?

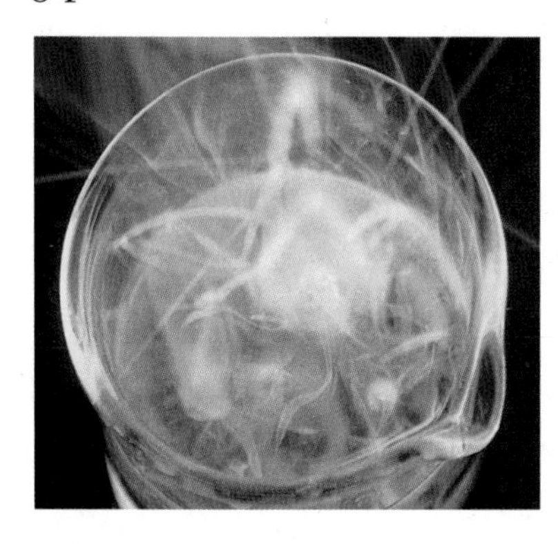

21. Use the energy diagram below to answer the questions that follow.

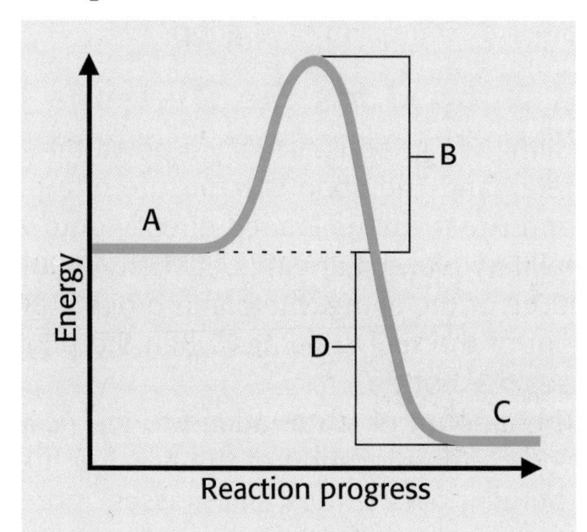

    a. Which letter represents the energy of the products?
    b. Which letter represents the activation energy of the reaction?
    c. Is energy given off or absorbed by this reaction?

NOW What Do You Think?

Take a minute to review your answers to the ScienceLog questions on page 373. Have your answers changed? If necessary, revise your answers based on what you have learned since you began this chapter.

EYE ON THE ENVIRONMENT

Slime That Fire!

Once a fire starts in the hard-to-reach mountains of the western United States, it is difficult to stop. Trees, grasses, and brush can provide an overwhelming supply of fuel. In order to stop a fire, firefighters make a fire line. This is an area where all the burnable materials are removed from the ground. How would you slow down a fire to give a ground crew more time to build a fire line? Would you suggest dropping water from a plane? That is not a bad idea, but what if you had something even better than water—like some slimy red goop?

▲ *This plane is dropping fire retardant on a forest fire.*

Red Goop Goes the Distance

The slimy red goop is actually a powerful fire retardant. The goop is a mixture of a powder and water that is loaded directly onto an old military plane. Carrying between 4,500 and 11,000 L of the slime, the plane drops it all in front of the raging flames when the pilot presses the button.

The amount of water added to the powder depends on the location of the fire. If a fire is burning over shrubs and grasses, more water is needed. In this form the goop actually rains down to the ground through the treetops. But if a fire is burning in tall trees, less water is used so the slime will glob onto the branches and ooze down very slowly.

Failed Flames

The burning of trees, grass, and brush is an exothermic reaction. A fire retardant slows or stops this self-feeding reaction. A fire retardant increases the activation energy for the materials it is applied to. Although a lot depends on how hot the fire is when it hits the area treated with the retardant and how much of the retardant is applied, firefighters on the ground can gain valuable time when a fire is slowed with a fire retardant. This extra time allows them to create a fire line that will ultimately stop the fire.

Neon Isn't Necessary

Once a fire is put out, the slimy red streaks left on the blackened ground can be an eyesore. To solve the problem, scientists have created special dyes for the retardant. These dyes make the goop neon colors when it is first applied, but after a few days in the sun, the goop turns a natural brown shade!

What Do They Study?

► Do some research to learn about a firefighter's training. What classes and exams are firefighters required to pass? How do they maintain their certifications once they become firefighters?

CAREERS

ARSON INVESTIGATOR

Once a fire dies down, you might see arson investigator **Lt. Larry McKee** on the scene. "After the fire is out, I can investigate the fire scene to determine where the fire started and how it started. If it was intentionally set and I'm successful at putting the arson case together, I can get a conviction. That's very satisfying," says Lt. McKee.

During a fire, fuel and oxygen combine in a chemical reaction called combustion. On the scene, Lt. Larry McKee questions witnesses and firefighters about what they saw. He knows, for example, that the color of the smoke can indicate certain chemicals.

McKee explains that fires usually burn "up and out, in a V shape." To find where the V begins, he says, "We work from the area with the least amount of damage to the one with the most damage. This normally leads us to the point of origin." Once the origin has been determined, it's time to call in the dogs!

An Accelerant-Sniffing Canine

"We have what we call an accelerant-sniffing canine. Our canine, Nikki, has been trained to detect approximately 11 different chemicals." When Nikki arrives on the scene, she sniffs for traces of chemicals, called accelerants, that may have been used to start the fire. When she finds one, she immediately starts to dig at it. At that point, McKee takes a sample from the area and sends it to the lab for analysis.

At the Lab

Once at the laboratory, the sample is treated so that any accelerants in it are dissolved in a liquid. A small amount of the liquid is then injected into an instrument called a *gas chromatograph.* The instrument heats the liquid, forming a mixture of gases. The gases then are passed through a flame. As each gas passes through the flame, it "causes a fluctuation in an electronic signal, which creates our graphs."

Solving the Case

If the laboratory report indicates that a suspicious accelerant has been found, McKee begins to search for arson suspects. By combining detective work with scientific evidence, fire investigators can successfully catch and convict arsonists.

Fascinating Fire Facts

▶ The temperature of a house fire can reach 980°C! At that temperature, aluminum window frames melt, and furniture goes up in flames. Do some research to discover three more facts about fires. Create a display with two or more classmates to illustrate some of your facts.

▲ *Nikki searches for traces of gasoline, kerosene, and other accelerants.*

CHAPTER 16

Chemical Compounds

Strange but True . . .

During World War II, the United States could not obtain natural rubber from Asian suppliers, who gathered it from rubber trees as shown below. Faced with a shortage of raw material, American scientists searched for other materials to use in truck tires and soldiers' boots.

James Wright, an engineer at General Electric, was working with silicone oil—a clear, gooey compound composed of silicon bonded to several other elements. By substituting silicon for carbon, the main element in rubber, Wright hoped to create a new compound with all the flexibility and bounce of rubber.

In 1943, Wright made a surprising discovery. He mixed boric acid with silicone oil in a test tube. Instead of forming the hard rubber material he sought, the compound remained slightly gooey to the touch. Disappointed with the results, Wright tossed a gob of the material from the test tube onto the floor. To his surprise, the gob bounced right back at him.

The new compound was very bouncy and could be stretched and pulled. However, it wasn't a good rubber substitute, so Wright and other General Electric scientists continued their search.

Seven years later, a toy seller named Peter Hodgson packaged some of Wright's creation in small plastic "eggs" and presented his new product at the 1950 International Toy Fair in New York. The material, called Silly Putty®, proved quite popular. Millions of eggs containing Silly Putty have been sold to kids of all ages since then.

Rubber and boric acid are substances with very different properties. In this chapter, you will learn about the properties that are used to classify many different compounds.

Natural rubber is collected from a rubber tree as it flows from cuts made in the bark.

Ionic Versus Covalent

Compounds are often classified based on similarities in their structure or properties. For some compounds, differences in properties are a result of the type of chemical bonding present. In this activity, you will investigate the properties of two substances and relate the properties to the type of bonding in the compounds.

What Do You Think?

In your ScienceLog, try to answer the following questions based on what you already know:

1. What is the difference between ionic compounds and covalent compounds?
2. What is an acid?
3. What elements make up a hydrocarbon?

Procedure

1. Place a small amount of **paraffin wax** into a **test tube.** Place an equal amount of **table salt** into a **second test tube.**
2. Fill a **plastic-foam cup** halfway with **hot water.**
3. Place the test tubes into the water. After 3 minutes, remove the test tubes from the water. Observe the contents of the test tubes, and record your observations in your ScienceLog.
4. Add 10 mL of **water** to each test tube using a **graduated cylinder.**
5. Stir each test tube with a **stirring rod.** Record your observations in your ScienceLog.

Analysis

6. Summarize the properties you observed for each type of compound.
7. Ionic bonding is present in many compounds that have a high melting point and that will dissolve in water. Covalent bonding is present in many compounds that have a low melting point and that will not dissolve in water. Identify the type of bonding present in paraffin wax and in table salt.

Section 1

Ionic and Covalent Compounds

NEW TERMS

ionic compounds
covalent compounds

OBJECTIVES

- Describe the properties of ionic and covalent compounds.
- Classify compounds as ionic or covalent based on their properties.

The world around you is made up of chemical compounds. Chemical compounds are pure substances composed of ions or molecules. There are millions of different kinds of compounds, so you can imagine how classifying them might be helpful. One simple way to classify compounds is by grouping them according to the type of bond they contain.

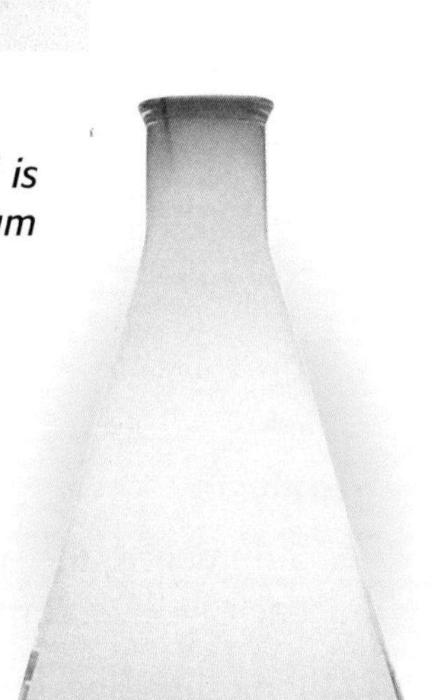

Figure 1 *An ionic compound is formed when the metal sodium reacts with the nonmetal chlorine. Sodium chloride is formed in the reaction, and energy is released as light and thermal energy.*

Ionic Compounds

Compounds that contain ionic bonds are called **ionic compounds.** Remember that an ionic bond is the force of attraction between two oppositely charged ions. Ionic compounds can be formed by the reaction of a metal with a nonmetal. Electrons are transferred from the metal atoms (which become positively charged ions) to the nonmetal atoms (which become negatively charged ions). For example, when sodium reacts with chlorine, as shown in **Figure 1,** the ionic compound sodium chloride, or ordinary table salt, is formed.

Properties of Ionic Compounds The forces acting between the ions that make up ionic compounds give these compounds certain properties. Ionic compounds tend to be brittle, as shown in **Figure 2.** The ions that make up an ionic compound are arranged in a repeating three-dimensional pattern called a crystal lattice. The ions that make up the crystal lattice are arranged as alternating positive and negative ions. Each ion in the lattice is surrounded by ions of the opposite charge, and each ion is bonded to the ions around it. When an ionic compound is struck with a hammer, the pattern of ions in the crystal lattice is shifted. Ions with the same charge line up and repel one another, causing the crystal to shatter.

Figure 2 *Ionic compounds tend to be brittle, and they will shatter when hit with a hammer.*

Because ionic bonds are very strong, ionic compounds have high melting points and are almost always solid at room temperature, as shown in **Figure 3.** An ionic compound will only melt at temperatures high enough to overcome the strong ionic bonds between the ions. Sodium chloride, for instance, must be heated to 801°C before it will melt. This temperature is much higher than you can produce in your kitchen or even your school laboratory.

Figure 3 *Each of these ionic compounds has a high melting point and is solid at room temperature.*

Another property shared by many ionic compounds is that they dissolve easily in water. Molecules of water attract each of the ions of an ionic compound and pull them away from one another. The solution created when an ionic compound dissolves in water can conduct an electric current, as shown in **Figure 4.** The dissolved ions are able to move past one another and allow the electric current to exist in the solution. In contrast, the ions in an undissolved crystal cannot move past one another, so a crystal of an ionic compound does not conduct an electric current.

Figure 4 *The beaker of pure water on the left does not conduct an electric current. As salt dissolves in the beaker of water on the right, an electric current can exist, and the bulb lights up.*

Covalent Compounds

Compounds composed of elements that are covalently bonded are called **covalent compounds.** Remember that a covalent bond is formed when two atoms share electrons. Gasoline, carbon dioxide, water, and sugar are well-known examples of covalent compounds.

BRAIN FOOD

Molecules of covalent compounds can have anywhere from two atoms to hundreds or thousands of atoms! Small, lightweight molecules, like water or carbon dioxide, tend to form liquids or gases at room temperature. Heavier molecules, such as sugar or plastics, tend to form solids at room temperature.

Properties of Covalent Compounds The properties of covalent compounds are quite different from those of ionic compounds. Covalent compounds exist as independent particles called molecules. The forces of attraction between molecules of covalent compounds are much weaker than the bonds between ions in a crystal lattice. Thus, covalent compounds have lower melting points than ionic compounds.

You have probably heard the phrase "oil and water don't mix." Oil, such as that used in salad dressing, is composed of covalent compounds. Many covalent compounds do not dissolve well in water. Water molecules have a stronger attraction for one another than they have for the molecules of most other covalent compounds. Thus, the molecules of the covalent compound get squeezed out as the water molecules pull together. Some covalent compounds do dissolve in water. Most of these solutions contain uncharged molecules dissolved in water and do not conduct an electric current, as shown in **Figure 5.** Some covalent compounds form ions when they dissolve in water. Solutions of these compounds, including compounds called acids, do conduct an electric current. You will learn more about acids in the next section.

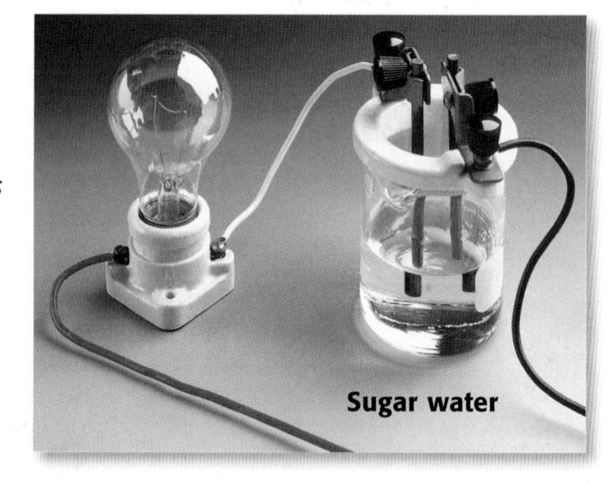

Figure 5 *This solution of sugar, a covalent compound, in water does not conduct an electric current because the individual molecules of sugar are not charged.*

REVIEW

1. List two properties of ionic compounds.
2. List two properties of covalent compounds.
3. Methane is a gas at room temperature. What type of compound is this most likely to be?
4. **Comparing Concepts** Compare ionic and covalent compounds based on the type of particle that makes up each.

Section 2

Acids, Bases, and Salts

NEW TERMS

acid
pH
base
salt

OBJECTIVES

- Describe the properties and uses of acids and bases.
- Explain the difference between strong acids and bases and weak acids and bases.
- Identify acids and bases using the pH scale.
- Describe the properties and uses of salts.

Have you ever noticed a change in your tea when you added lemon? When you squeeze lemon juice into tea, the color of the tea becomes lighter, as shown in **Figure 6.** Lemon juice contains a substance called an acid that changes the color of a substance in the tea. The ability to change the color of certain chemicals is one property used to classify substances as acids or bases. A third category of substances, called salts, will also be discussed in this section. Salts are formed by the reaction of an acid with a base.

Figure 6 *Acids, like those found in lemon juice, can change the color of tea.*

Acids

An **acid** is any compound that increases the number of hydrogen ions when dissolved in water, and whose solution tastes sour and can change the color of certain compounds.

NEVER touch or taste a concentrated solution of a strong acid.

Properties of Acids If you have ever had orange juice, you have experienced the sour taste of an acid. The taste of lemons, limes, and other citrus fruits is a result of citric acid. Taste, however, should NEVER be used as a test to identify an unknown chemical. Many acids are *corrosive,* meaning they destroy body tissue and clothing, and many are also poisonous.

In **Figure 7** you see the result of placing a piece of zinc into a hydrochloric acid solution. Acids react with some metals to produce hydrogen gas. Adding an acid to baking soda or limestone produces a different gas, carbon dioxide. Vinegar contains acetic acid. When vinegar is added to baking soda, bubbles of carbon dioxide are produced.

Figure 7 *Bubbles of hydrogen gas are produced when zinc metal reacts with hydrochloric acid.*

Solutions of acids conduct an electric current because acids break apart to form ions in water. Acids increase the number of hydrogen ions, H^+, in a solution. However, the hydrogen ion does not normally exist alone. In a water solution, the hydrogen ions strongly attract water molecules. Each hydrogen ion attaches to a water molecule to form a hydronium ion, H_3O^+.

Some hydrangea plants act as indicators. Leaves on the plants change from pink to blue as the soil becomes more acidic.

As mentioned earlier, a property of acids is their ability to change the color of a substance. An *indicator* is a substance that changes color in the presence of an acid or base. A commonly used indicator is litmus. Paper strips containing litmus are available in both blue and red. When an acid is added to blue litmus paper, the color of the litmus changes to red, as shown in **Figure 8.** (Red litmus paper is used to detect bases, as will be discussed shortly.) Many plant materials, such as red cabbage, contain compounds that are indicators.

Figure 8 *Vinegar turns blue litmus paper red because it contains acetic acid.*

Figure 9 *The label on this car battery warns you that sulfuric acid is found in the battery.*

Uses of Acids Acids are used in many areas of industry as well as in your home. Sulfuric acid is the most widely produced industrial chemical in the world. It is used in the production of metals, paper, paint, detergents, and fertilizers. It is also used in car batteries, as shown in **Figure 9.** Nitric acid is used to make fertilizers, rubber, and plastics. Hydrochloric acid is used in the production of metals and to help keep swimming pools free of algae. It is also found in your stomach, where it aids in digestion. Citric acid and ascorbic acid (vitamin C) are found in orange juice, while carbonic acid and phosphoric acid help give extra "bite" to soft drinks.

Strong Versus Weak As an acid dissolves in water, its molecules break apart and produce hydrogen ions. When all the molecules of an acid break apart in water to produce hydrogen ions, the acid is considered a strong acid. Strong acids include sulfuric acid, nitric acid, and hydrochloric acid.

When few molecules of an acid break apart in water to produce hydrogen ions, the acid is considered a weak acid. Acetic acid, citric acid, carbonic acid, and phosphoric acid are all weak acids.

Bases

A **base** is any compound that increases the number of hydroxide ions when dissolved in water, and whose solution tastes bitter, feels slippery, and can change the color of certain compounds.

To determine how acidic or basic a solution is, just use your head—of cabbage! Try it for yourself on page 582 of the LabBook.

Properties of Bases If you have ever accidentally tasted soap, then you have experienced the bitter taste of a base. Soap also demonstrates that a base feels slippery. However, NEVER use taste or touch as a test to identify an unknown chemical. Like acids, many bases are corrosive. If you are using a base in an experiment and your fingers begin to feel slippery, it might mean that some of the base got on your hands. You should immediately rinse your hands with large amounts of water.

Solutions of bases conduct an electric current because bases form ions in water. Bases increase the number of hydroxide ions, OH^-, in a solution. A hydroxide ion is actually a hydrogen atom and an oxygen atom bonded together. An extra electron gives the ion a negative charge.

Like acids, bases change the color of an indicator. Most indicators turn a different color for bases than they do for acids. For example, bases will change the color of red litmus paper to blue, as shown in **Figure 10.**

NEVER touch or taste a concentrated solution of a strong base.

Sodium Hydroxide NaOH

Figure 10 *Sodium hydroxide, a base, turns red litmus paper blue.*

Uses of Bases Like acids, bases have many uses. Sodium hydroxide is used to make soap and paper. You can find sodium hydroxide in your home in oven cleaners and in products that unclog your drain, as shown in **Figure 11.** Remember, bases can harm your skin, so carefully follow the safety instructions when using these products. Calcium hydroxide is used to make cement, mortar, and plaster. Ammonia is found in many household cleaners and is also used in the production of fertilizers. Magnesium hydroxide and aluminum hydroxide are used in antacids to treat heartburn.

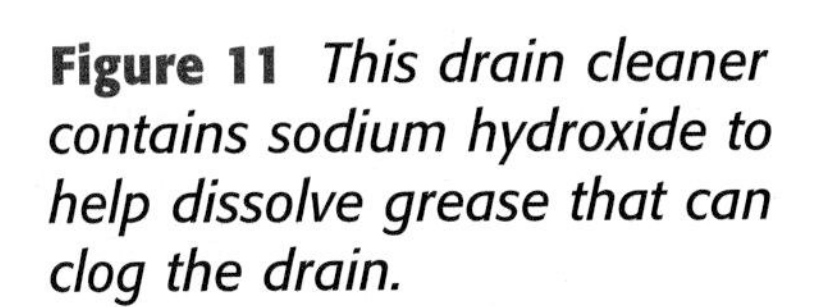

Figure 11 *This drain cleaner contains sodium hydroxide to help dissolve grease that can clog the drain.*

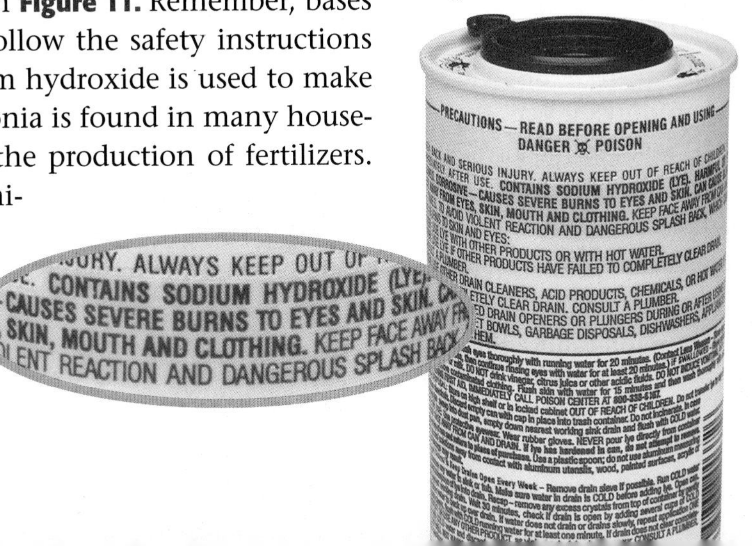

Strong Versus Weak When all the molecules of a base break apart in water to produce hydroxide ions, the base is called a strong base. Strong bases include sodium hydroxide, calcium hydroxide, and potassium hydroxide.

When only a few of the molecules of a base produce hydroxide ions in water, the base is called a weak base. Ammonia, magnesium hydroxide, and aluminum hydroxide are all weak bases.

Figure 12 *Have heartburn? Take an antacid! Antacid tablets contain a base that neutralizes the acid in your stomach.*

Acids and Bases Neutralize One Another

If you have ever suffered from an acid stomach, or heartburn, as shown in **Figure 12,** you might have taken an antacid. Antacids contain weak bases that soothe your heartburn by reacting with and neutralizing the acid in your stomach. Acids and bases neutralize one another because the H^+ of the acid and the OH^- of the base react to form water, H_2O. Other ions from the acid and base are also dissolved in the water. If the water is evaporated, these ions join to form a compound called a salt. You'll learn more about salts later in this section.

The pH Scale Indicators such as litmus can identify whether a solution contains an acid or base. To describe how acidic or basic a solution is, the pH scale is used. The **pH** of a solution is a measure of the hydronium ion concentration in the solution. By measuring the hydronium ion concentration, the pH is also a measure of the hydrogen ion concentration. On the scale, a solution that has a pH of 7 is neutral, meaning that it is neither acidic nor basic. Pure water has a pH of 7. Basic solutions have a pH greater than 7, and acidic solutions have a pH less than 7. Look at **Figure 13** to see the pH values for many common materials.

Figure 13 pH Values of Common Materials

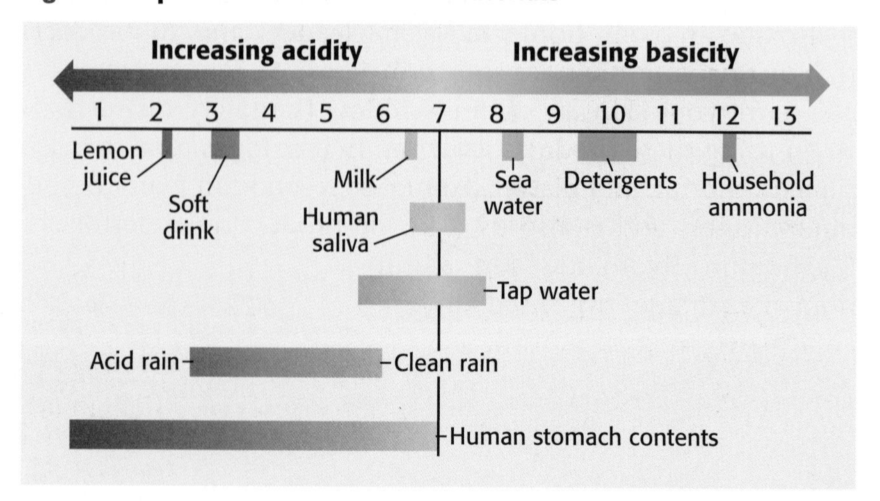

QuickLab

pHast Relief!

1. Fill a **small plastic cup** halfway with **vinegar.** Test the vinegar with **red** and **blue litmus paper.** Record your results in your ScienceLog.
2. Carefully crush an **antacid tablet,** and mix it with the vinegar. Test the mixture with litmus paper. Record your results in your ScienceLog.
3. Compare the acidity of the solution before and after the reaction.

Many indicators, including litmus, have only two colors. This allows you to determine if a solution is acidic or basic, but it does not identify its pH. A mixture of different indicators can be used to determine the pH of a solution. After determining the colors for this mixture at different pH values, the indicators can be used to determine the pH of an unknown solution, as shown in **Figure 14.** Indicators can be used as paper strips or solutions, and they are often used to test the pH of soil and of water in pools and aquariums. Another way to determine the acidity of a solution is to use an instrument called a pH meter, which can detect and measure hydrogen ions electronically.

Figure 14 *The paper strip contains several indicators. The pH of a solution is determined by comparing the color of the strip to the scale provided.*

pH and the Environment Living things depend on having a steady pH in their environment. Plants are known to have certain preferred growing conditions. Some plants, such as pine trees, prefer acidic soil with a pH between 4 and 6. Other plants, such as lettuce, require basic soil with a pH between 8 and 9. Fish require water near pH 7. As you can see in Figure 13, rainwater can have a pH as low as 3. This occurs in areas where compounds found in pollution react with water to make the strong acids sulfuric acid and nitric acid. As this acid precipitation collects in lakes, it can lower the pH to levels that may kill the fish and other organisms in the lake. To neutralize the acid and bring the pH closer to 7, a base can be added to the lakes, as shown in **Figure 15.**

Human blood has a pH of between 7.38 and 7.42. If the pH is above 7.8 or below 7, the body cannot function properly. Sudden changes in blood pH that are not quickly corrected can be fatal.

Figure 15 *This helicopter is adding a base to an acidic lake. Neutralizing the acid in the lake might help protect the organisms living in the lake.*

Self-Check

Which is more acidic, a soft drink or milk? (Hint: Refer to Figure 13 to find the pH values of these drinks.) *(See page 596 to check your answer.)*

Salts

When you hear the word *salt,* you probably think of the table salt you use to season your food. But the sodium chloride found in your salt shaker is only one example of a large group of compounds called salts. A **salt** is an ionic compound formed from the positive ion of a base and the negative ion of an acid. You may remember that a salt and water are produced when an acid neutralizes a base. However, salts can also be produced in other reactions, as shown in **Figure 16.**

Explore

Write the formula of the salt produced by each of the following pairs of acids and bases:

1. $HCl + LiOH$
2. $HBr + Ca(OH)_2$

Neutralization of an acid and a base:

$$HCl + KOH \longrightarrow H_2O + KCl$$

Reaction of a metal with an acid:

$$2K + 2HCl \longrightarrow 2KCl + H_2$$

Reaction of a metal and a nonmetal:

$$2K + Cl_2 \longrightarrow 2KCl$$

Figure 16 *The salt potassium chloride can be formed from several different reactions.*

Uses of Salts Salts have many uses in industry and in your home. You already know that sodium chloride is used to season foods. It is also used in the production of other compounds, including lye (sodium hydroxide), hydrochloric acid, and baking soda. The salt calcium sulfate is made into wallboard, or plasterboard, which is used in construction. Sodium nitrate is one of many salts used as a preservative in foods. Calcium carbonate is a salt that makes up limestone, chalk, and seashells. Another use of salts is shown in **Figure 17.**

Figure 17 *Salts are used during icy weather to help keep roads free of ice.*

REVIEW

1. What ion is present in all acid solutions?
2. What are two ways scientists can measure pH?
3. What products are formed when an acid and base react?
4. **Comparing Concepts** Compare the properties of acids and bases.
5. **Applying Concepts** Would you expect the pH of a solution of soap to be 4 or 9?

Section 3

Organic Compounds

NEW TERMS

organic compounds
biochemicals
carbohydrates
lipids
proteins
nucleic acids
hydrocarbons

OBJECTIVES

- Explain why so many organic compounds are possible.
- Describe the characteristics of carbohydrates, lipids, proteins, and nucleic acids and their functions in the body.
- Describe and identify saturated, unsaturated, and aromatic hydrocarbons.

Of all the known compounds, more than 90 percent are members of a group of compounds called organic compounds. **Organic compounds** are covalent compounds composed of carbon-based molecules. Sugar, starch, oil, protein, nucleic acid, and even cotton and plastic are organic compounds. How can there be so many different kinds of organic compounds? The huge variety of organic compounds is explained by examining the carbon atom.

Each Carbon Atom Forms Four Bonds

Carbon atoms form the backbone of organic compounds. Because each carbon atom has four valence electrons (electrons in the outermost energy level of an atom), each atom can make four bonds. Thus, a carbon atom can bond to one, two, or even three other carbon atoms and still have electrons remaining to bond to other atoms. Three types of carbon backbones on which many organic compounds are based are shown in the models in **Figure 18.**

Some organic compounds have hundreds or even thousands of carbon atoms making up their backbone! Although the elements hydrogen and oxygen, along with carbon, make up many of the organic compounds, sulfur, nitrogen, and phosphorus are also important—especially in forming the molecules that make up all living things.

Figure 18 *These models, called structural formulas, are used to show how atoms in a molecule are connected. Each line represents a pair of electrons shared in a covalent bond.*

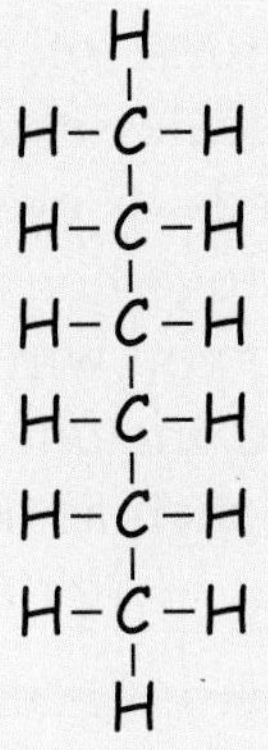

Straight Chain All carbon atoms are connected one after another in a line.

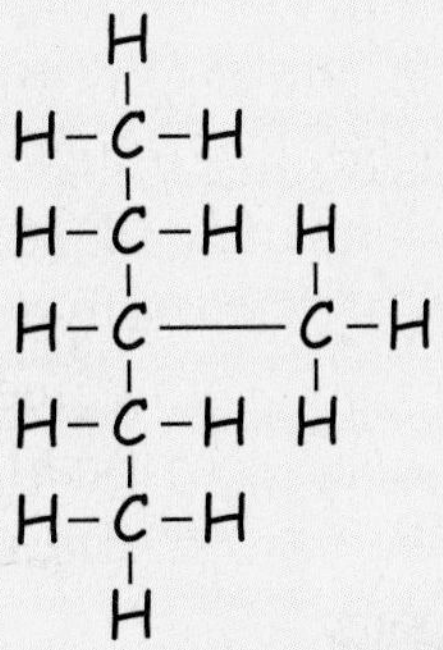

Branched Chain The chain of carbon atoms continues in more than one direction where a carbon atom bonds to three or more other carbon atoms.

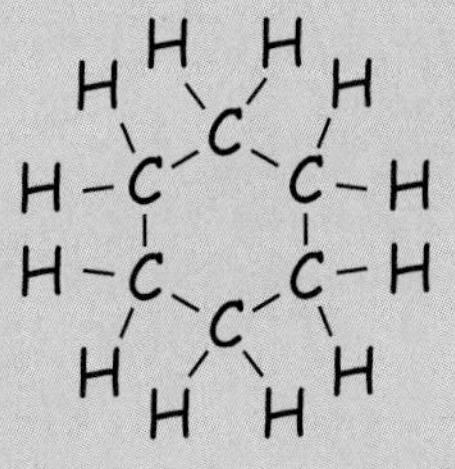

Ring The chain of carbon atoms forms a ring.

Biochemicals: The Compounds of Life

Many different compounds are found in living things. Some compounds are composed of very small, simple molecules that are not based on carbon. These compounds, including water and salt, are considered *inorganic.* Organic compounds made by living things are called **biochemicals.** The molecules of most biochemicals are very large. There are hundreds of thousands of different biochemicals, which can be divided into four categories: carbohydrates, lipids, proteins, and nucleic acids. Each type of biochemical has important functions in living organisms.

Carbohydrates Starch and cellulose are examples of carbohydrates. **Carbohydrates** are biochemicals that are composed of one or more simple sugars bonded together; they are used as a source of energy and for energy storage. The energy you get from these biochemicals is stored in the form of chemical bonds in the molecules. There are two types of carbohydrates: simple carbohydrates and complex carbohydrates. A single sugar molecule, represented using a hexagon, or a few sugar molecules bonded together are examples of simple carbohydrates, as illustrated in **Figure 19.** Glucose is a simple carbohydrate produced by plants through photosynthesis.

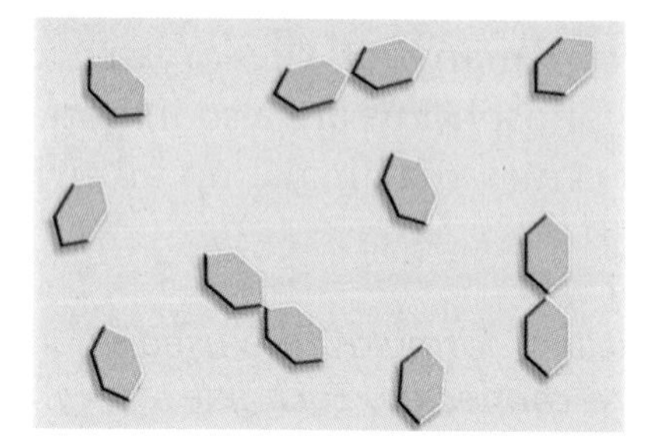

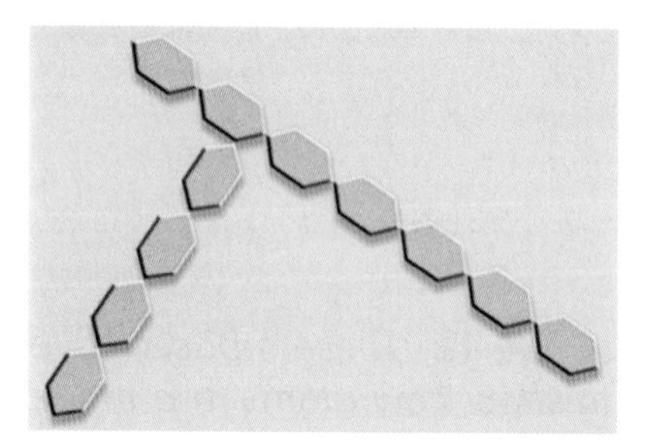

Figure 19 *The sugar molecules in the left image are simple carbohydrates. The starch in the right image is a complex carbohydrate because it is composed of many sugar molecules bonded together.*

When an organism has more sugar than it needs, its extra sugar may be stored for later use in the form of complex carbohydrates, as shown in Figure 19. Molecules of complex carbohydrates are composed of hundreds or even thousands of sugar molecules bonded together. Because carbohydrates are used to provide you with energy you need each day, you should include sources of carbohydrates in your diet, such as the foods shown in **Figure 20.**

Figure 20 *Simple carbohydrates include sugars found in fruits and honey. Complex carbohydrates, such as starches, are found in bread, cereal, and pasta.*

Lipids Fats, oils, waxes, and steroids are examples of lipids. **Lipids** are biochemicals that do not dissolve in water and have many different functions, including storing energy and making up cell membranes. Although too much fat in your diet can be unhealthy, some fat is extremely important to good health. The foods in **Figure 21** are sources of lipids.

Lipids store excess energy in the body. Animals tend to store lipids primarily as fats, while plants store lipids as oils. When an organism has used up most of its carbohydrates, it can obtain energy by breaking down lipids. Lipids are also used to store vitamins in the body. Vitamins that do not dissolve in water will often dissolve in fat.

Figure 21 *Vegetable oil, meat, cheese, nuts, and milk are sources of lipids in your diet.*

Lipids make up a structure called a cell membrane that surrounds each cell. Much of the cell membrane is formed from molecules of phospholipids. The structure of phospholipid molecules plays an important part in the phospholipid's role in the cell membrane. A phospholipid molecule has two regions with very different properties. The tail of a phospholipid molecule is a long, straight-chain carbon backbone composed only of carbon and hydrogen atoms. The tail is not attracted to water. In addition to carbon and hydrogen atoms, the head of a phospholipid molecule is composed of phosphorus, oxygen, and nitrogen atoms, which cause the head of the molecule to be attracted to water. When phospholipids are in water, the tails are forced together as water is attracted to the heads of the molecules. The result is the double layer of phospholipid molecules shown in the model in **Figure 22.** This arrangement of phospholipid molecules creates a barrier to help control the flow of chemicals into and out of the cell.

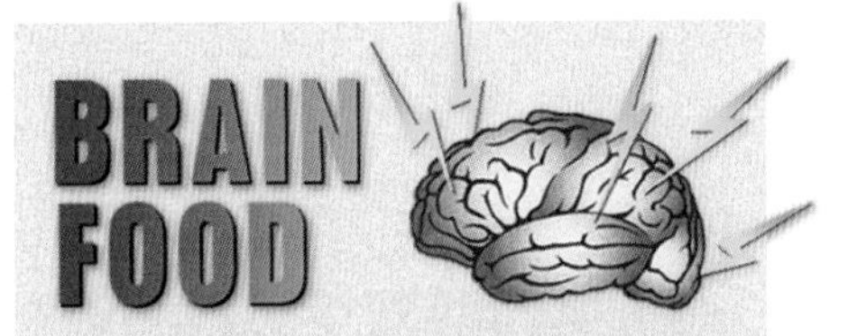

Deposits of the lipid cholesterol in the body have been linked to health problems such as heart disease. However, cholesterol is needed in nerve and brain tissue as well as to make certain hormones that regulate body processes such as growth.

Figure 22 *A cell membrane is composed primarily of two layers of phospholipid molecules.*

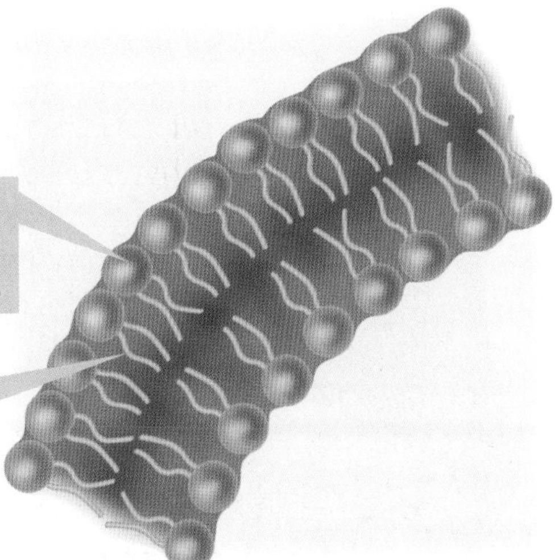

Figure 23 *Meat, fish, cheese, and beans contain proteins, which are broken down into amino acids as they are digested.*

Proteins Most of the biochemicals found in living things are proteins. In fact, after water, proteins are the most abundant molecules in your cells. **Proteins** are biochemicals that are composed of amino acids; they have many different functions, including regulating chemical activities, transporting and storing materials, and providing structural support.

Every protein is composed of small "building blocks" called *amino acids*. Amino acids are smaller molecules composed of carbon, hydrogen, oxygen, and nitrogen atoms. Some amino acids also include sulfur atoms. Amino acids chemically bond to form proteins of many different shapes and sizes, from short chains of only a few amino acids to large, twisted structures consisting of thousands of amino acids. The function of a protein depends on the shape that the bonded amino acids adopt. If even a single amino acid is missing or out of place, the protein may not function correctly or at all. The foods shown in **Figure 23** provide amino acids that your body needs to make new proteins.

Enzymes are proteins that regulate chemical reactions in the body by acting as catalysts to increase the rate at which the reactions occur. Some hormones that help control your bodily functions are proteins. Insulin, a hormone that helps regulate the level of sugar in your blood, is one of the smallest proteins, consisting of only 51 amino acids. Oxygen is carried by the protein hemoglobin, allowing red blood cells to deliver oxygen throughout your body. There are also large proteins that extend through cell membranes and help control the transport of materials into and out of cells. Proteins that provide structural support often form structures that are easy to see, like those in **Figure 24.**

life science CONNECTION

All the proteins in your body are made from just 20 amino acids. Nine of these amino acids are called essential amino acids because your body cannot make them. You must get them from the food you eat.

Figure 24 *Hair and spider webs are made up of proteins that are shaped like long fibers.*

Nucleic Acids The largest molecules made by living organisms are nucleic acids. **Nucleic acids** are biochemicals that store information and help to build proteins and other nucleic acids. Nucleic acids are sometimes called the "blueprints of life" because they contain all the information needed for the cell to make all of its proteins.

Like proteins, nucleic acids are long chains of smaller molecules joined together. These smaller molecules are composed of carbon, hydrogen, oxygen, nitrogen, and phosphorus atoms. Nucleic acids are much larger than proteins even though nucleic acids are composed of only five building blocks.

There are two types of nucleic acids: DNA and RNA. DNA (**d**eoxyribo**n**ucleic **a**cid), like that shown in **Figure 25,** is the genetic material of the cell. DNA molecules can store an enormous amount of information because of their length. If the DNA molecules in a single human cell were placed end to end and stretched out, their overall length would be about 2 m—that's over 6 ft long! When a cell needs to make a certain protein, it gets information from the DNA in the cell. The important part of the DNA molecule is copied. The information copied from DNA directs the order in which amino acids are bonded together to make that protein. DNA also contains information used to build the second type of nucleic acid, RNA (**r**ibo**n**ucleic **a**cid). RNA is involved in the actual building of proteins.

Nucleic acids store information—even about ancient peoples. Read more about these incredible biochemicals on page 418.

Figure 25 *The DNA from a fruit fly contains all of the instructions for making proteins, nucleic acids . . . in fact, for making everything in the organism!*

REVIEW

1. What are organic compounds?
2. What are the four categories of biochemicals?
3. What are two functions of proteins?
4. What biochemicals are used to provide energy?
5. **Inferring Relationships** Sickle-cell anemia is a condition that results from a change of one amino acid in the protein hemoglobin. Why is this condition a genetic disorder?

Hydrocarbons

Organic compounds that are composed of only carbon and hydrogen are called **hydrocarbons.** Hydrocarbons are an important group of organic compounds. Many fuels, including gasoline, methane, and propane, are hydrocarbons. Hydrocarbons can be divided into three categories: saturated, unsaturated, and aromatic.

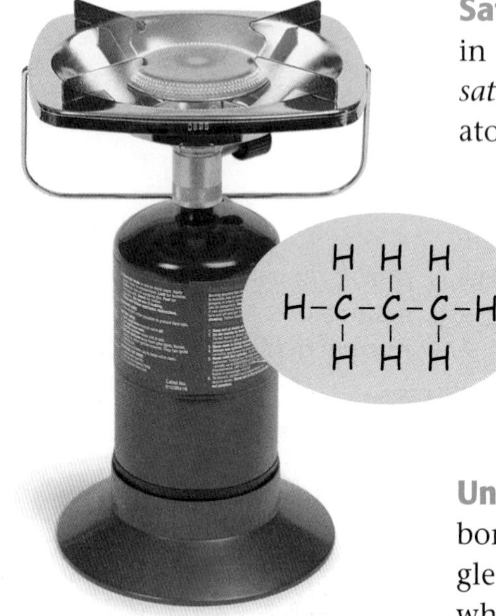

Figure 26 *The propane in this camping stove is a saturated hydrocarbon.*

Saturated Hydrocarbons Propane, like that used in the stove in **Figure 26,** is an example of a saturated hydrocarbon. A *saturated hydrocarbon* is a hydrocarbon in which each carbon atom in the molecule shares a single bond with each of four other atoms. A single bond is a covalent bond that consists of one pair of shared electrons. Hydrocarbons that contain carbon atoms connected only by single bonds are called saturated because no other atoms can be added without replacing an atom that is part of the molecule. Saturated hydrocarbons are also called *alkanes.*

Unsaturated Hydrocarbons Each carbon atom forms four bonds. However, these bonds do not always have to be single bonds. An *unsaturated hydrocarbon* is a hydrocarbon in which at least two carbon atoms share a double bond or a triple bond. A double bond is a covalent bond that consists of two pairs of shared electrons. Compounds that contain two carbon atoms connected by a double bond are called *alkenes.*

A triple bond is a covalent bond that consists of three pairs of shared electrons. Hydrocarbons that contain two carbon atoms connected by a triple bond are called *alkynes.*

Hydrocarbons that contain double or triple bonds are called unsaturated because the double or triple bond can be broken to allow more atoms to be added to the molecule. Examples of unsaturated hydrocarbons are shown in **Figure 27.**

Figure 27 *Fruits produce ethene, which helps ripen the fruit. Ethyne, better known as acetylene, is burned in this miner's lamp and is also used in welding.*

Aromatic Hydrocarbons Most aromatic compounds are based on benzene, the compound represented by the model in **Figure 28.** Look for this structure to help identify an aromatic hydrocarbon. As the name implies, aromatic hydrocarbons often have strong odors and are therefore used in such products as air fresheners and moth balls.

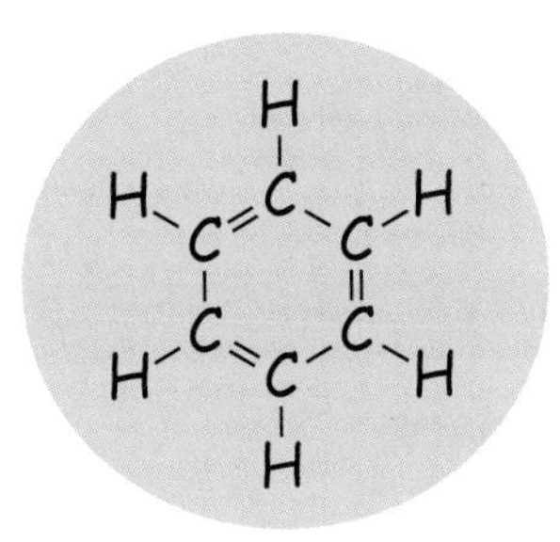

Figure 28 *Benzene has a ring of six carbons with alternating double and single bonds. Benzene is the starting material for manufacturing many products, including medicines.*

Other Organic Compounds

Many other types of organic compounds exist that have atoms of halogens, oxygen, sulfur, and phosphorus in their molecules. A few of these types of compounds and their uses are described in the chart below.

Types and Uses of Organic Compounds

Type of compound	Uses	Examples
Alkyl halide	starting material for Teflon refrigerant (freon)	chloromethane (CH_3Cl) bromoethane (C_2H_5Br)
Alcohols	rubbing alcohol gasoline additive antifreeze	methanol (CH_3OH) ethanol (C_2H_5OH)
Organic acids	food preservatives flavoring	ethanoic acid (CH_3COOH) propanoic acid (C_2H_5COOH)
Esters	flavorings fragrances clothing (polyester)	methyl ethanoate (CH_3COOCH_3) ethyl propanoate ($C_2H_5COOC_2H_5$)

REVIEW

1. What is a hydrocarbon?
2. How many electrons are shared in a double bond? a triple bond?
3. **Comparing Concepts** Compare saturated and unsaturated hydrocarbons.

Chapter Highlights

SECTION 1

Vocabulary

ionic compounds *(p. 398)*
covalent compounds *(p. 399)*

Section Notes

- Ionic compounds contain ionic bonds and are composed of oppositely charged ions arranged in a repeating pattern called a crystal lattice.
- Ionic compounds tend to be brittle, have high melting points, and dissolve in water to form solutions that conduct an electric current.
- Covalent compounds are composed of elements that are covalently bonded and consist of independent particles called molecules.
- Covalent compounds tend to have low melting points. Most do not dissolve well in water and do not form solutions that conduct an electric current.

SECTION 2

Vocabulary

acid *(p. 401)*
base *(p. 403)*
pH *(p. 404)*
salt *(p. 406)*

Section Notes

- An acid is a compound that increases the number of hydrogen ions in solution. Acids taste sour, turn blue litmus paper red, react with metals to produce hydrogen gas, and react with limestone or baking soda to produce carbon dioxide gas.
- A base is a compound that increases the number of hydroxide ions in solution. Bases taste bitter, feel slippery, and turn red litmus paper blue.
- When dissolved in water, every molecule of a strong acid or base breaks apart to form ions. Few molecules of weak acids and bases break apart to form ions.
- When combined, an acid and a base neutralize one another to produce water and a salt.
- pH is a measure of hydronium ion concentration in a solution. A pH of 7 indicates a neutral substance. A pH of less than 7 indicates an acidic substance. A pH of greater than 7 indicates a basic substance.
- A salt is an ionic compound formed from the positive ion of a base and the negative ion of an acid.

Labs

Cabbage Patch Indicators *(p. 582)*
Making Salt *(p. 584)*

☑ Skills Check

Visual Understanding

LITMUS PAPER You can use the ability of acids and bases to change the color of indicators to identify a chemical as an acid or base. Litmus is an indicator commonly used in schools. Review Figures 8 and 10, which show how the color of litmus paper is changed by an acid and by a base.

pH SCALE Knowing whether a substance is an acid or a base can help explain some of the properties of the substance. The pH scale shown in Figure 13 illustrates the pH ranges for many common substances.

DUAL-TINT® pH 1-12

SECTION 3

Vocabulary

organic compounds *(p. 407)*
biochemicals *(p. 408)*
carbohydrates *(p. 408)*
lipids *(p. 409)*
proteins *(p. 410)*
nucleic acids *(p. 411)*
hydrocarbons *(p. 412)*

Section Notes

- Organic compounds are covalent compounds composed of carbon-based molecules.
- Each carbon atom forms four bonds with other carbon atoms or with atoms of other elements to form straight chains, branched chains, or rings.
- Biochemicals are organic compounds made by living things.
- Carbohydrates are biochemicals that are composed of one or more simple sugars bonded together; they are used as a source of energy and for energy storage.
- Lipids are biochemicals that do not dissolve in water and have many functions, including storing energy and making up cell membranes.
- Proteins are biochemicals that are composed of amino acids and have many functions, including regulating chemical activities, transporting and storing materials, and providing structural support.
- Nucleic acids are biochemicals that store information and help to build proteins and other nucleic acids.
- Hydrocarbons are organic compounds composed of only carbon and hydrogen.
- In a saturated hydrocarbon, each carbon atom in the molecule shares a single bond with each of four other atoms.
- In an unsaturated hydrocarbon, at least two carbon atoms share a double bond or a triple bond.
- Many aromatic hydrocarbons are based on the six-carbon ring of benzene.
- Other organic compounds, including alkyl halides, alcohols, organic acids, and esters, are formed by adding atoms of other elements.

internet**connect**

GO TO: go.hrw.com

Visit the **HRW** Web site for a variety of learning tools related to this chapter. Just type in the keyword:

KEYWORD: HSTCMP

GO TO: www.scilinks.org

Visit the **National Science Teachers Association** on-line Web site for Internet resources related to this chapter. Just type in the ***sci*LINKS** number for more information about the topic:

TOPIC: Ionic Compounds — ***sci*LINKS NUMBER:** HSTP355
TOPIC: Covalent Compounds — ***sci*LINKS NUMBER:** HSTP360
TOPIC: Acids and Bases — ***sci*LINKS NUMBER:** HSTP365
TOPIC: Salts — ***sci*LINKS NUMBER:** HSTP370
TOPIC: Organic Compounds — ***sci*LINKS NUMBER:** HSTP375

Chapter Review

USING VOCABULARY

To complete the following sentences, choose the correct term from each pair of terms listed below:

1. Compounds that have low melting points and do not usually dissolve well in water are __?__. *(ionic compounds* or *covalent compounds)*

2. A(n) __?__ turns red litmus paper blue. *(acid* or *base)*

3. __?__ are composed of only carbon and hydrogen. *(Ionic compounds* or *Hydrocarbons)*

4. A biochemical composed of amino acids is a __?__. *(lipid* or *protein)*

5. A source of energy for living things can be found in __?__. *(nucleic acids* or *carbohydrates)*

UNDERSTANDING CONCEPTS

Multiple Choice

6. Which of the following describes lipids?
   a. used to store energy
   b. do not dissolve in water
   c. make up most of the cell membrane
   d. all of the above

7. An acid reacts to produce carbon dioxide when the acid is added to
   a. water.
   b. limestone.
   c. salt.
   d. sodium hydroxide.

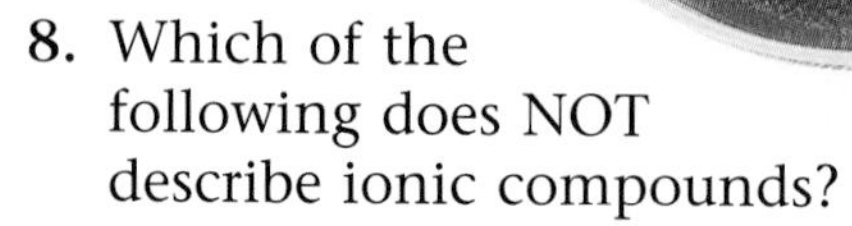

8. Which of the following does NOT describe ionic compounds?
   a. high melting point
   b. brittle
   c. do not conduct electric currents in water
   d. dissolve easily in water

9. An increase in the amount of hydrogen ions in solution __?__ the pH.
   a. raises
   b. lowers
   c. does not affect
   d. doubles

10. Which of the following compounds makes up the majority of cell membranes?
    a. lipids
    b. ionic compounds
    c. acids
    d. nucleic acids

11. The compounds that store information for building proteins are
    a. lipids.
    b. hydrocarbons.
    c. nucleic acids.
    d. carbohydrates.

Short Answer

12. What type of compound would you use to neutralize a solution of potassium hydroxide?

13. Explain why the reaction of an acid with a base is called *neutralization.*

14. What characteristic of carbon atoms helps to explain the wide variety of organic compounds?

15. Compare acids and bases based on the ion produced when each compound is dissolved in water.

Concept Mapping

16. Use the following terms to create a concept map: acid, base, salt, neutral, pH.

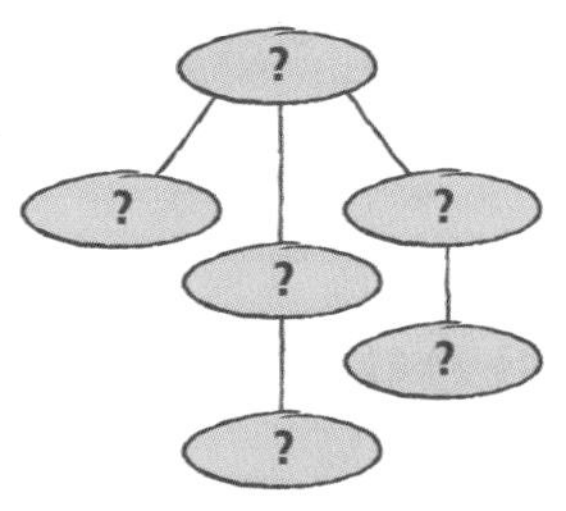

CRITICAL THINKING AND PROBLEM SOLVING

17. Fish give off the base ammonia, NH_3, as waste. How does the release of ammonia affect the pH of the water in the aquarium? What can be done to correct the problem?

18. Many insects, such as fire ants, inject formic acid, a weak acid, when they bite or sting. Describe the type of compound that should be used to treat the bite.

19. Organic compounds are also covalent compounds. What properties would you expect organic compounds to have as a result?

20. Farmers often can taste their soil to determine whether the soil has the correct acidity for their plants. How would taste help the farmer determine the acidity of the soil?

21. A diet that includes a high level of lipids is unhealthy. Why is a diet containing no lipids also unhealthy?

INTERPRETING GRAPHICS

Study the structural formulas below, and then answer the questions that follow.

a

b

c

d

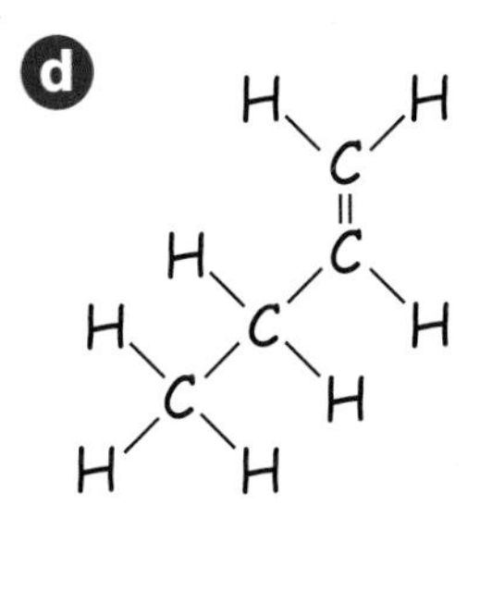

22. A saturated hydrocarbon is represented by which structural formula(s)?

23. An unsaturated hydrocarbon is represented by which structural formula(s)?

24. An aromatic hydrocarbon is represented by which structural formula(s)?

NOW What Do You Think?

Take a minute to review your answers to the ScienceLog questions on page 397. Have your answers changed? If necessary, revise your answers based on what you have learned since you began this chapter.

Unique Compounds

What makes you unique? Would you believe it's a complex pattern of information found on the deoxyribonucleic acid (DNA) in your cells? Well it is! And by analyzing how this information is arranged, scientists are finding clues about human ancestry.

Mummy Knows Best

If you compare the DNA from an older species and a more recent species, you can tell which traits were passed on. To consider the question of human evolution, scientists must use DNA from older humans—like mummies. Molecular archeologists study DNA from mummies in order to understand human evolution at a molecular level. Since well-preserved DNA fragments from mummies are scarce, you might be wondering why some ancient DNA fragments have been preserved better than others.

Neutralizing Acids

The condition of preserved DNA fragments depends on how the mummy was preserved. The tannic acid—commonly found in peat bogs—that is responsible for preserving mummies destroys DNA. But if there are limestone rocks nearby, the calcium carbonate from these rocks neutralizes the tannic acid, thereby preserving the DNA.

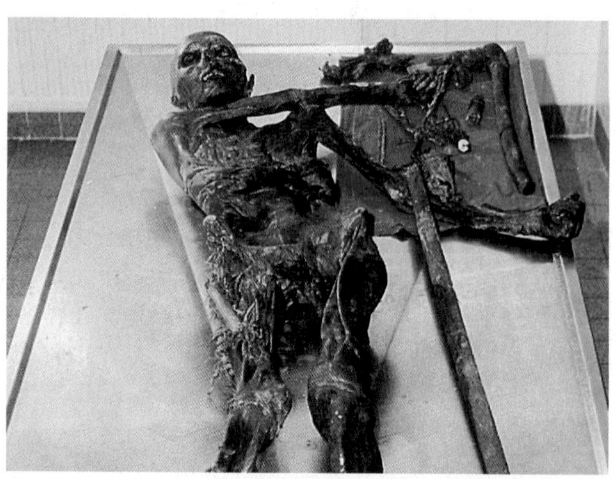

▲ *DNA from mummies like this one provides scientists with valuable information.*

Molecular Photocopying

When scientists find well-preserved DNA, they make copies of it by using a technique called polymerase chain reaction (PCR). PCR takes advantage of *polymerases* to generate copies of DNA fragments. Polymerases are found in all living things, and their job is to make strands of genetic material using existing strands as templates. That is why PCR is also called molecular photocopying. But researchers who use this technique risk contaminating the ancient DNA with their own DNA. If even one skin cell falls into the PCR mixture, the results are ruined.

Mysteries Solved?

PCR has been used to research ancient civilizations and peoples. For example, scientists found an 8,000-year-old human brain in Florida. This brain was preserved well enough for scientists to analyze the DNA and to conclude that today's Native Americans are not direct descendants of this group of people.

PCR has also been used to analyze the culture of ancient peoples. When archeologists tested the pigments used in 4,000-year-old paintings on rocks along the Pecos River, in Texas, they found DNA that was probably from bison. This was an unexpected discovery because there were no bison along the Pecos River when the paintings were made. The archeologists have concluded, therefore, that the artists must have tried very hard to find this specific pigment for their paint. This leads the archeologists to believe that the paintings must have had some spiritual meaning.

On Your Own

▶ Research ways PCR is being used to detect diseases and infections in humans and animals.

WEIRD SCIENCE

THE SECRETS OF SPIDER SILK

What is as strong as steel, more elastic than a rubber band, and able to stop a speeding bullet? Spider silk! Spiders make this silk to weave their delicate but deadly webs.

▲ *A golden orb-weaving spider on its web*

The Tangled Web We Weave

If you've seen a spider web, you've probably noticed that it resembles a bicycle wheel. The "spokes" of the web are made of a silk thread called *dragline silk.* The sticky, stretchy part of the web is called *capture silk* because that's what spiders use to capture their prey. Spider silk is made of proteins, and these proteins are made of blocks of amino acids.

There are 20 naturally occurring amino acids, but spider silk has only seven of them. Until recently, scientists knew what the silk is made of, but they didn't know how these amino acids were distributed throughout the protein chains.

Scientists used a technique called nuclear magnetic resonance (NMR) to see the structure of dragline silk. The silk fiber is made of two tough strands of alanine–rich protein embedded in a glycine–rich substance. If you look at this protein even closer, it looks like tangled spaghetti. Scientists believe that this tangled part makes the silk springy and a repeating sequence of five amino acids makes the protein stretchy.

Spinning Tails

Scientists think they have identified the piece of DNA needed to make spider silk. Synthetic silk can be made by copying a small part of this DNA and inserting it into the bacterium *Escherichia coli*. The bacteria read the gene and make liquid silk protein. Biologists at the University of Wyoming, in Laramie, have come up with a way to spin spider silk into threads by pushing the liquid protein through fine tubes.

What Do You Think?

► Scientists seem to think that there are many uses for synthetic spider silk. Make a list in your Science Log of as many things as you can think of that this material would be good for.

▲ ***Spiders use organs called spinnerets to spin their webs. This image of spinnerets was taken with a scanning electron microscope.***

TIMELINE

UNIT 6

Introduction to Astronomy

Have you ever looked up at the night sky and wondered, "What's out there?" For thousands of years, people have asked about and studied the lights in the night sky. In this unit, you will learn about our closest neighbors—the moon, the sun, and the planets that make up our solar system—as well as about galaxies and stars. This timeline includes some of the events leading to our current understanding of our solar system and our universe.

125 B.C.

Hipparchus, a Greek astronomer, divides the stars he can see into categories according to their brightness.

Around 1000

El Caracol, an early observatory, is constructed in Chichén Itzá, Mexico. The Maya use it to study the motion of the planets and stars.

1974

An X-ray source is discovered in the constellation Cygnus. It is believed to be a black hole.

1981

NASA successfully launches the first space shuttle, a manned space vehicle designed for reuse.

1610

Using the newly invented telescope, Galileo discovers the phases of Venus and four moons of Jupiter.

1781

William Herschel, an amateur astronomer, discovers the planet Uranus.

1801

An Italian astronomer named Giuseppe Piazzi discovers Ceres, the first known asteroid.

1961

Soviet cosmonaut Yuri Gagarin becomes the first person to orbit Earth.

1991

A total eclipse of the sun is visible from Hawaii and from Mexico.

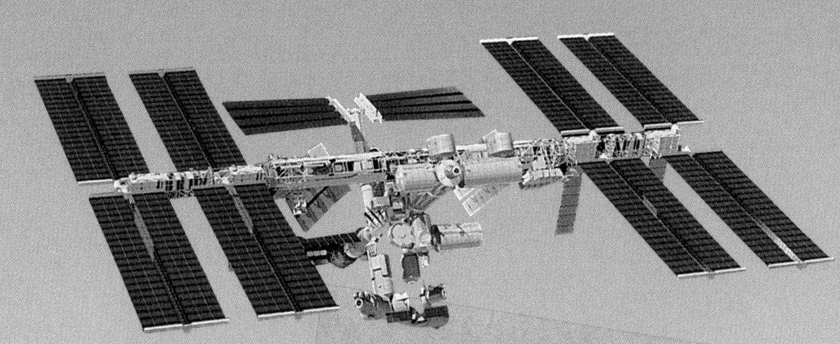

1994

Comet Shoemaker-Levy 9 crashes into Jupiter. The impact sites are visible to observers using telescopes and special imaging equipment.

1998

Russian cosmonauts and American astronauts begin construction of the International Space Station.

CHAPTER 17

Formation of the Solar System

Imagine . . .

Imagine you are in gym class. But this class has some really "far out" activities! You live inside a space station in orbit around the Earth. Unfortunately, you need gravity to get a good workout. The best the space station can offer is "artificial gravity" created by the station's slow rotation around its axis. This isn't really gravity, but it produces a force that is similar to the effect of the gravity you would feel on Earth's surface.

Think of whirling a pail of water attached to a string. If the bottom of the pail breaks, the water will fly off in a straight line. In the same way, the space-station floor pushes against your feet, keeping you inside the station. You feel this force as artificial gravity. The faster the station spins or the farther you are from the center, the heavier you feel.

Running on the station's track takes on a new meaning. To get a really good workout, you run in the direction of the station's rotation. The faster you run, the heavier you feel because your speed is added to that of the station. Of course, you can always run in the opposite direction to feel lighter!

Things get really interesting when you start lifting weights. If you lift a barbell off the floor, you will be lifting it closer to the center of the spinning station. The higher you lift the barbell, the lighter it gets because the distance to the center is getting shorter!

Today your class is heading toward the central axis of the station, where the artificial gravity drops to zero! On the way there, you'll stop to practice "flying" in the reduced gravity.

In this chapter you will explore gravity and pressure, the forces controlling the birth of our solar system. You will also explore energy production in the sun and the shape and structure of the Earth.

What Do You Think?

In your ScienceLog, try to answer the following questions based on what you already know:

1. What keeps the planets in their orbits?
2. Why does the sun shine?
3. Why is the Earth round?

Strange Gravity

More than 2,000 years ago, early Greek philosophers formed new ideas by simply making assumptions. They didn't see any value in testing their ideas with experiments. This changed, however, in the early 1600s when the Italian scientist Galileo Galilei started performing clever experiments to try to figure out how the world worked. Galileo's experiments helped later scientists understand how gravity works. In this activity you will experiment with how different objects behave under the pull of Earth's gravity.

Procedure

1. Select two pieces of identical **notebook paper.** Crumple one piece of paper into a ball.
2. Place the flat piece of paper on top of a **book** and the paper ball on top of the flat piece of paper.
3. Hold the book waist high, and then drop it to the floor.

Analysis

4. Which piece of paper reached the bottom first? Did either piece of paper fall slower than the book on which it rested? What does this mean about the way gravity pulls on objects of different mass? of equal mass? Record your observations in your ScienceLog.
5. Now hold the crumpled paper in one hand and the flat piece of paper in the other. Drop both pieces of paper at the same time. What do you observe? Does the result have anything to do with gravity? Why or why not? Record your observations in your ScienceLog, and share your ideas with your classmates.

Section 1

A Solar System Is Born

NEW TERMS

solar system
nebula
solar nebula
planetesimal
rotation
orbit
revolution
period of revolution
ellipse
astronomical unit

OBJECTIVES

- Explain the basic process of planet formation.
- Compare the inner planets with the outer planets.
- Describe the difference between rotation and revolution.
- Describe the shape of the orbits of the planets, and explain what keeps them in their orbits.

You probably know that Earth is not the only planet orbiting the sun. In fact, it has eight fellow travelers in its cosmic neighborhood. Together these nine planets and the sun are part of the solar system. The **solar system** is composed of the sun (a star) and the planets and other bodies that travel around the sun. (When talking about systems around other stars, the term *planetary system* is sometimes used.) But how did our solar system come to be?

The Solar Nebula

All the ingredients for building planets are found in the vast, seemingly empty regions between the stars. But these regions are not really empty. The "stuff between the stars" contains a mixture of gas and dust. The gas is mostly hydrogen and helium, while the dust is made up of tiny grains of elements such as carbon and iron. The dust and gas clump together in huge interstellar clouds called **nebulas** (or *nebulae*), which are so big that light takes many years to cross them! The nebulas are cold—only 10 Celsius degrees above absolute zero—and dark, as shown in **Figure 1.** Over time, light from nearby stars interacts with the dust and gas, and many new chemicals are formed. Chemicals, such as alcohol, and bits and pieces of complex molecules similar to those necessary for life are eventually formed deep within the nebulas. These clouds are the first ingredients of a new planetary system.

Figure 1 *The Horsehead nebula is a dark cloud of gas and dust as well as a possible site for future star formation.*

Gravity Pulls Matter Together Because these clouds of dust and gas consist of matter, they have mass. *Mass,* which is a measure of the amount of matter in an object, is affected by the force of gravity. But because the matter in a nebula is so spread out, the attraction between the dust and gas particles is very small. If a nebula's density were great enough, then the attraction between the particles might be strong enough to pull everything together into the center of the cloud. But even large clouds don't necessarily collapse toward the center because there is another effect, or force, that pushes in the opposite direction of gravity. You'll soon find out what that force is.

Pressure Pushes Matter Apart *Temperature* is a measure of how fast the particles in an object move around. If the gas molecules in a nebula move very slowly, the temperature is very low and the cloud is cold. If they move fast, the temperature is high and the cloud is warm. Because the cloud has a temperature that is above absolute zero, the gas molecules are moving. There is no particular structure in the cloud, and individual gas molecules can move in any direction. Sometimes they crash into each other. As shown in **Figure 2,** these collisions create a push, or *pressure,* away from the other gas particles. This pressure is what finally balances the gravity and keeps the cloud from collapsing.

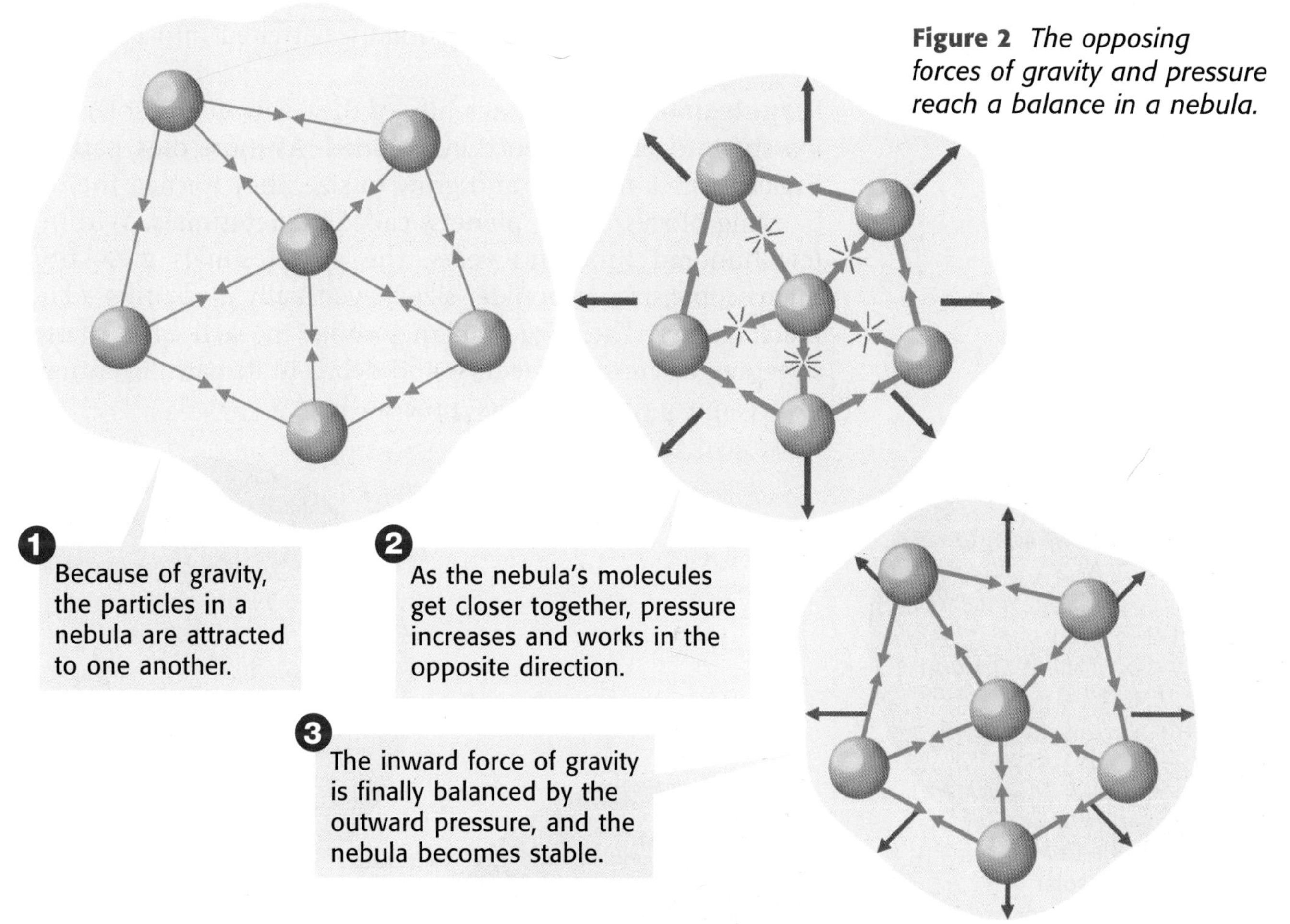

Figure 2 *The opposing forces of gravity and pressure reach a balance in a nebula.*

The Solar Nebula Forms Sometimes something happens to upset this balance. Two nebulas can crash into each other, for example, or a nearby star can explode, causing material from the star to crash into the cloud. These events compress small regions of the cloud so that gravity overcomes the pressure. Gravity then causes the cloud to collapse inward. At this point, the stage is set for the formation of a star and, as in the case of our sun, its planets. The **solar nebula** is the name of the nebula that formed into our own solar system.

Self-Check

What keeps a nebula from collapsing? *(See page 596 to check your answer.)*

From Planetesimals to Planets

Once the solar nebula started to collapse, things happened quickly, at least on a cosmic time scale. As the dark cloud collapsed, matter in the cloud got closer and closer together. This made the attraction between particles even stronger because gravity is stronger when things are closer together. The stronger attraction pulled the cloud together, and the gas and dust particles moved at a faster rate, increasing the temperature at the center of the cloud.

As things began to get crowded near the center of the solar nebula, particles of dust and gas in the cloud began to bump into other particles more often. Eventually much of the dust and gas began slowly rotating about the center of the cloud. The rotating solar nebula eventually flattened into a disk.

Planetesimals Sometimes bits of dust within the solar nebula stuck together when they collided. As more dust particles began to stick together and grow in size, they formed the tiny building blocks of the planets, called **planetesimals.** Within a few hundred thousand years, the planetesimals grew from microscopic sizes to boulder-sized, eventually measuring a kilometer across. The biggest planetesimal in each orbit started sweeping up most of the dust and debris in its path. Eventually it became a planet. This process is illustrated in **Figure 3.**

life science CONNECTION

Some of the complex molecules created in the cold, dark clouds that eventually form stars and planets contain amino acids. Amino acids are the building blocks of proteins and life. Scientists wonder if some of this material survives in the planetesimals that formed far from the sun—the comets. Could comets have brought life-forming molecules to Earth?

Figure 3 The Process of Solar System Formation

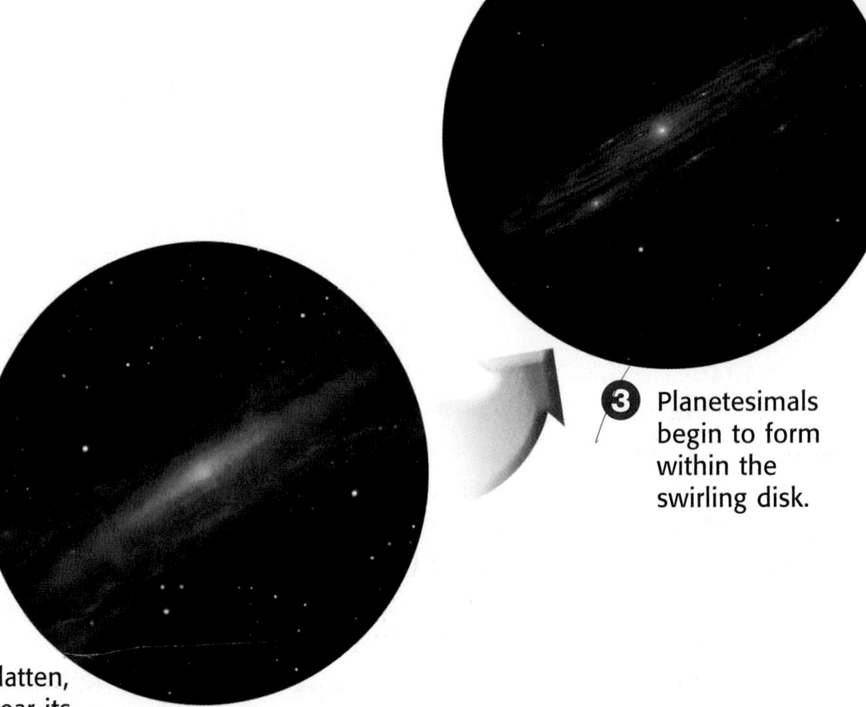

1. The young solar nebula begins to collapse due to gravity.
2. The solar nebula begins to rotate, flatten, and get warmer near its center.
3. Planetesimals begin to form within the swirling disk.

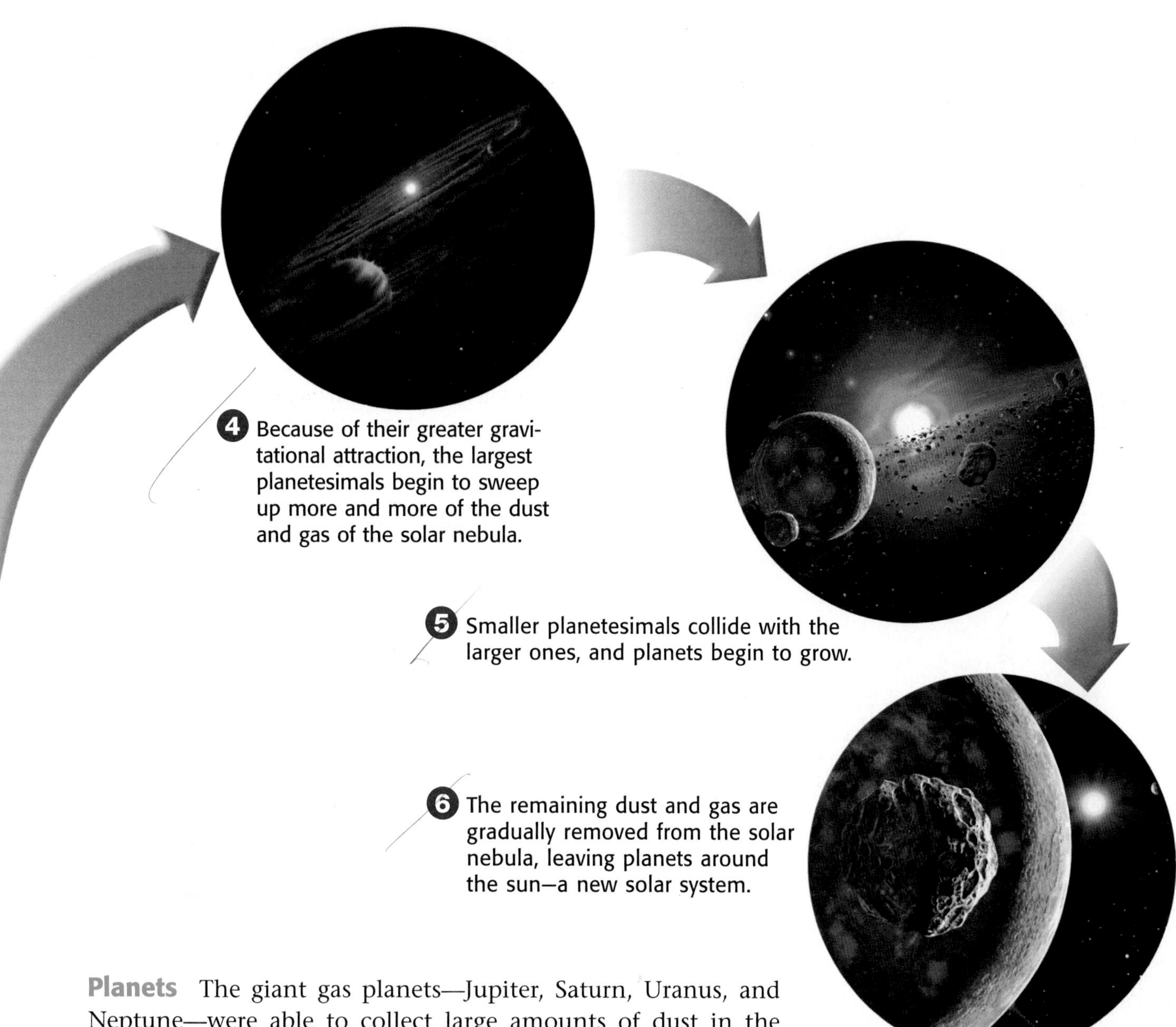

Planets The giant gas planets—Jupiter, Saturn, Uranus, and Neptune—were able to collect large amounts of dust in the cooler, outer solar nebula. Once they grew large enough, their gravity was strong enough to attract the nebula gases, hydrogen and helium. Far from the sun, it was cool enough for the giant planets to collect ices in addition to gases. Closer to the sun, it was too hot for the gases to remain, so these inner planets are made of rocky material.

Collisions with smaller planetesimals became more violent as pieces of debris became larger, leaving many craters on the surface of the rocky planets. We see evidence of this today particularly on Mercury and Mars as well as on our moon.

In the final steps of planet formation, the remaining planetesimals crashed down on the planets or got thrown to the outer edge of the solar nebula by the gravity of the larger planets—where they float in cold storage. Occasionally something, perhaps a passing star, sends them journeying toward the sun. If the planetesimal is icy, we see this visitor as a *comet.*

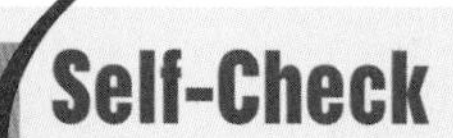

Self-Check

Why are the giant gas planets so large? *(See page 596 to check your answer.)*

Birth of a Star But what was happening at the middle of the solar nebula? The central part of the solar nebula contained so much mass and had become so hot—reaching temperatures of 10,000,000°C—that hydrogen fusion began. This created so much pressure at the center of the solar nebula that outward pressure balanced the inward force of gravity. At this point, the gas stopped collapsing.

As the sun was born, the remaining gas and dust of the nebula was blown into deep space by a strong solar wind, and the new solar system was complete. From the time the nebula first started to collapse, it took nearly 10 million years for the solar system to form.

Even though this was a fast process on a cosmic time scale, it is slow for us. So how do we know that our ideas of star and planet formation are correct? Powerful telescopes, such as the Hubble Space Telescope, are now able to show us some of the fine details inside distant nebulas. One such nebula is shown in **Figure 4.** For the first time, scientists can see disks of dust around stars that are in the process of forming.

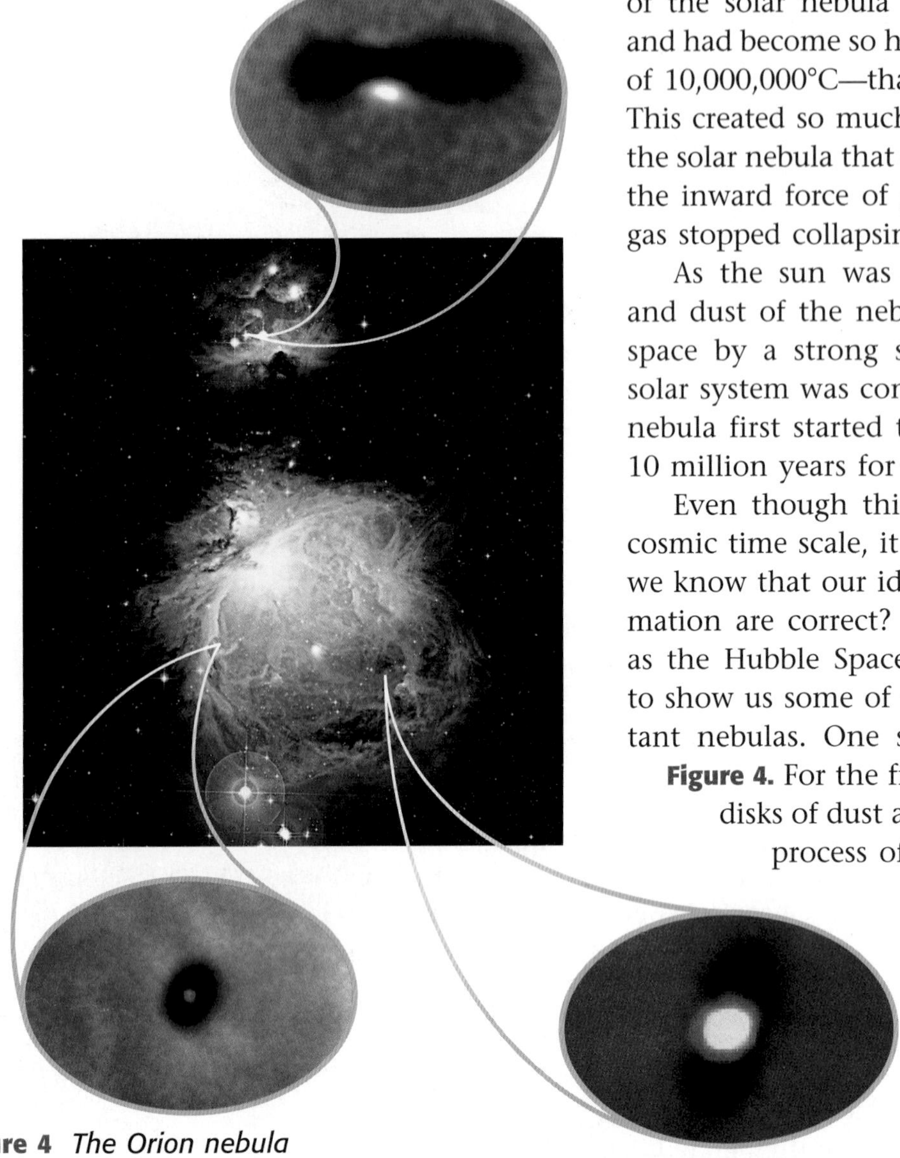

Figure 4 *The Orion nebula contains several disks of dust just beginning to form young stars.*

REVIEW

1. What two forces balance each other to keep a nebula of dust and gas from collapsing or flying apart?
2. Why does the composition of the giant gas planets differ from that of the rocky inner planets?
3. Explain why there is only one planet in each orbit around the sun.
4. **Making Inferences** Why do all the planets go around the sun in the same direction, and why do the planets all lie in a flat plane?

Planetary Motion

The solar system, which is now 4.6 billion years old, is not simply a collection of stationary planets and other bodies around the sun. Each one moves according to strict physical laws. The different ways these bodies move can have a variety of effects. The ways in which the Earth moves, for example, cause seasons and even day and night.

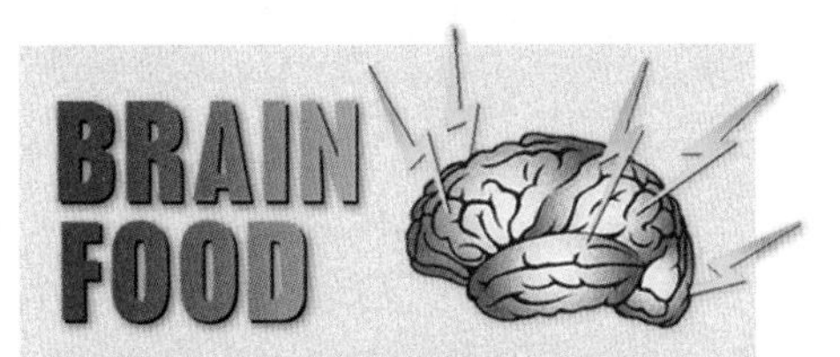

All planets *revolve* around the sun in the same direction. If you could look down on the solar system from above the sun's north pole, you would see all the planets revolving in a counterclockwise direction. Not all planets *rotate* in the same direction, however. Venus, and Pluto rotate backward compared with the rest of the planets, and Uranus rotates on its side.

Rotation and Revolution How does the motion of the Earth cause day and night? The answer has to do with the Earth's spinning on its axis, or **rotation.** Only one-half of the Earth faces the sun at any given time. As the Earth rotates, different parts of the Earth receive sunlight. The half facing the sun is light (day), and the half facing away from the sun is dark (night).

In addition to rotating on its axis, the Earth also travels around the sun in a path called an **orbit.** This motion around the sun along its orbit is called **revolution.** The other planets in our solar system also revolve around, or orbit, the sun. The amount of time it takes for a single trip around the sun is called a **period of revolution.** The period for the Earth to revolve around the sun is 365 days. Mercury orbits the sun in 88 days.

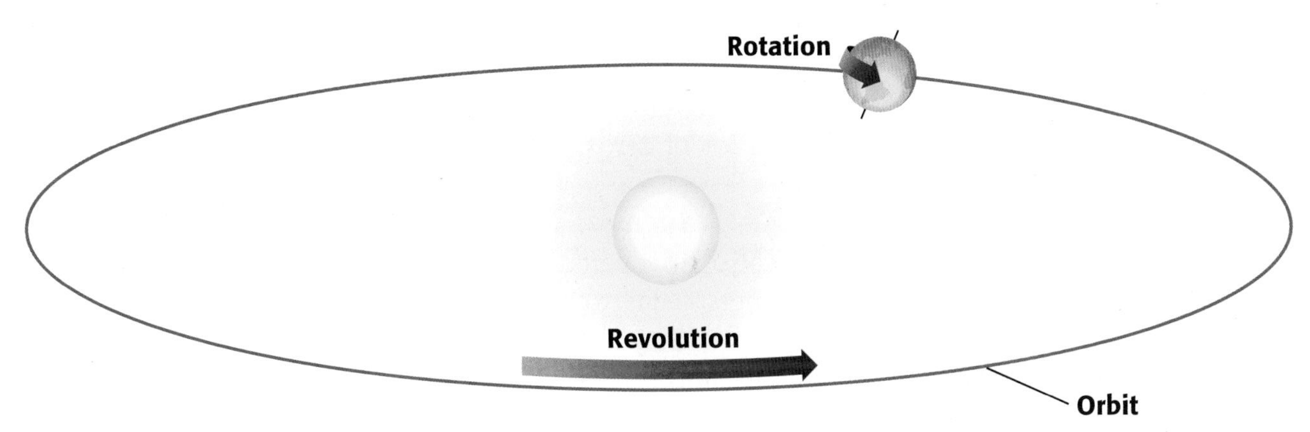

Figure 5 *A planet rotates on its own axis and revolves around the sun in a path called an orbit.*

Planetary Orbits But why do the planets continue to revolve around the sun? Does something hold them in their orbit? Why doesn't gravity pull the planets toward the sun? Or why don't they fly off into space? To answer these questions, we need to go back in time to look at the discoveries made by the scientists of the 1500s and 1600s.

Danish astronomer Tycho Brahe carefully observed the positions of the planets for over a quarter of a century. When he died, in 1601, a German astronomer named Johannes Kepler, inherited all of his records. Kepler set out to understand the motions of the planets and to make a simple description of the solar system.

*Quick*Lab

Staying in Focus

1. Take a short piece of **string,** and pin both ends to a **piece of paper** with two **thumbtacks.**
2. Keeping the string stretched tight at all times, use a **pencil** to trace out the path of an ellipse.
3. Change the distance between the thumbtacks to change the shape of the ellipse.
4. How does the position of the thumbtacks (foci) affect the ellipse?

Kepler's First Law of Motion Kepler's first discovery, or *first law of motion,* came from his careful study of the movement of the planet Mars. He discovered that the planet did not move in a circle around the sun, but in an elongated circle called an *ellipse.* An **ellipse** is a closed curve in which the sum of the distances from the edge of the curve to two points (called *foci*) inside the ellipse is always the same, as shown in **Figure 6.**

Figure 6 Parts of an Ellipse

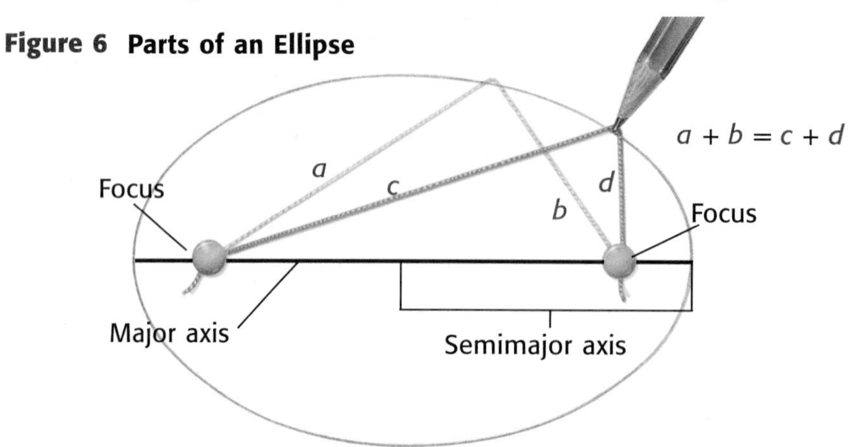

The maximum length of an ellipse is called its *major axis,* and half of this distance is the *semimajor axis,* which is usually used to give the size of an ellipse. The semimajor axis of Earth's orbit, for example, is 150 million kilometers. It represents the average distance between the Earth and the sun and is called one **astronomical unit,** or AU. Distances to other planets can be given in astronomical units rather than kilometers, saving a lot of zeros.

MATH BREAK

Kepler's Formula

Kepler's third law can be expressed with the formula

$$P^2 = a^3$$

where P is the period of revolution and a is the semimajor axis of an orbiting body. For example, Mars's period is 1.88 years, and its semimajor axis is 1.523 AU. Therefore, $1.88^2 = 1.523^3 = 3.53$. If astronomers know either the period or the distance, they can figure the other one out.

Kepler's Second Law Kepler also discovered that the planets seem to move faster when they are close to the sun and slower when they are farther away. To illustrate this, imagine that a planet is attached to the sun (which sits at one focus of the ellipse) by a string. The string will sweep out the same area in equal amounts of time. To keep the area of *A,* for example, equal to the area of *B,* the planet must move farther around its orbit in the same amount of time. This is Kepler's *second law of motion.*

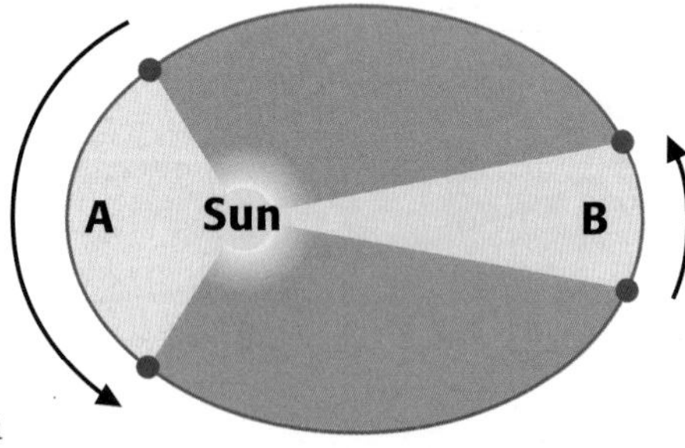

Kepler's Third Law Kepler's *third law of motion* compares the period of a planet's revolution with its semimajor axis. By doing some mathematical calculations, Kepler was able to demonstrate that by knowing a planet's period of revolution, the planet's distance from the sun can be calculated.

Newton's Law of Universal Gravitation

Kepler wondered if there was a reason that the planets closest to the sun move faster than the planets farther away, but he never got an answer. While his laws and the discoveries of other scientists formed the basis for understanding how the planets and the sun interact, it was Sir Isaac Newton (1642–1727) who finally put the puzzle together. He did this with his ideas about *gravity.* Newton didn't understand *why* gravity worked or what caused it. Even today, modern scientists do not fully understand gravity. But Newton was able to combine the work of earlier scientists to explain *how* the force of attraction between matter works.

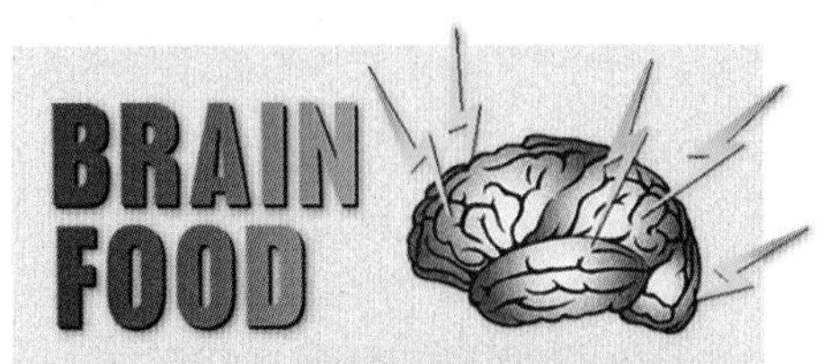

Legend has it that Newton got his ideas about gravity by watching an apple fall to the ground. Although it didn't hit him on the head, the apple may have inspired Newton to form his law of universal gravitation.

An Apple One Day Newton reasoned that small objects fall toward the Earth because they are attracted to each other by the force of gravity. But because the Earth has so much more mass than a small object, say an apple, only the object appears to move. Newton extended his idea to larger objects, realizing that the moon is also falling toward the Earth. But the moon is farther away from the Earth, so the effect is smaller.

Newton thus developed his *law of universal gravitation,* which states that the force of gravity depends on the product of the masses of the objects divided by the square of the distance between them. In other words, if two objects are moved twice as far apart, the gravitational attraction between them will decrease by a factor of $2 \times 2 = 4$, as shown in **Figure 7.** If the objects are moved 10 times as far apart, the gravitational attraction will decrease by a factor of $10 \times 10 = 100$.

Figure 7 *If two objects are moved twice as far apart, the gravitational attraction between them will be four times less.*

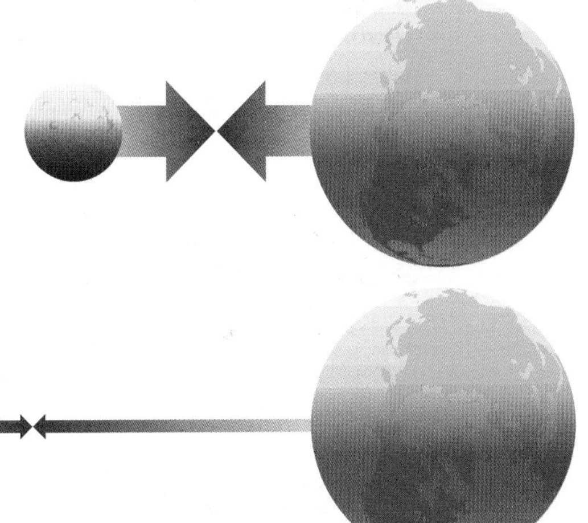

Space engineers that plan the paths of orbiting satellites must be able to calculate the height of the most appropriate orbit and the location of the satellite at each moment. To do this, they must take into account both Kepler's laws of motion and Newton's law of universal gravitation. Try this exercise: If the mass of the Earth were twice its actual mass, by how much would the gravity increase on a satellite in orbit around Earth? If the satellite were suddenly moved three times farther away, would Earth's gravitational pull on the satellite increase or decrease? By how much?

Explore

When the space shuttle is in orbit, we see the astronauts floating around as they work. Many people talk about this as a "zero-g" environment, meaning no gravity. Is this correct? Are shuttle astronauts affected by gravity? Do research to find out what happens when objects are in orbit around Earth.

Falling Down and Around How did Newton explain the orbit of the moon around the Earth? After all, according to gravity, the moon should come crashing into the Earth. And this is what the moon would do if it were not moving at a high velocity. In fact, if it were not for gravity, the moon would simply shoot off away from the Earth.

To understand this better, imagine twirling a ball on the end of a string. As long as you hold the string, the ball will orbit your hand. As soon as you let go of the string, the ball will fly off in a straight path. This same principle applies to the moon. But instead of a hand holding a string, gravity is keeping the moon from flying off in a straight path. **Figure 8** shows how this works. This same principle holds true for all bodies in orbit, including the Earth and other planets in our solar system.

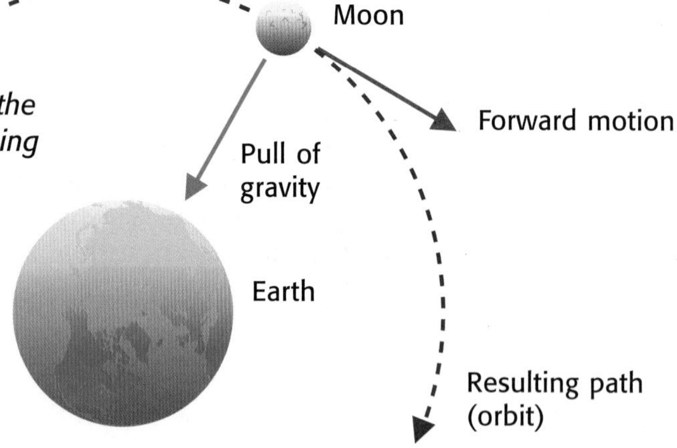

Figure 8 *Gravity is actually causing the moon to fall toward the Earth, changing what would be a straight-line path. The resulting path is a curved orbit.*

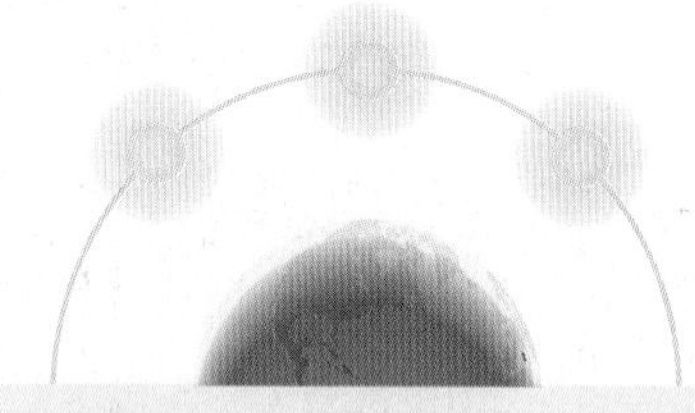

The Russians are planning to bounce sunlight from orbit to Earth. What's the scoop? Turn to page 449 to find out.

REVIEW

1. On what properties does the force of gravity between two objects depend?
2. Will a planet or comet be moving faster in its orbit when it is farther from or closer to the sun? Explain.
3. How does gravity keep a planet moving in an orbit around the sun?
4. **Applying Concepts** Suppose a certain planet had two moons, one of which was twice as far from the planet as the other. Which moon would complete one revolution of the planet first? Explain.

Section 2

The Sun: Our Very Own Star

NEW TERMS

corona
chromosphere
photosphere
convective zone
radiative zone
core
nuclear fusion
sunspot

OBJECTIVES

- Describe the basic structure and composition of the sun.
- Explain how the sun produces energy.
- Describe the surface activity of the sun, and name some of its effects on Earth.

There is nothing special about our sun, other than the fact that it is close enough to the Earth to give us light and warmth. Otherwise, the sun is similar to most of the other stars in our galaxy. It is basically a large ball of gas made mostly of hydrogen and helium held together by gravity. But let's take a closer look.

The Structure of the Sun

Although it may look like the sun has a solid surface, it does not. When we see a picture of the sun, we are really seeing through the sun's outer atmosphere, down to the point where the gas becomes so thick we cannot see through it anymore. As shown in **Figure 9,** the sun is composed of several layers.

Figure 9 Structure of the Sun and Its Atmosphere

a The **corona** forms the sun's outer atmosphere and can extend outward a distance equal to 10–12 times the diameter of the sun. The gases in the corona are so thin that it is visible only during a total solar eclipse. The corona can reach temperatures up to 2,000,000°C.

b The **chromosphere** is a thin region below the corona, only 3,000 km thick. Like the corona, the deep, red chromosphere is too faint to see unless there is a total solar eclipse. It ranges in temperature from 4,000°C to 50,000°C.

c The **photosphere** is where the gases get thick enough to see. The photosphere is what we know as the surface of the sun. It has a temperature of about 6,000°C and is only about 600 km thick.

d The **convective zone** is a region about 200,000 km thick where hot and cooler gases circulate in convection currents. Hot gases rise from the interior while cooler gases sink toward the interior. This is one way that the sun's energy reaches the surface.

e The **radiative zone** is a very dense region about 300,000 km thick. The atoms in this zone are so closely packed that light, which is absorbed and released by atoms along the way, takes millions of years to pass through this zone.

f The **core** is at the center of the sun. This is where the sun's energy is produced. The core has a radius of about 200,000 km and a temperature near 15,000,000°C.

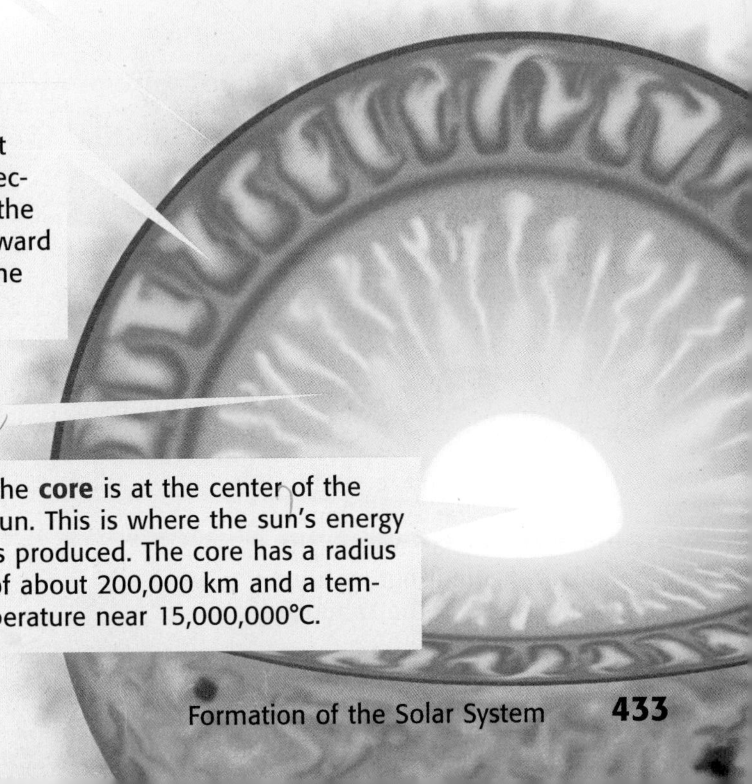

Energy Production in the Sun

BRAIN FOOD

Despite the large size of Jupiter and Saturn, the sun itself contains over 99 percent of all the matter in the solar system.

The sun has been shining on the Earth for about 4.6 billion years. How can it stay hot for so long? And what makes it shine? Over the years, several theories have been proposed to answer these questions. Because the sun is so bright and hot, many people thought that it was burning fuel to create the heat. But the amount of energy that is released during burning would not be enough to power the sun. If all of the matter in the sun were simply burned, the sun would last for only 10,000 years.

It eventually became clear that burning wouldn't last long enough to keep the sun shining. Scientists began to think that the sun was slowly shrinking due to gravity and that perhaps this would release enough energy to heat the sun. While the release of gravitational energy is more powerful than burning, it is still not enough to power the sun. If all of the sun's gravitational energy were released, the sun would last for only 45 million years. We know that dinosaurs roamed the Earth more than 65 million years ago, so this couldn't be the explanation. Something even more powerful was needed.

Some type of burning fuel was first thought to be the source of the sun's energy.

Figure 10 *Ideas about the source of the sun's energy have changed over time.*

A shrinking sun was another explanation for solar energy.

LabBook

The sun is difficult to study because it is far away from Earth. Just how far? You might be able to figure it out by turning to page 586 in the LabBook.

Nuclear Fusion At the beginning of the twentieth century, Albert Einstein demonstrated that matter and energy are interchangeable. Matter can be converted to energy according to his famous formula: $E = mc^2$, where E is energy, m is mass, and c is the speed of light. Because the speed of light is so large, even a small amount of matter can produce a large amount of energy. This idea paved the way for an understanding of a very powerful source of energy. **Nuclear fusion** is the process by which two or more nuclei with small masses (such as hydrogen) join together, or fuse, to form a larger, more massive nucleus (such as helium). During the process, energy is produced—a lot of it!

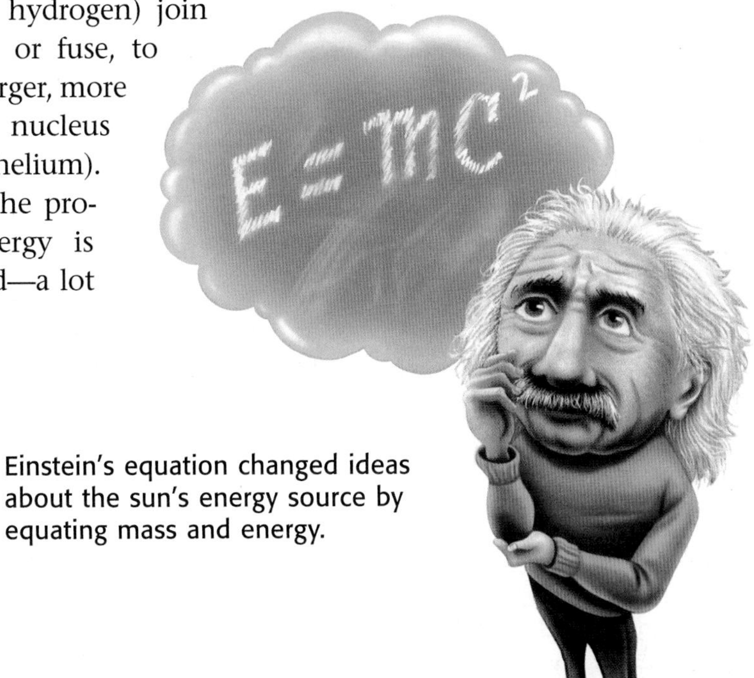

Einstein's equation changed ideas about the sun's energy source by equating mass and energy.

At the time Darwin introduced his theory of evolution, scientists thought that the sun was a few million years old at most. Some scientists argued that evolution—which takes place over billions of years—was therefore impossible because the sun could not have been shining that long. The nuclear fusion that fuels the sun, however, gives it a lifespan of about 10 billion years!

Atomic Review

Let's do a little review. *Atoms* are the smallest particles of matter that keep their chemical identity. A hydrogen atom and a helium atom are illustrated here. Notice that an atom consists of a *nucleus* surrounded by one or more *electrons,* which have a negative charge. A nucleus is made up of two types of smaller particles—a *proton,* with a positive charge, and a *neutron,* with no charge. The positively charged protons in the nucleus are balanced by an equal number of negatively charged electrons. The number of protons and electrons gives the atom its chemical identity.

Hydrogen

Proton(+)

Electron(–)

Neutron

Helium

Figure 11 *Like charges repel, just like similar poles on a pair of magnets.*

Under normal conditions, the nuclei of hydrogen atoms would never get close enough to combine because they are positively charged, and like charges repel each other, as shown in **Figure 11.** In the center of the sun, however, the temperature and pressure are very high because of the huge amount of matter within the core. This gives the hydrogen nuclei enough energy to overcome the repulsive force, allowing the conversion of hydrogen to helium. The conversion to helium occurs in three steps, as shown in **Figure 12.**

Figure 12 Fusion of Hydrogen in the Sun

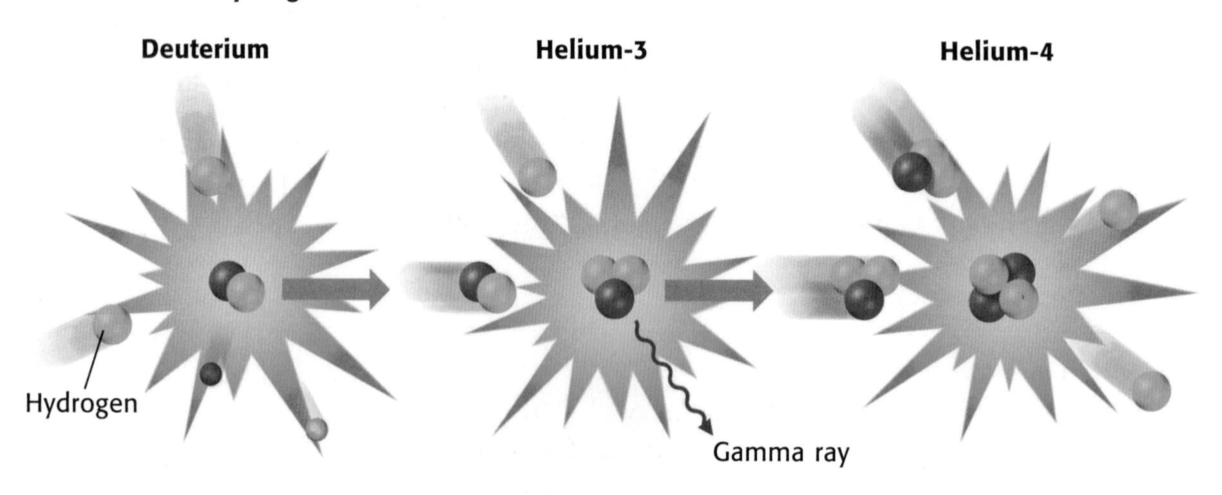

1. Two hydrogen nuclei (protons) collide. One proton emits particles and energy, then becomes a neutron. The proton and neutron combine to produce a heavy form of hydrogen called *deuterium.*
2. Deuterium combines with another hydrogen nucleus to form a variety of helium called helium-3. More energy is released, as well as gamma rays.
3. Two helium-3 atoms then combine to form ordinary helium-4, releasing more energy and a pair of hydrogen nuclei.

The energy produced in the core of the sun takes millions of years to reach the sun's surface. In the radiative zone just outside the core, the matter is so crowded that the light and energy keep getting blocked and sent off in different directions. Eventually the energy reaches the less crowded region of the convective zone, where hot gases carry it up to the photosphere relatively quickly. From there the energy leaves the sun as light, taking only 8.3 minutes to reach Earth.

The energy released during the nuclear fusion of 1 g of hydrogen is equal to about 100 tons of TNT! Each second, the sun converts about 5 million tons of matter into pure energy.

Activity on the Sun's Surface

The photosphere, or the visible surface of the sun, is a very dynamic place. As heat from the sun's interior reaches the surface, it causes the gas to boil and churn, a result of the rising and sinking of gases in the convective zone below. The surface, therefore, has a grainy appearance, though it can be seen only through special telescopes.

The circulation of the gases within the sun, in addition to the sun's own rotation, produces magnetic fields that reach out into space. But these magnetic fields also tend to slow down the activity in the convective zone. This causes areas on the photosphere above to be slightly cooler than surrounding areas. These cooler areas, which do not shine as brightly, show up as sunspots. **Sunspots** are cooler, dark spots on the sun, as shown in **Figure 13.**

The number of sunspots and their location on the sun change on a regular cycle. Records of the number of sunspots have been kept ever since the invention of the telescope. There may also be a connection between sunspot activity and Earth's long-term climate. In **Figure 14,** the cycle of sunspot numbers is shown, with the exception of the years 1645–1715, when sunspots were not observed. These years marked a much colder than average period for Europe and have been called the Little Ice Age.

The magnetic fields that cause sunspots also cause disturbances in the solar atmosphere. Giant storms on the surface of the sun, called *solar flares,* have temperatures of up to 5 million degrees Celsius. Solar flares send out huge streams of particles from the sun. These particles interact with the Earth's upper atmosphere, causing spectacular light shows called *auroras.*

Figure 13 *Sunspots mark cooler areas on the sun's surface. They are related to changes in the magnetic properties of the sun.*

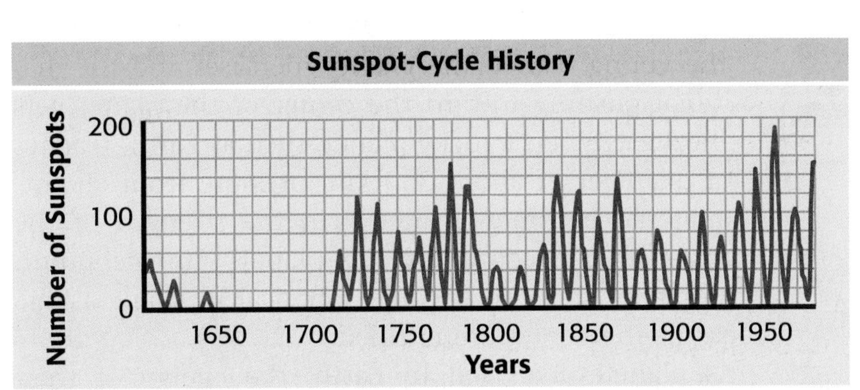

Figure 14 *This graph shows the number of sunspots that have occurred each year since Galileo's first observations, in 1610.*

REVIEW

1. According to modern understanding, what is the source of the sun's energy?
2. If nuclear fusion in the sun's core suddenly stopped today, would the sky be dark in the daytime tomorrow? Why?
3. **Interpreting Illustrations** In Figure 12, the nuclear fusion process ends up with one helium-4 nucleus and two free protons. What might happen to the two protons next?

Solar flares can interrupt radio communications on Earth. They can also affect satellites in orbit. The particles released during a solar flare can cause electronic circuits to fail. Scientists are trying to find ways to predict solar activity and give advanced warning of such events.

The Earth Takes Shape

Section 3

NEW TERMS

crust
mantle
core

OBJECTIVES

- Describe the shape and structure of the Earth.
- Explain how the Earth got its layered structure and how this process affects the appearance of Earth's surface.
- Explain the development of Earth's atmosphere and the influence of early life on the atmosphere.
- Describe how the Earth's oceans and continents were formed.

Investigating the early history of the Earth is not easy because no one was there to study it directly. Figuring out what the early Earth was like is similar to having a huge jigsaw puzzle with most of the pieces missing. Scientists develop ideas about what happened based on their knowledge of chemistry, biology, physics, geology, and other sciences. Astronomers are also gathering evidence from other stars where planets are forming to better understand how our own solar system formed. When new pieces of information come in, however, the pieces of the puzzle may have to be rearranged to make the new pieces fit.

The Solid Earth Takes Form

As scientists now understand it, the Earth was formed from the accumulation of planetesimals. The addition of planetesimals kept the Earth growing until it reached its present size. This happened within the first 10 million years of the collapse of the solar nebula.

When a young planet is still small, it can have an irregular shape. Bits can get broken off during collisions with planetesimals, and new material doesn't always collect on the surface evenly. As more matter builds up on the young planet, the gravity increases and the material pushing toward the center of the planet gets heavier. When a rocky planet, such as Earth, reaches a diameter of about 350 km, pressure from all this material becomes greater than the strength of the rocks in the center of the planet. At this point, the planet starts to become spherical in shape as the rocks in the center are crushed.

Figure 15 *The Earth has not always looked as inviting as it does today.*

As planetesimals fell to Earth, the energy of their motion was changed into heat. This energy made the Earth warmer. Once the Earth reached a certain size, the interior could not cool off as fast as its temperature rose, and the rocky material inside began to melt.

Self-Check

Why is the Earth spherical in shape, while most asteroids and comets are not? ***(See page 596 to check your answer.)***

The Earth and Its Layers Have you ever dropped pebbles into water or tried mixing oil and vinegar together for a salad? What happens? The heavier material (either solid or liquid) sinks, and the lighter material floats to the top. This is because of gravity. The material with a higher density experiences a stronger attraction and falls to the bottom. The same thing happened in the young Earth. As its rocks melted, the heavy elements, such as nickel and iron, sank to the center of the Earth, forming what we call the *core*. Lighter materials floated to the surface. This process is illustrated in **Figure 16.**

Figure 16 Earth's Materials Separate into Layers

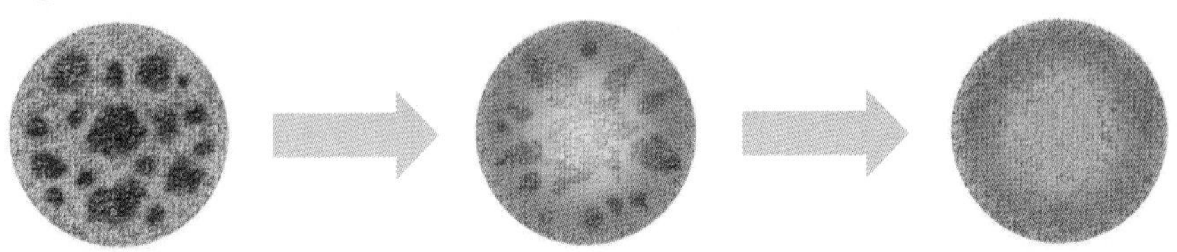

All materials in the early Earth are randomly mixed.

Rocks melt, and dense materials separate and sink.

Less-dense materials rise, and layers are formed.

The formation of the core started while the Earth was growing, and heat was added by the planetesimals and other material that fell to Earth. A second source of energy for heating the Earth was radioactive material, which was present in the solar nebula. Radioactive material radiates energy, and as it collected within the Earth, it also heated the planet.

*Quick*Lab

Mixing It Up

Have you ever mixed oil and water and watched what happened? Try this.

1. Pour 50 mL of **water** into a 150 mL **beaker.**
2. Add 50 mL of **cooking oil** to the water. Stir vigorously.
3. Let the mixture stand undisturbed for a few minutes.
4. What happens to the oil and water?
5. How does this relate to the interior of the early Earth?

The Earth's Interior The Earth is divided into three distinct layers according to the composition of its materials. These layers are shown in **Figure 17.** Geologists can map the interior of the Earth by measuring how sound waves pass through the planet during earthquakes and underground explosions.

The **crust** is the outermost layer of the Earth. It forms a thin skin over the entire planet, ranging from 5 km to 100 km thick.

The **mantle** lies below the crust, extending from about 100 km to about 2,900 km below the surface. The mantle contains denser rocks than the crust.

The **core,** at the center, contains the heaviest material (nickel and iron) and extends from the base of the mantle to the center of the Earth—almost 6,400 km below the surface.

Figure 17 *The interior of the Earth consists of three layers.*

The Atmosphere Evolves

Other than the presence of life, one of the biggest differences between the Earth of today and the Earth of 4.6 billion years ago is the character of its atmosphere. Earth's atmosphere today is composed of 21 percent oxygen, 78 percent nitrogen, and about 1 percent argon (with tiny amounts of many other gases). But it has not always been this way. Read on to discover how the Earth's atmosphere has changed through time.

chemistry CONNECTION

The Cassini Mission to Saturn (launched in October 1997) will study the chemistry of Saturn's moon Titan. Titan's atmosphere, like Earth's, is composed mostly of nitrogen, but it also contains many hydrogen-rich compounds. Scientists want to study how molecules essential to life may be formed in this atmosphere.

Earth's First Atmosphere Earth's early atmosphere was very different from the atmosphere of today. In the 1950s, laboratory experiments on the origins of life were based on the hypothesis that Earth's early atmosphere was largely made up of methane, ammonia, and water. And since the solar nebula was rich in hydrogen, many scientists thought that Earth's first atmosphere also contained a lot of hydrogen compounds.

New evidence is changing the way we think about Earth's first atmosphere. For one thing, 85 percent of the Earth's matter probably came from material similar to the *meteorites*—planetesimals made of rock. The other 15 percent probably came from the outer solar system in the form of *comets*—planetesimals made of ice.

During the final stages of formation, the Earth was hit many times by planetesimals, and the surface was very hot, even molten in places, as illustrated in **Figure 18.** The ground would have been venting large amounts of gas released from the heated minerals. The composition of meteorites tells us that much of that gas would have been water vapor and carbon dioxide. These two gases are also commonly released during volcanic eruptions when rocks turn to lava. Earth's first atmosphere was probably a steamy atmosphere made of water vapor and carbon dioxide.

Figure 18 *This is an artist's view of what Earth's surface may have looked like shortly after Earth's formation.*

Even though there was a lot of water vapor in the atmosphere, it probably didn't create our oceans right away. Planetesimal impacts may have helped release gases from the Earth, but they also helped to knock some of those gases back into space. Because planetesimals travel very fast, their impacts speed up the gas molecules in the atmosphere enough for them to overcome gravity and escape into space.

Heavier elements, such as iron, that were on the surface of the Earth also reacted chemically with water and gave off hydrogen—the lightest element. And because the early Earth was very warm, this hydrogen also had enough energy to escape.

Comets brought in a range of elements, such as carbon, hydrogen, oxygen, and nitrogen. They may also have brought water that eventually formed the oceans, as shown in **Figure 19.**

Figure 19 *Comets may have brought some of the water that formed the early Earth's oceans.*

Earth's Second Atmosphere After the Earth cooled off and the core formed, it was possible for the Earth's second atmosphere to take shape. This atmosphere was formed by gases contributed by both comets and volcanoes. Eruptions from Hawaiian volcanoes, like the one in **Figure 20,** show that a large amount of water vapor is produced, along with chlorine, nitrogen, sulfur, and large amounts of carbon dioxide. This carbon dioxide kept the planet much warmer than it is today.

Figure 20 *As this volcano in Hawaii shows, a large amount of gas is released during an eruption.*

environmental science CONNECTION

Because carbon dioxide is a very good *greenhouse gas*—one that traps heat—scientists have tried to estimate how much carbon dioxide the Earth must have had in its second atmosphere in order to keep it as warm as it was. For example, if all of the carbon dioxide that is now tied up in rocks and minerals of the ocean floor were released, it would make an atmosphere of carbon dioxide 60 times as thick as our present atmosphere.

Earth's Current Atmosphere How did this early atmosphere change to become the atmosphere we know today? It happened with the help of solar ultraviolet (UV) radiation, the very thing that we worry about now for its cancer-causing ability. Solar UV light is dangerous because it has a lot of energy and can break apart molecules in the air or in your skin. Today we are shielded from most of the sun's ultraviolet rays by Earth's protective ozone layer. But Earth's early atmosphere had no ozone, and many molecules were broken apart in the atmosphere. The pieces were later washed out into shallow seas and tide pools by rain. Eventually a rich supply of these pieces of molecules collected in protected areas, forming a rich organic solution that has sometimes been called a "primordial soup."

Although there was no ozone, a layer of water offers protection from the effects of ultraviolet radiation, and life is thought to have originated beneath the sea. The earliest forms of life may have been a primitive type of bacteria that lives near hydrothermal vents on the ocean floor. These primitive bacteria may have evolved into more-complex organisms such as blue-green algae, which were capable of photosynthesis and produced oxygen as a byproduct. These early life-forms are still around today, as shown in **Figure 21.**

Figure 21 *Fossilized algae (left) are among the earliest signs of life discovered. Today's stromatolites (right) are mats of bacteria thought to be similar to the first life on Earth.*

Oxygen didn't immediately build up in the atmosphere because it would have combined readily with minerals on the surface of the Earth such as iron. Evidence suggests that oxygen levels remained very low until about 2 billion years ago and then began to increase. The rate of this increase, however, is a subject of debate.

As plants began to cover the land, oxygen levels increased because plants produce oxygen during photosynthesis. Therefore, it was the emergence of life that completely changed our atmosphere into the one we have today.

Oceans and Continents

It is hard to say exactly when the first oceans appeared on Earth, but they probably formed early, as soon as the Earth was cool enough for rain to fall and remain on the surface. We know that Earth's secondary atmosphere had plenty of water vapor. After millions of years of rainfall, water began to cover the Earth, and by 4 billion years ago, a giant global ocean may have covered the planet. For the first few hundred million years of the Earth's history, there were probably no continents.

So how and when did the continents appear? Continental crust material is very light compared with material in the mantle. The composition of the granite and other rocks making up the continents tells geologists that the rocks of the crust have melted and cooled many times in the past. Each time the rocks melted, the heavier elements sank, leaving the lighter ones to rise to the surface. This process is illustrated in **Figure 22.**

After a while, some of the rocks were light enough that they no longer sank, and they began to pile up on the surface. This marked the beginning of the earliest continents. After gradually thickening, the continents slowly rose above the level of the seas. These scattered young continents didn't stay in the same place, however, because the slow convection in the mantle pushed the continents around. By about 3.5 billion years ago, less than 10 percent of the continents had been formed, but around 2.5 billion years ago, continents really started to grow. By 1.5 billion years ago, the upper mantle had cooled and become denser and heavier, so it was easier for the colder parts of it to sink. Then the real continental action, or *plate tectonics,* began.

Oxygen combines very quickly with other chemicals. Therefore, we would not expect to see oxygen in an atmosphere unless there were living organisms to keep making it. The discovery of oxygen in the atmosphere of another planet or moon would signal the presence of extraterrestrial life!

Figure 22 *The slow convective motion in the Earth's mantle causes rock to rise and sink.*

In each cycle, the rock partially melts as it rises, so that low-density materials can float to the top and solidify.

After cooling, some of the more dense material sinks back into Earth's interior. Lower density materials remain on the surface and are assembled into continents by plate motions

REVIEW

1. Why did the Earth separate into distinct layers?
2. How did the Earth's atmosphere change composition to become today's nitrogen and oxygen atmosphere?
3. Which are older, oceans or continents? Explain.
4. **Drawing Conclusions** If the Earth were not hot inside, would we have moving continents (plate tectonics)? Explain.

Chapter Highlights

SECTION 1

Vocabulary

solar system *(p. 424)*
nebula *(p. 424)*
solar nebula *(p. 425)*
planetesimal *(p. 426)*
rotation *(p. 429)*
orbit *(p. 429)*
revolution *(p. 429)*
period of revolution *(p. 429)*
ellipse *(p. 430)*
astronomical unit *(p. 430)*

Section Notes

- The solar system formed out of a vast cloud of cold gas and dust called a nebula.
- Gravity and pressure were balanced, keeping the cloud unchanging until something upset the balance. Then the nebula began to collapse.
- Collapse of the solar nebula caused heating in the center. As material crowded closer together, planetesimals began to form.
- The central mass of the nebula became the sun. Planets formed from the surrounding disk of material.
- It took about 10 million years for the solar system to form, and it is now 4.6 billion years old.
- The orbit of one body around another has the shape of an ellipse.
- Planets move faster in their orbits when they are closer to the sun.
- The square of the period of revolution of the planet is equal to the cube of its semimajor axis.
- Gravity depends on the masses of the interacting objects and the square of the distance between them.

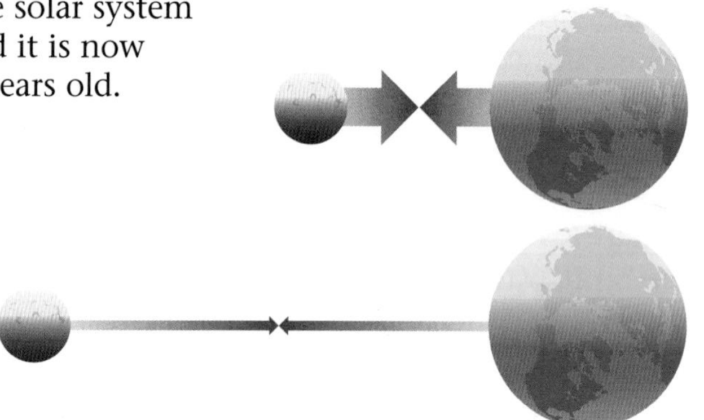

Skills Check

Math Concepts

SQUARES AND CUBES Let's take another look at Kepler's third law of motion. Expanding the formula $P^2 = a^3$ to $P \times P = a \times a \times a$ may be an easier way to consider the calculation. The period of Venus, for example, is 0.61 years, and its semimajor axis is 0.72 AU. Thus,

$$P^2 = a^3$$
$$P \times P = a \times a \times a$$
$$0.61 \times 0.61 = 0.72 \times 0.72 \times 0.72$$
$$0.37 = 0.37$$

Visual Understanding

LIKE AN ONION The sun is formed of six different layers of gas. From the inside out, the layers are the core, radiative zone, convective zone, photosphere, chromosphere, and corona. Look back at Figure 9 on page 433 to review the characteristics of each layer.

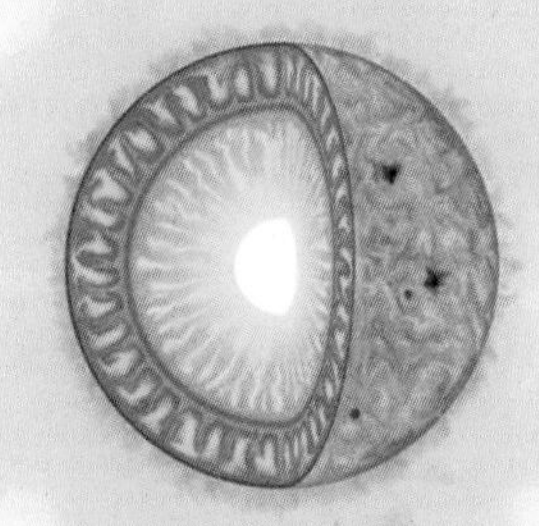

SECTION 2

Vocabulary

corona *(p. 433)*
chromosphere *(p. 433)*
photosphere *(p. 433)*
convective zone *(p. 433)*
radiative zone *(p. 433)*
core *(p. 433)*
nuclear fusion *(p. 435)*
sunspot *(p. 437)*

Section Notes

- The sun is a gaseous sphere made primarily of hydrogen and helium.
- The sun produces energy in its core by a process called nuclear fusion.
- Magnetic changes within the sun cause sunspots and solar flares.

Labs

How Far Is the Sun? *(p. 586)*

SECTION 3

Vocabulary

crust *(p. 439)*
mantle *(p. 439)*
core *(p. 439)*

Section Notes

- The Earth is divided into three main layers—crust, mantle, and core.
- Materials with different densities separarted because of melting inside Earth. Heavy elements sank to the center because of Earth's gravity.
- Earth's original atmosphere formed from the release of gases brought to Earth by meteorites and comets.
- Earth's second atmosphere arose from impacts by comets and volcanic eruptions. The composition was largely water and carbon dioxide.
- The presence of life dramatically changed Earth's atmosphere, adding free oxygen.
- Earth's oceans formed shortly after the Earth did, when it had cooled off enough for rain to fall.
- Continents were formed when lighter materials gathered on the surface and rose above sea level.

internet**connect**

GO TO: go.hrw.com

Visit the **HRW** Web site for a variety of learning tools related to this chapter. Just type in the keyword:

KEYWORD: HSTSOL

GO TO: www.scilinks.org

Visit the **National Science Teachers Association** on-line Web site for Internet resources related to this chapter. Just type in the ***sci*LINKS** number for more information about the topic:

TOPIC: The Planets — ***sci*LINKS NUMBER:** HSTP580
TOPIC: Kepler's Laws — ***sci*LINKS NUMBER:** HSTP585
TOPIC: The Sun — ***sci*LINKS NUMBER:** HSTP590
TOPIC: The Layers of the Earth — ***sci*LINKS NUMBER:** HSTP595
TOPIC: The Oceans — ***sci*LINKS NUMBER:** HSTP600

Chapter Review

USING VOCABULARY

For each pair of terms, explain the difference in their meanings.

1. rotation/revolution
2. ellipse/circle
3. solar system/solar nebula
4. planetesimal/planet
5. temperature/pressure
6. photosphere/corona

To complete the following sentences, choose the correct term from each pair of terms below.

7. It takes millions of years for light energy to travel through the sun's __?__. *(radiative zone* or *convective zone).*
8. __?__ of the Earth causes night and day. *(Rotation* or *Revolution)*
9. Convection in Earth's mantle causes __?__. *(plate tectonics* or *nuclear fusion)*

UNDERSTANDING CONCEPTS

Multiple Choice

10. Impacts in the early solar system
    a. brought new materials to the planets.
    b. released energy.
    c. dug craters.
    d. All of the above

11. Which type of planet will have a higher overall density?
    a. one that forms close to the sun
    b. one that forms far from the sun
12. Which process releases the most energy?
    a. nuclear fusion
    b. burning
    c. shrinking due to gravity
13. Which of the following planets has the shortest period of revolution?
    a. Pluto
    b. Earth
    c. Mercury
    d. Jupiter
14. Which gas in Earth's atmosphere tells us that there is life on Earth?
    a. hydrogen
    b. oxygen
    c. carbon dioxide
    d. nitrogen
15. Which layer of the Earth has the lowest density?
    a. the core
    b. the mantle
    c. the crust
16. What is the term for the speed of gas molecules?
    a. temperature
    b. pressure
    c. gravity
    d. force
17. Which of the following objects is least likely to have a spherical shape?
    a. a comet
    b. Venus
    c. the sun
    d. Jupiter

Short Answer

18. Why did the solar nebula begin to collapse to form the sun and planets if the forces of pressure and gravity were balanced?

19. How is the period of revolution related to the semimajor axis of an orbit? Draw an ellipse and label the semimajor axis.

20. How did our understanding of the sun's energy change over time?

Concept Mapping

21. Use the following terms to create a concept map: solar nebula, solar system, planetesimals, sun, photosphere, core, nuclear fusion, planets, Earth.

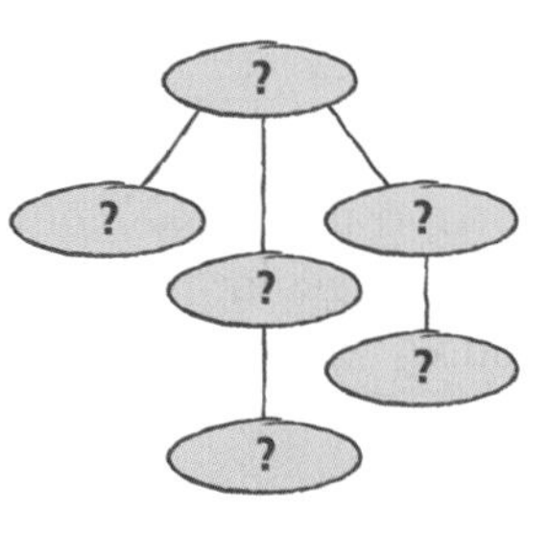

CRITICAL THINKING AND PROBLEM SOLVING

22. Explain why nuclear fusion works inside the sun but not inside Jupiter, which is also made mostly of hydrogen and helium.

23. Why is it less expensive to launch an interplanetary spacecraft from the international space station in Earth's orbit than from Earth itself?

24. Soon after the formation of the universe, there was only hydrogen and helium. Heavier elements, such as carbon, oxygen, silicon, and all the matter that makes up the heavier minerals and rocks in the solar system, were made inside an earlier generation of stars. Do you think the first generation of stars had any planets like Earth, Venus, Mercury, and Mars? Explain.

MATH IN SCIENCE

25. Suppose astronomers discover a new planet orbiting our sun. The orbit has a semimajor axis of 2.52 AU. What is the planet's period of revolution?

26. If the planet in the previous question is twice as massive as the Earth but is the same size, how much would a person who weighs 100 lb on Earth weigh on this planet?

INTERPRETING GRAPHICS

Examine the illustration below, and answer the questions that follow.

27. Do you think this is a rocky, inner planet or a gas giant?

28. Did this planet form close to the sun or far from the sun? Explain.

29. Does this planet have an atmosphere? Why or why not?

NOW What Do You Think?

Take a minute to review your answers to the ScienceLog questions on page 423. Have your answers changed? If necessary, revise your answers based on what you have learned since you began this chapter.

Science, Technology, and Society

Don't Look at the Sun!

You know you are not supposed to look at the sun, right? But how can we learn anything about the sun if we can't look at it? By using a solar telescope, of course! Where would you find one of these, you ask? Well, if you travel about 70 km southwest of Tucson, Arizona, you will arrive at Kitt Peak National Observatory, where you will find three of them. One telescope in particular has gone to great lengths to give astronomers extraordinary views of the sun!

Top Selection

In 1958, Kitt Peak was chosen from more than 150 mountain ranges to be the site for a national observatory. Located in the Sonoran Desert, Kitt Peak is a part of lands belonging to the Tohono O'odham nation. The McMath-Pierce Facility houses the three largest solar telescopes in the world. Astronomers come from around the globe to use these telescopes. The largest of the three, called the McMath-Pierce telescope, creates an image of the sun that is almost 1 m wide!

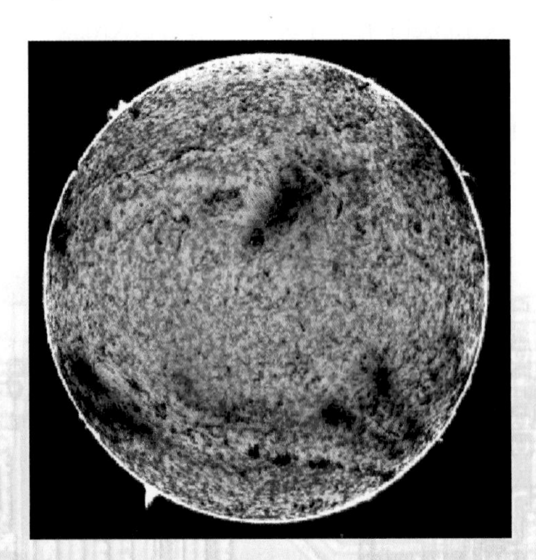

▲ *This is an image of the sun as viewed through the McMath-Pierce solar telescope.*

Too Hot to Handle

Have you ever caught a piece of paper on fire using only a magnifying glass and the rays from the sun? Sunlight that has been focused can produce a great amount of thermal energy—enough to start a fire. Now imagine a magnifying glass 1.6 m in diameter focusing the sun's rays. The resulting heat could melt metal. This is what would happen to a conventional telescope if it were pointed directly at the sun.

To avoid a meltdown, the McMath-Pierce solar telescope uses a mirror that produces a large image of the sun. This mirror directs the sun's rays down a diagonal shaft to another mirror 50 m underground. This mirror is adjustable to focus the sunlight. The sunlight is then directed to a third mirror, which directs the light to an observing room and instrument shaft.

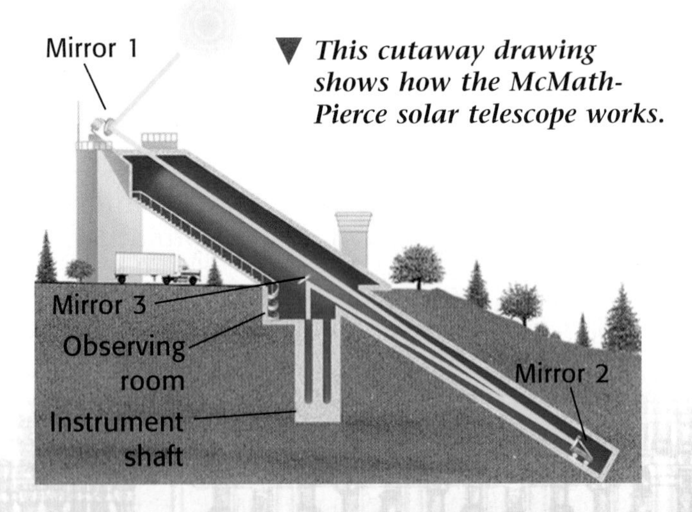

▼ *This cutaway drawing shows how the McMath-Pierce solar telescope works.*

Scope It Out

► Kitt Peak Observatory also has optical telescopes, which differ from solar telescopes. Do some research to find out how optical telescopes work and what the ones at Kitt Peak are used for.

SCIENTIFICDEBATE

Mirrors in Space

People who live in areas that do not get much sunshine are more prone to health problems such as depression and alcoholism. The people of Siberia, Russia, experience a shortage of sunshine during the winter, when the sun shines only 6 hours on certain days. Could there be a solution to this problem?

A Mirror From *Mir*

In February 1999, the crew of the space station *Mir* was scheduled to launch a large, umbrella-like mirror into orbit. The mirror was designed to reflect sunlight to Siberia. Once placed into orbit, however, problems arose and the crew was unable to unfold the mirror. Had things gone as planned, the beam of reflected sunlight was expected to be 5 to 10 times brighter than the light from the moon. If the first mirror had worked, this would have opened the door for Russia to build many more mirrors that are larger in diameter. These larger mirrors would have been launched into space to lengthen winter days, provide additional heat, and even reduce the amount of electricity used for lighting. The idea of placing mirrors in space, however, caused some serious concerns about the effects it could have.

▲ *The end of a winter day in Siberia*

Overcrowding

The first mirror was about 30 m in diameter. Because it was launched in the Low Earth Orbit (LEO), the light beam would be obstructed by the Earth's horizon as the mirror made its orbit. As a result, it would only reflect light on a single area for about 30 seconds. In order to shine light on Siberia on a large scale, hundreds of larger mirrors would have to be used. But using this many mirrors could result in collisions with satellites that share the LEO.

Damage to Ecosystems

It is very difficult to determine what effects extra daylight would have on Siberian ecosystems. Many plants and animals have cycles for various biological functions, such as feeding, sleeping, moving, and reproducing. Extra light and increased temperatures could adversely affect these cycles. Birds might migrate so late that they wouldn't survive the trip across the colder climates because food would be scarce. Plants might sprout too soon and freeze. Arctic ice might melt and cause flooding.

Light Pollution

Astronomers may also be affected by orbiting mirrors. Already astronomers must plan their viewing times to avoid the passing of bright planets and satellites. More sunlight directed toward the Earth would increase light pollution and could make seeing into space more difficult. A string of several hundred mirrors shining light toward the Earth would likely cause additional light pollution in certain locations as the mirrors passed overhead.

What's the Current Status?

▶ Find out more about the Russian project and where it stands now. If you had to decide whether to pursue this project, what would you decide? Why?

CHAPTER 18

A Family of Planets

Jupiter
142,984 km

Sun
1,391,960 km

Mars
6,794 km

Earth
12,756 km

Venus
12,104 km

Mercury
4,878 km

These images show the relative sizes and the diameters of the planets and the sun.

Imagine . . .

Imagine that it is 200 B.C. and you are an apprentice to a Greek astronomer. After years of observing the sky, he knows all of the constellations as well as you know the back of your hand. He shows you how the stars all move together—the whole sky spins slowly as the night goes on.

He also shows you that among the thousands of stars in the sky, some of the brighter ones slowly change their position relative to the other stars. These he names *planetai,* the Greek word for "wanderers." Building on the observations of the ancient Greeks, we now know that the *planetai* are actually planets, not wandering stars.

In this chapter you will learn about all of the bodies of the solar system—the planets and their moons, asteroids, comets, and meteoroids. You will also learn how space missions are making discoveries about our solar system that expand our knowledge of the solar neighborhood in which we live.

Interplanetary Distances		
Planet	**Distance from sun in AU**	**Scaled distance in yards**
Mercury	0.39	1
Venus	0.72	1.8
Earth	1.00	2.5
Mars	1.52	4
Jupiter	5.20	13
Saturn	9.54	24
Uranus	19.19	49
Neptune	30.06	76
Pluto	39.53	100

Investigate!

Measuring Space

The Earth is about 150 million kilometers from the sun. Another way of describing this distance is 1 AU. AU stands for *astronomical unit,* which is the average distance between the Earth and the sun. Using this unit makes it easier for scientists to describe the large distances between planets in the solar system. Do the following exercise to get a better idea of your solar neighborhood.

Procedure

1. Use **10 stakes** as markers to map the distances between the planets. Attach a **flag** to the top of each stake.
2. Plant a stake at the goal line of a **football field**—this stake represents the sun. Then use the table to plant stakes to represent the position of each planet relative to the sun.

Analysis

3. After you have placed all the "planets" at their relative distances, what do you notice about how the planets are spaced?
4. Besides the way they are spaced, in what other way are most of the outer planets different from the inner planets?

What Do You Think?

In your ScienceLog, try to answer the following questions based on what you already know:

1. What are the differences between planets, moons, asteroids, comets, and meteoroids?
2. How can surface features tell us about a planet's history?

Section 1

The Nine Planets

NEW TERMS

astronomical unit (AU)
terrestrial planets
prograde rotation
retrograde rotation
gas giants

OBJECTIVES

- List the names of the planets in the order they orbit the sun.
- Describe three ways in which the inner and outer planets are different from each other.

Ancient people knew about the existence of planets and could predict their motions. But it wasn't until the seventeenth century, when Galileo used the telescope to study planets and stars, that we began our first exploration of these alien worlds. Since the former Soviet Union launched *Sputnik*—the first artificial satellite—in 1957, over 150 successful missions have been launched to moons, planets, comets, and asteroids. **Figure 1** shows how far we have come since Galileo's time.

Galileo Galilei

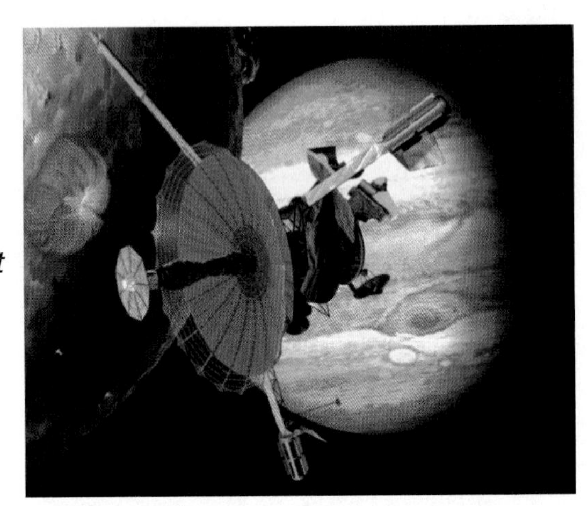

Figure 1 *Galileo Galilei (shown at left) discovered Jupiter's four largest moons using the newly invented telescope in 1610. The* Galileo *spacecraft (shown at right) arrived at Jupiter on December 7, 1995.*

Measuring Interplanetary Distances

As you have seen, one way scientists measure distances in space is by using the astronomical unit. The **astronomical unit (AU)** is the average distance between the Earth and the sun. Another way to measure distances in space is by the distance light travels in a given amount of time. Light travels at about 300,000 km per second in space. This means that in 1 second, light travels a distance of 300,000 km—or about the distance you would cover if you traveled around Earth 7.5 times.

In 1 minute, light travels nearly 18,000,000 km! This distance is also called 1 *light-minute*. For example, it takes light from the sun 8.3 minutes to reach Earth, so the distance from the Earth to the sun is 8.3 light-minutes. Distances within the solar system can be measured in light-minutes and light-hours, but the distances between stars are measured in light-years!

Figure 2 *One astronomical unit equals about 8.3 light-minutes.*

Sun
1 Light-minute
1 Astronomical unit
Earth

The Inner Planets

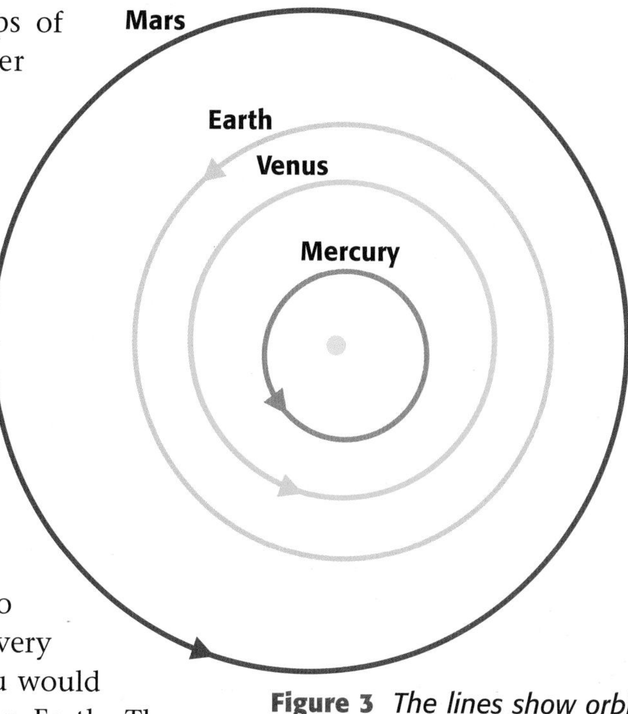

Figure 3 *The lines show orbits of the inner planets. The arrows indicate the direction of motion and the location of each planet on January 1, 2005.*

The solar system is divided into two groups of planets—the inner planets and the outer planets. Planets and their moons shine because they reflect sunlight. As you learned from the Investigate, the inner planets are more closely spaced than the outer planets. Other differences between the inner and outer planets are their sizes and the materials of which they are made. The inner planets are called **terrestrial planets** because they are like Earth—small, dense, and rocky. The outer planets, except for icy Pluto, are much larger and are made mostly of gases.

Mercury—Closest to the Sun If you were to visit the planet Mercury, you would find a very strange world. For one thing, on Mercury you would weigh only 38 percent of what you weigh on Earth. The weight you experience on Earth is due to *surface gravity,* which is less on less massive planets. Also, a day on Mercury would be quite different from a day on Earth. A day on Mercury is almost 59 Earth days long! This is because Mercury spins on its axis much more slowly than Earth does. The spin of an object in space is called *rotation.* The amount of time it takes for an object to rotate once is called its *period of rotation.*

Another curious thing about Mercury is that its year is only 88 Earth days long. As you know, a year is the time it takes for a planet to go around the sun once. The motion of a body as it *orbits* another body in space is called *revolution.* The time it takes for an object to revolve around the sun once is called its *period of revolution.* Every 88 Earth days, or 1.5 Mercurian days, Mercury completes one revolution around the sun.

Figure 4 *This image of Mercury was put together from a series of pictures taken by the* Mariner 10 *spacecraft on March 24, 1974, from a distance of 5,380,000 km.*

Mercury Statistics	
Distance from sun	**3.2** light-minutes
Period of rotation	**58** days, **16** hours
Period of revolution	**88** days
Diameter	**4,878** km
Density	**5.43** g/cm^3
Surface temperature	**−173** to **427**°C
Surface gravity	**38**% of Earth's

Figure 5 *This image of Venus was made from a series of images taken by* Mariner 10 *on February 5, 1974. The uppermost layer of clouds that blankets the planet consists of sulfuric acid.*

Venus Statistics	
Distance from sun	**6.0** light-minutes
Period of rotation	**243** days, (R)*
Period of revolution	**224** days, **17** hours
Diameter	**12,104** km
Density	**5.24** g/cm^3
Surface temperature	**464**°C
Surface gravity	**91**% of Earth's

*R = retrograde rotation

Venus—Earth's Twin? In many ways Venus is more similar to Earth than is any other planet—they have about the same size, mass, and density. But in other ways Venus is very different from Earth. Unlike on Earth, on Venus the sun rises in the west and sets in the east. This is because Venus rotates in the opposite direction that Earth rotates. Earth is said to have **prograde rotation,** because when viewed from above its north pole, Earth appears to spin in a *counterclockwise* direction. If a planet spins in a *clockwise* direction, it is said to have **retrograde rotation.** Venus also rotates much more slowly than Earth. As you can see in the table above, on Venus, a day is longer than an entire year!

At 90 times the pressure of Earth's atmosphere, the atmosphere of Venus is the densest of the terrestrial planets. It consists mostly of carbon dioxide, but it also contains some of the most corrosive acids known. The carbon dioxide in the atmosphere traps heat from sunlight in a process known as the *greenhouse effect.* This is why the surface temperature is so high. With an average temperature of 464°C, Venus has the hottest surface of any planet in the solar system. Because of the extreme pressure, heat, and acidity, even the hardiest of the former Soviet Union's Venera landers lasted little more than 2 hours on the surface!

Between 1990 and 1992, the *Magellan* spacecraft mapped the surface of Venus by using radar waves. The radar waves traveled through the clouds and bounced off the planet's surface. The radar image in **Figure 6** shows that, like Earth, Venus has an active surface.

Figure 6 *This false-color, three-dimensional image of the volcano Maat Mons, on the surface of Venus, was made with radar data gathered by the* Magellan *spacecraft. Bright areas indicate massive lava flows.*

Earth—An Oasis in Space As viewed from space, Earth is like a sparkling blue oasis suspended in a black sea. Constantly changing weather patterns create the swirls of clouds that blanket the blue and brown sphere we call home. Why did Earth have such good fortune while its two nearest neighbors, Venus and Mars, are unsuitable for life?

Earth is fortunate enough to have formed at just the right distance from the sun. The temperatures are warm enough to prevent most of its water from freezing but cool enough to keep it from boiling away. Liquid water was the key to the development of life on Earth. Water provides a means for much of the chemistry that living things depend on for survival.

You might think the only goal of space exploration is to get away from Earth, but NASA has a program to study Earth using satellites—just as we study other planets. The goal of this project, called the Earth Science Enterprise, is to study the Earth as a system and to determine the effects humans have in changing the global environment. For example, because humans can affect the global environment, does this mean that we can create conditions that would result in Earth's becoming more like Venus or Mars? By studying Earth from space, we hope to understand how different parts of the global system—such as weather, climate, and pollution—interact.

Figure 7 *Earth is the only planet we know of that supports life.*

Figure 8 *This image of Earth was taken on December 7, 1972, by the crew of the* Apollo 17 *spacecraft on the way to the moon. The photograph shows the African and Antarctic continents as well as the Atlantic and Indian Oceans.*

Earth Statistics	
Distance from sun	**8.3** light-minutes
Period of rotation	**23** hours, **56** minutes
Period of revolution	**365** days, **6** hours
Diameter	**12,756** km
Density	**5.52** g/cm^3
Surface temperature	**−13** to **37**°C
Surface gravity	**100%** of Earth's

Mars Statistics	
Distance from sun	**12.7** light-minutes
Period of rotation	**24** hours, **37** minutes
Period of revolution	**1** year, **322** days
Diameter	**6,794** km
Density	**3.93** g/cm^3
Surface temperature	**−123** to **37°C**
Surface gravity	**38%** of Earth's

Figure 9 *This* Viking *orbiter image of Mars was made from a series of images. The large circular feature in the center is the impact crater Schiaparelli, with a diameter of 450 km. The southern icecap is just visible at the lower right part of the image.*

Mars—The Red Planet Other than Earth, Mars is perhaps the most studied planet in the solar system. Images from ground-based telescopes and space probes indicate that the surface of Mars has a rich geologic history. Much of our knowledge of Mars has come from information gathered by the *Viking 1* and *Viking 2* spacecraft that landed on Mars in 1976 and from the *Pathfinder* spacecraft that landed on Mars in 1997.

Because of its thin atmosphere and its great distance from the sun, Mars is a cold planet. Mid-summer temperatures recorded by the *Pathfinder* lander ranged from –13°C to –77°C. The atmosphere of Mars is so thin that the air pressure at the planet's surface is roughly equal to the pressure 30 km above Earth's surface—about three times higher than most planes fly. The pressure of Mars's thin atmosphere is so low that any liquid water would quickly boil away. The only water you'll find on Mars is in the form of ice.

Even though liquid water cannot exist on Mars's surface today, there is strong evidence that it did exist there in the past! **Figure 10** shows a region on Mars with features that look like dry river beds on Earth. This means that in the past Mars might have been a warmer place with a thicker atmosphere. Where is the water now?

Figure 10 *This* Viking *orbiter image shows a drainage system on Mars formed by running water.*

Mars has two polar icecaps that contain both frozen water and frozen carbon dioxide, but this cannot account for all the water. Looking closely at the walls of some Martian craters, scientists have found that the debris surrounding the craters looks as if it were made by a mud flow rather than by the movement of dry material. Where does this suggest some of the "lost" Martian water went? Many scientists think it is frozen beneath the Martian soil.

Mars has a rich volcanic history. Unlike on Earth, where volcanoes occur in many places, Mars has only two large volcanic systems. The largest, the Tharsis region, stretches 8,000 km across the planet. The largest mountain in the solar system, Olympus Mons, is an extinct shield volcano similar to Mauna Kea, on the island of Hawaii.

How did a small planet like Mars get such enormous volcanoes? The answer may lie in the fact that Mars formed farther from the sun than Earth did. Mars not only is smaller and cooler than Earth but also has a slightly different chemical composition. Those factors may have prevented the Martian crust from moving around as Earth's crust has, so the volcanoes kept building up in the same spots. Images and data sent back by probes like the *Sojourner* rover, shown in **Figure 11,** are helping to explain Mars's mysterious past.

physical science CONNECTION

At sea level on Earth's surface, water boils at 100°C, but if you try to boil water on top of a high mountain, you will find that the boiling point is lower than 100°C. This is because the atmospheric pressure is less at high altitude. The atmospheric pressure on the surface of Mars is so low that liquid water can't exist at all!

REVIEW

1. What three characteristics do the inner planets have in common?
2. List three differences and three similarities between Venus and Earth.
3. **Analyzing Relationships** Mercury is closest to the sun, yet Venus has a higher surface temperature. Explain why this is so.

Figure 11 *The* Sojourner *rover, part of the Pathfinder mission, is shown here creeping up to a rock named Yogi to measure its composition. The dark panel on top of the rover collected the solar energy used to power its motor.*

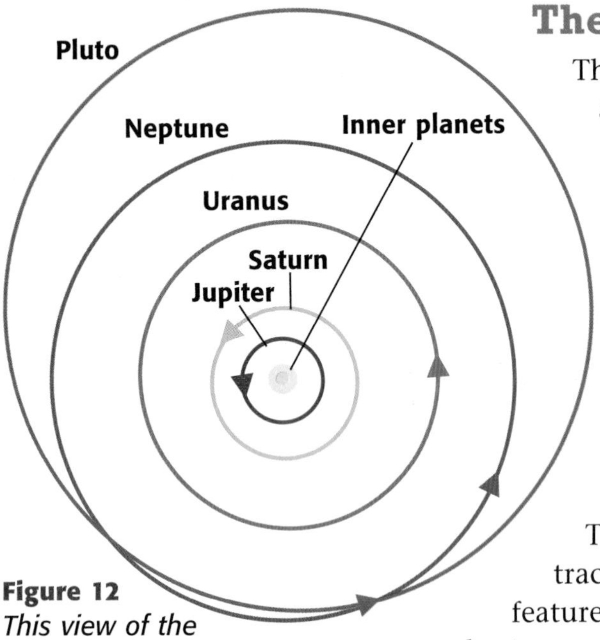

The Outer Planets

The outer planets differ significantly in composition and size from the inner planets. All of the outer planets, except for Pluto, are gas giants. **Gas giants** are very large planets that don't have any known solid surfaces—their atmospheres blend smoothly into the denser layers of their interiors, very deep beneath the outer layers.

Pluto
Neptune
Inner planets
Uranus
Saturn
Jupiter

Figure 12 *This view of the solar system shows the orbits and positions of the outer planets on January 1, 2005. The circle in the center represents the region of the inner planets.*

Jupiter—A Giant Among Giants Like the sun, Jupiter is made primarily of hydrogen and helium. The outer part of Jupiter's atmosphere is made of layered clouds of water, methane, and ammonia. The beautiful colors in **Figure 13** are probably due to trace amounts of organic compounds. Another striking feature of Jupiter is the Great Red Spot, which is a long-lasting storm system that has a diameter of about one and a half times that of Earth! At a depth of about 10,000 km, the pressure is high enough to change hydrogen gas into a liquid. Deeper still, the pressure changes the liquid hydrogen into a metallic liquid state.

Unlike most planets, Jupiter radiates much more heat into space than it receives from the sun. This is because heat is continuously transported from Jupiter's interior to its outer atmospheric layers, where it is radiated into space.

There have been five NASA missions to Jupiter—two Pioneer missions, two Voyager missions, and the recent Galileo mission. The *Voyager 1* and *Voyager 2* spacecraft sent back images that revealed a thin faint ring around the planet, as well as the first detailed images of its moons. The *Galileo* spacecraft reached Jupiter in 1995 and released a probe that plunged into Jupiter's atmosphere. The probe sent back data on the atmosphere's composition, temperature, and pressure. The mission found that the relative amounts of hydrogen and helium in Jupiter are very similar to those in the sun.

Figure 13 *This* Voyager 2 *image of Jupiter was taken at a distance of 28.4 million kilometers. Io, one of Jupiter's 16 known moons, can also be seen in this image.*

Jupiter Statistics	
Distance from sun	**43.3** light-minutes
Period of rotation	**9** hours, **50** minutes
Period of revolution	**11** years, **313** days
Diameter	**142,984** km
Density	**1.32** g/cm^3
Temperature	**−153°C**
Gravity	**236%** of Earth's

Saturn Statistics	
Distance from sun	**1.3** light-hours
Period of rotation	**10** hours, **30** minutes
Period of revolution	**29** years, **155** days
Diameter	**120,536** km
Density	**0.69** g/cm^3
Temperature	**−185**°C
Gravity	**92**% of Earth's

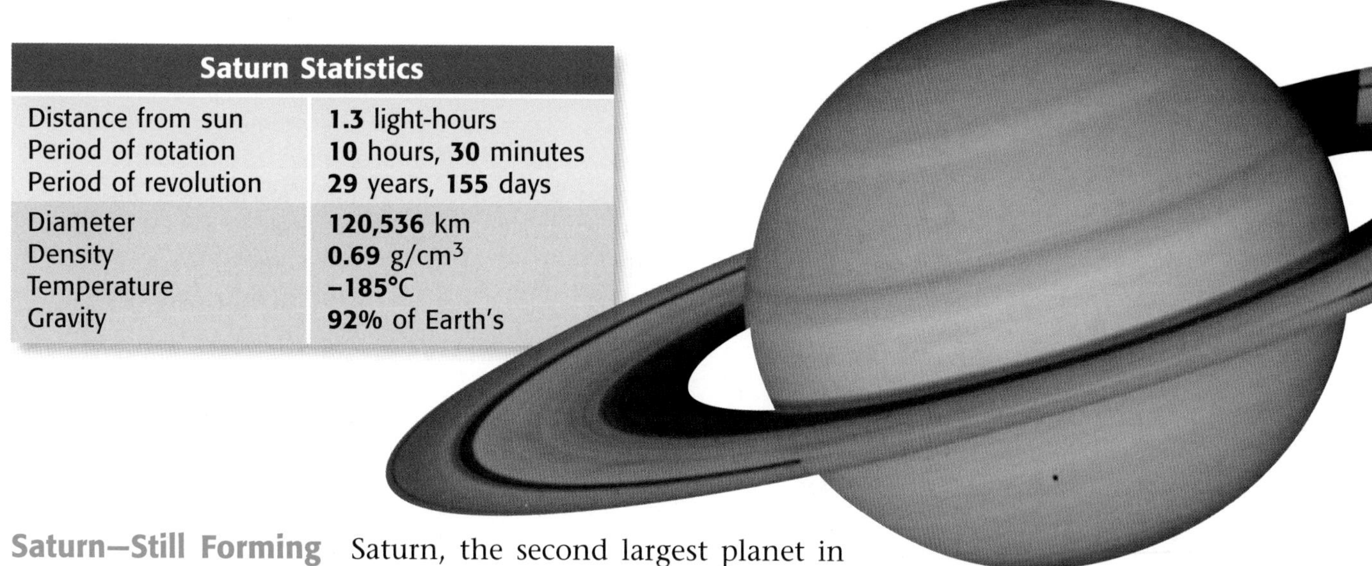

Figure 14 *This* Voyager 2 *image of Saturn was taken from 21 million kilometers away. The dot you see below the rings is the shadow of Tethys, one of Saturn's moons.*

Saturn—Still Forming Saturn, the second largest planet in the solar system, has roughly 764 times the volume of Earth and is 95 times more massive. Its overall composition, like Jupiter's, is mostly hydrogen and helium, with methane, ammonia, and ethane in the upper atmosphere. The colors in the atmosphere are not as brilliant as on Jupiter, and it is harder to see the features. This is because Saturn is colder, and a layer of white ammonia clouds blocks our view. Saturn's interior is probably very similar to that of Jupiter.

Like Jupiter, Saturn gives off a lot more heat than it receives from the sun. Scientists believe that this is because the helium, which is heavier than hydrogen, is raining out of the atmosphere and sinking to the core, releasing heat. In essence, Saturn is still forming!

Although all of the gas giants have rings, Saturn's rings are the largest. Saturn's rings start near the top of Saturn's atmosphere and extend out 136,000 km, yet they are only a few hundred meters thick. The rings consist of icy particles that range in size from a few centimeters to several meters across. **Figure 15** shows a close-up view of Saturn's rings.

More than 300,000 color images of Saturn, its moon Titan, its rings, and its other moons will be sent to Earth during the Cassini mission.

The most recent NASA mission to Saturn, called the Cassini mission, was launched in October 1997. The goals of the Cassini mission are to study Saturn's rings, to get close-up images of its moons, and to study its atmosphere.

Figure 15 *This false-color image of Saturn's rings was made from data gathered by* Voyager 2 *in 1981. The different colors show differences in the chemical composition of the rings.*

Figure 16 *This image of Uranus was taken by* Voyager 2 *at a distance of 9.1 million kilometers.*

Uranus Statistics	
Distance from sun	**2.7** light-hours
Period of rotation	**17** hours, **14** minutes (R)*
Period of revolution	**83** years, **274** days
Diameter	**51,118** km
Density	**1.32** g/cm^3
Temperature	**–214**°C
Gravity	**89%** of Earth's

*R = retrograde rotation

Uranus—A Giant Impact Uranus (YOOR uh nuhs) has about 63 times the volume of Earth and is nearly 15 times as massive. One especially unusual quality of Uranus is that it is tipped over on its side—the axis of rotation is tilted by almost 90° and lies almost in the plane of its orbit. **Figure 17** shows how far Uranus's axis is inclined. For part of a Uranus year, one pole points toward the sun while the other pole is in darkness. At the other end of Uranus's orbit the poles are reversed.

The moons and the thin rings of Uranus all lie in a disk that is in the same plane as the equator of Uranus. In essence, the orbits of Uranus's moons are all tilted out of the plane of the solar system. Scientists suggest that early in its history, Uranus got hit by a massive object that tipped the planet over.

Uranus, Neptune, and Pluto were not known to ancient people because they were too faint to see with the naked eye. (Actually, if you have good eyesight and you know where to look, you can see Uranus—just barely.) Uranus was discovered by the English amateur astronomer William Herschel in 1781. Viewed through a telescope, Uranus looks like a featureless blue-green disk. The atmosphere is mainly hydrogen and methane gas, which absorbs the red part of sunlight very strongly. Uranus and Neptune are much smaller than Jupiter and Saturn, and yet they have similar densities. This suggests that they have lower percentages of light elements and more water in their interiors.

Figure 17 *Uranus's axis of rotation is tilted so that it is nearly parallel to the plane of Uranus's orbit. In contrast, the axes of most other planets are closer to being perpendicular to the plane of their orbits.*

Imagine that it is the year 2120 and you are the pilot of an interplanetary spacecraft on your way to explore Pluto. In the middle of your journey, your navigation system malfunctions, giving you only one chance to land safely. You will not be able to make it to your original destination or back to Earth, so you must choose one of the other planets to land on. Your equipment includes two years' supply of food, water, and air. You will be stranded on the planet you choose until a rescue mission can be launched from Earth. Which planet will you choose to land on? How would your choice of this planet increase your chances of survival? Explain why you did not choose each of the other planets.

Neptune—The Blue World Irregularities in the orbit of Uranus suggested to early astronomers that there must be another planet beyond Uranus whose gravitational force causes Uranus to move off its predicted path. By using the predictions of the new planet's orbit, astronomers discovered the planet Neptune in 1846. Galileo saw Neptune in 1613 while observing Jupiter, but he failed to realize that it was a planet, so the discovery of Neptune did not occur for another 200 years!

The *Voyager 2* spacecraft sent back images that gave us much new information about the nature of Neptune's atmosphere. Although the composition of Neptune's atmosphere is nearly the same as that of Uranus's atmosphere, Neptune's atmosphere contains belts of clouds. At the time of *Voyager 2*'s visit, Neptune had a Great Dark Spot, similar to the Great Red Spot on Jupiter. And like the interiors of Jupiter and Saturn, Neptune's interior releases heat to its outer layers. This helps the warm gases rise and the cool gases sink, setting up the wind patterns in the atmosphere that create the belts of clouds. *Voyager 2* images also revealed that Neptune has a set of very narrow rings.

Figure 18 *This* Voyager 2 *image of Neptune, taken while the spacecraft was more than 7 million kilometers away, shows the Great Dark Spot as well as some bright cloud bands.*

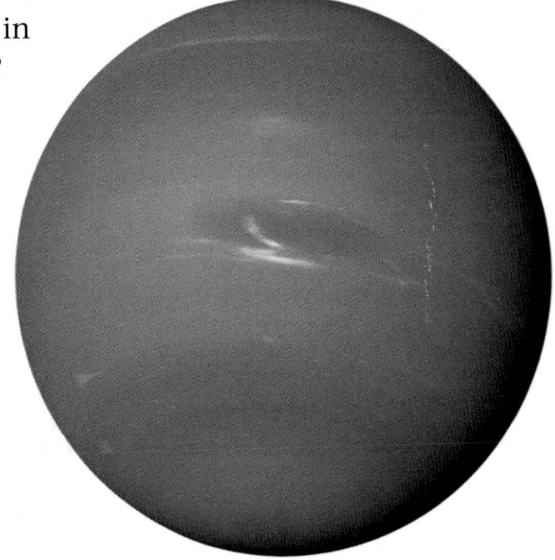

Neptune Statistics	
Distance from sun	**4.2** light-hours
Period of rotation	**16** hours, **7** minutes
Period of revolution	**163** years, **263** days
Diameter	**49,528** km
Density	**1.64** g/cm^3
Temperature	**–225**°C
Gravity	**112**% of Earth's

Figure 19 *This Hubble Space Telescope image is one of the clearest ever taken of Pluto and its moon, Charon.*

Pluto Statistics	
Distance from sun	**5.5** light-hours
Period of rotation	**6** days, **10** hours (R)*
Period of revolution	**248** years
Diameter	**2,390** km
Density	**2.05** g/cm^3
Surface temperature	**−236°**C
Surface gravity	**7%** of Earth's

*R = retrograde rotation

Pluto—A Double Planet? Pluto is the farthest planet from the sun. It is also the smallest planet—less than half the size of Mercury. Another reason Pluto is unusual is that its moon, Charon (KER uhn), is more than half its size! In fact, Charon is the largest satellite relative to its planet in the solar system.

From Earth, it is hard to separate the images of Pluto and Charon because they are so far away. **Figure 20** shows just how far away from the sun Pluto and Charon really are—from the surface of Pluto the sun appears to be only a very distant, bright star.

From Pluto's density, we know that it must be made of rock and ice. A very thin atmosphere of methane has been detected. While Pluto is covered by nitrogen ice, Charon is covered by water ice. Pluto is the only planet that has not been visited by a NASA mission, but plans are underway to finally visit this world and its moon in 2010.

Figure 20 *An artist's view of the sun and Charon from Pluto shows just how little light and heat Pluto receives from the sun.*

Is Pluto Really a Planet?
Pluto is neither a terrestrial planet nor a gas giant, so why do we call it a planet? Turn to page 480 to find out more.

REVIEW

1. How are the gas giants different from the terrestrial planets?
2. What is so unusual about Uranus's axis of rotation?
3. What conclusion can you draw about a planet's properties just by knowing how far it is from the sun?
4. **Applying Concepts** Why is the word *surface* not included in the statistics for the gas giants?

Section 2

Moons

NEW TERMS
satellite
phases
eclipse

OBJECTIVES
- Describe the current theory for the origin of Earth's moon.
- Describe what causes the phases of Earth's moon.
- Explain the difference between a solar eclipse and a lunar eclipse.

Satellites are natural or artificial bodies that revolve around larger bodies like planets. Except for Mercury and Venus, all of the planets have natural satellites called *moons.*

Luna: The Moon of Earth

We know that Earth's moon—also called *Luna*—has a different composition from the Earth because its density is much less than Earth's. This tells us that the moon has a lower percentage of heavy elements than the Earth has. The composition of lunar rocks brought back by Apollo astronauts suggests that the overall composition of the moon is similar to that of the Earth's mantle.

The Surface of the Moon The explorations of the moon's surface by the Apollo astronauts have given us insights about other planets and moons of the solar system. For example, the ages of lunar rocks brought back during the Apollo missions of the 1960s and 1970s were measured using radiometric-dating techniques. The oldest lunar rocks were found to be about 4.6 billion years old. Because these rocks have hardly changed since they formed, we know the solar system itself is about 4.6 billion years old.

Figure 21 Apollo 17 *astronaut Harrison Schmidt—the first geologist to walk on the moon—samples the lunar soil.*

In addition, we know that the surfaces of bodies that have no atmospheres preserve a record of almost all the impacts they have had with other objects. As shown in **Figure 22,** the moon's history is written on its face! Because we now know the age of the moon, we can count the number of impact craters on the moon and use that number to calculate the rate of cratering that has occurred since the birth of our solar system. By knowing the rate of cratering, scientists are able to use the number of craters on the surface of any body to estimate how old its surface is—without having to bring back rock samples!

Figure 22 *This image of the moon was taken by the* Galileo *spacecraft while on its way to Jupiter. The large dark areas are lava plains called* maria.

Moon Statistics	
Period of rotation	27 days, 8 hours
Period of revolution	27 days, 8 hours
Diameter	3,476 km
Density	3.34 g/cm^3
Surface temperature	−170 to 134°C
Surface gravity	17% of Earth's

physics CONNECTION

Did you know that the moon is falling? It's true. Because of gravity, every object in orbit around Earth is falling toward the planet. But the moon is also moving forward at the same time it is falling. In fact, it is the Earth's gravity that keeps the moon from flying off in a straight line. The combination of the moon's forward motion and its falling motion results in the moon's curved orbit around Earth.

Lunar Origins Before rock samples from the Apollo missions confirmed the composition of the moon, there were three popular explanations for the formation of the moon: (1) it was a separate body captured by Earth's gravity, (2) it formed at the same time and from the same materials as the Earth, and (3) the newly formed Earth was spinning so fast that a piece flew off and became the moon. Each idea had problems. If the moon were captured by Earth's gravity, it would have a completely different composition from that of Earth, which is not the case. On the other hand, if the moon formed at the same time as the Earth or as a spin off of the Earth, the moon would have exactly the same composition as Earth, which it doesn't.

The current theory is that a large, Mars-sized object collided with Earth while the Earth was still forming. The collision was so violent that part of the Earth's mantle was blasted into orbit around Earth. Once in orbit, part of the Earth's mantle material and debris from the impacting body eventually joined

Formation of the Moon

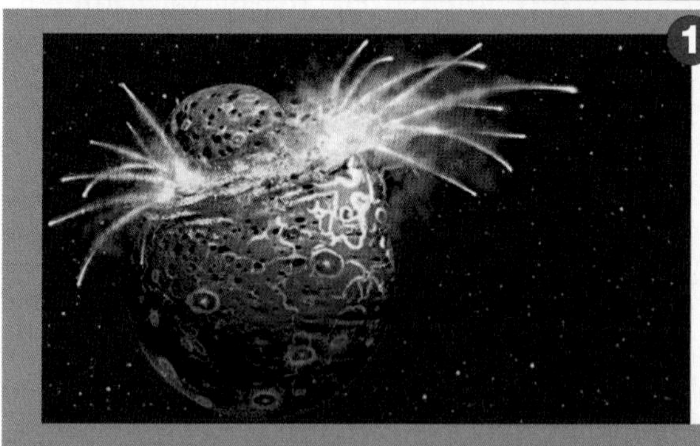

1 Impact
About 4.6 billion years ago, when Earth was still mostly molten, a large body collided with Earth. Scientists reason that the object must have been large enough to blast part of Earth's mantle into space, because the composition of the moon is similar to Earth's mantle.

2 Ejection
The resulting debris began to revolve around the Earth within a few hours of the impact. This debris consisted of mantle material from Earth and the impacting body as well as part of the iron core of the impacting body.

3 Formation
Soon after the giant impact, the clumps of material ejected into orbit around Earth began to join together to form the moon. Much later, as the moon cooled, additional impacts created deep basins and fractured the moon's surface. Lunar lava flowed from those cracks and flooded the basins to form the lunar maria we see today.

to form the moon. The moon would then be a combination of Earth's mantle and the impacting body. This theory is consistent with the composition of the lunar rocks brought back by the Apollo missions and is generally accepted today.

The moon's appearance changes every night, To find out how this occurs, turn to page 591 in your LabBook.

Phases of the Moon From Earth, one of the most noticeable aspects of the moon is its continually changing appearance. Within a month, its Earthward face changes from a fully lit circle to a thin crescent and then back to a circle. These different appearances of the moon result from its changing position with respect to the Earth and the sun. As the moon revolves around the Earth, the amount of sunlight on the side of the moon that faces the Earth changes. The different appearances of the moon due to its changing position are called **phases.** The phases of the moon are shown in **Figure 23.**

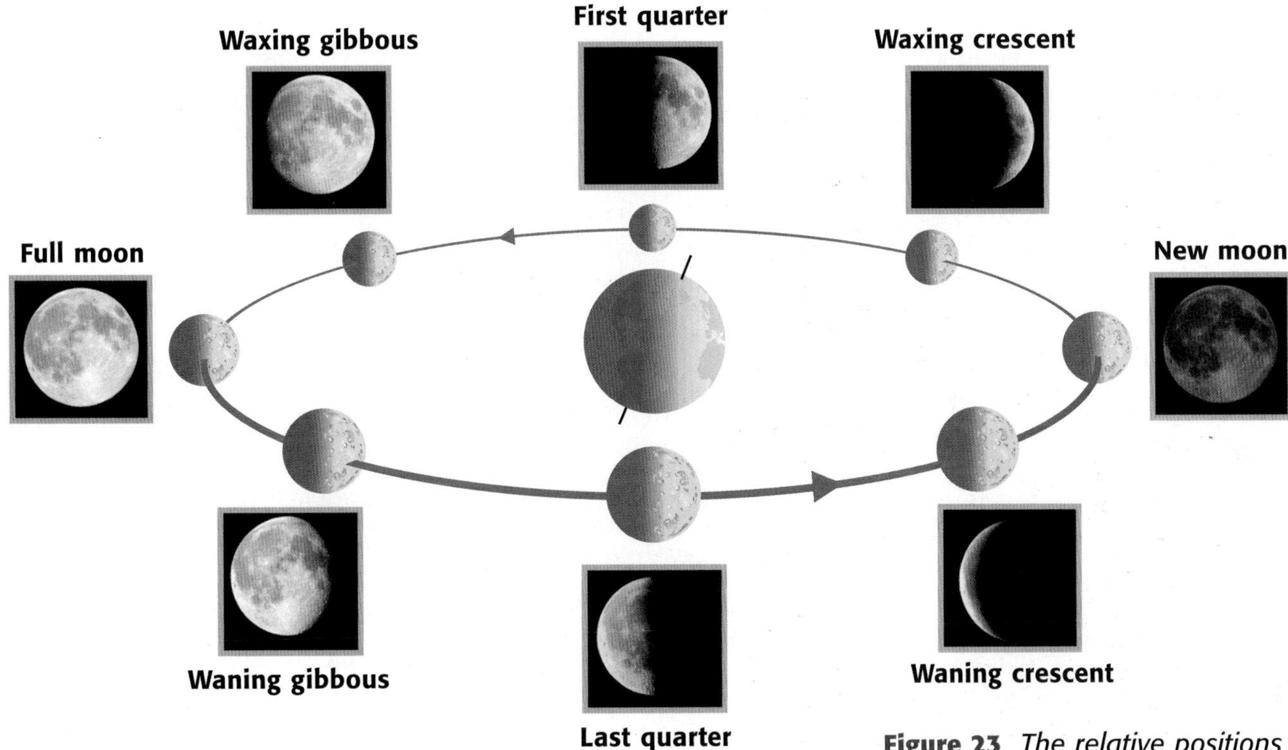

Figure 23 *The relative positions of the moon, sun, and Earth determine which phase the moon is in. The photo insets show how the moon looks from Earth at each phase.*

When the moon is waxing, it means that the sunlit fraction we can see from Earth is getting larger. When it is waning, the sunlit fraction is getting smaller. Notice in Figure 23 that even as the phases of the moon change, the total amount of sunlight the moon gets remains the same. Half the moon is always in sunlight, just as half the Earth is always in sunlight. But because the period of rotation for the moon is the same as its period of revolution, on Earth we always see the same side of the moon. If you lived on the far side of the moon, you would see the sun for half of each lunar day, but you would never see the Earth!

*Quick*Lab

Clever Insight

Pythagoras (540–510 B.C.) and Aristotle (384–322 B.C.) used observations of lunar eclipses and a little logic to figure out that Earth is a sphere. Can you?

1. Cut out a circle of **heavy white paper.** This will represent Earth.
2. Find **two spherical objects** and several other **objects** with different shapes.
3. Hold each object up in front of a **lamp** (representing the sun) so that its shadow falls on the white paper circle.
4. Rotate your objects in all directions, and record the shapes of the shadows they make.
5. Which objects always cast a curved shadow?

NEVER look directly at the sun! You can permanently damage your eyes.

Eclipses An **eclipse** occurs when the shadow of one celestial body falls on another. A *lunar eclipse* happens when the Earth comes between the sun and the moon, and the shadow of the Earth falls on the moon. A *solar eclipse* happens when the moon comes between the Earth and the sun, and the shadow of the moon falls on part of Earth.

By a remarkable coincidence, the moon in the sky appears to be nearly the same size as the sun. Even though the moon is much smaller than the sun, it appears to be the same size because it is so much closer. So during a solar eclipse, the disk of the moon almost always covers the disk of the sun. However, because the moon's orbit is not completely circular, sometimes the moon is farther away from the Earth, and a thin ring of sunlight shows around the outer edge of the moon. This type of solar eclipse is called an *annular eclipse*. **Figure 24** illustrates the position of the Earth and the moon during a solar eclipse.

Figure 24 *Because the shadow of the moon on Earth is small, a solar eclipse can be viewed from only a few locations.*

Figure 25 *This is an image of the sun's corona during the February 26, 1998, eclipse in the Caribbean. The solar corona is visible only when the entire disk of the sun is blocked by the moon.*

As you can see in **Figure 26,** the view during a lunar eclipse is also spectacular. During the hours of a total lunar eclipse, the moon often appears to turn a deep red color. Earth's atmosphere acts like a lens and bends some of the sunlight into the Earth's shadow, and the interaction of the sunlight with the molecules in the atmosphere filters out the blue light. With the blue part of the light removed, most of the remaining light that illuminates the moon is red.

Figure 26 *Because of atmospheric effects on Earth, the moon can have a reddish color during a lunar eclipse.*

Figure 27 *During a lunar eclipse, the moon passes within the Earth's shadow.*

From our discussion of the moon's phases, you might now be asking the question, "Why don't we see solar and lunar eclipses every month?" The answer is that the moon's orbit around the Earth is tilted—by about 5°—with respect to the orbit of the Earth around the sun. This tilt of the moon's orbit is enough to place the moon out of Earth's shadow for most full moons and the Earth out of the moon's shadow for most new moons.

REVIEW

1. What evidence suggests that Earth's moon formed from a giant impact?
2. Why do we always see the same side of the moon?
3. How are lunar eclipses different from solar eclipses?
4. **Analyzing Methods** How does knowing the age of a lunar rock help astronomers estimate the age of the surface of a planet like Mercury?

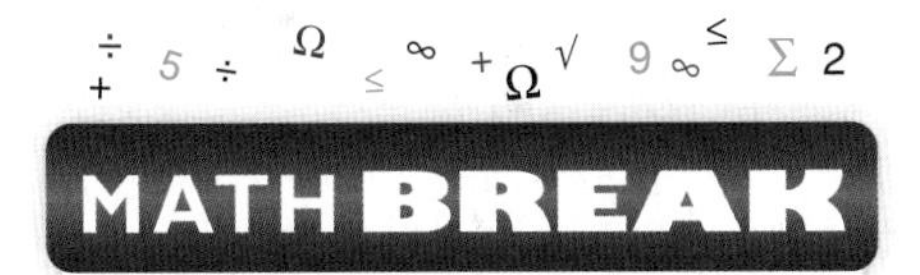

Orbits Within Orbits

The average distance between the Earth and the moon is about 384,400 km. As you have read, the average distance between the Earth and the sun is 1 AU, or about 150,000,000 km. Assume that the orbit of the Earth around the sun and the orbit of the moon around the Earth are perfectly circular. Using the distances given above, calculate the maximum and minimum distances between the moon and the sun.

The Moons of Other Planets

The moons of the other planets range in size from very small to as large as terrestrial planets. All of the gas giants have multiple moons, and scientists are still discovering new moons. Some moons have very elongated, or elliptical, orbits, and some even revolve around their planet backward! Many of the very small moons may be captured asteroids. As we are learning from recent space missions, moons can be some of the most bizarre and interesting places in the solar system!

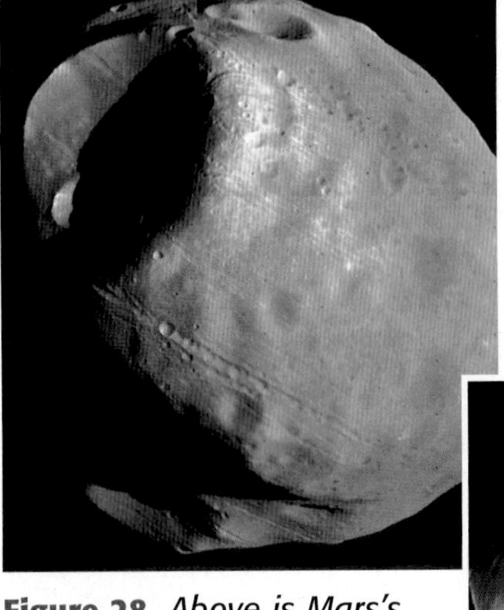

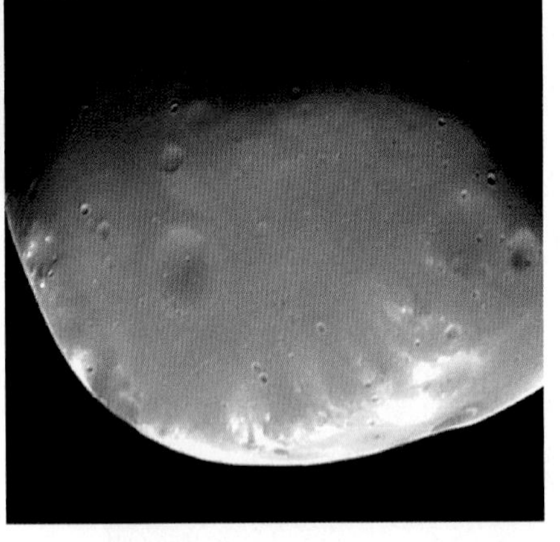

Figure 28 *Above is Mars's largest moon, Phobos, which is 28 km long. At right is the smaller moon, Deimos, which is 16 km long.*

The Moons of Mars Mars's two moons, Phobos and Deimos, are both small satellites that have irregular shapes. The two moons have very dark surfaces that reflect even less light than asphalt does. The surface materials are very similar to those found in asteroids, and scientists speculate that these two moons are probably captured asteroids.

Could you possibly live on Io? To find out how a person could learn to adapt to this harsh environment, turn to page 481.

The Moons of Jupiter Jupiter has a total of 17 known moons. The four largest—Ganymede, Callisto, Io, and Europa—were discovered in 1610 by Galileo and are known as the Galilean satellites. The largest moon, Ganymede, is even larger than the planet Mercury! The Galilean satellites are accompanied by at least 12 smaller satellites. These satellites are probably captured asteroids.

Moving outward from Jupiter, the first Galilean satellite is Io (IE oh), a truly bizarre world. Io is caught in a gravitational tug of war between Jupiter and Io's nearest neighbor, the moon Europa. This constant tugging stretches Io a little, causing it to heat up. Because of this, Io is the most volcanically active body in the solar system!

Recent pictures of the moon Europa support the idea that liquid water may lie beneath the moon's icy surface. This has many scientists wondering if life could have evolved in the subterranean oceans of Europa.

Figure 29 *At left is a* Galileo *image of Jupiter's innermost moon, Io. At right is a* Galileo *image of Jupiter's fourth largest moon, Europa.*

The Moons of Saturn Saturn has a total of 22 known moons. Most of these moons are small bodies made mostly of water ice with some rocky material. The largest satellite, Titan, was discovered in 1655 by Christiaan Huygens. In 1980, the *Voyager 1* spacecraft flew past Titan and discovered a hazy orange atmosphere, similar to what Earth's atmosphere may have been like before life began to evolve. In 1997, NASA launched the *Cassini* spacecraft to study Saturn and its moons, including Titan. The Cassini mission includes a probe, called the *Huygens* probe, that will parachute through Titan's thick atmosphere and land on its frozen surface. By studying the "primordial soup" of hydrocarbons on Titan, scientists hope to answer some of the questions about how life began on Earth.

Figure 30 *Titan is one of only two moons that have a thick atmosphere—in fact, its atmosphere is thicker than Earth's! This hazy orange atmosphere is made of nitrogen plus several other gases, such as methane.*

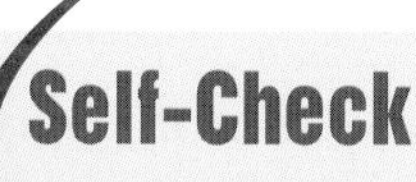

Self-Check

What is one major difference between Titan and the early Earth that would suggest that there probably isn't life on Titan right now? *(See page 596 to check your answer.)*

The Moons of Uranus Uranus has 21 moons, three of which were just discovered by ground-based telescopes during the summer of 1999. Like the moons of Saturn, the four largest moons are made of ice and rock and are heavily cratered. The little moon Miranda, shown in **Figure 31,** has some of the most unusual features in the solar system. Miranda's surface includes smooth, cratered plains as well as regions with grooves and cliffs up to 20 km high. Current ideas suggest that Miranda may have been hit and broken apart in the past but was able to come together again, leaving a patchwork surface.

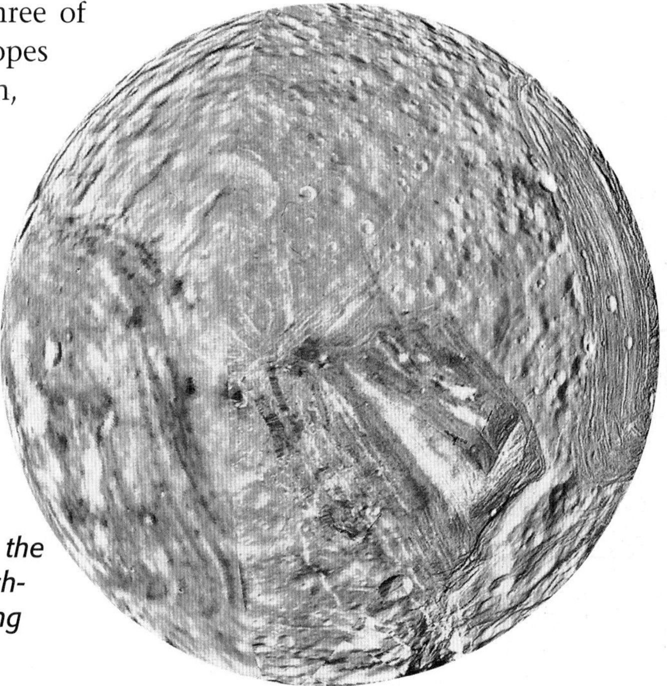

Figure 31 *This* Voyager 2 *image shows Miranda, the most unusual moon of Uranus. Miranda is a patchwork of several different types of terrain, indicating that this small moon has a violent history.*

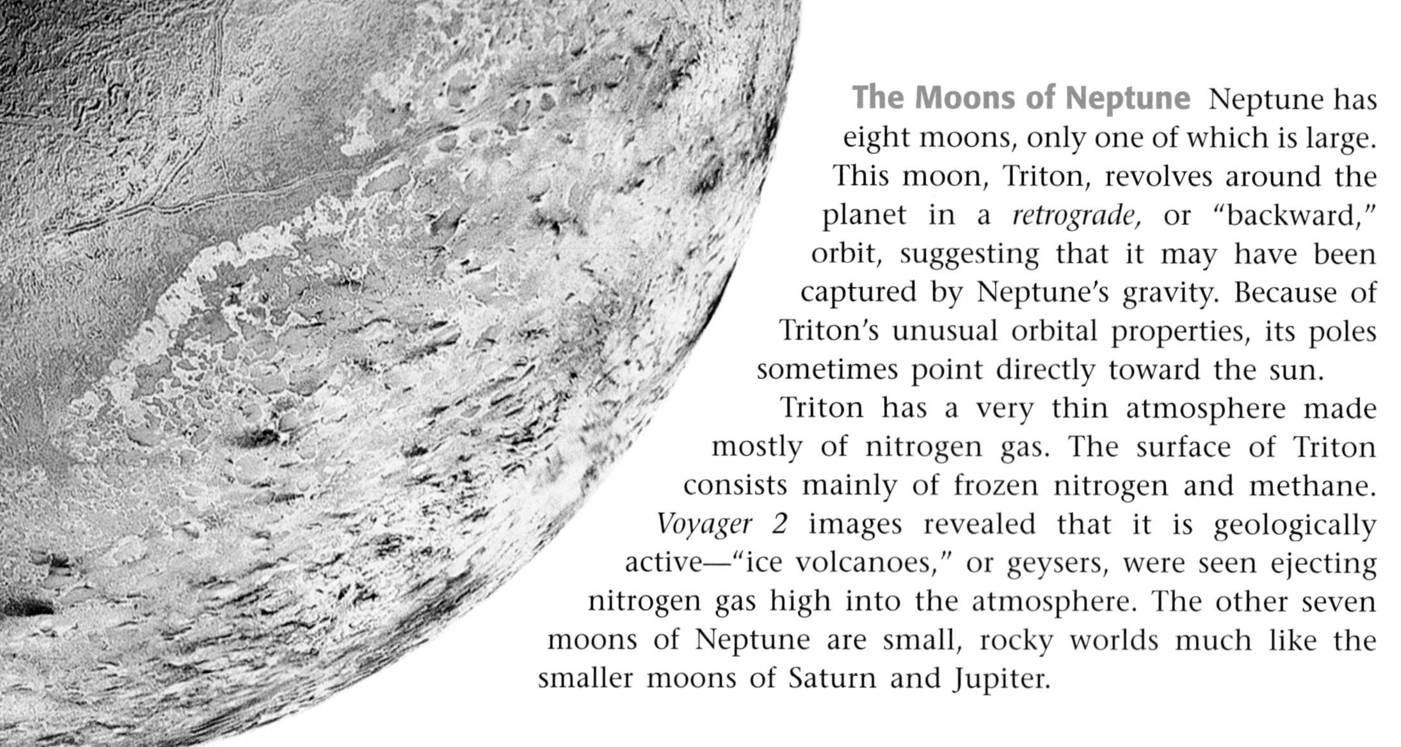

The Moons of Neptune Neptune has eight moons, only one of which is large. This moon, Triton, revolves around the planet in a *retrograde,* or "backward," orbit, suggesting that it may have been captured by Neptune's gravity. Because of Triton's unusual orbital properties, its poles sometimes point directly toward the sun.

Triton has a very thin atmosphere made mostly of nitrogen gas. The surface of Triton consists mainly of frozen nitrogen and methane. *Voyager 2* images revealed that it is geologically active—"ice volcanoes," or geysers, were seen ejecting nitrogen gas high into the atmosphere. The other seven moons of Neptune are small, rocky worlds much like the smaller moons of Saturn and Jupiter.

Figure 32 *This high-resolution* Voyager 2 *image shows Neptune's largest moon, Triton. The polar icecap currently facing the sun may have a slowly evaporating layer of nitrogen ice, adding to Triton's thin atmosphere.*

The Moon of Pluto Pluto's only moon, Charon, was discovered in 1978. Charon's period of revolution is the same as Pluto's period of rotation—about 6.4 days. This means that one side of Pluto always faces Charon. In other words, if you stood on the surface of Pluto, Charon would always occupy the same place in the sky. Imagine Earth's moon staying in the same place every night! Because Charon's orbit around Pluto is tilted with respect to Pluto's orbit around the sun, as seen from Earth, Pluto is sometimes eclipsed by Charon. But don't hold your breath; this happens only once every 120 years!

While scientists continue to gather data to determine whether global warming is happening on Earth, an MIT researcher has discovered that this process is actually occurring on Neptune's largest moon, Triton. Triton is entering an unusually warm summer in which its southern polar icecap is receiving more direct sunlight than usual. This extra sunlight causes some of the ice to vaporize and become part of Triton's atmosphere. The thicker its atmosphere, the warmer Triton gets. Scientists say that the surface temperature of Triton has risen by about 5 percent in the last 10 years.

REVIEW

1. What makes Io the most volcanically active body in the solar system?
2. Why is Saturn's moon Titan of so much interest to scientists studying the origins of life on Earth?
3. What two properties of Neptune's moon Triton make it unusual?
4. **Identifying Relationships** Charon always stays in the same place in Pluto's sky, but the moon always moves across Earth's sky. What causes this difference?

Section 3

Small Bodies in the Solar System

NEW TERMS

comet
perihelion
aphelion
asteroid
asteroid belt
meteoroid
meteorite
meteor

OBJECTIVES

- Explain why comets, asteroids, and meteoroids are important to the study of the formation of the solar system.
- Compare the different types of asteroids with the different types of meteoroids.
- Describe the risks to life on Earth from cosmic impacts.

In addition to planets and moons, the solar system contains many other types of objects, including comets, asteroids, and meteoroids. Some of these relatively small objects have orbits that bring them very close to Earth as they revolve around the sun. As you will see, these objects play an important role in the study of the origins of the solar system.

Comets

A **comet** is a small body of ice, rock, and cosmic dust loosely packed together. Because of their composition, some scientists refer to comets as "dirty snowballs." Unlike the planets, comets are very small and originate from the cold, outer solar system. Nothing much has happened to them since the birth of the solar system some 4.6 billion years ago. Comets are probably the leftovers from the process of planet formation. Each comet is a sample of the early solar system. Scientists want to learn more about comets in order to piece together the chemical and physical history of the solar system.

When a comet passes close enough to the sun, solar radiation heats the water ice so that the comet gives off gas and dust in the form of a long tail, as shown in **Figure 33.** As the gases flow away from the comet they can cause the cosmic dust to escape into space too. Sometimes this process can give a comet two tails—an *ion tail* and a *dust tail*. The ion tail consists of electrically charged particles that are blown directly away from the sun by the solar wind, which also consists of charged particles. The solid center of a comet is called its *nucleus*. Comet nuclei can range in size from less than half a kilometer to more than 100 km in diameter. **Figure 34** shows the different features a comet may have when it passes close to the sun.

Figure 33 *Comet Hale-Bopp appeared in North American skies in the spring of 1997.*

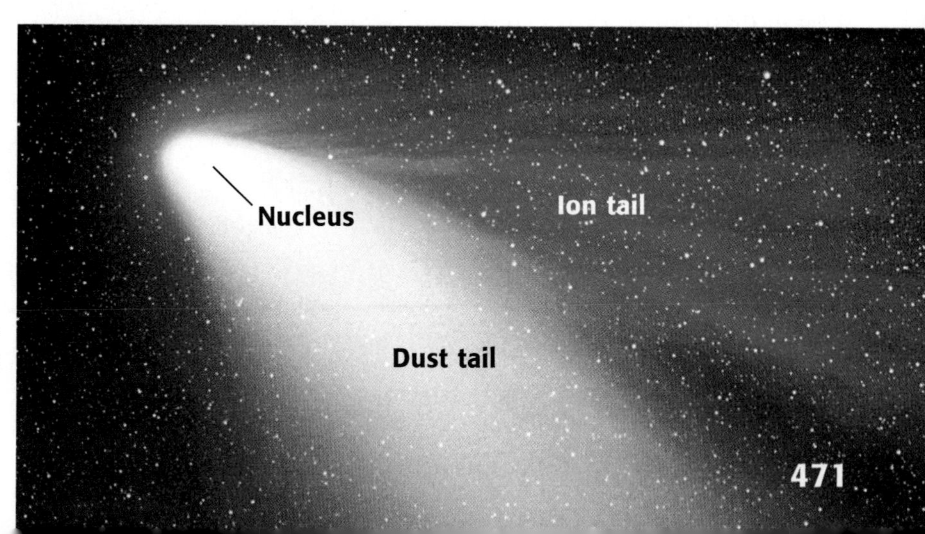

Figure 34 *The image at right shows the physical features of a comet when it is close to the sun. The nucleus of a comet is hidden by brightly lit gases and dust.*

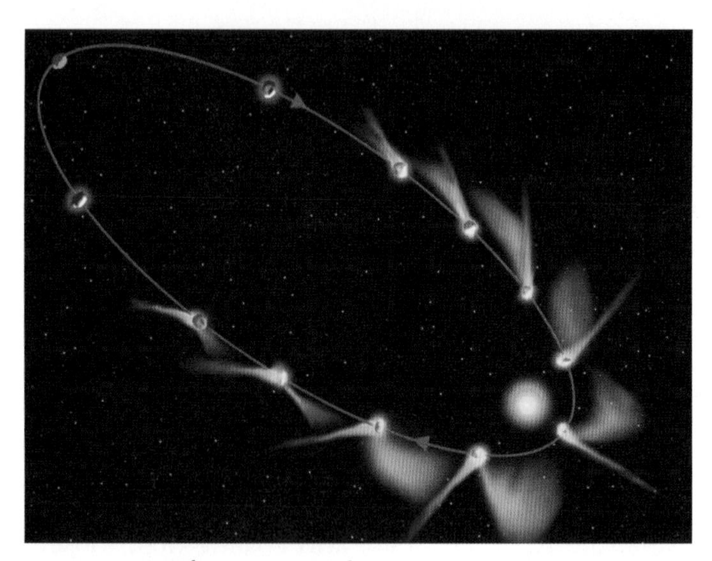

Figure 35 *When a comet's highly elliptical orbit carries it close to the sun, it can develop one or two tails. As shown here, the ion tail is blue and the dust tail is yellow.*

All orbits are *ellipses*—circles that are somewhat stretched out of shape. Whereas the orbits of most planets are nearly circular, comet orbits are highly elliptical—they are very elongated. When a body, such as a comet, is at the point in its orbit closest to the sun, it is said to be at **perihelion** (PER i HEE lee uhn). The point in an orbit farthest from the sun is called the **aphelion** (uh FEE lee uhn). When a comet is at perihelion its tail can extend millions of kilometers through space!

Notice in **Figure 35** that a comet's ion tail always points directly away from the sun. This is because the ion tail is blown away from the sun by the solar wind. The dust tail tends to follow the comet's orbit around the sun and does not always point away from the sun.

Where do comets come from? Many scientists think they may come from a spherical region, called the *Oort* (ohrt) *cloud,* that surrounds the solar system. When the gravity of a passing planet or star disturbs part of this cloud, comets can be pulled in toward the sun. These distant members of the solar system then assume a smaller orbit around the sun. Another recently discovered region where comets exist is called the *Kuiper* (KIE per) *belt,* which is the region outside the orbit of Neptune. These two regions where comets orbit are shown in **Figure 36.**

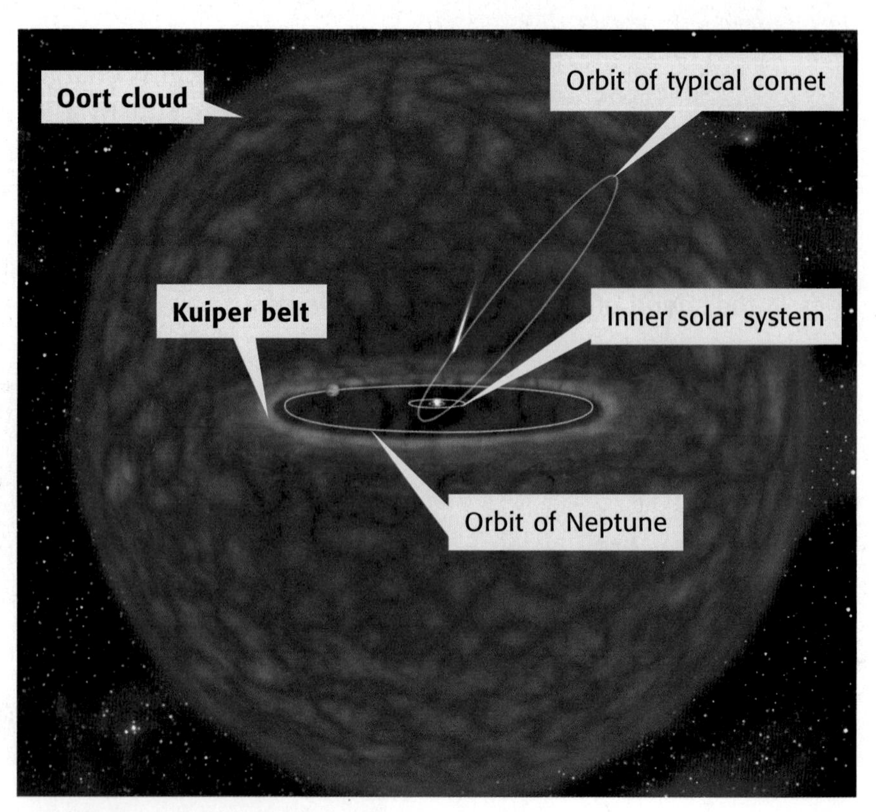

Figure 36 *The Kuiper belt is a disk-shaped region that extends outward from the orbit of Neptune. The Oort cloud is a spherical region far beyond the orbit of Pluto.*

Asteroids

Asteroids are small, rocky bodies in orbit around the sun. They range in size from a few meters to more than 900 km in diameter. Asteroids have irregular shapes, although some of the larger ones are spherical. Most asteroids orbit the sun in a wide region between the orbits of Mars and Jupiter, called the **asteroid belt.**

Asteroids can have a variety of compositions, depending on where they are located within the asteroid belt. In the outermost region of the asteroid belt, asteroids have dark reddish brown to black surfaces, which may indicate that they are rich in organic material. A little closer to the sun, asteroids have dark gray surfaces, indicating that they are rich in carbon. In the innermost part of the asteroid belt are light gray asteroids that have either a stony or metallic composition. **Figure 38** shows some examples of what some of the asteroids may look like.

Like comets, asteroids are thought to be material left over from the formation of the solar system. NASA's NEAR (Near Earth Asteroid Rendezvous) mission was the first in a series of missions to send spacecraft to study asteroids. Scientists hope data gathered on these missions will help us better understand the way our solar system formed.

Figure 37 *The asteroid Ida has a small companion asteroid that orbits it called Dactyl. Ida is about 52 km long.*

Figure 38 *The asteroid belt is a disk-shaped region located between the orbits of Mars and Jupiter.*

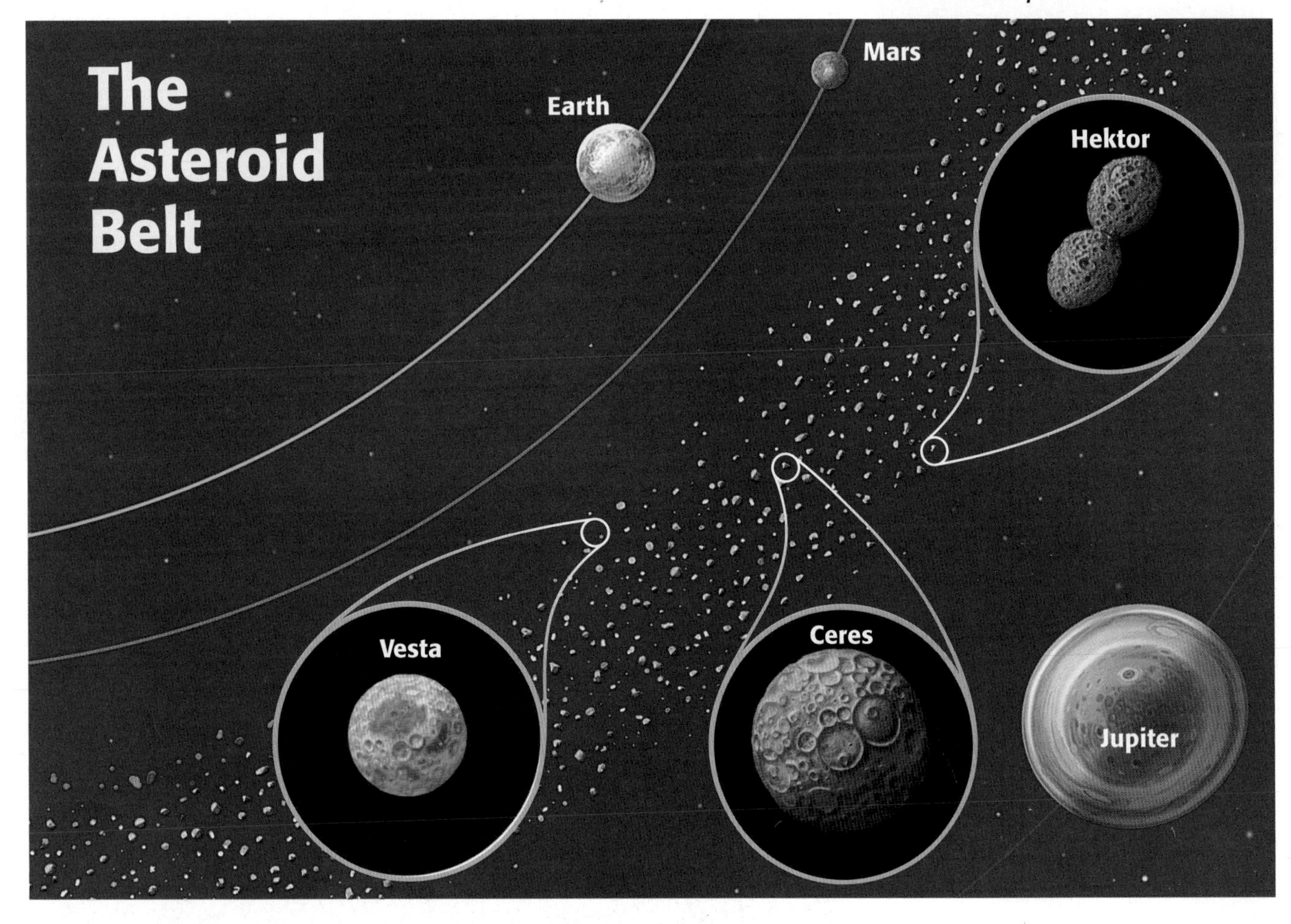

Meteoroids

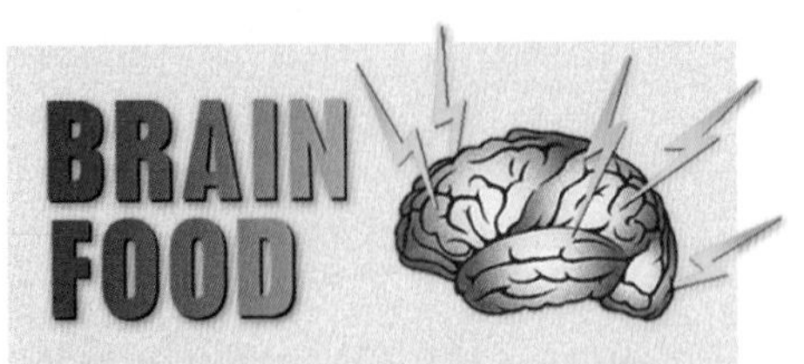

The total mass of meteorites that fall to Earth each year is between 10,000 and 1 million metric tons!

A **meteoroid** is a small, rocky body orbiting the sun. Meteoroids are similar to asteroids, but they are much smaller. In fact, most meteoroids probably come from asteroids. If a meteoroid enters Earth's atmosphere and strikes the ground, it is then called a **meteorite.** When a meteoroid falls into Earth's atmosphere, it is usually traveling at such a high speed that its surface heats up and melts. As it burns up, the meteoroid glows and gives off an enormous amount of light and heat. From the ground, we see a spectacular streak of light, or a shooting star. The bright streak of light caused by a meteoroid or comet dust burning up in the atmosphere is called a **meteor.**

Many of the meteors that we see come from very small (dust-sized to pebble-sized) rocks and can be seen on almost any night if you are far enough away from the city to avoid the glare of its lights. At certain times of the year, you can see large numbers of meteors, as shown in **Figure 39.** These events are called *meteor showers*. Meteor showers occur when Earth passes through the dusty debris left behind in the orbit of a comet. During some meteor showers, alert observers can see up to several thousand meteors per hour!

Figure 39 *Meteors are the streaks of light caused by meteoroids as they burn up in Earth's atmosphere.*

Some kinds of meteorites are easy to recognize because they are very dense. If you pick up a rock that is especially heavy for its size, it might be a meteorite. Like their relatives the asteroids, meteorites have a variety of compositions. The three major types of meteorites—stony, metallic, and stony-iron—are shown in **Figure 40.** Many of the stony meteorites probably come from carbon-rich asteroids and may contain organic materials and water. Scientists use meteorites to study the early solar system. Like comets and asteroids, meteoroids are some of the building blocks of planets.

Figure 40 *There are three major types of meteorites.*

Stony meteorite
rocky material

Metallic meteorite
iron and nickel

Stony-iron meteorite
rocky material, iron and nickel

The Role of Impacts in the Solar System

Planets and moons that have no atmosphere have many more impact craters than those that do have atmospheres. Look at **Figure 41.** The Earth's moon has many more impact craters than the Earth because it has no atmosphere or tectonic activity. Fewer objects land on Earth because Earth's atmosphere acts like a shield. Smaller bodies burn up before they ever reach the surface. On the moon, there is nothing to stop them! Also, most craters left on Earth have been erased due to weathering, erosion, and tectonic activity.

Objects smaller than about 10 m across usually burn up in the atmosphere, causing a meteor. Larger objects are more likely to strike Earth's surface. In order to estimate the risk of cosmic impacts, we need to consider how often large impacts occur. The solar system still contains a large amount of small debris—most of which we enjoy as meteor showers. The number of large objects that could collide with Earth is relatively small. Scientists estimate that impacts powerful enough to cause a natural disaster might occur once every few thousand years. An impact large enough to cause a global catastrophe—such as the extinction of the dinosaurs—is estimated to occur once every 30 million to 50 million years on average.

Figure 41 *The Earth and the moon are about the same age, but unlike the Earth's surface, shown here covered with snow, the surface of the moon preserves a record of billions of years of cosmic impacts.*

REVIEW

1. Why is the study of comets, asteroids, and meteoroids important in understanding the formation of the solar system?
2. Why do a comet's two tails often point in different directions?
3. What is the difference between an asteroid and a meteoroid?
4. Describe one reason asteroids may become a natural resource in the future.
5. **Analyzing Viewpoints** Do you think the government should spend money on programs to search for asteroids and comets with Earth-crossing orbits? Discuss why.

Chapter Highlights

SECTION 1

Vocabulary

astronomical unit (AU) *(p. 452)*
terrestrial planets *(p. 453)*
prograde rotation *(p. 454)*
retrograde rotation *(p. 454)*
gas giants *(p. 458)*

Section Notes

- The solar system has nine planets.
- Distances within the solar system can be expressed in astronomical units (AU) or in light-minutes.
- The inner four planets, called the terrestrial planets, are small and rocky.
- The outer planets, with the exception of Pluto, are gas giants.
- By learning about the properties of the planets, we get a better understanding of global processes on Earth.

Labs

Why Do They Wander? *(p. 588)*

SECTION 2

Vocabulary

satellite *(p. 463)*
phases *(p. 465)*
eclipse *(p. 466)*

Section Notes

- Earth's moon probably formed from a giant impact on Earth.
- The moon's phases are caused by the moon's orbit around the Earth. At different times of the month, we view different amounts of sunlight on the moon because of the moon's position relative to the sun and the Earth.
- Lunar eclipses occur when the Earth's shadow falls on the moon.

☑ Skills Check

Math Concepts

INTERPLANETARY DISTANCES The distances between planets are so vast that scientists have invented new units of measurement to describe them. One of these units is the astronomical unit (AU). One AU is equal to the average distance between the Earth and the sun—about 150 million kilometers. If you wanted to get to the sun from the Earth in 10 hours, you would have to travel at a rate of 15,000,000 km/h!

$$\frac{\text{150 million kilometers}}{\text{15 million kilometers/hour}} = \text{10 hours}$$

Visual Understanding

AXIAL TILT A planet's axis of rotation is an imaginary line that runs through the center of the planet and comes out its north and south poles. The tilt of a planet's axis is the angle between the planet's axis and the plane of the planet's orbit around the sun.

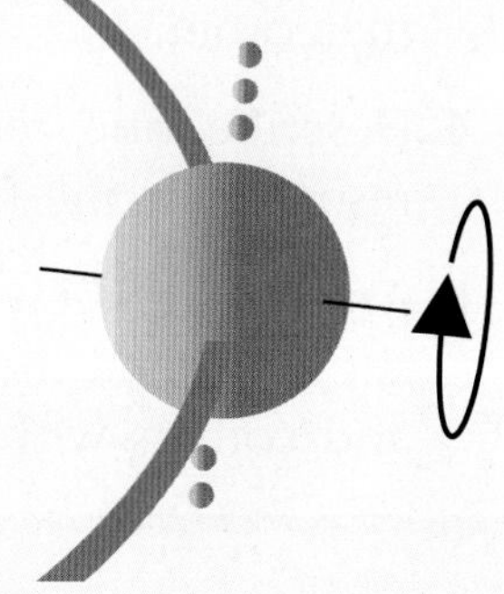

SECTION 2

- Solar eclipses occur when the moon is between the sun and the Earth, causing the moon's shadow to fall on the Earth.
- The plane of the moon's orbit around the Earth is tilted by 5° relative to the plane of the Earth's orbit around the sun.

Labs

Eclipses *(p. 590)*
Phases of the Moon *(p. 591)*

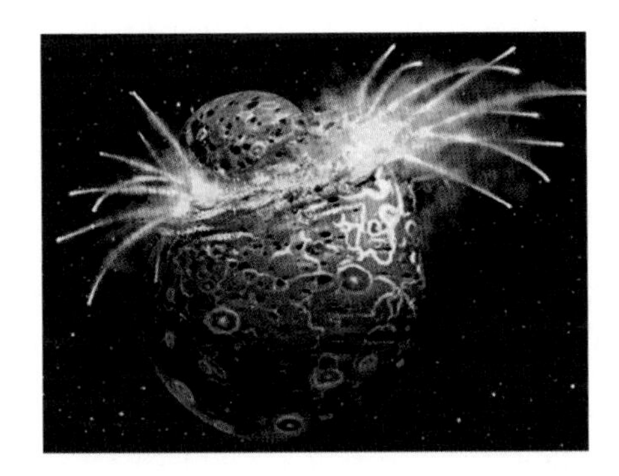

SECTION 3

Vocabulary

comet *(p. 471)*
perihelion *(p. 472)*
aphelion *(p. 472)*
asteroid *(p. 473)*
asteroid belt *(p. 473)*
meteoroid *(p. 474)*
meteorite *(p. 474)*
meteor *(p. 474)*

Section Notes

- Comets are small bodies of water ice and cosmic dust left over from the formation of the solar system.
- When a comet is heated by the sun, the ices convert to gases that leave the nucleus and form an ion tail. Dust also comes off a comet to form a second kind of tail called a dust tail.
- All orbits are ellipses—circles that have been stretched out. The point on an orbit closest to the sun is called the perihelion. The point on an orbit farthest from the sun is the aphelion.
- Asteroids are small, rocky bodies that orbit the sun between the orbits of Mars and Jupiter.
- Meteoroids are small, rocky bodies that probably come from asteroids.
- Meteor showers occur when Earth passes through the dusty debris along a comet's orbit.
- Impacts that cause natural disasters occur once every few thousand years, but impacts large enough to cause global extinctions occur once every 30 million to 50 million years.

internetconnect

GO TO: go.hrw.com

Visit the **HRW** Web site for a variety of learning tools related to this chapter. Just type in the keyword:

KEYWORD: HSTFAM

GO TO: www.scilinks.org

Visit the **National Science Teachers Association** on-line Web site for Internet resources related to this chapter. Just type in the ***sci*LINKS** number for more information about the topic:

TOPIC: The Nine Planets — ***sci*LINKS NUMBER:** HSTP605
TOPIC: Studying Earth from Space — ***sci*LINKS NUMBER:** HSTP610
TOPIC: The Earth's Moon — ***sci*LINKS NUMBER:** HSTP615
TOPIC: The Moons of Other Planets — ***sci*LINKS NUMBER:** HSTP620
TOPIC: Comets, Asteroids, and Meteoroids — ***sci*LINKS NUMBER:** HSTP625

Chapter Review

USING VOCABULARY

For each pair of terms, explain the difference in their meaning.

1. aphelion/perihelion
2. asteroid/comet
3. meteor/meteorite
4. satellite/moon
5. Kuiper belt/Oort cloud

To complete the following sentences, choose the correct term from each pair of terms listed below:

6. The average distance between the sun and the Earth is 1 __?__. *(light-minute,* or *AU)*
7. A small rock in space is called a __?__. *(meteorite, meteor,* or *meteoroid)*
8. The time it takes for the Earth to __?__ around the sun is one year. *(rotate* or *revolve)*
9. Most lunar craters are the result of __?__. *(volcanoes* or *impacts)*

UNDERSTANDING CONCEPTS

Multiple Choice

10. When do annular eclipses occur?
    a. every solar eclipse
    b. when the moon is closest to the Earth
    c. only during full moon
    d. when the moon is farthest from the Earth

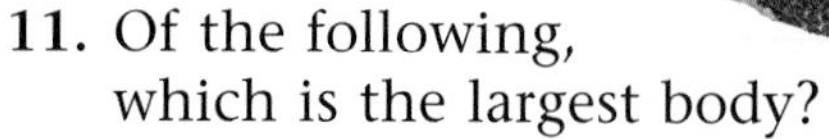

11. Of the following, which is the largest body?
    a. the moon
    b. Pluto
    c. Mercury
    d. Ganymede

12. Which is not true about impacts?
    a. They are very destructive.
    b. They can bring water to dry worlds.
    c. They only occurred as the solar system formed.
    d. They can help us do remote geology.

13. Which of these planets does not have any moons?
    a. Mercury
    b. Mars
    c. Uranus
    d. none of the above

14. What is the most current theory for the formation of Earth's moon?
    a. The moon formed from a collision between another body and the Earth.
    b. The moon was captured by the Earth.
    c. The moon formed at the same time as the Earth.
    d. The moon formed by spinning off from the Earth early in its history.

15. Liquid water cannot exist on the surface of Mars because
    a. the temperature is too hot.
    b. liquid water once existed there.
    c. the gravity of Mars is too weak.
    d. the atmospheric pressure is too low.

16. Which of the following planets is not a terrestrial planet?
    a. Mercury
    b. Mars
    c. Earth
    d. Pluto

17. All of the gas giants have ring systems.
    a. true
    b. false

18. A comet's ion tail consists of
    a. dust.
    b. electrically charged particles of gas.
    c. light rays.
    d. comet nuclei.

Short Answer

19. Do solar eclipses occur at the full moon or at the new moon? Explain why.
20. How do we know there are small meteoroids and dust in space?
21. Which planets have retrograde rotation?

Concept Mapping

22. Use the following terms to create a concept map: solar system, terrestrial planets, gas giants, moons, comets, asteroids, meteoroids.

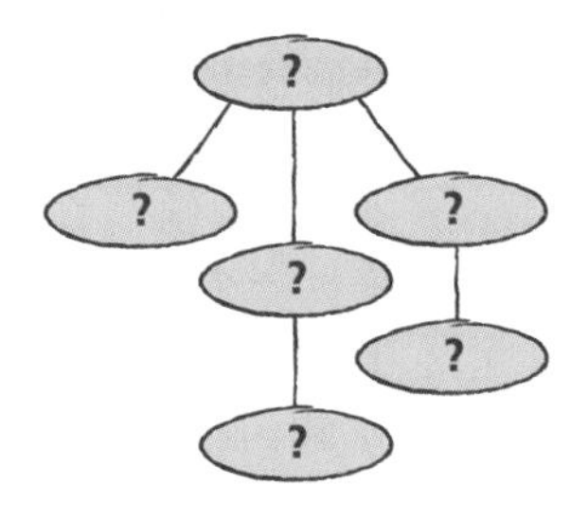

CRITICAL THINKING AND PROBLEM SOLVING

23. Even though we haven't yet retrieved any rock samples from Mercury's surface for radiometric dating, we know that the surface of Mercury is much older than that of Earth. How do we know this?
24. Where in the solar system might we search for life, and why?
25. Is the far side of the moon always dark? Explain your answer.
26. If we could somehow bring Europa as close to the sun as the Earth is, 1 AU, what do you think would happen?

MATH IN SCIENCE

27. Suppose you have an object that weighs 200 N (45 lbs.) on Earth. How much would that same object weigh on each of the other terrestrial planets?

INTERPRETING GRAPHICS

The graph below shows density versus mass for Earth, Uranus, and Neptune. Mass is given in Earth masses—the mass of Earth equals one. The relative volumes for the planets are shown by the size of each circle.

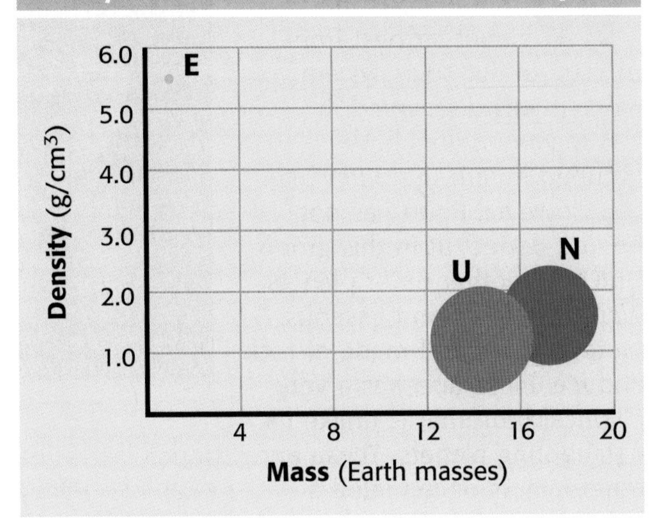

28. Which planet is denser, Uranus or Neptune? How can you tell?
29. You can see that although Earth has the smallest mass, it has the highest density. How can Earth be the densest of the three when Uranus and Neptune have so much more mass?

NOW What Do You Think?

Take a minute to review your answers to the ScienceLog questions on page 451. Have your answers changed? If necessary, revise your answers based on what you have learned since you began this chapter.

SCIENTIFICDEBATE

Is Pluto Really a Planet?

We have all learned that Pluto is the planet farthest from the sun in our solar system. Since it was discovered in 1930, astronomers have grouped it with the outer planets. However, Pluto has not been a perfect fit in this group. Unlike the other outer planets, which are large and gaseous, Pluto is small and made of rock and ice. Pluto also has a very elliptical orbit that is unlike its neighboring planets. These and other factors once fueled a debate as to whether Pluto really is a planet and how it should be classified.

◀ *A composite drawing of Pluto, Charon, Triton, and Halley's comet*

Kuiper Belt

In 1997, astronomers discovered a belt of comets outside the orbit of Neptune. The belt was named the Kuiper Belt in honor of Gerard Kuiper, a Dutch-born American astronomer. So what does this belt have to do with Pluto? Given its proximity to Pluto, some astronomers thought Pluto may actually be a large comet that escaped the Kuiper Belt.

Comet?

Comets are basically dirty snowballs made of ice and cosmic dust. Pluto is about 30 percent ice and 70 percent rock. This is much more rock than is in a normal comet. Also, at 2,390 km in diameter, Pluto is much larger than a comet. For example, Halley's comet is only about 20 km in diameter. Even so, Pluto's orbit is very similar to that of a comet. Both have orbits that are very elliptical.

Escaped Moon?

Pluto and its moon, Charon, have much in common with Neptune's moon, Triton. All three have atmospheres made of nitrogen and methane, which suggests that they share a similar origin. And since Triton has a "backward" orbit compared with Neptune's other moons, it may have been captured by Neptune's gravity. Some astronomers thought Pluto might also have been captured by Neptune but broke free by some cataclysmic event.

New Category of Planet?

Some astronomers suggested that perhaps we should create a new subclass of planets, such as the ice planets, to add to the Jovian and terrestrial classification we currently use. Pluto would be the only planet in this class, but scientists think we are likely to find others.

As there are more new discoveries, astronomers will likely continue to debate the issues. To date, however, Pluto is still officially considered a planet. This decision is firmly grounded by the fact that Pluto has been called a planet since its discovery.

You Decide

▶ Do some additional research about Pluto, the Kuiper Belt, and comets. What do you think Pluto should be called?

Science Fiction

"The Mad Moon"

by Stanley Weinbaum

The third largest satellite of Jupiter, called Io, can be a hard place to live. Although living comfortably is possible in the small cities at the polar regions, most of the moon is hot, humid, and jungle-like. There is also *blancha,* a kind of tropical fever that causes hallucinations, weakness, and vicious headaches. Without proper medication a person with *blancha* can go mad or even die.

Just 2 years ago, Grant Calthorpe was a wealthy hunter and famous sportsman. Then the gold market crashed, and he lost his entire fortune. What better way for an experienced hunter and explorer to get a fresh start than to set out for a little space travel? The opportunity to rekindle his fortune by gathering ferva leaves so that they can be converted into useful human medications lures Calthorpe to Io.

There he meets the loonies—creatures with balloon heads and silly grins atop *really* long necks. The parcat Oliver quickly becomes Calthorpe's pet and helps him cope with the loneliness and the slinkers. The slinkers, well, they would just as soon *not* have Calthorpe around at all, but they are pretty good at making even this famous outdoorsman wonder why he ever took this job.

In "The Mad Moon," you'll discover a dozen adventures with Grant Calthorpe as he struggles to stay alive—and sane. Read Stanley Weinbaum's story "The Mad Moon" in the *Holt Anthology of Science Fiction.* Enjoy your trip!

CHAPTER 19

The Universe Beyond

Imagine . . .

Suppose you are the director of the Hubble Space Telescope Science Institute and you are allowed some of the precious observing time on the Hubble Space Telescope (HST) for any project you want. While other astronomers have to write long requests to use this special telescope, you get to use it without doing all the paperwork. You can look at *anything* in the universe! If you were to ask your friends for suggestions, what would they say? What kinds of things are *you* curious about? Would you choose a planet, like Jupiter, to look at, or would you focus on a strangely shaped cloud of gas or perhaps a galaxy? Would you look at a big part of the sky or just a tiny piece?

In 1995, Robert Williams was the director of the institute, and he had this very problem. He finally decided to look at what seemed like an empty piece of sky—but to look at it longer than any one else had done with the HST. Over a 10-day period, the HST took 342 pictures of the same small part of the sky. Later, computers combined the pictures to get a single image called the Hubble Deep Field. The Hubble Deep Field shows almost 2,000 galaxies in that one spot of sky. As you can see in the photo above, the galaxies have different shapes, sizes, and colors. Some are even colliding with each other. In this chapter, you will learn about galaxies and the stars they are made of.

To get an idea of how much area the Hubble Deep Field covers, hold a grain of sand at arm's length while looking up at the sky. The sand grain covers an area about the same size as the one that Robert Williams studied.

What Do You Think?

In your ScienceLog, try to answer the following questions based on what you already know:

1. Why do stars shine?
2. What is a galaxy?
3. How did the universe begin, and how will it end? or will it?

Exploring Galaxies

Galaxies are groups of stars and other material floating like islands in the sea of space. Each galaxy contains billions and billions of stars. But not all galaxies are the same. As you saw in the Hubble Deep Field, they come in different sizes and shapes. Let's explore some of these differences.

Procedure

1. Look at the different galaxies in the Hubble Deep Field image on the previous page. (The two bright spots with spikes are stars that are much closer to Earth; you can ignore them.)
2. Can you find different types of galaxies? Look for different shapes and colors. In your ScienceLog, make sketches of at least three different types. Make up a name that describes each type of galaxy.
3. With a **metric ruler,** measure the size of at least four galaxies of each type in millimeters. Record your measurements in your ScienceLog.

Analysis

4. Why did you classify the galaxies the way you did?
5. Compare your types of galaxies with those of your classmates. Are there similarities? Did you give them similar names?
6. What conclusions can you draw about galaxies from the measurements you took in step 3?

Section 1

Stars

NEW TERMS

spectrum
apparent magnitude
absolute magnitude
light-year
parallax

OBJECTIVES

- Describe how color indicates temperature.
- Compare absolute magnitude with apparent magnitude, and discuss how each measures brightness.
- Describe the difference between the apparent motion of stars and the real motion of stars.

Most stars look like faint dots of light in the night sky. But stars are actually huge, hot, brilliant balls of gas trillions of kilometers away from Earth. How do astronomers learn about stars when they are too far away to visit? They study starlight!

Color of Stars

Look closely at the flames on the candle and the Bunsen burner in **Figure 1.** Which one has the hotter flame? How can you tell? If you draw the colors in the candle flame, in what order are they? Although artists may speak of *red* as a "hot" color, to a scientist, *red* is a "cool" color. The blue flame of the Bunsen burner is much hotter than the yellow flame of the candle. The candle's flame, however, would be hotter than the red glowing embers of a campfire.

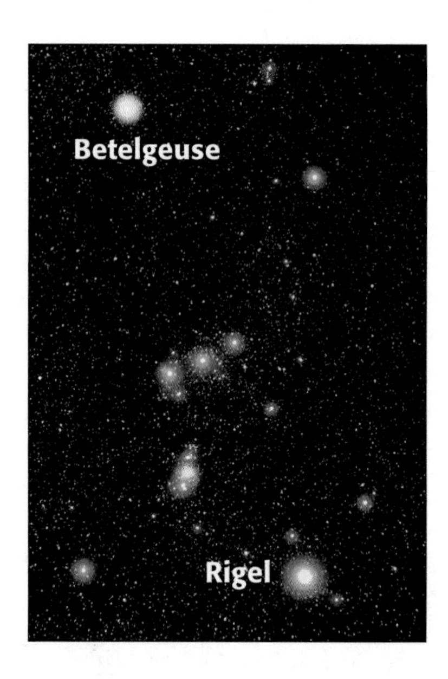

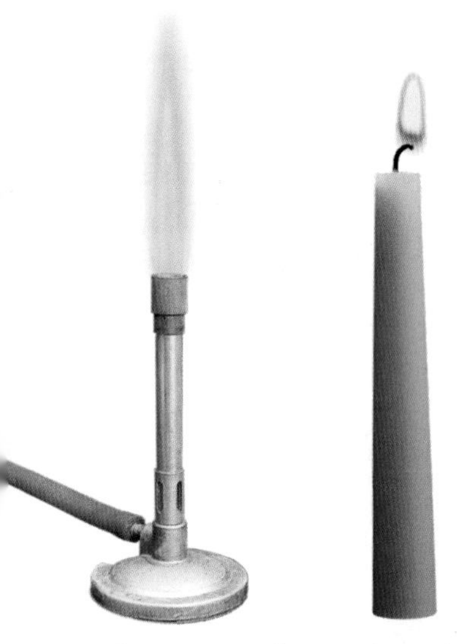

Figure 1 *What colors do you see when you examine the flames of a Bunsen burner and an ordinary candle?*

If you allow time for your eyes to adjust to the night sky and look carefully, you might notice the different colors of some familiar stars. Betelgeuse, which is red, and Rigel, which is blue, are the stars that form the top left and bottom right corners of the constellation Orion, as shown above. This constellation is easy to see in the evenings during the winter months. Because stars are different colors, we can infer that they have different temperatures.

Composition of Stars

When you look at white light through a glass prism, you see a rainbow. This rainbow of colors is called a **spectrum.** The spectrum contains the colors we recognize as red, orange, yellow, green, blue, indigo, and violet. A hot solid object, like the glowing wire inside a light bulb or a piece of molten metal, gives off a *continuous spectrum*—one that shows all the colors. Astronomers use an instrument called a *spectrograph* to spread starlight out into its colors, just as you might use a prism to spread sunlight. Stars, however, don't have continuous spectra. Because they are not solid objects, stars give off spectra that are different from those of light bulbs.

Making an ID Stars are made of various gases that are so dense, they act like a hot solid. For this reason, the "surface" of a star, or the part that we see, gives off a continuous spectrum. But the light we see passes through the star's "atmosphere," which is made of somewhat cooler gases than the star itself is made of. A star therefore produces a spectrum with various lines in it. To understand what these lines are, let's look at something you might be more familiar with than stars.

Many restaurants use neon signs to attract customers. The gas in a neon sign glows orange-red when an electric current flows through it. If we were to look at the sign with an astronomer's spectrograph, we would not see a continuous spectrum. Instead we would see *emission lines*. Emission lines are bright lines that are made when certain wavelengths of light are given off, or emitted, by hot gases. Only some colors in the spectrum show up, while all the other colors that make up white light are missing. Every tube of neon gas, for example, emits light with the same emission lines. Every other element has its own set of emission lines. Emission lines are like fingerprints for the elements. You can see some of these "fingerprints" in **Figure 2.**

Police use spectrographs to "fingerprint" cars. Automobile manufacturers put trace elements in the paint of cars. Each make of car has its own special paint and therefore its own trace element. When a car is involved in a hit-and-run accident, the police can identify the make of the car by the paint that is left behind.

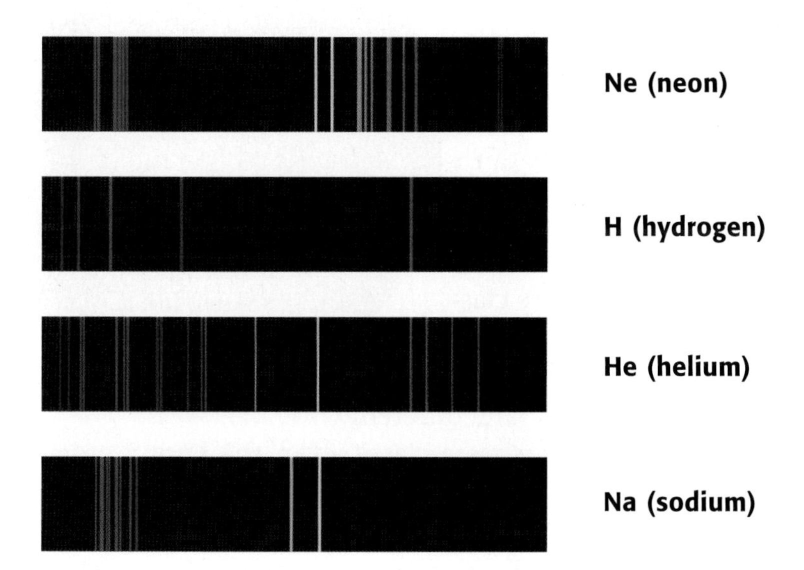

Figure 2 *Neon gas produces its own characteristic pattern of emission lines, as do hydrogen, helium, and sodium.*

One More Thing If we could look at *just* the gas in a star's atmosphere, that gas would produce emission lines. But we cannot see the star's atmosphere without also seeing the star behind it, which makes a continuous spectrum. One more thing to learn—cool gases behave differently from hot gases. The relatively cooler gases in a star's atmosphere absorb light and remove certain colors of light from the continuous spectrum of the hot star. In fact, the colors that the atmosphere absorbs are the same colors it would emit if heated.

To learn more about the color and temperature of stars, turn to page 592 in the LabBook.

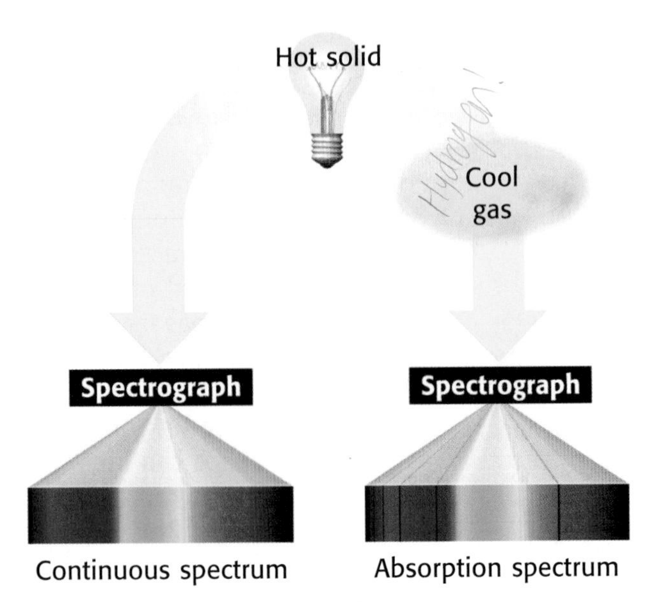

Figure 3 *An absorption spectrum (right) is produced when light passes through a cooler gas. Notice the dark lines in the spectrum.*

If light from a hot solid passes through a cooler gas, it produces an *absorption spectrum*—a continuous spectrum with dark lines where less light gets through. Take a look at **Figure 3.** Can you identify the element in the gas by comparing the position of the dark lines in its spectrum with the bright lines in Figure 2?

An astronomer's spectrum of a star shows an absorption spectrum. The pattern of lines shows some of the elements that are in the star's atmosphere. If a star were made of just one element, it would be simple to identify the element. But stars are a mixture of things, and all the different sets of lines for its elements appear together in a star's spectrum. Sorting out the patterns is often a puzzle.

Classifying Stars

In the 1800s, the first people to look at spectra found that different stars had different spectra. They started to collect the spectra of lots of stars and tried to classify them. At first, letters were assigned to each type of spectra. Stars with spectra that had very noticeable hydrogen patterns were classified as A type stars. Other stars were classified as B, and so on.

Later, when scientists finally understood why spectra are different, they realized that the stars were classified in the wrong order. Stars are now classified by how hot they are. The main differences in the spectra of stars are related to the temperature of the stars. We see the temperature differences as colors. The original class O stars are blue—they are very hot, the hottest of all stars. If you arrange the letters in order of temperature, they are no longer in alphabetical order. The resulting order of star classes—OBAFGKM—is shown in the table on the next page.

If you see a certain pattern of absorption lines in a star, you know that a certain element or molecule is in the star or at least in its atmosphere. But the absence of a pattern doesn't mean the element isn't there; the temperature might not be high enough or low enough for absorption lines to be produced.

life science CONNECTION

Imagine you are almost asleep in a darkened room when your grandmother comes in to say goodnight. You can make out her shape, but you can't tell what color her dress is.

Our eyes are not sensitive to colors when light levels are low. There are two types of light-sensitive cells inside the eye: rods and cones. Rods are good at distinguishing shades of light and dark as well as shape and movement. Cones are good for distinguishing colors. Cones, however, do not work well in low light. This is why it is hard to distinguish between star colors.

Types of Stars				
Class	Color	Surface temperature (°C)	Elements detected	Examples of stars
O	blue	above 30,000	helium	10 Lacertae
B	blue-white	10,000–30,000	helium and hydrogen	Rigel, Spica
A	blue-white	7,500–10,000	hydrogen	Vega, Sirius
F	yellow-white	6,000–7,500	hydrogen and heavier elements	Canopus, Procyon
G	yellow	5,000–6,000	calcium and other metals	the sun, Capella
K	orange	3,500–5,000	calcium and molecules	Arcturus, Aldebaran
M	red	less than 3,500	molecules	Betelgeuse, Antares

How Bright Is That Star?

When you look up at the sky on a dark night, you might see lots of stars. It is easy to see that some stars are bright and some are faint. Look at **Figure 4.** It shows the constellation Ursa Major (Big Bear), which includes the Big Dipper. Ancient astronomers saw the same stars that you see today. They called the brightest stars in the sky *first magnitude* stars and the faintest stars *sixth magnitude* stars. *Magnitude* means size, or in this case brightness. A first magnitude star is 100 times brighter than a sixth magnitude star. Notice that a smaller number means a brighter star.

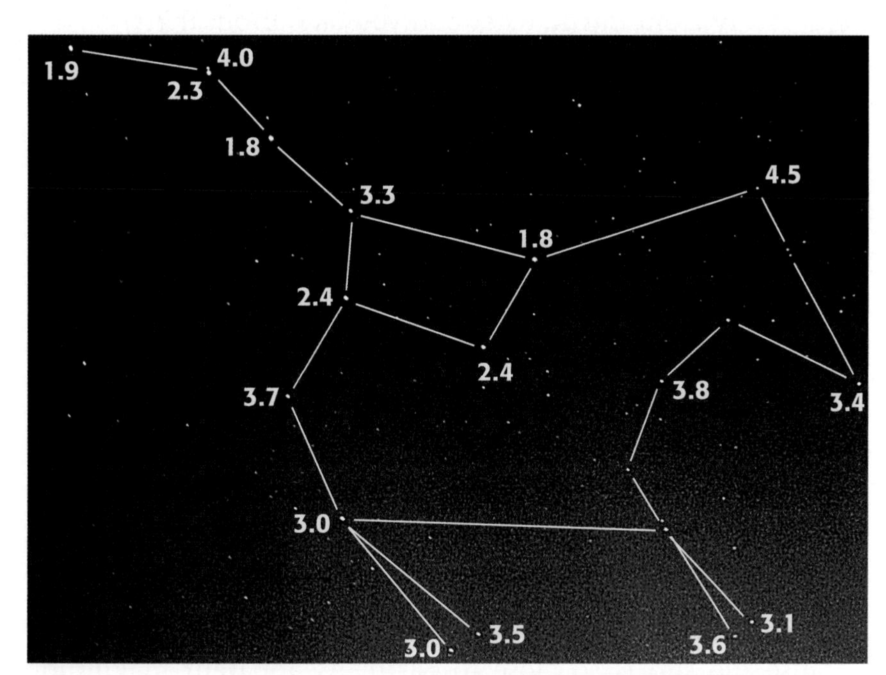

Figure 4 *Although you may recognize this constellation by its bright stars, it has many fainter stars as well. Numbers indicate their relative brightness.*

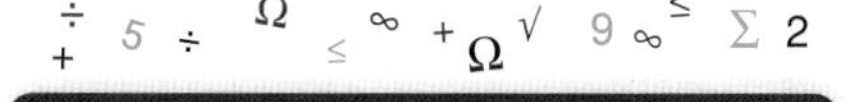

MATH BREAK

Starlight, Star Bright

Magnitude is used to indicate how bright one object is compared with another. Five magnitudes equal a factor of 100 times in brightness. The brightest blue stars, for example, have an absolute magnitude of –10. The sun is about +5. How much brighter is a blue star than the sun? Since each five magnitudes is a factor of 100 and the blue star and the sun are 15 magnitudes different, the blue star must be 100 × 100 × 100 times brighter than the sun. This is 1,000,000 (one million) times!

Figure 5 *You can estimate how far away each street light is by looking at its apparent brightness. Does this work with stars?*

The ancient astronomers probably thought they could see all the stars that existed. But in 1609, the telescope was invented. With a telescope, astronomers could see many more stars. But no one wanted to change the system of magnitudes that was already in place. Instead, modern astronomers simply added higher numbers for fainter stars. Now, with large telescopes, very faint stars of 29th magnitude have been found. Stars brighter than the original first magnitude stars are now given negative numbers. Sirius, the brightest star in the night sky, has a magnitude of –1.4.

Apparent Magnitude If you look at a row of street lights along a highway, like those shown in **Figure 5,** do they all look exactly the same? Does the light you are standing under look the same as a light several blocks away? Of course not! The nearest ones look bright, and the farthest ones look dim.

How bright a light looks, or appears, is called **apparent magnitude.** If you measure the brightness of a street light with a light meter, you will find that its brightness depends on the square of the distance between them. For example, a light that is 10 m away will appear four (2×2 or 2^2) times as bright as a light that is 20 m away. The same light will appear nine (3×3 or 3^2) times as bright as a light that is 30 m away.

Self-Check

If two identical stars are located the same distance away from Earth, what can you say about their apparent magnitudes? *(See page 596 to check your answer.)*

But unlike street lights, some stars are brighter than others because of their size or energy output, not their distance from Earth. So how can you tell the difference?

Absolute Magnitude Astronomers use a star's apparent magnitude (how bright it seems to be) and its distance from Earth to calculate its absolute magnitude. **Absolute magnitude** is the actual brightness of a star. In other words, if all stars could be placed the same distance away, their absolute magnitudes would be the same as their apparent magnitudes and the brighter stars would look brighter. The sun, for example, has an absolute magnitude of +4.8—pretty ordinary for a star. But because the sun is so close to Earth, its apparent magnitude is –26.8, making it the brightest object in the sky.

And speaking of street lights . . . Cities have many street lights and other lights from buildings and homes. Because of this, someone looking at the night sky in a city would not see as many stars as someone looking at the sky in the country. Light pollution is a big problem for astronomers and backyard stargazers alike. Certain types of lighting can help reduce glare, but there will continue to be a conflict between lighting buildings at night and seeing the stars.

Distance to the Stars

Because they are so far away, astronomers use light-years to give the distances to the stars. A **light-year** is the distance that light travels in one year. Because the speed of light is about 300,000 km/s, it travels almost 9.5 trillion kilometers in one year. Obviously it would be easier to give the distance to the North Star as 431 light-years than 4,080,000,000,000,000 km. But how do astronomers measure a star's distance?

To get a clue, take a look at the QuickLab at right. Just as your thumb appeared to move, stars near the Earth seem to move compared with more-distant stars as Earth revolves around the sun, as shown in **Figure 6.** This apparent shift in position is called **parallax.** While this shift can be seen only through telescopes, using parallax and simple trigonometry (a type of math), astronomers can find the actual distance to stars that are close to Earth. Farther stars, however, are measured in more-complicated ways.

Quick Lab

Not All Thumbs!

1. Hold your **thumb** in front of your face at arm's length.
2. Close one **eye** and focus on an **object** some distance behind your thumb.
3. Slowly move your **head** back and forth a small amount, and notice how your thumb seems to be moving compared with the background you are looking at.
4. Now move your thumb in close to your face and move your head the same amount. Notice how much more your thumb moves.

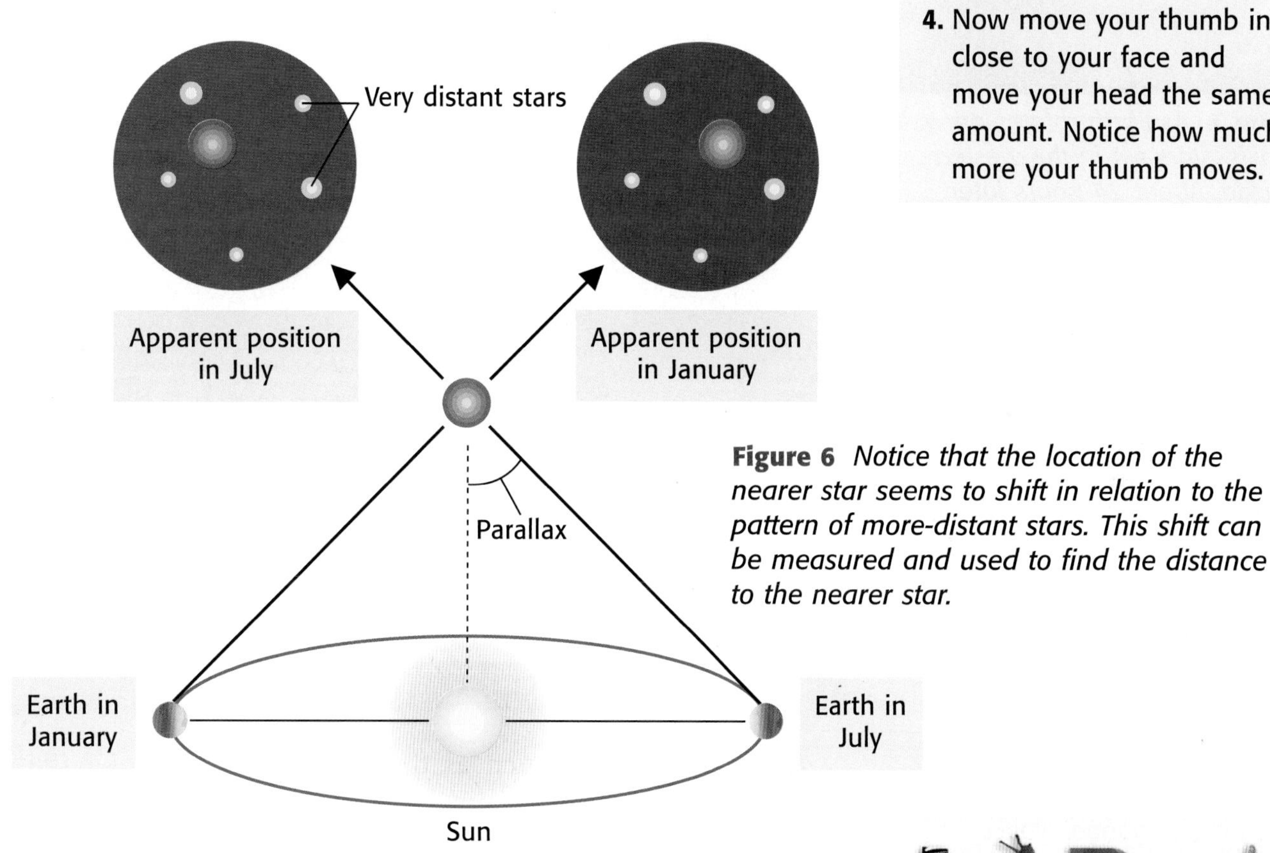

Figure 6 *Notice that the location of the nearer star seems to shift in relation to the pattern of more-distant stars. This shift can be measured and used to find the distance to the nearer star.*

Motions of Stars

As you know, the Earth rotates on its axis. As the Earth turns, different parts of its surface face the sun. This is why we have days and nights. The Earth also revolves around the sun. At different times of the year, you see different stars in the night sky. This is because the side of Earth that is away from the sun at night faces a different part of the universe.

To learn more about parallax, turn to page 594 in the LabBook.

Figure 7 *As Earth rotates on its axis, stars set in the western horizon. About 24 hours later, the stars are in the same position.*

Apparent Motion Because of our location on the Earth's surface, the sun appears to rise in the east and set in the west. Stars also seem to rise and set, as shown in **Figure 7.** During the day the atmosphere scatters light from the sun, and we can't see the stars because the sky is too bright.

Actual Motion You now know that the rising and setting of the sun and stars that we see is due to Earth's rotation. But each star is also really moving in space. Because the stars are so distant, though, their motion is hard for us to measure. Most of the stars nearest the sun are traveling in the same direction as the sun. This is like driving on a highway while all the cars are going about the same speed and direction. It would be difficult to measure the speed of the cars in the lane next to you by just watching them. If you could watch stars over thousands of years, their movement would be obvious. Because the stars in constellations only look like they belong together, the constellations will slowly change shape. This is shown in **Figure 8.**

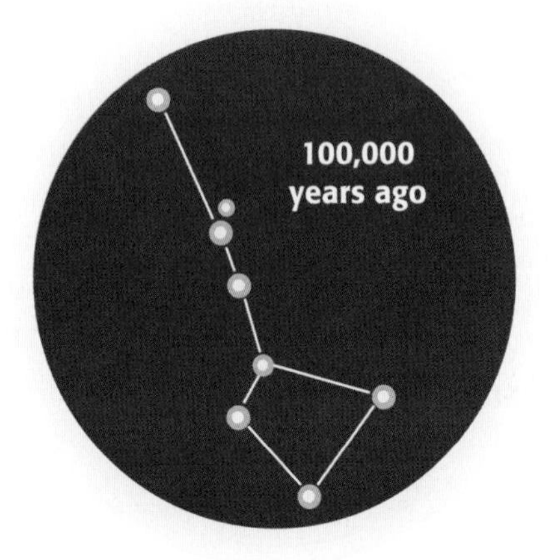

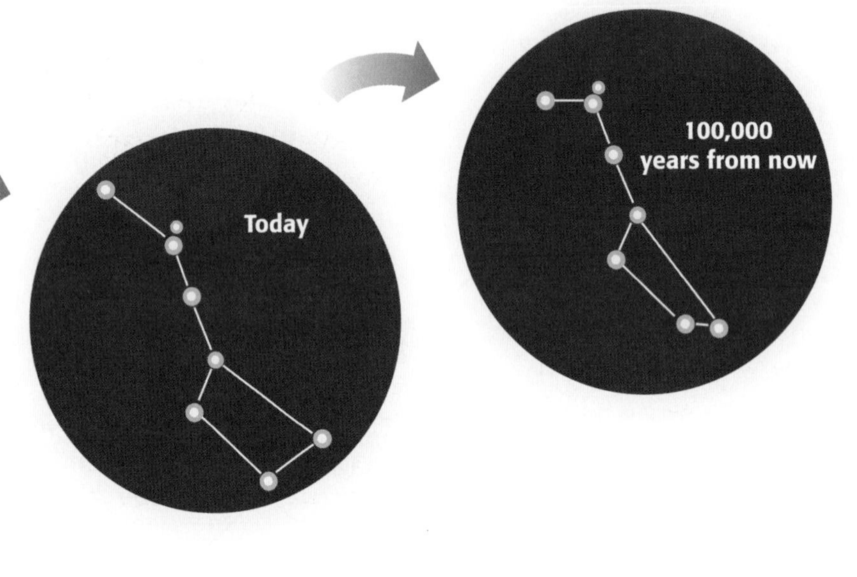

Figure 8 *Over time, the shapes of the constellations and other star groups change. Notice how the individual motion of stars will cause the Big Dipper to change its shape over 200,000 years.*

REVIEW

1. Is a yellow star, such as the sun, hotter or cooler than an orange star? Explain.
2. Suppose you see two stars that have the same apparent magnitude. If one star is actually four times as far away as the other, how much brighter would the farther star really be?
3. **Interpreting Illustrations** Look back at Figure 7. How many hours passed between the first image and the second image? How can you tell?

Section 2

The Life Cycle of Stars

NEW TERMS

H-R diagram
main sequence
white dwarf
red giant
supernova
neutron star
pulsar
black hole

OBJECTIVES

- Describe the quantities that are plotted in the H-R diagram.
- Explain how stars at different stages in their life cycle appear on different parts of the H-R diagram.

Explore

If you wanted to compare people, you might consider their height, age, or shirt size. You could then graph one variable against another to see what patterns show up.

If you graphed the heights of your school's basketball team, you might get a different result than if you graphed the heights of your classmates. In your classroom, ages might be almost the same. So choosing *what* to graph is important.

Think of two variables about your classmates, gather the data, and plot a graph. What does the graph tell you?

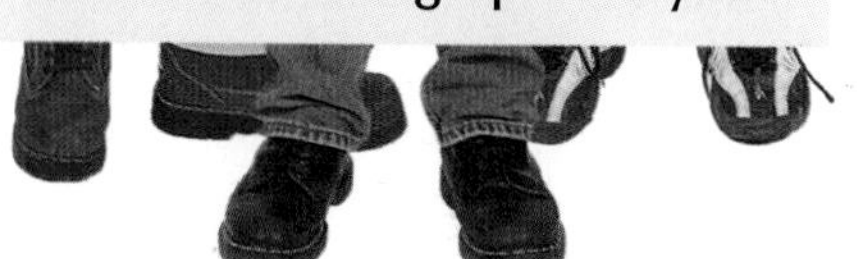

Just like people, stars are born, grow old, and eventually die. But unlike people, stars exist for billions of years. They are born when clouds of gas and dust come together and become very hot and dense. As stars get older, they lose some of their material. Usually this is a gradual change, but sometimes it happens in a big explosion. Either way, when a star dies, much of its material returns to space. There some of it combines with more gas and dust to form new stars. How do scientists know these things about stars? Read on to find out.

The Diagram That Did It!

In 1911 a Danish astronomer named Ejnar Hertzsprung plotted the temperature and brightness of stars on a graph. Two years later, American astronomer Henry Norris Russell made some graphs of his own. Although they used different data, these astronomers had similar results. By plotting the temperature of a star against its absolute magnitude, both came up with an interesting graph. The combination of their ideas is now called the *Hertzsprung-Russell,* or *H-R, diagram.* The **H-R diagram** is a graph showing the relationship between a star's surface temperature and its absolute magnitude. Russell's original diagram is shown in **Figure 9.**

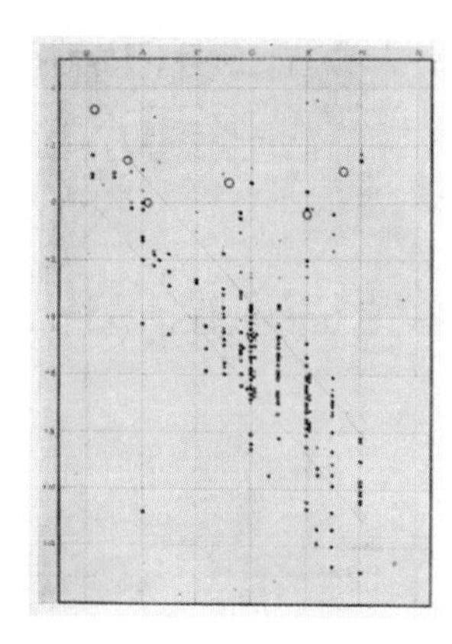

Figure 9 *Notice that a pattern begins to appear from the lower right to the upper left of the graph. Although it may not look like much, this graph began a revolution in astronomy.*

Some of the data in Russell's diagram seemed to show a trend, but the two astronomers didn't have very much data to go on because measuring the distance to stars is difficult. And as you know, distance is needed to calculate absolute magnitude.

Over the years, the H-R diagram has become a tool for studying the nature of stars. It not only shows how stars are classified by temperature and brightness but also is a good way to illustrate how stars change over time. Turn the page to see a modern version of this diagram.

The H-R Diagram

Look closely at the diagram on these two pages. Temperature is given along the bottom of the diagram. Absolute magnitude, or brightness, is given along the left side. Hot (blue) stars are located on the left, and cool (red) stars are on the right. Bright stars are at the top, and faint stars are at the bottom. The brightest stars are a million times brighter than the sun. The faintest are 1/10,000 as bright as the sun. As you can see, there seems to be a band of stars going from the top left to the bottom right corner. This diagonal pattern of stars is called the **main sequence.** A star spends most of its lifetime as a main-sequence star and then changes into one of the other types of stars shown here.

Main sequence
Stars in the main sequence form a band that runs along the middle of the H-R diagram. The sun is a main-sequence star. Stars similar to the sun are called *dwarfs.* The sun has been shining for about 5 billion years. Scientists think the sun is in midlife and that it will shine for another 5 billion years.

Absolute magnitude is measured upside down. That means the larger the number, the dimmer the star. At +5, the sun is not as bright as a –7 star.

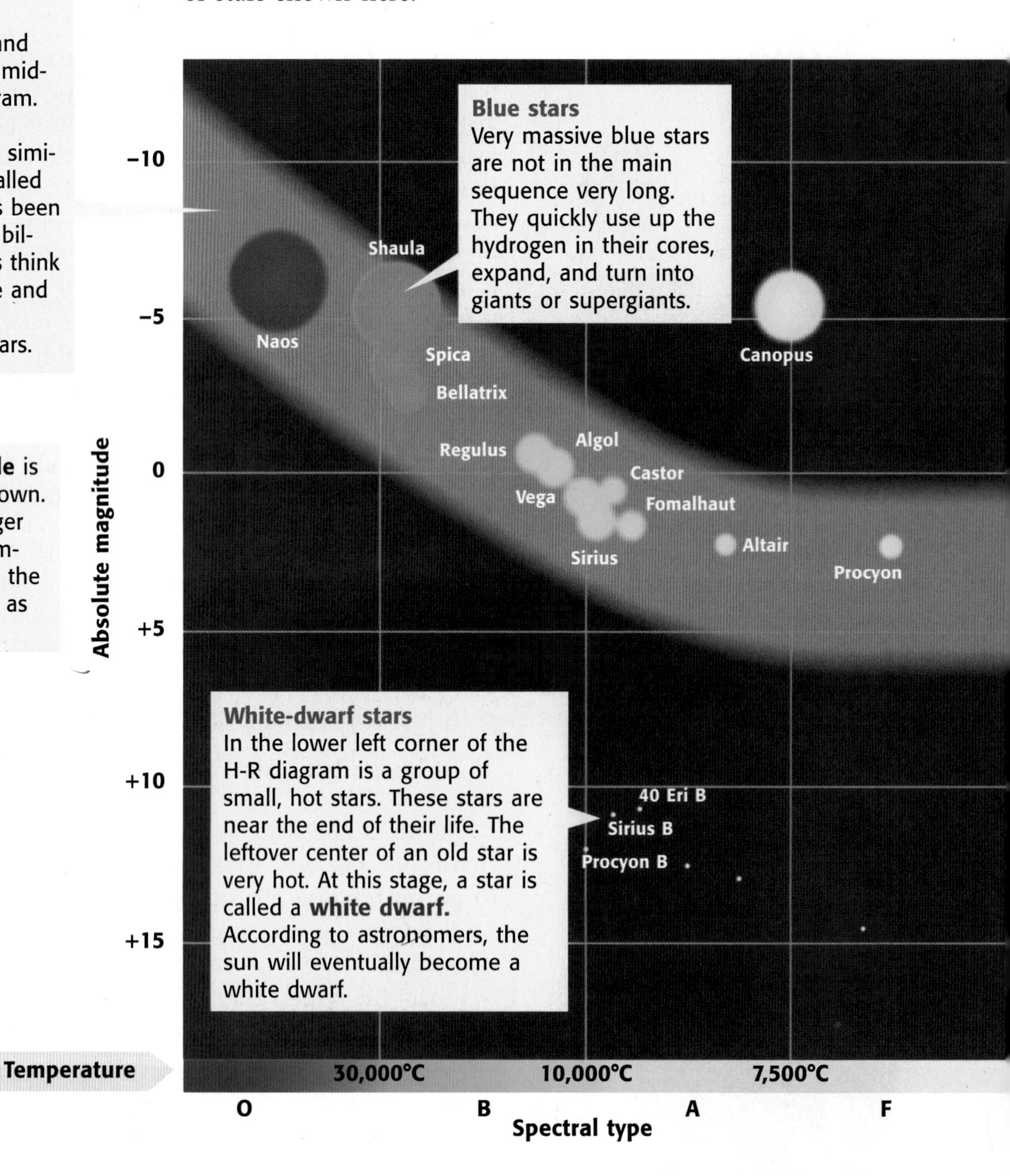

All stars begin as a ball of gas and dust in space. Gravity pulls the gas and dust together. The gas becomes hotter as it becomes more dense. When it is hot enough in the center, hydrogen turns into helium in a process called *nuclear fusion,* and lots of energy is given off. A star is born.

Most stars can be plotted on the main sequence. Small-mass stars tend to be located at the lower right end of the main sequence; larger stars are found at the left end. As main-sequence stars age, they move up and to the right on the H-R diagram to become giants or supergiants. Such stars can then lose their atmospheres, leaving small cores behind, which end up in the lower left corner of the diagram as white dwarfs.

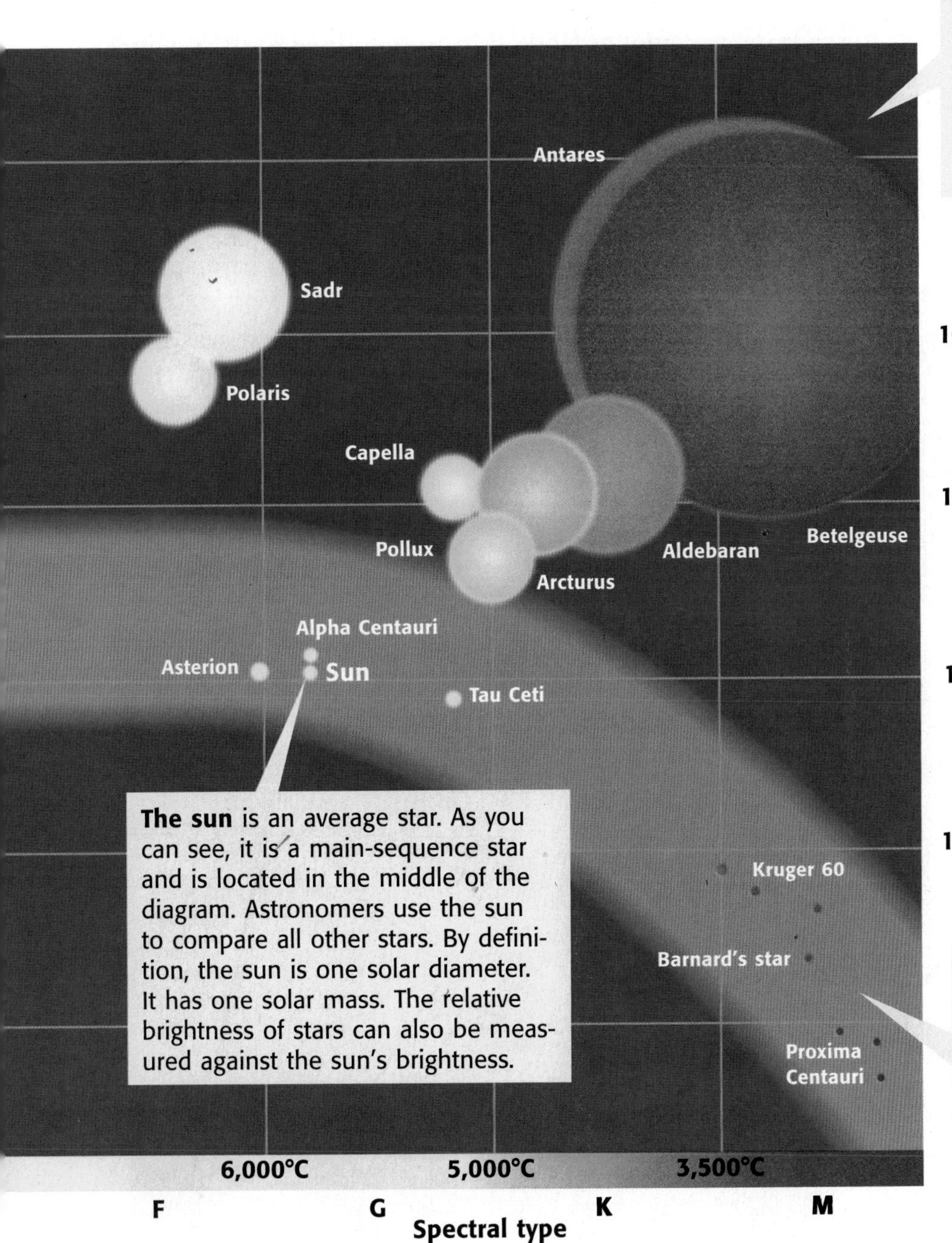

Giants and supergiants
When a star runs out of hydrogen in its core, the center of the star shrinks and the outer parts expand outward. In a star the size of our sun, the atmosphere will grow very large and cool. When this happens, the star becomes a **red giant.** If the star is very massive, it becomes a *supergiant.*

Red-dwarf stars
At the lower end of the main sequence are the red-dwarf stars. Red dwarfs are low-mass stars. Low-mass stars remain on the main sequence a long time. The lowest-mass stars may be some of the oldest stars in the galaxy.

When Stars Get Old

While stars may stay on the main sequence for a long time, they don't stay there forever. You have already seen that average stars, such as the sun, turn into red giants and then white dwarfs. But when massive stars get old, they may leave the main sequence in a more spectacular fashion. Stars much larger than the sun may explode with such violence that they turn into a variety of strange new objects. Let's take a look at some of these objects.

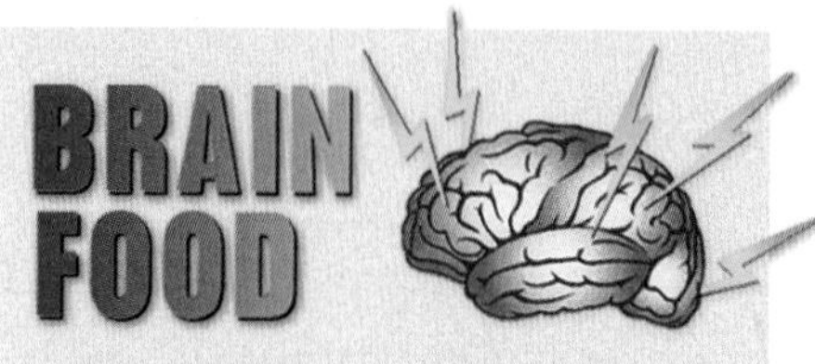

Many of the elements in your body were made during supernova explosions. In other words, you are made of "starstuff"!

Supernovas Large blue stars use up their hydrogen much faster than stars like the sun. This means they make a lot more energy, which makes them very hot and therefore blue! And compared with other stars, they don't last long. At the end of its life, a blue star may explode in a tremendous flash of light called a *supernova.* A **supernova** is basically the death of a large star by explosion. A supernova explosion is so powerful it can be brighter than an entire galaxy. It may shine for several days after the initial explosion and then gradually dim. Heavy elements, such as silver, gold, and lead, are made during a supernova explosion and are then scattered into space.

The ringed structure shown in **Figure 10** is the result of a supernova explosion that was first observed on February 23, 1987. In the years that followed, parts of the star were thrown out to form a double ring of gas and dust around the remains of the original star. The star, located in a nearby galaxy, actually exploded before civilization began here on Earth, but it took 169,000 years for the light from the explosion to reach our planet.

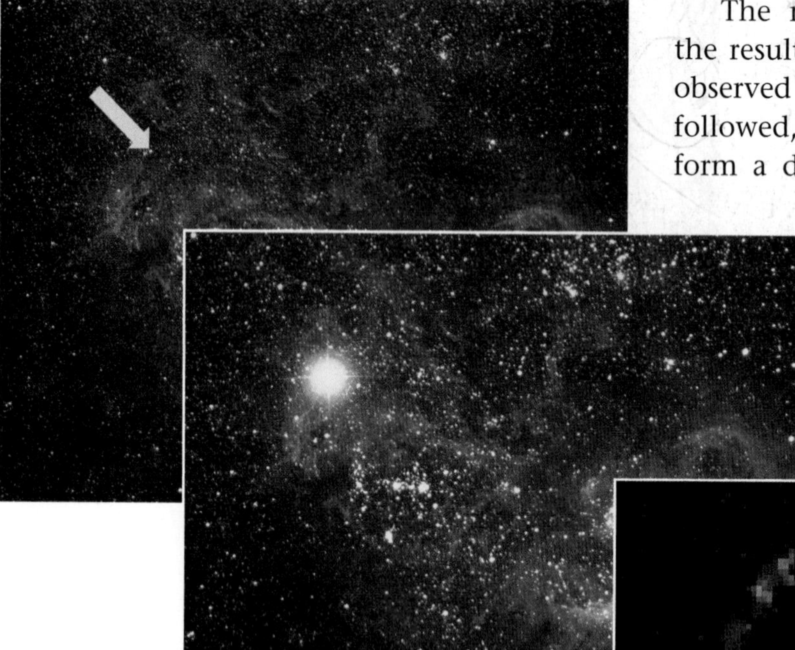

Figure 10 *Supernova 1987A, before and after (above), was the first supernova visible to the unaided eye in 400 years. Today its remains form a double ring of gas and dust, as shown in the highly magnified image at right.*

Figure 11 *A pulsar is a swiftly rotating neutron star. Pulsars can be detected on Earth only if their beams of radiation sweep past the Earth.*

Neutron Stars and Pulsars So what happens to a star that becomes a supernova? The leftover materials in the center of a supernova are squeezed together to form a star of about two solar masses. But all the material is found in a sphere only about 20 km in diameter. The particles inside the star become neutrons, so this star is called a **neutron star.** The squeezed material in a neutron star is so dense, a teaspoon of neutron star matter brought back to Earth would weigh a billion tons.

If a neutron star is also spinning, it is called a **pulsar.** A pulsar sends out beams of radiation that also spin around very rapidly. These beams, shown in **Figure 11,** are much like the beams from a lighthouse. The beams are detected as rapid clicks or pulses by radio telescopes.

Black Holes Sometimes the leftovers of a supernova are so massive that they collapse to form a *black hole.* A **black hole** is an object with more than three solar masses squeezed into a ball only 10 km across—100 football fields long. A black hole is so small and massive and its gravity is so strong that not even light can escape. That is why it is called a *black* hole. Contrary to some movie depictions, a black hole doesn't gobble up other stars. But if a star is nearby, some gas or dust from the star will spiral into the black hole, as shown in **Figure 12,** giving off X rays. It is by these X rays that astronomers can detect the existence of black holes.

Figure 12 *A black hole's gravity is so strong that it can pull in material from a nearby star, as shown in this artist's drawing.*

REVIEW

1. Are blue stars young or old? How can you tell?
2. In main-sequence stars, what is the relationship between brightness and temperature?
3. Arrange the following in order of their appearance in the life cycle of a star: white dwarf, red giant, main-sequence star. Explain your answer.
4. **Applying Concepts** Given that there are more low-mass stars than high-mass stars in the universe, do you think there are more white dwarfs or more black holes? Explain.

As a famous astronomer once said, "The black hole seems much more at home in science fiction or in ancient myth than in the real universe." Turn to page 508 to learn more about these mysterious objects in space.

Section 3

Galaxies

NEW TERMS

galaxy
spiral galaxy
elliptical galaxy
irregular galaxy
nebula
open cluster
globular cluster
quasar

OBJECTIVES

- Identify the various types of galaxies from pictures.
- Describe the contents of galaxies.
- Summarize one theory of the origin of galaxies.

Stars don't exist alone in space. They belong to larger groups that are held together by the attraction of gravity. The most common groupings are galaxies. **Galaxies** are large groupings of stars in space. Galaxies come in a variety of sizes and shapes. The largest galaxies contain more than a trillion stars. Some of the smaller ones have only a few million. Astronomers don't count the stars, of course; they estimate from the size and brightness of the galaxy how many sun-sized stars the galaxy might have.

Types of Galaxies

Look again at the Hubble Deep Field image at the beginning of this chapter. You'll notice many different types of *galaxies*. Edwin Hubble, the astronomer for whom the Hubble Space Telescope is named, began to classify galaxies in the 1920s, mostly by their shapes. We still use the galaxy names that Hubble originally assigned.

Spiral Galaxies Spiral galaxies are what most people think of when you say *galaxy*. **Spiral galaxies** have a bulge at the center and very distinctive spiral arms. When the center has a bar shape, the galaxy is called a *barred spiral*. Hot blue stars in the spiral arms make the arms in spiral galaxies appear blue. The central region, or *nuclear bulge*, appears yellow because it contains cooler stars. **Figure 13** shows a spiral galaxy tilted so you can see its pinwheel shape. Other spiral galaxies appear to be "edge-on." These galaxies are harder to classify. With them, the relative shape and size of the nuclear bulge is an important clue to determining the type of galaxy it is.

Figure 13 *The Milky Way galaxy is thought to be a spiral galaxy similar to the galaxy in Andromeda, shown here.*

Our sun is located about two-thirds the distance from the center to the outer edge of our galaxy, the *Milky Way.* It is hard to tell what type of galaxy we are in because the gas, dust, and stars keep us from having a good view. It is like trying to figure out what pattern a marching band is making while you are in the band, as shown in **Figure 14.** Observing other galaxies and making measurements inside our galaxy lead astronomers to think that Earth is in a spiral galaxy.

Figure 14 *The members of a marching band cannot see the formation the band is making while they are in the formation. We have the same problem inside the Milky Way.*

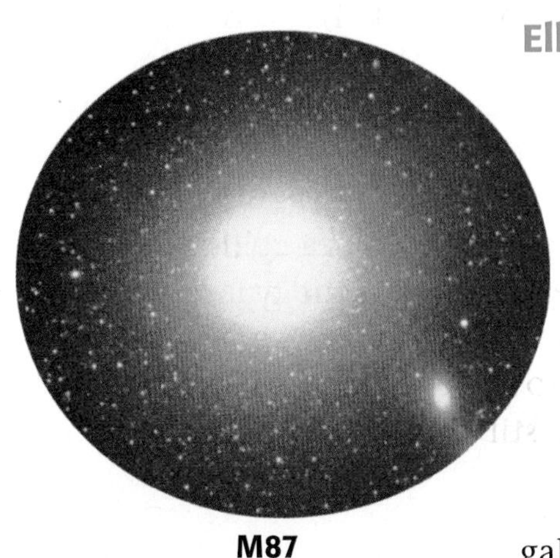

M87

Elliptical Galaxies About one-third of all galaxies are simply massive blobs of stars. Many look like spheres, while others are more elongated. Because we don't know how they are oriented, these galaxies could be cucumber shaped, with the round end facing us. These galaxies are called *elliptical galaxies.* **Elliptical galaxies** have very bright centers and very little dust and gas. Because there is so little gas, there are no new stars forming, and therefore elliptical galaxies contain only old stars. Some elliptical galaxies, like M87, above, are huge and are therefore called *giant elliptical galaxies.* Others are much smaller than the Milky Way and are called *dwarf elliptical galaxies.* There are probably lots of dwarf ellipticals, but because they are small and faint, they are very hard to detect. Astronomers have only recently begun intense searches to find more dwarf elliptical galaxies.

Irregular Galaxies When Hubble first classified galaxies, he had a group of leftovers. He named them "irregulars." **Irregular galaxies** are galaxies that don't fit into any other class. As their name suggests, their shape is irregular. Many of these galaxies, such as the Large Magellanic Cloud, shown at right, are close companions of large spiral galaxies, whose gravity may be distorting the shape of their smaller neighbors.

Large Magellanic Cloud

Explore

Now that you know the names Edwin Hubble gave to different shapes of galaxies, look at the names you gave the galaxies in the Hubble Deep Field activity at the beginning of this chapter. Rename your types with the Hubble names. Look for examples of spirals, ellipticals, and irregular galaxies.

Contents of Galaxies

Galaxies are composed of billions and billions of stars. But besides the stars and the planetary systems many of them probably have, there are larger features within galaxies that are made up of stars or the material of stars. Among these are gas clouds and star clusters.

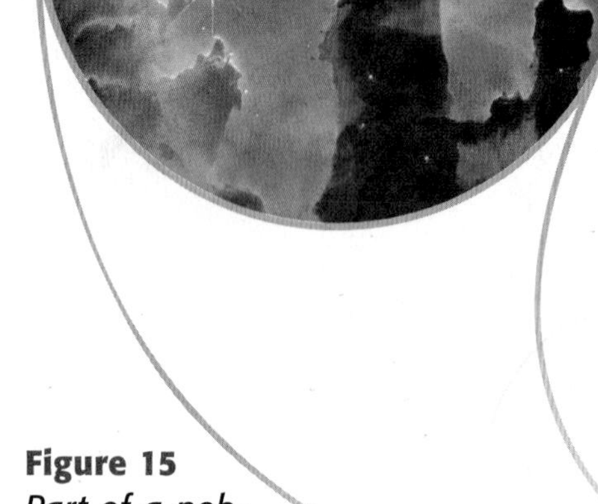

Gas Clouds The Latin word for "cloud" is *nebula.* In space, **nebulas** (or *nebulae*) are giant clouds of gas and dust. Some types of nebulas glow by themselves, while others absorb light and hide stars. Still others reflect starlight, producing some amazing images. Some nebulas are regions where new stars are formed. **Figure 15** shows part of the Eagle nebula. Spiral galaxies generally contain nebulas, but elliptical galaxies don't. When two galaxies collide, new stars may form where the gas and dust from the galaxies mix.

Figure 15 *Part of a nebula in which stars are born is shown above. The finger-like shape to the left of the bright star is slightly wider than our solar system.*

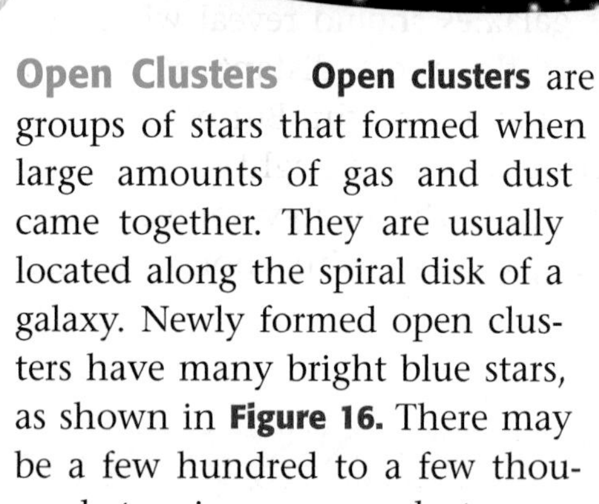

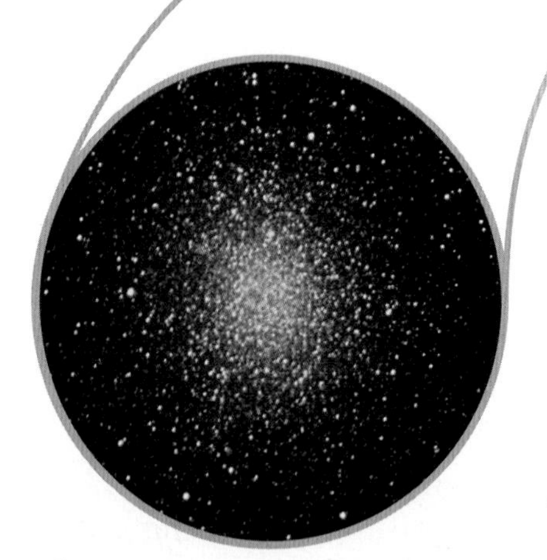

Open Clusters **Open clusters** are groups of stars that formed when large amounts of gas and dust came together. They are usually located along the spiral disk of a galaxy. Newly formed open clusters have many bright blue stars, as shown in **Figure 16.** There may be a few hundred to a few thousand stars in an open cluster.

Figure 17 *Omega Centauri is the largest globular cluster in the Milky Way. It contains 5 to 10 million stars. The Milky Way galaxy is surrounded by globular clusters.*

Figure 16 *The open cluster Pleiades is just visible without a telescope.*

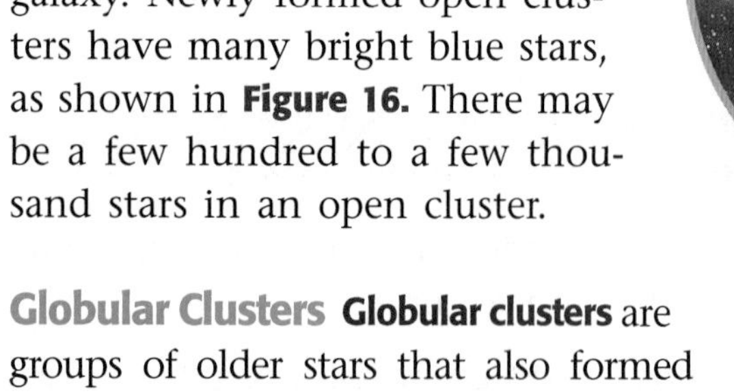

Globular Clusters **Globular clusters** are groups of older stars that also formed from a large gas cloud. A globular cluster looks like a ball of stars, as shown in **Figure 17.** There may be 20,000 to 100,000 stars in an average globular cluster. Globular clusters are located in a spherical *halo* that surrounds spiral galaxies such as the Milky Way. Globular clusters are also common around giant elliptical galaxies.

Origin of Galaxies

Astronomers do not know for sure how galaxies form. One theory is that galaxies form from collapsing clouds of gas and dust. If the cloud is rotating, a spiral galaxy will form. Some of the material will form stars during the collapse to become the galaxy's bulge. The rest of the material will collapse into a disk and form the spiral arms, where new stars will continue to form. If the cloud is not rotating fast enough, an elliptical galaxy will form. Because most of the gas and dust will be used up during the collapse, stars will no longer form once the galaxy is complete. See **Figure 18.**

Another theory is that most galaxies form as spirals. Some of these spiral galaxies then collide and merge their stars together, causing any remaining gas and dust to form stars. The rotations of the merging galaxies are affected by the collision, and once they merge completely, they become an elliptical galaxy.

Figure 18 *The formation of galaxies can follow one of two paths, according to one theory.*

Back in Time Because it takes time for light to travel through space, looking through a telescope is like looking back in time. The farther out one looks, the further back in time one travels. Because of this, distant galaxies should reveal what early galaxies looked like. Among the most distant objects are **quasars,** which look like tiny points of light. But because they are very far away, they must be extremely bright for their size. Quasars are among the most powerful energy sources in the universe. They may be young galaxies with enormous black holes at their centers.

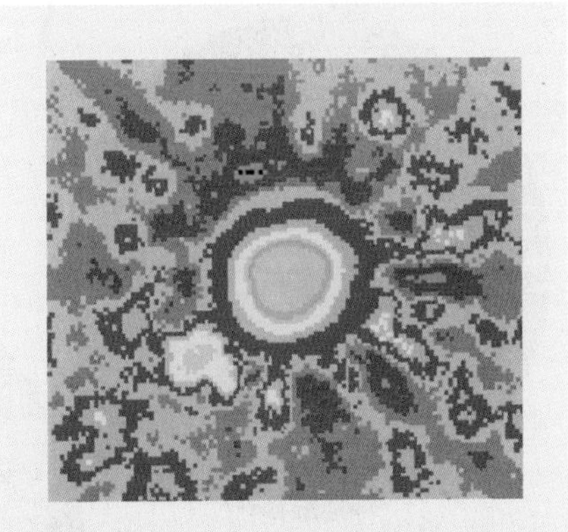

What happens when you're looking for quasars and find something else? You get famous! Turn to page 509 and see who it is.

REVIEW

1. Arrange these galaxies in order of decreasing size: spiral, giant elliptical, dwarf elliptical, irregular.
2. Describe the difference between an elliptical galaxy and a globular cluster.
3. **Analyzing Relationships** If you could observe very distant galaxies just as they begin to form stars, would you expect them to have more red or more blue stars? Why?

Section 4

Formation of the Universe

NEW TERMS

cosmology
big bang theory
cosmic background radiation

OBJECTIVES

- Describe the big bang theory.
- Explain evidence used to show support for the big bang theory.
- Explain how the expansion of the universe is explained by the big bang theory.

So far you've learned about the contents of the universe. But what about its history? How did the universe begin? How might it end? Questions like these are a special part of astronomy called *cosmology*. **Cosmology** is the study of the origin and future of the universe. Like other scientific theories, theories about the beginning and end of the universe must be tested by observations or experiments. Because the universe is so large, theories about its origin are difficult to test.

The Big Bang Theory

One of the most important theories in cosmology is the big bang theory. The **big bang theory** states that the universe began with a tremendous explosion. According to the theory, 12 billion to 15 billion years ago, all the contents of the universe were gathered together under extreme pressure, temperature, and density in a very tiny spot. Then, for some reason, it rapidly expanded outward. In the early moments of the universe, some of the expanding energy turned into matter that eventually became the galaxies, as shown in **Figure 19.**

As the galaxies move apart, they get older and eventually stop forming stars. What happens next depends on how much matter is contained in the universe. If there is enough matter, gravity will slow and eventually stop the expansion of the universe. The universe may even start collapsing, causing a "big crunch." If there is not enough matter to stop the expansion of the universe, galaxies will become more widely separated. Then as stars age and die, the universe will eventually become cold and dark. Recent observations suggest that there may not be enough matter to stop the universe from expanding forever, but the answer is still uncertain.

Figure 19 *The big bang caused the universe to expand in all directions.*

Supporting the Theory So how do we know if the big bang really happened? In 1964, two scientists, using the antenna shown in **Figure 20**, accidentally found radiation coming from all directions in space. One explanation for this radiation is that it is **cosmic background radiation** left over from the big bang. Think about what happens when an oven door is left open after the oven has been used. The heat spreads out through the kitchen as the oven cools. Eventually the room and the oven are the same temperature. According to the big bang theory, the heat from the original explosion became scattered as it traveled outward. It now fills all of space at the same temperature—a chilly –270°C, or three Celsius degrees above absolute zero.

Figure 20 *Arno Penzias (left) and Robert Wilson (right) used this strange-looking horn antenna to discover the cosmic background radiation, giving a big boost to the big bang theory.*

Today the big bang theory is widely accepted by astronomers. It seems to explain all of their observations so far. However, it is possible that new observations may not fit this theory. Scientists must remain open to new theories that might give alternative explanations for the cosmic background radiation.

Universal Expansion

But where did the idea of a big bang come from? The answer is found in deep space. No matter what direction we look, the galaxies are moving away from us, as shown in **Figure 21.** Distant galaxies are moving faster than nearby galaxies. When scientists first measured the speeds of galaxies, they observed that almost everything in the universe is moving away from our galaxy. There doesn't seem to be anything special about our galaxy, except for the fact that we live here. Are we the center of the universe? Well, not quite. A close measurement of speed and distances has shown that all galaxies are moving away from all other galaxies. But what does this mean?

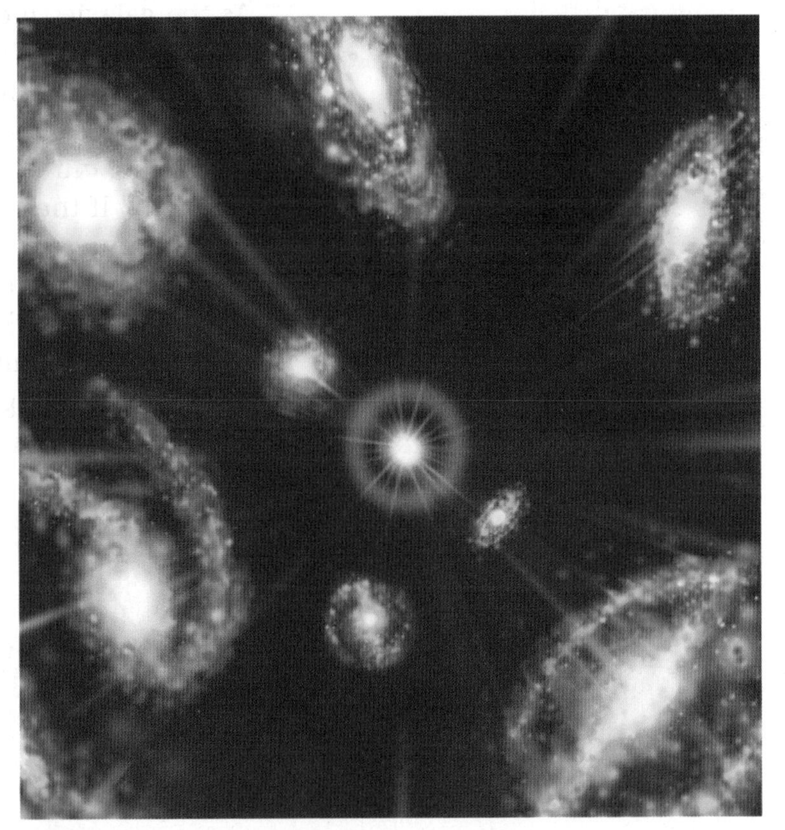

Figure 21 *The big bang theory explains the expansion of the universe we observe as galaxies move outward in all directions.*

Imagine a loaf of raisin bread before it is baked. Inside the dough, each raisin is a certain distance from each other raisin. None are moving. As the dough gets warm and rises, however, all the raisins move away from each other. No matter which raisin you choose, the others are farther from it at each moment of time. The most distant raisins seem to move fastest. In a similar way, almost all galaxies are moving away from each other. Our Milky Way galaxy, for example, is like one of the raisins. All other distant galaxies are moving away from us. And the same would be true for observers in any other galaxy. In other words, there isn't any way to find the "center" of the universe.

BRAIN FOOD

Objects in very distant space look "younger" than they really are. In fact, we cannot even be sure they still exist. If a distant galaxy disappeared, for example, people on Earth wouldn't know about it for billions of years.

Some Puzzling Questions While astronomers have learned a lot about the universe, there are still some unanswered questions. For one, we are not certain about the age of the universe. Remember, since light travels at a certain speed, looking at very distant parts of the universe is like looking back in time. Because the light from distant galaxies took a long time to travel to us, we see the galaxies as they were a long time ago. Scientists can measure how old the universe is by determining the distance of these early galaxies. But because measurements of distance to galaxies are uncertain, our calculation of the age of the universe is also uncertain.

Suppose you decide to make some raisin bread. You would form a lump of dough, as shown in the top image. The lower image represents dough that has been rising for 2 hours. Look at raisin **B** in the top image. Measure how far it is from each of the other raisins—**A, C, D, E, F,** and **G**—in millimeters. Now measure how far each raisin has moved away from **B** in the lower image. Make a graph of speed (in units of mm/h) versus original distance (in mm). Remember that speed equals distance divided by time. For example, if raisin **E** was originally 15 mm from raisin **B** and is now 30 mm away, it moved 15 mm in 2 hours. Its speed is therefore 7.5 mm/h. Repeat the procedure, starting with raisin **D.** Plot your results on the same graph, and compare the two results. What can you conclude from the information you graphed?

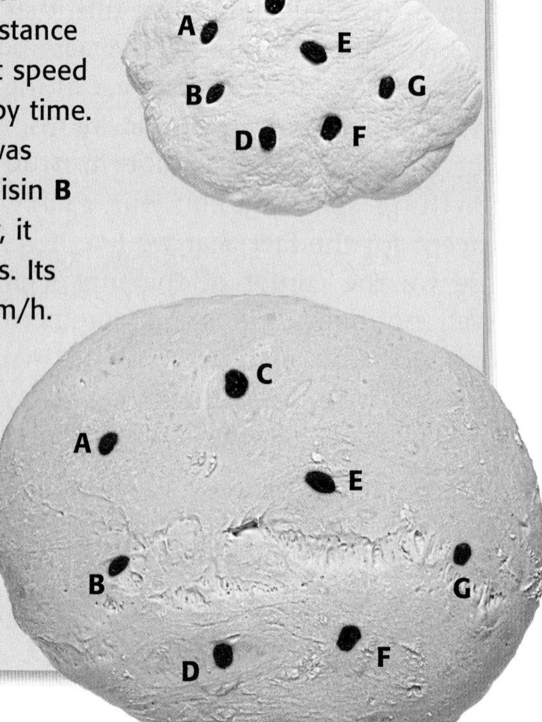

Other scientists have calculated the ages of stars by comparing theories of how stars change over time with observations of stars in and around our galaxy. Since the universe must be at least as old as the oldest stars it contains, their ages might provide a clue to the age of the universe. However, these calculations indicate that some stars are older than the universe itself! This is clearly a contradiction. Is the problem with calculating stellar ages or with measuring galaxy distances? Astronomers continue to search for scientific answers to these questions.

Structure of the Universe

The universe is an amazing place. From our home on planet Earth, it stretches out farther than we can see with our most sensitive instruments. It contains a variety of objects, some of which you have just learned about. But these objects are not simply scattered through the universe at random.

The universe has a structure that is repeated over and over again. You already know that the Earth is a planet. But planets are part of planetary systems. Our solar system is the only example we actually know about, but other stars with planets around them have been identified. Stars sprinkle the sky. But stars are grouped in larger systems, ranging from star clusters to galaxies. Galaxies themselves are arranged in groups bound together by gravity. Even galaxy groups form galaxy clusters and superclusters, as shown in **Figure 22.**

Farther than the eye can see, the universe continues with this pattern, with great collections of galaxy clusters and vast empty regions of space in between. But is the universe itself alone? Some cosmologists think that our universe is only one of a great many other universes, perhaps similar to ours or perhaps not. At present, we cannot observe other universes. But someday, who knows? Maybe students in future classrooms will have much more to study!

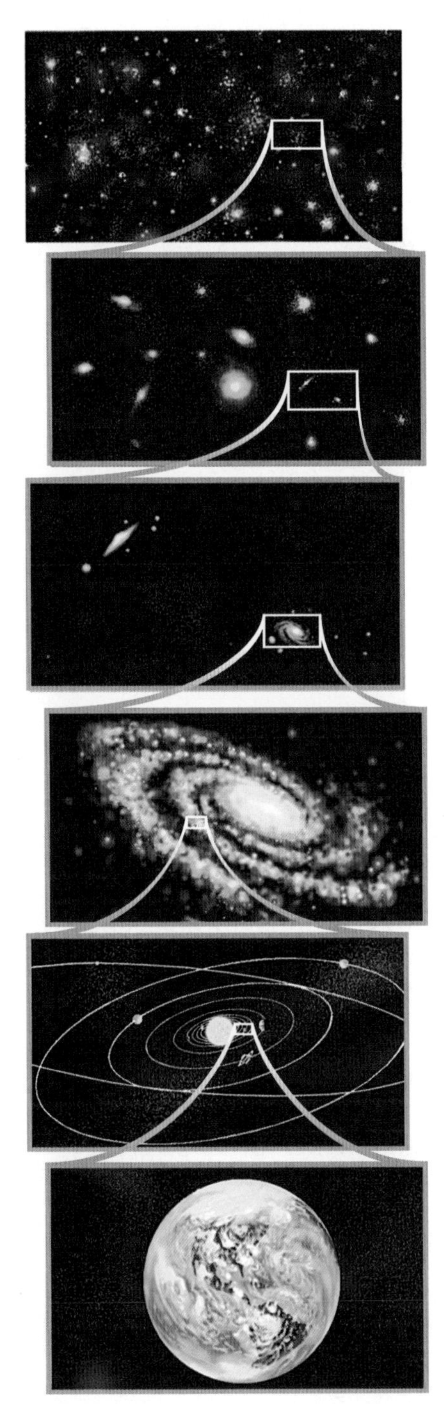

Figure 22 *The Earth is only part of a vast system of matter.*

REVIEW

1. Name one observation that supports the big bang theory.
2. How does the big bang theory explain the observed expansion of the universe?
3. **Understanding Technology** Large telescopes gather more light than small telescopes gather. Why are large telescopes used to study very distant galaxies?

Chapter Highlights

SECTION 1

Vocabulary

spectrum *(p. 484)*
apparent magnitude *(p. 488)*
absolute magnitude *(p. 488)*
light-year *(p. 489)*
parallax *(p. 489)*

Section Notes

- The color of a star depends on its temperature. Hot stars are blue. Cool stars are red.
- The spectra of stars indicate their composition. Spectra are also used to classify stars.
- The magnitude of a star is a measure of its brightness.
- Apparent magnitude is how bright a star appears from Earth.
- Absolute magnitude is how bright a star actually is. Lower absolute magnitude numbers indicate brighter stars.
- Distance to nearby stars can be measured by their movement relative to stars farther away.

Labs

Red Hot, or Not? *(p. 592)*
I See the Light! *(p. 594)*

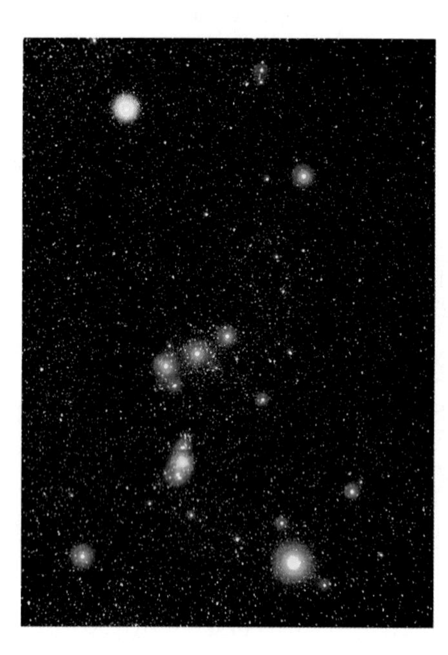

SECTION 2

Vocabulary

H-R diagram *(p. 491)*
main sequence *(p. 492)*
white dwarf *(p. 492)*
red giant *(p. 493)*
supernova *(p. 494)*
neutron star *(p. 495)*
pulsar *(p. 495)*
black hole *(p. 495)*

Section Notes

- New stars form from the material of old stars that have gone through their life cycles.
- The H-R diagram relates the temperature and brightness of a star. It also illustrates the life cycle of stars.
- Most stars are main-sequence stars. Red giants and white dwarfs are later stages in a star's life cycle.
- Massive stars become supernovas. Their cores turn into neutron stars or black holes.

☑ Skills Check

Math Concepts

SQUARING THE DIFFERENCE The difference in brightness (apparent magnitude) between a pair of similar stars depends on the difference in their distances from Earth. Compare a star that is 10 light-years away with a star that is 5 light-years away. One star is twice as close, so it is $2 \times 2 = 4$ times brighter than the other star. The star that is 5 light-years away is also 3^2, or 9, times brighter than one that is 15 light-years away.

Visual Understanding

READING BETWEEN THE LINES The composition of a star is determined by the absorption spectra it displays. Dark lines in the spectrum of a star indicate which elements are present. Look back at Figure 3 to review.

SECTION 3

Vocabulary

galaxy *(p. 496)*
spiral galaxy *(p. 496)*
elliptical galaxy *(p. 497)*
irregular galaxy *(p. 497)*
nebula *(p. 498)*
open cluster *(p. 498)*
globular cluster *(p. 498)*
quasar *(p. 499)*

Section Notes

- Edwin Hubble classified galaxies according to their shape. Major types include spiral, elliptical, and irregular galaxies.
- The Milky Way is our galaxy. Our sun is located in a spiral arm about two-thirds the distance from the galaxy's center to its edge.
- A nebula is a cloud of gas and dust. New stars are born in some nebulas.
- Open clusters are groups of stars located along the spiral disk of a galaxy. Globular star clusters are found in the halos of spiral galaxies and in elliptical galaxies.

SECTION 4

Vocabulary

cosmology *(p. 500)*
big bang theory *(p. 500)*
cosmic background radiation *(p. 501)*

Section Notes

- The big bang theory states that the universe began with an explosion about 12 billion to 15 billion years ago.
- Cosmic background radiation fills the universe with radiation that is left over from the big bang. It is the primary evidence supporting the big bang theory.
- Observations show that the universe is expanding outward. There is no measurable center and no apparent edge.
- All matter in the universe is a part of larger systems, from planets to superclusters of galaxies.

internet**connect**

GO TO: go.hrw.com

Visit the **HRW** Web site for a variety of learning tools related to this chapter. Just type in the keyword:

KEYWORD: HSTUNV

GO TO: www.scilinks.org

Visit the **National Science Teachers Association** on-line Web site for Internet resources related to this chapter. Just type in the ***sci*LINKS** number for more information about the topic:

TOPIC: The Hubble Space Telescope — ***sci*LINKS NUMBER:** HSTP630
TOPIC: Stars — ***sci*LINKS NUMBER:** HSTP635
TOPIC: Supernovas — ***sci*LINKS NUMBER:** HSTP640
TOPIC: Galaxies — ***sci*LINKS NUMBER:** HSTP645

Chapter Review

USING VOCABULARY

For each pair of terms, explain the difference in their meanings.

1. absolute magnitude/apparent magnitude
2. spectrum/parallax
3. main-sequence star/red giant
4. white dwarf/black hole
5. elliptical galaxy/spiral galaxy
6. big bang/cosmic background radiation

UNDERSTANDING CONCEPTS

Multiple Choice

7. The majority of stars in our galaxy are
   a. blue.
   b. white dwarfs.
   c. main-sequence stars.
   d. red giants.

8. Which would be seen as the brightest star in the following group?
   a. Alcyone—apparent magnitude of 3
   b. Alpheratz—apparent magnitude of 2
   c. Deneb—apparent magnitude of 1
   d. Rigel—apparent magnitude of 0

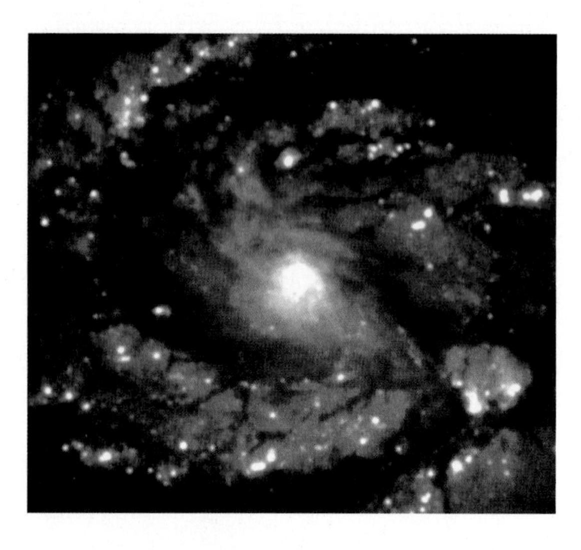

9. A cluster of stars forms in a nebula. There are red stars, blue stars, yellow stars, and white stars. Which stars are most like the sun?
   a. red
   b. yellow
   c. blue
   d. white

10. Individual stars are moving in space. How long will it take to see a noticeable difference without using a telescope?
    a. 24 hours
    b. 1 year
    c. 100 years
    d. 100,000 years

11. You visited an observatory and looked through the telescope. You saw a ball of stars through the telescope. What type of object did you see?
    a. a spiral galaxy
    b. an open cluster
    c. a globular cluster
    d. a barred spiral galaxy

12. In which part of a spiral galaxy do you expect to find nebulas?
    a. the spiral arms
    b. the nuclear bulge
    c. the halo
    d. all parts of the galaxy

13. Which statement about the big bang theory is accurate?
    a. The universe will never end.
    b. New matter is being continuously created in the universe.
    c. The universe is filled with radiation coming from all directions in space.
    d. We can locate the center of the universe.

Short Answer

14. Describe how the apparent magnitude of a star varies with its distance from Earth.

15. Name six types of astronomical objects in the universe. Arrange them by size.

16. What happens when two spiral galaxies collide with each other?

17. What does the big bang theory have to say about how the universe will end?

Concept Mapping

18. Use the following terms to create a concept map: black hole, neutron star, main-sequence star, red giant, nebula, white dwarf.

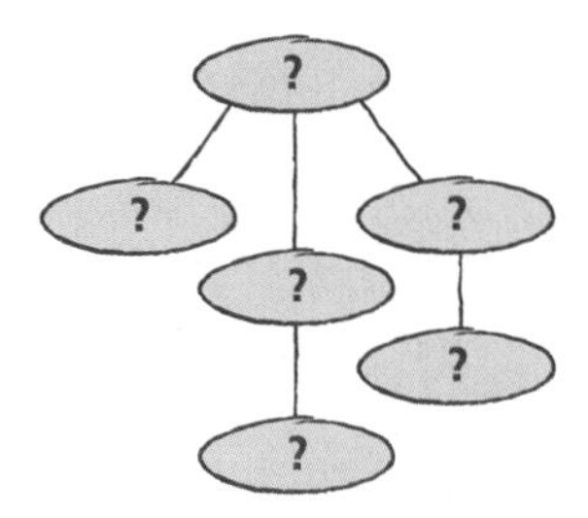

CRITICAL THINKING AND PROBLEM SOLVING

Write one or two sentences to answer the following questions:

19. If a certain star displayed a large parallax, what could you say about its distance from Earth?

20. Two M-type stars have the same apparent magnitude. Their spectra show that one is a red giant and the other is a red-dwarf star. Which one is farther from Earth? Explain your answer.

21. Look back at the H-R diagram in Section 2. Why do astronomers use absolute magnitudes to plot the stars? Why don't they use apparent magnitudes?

22. While looking at a galaxy through a nearby university's telescope, you notice that there are no blue stars present. What kind of galaxy is it most likely to be?

MATH IN SCIENCE

23. An astronomer observes two stars of about the same temperature and size. Alpha Centauri B is about 4 light-years away, and sigma[2] Eridani A is about 16 light-years away. How much brighter does Alpha Centauri B appear?

INTERPRETING GRAPHICS

The following graph illustrates the Hubble law relating the distances of galaxies and their speed away from us.

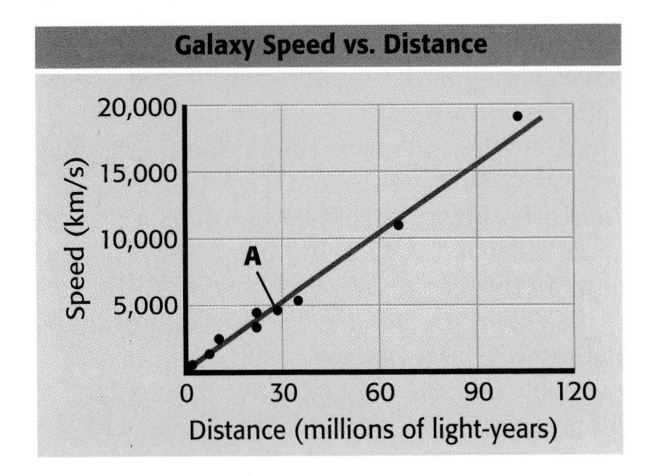

24. Look at the galaxy marked **A** in the graph. What is its speed and distance?

25. If a new galaxy with a speed of 15,000 km/s were found, at what distance would you expect it to be?

NOW What Do You Think?

Take a minute to review your answers to the ScienceLog questions on page 483. Have your answers changed? If necessary, revise your answers based on what you have learned since you began this chapter.

WEIRD SCIENCE

HOLES WHERE STARS ONCE WERE

An invisible phantom lurks in outer space, ready to swallow up everything that comes near it. Once trapped in its grasp, matter is stretched, torn, and crushed into oblivion. Does this sound like a horror story? Guess again! Scientists call it a black hole.

Born of a Collapsing Star

As a star runs out of fuel, it cools and eventually collapses under the force of its own gravity. If the collapsing star is large enough, it may shrink enough to become a black hole. One spoonful of the matter in a black hole would be heavier than the Earth! The resulting gravitational attraction is so enormous that even light cannot escape.

Scientists predict that at the center of the black hole is a *singularity,* a tiny point of incredible density, temperature, and pressure. The area around the singularity is called the *event horizon.* The event horizon represents the boundary of the black hole. Anything that crosses the event horizon, including light, will eventually be pulled into the black hole. As matter comes near the event horizon, the matter begins to swirl in much the same way that water swirls down a drain.

▲ *The Hubble Space Telescope*

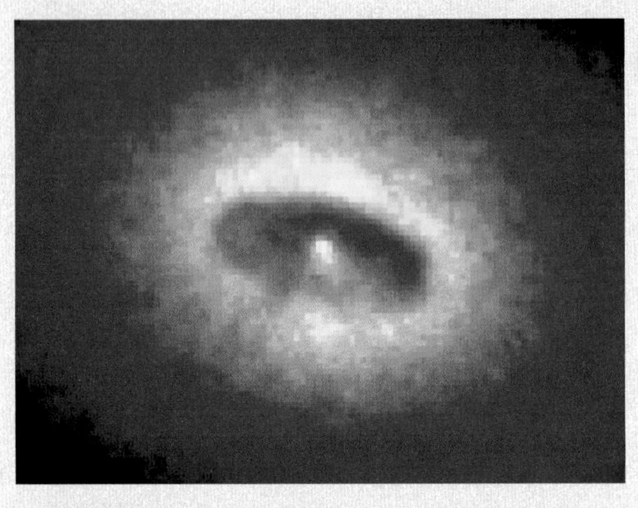

▲ *This photograph of M87 was taken by the Hubble Space Telescope.*

The Story of M87

For years, scientists had theorized about black holes but hadn't actually found one. Then in 1994, scientists found something strange at the core of a galaxy called M87. Scientists detected a disk-shaped cloud of gas with a diameter of 60 light-years, rotating at about 2 million kilometers per hour. When scientists realized that a mass more than 2 billion times that of the sun was crammed into a space no bigger than our solar system, they knew that something was pulling in the gases at the center of the galaxy.

Many astronomers think that black holes, such as the one in M87, lie at the heart of many galaxies. Some scientists suggest that there is a giant black hole at the center of our own Milky Way galaxy. But don't worry. The Earth is too far away to be caught.

Modeling a Black Hole

► Make a model to show how a black hole pulls in the matter surrounding it. Indicate the singularity and event horizon.

CAREERS

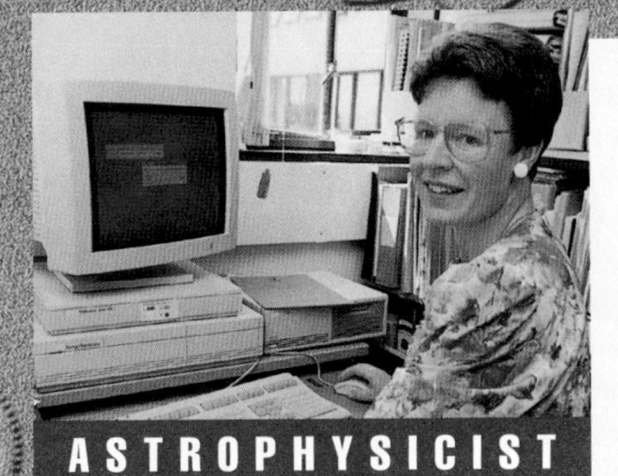

ASTROPHYSICIST

Jocelyn Bell-Burnell became fascinated with astronomy at an early age. As a research student at Cambridge University, Bell-Burnell discovered pulsars, celestial objects that emit radio waves at short and regular intervals. Today Bell-Burnell is a leading expert in the field of astrophysics and the study of stars. She is currently head of the physics department at the Open University, in Milton Keynes, England.

At Cambridge University in 1967, Bell-Burnell and her adviser, Antony Hewish, completed work on a gigantic radio telescope designed to pick up signals from quasars. Bell-Burnell's job was to operate the telescope and analyze the "chart paper" recordings of the telescope on a graph. Each day, the telescope recorded 29.2 m of chart paper! After a month of operating the telescope, Bell-Burnell noticed a few "bits of scruff" that she could not explain—they were very short, pulsating radio signals. The signals were only 6.3 mm long, and they occurred only once every 4 days. What Bell-Burnell had accidentally found was a needle in a cosmic haystack!

LGM 1

Bell-Burnell and Hewish struggled to find the source of this mysterious new signal. They double-checked the equipment and began eliminating all of the possible sources of the signal, such as satellites, television, and radar. Because they could not rule out that the signal was coming from aliens, Bell-Burnell and Hewish called it LGM 1. Can you guess why? LGM stood for "Little Green Men"!

The Answer: Neutron Stars

Shortly after finding the first signal, Bell-Burnell discovered yet another strange, pulsing signal within the vast quantity of chart paper. This signal was similar to the first, except that it came from the other side of the sky. To Bell-Burnell, this second signal was exciting because it meant that her first signal was not of local origin and that she had stumbled on a new and unknown signal from space! By January 1968, Bell-Burnell had discovered two more pulsating signals. In March of that year, her findings were published, to the amazement of the scientific community. The scientific press coined the term *pulsars,* from pulsating radio stars. Bell-Burnell and other scientists reached the conclusion that her "bits of scruff" were caused by rapidly spinning neutron stars!

Star Tracking

▶ Pick out a bright star in the sky, and keep a record of its position in relation to a reference point, such as a tree or building. Each night, record what time the star appears at this point in the sky. Do you notice a pattern?

▲ *An artist's depiction of a pulsar*

LabBook

Contents

Exploring, inventing, and investigating are essential to the study of science. However, these activities can also be dangerous. To make sure that your experiments and explorations are safe, you must be aware of a variety of safety guidelines.

You have probably heard of the saying, "It is better to be safe than sorry." This is particularly true in a science classroom where experiments and explorations are being performed. Being uninformed and careless can result in serious injuries. Don't take chances with your own safety or with anyone else's.

Following are important guidelines for staying safe in the science classroom. Your teacher may also have safety guidelines and tips that are specific to your classroom and laboratory. Take the time to be safe.

Safety Rules!

Start Out Right

Always get your teacher's permission before attempting any laboratory exploration. Read the procedures carefully, and pay particular attention to safety information and caution statements. If you are unsure about what a safety symbol means, look it up or ask your teacher. You cannot be too careful when it comes to safety. If an accident does occur, inform your teacher immediately, regardless of how minor you think the accident is.

If you are instructed to note the odor of a substance, wave the fumes toward your nose with your hand. Never put your nose close to the source.

Safety Symbols

All of the experiments and investigations in this book and their related worksheets include important safety symbols to alert you to particular safety concerns. Become familiar with these symbols so that when you see them, you will know what they mean and what to do. It is important that you read this entire safety section to learn about specific dangers in the laboratory.

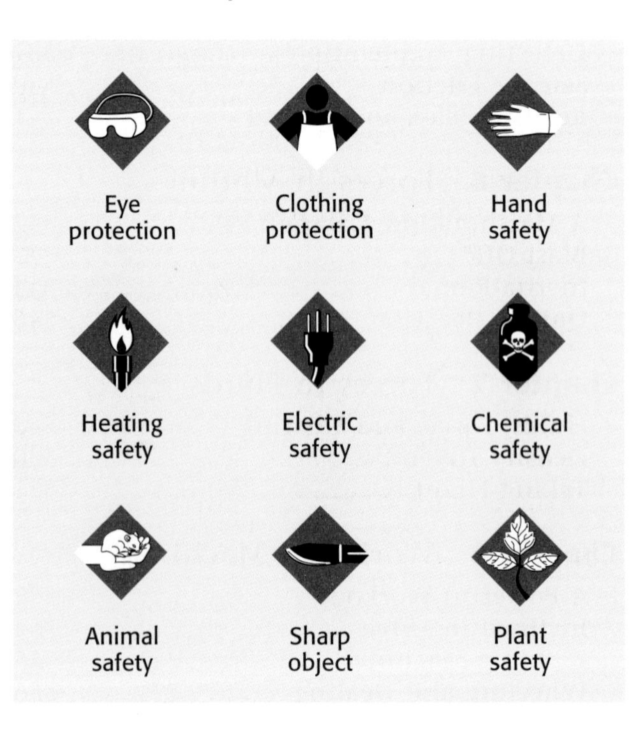

Eye Safety 

Wear safety goggles when working around chemicals, acids, bases, or any type of flame or heating device. Wear safety goggles any time there is even the slightest chance that harm could come to your eyes. If any substance gets into your eyes, notify your teacher immediately, and flush your eyes with running water for at least 15 minutes. Treat any unknown chemical as if it were a dangerous chemical. Never look directly into the sun. Doing so could cause permanent blindness.

Avoid wearing contact lenses in a laboratory situation. Even if you are wearing safety goggles, chemicals can get between the contact lenses and your eyes. If your doctor requires that you wear contact lenses instead of glasses, wear eye-cup safety goggles in the lab.

Safety Equipment

Know the locations of the nearest fire alarms and any other safety equipment, such as fire blankets and eyewash fountains, as identified by your teacher, and know the procedures for using them.

Be extra careful when using any glassware. When adding a heavy object to a graduated cylinder, tilt the cylinder so the object slides slowly to the bottom.

Neatness

Keep your work area free of all unnecessary books and papers. Tie back long hair, and secure loose sleeves or other loose articles of clothing, such as ties and bows. Remove dangling jewelry. Don't wear open-toed shoes or sandals in the laboratory. Never eat, drink, or apply cosmetics in a laboratory setting. Food, drink, and cosmetics can easily become contaminated with dangerous materials.

Certain hair products (such as aerosol hair spray) are flammable and should not be worn while working near an open flame. Avoid wearing hair spray or hair gel on lab days.

Sharp/Pointed Objects 

Use knives and other sharp instruments with extreme care. Never cut objects while holding them in your hands. Place objects on a suitable work surface for cutting.

Heat

Wear safety goggles when using a heating device or a flame. Whenever possible, use an electric hot plate as a heat source instead of an open flame. When heating materials in a test tube, always angle the test tube away from yourself and others. In order to avoid burns, wear heat-resistant gloves whenever instructed to do so.

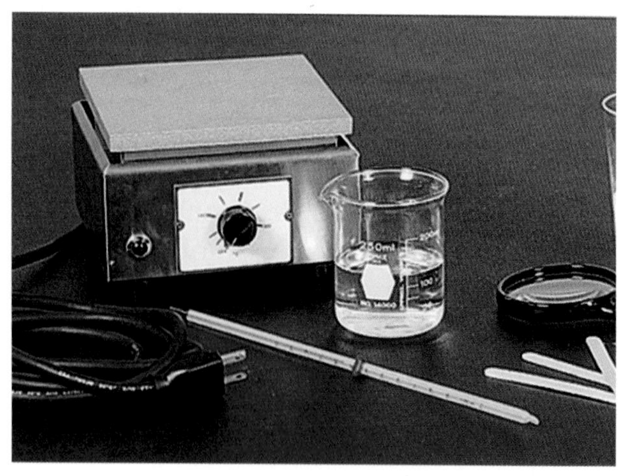

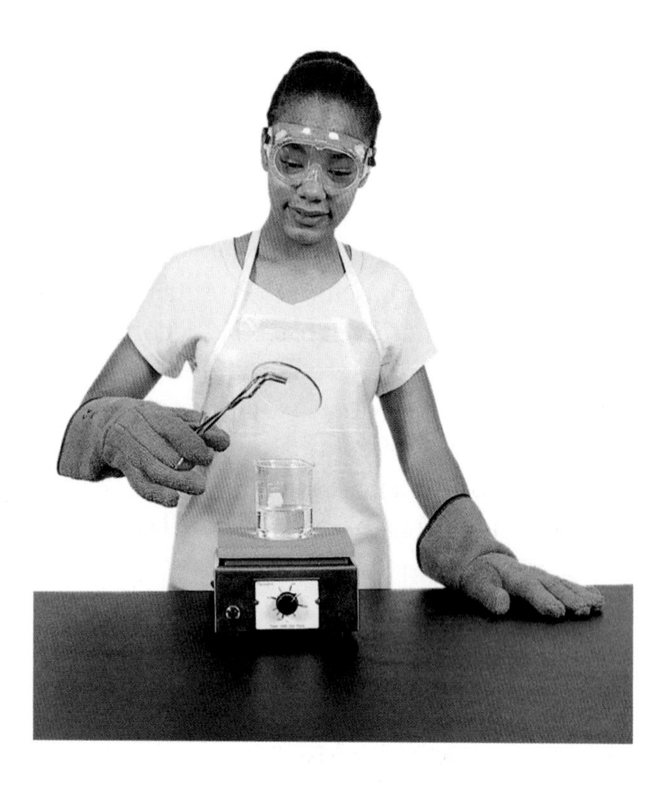

Chemicals    

Wear safety goggles when handling any potentially dangerous chemicals, acids, or bases. If a chemical is unknown, handle it as you would a dangerous chemical. Wear an apron and safety gloves when working with acids or bases or whenever you are told to do so. If a spill gets on your skin or clothing, rinse it off immediately with water for at least 5 minutes while calling to your teacher.

Never mix chemicals unless your teacher tells you to do so. Never taste, touch, or smell chemicals unless you are specifically directed to do so. Before working with a flammable liquid or gas, check for the presence of any source of flame, spark, or heat.

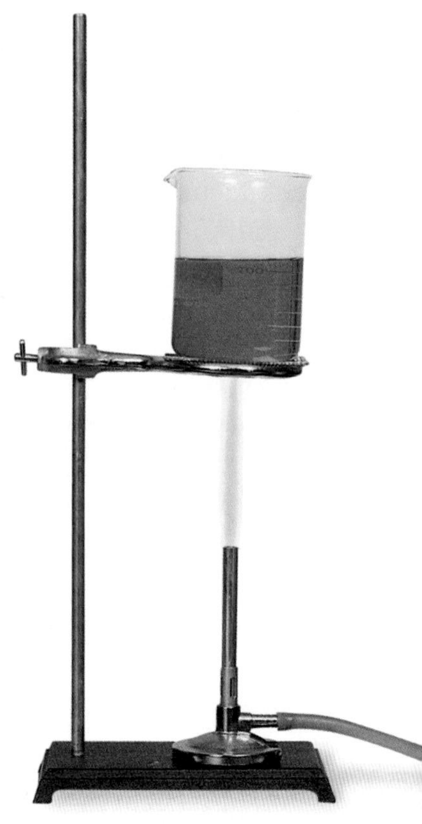

Electricity

Be careful with electrical cords. When using a microscope with a lamp, do not place the cord where it could trip someone. Do not let cords hang over a table edge in a way that could cause equipment to fall if the cord is accidentally pulled. Do not use equipment with damaged cords. Be sure your hands are dry and that the electrical equipment is in the "off" position before plugging it in. Turn off and unplug electrical equipment when you are finished.

Animal Safety 

Always obtain your teacher's permission before bringing any animal into the school building. Handle animals only as your teacher directs. Always treat animals carefully and with respect. Wash your hands thoroughly after handling any animal.

Plant Safety 

Do not eat any part of a plant or plant seed used in the laboratory. Wash hands thoroughly after handling any part of a plant. When in nature, do not pick any wild plants unless your teacher instructs you to do so.

Glassware

Examine all glassware before use. Be sure that glassware is clean and free of chips and cracks. Report damaged glassware to your teacher. Glass containers used for heating should be made of heat-resistant glass.

Exploring the Unseen

Your teacher will give you a box in which a special divider has been created. Your task is to describe this divider as precisely as possible—without opening the box! Your only aid is a marble that is also inside the box. This task will allow you to demonstrate your understanding of the scientific method. Good luck!

Materials

- a sealed mystery box

Ask a Question

1. In your ScienceLog, record the question that you are trying to answer by doing this experiment. (Hint: Read the introductory paragraph again if you are not sure what your task is.)

Form a Hypothesis

2. Before you begin the experiment, think about what's required. Do you think you will be able to easily determine the shape of the divider? What about its texture? its color? In your ScienceLog, write a hypothesis that states how much you think you will be able to determine about the divider during the experiment. (Remember, you can't open the box!)

Test the Hypothesis

3. Using all the methods you can think of (except opening the box), test your hypothesis. Make careful notes about your testing and observations in your ScienceLog.

Analyze the Results

4. What characteristics of the divider were you able to identify? Draw or write your best description of the interior of the box.
5. Do your observations support your hypothesis? Explain. If your results do not support your hypothesis, write a new hypothesis and test it.
6. With your teacher's permission, open the box and look inside. Record your observations in your ScienceLog.

Communicate Results

7. Write a paragraph summarizing your experiment. Be sure to include your methods, whether your results supported your hypothesis, and how you could improve your methods.

Off to the Races!

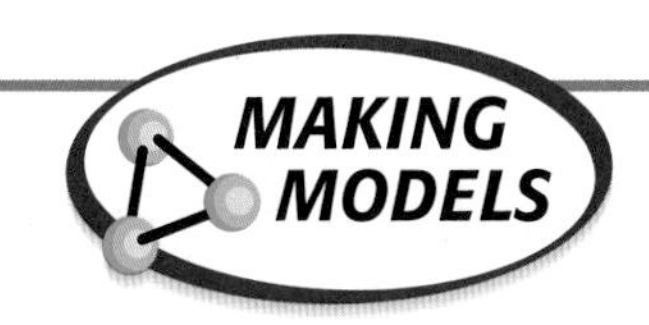

Scientists often use models—representations of objects or systems. Physical models, such as a model airplane, are generally a different size than the objects they represent. In this lab you will build a model car, test its design, and then try to improve the design.

Materials

- 2 sheets of typing paper
- glue
- 16 cm clothes-hanger wire
- pliers or wire cutters
- metric ruler
- rubber eraser or wooden block
- ramp (board and text-books)
- stopwatch

Procedure

1. Using the materials listed, design and build a car that will carry the load (the eraser or block of wood) down the ramp as quickly as possible. Your car must be no wider than 8 cm, it must have room to carry the load, and it must roll.
2. As you test your design, do not be afraid to rebuild or redesign your car. Improving your methods is an important part of scientific progress.
3. When you have a design that works well, measure the time required for your car to roll down the ramp. Record this time in your ScienceLog. Repeat this step several times.
4. Try to improve your model. Find one thing that you can change to make your model car roll faster down the ramp. In your ScienceLog, write a description of the change.
5. Repeat step 3.

Analysis

6. Why is it important to have room in the model car for the eraser or wood block? (Hint: Think about the function of a real car.)
7. Before you built the model car, you created a design for it. Do you think this design is also a model? Explain.
8. Based on your observations in this lab, list three reasons why it is helpful for automobile designers to build and test small model cars rather than immediately building a full-size car.
9. In this lab you built a model that was smaller than the object it represented. Some models are larger than the objects they represent. List three examples of larger models that are used to represent objects. Why is it helpful to use a larger model in these cases?

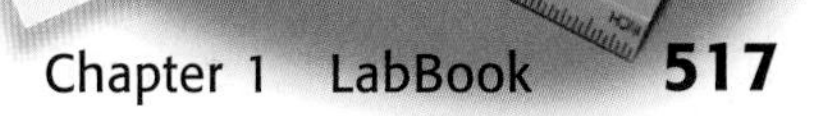

Measuring Liquid Volume

In this lab you will practice your science skills by using a graduated cylinder to measure and transfer precise amounts of liquids. Remember, in order to accurately measure liquids in a graduated cylinder, you should read the level at the bottom of the meniscus, the curved surface of the liquid.

Materials

- masking tape
- marker
- 6 large test tubes
- test-tube rack
- 10 mL graduated cylinder
- 3 beakers filled with colored liquid
- small funnel

Procedure

1. Using the masking tape and marker, label the test tubes A, B, C, D, E, and F. Place them in the test-tube rack. Be careful not to confuse the test tubes.
2. Using the 10 mL graduated cylinder, pour 14 mL of the red liquid into test tube A. To do this, first pour 10 mL of the liquid into the test tube and then add 4 mL of liquid. (Pouring the liquid through a funnel makes the task easier.)
3. Rinse the graduated cylinder and funnel before each use.
4. Using the same technique as you did in step 2, measure 13 mL of the yellow liquid, and pour it into test tube C. Then measure 13 mL of the blue liquid, and pour it into test tube E.
5. Transfer 4 mL of liquid from test tube C into test tube D. Transfer 7 mL of liquid from test tube E into test tube D.
6. Measure 4 mL of blue liquid from the beaker, and pour it into test tube F. Measure 7 mL of red liquid from the beaker, and pour it into test tube F.
7. Transfer 8 mL of liquid from test tube A into test tube B. Transfer 3 mL of liquid from test tube C into test tube B.

Collect Data

8. Make a data table in your ScienceLog, and record the color of the liquid in each test tube.
9. Use the graduated cylinder to measure the volume of liquid in each test tube, and record the volumes in your data table. Be sure to include the proper units.
10. On an overhead transparency or the board, record your color observations in a table of class data prepared by your teacher.
11. Make another data table in your ScienceLog, and record all the color observations for your class.

Analysis

12. Did all of the groups report the same colors? Explain why the colors were the same or different.
13. Why should you not fill the graduated cylinder to the very top?

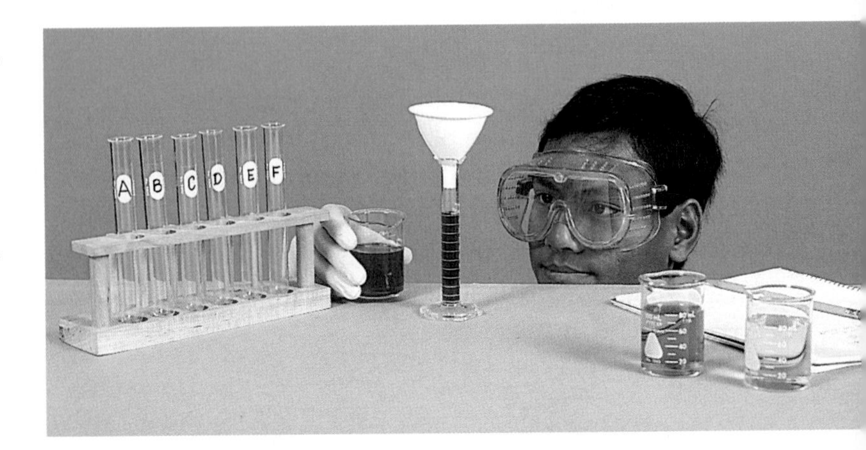

Coin Operated

All pennies are exactly the same, right? Probably not! After all, each penny was made in a certain year at a specific mint, and each has traveled a unique path to reach your classroom. But all pennies *are* similar. In this lab you will investigate differences and similarities among a group of pennies.

Materials

- 10 pennies
- metric balance
- few sheets of paper
- 100 mL graduated cylinder
- water
- paper towels

Procedure

1. Write the numbers 1 through 10 on a page in your ScienceLog, and place a penny next to each number.
2. Use the metric balance to find the mass of each penny. Round off the values to the nearest 0.1 g. Record the mass of each penny in your ScienceLog.
3. On an overhead projector or the board, your teacher will make a table that is marked in 0.1 g units. Make a mark in the correct column of the table for each penny you measured.
4. Separate your pennies into piles based on the class data. Place each pile on its own sheet of paper.
5. Measure and record the mass of each pile. Write the mass on the paper you are using to identify the pile.
6. Fill a graduated cylinder about halfway with water, and determine the volume as precisely as possible. Record the volume of the water.
7. Carefully place the pennies from one pile in the graduated cylinder. Measure and record the new volume.
8. Carefully remove the pennies from the graduated cylinder, and dry them off.
9. Repeat steps 6 through 8 for each pile of pennies.
10. Determine the volume of the displaced water by subtracting the initial volume from the final volume. This amount is equal to the volume of the pennies. Record the volume of each pile of pennies.
11. Calculate the density of the pile. To do this, divide the total mass of the pennies by the volume of the pennies. Record the density in the table on the overhead projector or chalkboard.

Analysis

12. How is it possible for the pennies to have different densities?
13. What clues might allow you to separate the pennies into the same groups without experimentation? Explain.

Volumania!

You have learned how to measure the volume of a solid object that has square or rectangular sides. But there are lots of objects in the world that have irregular shapes. In this lab activity, you'll learn some ways to find the volume of objects that have irregular shapes.

Materials

Part A

- graduated cylinder
- water
- various small objects supplied by your teacher

Part B

- bottom half of a 2 L plastic bottle or similar container
- water
- aluminum pie pan
- paper towels
- funnel
- graduated cylinder

Part A: Finding the Volume of Small Objects

Procedure

1. Fill a graduated cylinder half full with water. Read the volume of the water, and record it in your ScienceLog. Be sure to look at the surface of the water at eye level and to read the volume at the bottom of the meniscus, as shown below.

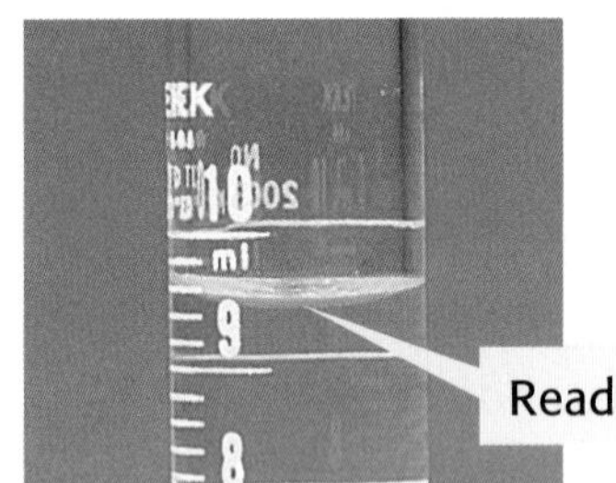

2. Carefully slide one of the objects into the tilted graduated cylinder, as shown below.
3. Read the new volume, and record it in your ScienceLog.
4. Subtract the old volume from the new volume. The resulting amount is equal to the volume of the solid object.
5. Use the same method to find the volume of the other objects. Record your results in your ScienceLog.

Analysis

6. What changes do you have to make to the volumes you determine in order to express them correctly?
7. Do the heaviest objects always have the largest volumes? Why or why not?

Part B: Finding the Volume of Your Hand

Procedure

8. Completely fill the container with water. Put the container in the center of the pie pan. Be sure not to spill any of the water into the pie pan.
9. Make a fist, and put your hand into the container up to your wrist.
10. Remove your hand, and let the excess water drip into the container, not the pie pan. Dry your hand with a paper towel.
11. Use the funnel to pour the overflow water into the graduated cylinder. Measure the volume. This is the volume of your hand. Record the volume in your ScienceLog. (Remember to use the correct unit of volume for a solid object.)
12. Repeat this procedure with your other hand.

Analysis

13. Was the volume the same for both of your hands? If not, were you surprised? What might account for a person's hands having different volumes?
14. Would it have made a difference if you had placed your open hand into the container instead of your fist? Explain your reasoning.
15. Compare the volume of your right hand with the volume of your classmates' right hands. Create a class graph of right-hand volumes. What is the average right-hand volume for your class?

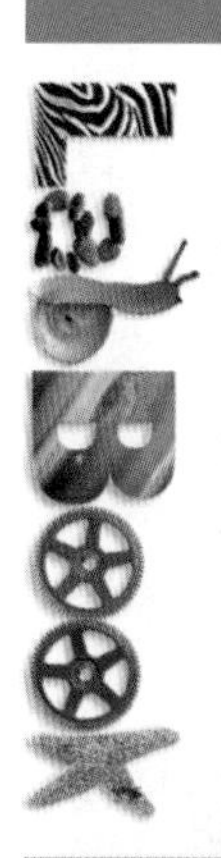

Going Further

- Design an experiment to determine the volume of a person's body. In your plans, be sure to include the materials needed for the experiment and the procedures that must be followed. Include a sketch that shows how your materials and methods would be used in this experiment.
- Using an encyclopedia, the Internet, or other reference materials, find out how the volumes of very large samples of matter—such as an entire planet—are determined.

Determining Density

The density of an object is its mass divided by its volume. But how does the density of a small amount of a substance relate to the density of a larger amount of the same substance? In this lab, you will calculate the density of one marble and of a group of marbles. Then you will confirm the relationship between the mass and volume of a substance.

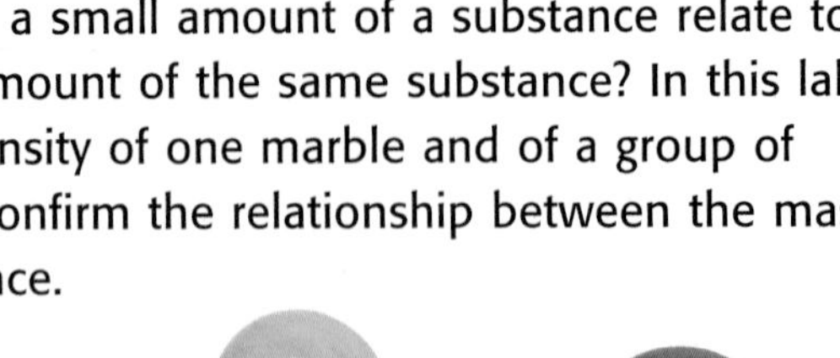

Materials

- 100 mL graduated cylinder
- water
- paper towels
- 8 to 10 glass marbles
- metric balance
- graph paper

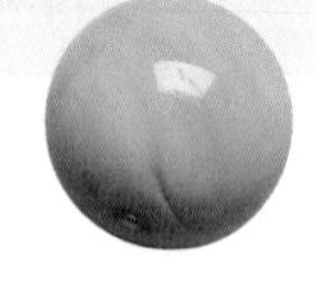

Collect Data

1. Copy the table below in your ScienceLog. Include one row for each marble.

Mass of marble, g	Total mass of marbles, g	Total volume, mL	Volume of marbles, mL (total volume minus 50.0 mL)	Density of marbles, g/mL (total mass of marbles divided by volume of marbles)
DO NOT WRITE IN BOOK			DO NOT WRITE IN BOOK	

2. Fill the graduated cylinder with 50.0 mL of water. If you put in too much water, twist one of the paper towels and use its end to absorb excess water.
3. Measure the mass of a marble as accurately as you can (to at least one-tenth of a gram). Record the marble's mass in the table.
4. Carefully drop the marble in the tilted cylinder, and measure the total volume. Record the volume in the third column.
5. Measure and record the mass of another marble. Add the masses of the marbles together, and record this value in the second column of the table.
6. Carefully drop the second marble in the graduated cylinder. Complete the row of information in the table.
7. Repeat steps 5 and 6, adding one marble at a time. Stop when you run out of marbles, the water no longer completely covers the marbles, or the graduated cylinder is full.

Analyze the Results

8. Examine the data in your table. As the number of marbles increases, what happens to the total mass of the marbles? What happens to the volume of the marbles? What happens to the density of the marbles?
9. Graph the total mass of the marbles (*y*-axis) versus the volume of the marbles (*x*-axis). Is the graph a straight line or a curved line?

Draw Conclusions

10. Does the density of a substance depend on the amount of substance present? Explain how your results support your answer.

Going Further

Calculate the slope of the graph. How does the slope compare with the values in the column titled "Density of marbles"? Explain.

Layering Liquids

You have learned that liquids form layers according to their densities. In this lab, you'll discover whether it matters in which order you add the liquids.

Materials

- liquid A
- liquid B
- liquid C
- beaker or other small, clear container
- 10 mL graduated cylinders (3)
- 3 funnels

Make a Prediction

1. Does the order in which you add liquids of different densities to a container affect the order of the layers formed by those liquids?

Conduct an Experiment

2. Using the graduated cylinders, add 10 mL of each liquid to the clear container. Remember to read the volume at the bottom of the meniscus, as shown below. In your ScienceLog, record the order in which you added the liquids.
3. Observe the liquids in the container. In your ScienceLog, sketch what you see. Be sure to label the layers and the colors.
4. Add 10 mL more of liquid C. Observe what happens, and write your observations in your ScienceLog.
5. Add 20 mL more of liquid A. Observe what happens, and write your observations in your ScienceLog.

Analyze Your Results

6. Which of the liquids has the greatest density? Which has the least density? How can you tell?
7. Did the layers change position when you added more of liquid C? Explain your answer.
8. Did the layers change position when you added more of liquid A? Explain your answer.

Communicate Your Results

9. Find out in what order your classmates added the liquids to the container. Compare your results with those of a classmate who added the liquids in a different order. Were your results different? In your ScienceLog, explain why or why not.

Draw Conclusions

10. Based on your results, evaluate your prediction from step 1.

White Before Your Eyes

You have learned how to describe matter based on its physical and chemical properties. You have also learned some clues that can help you determine whether a change in matter is a physical change or a chemical change. In this lab, you'll use what you have learned to describe four substances based on their properties and the changes they undergo.

Materials

- 4 spatulas
- baking powder
- plastic-foam egg carton
- 3 eyedroppers
- water
- stirring rod
- vinegar
- iodine solution
- baking soda
- cornstarch
- sugar

Procedure

1. Copy Table 1 and Table 2, shown on the next page, into your ScienceLog. Be sure to leave plenty of room in each box to write down your observations.
2. Use a spatula to place a small amount (just enough to cover the bottom of the cup) of baking powder into three cups of your egg carton. Look closely at the baking powder, and record observations of its color, texture, etc., in the column of Table 1 titled "Unmixed."
3. Use an eyedropper to add 60 drops of water to the baking powder in the first cup, as shown below. Stir with the stirring rod. Record your observations in Table 1 in the column titled "Mixed with water." Clean your stirring rod.
4. Use a clean dropper to add 20 drops of vinegar to the second cup of baking powder. Stir. Record your observations in the column titled "Mixed with vinegar." Clean your stirring rod.

5. Use a clean dropper to add five drops of iodine solution to the third cup of baking powder. Stir. Record your observations in the column in Table 1 titled "Mixed with iodine solution." Clean your stirring rod.
   Caution: Be careful when using iodine. Iodine will stain your skin and clothes.
6. Repeat steps 2–5 for each of the other substances. Use a clean spatula for each substance.

Analysis

7. In Table 2, write the type of change you observed, and state the property that the change demonstrates.
8. What clues did you use to identify when a chemical change happened?

Table 1 Observations

Substance	Unmixed	Mixed with water	Mixed with vinegar	Mixed with iodine solution
Baking powder				
Baking soda				
Cornstarch				
Sugar				

Table 2 Changes and Properties

	Mixed with water		Mixed with vinegar		Mixed with iodine solution	
Substance	**Change**	**Property**	**Change**	**Property**	**Change**	**Property**
Baking powder						
Baking soda						
Cornstarch						
Sugar						

Full of Hot Air!

Why do hot-air balloons float gracefully above Earth, while balloons you blow up fall to the ground? The answer has to do with the density of the air inside the balloon. Density is mass per unit volume, and volume is affected by changes in temperature. In this experiment, you will investigate the relationship between the temperature of a gas and its volume. Then you will be able to determine how the temperature of a gas affects its density.

Materials

- 2 aluminum pans
- water
- metric ruler
- hot plate
- ice water
- balloon
- 500 mL beaker
- heat-resistant gloves

Form a Hypothesis

1. How does an increase or decrease in temperature affect the volume of a balloon? Write your hypothesis in your ScienceLog.

Test the Hypothesis

2. Fill an aluminum pan with water about 4 to 5 cm deep. Put the pan on the hot plate, and turn the hot plate on.
3. While the water is heating, fill the other pan 4 to 5 cm deep with ice water.
4. Blow up a balloon inside the 500 mL beaker, as shown. The balloon should fill the beaker but should not extend outside the beaker. Tie the balloon at its opening.

5. Place the beaker and balloon in the ice water. Observe what happens. Record your observations in your ScienceLog.
6. Remove the balloon and beaker from the ice water. Observe the balloon for several minutes. Record any changes.
7. Put on heat-resistant gloves. When the hot water begins to boil, put the beaker and balloon in the hot water. Observe the balloon for several minutes, and record your observations.
8. Turn off the hot plate. When the water has cooled, carefully pour it into a sink.

Analyze the Results

9. Summarize your observations of the balloon. Relate your observations to Charles's law.
10. Was your hypothesis for step 1 supported? If not, revise your hypothesis.

Draw Conclusions

11. Based on your observations, how is the density of a gas affected by an increase or decrease in temperature?
12. Explain in terms of density and Charles's law why heating the air allows a hot-air balloon to float.

Can Crusher

Condensation can occur when gas particles come near the surface of a liquid. The gas particles slow down because they are attracted to the liquid. This reduction in speed causes the gas particles to condense into a liquid. In this lab, you'll see that particles that have condensed into a liquid don't take up as much space and therefore don't exert as much pressure as they did in the gaseous state.

Materials

- water
- 2 empty aluminum cans
- heat-resistant gloves
- hot plate
- tongs
- 1 L beaker

Conduct an Experiment

1. Place just enough water in an aluminum can to slightly cover the bottom.
2. Put on heat-resistant gloves. Place the aluminum can on a hot plate turned to the highest temperature setting.
3. Heat the can until the water is boiling. Steam should be rising vigorously from the top of the can.
4. Using tongs, quickly pick up the can and place the top 2 cm of the can upside down in the 1 L beaker filled with room-temperature water.
5. Describe your observations in your ScienceLog.

Analyze the Results

6. The can was crushed because the atmospheric pressure outside the can became greater than the pressure inside the can. Explain what happened inside the can to cause this.

Draw Conclusions

7. Inside every popcorn kernel is a small amount of water. When you make popcorn, the water inside the kernels is heated until it becomes steam. Explain how the popping of the kernels is the opposite of what you saw in this lab. Be sure to address the effects of pressure in your explanation.

Going Further

Try the experiment again, but use ice water instead of room-temperature water. Explain your results in terms of the effects of temperature.

A Hot and Cool Lab

When you add energy to a substance through heating, does the substance's temperature always go up? When you remove energy from a substance through cooling, does the substance's temperature always go down? In this lab you'll investigate these important questions with a very common substance—water.

Materials

Part A

- 250 or 400 mL beaker
- water
- heat-resistant gloves
- hot plate
- thermometer
- stopwatch
- graph paper

Part B

- 100 mL graduated cylinder
- water
- large coffee can
- crushed ice
- rock salt
- thermometer
- wire-loop stirring device
- stopwatch
- graph paper

Part A: Boiling Water

Make a Prediction

1. What happens to the temperature of boiling water when you continue to add energy through heating?

Procedure

2. Fill the beaker about one-third to one-half full with water.
3. Put on heat-resistant gloves. Turn on the hot plate, and put the beaker on the burner. Put the thermometer in the beaker. **Caution:** Be careful not to touch the burner.

Collect Data

4. In a table like the one below, record the temperature of the water every 30 seconds. Continue doing this until about one-fourth of the water boils away. Note the first temperature reading at which the water is steadily boiling.

Time (s)	30	60	90	120	150	180	210	etc.
Temperature (°C)	DO NOT WRITE IN BOOK							

5. Turn off the hot plate.
6. While the beaker is cooling, make a graph of temperature (*y*-axis) versus time (*x*-axis). Draw an arrow pointing to the first temperature at which the water was steadily boiling.
7. After you finish the graph, use heat-resistant gloves to pick up the beaker. Pour the warm water out, and rinse the warm beaker with cool water. **Caution:** Even after cooling, the beaker is still too warm to handle without gloves.

Part B: Freezing Water

Make Another Prediction

8. What happens to the temperature of freezing water when you continue to remove energy through cooling?

Procedure

9. Put approximately 20 mL of water in the graduated cylinder.
10. Put the graduated cylinder in the coffee can, and fill in around the graduated cylinder with crushed ice. Pour rock salt on the ice around the graduated cylinder. Place the thermometer and the wire-loop stirring device in the graduated cylinder.
11. As the ice melts and mixes with the rock salt, the level of ice will decrease. Add ice and rock salt to the can as needed.

Collect Data

12. In a new table, record the temperature of the water in the graduated cylinder every 30 seconds. Stir the water with the stirring device. **Caution:** Do not stir with the thermometer.
13. Once the water begins to freeze, stop stirring. Do not try to pull the thermometer out of the solid ice in the cylinder.
14. Note the temperature when you first notice ice crystals forming in the water. Continue taking readings until the water in the graduated cylinder is completely frozen.
15. Make a graph of temperature (*y*-axis) versus time (*x*-axis). Draw an arrow to the temperature reading at which the first ice crystals form in the water in the graduated cylinder.

Analyze the Results (Parts A and B)

16. What does the slope of each graph represent?
17. How does the slope when the water is boiling compare with the slope before the water starts to boil? Why is the slope different for the two periods?
18. How does the slope when the water is freezing compare with the slope before the water starts to freeze? Why is the slope different for the two periods?

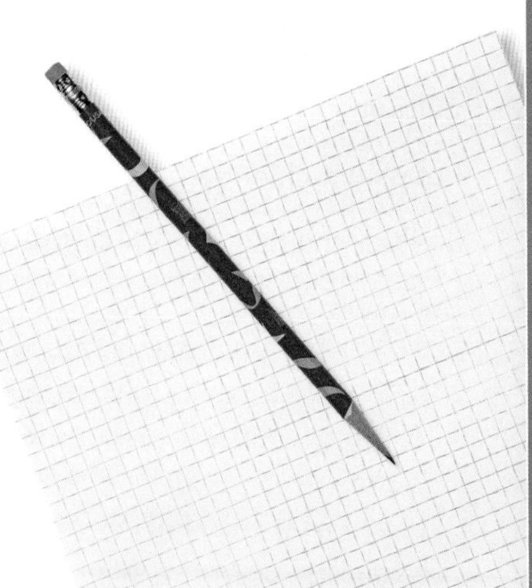

Draw Conclusions (Parts A and B)

19. Addition or subtraction of energy leads to changes in the movement of particles that make up solids, liquids, and gases. Use this idea to explain why the temperature graphs of the two experiments look the way they do.

Flame Tests

Fireworks produce fantastic combinations of color when they are ignited. The different colors are the results of burning different compounds. Imagine that you are the lead chemist for a fireworks company. The label has fallen off one box filled with a compound, and you must identify the unknown compound so that it may be used in the correct fireworks display. To identify the compound, you will use your knowledge that every compound has a unique set of properties.

Materials

- 4 small test tubes
- test-tube rack
- masking tape
- 4 chloride test solutions
- spark igniter
- Bunsen burner
- wire and holder
- dilute hydrochloric acid in a small beaker
- distilled water in a small beaker

Make a Prediction

1. Can you identify the unknown compound by heating it in a flame? Explain.

Conduct an Experiment

Caution: Be very careful in handling all chemicals. Tell your teacher immediately if you spill a chemical.

2. Arrange the test tubes in the test-tube rack. Use masking tape to label the tubes with the following names: calcium chloride, potassium chloride, sodium chloride, and unknown.
3. Copy the table below into your ScienceLog. Then ask your teacher for your portions of the solutions.

Test Results	
Compound	**Color of flame**
Calcium chloride	
Potassium chloride	
Sodium chloride	DO NOT WRITE IN BOOK
Unknown	

4. Light the burner. Clean the wire by dipping it into the dilute hydrochloric acid and then into distilled water. Holding the wooden handle, heat the wire in the blue flame of the burner until the wire is glowing and it no longer colors the flame. **Caution:** Use extreme care around an open flame.

Collect Data

5. Dip the clean wire into the first test solution. Hold the wire at the tip of the inner cone of the burner flame. In the table, record the color given to the flame.
6. Clean the wire by repeating step 4.
7. Repeat steps 5 and 6 for the other solutions.
8. Follow your teacher's instructions for cleanup and disposal.

Analyze the Results

9. Is the flame color a test for the metal or for the chloride in each compound? Explain your answer.
10. What is the identity of your unknown solution? How do you know?

Draw Conclusions

11. Why is it necessary to carefully clean the wire before testing each solution?
12. Would you expect the compound sodium fluoride to produce the same color as sodium chloride in a flame test? Why or why not?
13. Each of the compounds you tested is made from chlorine, which is a poisonous gas at room temperature. Why is it safe to use these compounds without a gas mask?

A Sugar Cube Race!

If you drop a sugar cube into a glass of water, how long will it take to dissolve? Will it take 5 minutes, 10 minutes, or longer? What can you do to speed up the rate at which it dissolves? Should you change something about the water, the sugar cube, or the process? In other words, what variable should you change? Before reading further, make a list of variables that could be changed in this situation. Record your list in your ScienceLog.

Materials

- water
- graduated cylinder
- 2 sugar cubes
- 2 beakers or other clear containers
- clock or stopwatch
- other materials approved by your teacher

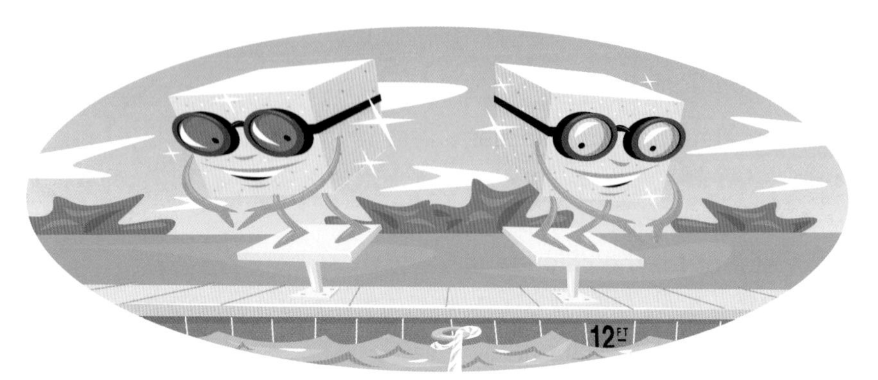

Make a Prediction

1. Choose one variable to test. In your ScienceLog, record your choice, and predict how changing your variable will affect the rate of dissolving.

Conduct an Experiment

2. Pour 150 mL of water into one of the beakers. Add one sugar cube, and use the stopwatch to measure how long it takes for the sugar cube to dissolve. You must not disturb the sugar cube in any way! Record this time in your ScienceLog.
3. Tell your teacher how you wish to test the variable. Do not proceed without his or her approval. You may need additional equipment.
4. Prepare your materials to test the variable you have picked. When you are ready, start your procedure for speeding up the dissolving of the sugar cube. Use the stopwatch to measure the time. Record this time in your ScienceLog.

Analyze the Results

5. Compare your results with the results obtained in step 2. Was your prediction correct? Why or why not?

Draw Conclusions

6. Why was it necessary to observe the sugar cube dissolving on its own before you tested the variable?
7. Do you think that changing more than one variable would speed up the rate of dissolving even more? Explain your reasoning.

Communicate Results

8. Discuss your results with a group that tested a different variable. Which variable had a greater effect on the rate of dissolving?

Making Butter

A colloid is an interesting substance. It has properties of both solutions and suspensions. Colloidal particles are not heavy enough to settle out, so they remain evenly dispersed throughout the mixture. In this activity, you will make butter—a very familiar colloid—and observe the characteristics that classify butter as a colloid.

Materials

- marble
- small, clear container with lid
- heavy cream
- clock or stopwatch

Procedure

1. Place a marble inside the container, and fill the container with heavy cream. Put the lid tightly on the container.
2. Take turns shaking the container vigorously and constantly for 10 minutes. Record the time when you begin shaking in your ScienceLog. Every minute, stop shaking the container and hold it up to the light. Record your observations.
3. Continue shaking the container, taking turns if necessary. When you see, hear, or feel any changes inside the container, note the time and change in your ScienceLog.
4. After 10 minutes of shaking, you should have a lump of "butter" surrounded by liquid inside the container. Describe both the butter and the liquid in detail in your ScienceLog.
5. Let the container sit for about 10 minutes. Observe the butter and liquid again, and record your observations in your ScienceLog.

Analysis

6. When you noticed the change in the container, what did you think was happening at that point?
7. Based on your observations, explain why butter is classified as a colloid.
8. What kind of mixture is the liquid that is left behind? Explain.

Unpolluting Water

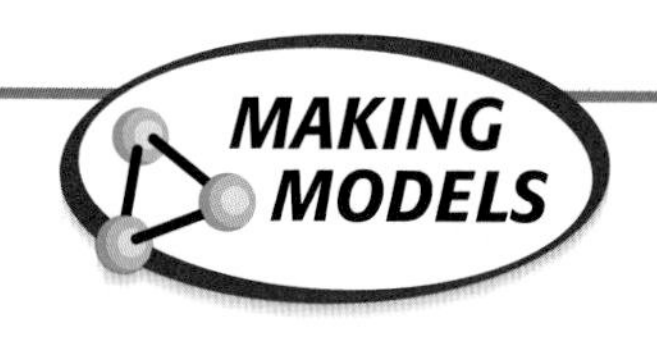

In many cities, the water supply comes from a river, lake, or reservoir. This water may include several mixtures, including suspensions (with suspended dirt, oil, or living organisms) and solutions (with dissolved chemicals). To make the water safe to drink, your city's water supplier must remove impurities. In this lab, you will model the procedures used in real water-treatment plants.

Materials

- "polluted" water
- graduated cylinder
- 250 mL beakers (4)
- 2 plastic spoons
- small nail
- 8 oz plastic-foam cup (2)
- scissors
- 2 pieces of filter paper
- washed fine sand
- metric ruler
- washed activated charcoal
- rubber band

Part A: Untreated Water

Procedure

1. Measure 100 mL of "polluted" water into a graduated cylinder. Be sure to shake the bottle of water before you pour so your sample will include all the impurities.
2. Pour the contents of the graduated cylinder into one of the beakers.
3. Copy the table below into your ScienceLog, and record your observations of the water in the "Before treatment" row.

Observations						
	Color	Clearness	Odor	Any layers?	Any solids?	Water volume
Before treatment						
After oil separation						
After sand filtration						
After charcoal						

Part B: Settling In

If a suspension is left standing, the suspended particles will settle to the top or bottom. You should see a layer of oil at the top.

Procedure

4. Separate the oil by carefully pouring the oil into another beaker. You can use a plastic spoon to get the last bit of oil from the water. Record your observations.

Part C: Filtration

Cloudy water can be a sign of small particles still in suspension. These particles can usually be removed by filtering. Water-treatment plants use sand and gravel as filters.

Procedure

5. Make a filter as follows:
   a. Use the nail to poke 5 to 10 small holes in the bottom of one of the cups.
   b. Cut a circle of filter paper to fit inside the bottom of the cup. (This will keep the sand in the cup.)
   c. Fill the cup to 2 cm below the rim with wet sand. Pack the sand tightly.
   d. Set the cup inside an empty beaker.
6. Pour the polluted water on top of the sand, and let it filter through. Do not pour any of the settled mud onto the sand. (Dispose of the mud as instructed by your teacher.) In your table, record your observations of the water collected in the beaker.

Part D: Separating Solutions

Something that has been dissolved in a solvent cannot be separated using filters. Water-treatment plants use activated charcoal to absorb many dissolved chemicals.

Procedure

7. Place activated charcoal about 3 cm deep in the unused cup. Pour the water collected from the sand filtration into the cup, and stir for a minute with a spoon.
8. Place a piece of filter paper over the top of the cup, and fasten it in place with a rubber band. With the paper securely in place, pour the water through the filter paper and back into a clean beaker. Record your observations in your table.

Analysis (Parts A–D)

9. Is your unpolluted water safe to drink? Why or why not?
10. When you treat a sample of water, do you get out exactly the same amount of water that you put in? Explain your answer.
11. Some groups may still have cloudy water when they finish. Explain a possible cause for this.

DESIGN YOUR OWN

Built for Speed

Imagine that you are an engineer at GoCarCo, a toy-vehicle company. GoCarCo is trying to beat the competition by building a new toy vehicle. Several new designs are being tested. Your boss has given you one of the new toy vehicles and instructed you to measure its speed as accurately as possible with the tools you have. Other engineers (your classmates) are testing the other designs. Your results could decide the fate of the company!

Materials

- toy vehicle
- meterstick
- masking tape
- stopwatch

Procedure

1. How will you accomplish your goal? Write a paragraph in your ScienceLog to describe your goal and your procedure for this experiment. Be sure that your procedure includes several trials.
2. Show your plan to your boss (teacher). Get his or her approval to carry out your procedure.
3. Perform your stated procedure. Record all data in your ScienceLog. Be sure to express all data in the correct units.

Analysis

4. What was the average speed of your vehicle? How does your result compare with the results of the other engineers?
5. Compare your technique for determining the speed of your vehicle with the techniques of the other engineers. Which technique do you think is the most effective?
6. Was your toy vehicle the fastest? Explain why or why not.

Going Further

Think of several conditions that could affect your vehicle's speed. Design an experiment to test your vehicle under one of those conditions. Write a paragraph in your ScienceLog to explain your procedure. Be sure to include an explanation of how that condition changes your vehicle's speed.

Detecting Acceleration

Have you ever noticed that you can "feel" acceleration? In a car or in an elevator you notice the change in speed or direction—even with your eyes closed! Inside your ears are tiny hair cells. These cells can detect the movement of fluid in your inner ear. When you accelerate, the fluid does, too. The hair cells detect this acceleration in the fluid and send a message to your brain. This allows you to sense acceleration.

In this activity you will build a device that detects acceleration. Even though this device is made with simple materials, it is very sensitive. It registers acceleration only briefly. You will have to be very observant when using this device.

Materials

- scissors
- string
- 1 L container with watertight lid
- pushpin
- small cork or plastic-foam ball
- modeling clay
- water

Procedure

1. Cut a piece of string that is just long enough to reach three quarters of the way inside the container.
2. Use a pushpin to attach one end of the string to the cork or plastic-foam ball.
3. Use modeling clay to attach the other end of the string to the center of the *inside* of the container lid. Be careful not to use too much string—the cork (or ball) should hang no farther than three-quarters of the way into the container.
4. Fill the container to the top with water.
5. Put the lid tightly on the container with the string and cork (or ball) on the inside.
6. Turn the container upside down (lid on the bottom). The cork should float about three-quarters of the way up inside the container, as shown at right. You are now ready to use your accelerometer to detect acceleration by following the steps on the next page.

7. Put the accelerometer lid side down on a tabletop. Notice that the cork floats straight up in the water.
8. Now gently start pushing the accelerometer across the table at a constant speed. Notice that the cork quickly moves in the direction you are pushing then swings backward. If you did not see this happen, try the same thing again until you are sure you can see the first movement of the cork.
9. Once you are familiar with how to use your accelerometer, try the following changes in motion, and record your observations of the cork's first motion for each change in your ScienceLog.
   a. While moving the device across the table, push a little faster.
   b. While moving the device across the table, slow down.
   c. While moving the device across the table, change the direction that you are pushing. (Try changing both to the left and to the right.)
   d. Make any other changes in motion you can think of. You should only change one part of the motion at a time.

Analysis

10. The cork moves forward (in the direction you were pushing the bottle) when you speed up but backward when you slow down. Why? (Hint: Think about the direction of acceleration.)

11. When you push the bottle at a constant speed, why does the cork quickly swing back after it shows you the direction of acceleration?

12. Imagine you are standing on a corner, watching a car that is waiting at a stoplight. A passenger inside the car is holding some helium balloons. Based on what you observed with your accelerometer, what do you think will happen to the balloons when the car begins moving?

Going Further

If you move the bottle in a circle at a constant speed, what do you predict the cork will do? Try it, and check your answer.

LabBook

Science Friction

In this experiment, you will investigate three types of friction—static, sliding, and rolling—to determine which is the largest force and which is the smallest force.

Materials

- scissors
- string
- textbook (covered)
- spring scale (force meter)
- 3 to 4 wooden or metal rods

Ask a Question

1. Which type of friction is the largest force—static, sliding, or rolling? Which is the smallest?

Form a Hypothesis

2. In your ScienceLog, write a statement or statements that answer the questions above. Explain your reasoning.

Test the Hypothesis/Collect Data

3. Cut a piece of string, and tie it in a loop that fits in the textbook, as shown below. Hook the string to the spring scale.
4. Practice the next three steps several times before you collect data.
5. To measure the static friction between the book and the table, pull the spring scale very slowly. Record the largest force on the scale before the book starts to move.
6. After the book begins to move, you can determine the sliding friction. Record the force required to keep the book sliding at a slow, constant speed.
7. Place two or three rods under the book to act as rollers. Make sure the rollers are evenly spaced. Place another roller in front of the book so that the book will roll onto it. Pull the force meter slowly. Measure the force needed to keep the book rolling at a constant speed.

Analyze the Results

8. Which type of friction was the largest? Which was the smallest?
9. Do the results support your hypothesis? If not, how would you revise or retest your hypothesis?

Communicate Results

10. Compare your results with those of another group. Are there any differences? Working together, design a way to improve the experiment and resolve possible differences.

Relating Mass and Weight

Why do objects with more mass weigh more than objects with less mass? All objects have weight on Earth because their mass is affected by Earth's gravitational force. Because the mass of an object on Earth is constant, the relationship between the mass of an object and its weight is also constant. You will measure the mass and weight of several objects to verify the relationship between mass and weight on the surface of Earth.

Materials

- metric balance
- small classroom objects
- spring scale (force meter)
- string
- scissors
- graph paper

Collect Data

1. Copy the table below into your ScienceLog.

Mass and Weight Measurements

Object	Mass (g)	Weight (N)

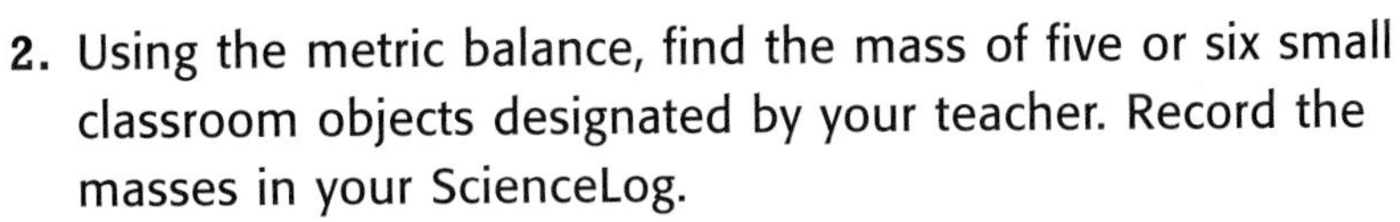

2. Using the metric balance, find the mass of five or six small classroom objects designated by your teacher. Record the masses in your ScienceLog.
3. Using the spring scale, find the weight of each object. Record the weights in your ScienceLog. (You may need to use the string to create a hook with which to hang some objects from the spring scale, as shown at right.)

Analyze the Results

4. Using your data, construct a graph of weight (*y*-axis) versus mass (*x*-axis). Draw a line that best fits all your data points.
5. Does the graph confirm the relationship between mass and weight on Earth? Explain your answer.

A Marshmallow Catapult

Catapults use projectile motion to launch objects across distances. A variety of factors can affect the distance an object can be launched, such as the weight of the object, how far the catapult is pulled back, and the catapult's strength. In this lab, you will build a simple catapult and determine the angle at which the catapult will launch an object the farthest.

Materials

- plastic spoon
- block of wood, 3.5 cm × 3.5 cm × 1 cm
- duct tape
- miniature marshmallows
- protractor
- meterstick

Form a Hypothesis

1. At what angle, from 10° to 90°, will a catapult launch a marshmallow the farthest?

Test the Hypothesis

2. Copy the table below into your ScienceLog. In your table, add one row each for 20°, 30°, 40°, 50°, 60°, 70°, 80°, and 90° angles.

Angle	Distance 1 (cm)	Distance 2 (cm)	Average distance (cm)
10°	DO NOT WRITE IN BOOK		

3. Attach the plastic spoon to the 1 cm side of the block with duct tape. Use enough tape so that the spoon is attached securely.
4. Place one marshmallow in the center of the spoon, and tape it to the spoon. This serves as a ledge to hold the marshmallow that will be launched.
5. Line up the bottom corner of the block with the bottom center of the protractor. Start with the block at 10° as shown in the photograph.
6. Place a marshmallow in the spoon, on top of the taped marshmallow. Pull back lightly, and let go. Measure and record the distance from the catapult that the marshmallow lands. Repeat the measurement, and calculate an average.
7. Repeat step 6 for each angle up to 90°.

Analyze the Results

8. At what angle did the catapult launch the marshmallow the farthest? Compare this with your hypothesis. Explain any differences.

Draw Conclusions

9. Does the path of an object's projectile motion depend on the catapult's angle? Support your answer with your data.
10. At what angle should you throw a ball or shoot an arrow so that it will fly the farthest? Why? Support your answer with your data.

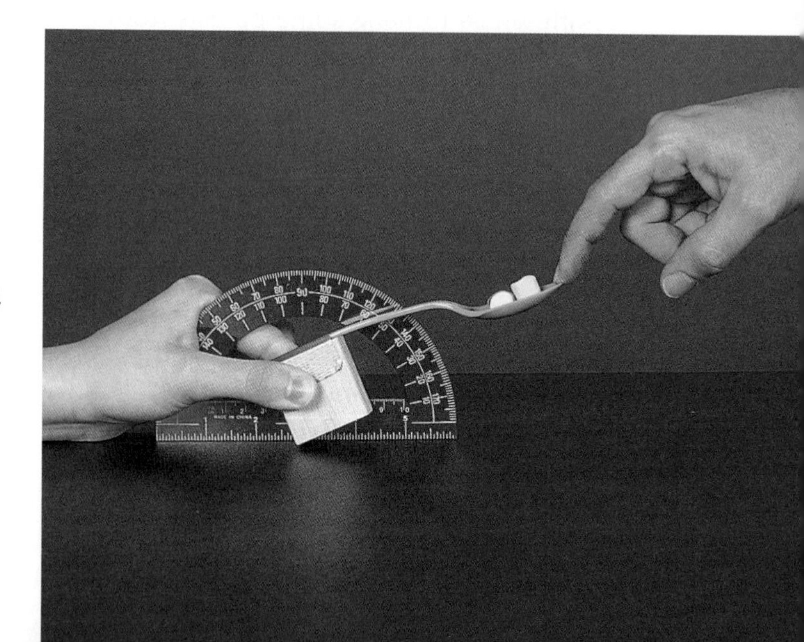

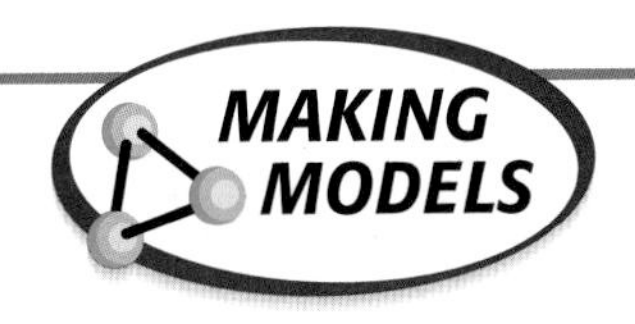

Blast Off!

You have been hired as a rocket scientist for NASA. Your job is to design a rocket that will have a controlled flight while carrying a payload. Keep in mind that Newton's laws will have a powerful influence on your rocket.

Materials

- tape
- 3 m fishing line
- sharpened pencil
- small paper cup
- 15 cm pieces of string (2)
- long, thin balloon
- twist tie
- drinking straw
- meterstick
- pennies

Procedure

1. When you begin your experiment, your teacher will tape one end of the fishing line to the ceiling.
2. Use a sharpened pencil to poke a small hole in one side of the cup near the top. Place a 15 cm piece of string through the hole, and tape down the end inside. Do the same on the other side of the cup.
3. Inflate the balloon, and use the twist tie to hold it closed.
4. Tape the free ends of the strings to the sides of the balloon about 4 cm from the bottom. The cup should hang below the balloon. Your model rocket should look similar to a hot air balloon.
5. Thread the fishing line that is hanging from the ceiling through the straw. Tape the balloon securely to the straw.
6. Tape the loose end of the fishing line to the floor.
7. Untie the twist tie while holding the end of the balloon closed. When you are ready, release the end of the balloon. Mark and record the maximum height of the rocket.
8. Repeat the procedure several times, adding a penny to the cup each time. Continue until your rocket cannot lift any more pennies.

Analysis

9. In a paragraph, describe how all three of Newton's laws influenced the flight of your rocket.
10. Draw a diagram of your rocket, and label the action and reaction forces on it.

Going Further

Brainstorm about how to modify your rocket so that it will carry the maximum number of pennies to the maximum height. Select the best design. When your teacher has approved all the designs, each team will build and launch their rocket. Which variable did you modify? How did this variable affect the flight of the rocket?

Inertia-Rama!

Inertia is a property of all matter, from small particles of dust to enormous planets and stars. In this lab, you will investigate the inertia of various shapes and types of matter. Keep in mind that each investigation requires you to either overcome or use the object's inertia.

Materials

Station 1

- hard-boiled egg
- raw egg

Station 2

- coin
- index card
- cup

Station 3

- spool of thread
- suspended mass
- scissors
- meterstick

Station 1: Magic Eggs

Procedure

1. There are two eggs at this station—one is hard-boiled (solid all the way through) and the other is raw (liquid inside). The masses of the two eggs are about the same. The eggs are marked so that you can tell them apart.
2. You will spin each egg and then stop it from spinning by placing a finger on its center. Before you do anything to either egg, write some predictions in your ScienceLog: Which egg will be the easiest to spin? Which egg will be the easiest to stop?
3. Spin the hard-boiled egg. Then place your finger on it to make it stop spinning. Record your observations in your ScienceLog.
4. Repeat step 3 with the raw egg.
5. Compare your predictions with your observations. (Repeat steps 3 and 4 if necessary.)

Analysis

6. Explain why the eggs behave differently when you spin them even though they should have the same inertia. (Hint: Think about what happens to the liquid inside the raw egg.)
7. In terms of inertia, explain why the eggs react differently when you try to stop them.

Station 2: Coin in a Cup

Procedure

8. At this station, you will find a coin, an index card, and a cup. Place the card over the cup. Then place the coin on the card over the center of the cup, as shown at right.

9. In your ScienceLog, write down a method for getting the coin into the cup without touching the coin and without lifting the card.
10. Try your method. If it doesn't work, try again until you find a method that does work. When you are done, place the card and coin on the table for the next group.

Analysis

11. Use Newton's first law of motion to explain why the coin falls into the cup if you remove the card quickly.
12. Explain why pulling on the card slowly will not work, even though the coin has inertia. (Hint: Friction is a force.)

Station 3: The Magic Thread

Procedure

13. At this station, you will find a spool of thread and a mass hanging from a strong string. Cut a piece of thread about 40 cm long. Tie the thread around the bottom of the mass, as shown at right.
14. Pull gently on the end of the thread. Observe what happens, and record your observations in your ScienceLog.
15. Stop the mass from moving. Now hold the end of the thread so that there is a lot of slack between your fingers and the mass.
16. Give the thread a quick, hard pull. You should observe a very different event. Record your observations in your ScienceLog. Throw away the thread.

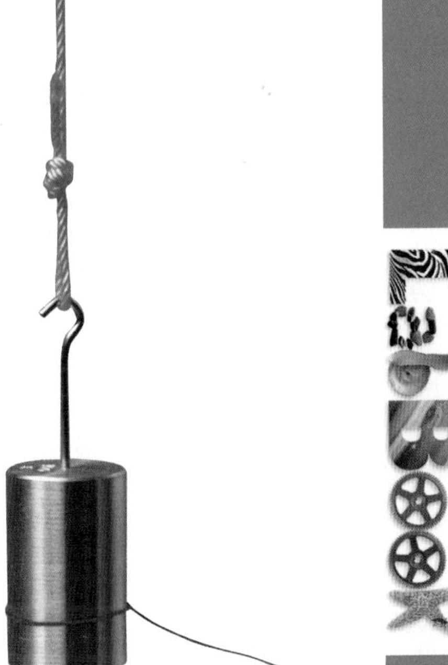

Analysis

17. Use Newton's first law of motion to explain why the results of a gentle pull are so different from the results of a hard pull.

Draw Conclusions

18. Remember that both moving and nonmoving objects have inertia. Explain why it is hard to throw a bowling ball and why it is hard to catch a thrown bowling ball.
19. Why is it harder to run with a backpack full of books than with an empty backpack?

Quite a Reaction

Catapults have been used for centuries to throw objects great distances. You may already be familiar with catapults after doing the marshmallow catapult lab. According to Newton's third law of motion (whenever one object exerts a force on a second object, the second object exerts an equal and opposite force on the first), when an object is launched, something must also happen to the catapult. In this activity, you will build a kind of catapult that will allow you to observe the effects of Newton's third law of motion and the law of conservation of momentum.

Materials

- glue
- 10 cm × 15 cm rectangles of cardboard (3)
- 3 pushpins
- string
- rubber band
- 6 plastic straws
- marble
- scissors
- meterstick

Conduct an Experiment

1. Glue the cardboard rectangles together to make a stack of three.
2. Push two of the pushpins into the cardboard stack near the corners at one end, as shown below. These will be the anchors for the rubber band.
3. Make a small loop of string.
4. Put the rubber band through the loop of string, and then place the rubber band over the two pushpin anchors. The rubber band should be stretched between the two anchors with the string loop in the middle.
5. Pull the string loop toward the end of the cardboard stack opposite the end with the anchors, and fasten the loop in place with the third pushpin.
6. Place the six straws about 1 cm apart on a tabletop or on the floor. Then carefully center the catapult on top of the straws.
7. Put the marble in the closed end of the V formed by the rubber band.

8. Use scissors to cut the string holding the rubber band, and observe what happens. (Be careful not to let the scissors touch the cardboard catapult when you cut the string.)
9. Reset the catapult with a new piece of string. Try launching the marble several times to be sure that you have observed everything that happens during a launch. Record all your observations in your ScienceLog.

Analyze the Results

10. Which has more mass, the marble or the catapult?
11. What happened to the catapult when the marble was launched?
12. How far did the marble fly before it landed?
13. Did the catapult move as far as the marble did?

Draw Conclusions

14. Explain why the catapult moved backward.
15. If the forces that made the marble and the catapult move apart are equal, why didn't the marble and the catapult move apart the same distance? (Hint: The fact that the marble can roll after it lands is not the answer.)
16. The momentum of an object depends on the mass and velocity of the object. What is the momentum of the marble before it is launched? What is the momentum of the catapult? Explain your answers.
17. Using the law of conservation of momentum, explain why the marble and the catapult move in opposite directions after the launch.

Going Further

How would you modify the catapult if you wanted to keep it from moving backward as far as it did? (It still has to rest on the straws.) Using items that you can find in the classroom, design a catapult that will move backward less than the original design.

Fluids, Force, and Floating

Why do some objects sink in fluids but others float? In this lab, you'll get a sinking feeling as you determine that an object floats when its weight is less than the buoyant force exerted by the surrounding fluid.

Materials

- large rectangular tank or plastic tub
- water
- metric ruler
- small rectangular baking pan
- labeled masses
- metric balance
- paper towels

Procedure

1. Copy the table below into your ScienceLog.

Measurement	Trial 1	Trial 2
Length (l), cm		
Width (w), cm		
Initial height (h_1), cm		
Initial volume (V_1), cm^3 $V_1 = l \times w \times h_1$		
New height (h_2), cm		
New volume (V_2), cm^3 $V_2 = l \times w \times h_2$		
Displaced volume (ΔV), cm^3 $\Delta V = V_2 - V_1$		
Mass of displaced water, g $m = \Delta V \times 1 \text{ g/cm}^3$		
Weight of displaced water, N (buoyant force)		
Weight of pan and masses, N		

2. Fill the tank or tub half full with water.
3. Measure (in centimeters) the length, width, and initial height of the water. Record your measurements in the table.
4. Using the equation given in the table, determine the initial volume of water in the tank. Record your results in the table.
5. Place the pan in the water, and place masses in the pan, as shown on the next page. Keep adding masses until the pan sinks to about three-quarters of its height. This will cause the water level in the tank to rise. Record the new height of the water in the table. Then use this value to determine and record the new volume of water.

6. Determine the volume of the water that was displaced by the pan and masses, and record this value in the table. The displaced volume is equal to the new volume minus the initial volume.
7. Determine the mass of the displaced water by multiplying the displaced volume by its density (1 g/cm^3). Record the mass in the table.
8. Divide the mass by 100. The value you get is the weight of the displaced water in newtons (N). This is equal to the buoyant force. Record the weight of the displaced water in the table.
9. Remove the pan and masses, and determine their total mass (in grams) using the balance. Convert the mass to weight (N), as you did in step 8. Record the weight of the masses and pan in the table.
10. Place the empty pan back in the tank. Perform a second trial by repeating steps 5–9. This time add masses until the pan is just about to sink.

Analysis

11. In your ScienceLog, compare the buoyant force (the weight of the displaced water) with the weight of the pan and masses for both trials.
12. How did the buoyant force differ between the two trials? Explain.

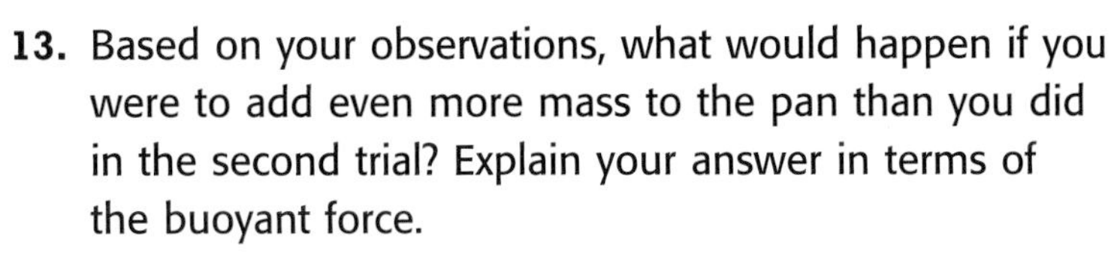

13. Based on your observations, what would happen if you were to add even more mass to the pan than you did in the second trial? Explain your answer in terms of the buoyant force.
14. What would happen if you put the masses in the water without the pan? What difference does the pan's shape make?

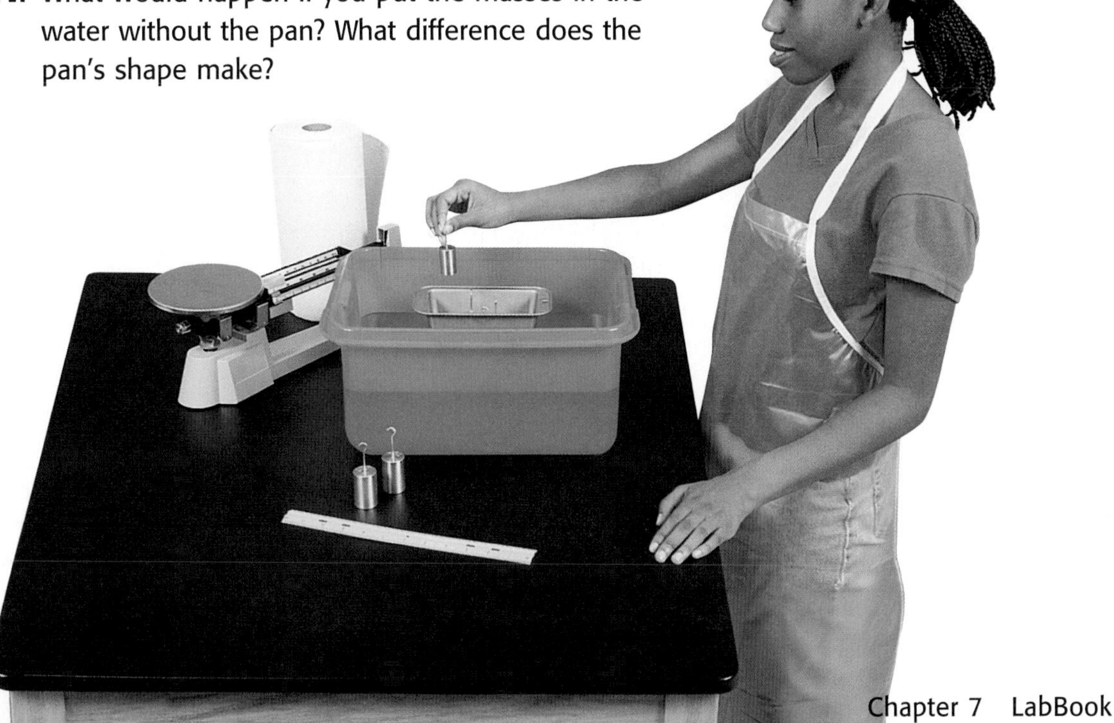

Density Diver

Crew members of a submarine can control the submarine's density underwater by allowing water to flow into and out of special tanks. These changes in density affect the submarine's position in the water. In this lab, you'll control a "density diver" to learn for yourself how the density of an object affects its position in a fluid.

Materials

- 2 L plastic bottle with screw-on cap
- water
- medicine dropper

Form a Hypothesis

1. How does the density of an object determine whether the object floats, sinks, or maintains its position in a fluid? Write your hypothesis in your ScienceLog.

Test the Hypothesis

2. Completely fill the 2 L plastic bottle with water.
3. Fill the diver (medicine dropper) approximately halfway with water, and place it in the bottle. The diver should float with only part of the rubber bulb above the surface of the water. If the diver floats too high, carefully remove it from the bottle and add a small amount of water to the diver. Place the diver back in the bottle. If you add too much water and the diver sinks, empty out the bottle and diver and go back to step 2.
4. Put the cap on the bottle tightly so that no water leaks out.
5. Apply various pressures to the bottle. Carefully watch the water level inside the diver as you squeeze and release the bottle. Record what happens in your ScienceLog.
6. Try to make the diver rise, sink, or stop at any level. Record your technique and your results.

Analyze the Results

7. How do the changes inside the diver affect its position in the surrounding fluid?
8. What is the relationship between the water level inside the diver and the diver's density? Explain.

Draw Conclusions

9. What relationship did you observe between the diver's density and the diver's position in the fluid?
10. Explain how your density diver is like a submarine.
11. Explain how pressure on the bottle is related to the diver's density. Be sure to include Pascal's principle in your explanation.
12. What was the variable in this experiment? What factors were controlled?

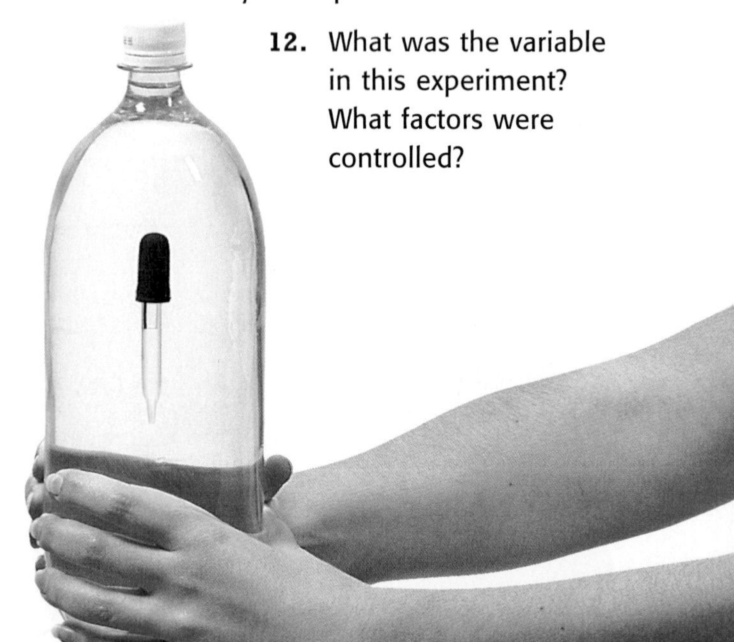

Taking Flight

MAKING MODELS

When air moves above and below the wing of an airplane, the air pressure below the wing is higher than the air pressure above the wing. This creates lift. In this activity, you will build a model airplane to help you identify how wing size and thrust (forward force provided by the engine) can affect the lift that is necessary for flight.

Materials

- sheet of paper

Procedure

1. Fold the paper in half lengthwise and open it again, as shown at right. Make sure to crease all folds well.
2. Fold the right- and left-hand corners toward the center crease.
3. Fold the entire sheet in half along the center crease.
4. With the plane lying on its side, fold the top front edge down so that it meets the bottom edge, as shown.
5. Fold the top wing down again, bringing the top edge to the bottom edge.
6. Turn the plane over, and repeat steps 4 and 5 for the other wing.
7. Raise both wings away from the body to a position slightly above horizontal. Your plane is ready!
8. Point the plane slightly upward. Gently throw the plane overhand. Repeat several times. Describe your observations in your ScienceLog.
   Caution: Be sure to point the plane away from people.
9. Make the wings smaller by folding them one more time. Gently throw the plane overhand. Repeat several times. Describe your observations in your ScienceLog.
10. Try to achieve the same flight path you saw when the plane's wings were bigger. Record your technique.

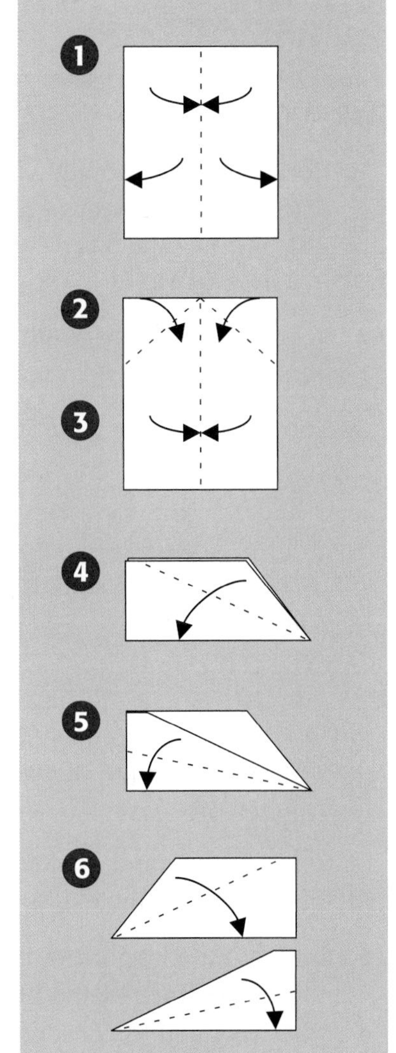

Analysis

11. What happened to the plane's flight when you reduced the size of its wings? Explain.
12. What provided your airplane's thrust?
13. Based on your observations, how does changing the thrust affect the plane's lift?

A Powerful Workout

Does the amount of work you do depend on how fast you do it? No! But doing work in a shorter amount of time does affect your power—the rate at which work is done. In this lab, you'll calculate your work and power when climbing a flight of stairs at different speeds. Then you'll compare your power with that of an ordinary household object—a 100 W light bulb.

Materials

- flight of stairs
- metric ruler
- stopwatch

Ask a Question

1. How does your power when climbing a flight of stairs compare with the power of a 100 W light bulb?

Form a Hypothesis

2. In your ScienceLog, write a hypothesis that answers the question in step 1. Explain your reasoning.
3. Copy Table 1 into your ScienceLog.

Table 1 Data Collection

Height of step (cm)	Number of steps	Height of stairs (m)	Time for slow walk (s)	Time for quick walk (s)
	DO NOT WRITE IN BOOK			

4. Measure the height of one stair step. Record the measurement in Table 1.
5. Count the number of stairs, including the top step, and record this number in Table 1.
6. Calculate the height (in meters) of the stairs by multiplying the number of steps by the height of one step. Record your answer. (You will need to convert from centimeters to meters.)
7. Using a stopwatch, measure how many seconds it takes you to walk slowly up the flight of stairs. Record your measurement in Table 1.
8. Now measure how many seconds it takes you to walk quickly up the flight of stairs. Be careful not to overexert yourself.

Analyze the Results

9. Copy Table 2 into your ScienceLog.

Table 2 Work and Power Calculations			
Weight (N)	Work (J)	Power for slow walk (W)	Power for quick walk (W)
	DO NOT WRITE IN BOOK		

10. Determine your weight in newtons, and record it in Table 2. Your weight in newtons is your weight in pounds (lb) multiplied by 4.45 N/lb.
11. Calculate and record your work done to climb the stairs using the following equation:

$$\text{Work} = \text{Force} \times \text{distance}$$

Remember that 1 N•m is 1 J. (Hint: When determining the force exerted, remember that force is expressed in newtons.)

12. Calculate and record your power for each trial (the slow walk and the quick walk) using the following equation:

$$\text{Power} = \frac{\text{Work}}{\text{time}}$$

Remember that the unit for power is the watt (1 W = 1 J/s).

Draw Conclusions

13. In step 11 you were asked to calculate your work done in climbing the stairs. Why weren't you asked to calculate your work for each trial?
14. Look at your hypothesis in step 2. Was your hypothesis supported? Write a statement in your ScienceLog that describes how your power in each trial compares with the power of a 100 W light bulb.
15. The work done to move one electron in a light bulb is very small. Write down two reasons why the power is large. (Hint 1: How many electrons are in the filament of a light bulb? Hint 2: How did you use more power in your second trial?)

Communicate Results

16. Your teacher will provide a class data table on the board. Write your average power in the table. Then calculate the average power from the class data. How many students would it take to equal the power of a 100 W bulb?

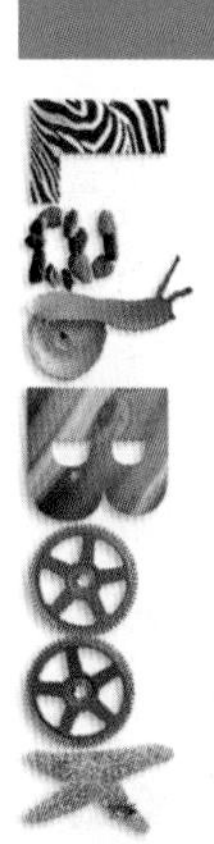

Where is work done in a light bulb? Electrons in the filament move back and forth very quickly. These moving electrons do work by heating up the filament and making it glow.

Inclined to Move

In this lab, you will examine a simple machine—an inclined plane. Your task is to compare the work done with and without the inclined plane and to analyze the effects of friction.

Materials

- string
- small book
- spring scale
- meterstick
- wooden board
- blocks
- graph paper

Collect Data

1. Copy the table below into your ScienceLog.
2. Tie a piece of string around a book. Attach the spring scale to the string. Use the spring scale to slowly lift the book to a height of 50 cm. Record the output force (the force needed to lift the book). The output force is constant throughout the lab.
3. Create a ramp using the board and blocks. The ramp should be 10 cm high at the highest point. Measure and record the ramp length.
4. Keeping the spring scale and string parallel to the ramp, as shown below, slowly pull the book up the ramp. Record the input force (the force needed to pull the book up the ramp).
5. Increase the height of the ramp by 10 cm. Repeat step 4. Repeat this step for each ramp height up to 50 cm.

Force vs. Height

Ramp height (cm)	Output force (N)	Ramp length (cm)	Input force (N)
10			
20			
30			
40			
50			

Analyze the Results

6. The *real* amount of work done includes the work done to overcome friction while sliding the book up the inclined plane. Calculate the real work at each height by multiplying the ramp length by the input force (remember to convert centimeters to meters). Graph your results, plotting work (y-axis) versus height (x-axis).
7. The *ideal* amount of work is the work you would do if there were no friction. Calculate the ideal work at each height by multiplying the ramp height by the output force. Plot the data on your graph.

Draw Conclusions

8. Does it require more or less force and work to raise the book using the ramp? Explain, using your calculations and graphs.
9. What is the relationship between the height of the inclined plane and the input force?
10. Write a statement that summarizes why the slopes of the two graphs are different.

Building Machines

The world around you is filled with machines. Some are simple machines, such as ramps for wheelchair access to a building. Others, like elevators and escalators, are compound machines. A compound machine is made of two or more simple machines. In this lab, you will design and build several simple machines and a compound machine.

Materials

- bottle caps
- cardboard
- craft sticks
- empty thread spools
- glue
- modeling clay
- paper
- pencils
- rubber bands
- scissors
- shoe boxes
- stones
- straws
- string
- tape
- other materials available in your classroom that are approved by your teacher

Procedure

1. Using any of the listed materials, build a model of each type of simple machine: inclined plane, lever, wheel and axle, pulley, screw, and wedge. Describe and draw each of your machines in your ScienceLog.
2. In your ScienceLog, design a compound machine that you can build out of the materials listed. Try to be creative. You may design a machine that already exists, or you may invent your own machine.
3. After your teacher approves your design, build your compound machine.

Analysis

4. List a possible use for each of your simple machines.
5. Compare your simple machines with those created by your classmates. How are they different? How are they similar?
6. How many simple machines are in your compound machine? List them.
7. Compare your compound machine with those created by your classmates. How is yours different? How is yours similar?
8. What is a possible use for your compound machine? Why did you design it as you did?
9. A compound machine is listed in the Materials list. What is it?

Going Further

Design a compound machine that has all the simple machines in it. Explain what the machine will do and how it will make work easier. With your teacher's approval, build your machine.

Wheeling and Dealing

A wheel and axle is one type of simple machine. One kind of wheel and axle you may have seen is a crank handle. Pencil sharpeners, ice-cream makers, and water wells all utilize a wheel and axle in this manner. In this lab, you will use a crank handle to find out how a wheel and axle helps you do work. You will use a large dowel in a pipe as an axle and smaller dowels as handles of different lengths. You will determine what effect the length of the handle has on the operation of the machine.

Materials

- wheel and axle assembly
- meterstick
- large mass
- spring scale
- handles
- 0.5 m string
- 2 C-clamps

Collect Data

1. Copy Table 1 into your ScienceLog.
2. Measure the radius (in meters) of the large dowel in the wheel and axle assembly. This is the axle radius, which remains constant throughout the lab. It may be easier to measure the diameter and divide by two. Record the axle radius in Table 1.
3. Lift the large mass with the spring scale, and determine the force required. This is the output force, and it remains constant throughout the lab. Record the output force in Table 1.
4. Measure the length (in meters) of handle 1. This is the radius of the wheel. Record the wheel radius in Table 1.
5. Insert the handle into the hole in the axle. Attach one end of the string to the large mass to be lifted. Tie the other end of the string to the screw in the axle near where the handle is inserted.
6. Hang the wheel and axle assembly over the table so that the mass is hanging down and the handle can turn freely. Use two C-clamps to secure the assembly to the table, as shown at right.

Table 1 Data Collection

Handle	Axle radius (m)	Output force (N)	Wheel radius (m)	Input force (N)
1				
2				
3				
4				

7. Turn the handle several times to lift the mass off the floor. Hook the end of the spring scale to the end of the handle. Hold the spring scale upside down so that the handle is pulling up on the spring scale, and record the force (in newtons) exerted on the handle. This is the input force.

8. Remove the spring scale, and lower the mass to the floor. Remove the handle.

9. Repeat steps 4 through 8 with the other three handles. Record all data in Table 1.

Analyze the Results

10. Copy Table 2 into your ScienceLog.

Table 2 Calculations						
Handle	Axle distance (m)	Wheel distance (m)	Work input (J)	Work output (J)	Mechanical efficiency (%)	Mechanical advantage
1						
2						
3						
4						

11. For each handle, calculate the distance the axle rotates and the distance the wheel rotates using the following equation:

$$\text{Distance} = 2\pi r$$

Use the data from Table 1 for the axle radius and wheel radius. Use 3.14 for the value of π. Record your answer in Table 2.

12. Calculate and record the work input (input force × wheel distance) and work output (output force × axle distance).

13. Using the equation below, calculate and record the mechanical efficiency of the wheel and axle for each trial.

$$\text{Mechanical efficiency} = \frac{\text{work output}}{\text{work input}} \times 100$$

14. Using the equation below, calculate and record the mechanical advantage of the wheel and axle for each trial.

$$\text{Mechanical advantage} = \frac{\text{wheel radius}}{\text{axle radius}}$$

Draw Conclusions

15. What happens to work output and work input as the handle length increases? Why?

16. What happens to mechanical efficiency as the handle length increases? Why?

17. What happens to mechanical advantage as the handle length increases? Why?

18. What will happen to mechanical advantage if the handle length is kept constant and the axle radius gets larger?

19. What factors were controlled in this experiment? What was the variable?

Finding Energy

When you are coasting down a big hill on a bike or skateboard, you may notice that you pick up speed, going faster and faster. Because you are moving, you have kinetic energy—the energy of motion. Where does that energy come from? Sometimes you pedal the bike or push yourself along on a skateboard, but you can also roll down a hill without exerting yourself at all! The kinetic energy must come from somewhere, and in this lab you will find out where.

Materials

- 2 or 3 books
- wooden board
- masking tape
- meterstick
- metric balance
- rolling cart
- stopwatch

Form a Hypothesis

1. Where does the kinetic energy come from when you roll down a hill? Write your hypothesis in your ScienceLog.

Test the Hypothesis/Collect Data

2. Copy Table 1 into your ScienceLog.

Table 1 Data Collection

Height of ramp (m)	Length of ramp (m)	Mass of cart (kg)	Weight of cart (N)	Time of trial (s)			Average time (s)
				1	2	3	
		DO NOT WRITE IN BOOK					

3. Make a ramp with the books and board.
4. Use masking tape to make a starting line. Be sure the starting line is far enough from the top so the cart can be placed behind the line.
5. Place a strip of masking tape at the bottom of the ramp to mark the finish line.
6. Determine the height of the ramp by measuring the height of the starting line and subtracting the height of the finish line. Record the height of the ramp in meters in Table 1.
7. Measure the distance in meters between the starting and the finish lines. Record this distance as the length of the ramp in Table 1.
8. Use the metric balance to find the mass of the cart in grams. Convert this to kilograms by dividing by 1,000. Record the mass in kilograms in Table 1.

9. Multiply the mass by 10. The number you get is the weight of the cart in newtons. Record the weight in Table 1.

10. Set the cart so that it is behind the starting line, and release it. Use the stopwatch to time how long it takes for the cart to reach the finish line. Record the time in Table 1.

11. Repeat step 10 twice more, and average the results. Record the average time in Table 1.

Analyze the Results

12. Copy Table 2 into your ScienceLog.

Table 2 Calculations

Average speed (m/s)	Final speed (m/s)	Kinetic energy at bottom (J)	Gravitational potential energy at top (J)

13. Find and record the average speed of the cart (at the bottom of the ramp) by dividing the distance traveled down the ramp by the average time taken to travel that distance.

14. Find and record the final speed of the cart. The cart is accelerating smoothly from zero speed, so the final speed is twice the average speed.

15. Calculate the kinetic energy of the cart at the bottom of the ramp using the following equation:

$$\text{kinetic energy} = \frac{mv^2}{2}$$

where m equals the mass of the cart and v equals the final speed of the cart. Record the kinetic energy in joules. (Remember that 1 J = 1 kg•m²/s/s.)

16. Calculate the gravitational potential energy of the cart at the top of the ramp using the following equation:

$$\text{GPE} = \text{weight} \times \text{height}$$

Record the gravitational potential energy in joules. (Remember that 1 N = 1 kg•m/s/s, so 1 N × 1 m = 1 kg•m²/s/s = 1 J.)

Draw Conclusions

17. How does the cart's gravitational potential energy at the top of the ramp compare with its kinetic energy at the bottom? Does this support your hypothesis? Explain your answer.

18. You probably found that the gravitational potential energy of the cart at the top of the ramp was close but not exactly equal to the kinetic energy of the cart at the bottom. Explain this finding.

19. While riding your bike, you coast down both a small hill and a large hill. Compare your final speed at the bottom of the small hill with your final speed at the bottom of the large hill. Explain your answer.

Energy of a Pendulum

SKILL BUILDER

A pendulum clock is a compound machine that uses stored energy to do work, that is, to move the hands of the clock. A pendulum is used to store energy. With each swing of the pendulum, some of that stored energy is used to move the hands of the clock. A pendulum is an example of a device that converts potential energy into kinetic energy and back. In this lab you will take a close look at the energy conversions that occur as a pendulum swings.

Materials

- 1 m of string
- 100 g hooked mass
- marker
- meterstick

Procedure

1. Make a pendulum by tying the string around the hook of the mass. Use the marker and the meterstick to mark points on the string that are 50 cm, 70 cm, and 90 cm away from the mass.
2. Hold the string at the 50 cm mark. To swing the pendulum, gently pull the mass to the side, and release the mass without pushing it. Observe at least 10 swings of the pendulum.
3. In your ScienceLog, record your observations about the motion of the pendulum. Be sure to consider how fast and how high the pendulum swings.
4. Repeat steps 2 and 3 while holding the string at the 70 cm mark and again while holding the string at the 90 cm mark.

Analysis

5. In your ScienceLog, list similarities and differences in the motion of the pendulum during all three trials.
6. Based on your observations of each trial, at which point (or points) was the pendulum moving the slowest?
7. At which point (or points) of the swing was the pendulum moving the fastest?
8. In each trial, at which point (or points) of the swing did the pendulum have the greatest potential energy? the smallest potential energy? (Hint: Think about your answers to questions 6 and 7.)
9. At which point (or points) of the swing did the pendulum have the greatest kinetic energy? the smallest kinetic energy? Explain your answers.
10. Does the pendulum's kinetic energy increase or decrease on its way down? Explain.
11. Does the pendulum's potential energy increase or decrease on its way down? Explain.
12. What improvements might reduce the amount of energy used to overcome friction so that the pendulum would swing for a longer period of time?

Eggstremely Fragile

All moving objects have kinetic energy. The faster an object is moving, the more kinetic energy it has. When a falling object hits something that stops its motion (for example, the floor), the law of conservation of energy requires that the energy can't just disappear. It must be transferred or changed to another form of energy.

When an unprotected egg hits the ground from a height of 1 m, most of the kinetic energy is transferred into the shell—with messy results. A fall of 2 m means two times as much egg-shattering energy. Obviously, an egg needs protection if it might fall any distance. In this lab you will design your own protection system for an egg.

Materials

- raw egg
- empty half-pint milk carton
- assorted materials provided by your teacher

Procedure

1. Using the materials provided by your teacher, design a protection system that will prevent the egg from breaking when it is dropped from heights of 1, 2, and 3 m. The job of the protection system is to transfer the kinetic energy of falling somewhere besides the eggshell or to somehow absorb the kinetic energy. Keep the following points in mind while developing your egg-protection system:
   - **a.** The egg and its protective materials must be small enough to fit inside the closed milk carton. (Note: This is for size only—the milk carton will not be dropped with the egg.)
   - **b.** The protective materials don't have to be soft (to absorb the impact); they can be hard if they are used properly.
   - **c.** The protective materials can surround the egg or can be attached to the egg at various points.
2. In your ScienceLog, explain why you chose the materials you did.
3. You and your classmates will perform the three dropping trials at a time and location specified by your teacher. Record your results for each trial in your ScienceLog.

Analysis

4. Did your egg survive all three trials? If it did not, why did your egg-protection system fail? If your egg did survive, what features of your egg-protecting system transferred or absorbed the energy?
5. How do egg cartons like those you find in a grocery store protect eggs from mishandling?

Wave Energy and Speed

If you threw a rock into a pond, waves would carry energy away from the point of origin. But if you threw a large rock into a pond, would the waves carry more energy away from the point of origin than waves created by a small rock? And would a large rock create waves that move faster than waves created by a small rock? In this lab you'll answer these questions using pencils instead of rocks.

Materials

- shallow pan, approximately 20 × 30 cm
- newspaper
- small beaker
- water
- 2 pencils
- stopwatch

Ask a Question

1. In this lab you will answer the following questions: Do waves created by a large disturbance carry more energy than waves created by a small disturbance? Do waves created by a large disturbance travel faster than waves created by a small disturbance?

Form a Hypothesis

2. In your ScienceLog, write a few brief sentences stating what you think the outcome of this experiment will be. Your hypothesis should answer the questions in step 1.

Test the Hypothesis

3. Place the pan on a few sheets of newspaper. Using the small beaker, fill the pan with water.
4. Make sure that the water is still. Tap the surface of the water near one end of the pan with the eraser end of one pencil. This represents the small disturbance. In your ScienceLog, record your observations about the size of the waves that are created and the path they take.
5. Repeat this procedure after the waves have disappeared. This time, use the stopwatch to record the amount of time it takes for one of the waves to reach the other side of the pan. Record your data in your ScienceLog.
6. Repeat steps 4 and 5 using two pencils at once. This represents the large disturbance. (Try to use the same amount of force to tap the water as you did with just one pencil.) Observe, and record your results.

Analyze the Results

7. Compare the appearance of the waves created by one pencil with that of the waves created by two pencils. Were there any differences in amplitude (wave height)? Record your observations in your ScienceLog.

8. Compare the amount of time required for the waves to reach the side of the pan. Did the waves travel faster when two pencils were used? Record your answer in your ScienceLog.

Draw Conclusions

9. Do waves created by a large disturbance carry more energy than waves created by a small disturbance? Explain your answer. Be sure to discuss how your results support your answer. (Hint: Remember the relationship between amplitude and energy.)

10. Do waves created by a large disturbance travel faster than waves created by a small disturbance? Explain your answer. Be sure to discuss how your results support your answer.

11. Based on your results, explain whether you think the following statements are true:
    a. The energy carried by a wave does not depend on what caused the wave; the energy depends only on the medium that the wave is traveling through.
    b. The speed of a wave does not depend on what caused the wave; the speed depends only on the medium that the wave is traveling through.

12. A tsunami is a giant ocean wave; tsunamis can reach a height of 30 m. Tsunamis that reach land can cause injury and enormous property damage. Fortunately, scientists can often predict when tsunamis will reach land and can warn people in threatened areas. Using what you learned in this lab about wave energy and speed, explain why tsunamis are so dangerous and how scientists can forecast when they will reach land.

Wave Speed, Frequency, and Wavelength

Wave speed, frequency, and wavelength are three related properties of waves. In this lab you will make observations and collect data to determine the relationship among these properties.

Materials

- coiled spring toy
- meterstick
- stopwatch

Part A—Wave Speed

Procedure

1. Copy Table 1 into your ScienceLog.

Table 1 Wave Speed Data			
Trial	Length of spring (m)	Time for wave (s)	Speed of wave (m/s)
1			
2			
3			
Average			

2. On the floor or a table, two group members should stretch the spring between them so that the spring is 2 to 4 m long. A third member should measure the length of the spring. Record the length in Table 1.
3. One of the students holding the spring should pull part of the spring sideways with one hand, as shown at right. The student should then release the pulled-back portion, which will cause a wave to travel down the spring.
4. Using a stopwatch, the third group member should measure how long it takes for the wave to travel down the length of the spring and back. Record this time in the data table.
5. Repeat steps 3 and 4 two more times.
6. Calculate and record the wave speed for each trial. (Hint: Speed equals distance divided by time.)
7. Calculate and record the average time it takes for the waves to travel and the average wave speed.

Part B—Wavelength and Frequency

Procedure

8. Keep the spring the same length that you used in Part A.
9. Copy Table 2 into your ScienceLog.

Table 2 Wavelength and Frequency Data				
Trial	Length of spring (m)	Time for 10 cycles (s)	Wave frequency (Hz)	Wavelength (m)
1				
2				
3				
Average				

10. One of the two students holding the spring should start shaking the spring from side to side until a wave pattern appears that resembles one of those shown below.
11. Using the stopwatch, the third group member should measure and record how long it takes for 10 cycles of the wave pattern to occur. (One back-and-forth shake is one cycle.) Keep the pattern going so that measurements for three trials can be made.
12. Calculate the frequency (Hz) for each trial. To do this, divide the number of cycles (10) by the number of seconds that elapsed during the 10 cycles. Record the answers in Table 2.
13. Determine the wavelength (m) of your wave by using the equation given with the pattern at right that matches your wave pattern. Record your answer in Table 2.
14. Calculate and record the average time and frequency.

Wave Patterns

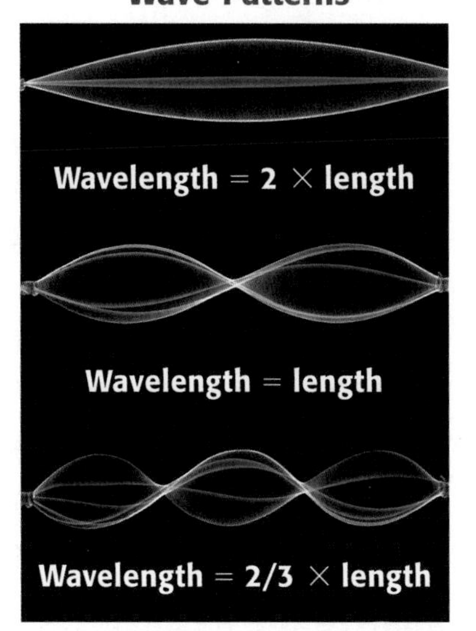

Analysis—Parts A and B

15. To discover the relationship among speed, wavelength, and frequency, try multiplying or dividing any two of them to see if the result equals the third. (Use the average speed, wavelength, and average frequency from your data tables.) In your ScienceLog, write the equation that shows the relationship.
16. Reread the definitions for *frequency* and *wavelength* in the chapter titled "The Energy of Waves." Use these definitions to explain the relationship that you discovered.

Stop the Static Electricity!

Imagine this scenario: Some of your clothes cling together when they come out of the dryer. This annoying problem is caused by static electricity—the buildup of electric charges on an object. In this lab, you'll discover how this buildup occurs.

Materials

- 30 cm thread
- plastic-foam packing peanut
- tape
- rubber rod
- wool cloth
- glass rod
- silk cloth

Ask a Question

1. How do electric charges build up on clothes in a dryer?

Form a Hypothesis

2. Write a statement that answers the question above. Explain your reasoning.

Test the Hypothesis

3. Tie a piece of thread approximately 30 cm in length to a packing peanut. Hang the peanut by the thread from the edge of a table. Tape the thread to the table.
4. Rub the rubber rod with the wool cloth for 10–15 seconds. Bring the rod near, but not touching, the peanut. Observe the peanut and record your observations. If nothing happens, repeat this step.
5. Touch the peanut with the rubber rod. Pull the rod away from the peanut, and then bring it near again. Record your observations.
6. Repeat steps 4 and 5 with the glass rod and silk cloth.
7. Now rub the rubber rod with the wool cloth, and bring the rod near the peanut again. Record your observations.

Analyze the Results

8. What caused the peanut to act differently in steps 4 and 5?
9. Did the glass rod have the same effect on the peanut as the rubber rod did? Explain how the peanut reacted in each case.
10. Was the reaction of the peanut the same in steps 5 and 7? Explain.

Draw Conclusions

11. Based on your results, how do you think electric charges build up on clothes in a dryer?
12. Was your hypothesis correct? Explain your answer, and write a new statement if necessary.
13. Explain why the rubber rod and the glass rod affected the peanut.

Going Further

Do some research to find out how a dryer sheet helps stop the buildup of electric charges in the dryer.

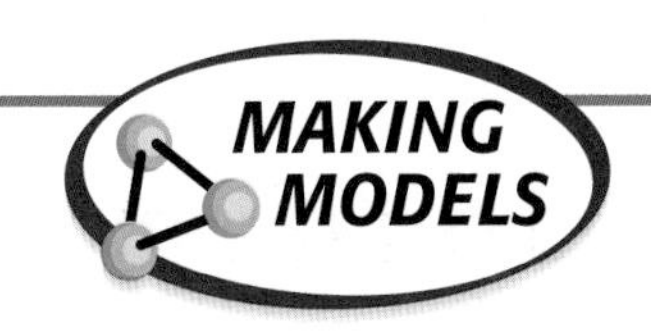

Potato Power

Have you ever wanted to look inside a D cell from a flashlight or an AA cell from a portable radio? All cells include the same basic components, as shown below. There is a metal "bucket," some electrolyte (a paste), and a rod of some other metal (or solid) in the middle. Even though the construction is simple, companies that manufacture cells are always trying to make a product with the highest voltage possible from the least expensive materials. Sometimes they try different pastes, and sometimes they try different combinations of metals. In this lab, you will make your own cell. Using inexpensive materials, you will try to produce the highest voltage you can.

Materials

- labeled metal strips
- potato
- metric ruler
- voltmeter

Procedure

1. Choose two metal strips. Carefully push one of the strips into the potato at least 2 cm deep. Insert the second strip the same way, and measure how far apart the two strips are. (If one of your metal strips is too soft to push into the potato, push a harder strip in first, remove it, and then push the soft strip into the slit.) Record the two metals you have used and the distance between them in your ScienceLog. **Caution:** The strips of metal may have sharp edges.
2. Connect the voltmeter to the two strips, and record the voltage.
3. Move one of the strips closer to or farther from the other. Measure the new distance and voltage. Record your results.
4. Repeat steps 1 through 3, using different combinations of metal strips and distances until you find the combination that produces the highest voltage.

Analysis

5. What combination of metals and distance produced the highest voltage?
6. If you change only the distance but use the same metal strips, what is the effect on the voltage?

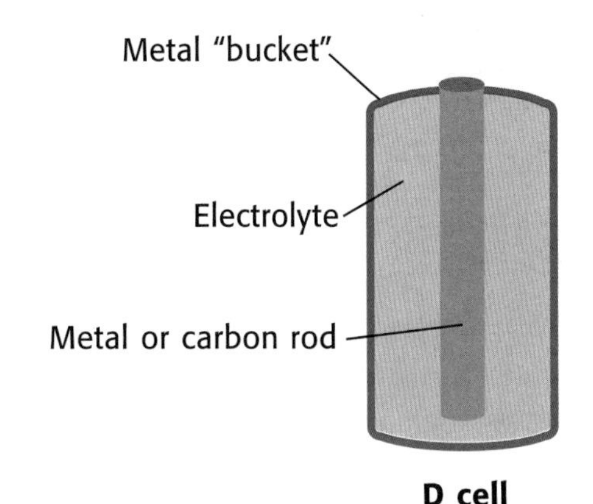

D cell

7. One of the metal strips tends to lose electrons, while the other tends to gain electrons. What do you think would happen if you used two strips of the same metal?

Circuitry 101

You have learned that there are two basic types of electrical circuits. A series circuit connects all the parts in a single loop, and a parallel circuit connects each of the parts on separate branches to the power source. If you want to control the whole circuit, the loads and the switch must be wired in series. If you want parts of the circuit to operate independently, the loads must be wired in parallel.

No matter how simple or complicated a circuit may be, Ohm's law (current equals voltage divided by resistance) applies. In this lab, you will construct both a series circuit and a parallel circuit. You will use an ammeter to measure current and a voltmeter to measure voltage. With each circuit, you will test and apply Ohm's law.

Materials

- power source–dry cell(s)
- switch
- 3 light-bulb holders
- 3 light bulbs
- insulated wire, cut into 15 cm lengths with both ends stripped
- ammeter
- voltmeter

Part A—Series Circuit

Procedure

1. Construct a series circuit with a power source, a switch, and three light bulbs. **Caution:** Always leave the switch open when constructing or changing the circuit. Close the switch only when you are testing or taking a reading.
2. Draw a diagram of your circuit in your ScienceLog.
3. Test your circuit. Do all three bulbs light up? Are they all the same brightness? What happens if you carefully unscrew one light bulb? Does it make any difference which bulb you unscrew? Record your observations in your ScienceLog.
4. Connect the ammeter between the power source and the switch. Close the switch, and record the current with a label on your diagram in your ScienceLog. Be sure to show where you measured the current and what the value was.
5. Reconnect the circuit so the ammeter is between the first and second bulbs. Record the current, as you did in step 4.
6. Move the ammeter so it is between the second and third bulbs, and record the current again.
7. Remove the ammeter from the circuit, and connect the voltmeter to the two ends of the power source. Record the voltage with a label on your diagram.
8. Use the voltmeter to measure the voltage across each bulb. Label the voltage across each bulb on your diagram.

Part B—Parallel Circuit

Procedure

9. Take apart your series circuit, and reassemble the same power source, switch, and three light bulbs so that the bulbs are wired in parallel. (Note: The switch must remain in series with the power source to be able to control the whole circuit.)

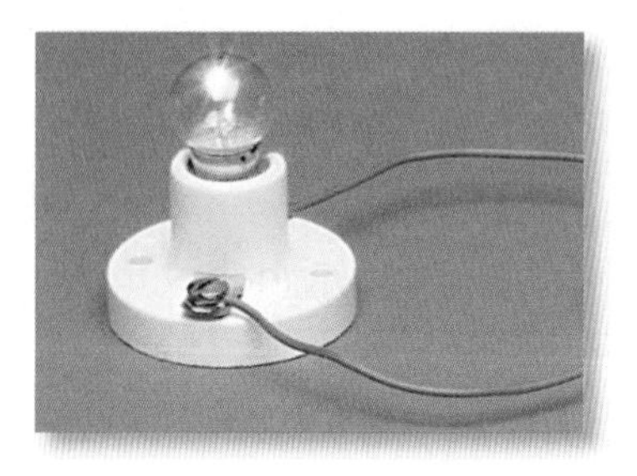

10. Draw a diagram of your parallel circuit in your ScienceLog.
11. Test your circuit, and record your observations, as you did in step 3.
12. Connect the ammeter between the power source and the switch. Record the reading on your diagram.
13. Reconnect the circuit so that the ammeter is right next to one of the three bulbs. Record the current on your diagram.
14. Repeat step 13 for the two remaining bulbs.
15. Remove the ammeter from your circuit, and connect the voltmeter to the two ends of the power source. Record this voltage.
16. Measure and record the voltage across each light bulb.

Analysis—Parts A and B

17. Was the current the same at all places in the series circuit? Was it the same everywhere in the parallel circuit?
18. For each circuit, compare the voltage at each light bulb with the power source.
19. What is the relationship between the voltage at the power source and the voltages at the light bulbs in a series circuit?
20. Use Ohm's law and the readings for current (*I*) and voltage (*V*) at the power source for both circuits to calculate the total resistance (*R*) in both the series and parallel circuits.
21. Was the total resistance for both circuits the same? Explain your answer.
22. Why did the bulbs differ in brightness?
23. Based on your results, what do you think might happen if too many electric appliances are plugged into the same series circuit? the same parallel circuit?

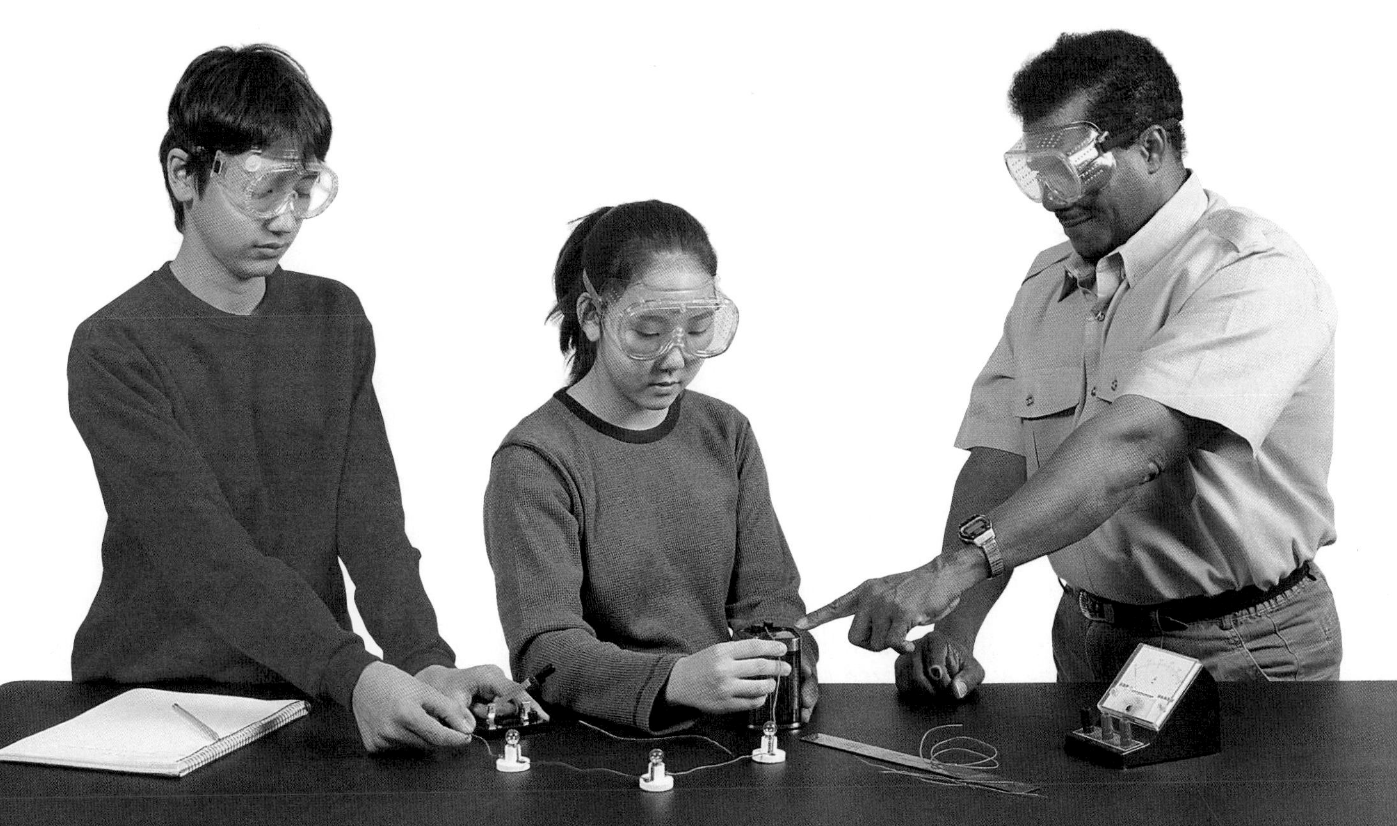

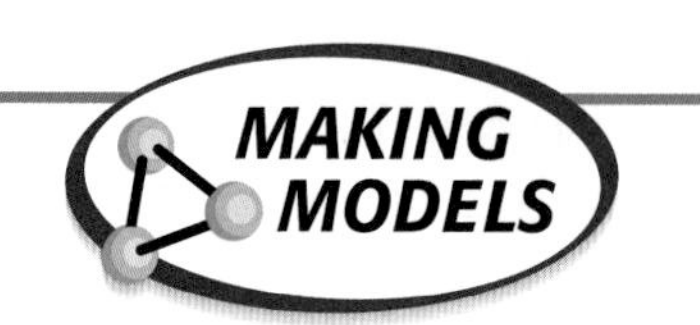

Made to Order

Imagine that you are a new employee at the Elements-4-U Company, which custom builds elements. Your job is to construct the atomic nucleus for each element ordered by your clients. You were hired for the position because of your knowledge about what a nucleus is made of and your understanding of how isotopes of an element differ from each other. Now it's time to put that knowledge to work!

Materials

- 4 protons (white plastic-foam balls, 2–3 cm in diameter)
- 6 neutrons (blue plastic-foam balls, 2–3 cm in diameter)
- 20 strong-force connectors (toothpicks)
- periodic table

Procedure

1. Copy the table below into your ScienceLog. Be sure to leave room to expand the table to include more elements.

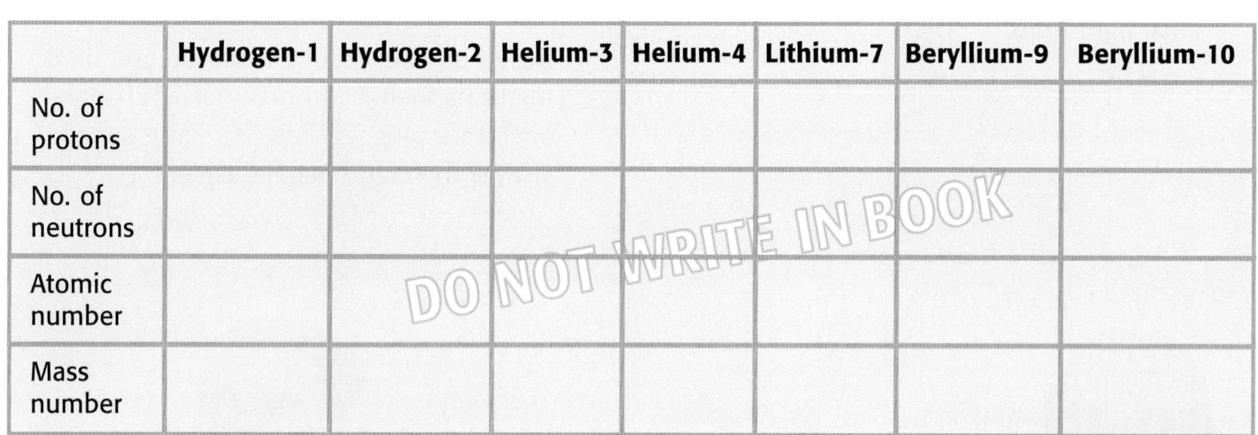

	Hydrogen-1	Hydrogen-2	Helium-3	Helium-4	Lithium-7	Beryllium-9	Beryllium-10
No. of protons							
No. of neutrons							
Atomic number							
Mass number							

2. Your first assignment: the nucleus of hydrogen-1. Pick up one proton (a white plastic-foam ball). Congratulations! You have just built a hydrogen-1 nucleus, the simplest nucleus possible.
3. Count the number of protons and neutrons in the nucleus, and fill in rows 2 and 3 for this element in the table.
4. Use the information in rows 2 and 3 to determine the atomic number and mass number of the element. Record this information in the table.
5. Draw a picture of your model in your ScienceLog.
6. Hydrogen-2 is an isotope of hydrogen that has one proton and one neutron. Using a strong-force connector, add a neutron to your hydrogen-1 nucleus. (Remember that in a nucleus, the protons and neutrons are held together by the strong force, which is represented in this activity by the toothpicks.) Repeat steps 3–5.

7. Helium-3 is an isotope of helium that has two protons and one neutron. Add one proton to your hydrogen-2 nucleus to create a helium-3 nucleus. Each particle should be connected to the other two particles so they make a triangle, not a line. Protons and neutrons always form the smallest arrangement possible because the strong force pulls them together. Repeat steps 3–5.

8. For the next part of the lab, you will need to use information from the periodic table of the elements. Look at the illustration at right. It shows the periodic table entry for carbon, one of the most abundant elements on Earth. For your job, the most important information in the periodic table is the atomic number. You can find the atomic number of any element at the top of its entry on the table. In the example, the atomic number of carbon is 6.

9. Use the information in the periodic table to build models of the following isotopes of elements: helium-4, lithium-7, beryllium-9, and beryllium-10. Remember to put the protons and neutrons as close together as possible—each particle should attach to at least two others. Repeat steps 3–5 for each isotope.

Analyze the Results

10. What is the relationship between the number of protons and the atomic number?

11. If you know the atomic number and the mass number of an isotope, how could you figure out the number of neutrons in its nucleus?

12. Look up uranium on the periodic table.
    a. What is the atomic number of uranium?
    b. How many neutrons does the isotope uranium-235 have?

Communicate Results

13. Compare your model with the models of other groups. How are they similar? How are they different?

Going Further
Working with another group, combine your models. Identify the element (and isotope) you have created.

Create a Periodic Table

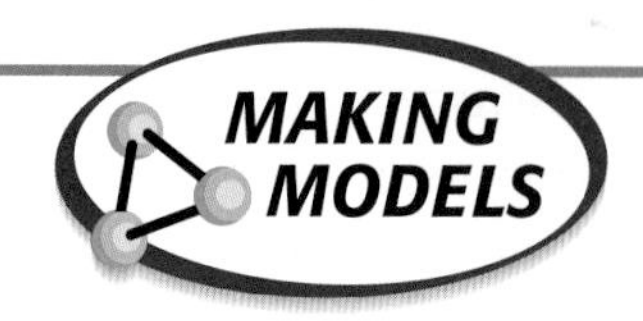

You probably have classification systems for many things in your life, such as your clothes, your books, and your CDs. You see many other classification systems every day—in the library, the grocery store, and the video rental store. There are many classification systems in science, and one of the most important is the periodic table of the elements. In this lab you will develop your own classification system for a collection of ordinary objects. You will analyze trends in your system and compare your system with the periodic table of the elements.

Materials

- bag of objects
- 20 squares of paper, each 3 × 3 cm
- metric balance
- metric ruler
- 2 sheets of graph paper

Procedure

1. Your teacher will give you a bag of objects. Your bag is missing one item. Examine the items carefully, and identify the missing object. Describe the object in as many ways as you can imagine in your ScienceLog. Be sure to include the reasons why you think the missing object has these characteristics.
2. Lay the paper squares out on your desk or table so that you have a grid of five rows of four squares each.
3. Arrange your objects on the grid in a logical order. (You must decide what order is logical!) You should end up with one blank square for the missing object.
4. In your ScienceLog, describe the basis for your arrangement.
5. Measure the mass (g) and diameter (mm) of each object, and record your results in the appropriate square. Each square (except the empty one) should have one object and two written measurements on it.

6. Examine your pattern again. Does the order in which your objects are arranged still make sense? Explain.

7. Rearrange the squares and their objects if necessary to improve your arrangement. Describe the basis for the new arrangement in your ScienceLog.

8. Working across the rows, number the squares 1 to 20. When you get to the end of a row, continue numbering in the first square of the next row.

9. Copy your grid into your ScienceLog. In each square, be sure to list the type of object and label all measurements with appropriate units.

Analyze the Results

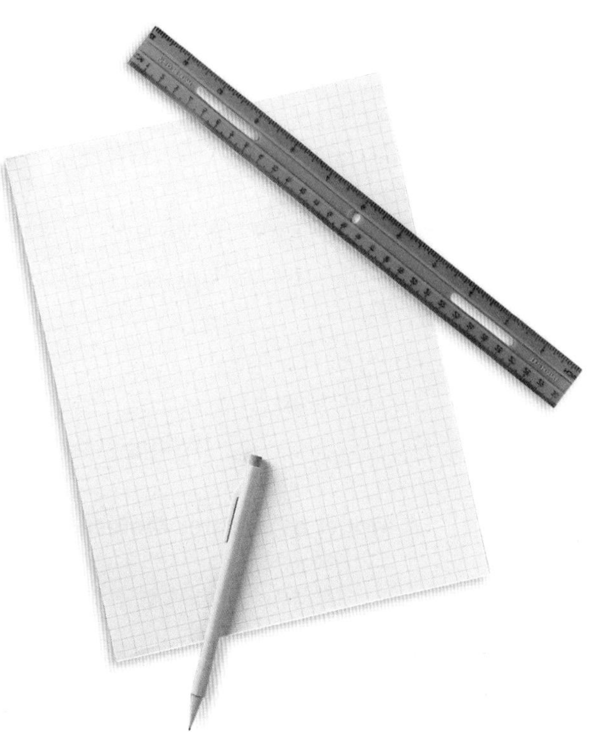

10. Make a graph of mass (*y*-axis) versus object number (*x*-axis). Label each axis, and put a title on the graph.

11. Discuss the graph with your classmates. Try to identify any important features of the graph. For example, does the graph form a line or a curve? Is there anything unusual about the graph? What do these features tell you? Write your answers in your ScienceLog.

12. Now make a graph of diameter (*y*-axis) versus object number (*x*-axis).

13. Repeat step 11.

Draw Conclusions

14. How is your arrangement of objects similar to the periodic table of the elements found in this textbook? How is your arrangement different from that periodic table?

15. Look back at your prediction about the missing object. Do you think it is still accurate? Try to improve your description by estimating the mass and diameter of the missing object. Record your estimates in your ScienceLog.

16. Mendeleev created a periodic table of elements and predicted characteristics of missing elements. How is your experiment similar to Mendeleev's work?

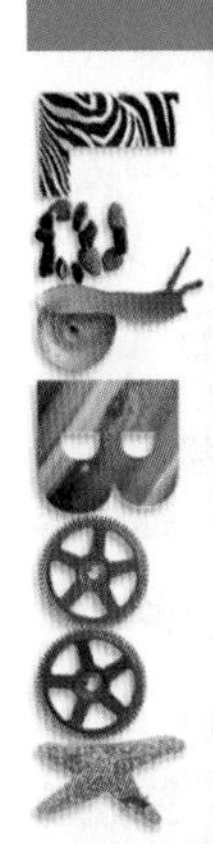

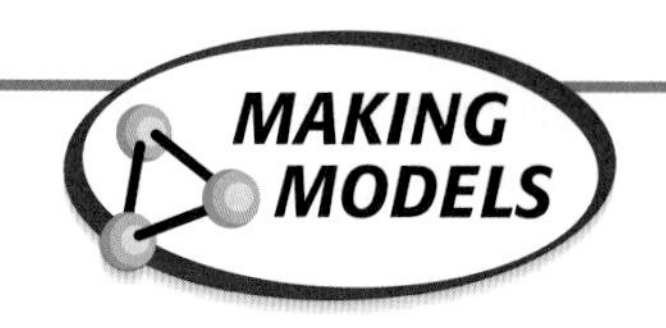

Covalent Marshmallows

A hydrogen atom has one electron in its outer energy level, but two electrons are required to fill its outer level. An oxygen atom has six electrons in its outer energy level, but eight electrons are required to fill its outer level. In order to fill their outer energy levels, two atoms of hydrogen and one atom of oxygen can share electrons, as shown below. Such a sharing of electrons to fill the outer level of atoms is called covalent bonding. When hydrogen and oxygen bond in this manner, a molecule of water is formed. In this lab you will build a three-dimensional model of water in order to better understand the covalent bonds formed in a water molecule.

Materials

- marshmallows (2 of one color, 1 of another color)
- toothpicks

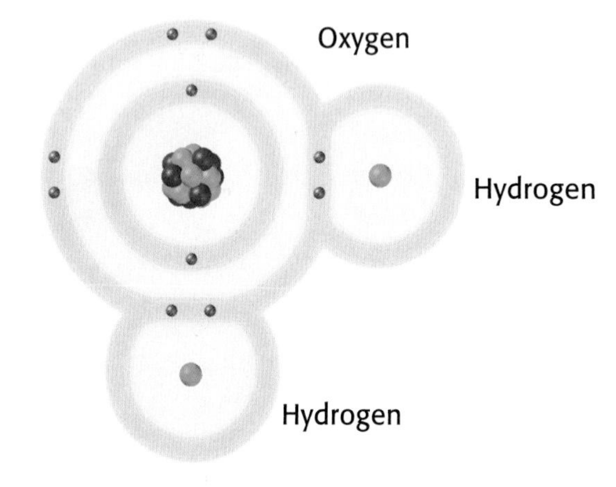

A Model of a Water Molecule

Procedure

1. Using the marshmallows and toothpicks, create a model of a water molecule. Use the diagram above for guidance in building your model.
2. Draw a sketch of your model in your ScienceLog. Be sure to label the hydrogen and oxygen atoms on your sketch.
3. Draw an electron-dot diagram of the water molecule in your ScienceLog. (Refer to the chapter text if you need help drawing an electron-dot diagram.)

Analysis

4. What do the marshmallows represent? What do the toothpicks represent?
5. Why are the marshmallows different colors?
6. Compare your model with the picture above. How might your model be improved to more accurately represent a water molecule?

7. Hydrogen in nature can covalently bond to form hydrogen (H_2) molecules. How could you model this using the marshmallows and toothpicks?
8. Draw an electron-dot diagram of an H_2 molecule in your ScienceLog.
9. Which do you think would be more difficult to create—a model of an ionic bond or a model of a covalent bond? Explain your answer.

Going Further

Create a model of a carbon dioxide molecule, which consists of two oxygen atoms and one carbon atom. The structure is similar to the structure of water, although the three atoms bond in a straight line instead of at angles. The bond between each oxygen atom and the carbon atom in a carbon dioxide molecule is a "double bond," so use two connections. Do the double bonds in carbon dioxide appear stronger or weaker than the single bonds in water? Explain your answer.

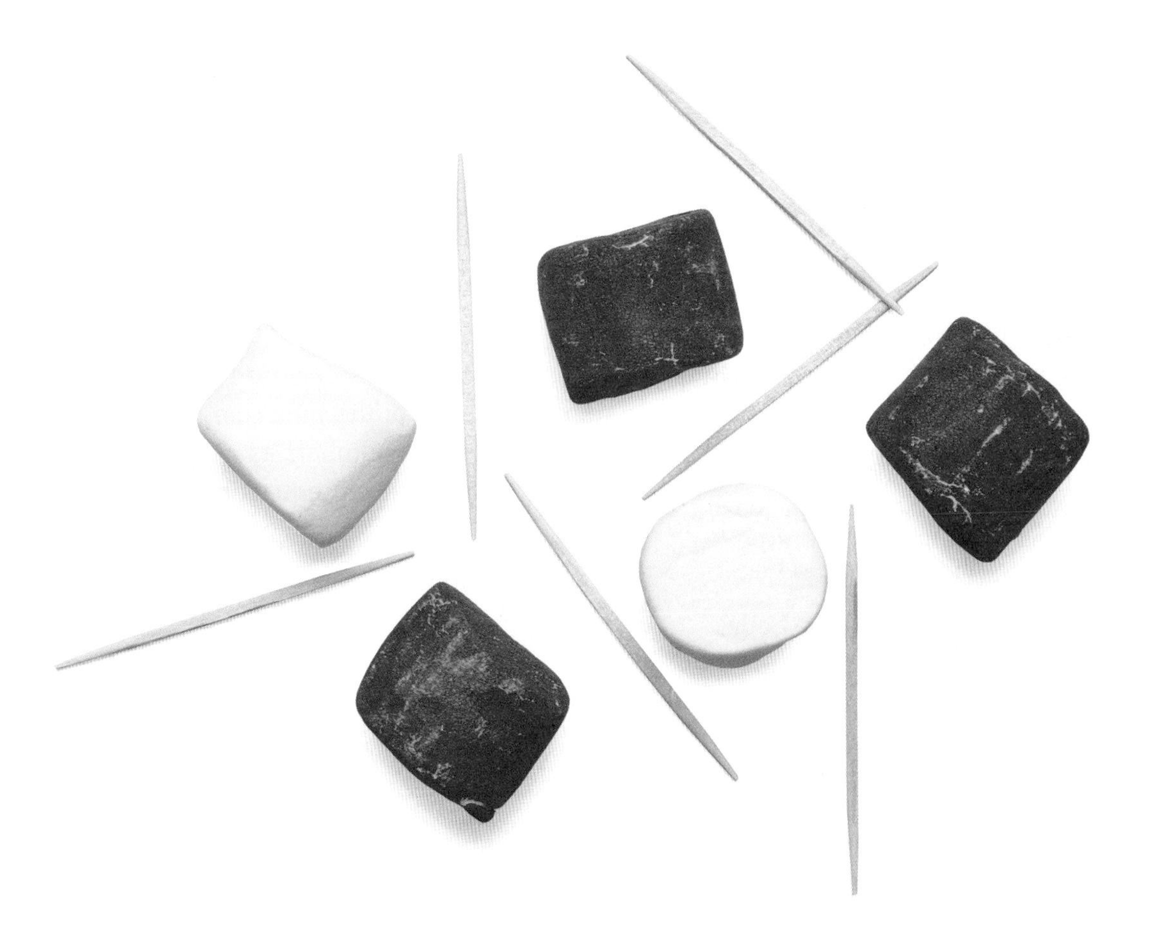

Finding a Balance

Usually, balancing a chemical equation involves just writing in your ScienceLog. But in this activity, you will use models to practice balancing chemical equations, as shown below. By following the rules, you will soon become an expert equation balancer!

Materials

- envelopes, each labeled with an unbalanced equation

Example

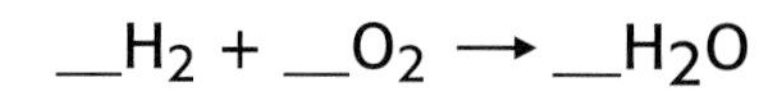

$\_H_2 + \_O_2 \rightarrow \_H_2O$

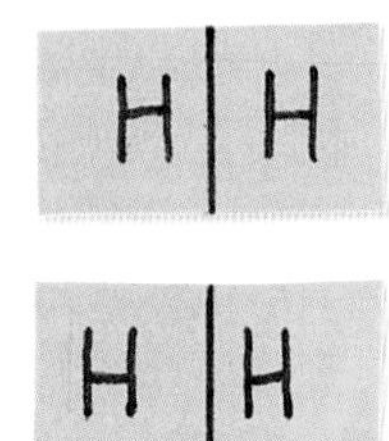

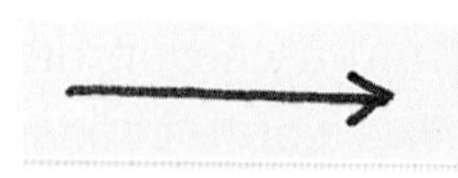

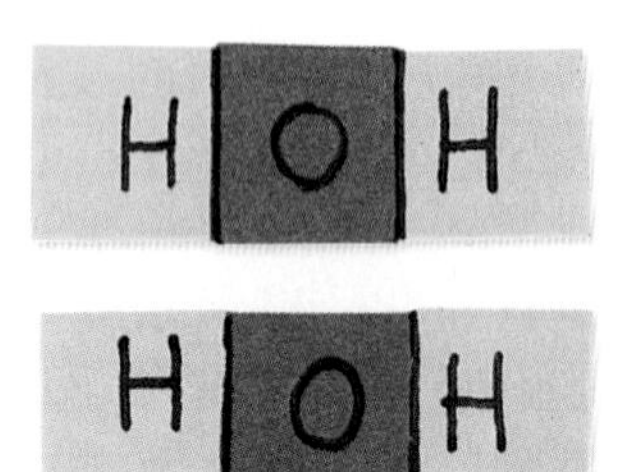

Balanced Equation

$2H_2 + O_2 \rightarrow 2H_2O$

Procedure

1. The rules:
   - **a.** Reactant-molecule models may be placed only to the left of the arrow.
   - **b.** Product-molecule models may be placed only to the right of the arrow.
   - **c.** You may use only complete molecule models.
   - **d.** At least one of each of the reactant and product molecules shown in the equation must be included in the model when you are finished.
2. Select one of the labeled envelopes. Copy the unbalanced equation written on the envelope into your ScienceLog.
3. Open the envelope, and pull out the molecule models and the arrow. Place the arrow in the center of your work area.
4. Put one model of each molecule that is a reactant on the left side of the arrow and one model of each product on the right side.
5. Add one reactant-molecule or product-molecule model at a time until the number of each of the different-colored squares on each side of the arrow is the same. Remember to follow the rules.
6. When the equation is balanced, count the number of each of the molecule models you used. Write these numbers as coefficients, as shown in the balanced equation above.
7. Select another envelope, and repeat the steps until you have balanced all of the equations.

Analysis

8. The rules specify that you are only allowed to use complete molecule models. How is this similar to what occurs in a real chemical reaction?
9. In chemical reactions, energy is either released or absorbed. In your ScienceLog, devise a way to improve the model to show energy being released or absorbed.

Cata-what? Catalyst!

Catalysts increase the rate of a chemical reaction without being changed during the reaction. In this experiment, hydrogen peroxide, H_2O_2, decomposes into oxygen, O_2, and water, H_2O. An enzyme present in liver cells acts as a catalyst for this reaction. You will investigate the relationship between the amount of the catalyst and the rate of the decomposition reaction.

Materials

- 10 mL test tubes (3)
- masking tape
- 600 mL beaker
- hot water
- funnel
- 10 mL graduated cylinder
- hydrogen peroxide solution
- 2 small liver cubes
- mortar and pestle
- tweezers

Ask a Question

1. How does the amount of a catalyst affect reaction rate?

Form a Hypothesis

2. In your ScienceLog, write a statement that answers the question above. Explain your reasoning.

Test the Hypothesis

3. Put a small piece of masking tape near the top of each test tube, and label the tubes 1, 2, and 3.
4. Create a hot-water bath by filling the beaker half-full with hot water.
5. Using the funnel and graduated cylinder, measure 5 mL of the hydrogen peroxide solution into each test tube. Place the test tubes in the hot-water bath for 5 minutes.
6. While the test tubes warm up, grind one liver cube with the mortar and pestle.
7. After 5 minutes, use the tweezers to place the cube of liver in test tube 1. Place the ground liver in test tube 2. Leave test tube 3 alone.

Make Observations

8. Observe the reaction rate (the amount of bubbling) in all three test tubes, and record your observations in your ScienceLog.

Analyze Your Results

9. Does liver appear to be a catalyst? Explain your answer.
10. Which type of liver (whole or ground) produced a faster reaction? Why?
11. What is the purpose of test tube 3?

Draw Conclusions

12. How do your results support or disprove your hypothesis?
13. Why was a hot-water bath used? (Hint: Look in your book for a definition of activation energy.)

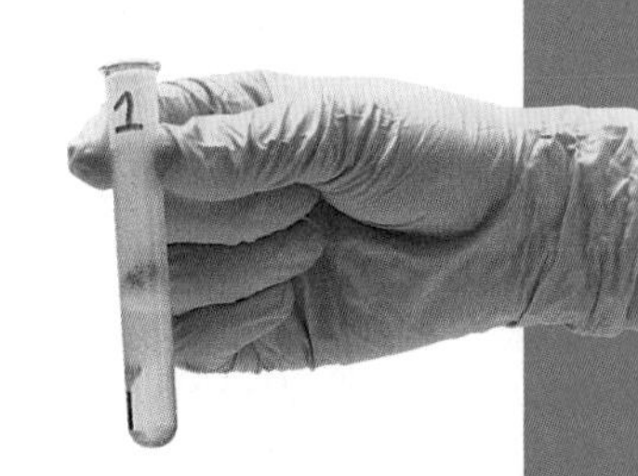

Putting Elements Together

A synthesis reaction is a reaction in which two or more substances combine to form a single compound. The resulting compound has different chemical and physical properties than the substances from which it is composed. In this activity, you will synthesize, or create, copper(II) oxide from the elements copper and oxygen.

Materials

- metric balance
- evaporating dish
- weighing paper
- copper powder
- ring stand and ring
- wire gauze
- Bunsen burner or portable burner
- spark igniter
- tongs

Conduct an Experiment/Collect Data

1. Copy the table below into your ScienceLog.

Data Collection Table	
Object	**Mass (g)**
Evaporating dish	
Copper powder	
Copper + evaporating dish after heating	
Copper(II) oxide	

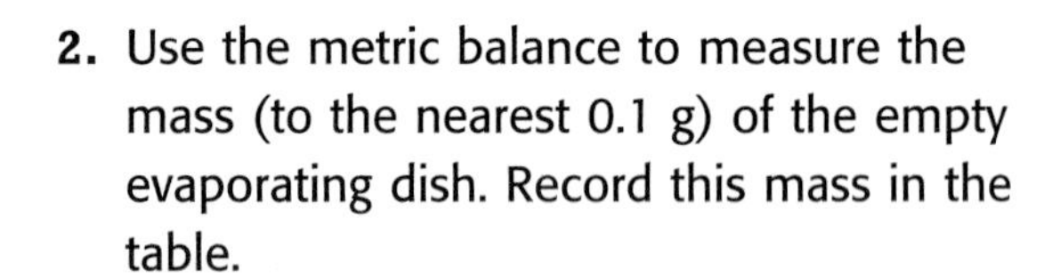

2. Use the metric balance to measure the mass (to the nearest 0.1 g) of the empty evaporating dish. Record this mass in the table.
3. Place a piece of weighing paper on the metric balance, and measure approximately 10 g of copper powder. Record the mass (to the nearest 0.1 g) in the table.
   Caution: Wear protective gloves when working with copper powder.
4. Use the weighing paper to place the copper powder in the evaporating dish. Spread the powder over the bottom and up the sides as much as possible. Discard the weighing paper.

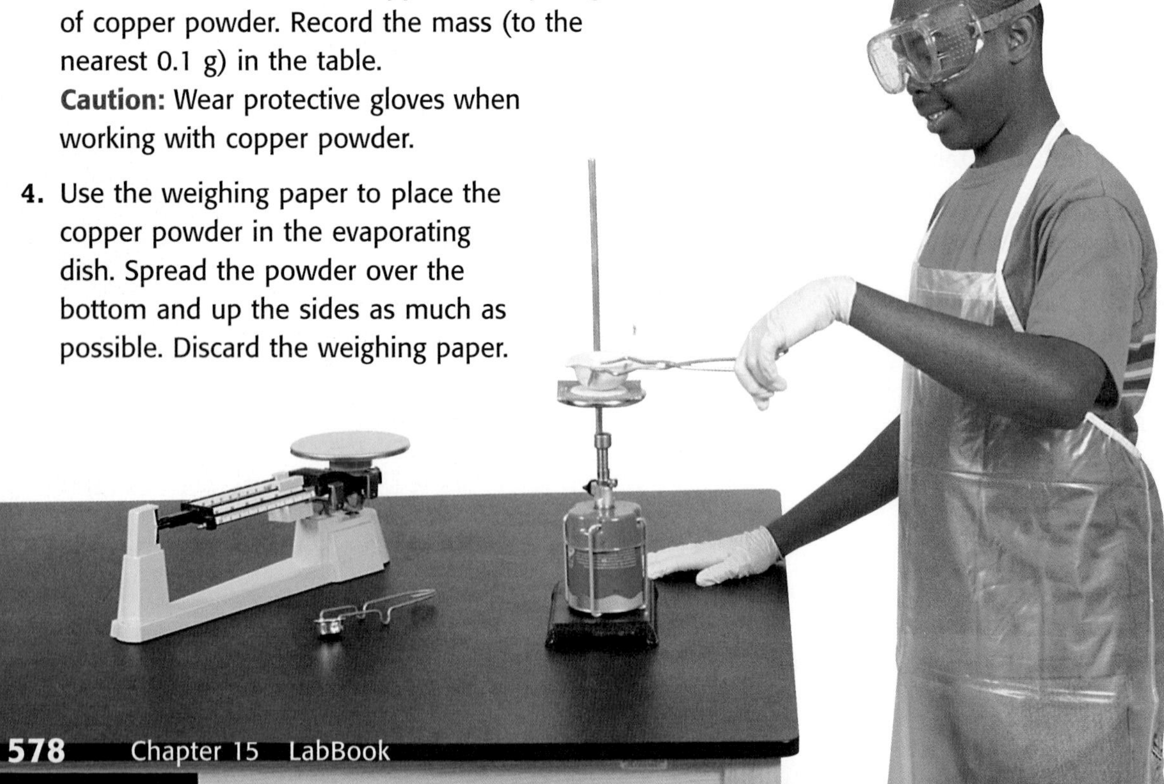

5. Set up the ring stand and ring. Place the wire gauze on top of the ring. Carefully place the evaporating dish on the wire gauze.
6. Place the Bunsen burner under the ring and wire gauze. Use the spark igniter to light the Bunsen burner.
   Caution: Use extreme care when working near an open flame.
7. Heat the evaporating dish for 10 minutes.
8. Turn off the burner, and allow the evaporating dish to cool for 10 minutes. Use tongs to remove the evaporating dish and place it on the balance to determine the mass. Record the mass in the table.
9. Determine the mass of the reaction product—copper(II) oxide—by subtracting the mass of the evaporating dish from the mass of the evaporating dish and copper powder after heating. Record this mass in the table.

Analyze the Results

10. What evidence of a chemical reaction did you observe after the copper was heated?
11. Explain why a change in mass occurred.
12. How does the change in mass support the idea that this is a synthesis reaction?

Draw Conclusions

13. Why was powdered copper used rather than a small piece of copper? (Hint: How does surface area affect the rate of the reaction?)
14. Why was the copper heated? (Hint: Look in your book for the discussion of *activation energy*.)
15. Sometimes the copper bottoms of cooking pots turn black after being used. How is that similar to the results you obtained in this lab?
16. Rust, shown below, is iron(III) oxide—the product of a synthesis reaction between iron and oxygen. How does painting a car help prevent this type of reaction?

Speed Control

The reaction rate (how fast a chemical reaction happens) is an important factor that people often want to change. Sometimes you want a reaction to take place rapidly, such as when you are removing tarnish from a metal surface. Other times you want a reaction to happen very slowly, such as when you are depending on a battery as a power source. In this lab, you will use aluminum and hydrochloric acid to discover how changing the surface area and concentration of the reactants affects reaction rate. Because metals react with acids to produce hydrogen gas, you can estimate the rate of reaction by observing how fast bubbles form.

Materials

- 30 mL test tubes (6)
- 6 strips of aluminum, approximately 5 × 1 cm each
- test-tube rack
- scissors
- 2 funnels
- 10 mL graduated cylinders (2)
- acid A
- acid B

Part A—Surface Area

Ask a Question

1. How does changing the surface area of a metal affect reaction rate?

Form a Hypothesis

2. In your ScienceLog, write a statement that answers the question above. Explain your reasoning.

Test the Hypothesis

3. You will use three identical strips of aluminum. Put one strip into a test tube without changing its shape. Place the test tube in the test-tube rack.
   Caution: The strips of metal may have sharp edges.
4. Carefully fold a second strip in half the long way and then in half again. Use a text book or other large object to flatten the folded strip as much as possible. Place the flattened strip in a second test tube, and place the test tube in the test-tube rack.
5. Use scissors to cut a third strip of aluminum into the smallest possible pieces. Place all of the pieces into a third test tube, and place the test tube in the test-tube rack.
6. Use a funnel and a graduated cylinder to pour 10 mL of acid A into each of the three test tubes.
   Caution: Hydrochloric acid is corrosive. If any acid should spill on you, immediately flush the area with water and notify your teacher.

Make Observations

7. Observe the rate of bubble formation in each test tube. Record your observations in your ScienceLog.

Analyze the Results

8. Which form of aluminum had the greatest surface area? Which had the smallest?
9. In the three test tubes, the amount of aluminum and the amount of acid were the same. Which form of the aluminum seemed to react the fastest? Which form reacted the slowest? Explain your answers.
10. Do your results support the hypothesis you made in step 2? Explain.

Draw Conclusions

11. Would powdered aluminum react faster or slower than the forms of aluminum you used? Explain your answer.

Part B—Concentration

Ask a Question

12. How does changing the concentration of acid affect the reaction rate?

Form a Hypothesis

13. In your ScienceLog, write a statement that answers the question above. Explain your reasoning.

Test the Hypothesis

14. Place one of the three remaining aluminum strips in each of the three clean test tubes. (Note: Do not alter the strips.) Place the test tubes in the test-tube rack.

15. Using the second funnel and graduated cylinder, pour 10 mL of water into one of the test tubes. Pour 10 mL of acid B into the second test tube. Pour 10 mL of acid A into the third test tube.

Make Observations

16. Observe the rate of bubble formation in the three test tubes. Record your observations in your ScienceLog.

Analyze the Results

17. In this set of test tubes, the strips of aluminum were the same, but the concentration of the acid was different. Acid A is more concentrated than acid B. Was there a difference between the test tube with water and the test tubes with acid? Which test tube formed bubbles the fastest? Explain your answers.

18. Do your results support the hypothesis you made in step 13? Explain.

Draw Conclusions

19. Explain why spilled hydrochloric acid should be diluted with water before it is wiped up.

Cabbage Patch Indicators

Indicators are weak acids or bases that change color due to the pH of the substance to which they are added. Red cabbage contains a natural indicator that turns specific colors at specific pHs. In this lab you will extract the indicator from red cabbage and use it to determine the pH of several liquids.

Materials

- distilled water
- 250 mL beaker
- red cabbage leaf
- hot plate
- beaker tongs
- masking tape
- test tubes
- test-tube rack
- eyedropper
- sample liquids provided by teacher
- litmus paper

Procedure

1. Copy the table below into your ScienceLog. Be sure to include one line for each sample liquid.

Data Collection Table			
Liquid	**Color with indicator**	**pH**	**Effect on litmus paper**
Control			

2. Put on protective gloves. Place 100 mL of distilled water in the beaker. Tear the cabbage leaf into small pieces, and place the pieces in the beaker.
3. Use the hot plate to heat the cabbage and water to boiling. Continue boiling until the water is deep blue.
   Caution: Use extreme care when working near a hot plate.
4. Use tongs to remove the beaker from the hot plate, and turn the hot plate off. Allow the solution to cool for 5–10 minutes.

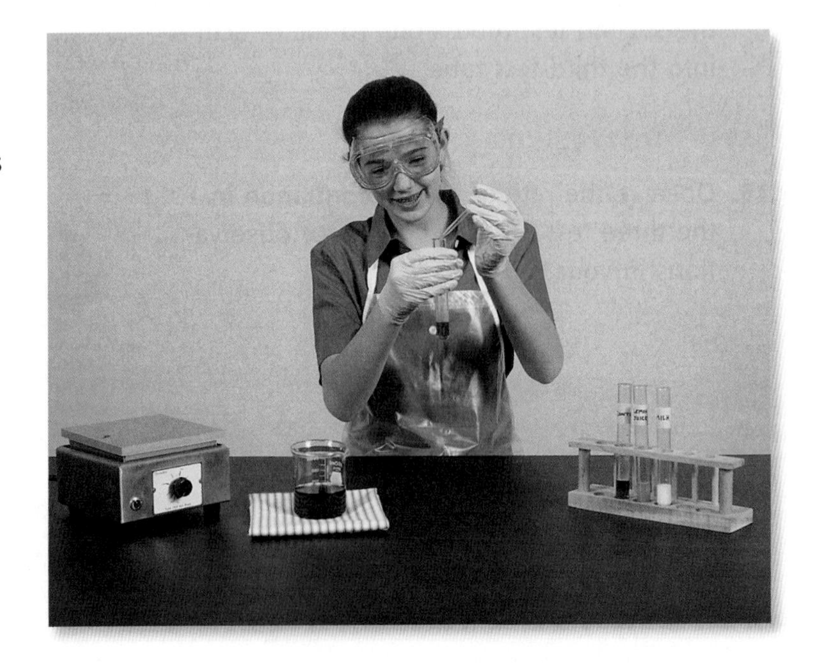

5. While the solution is cooling, use masking tape and a pen to label the test tubes for each sample liquid. Label one test tube as the control. Place the tubes in the rack.

6. Use the eyedropper to place a small amount (about 5 mL) of the indicator (cabbage juice) in the test tube labeled as the control. Pour a small amount (about 5 mL) of each sample liquid into the appropriate test tube.

7. Using the eyedropper, place several drops of the indicator into each test tube and swirl gently. Record the color of each liquid in the table. Use the chart below to determine and record the pH for each sample.

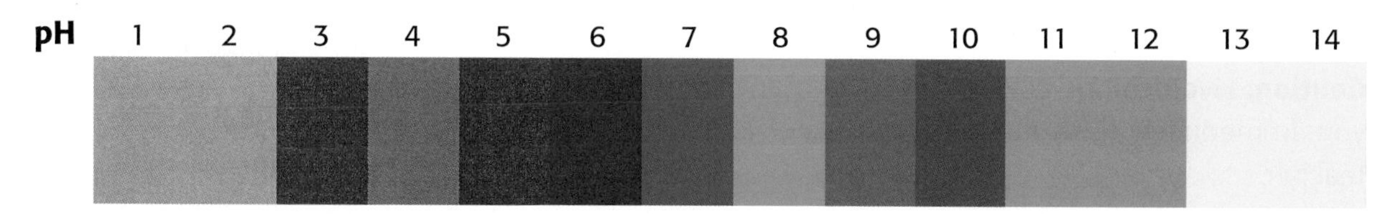

8. Litmus paper is an indicator that turns red in an acid and blue in a base. Test each liquid with a strip of litmus paper, and record the results.

Analysis

9. What purpose does the control serve? What is the pH of the control?

10. What colors are associated with acids? with bases?

11. Why is red cabbage juice considered a good indicator?

12. Which do you think would be more useful to help identify an unknown liquid—litmus paper or red cabbage juice? Why?

Going Further

Unlike distilled water, rainwater has some carbon dioxide dissolved in it. Is rainwater acidic, basic, or neutral? To find out, place a small amount of the cabbage juice indicator (which is water-based) in a clean test tube. Use a straw to gently blow bubbles in the indicator. Continue blowing bubbles until you see a color change. What can you conclude about the pH of your "rainwater?" What is the purpose of blowing bubbles in the cabbage juice?

Making Salt

A neutralization reaction between an acid and a base produces water and a salt. In this lab, you will react an acid with a base and then let the water evaporate. You will then examine what is left for properties that tell you that it is indeed a salt.

Materials

- hydrochloric acid
- 50 mL graduated cylinder
- 100 mL beaker
- distilled water
- phenolphthalein solution in a dropper bottle
- sodium hydroxide
- glass stirring rod
- 2 eyedroppers
- evaporating dish
- magnifying lens

Procedure

1. Put on protective gloves. Carefully measure 25 mL of hydrochloric acid in a graduated cylinder, then pour it into the beaker. Carefully rinse the graduated cylinder with distilled water to clean out any leftover acid.
   Caution: Hydrochloric acid is corrosive. If any should spill on you, immediately flush the area with water and notify your teacher.
2. Add three drops of phenolphthalein indicator to the acid in the beaker. You will not see anything happen yet because this indicator won't show its color unless too much base is present.
3. Measure 20 mL of sodium hydroxide (base) in the graduated cylinder, and add it slowly to the beaker with the acid. Use the stirring rod to mix the substances completely.
   Caution: Sodium hydroxide is also corrosive. If any should spill on you, immediately flush the area with water and notify your teacher.
4. Use an eyedropper to add more base to the acid-base mixture in the beaker a few drops at a time. Be sure to stir the mixture after each few drops. Continue adding drops of base until the mixture remains colored after stirring.
5. Use another eyedropper to add acid to the beaker, one drop at a time, until the color just disappears after stirring.

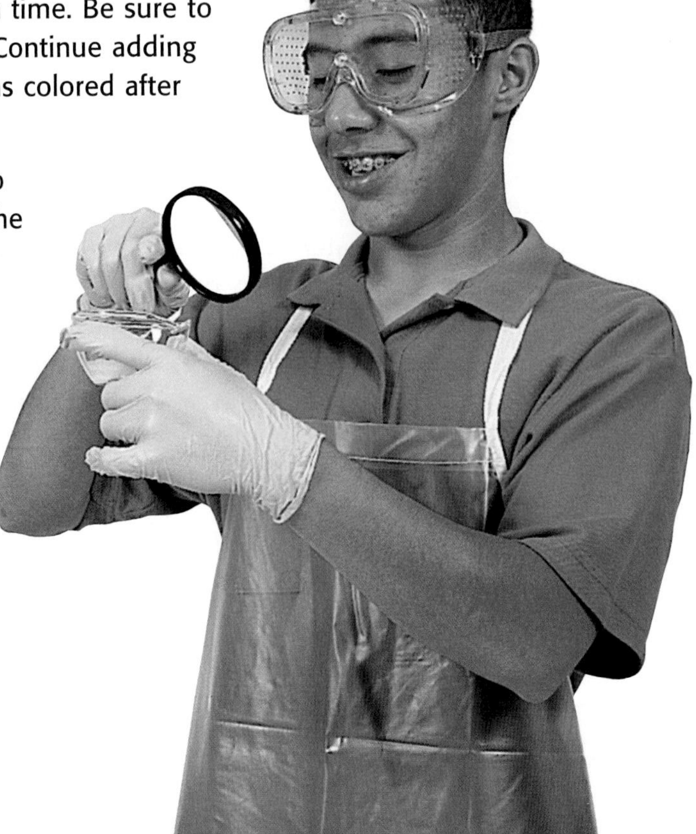

6. Pour the mixture carefully into an evaporating dish, and place the dish where your teacher tells you to allow the water to evaporate overnight.

7. The next day, examine your evaporating dish and study the crystals that were left with a magnifying lens. Identify the color, shape, and other properties that the crystals have.

Analysis

8. The equation for the reaction above is:

$$HCl + NaOH \longrightarrow H_2O + NaCl.$$

NaCl is ordinary table salt and forms very regular cubic crystals that are white. Did you find white cubic crystals?

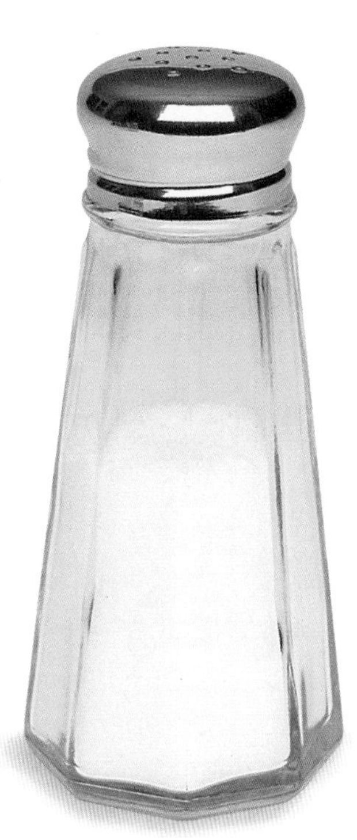

9. The phenolphthalein indicator changes color in the presence of a base. Why did you add more acid in step 5 until the color disappeared?

Going Further

Another neutralization reaction occurs between hydrochloric acid and potassium hydroxide, KOH. The equation for this reaction is as follows:

$$HCl + KOH \longrightarrow H_2O + KCl$$

What are the products of this neutralization reaction? How do they compare with those you discovered in this experiment?

DISCOVERY LAB

How Far Is the Sun?

It doesn't slice, it doesn't dice, but it can give you an idea of how big our universe is! You can build your very own stellar-distance measuring device from household items. Amaze your friends by figuring out how many metersticks can be placed between Earth and the sun.

Materials

- poster board
- scissors
- square of aluminum foil
- thumbtack
- masking tape
- index card
- meterstick
- metric ruler

Ask a Question

1. If it were possible, how many metersticks could I place between the sun and the Earth?

Conduct an Experiment

2. Measure and cut a 4×4 cm square from the middle of the poster board. Tape the foil square in the center of the poster board.
3. Carefully prick the foil with a thumbtack to form a tiny hole in the center. Congratulations—you have just constructed your very own stellar-distance measuring device!
4. Tape the device to a window facing the sun so that sunlight shines directly through the pinhole.
   Caution: Do not look directly into the sun.

5. Place one end of the meterstick against the window and beneath the foil square, and steady the meterstick with one hand.
6. With the other hand, hold the index card close to the pinhole. You should be able to see a circular image on the card. This is an image of the sun.
7. Move the card back until the image is large enough to measure. Be sure to keep the image on the card sharply focused. Reposition the meterstick so that it touches the bottom of the card.
8. Ask your partner to measure the diameter of the image on the card with the metric ruler. Record the diameter of the image in your ScienceLog.
9. Record the distance between the window and the index card by reading the point at which the card rests on the meterstick.
10. Calculate the distance between the Earth and the sun using the following formula:

$$\text{Distance between the sun and Earth} = \text{sun's diameter} \times \frac{\text{distance to the image}}{\text{image's diameter}}$$

1 cm = 10 mm
1 m = 100 cm
1 km = 1,000 m

Hint: The sun's diameter is 1,392,000,000 m.

Analyze the Results

11. According to your calculations, how far is the sun from the Earth? Don't forget to convert your measurements to meters.

Draw Conclusions

12. You could put 150 billion metersticks between the Earth and the sun. Compare this with your result in step 10. Do you think that this is a good way to measure the Earth's distance from the sun? Support your answer.

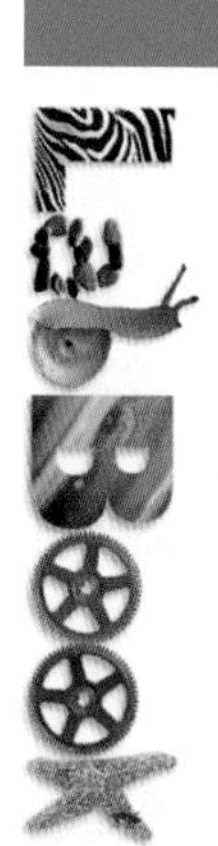

Why Do They Wander?

Before the discoveries of Nicolaus Copernicus in the early 1500s, most people thought that the planets and the sun revolved around the Earth and that the Earth was the center of the solar system. But Copernicus observed that the sun is the center of the solar system and that all the planets, including Earth, revolve around the sun. He also explained a puzzling aspect of the movement of planets across the night sky. If you watch a planet every night for several months, you'll notice that it appears to "wander" among the stars. While the stars remain in fixed positions relative to each other, the planets appear to move independently of the stars. First Mars travels to the left, then it goes back to the right a little, and finally it reverses direction and travels again to the left. No wonder the early Greeks called the planets wanderers! In this lab you will make your own model of part of the solar system to find out how Copernicus's model of the solar system explained this zigzag motion of the planets.

Materials

- drawing compass
- white paper
- metric ruler
- colored pencils

Ask a Question

1. Why do the planets appear to move back and forth in the Earth's night sky?

Conduct an Experiment

2. Use the compass to draw a circle with a diameter of 9 cm on the paper. This circle will represent the orbit of the Earth around the sun. (Note: The orbits of the planets are actually slightly elliptical, but circles will work for this activity.)
3. Using the same center point, draw a circle with a diameter of 12 cm. This circle will represent the orbit of Mars.
4. Using a blue pencil, draw three parallel lines in a diagonal across one end of your paper, as shown at right. These lines will help you plot the path Mars appears to travel in Earth's night sky. Turn your paper so that the diagonal lines are at the top of the page.
5. Place 11 dots on your Earth orbit, as shown on the next page, and number them 1 through 11. These dots will represent Earth's position from month to month.

6. Now place 11 dots along the top of your Mars orbit, as shown below. Number the dots as shown. These dots will represent the position of Mars at the same time intervals. Notice that Mars travels slower than Earth.

7. Use a green line to connect the first dot on Earth's orbit to the first dot on Mars's orbit, and extend the line all the way to the first diagonal line at the top of your paper. Place a green dot where this green line meets the first blue diagonal line, and label the green dot *1.*

8. Now connect the second dot on Earth's orbit to the second dot on Mars's orbit, and extend the line all the way to the first diagonal at the top of your paper. Place a green dot where this line meets the first blue diagonal line, and label this dot *2.*

9. Continue drawing green lines from Earth's orbit through Mars's orbit and finally to the blue diagonal lines. Pay attention to the pattern of dots you are adding to the diagonal lines. When the direction of the dots changes, extend the green line to the next diagonal, and add the dots to that line instead.

10. When you are finished adding green lines, draw a red line to connect all the dots on the blue diagonal lines in the order you drew them.

Analyze the Results

11. What do the green lines connecting points along Earth's orbit and Mars's orbit represent?

12. What does the red line connecting the dots along the diagonal lines look like? How can you explain this?

Draw Conclusions

13. What does this demonstration show about the motion of Mars?

14. Why do planets appear to move back and forth across the sky?

15. Were the Greeks justified in calling the planets wanderers? Explain.

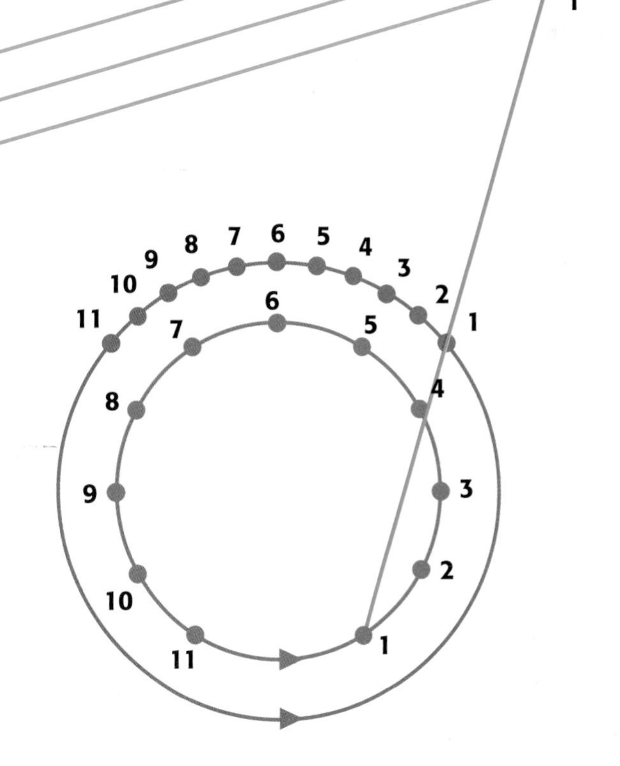

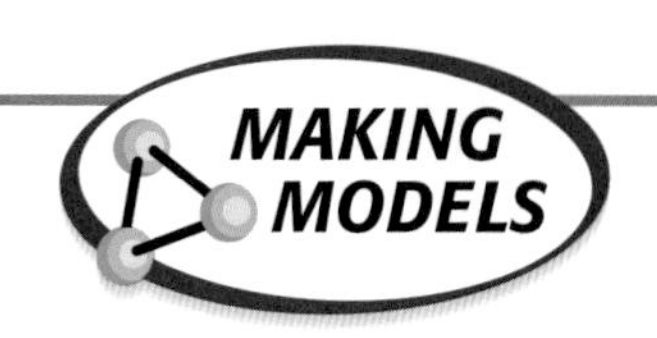

Eclipses

As the Earth and the moon revolve around the sun, they both cast shadows into space. An eclipse occurs when one planetary body passes through the shadow of another. You can demonstrate how an eclipse occurs by using clay models of planetary bodies.

Materials

- modeling clay
- metric ruler
- sheet of notebook paper
- small flashlight

Procedure

1. Make two balls out of the modeling clay. One ball should have a diameter of about 4 cm and will represent the Earth. The other should have a diameter of about 1 cm and will represent the moon.
2. Place the two balls about 15 cm apart on the sheet of paper. (You may want to prop the smaller ball up on folded paper or on clay so that the centers of the two balls are at the same level.)
3. Hold the flashlight approximately 15 cm away from the large ball. The flashlight and the two balls should be in a straight line. Keep the flashlight at about the same level as the clay. When the whole class is ready, your teacher will turn off the lights.
4. Turn on your flashlight. Shine the light on the larger ball, and sketch your model in your ScienceLog. Include the beam of light in your drawing.
5. Move the flashlight to the opposite side of the paper. The flashlight should now be approximately 15 cm away from the smaller clay ball. Repeat step 4.

Analysis

6. What does the flashlight in your model represent?
7. As viewed from Earth, what event did your model represent in step 4?
8. As viewed from the moon, what event did your model represent in step 4?
9. As viewed from Earth, what event did your model represent in step 5?
10. As viewed from the moon, what event did your model represent in step 5?
11. According to your model, how often would solar and lunar eclipses occur? Is this accurate? Explain.

Phases of the Moon

It's easy to see when the moon is full. But you may have wondered exactly what happens when the moon appears as a crescent or when you cannot see the moon at all. Does the Earth cast its shadow on the moon? In this activity, you will discover how and why the moon appears as it does in each phase.

Materials

- globe
- light source
- plastic-foam ball

Procedure

1. Place your globe near the light source. Be sure that the north pole is tilted toward the light. Rotate the globe so that your part of the world faces the light.
2. Using the ball as your model of the moon, move the moon between the Earth (the globe) and the sun (the light). The side of the moon that faces the Earth will be in darkness. Write your observations of this new-moon phase in your ScienceLog.
3. Continue to move the moon in its orbit around the Earth. When part of the moon is illuminated by the light, as viewed from Earth, the moon is in the crescent phase. Add your observations to your ScienceLog.
4. If you have time, you may draw your own moon-phase diagram.

Analysis

5. About 2 weeks after the new moon appears, the entire moon is visible in the sky. Move the ball to show this event.
6. What other phases can you add to your diagram? For example, when does the quarter moon appear?
7. Explain why the moon sometimes appears as a crescent to viewers on Earth.

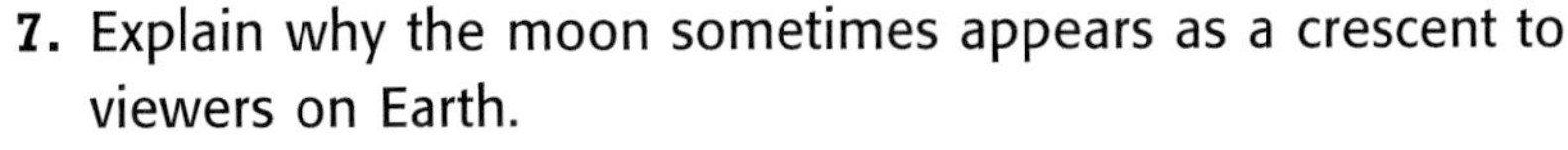

Red Hot, or Not?

When you look at the night sky, some stars are brighter than others. Some are even different colors from what you might expect. For example, one star in the constellation Orion glows red; and Sirius, the brightest star in the sky, glows a bluish white. Astronomers use these colors to estimate the temperature of the stars. In this activity, you will experiment with a light bulb and some batteries to discover what the color of a glowing object reveals about the temperature of the object.

Materials

- electrical tape
- 2 conducting wires
- weak D cell
- flashlight bulb
- 2 fresh D cells

Procedure

1. Tape one end of a conducting wire to the positive pole of the weak D cell. Tape one end of the second conducting wire to the negative pole.
2. Touch the free end of each wire to the light bulb. Hold one of the wires against the bottom tip of the light bulb. Hold the second wire against the side of the metal portion of the bulb. The bulb should light.
3. In your ScienceLog, record the color of the filament in the light bulb. Carefully touch your hand to the bulb. Observe the temperature of the bulb. Record your observations in your ScienceLog.

4. Repeat steps 1–3 with one of the two fresh D cells.
5. Use the electrical tape to connect the two fresh D cells in a continuous circuit so that the positive pole of the first cell is connected to the negative pole of the second cell.
6. Repeat steps 1–3 with both fresh D cells in combination.

Analysis

7. How did the color of the filament change in the three trials? How did the temperature change?
8. What information does the color of a star provide?
9. What color are stars with relatively high surface temperatures? What color are stars with relatively low surface temperatures?
10. Arrange the following stars in order from highest to lowest surface temperature: Sirius is bluish white. Aldebaran is orange. Procyon is yellow-white. Capella is yellow. Betelgeuse (BET uhl jooz) is red.

I See the Light!

How do you find the distance to an object you can't reach? You can do it by measuring something you can reach, finding a few angles, and using mathematics. In this activity, you'll practice measuring the distances of objects here on Earth. When you get used to it, you can take your skills to the stars!

Materials

- 16 × 16 cm piece of poster board
- metric ruler
- protractor
- scissors
- sharp pencil
- 30 cm string
- transparent tape
- meterstick
- metric measuring tape
- scientific calculator

Procedure

1. Draw a line 4 cm away from the edge of one side of the piece of poster board. Fold the poster board along this line.
2. Tape the protractor to the poster board with its flat edge against the fold, as shown in the photo below.
3. Use a sharp pencil to carefully punch a hole through the poster board along its folded edge at the center of the protractor.
4. Thread the string through the hole, and tape one end to the underside of the poster board. The other end should be long enough to hang off the far end of the poster board.
5. Carefully punch a second hole in the smaller area of the poster board halfway between its short sides. The hole should be directly above the first hole and should be large enough for the pencil to fit through. This is the viewing hole of your new parallax device. This device will allow you to measure the distance of faraway objects.
6. Find a location outside that is at least 50 steps away from a tall, narrow object, such as the school's flagpole or a tall tree. (This object will represent background stars.) Set the meterstick on the ground with one of its long edges facing the flagpole.
7. Ask your partner, who represents a nearby star, to take 10 steps toward the flagpole, starting at the left end of the meterstick. You will be the observer. When you stand at the left end of the meterstick, which represents the location of the sun, your partner's nose should be lined up with the flag pole.

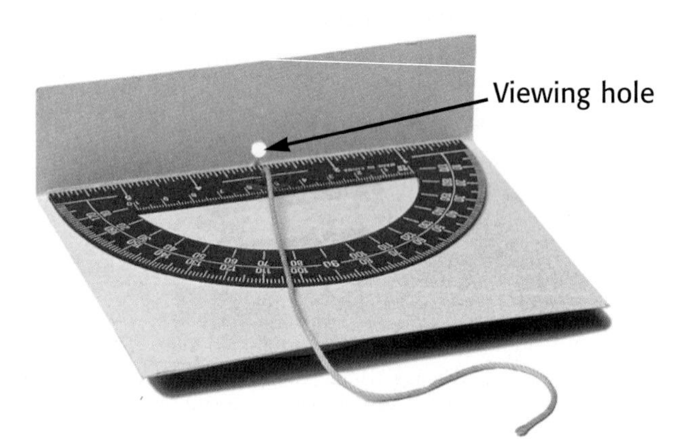

8. Move to the other end of the meterstick, which represents the location of Earth. Does your partner appear to the left or right of the flagpole? Record your observation in your ScienceLog.
9. Hold the string so that it runs straight from the viewing hole to the 90° mark on the protractor. Using one eye, look through the viewing hole along the string and point the device at your partner's nose.

10. Holding the device still, slowly move your head until you can see the flagpole through the viewing hole. Move the string so that it lines up between your eye and the flagpole. Make sure the string is taut, and hold it tightly against the protractor.
11. Read and record the angle made by the string and the string's original position at 90° (count the number of degrees between 90° and the string's new position).
12. Use the measuring tape to find the distance from the left end of the meterstick to your partner's nose. Record this distance in your ScienceLog.
13. Now find a place outside that is at least 100 steps away from the flagpole. Set the meterstick on the ground as before.
14. Repeat steps 7–12.

Analysis

15. The angle you recorded in step 11 is called the *parallax angle.* The distance from one end of the meterstick to the other is called the *baseline.* With this angle and the length of your baseline, you can calculate the distance to your partner.
16. To calculate the distance (d) to your partner, use the following equation:

$$d = b/\tan A$$

In this equation, A is the parallax angle and b is the length of the baseline (1 m). (Tan A means the tangent of angle A, which you will learn more about in high school math classes.)

17. To find d, enter 1 (the length of your baseline in meters) into the calculator, press the "divide" key, enter the value of A (the parallax angle you recorded), then press the "tan" key. Finally, press the "equals" key.
18. Record this result in your ScienceLog. It is the distance in meters between the left end of the meterstick and your partner. You may want to use a table like the one shown at right.
19. How close is this calculated distance to the distance you measured in step 12?
20. Repeat steps 16–18 using the angle you found when the flagpole was 100 steps away.

Conclusions

21. At which position, 50 steps or 100 steps from the flagpole, did your calculated distance better match the actual distance as measured in step 12?
22. What do you think would happen if you were even farther from the flagpole?
23. When astronomers use parallax, their "flagpole" is the very distant stars. How might this affect the accuracy of their parallax readings?

Distance by Parallax Versus Measuring Tape

	At 50 steps	At 100 steps
Parallax angle		
Distance (calculated)		
Distance (measured)		

SELF-CHECK

Self-Check Answers

Chapter 1—The World of Physical Science

Page 16: flapping rate

Chapter 2—The Properties of Matter

Page 41: approximately 30 N

Chapter 3—States of Matter

Page 64: The pressure would increase.

Page 70: endothermic

Chapter 4—Elements, Compounds, and Mixtures

Page 87: No, the properties of pure water are the same no matter what its source is.

Page 93: Copper and silver are solutes. Gold is the solvent.

Chapter 5—Matter in Motion

Page 110: Numbers 1 and 3 are examples of velocity.

Page 117: 2 N north

Page 122: sliding friction

Page 126: Gravity is a force of attraction between objects that is due to the masses of the objects.

Chapter 6—Forces in Motion

Page 140: A leaf is more affected by air resistance.

Page 147: This can be answered in terms of either Newton's first law or inertia.

Newton's first law: When the bus is still, both you and the bus are at rest. The bus started moving, but no unbalanced force acted on your body, so your body stayed at rest.

Inertia: You have inertia, and that makes you difficult to move, so when the bus started to move, you didn't move with it.

Chapter 7—Forces in Fluids

Page 175: Air travels faster over the top of a wing.

Chapter 8—Work and Machines

Page 189: Pulling a wheeled suitcase is doing work because the force applied and the motion of the suitcase are in the same direction.

Chapter 9—Energy and Energy Resources

Page 223: A roller coaster has the greatest potential energy at the top of the highest hill (usually the first hill) and the greatest kinetic energy at the bottom of the highest hill.

Chapter 10—The Energy of Waves

Page 248: Mechanical waves require a medium; electromagnetic waves do not.

Page 257: A light wave will not refract if it enters a new medium perpendicular to the surface because the entire wave enters the new medium at the same time.

Chapter 11—Introduction to Electricity

Page 273: Plastic wrap is charged by friction as it is pulled off the roll.

Page 286: $E = P \times t$; $E = 200\text{ W} \times 2\text{ h} = 400\text{ Wh}$

Students will have to use data from the table on page 286 to answer this question.

Page 289: Yes; a microwave is an example of a load because it uses electrical energy to do work.

Chapter 12—Introduction to Atoms

Page 309: The particles Thomson discovered had negative charges. Because an atom has no charge, it must contain positively charged particles to cancel the negative charges.

Chapter 13—The Periodic Table

Page 336: It is easier for atoms of alkali metals to lose one electron than for atoms of alkaline-earth metals to lose two electrons. Therefore, alkali metals are more reactive than alkaline-earth metals.

Chapter 14—Chemical Bonding

Page 357: neon

Page 361: **1.** 6 **2.** In a covalent bond, electrons are shared between atoms. In an ionic bond, electrons are transferred from one atom to another.

Chapter 15—Chemical Reactions

Page 377: 2 sodium atoms, 1 sulfur atom, and 4 oxygen atoms

Page 379: $CaBr_2 + Cl_2 \rightarrow Br_2 + CaCl_2$
reactants: $CaBr_2$ and Cl_2
products: Br_2 and $CaCl_2$

Chapter 16—Chemical Compounds

Page 405: a soft drink

Chapter 17—Formation of the Solar System

Page 425: The balance between gravity and pressure keeps a nebula from collapsing.

Page 427: Gas giants were formed far out in the solar nebula, where it was cold and there was more solid material (dust and ices) for the planets to collect.

Page 438: Earth has enough mass that gravitational pressure crushed and melted rocks during its formation. The force of gravity pulled this material toward the center, forming a sphere. Asteroids are not massive enough for their interiors to be crushed or melted.

Chapter 18—A Family of Planets

Page 469: The surface of Titan is much colder than that of the Earth. In fact, the temperature on Titan is close to –178°C!

Chapter 19—The Universe Beyond

Page 488: The two stars would have the same apparent magnitude.

CONTENTS

Inch
Yard
Fathom
Foot

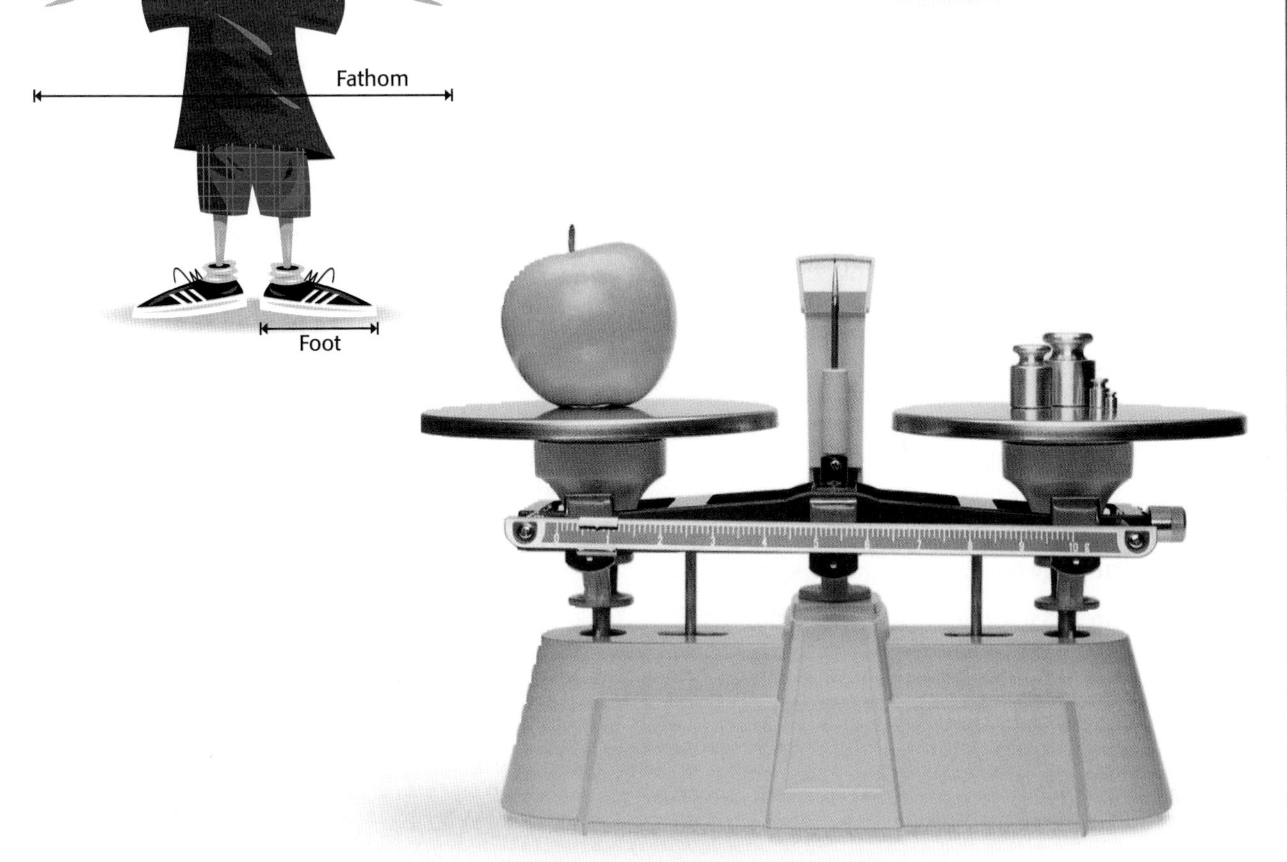

Concept Mapping: A Way to Bring Ideas Together

What Is a Concept Map?

Have you ever tried to tell someone about a book or a chapter you've just read and found that you can remember only a few isolated words and ideas? Or maybe you've memorized facts for a test, and then weeks later discover you're not even sure what topics those facts cover.

In both cases, you may have understood the ideas or concepts by themselves but not in relation to one another. If you could somehow link the ideas together, you would probably understand them better and remember them longer. This is something a concept map can help you do. A concept map is a way to see how ideas or concepts fit together. It can help you see the "big picture."

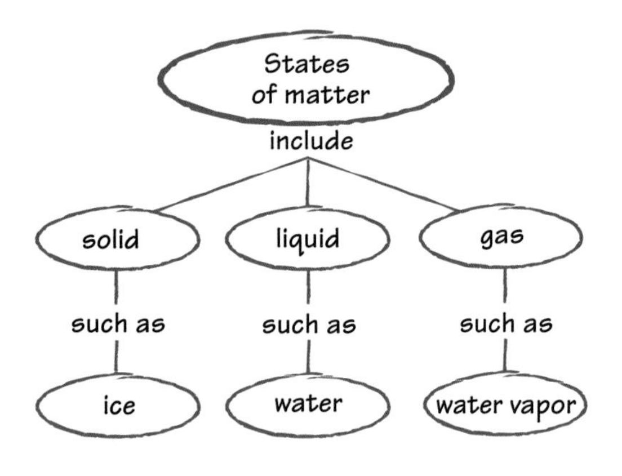

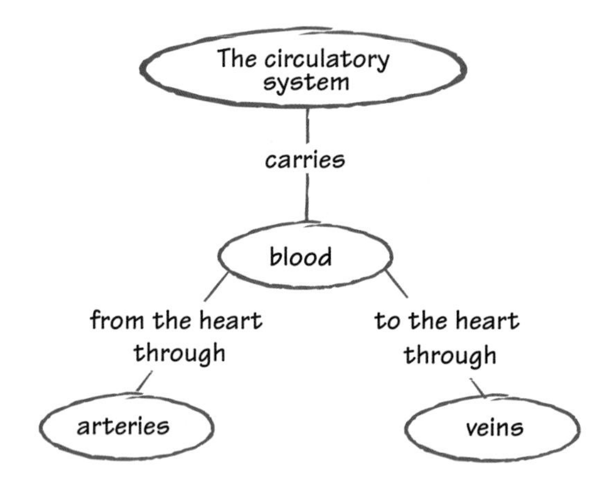

How to Make a Concept Map

1. **Make a list of the main ideas or concepts.**

   It might help to write each concept on its own slip of paper. This will make it easier to rearrange the concepts as many times as necessary to make sense of how the concepts are connected. After you've made a few concept maps this way, you can go directly from writing your list to actually making the map.

2. **Spread out the slips on a sheet of paper, and arrange the concepts in order from the most general to the most specific.**

   Put the most general concept at the top and circle it. Ask yourself, "How does this concept relate to the remaining concepts?" As you see the relationships, arrange the concepts in order from general to specific.

3. **Connect the related concepts with lines.**

4. **On each line, write an action word or short phrase that shows how the concepts are related.**

   Look at the concept maps on this page, and then see if you can make one for the following terms:

   plants, water, photosynthesis, carbon dioxide, sun's energy

   One possible answer is provided at right, but don't look at it until you try the concept map yourself.

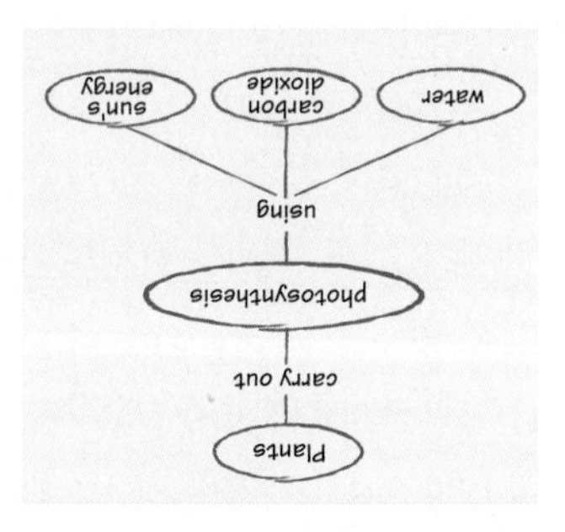

SI Measurement

The International System of Units, or SI, is the standard measuring system for many scientists. Using the same standards of measurement makes it easier for scientists to communicate with one another.

SI works by combining prefixes and base units. Each base unit can be used with different prefixes to define smaller and larger quantities. The table below lists common SI prefixes.

SI Prefixes			
Prefix	**Abbreviation**	**Factor**	**Example**
kilo-	k	1,000	kilogram, 1 kg = 1,000 g
hecto-	h	100	hectoliter, 1 hL = 100 L
deka-	da	10	dekameter, 1 dam = 10 m
		1	meter, liter
deci-	d	0.1	decigram, 1 dg = 0.1 g
centi-	c	0.01	centimeter, 1 cm = 0.01 m
milli-	m	0.001	milliliter, 1 mL = 0.001 L
micro-	µ	0.000001	micrometer, 1 µm = 0.000 001 m

SI Conversion Table		
SI units	**From SI to English**	**From English to SI**
Length		
kilometer (km) = 1,000 m	1 km = 0.621 mi	1 mi = 1.609 km
meter (m) = 100 cm	1 m = 3.281 ft	1 ft = 0.305 m
centimeter (cm) = 0.01 m	1 cm = 0.394 in.	1 in. = 2.540 cm
millimeter (mm) = 0.001 m	1 mm = 0.039 in.	
micrometer (µm) = 0.000 001 m		
nanometer (nm) = 0.000 000 001 m		
Area		
square kilometer (km^2) = 100 hectares	1 km^2 = 0.386 mi^2	1 mi^2 = 2.590 km^2
hectare (ha) = 10,000 m^2	1 ha = 2.471 acres	1 acre = 0.405 ha
square meter (m^2) = 10,000 cm^2	1 m^2 = 10.765 ft^2	1 ft^2 = 0.093 m^2
square centimeter (cm^2) = 100 mm^2	1 cm^2 = 0.155 $in.^2$	1 $in.^2$ = 6.452 cm^2
Volume		
liter (L) = 1,000 mL = 1 dm^3	1 L = 1.057 fl qt	1 fl qt = 0.946 L
milliliter (mL) = 0.001 L = 1 cm^3	1 mL = 0.034 fl oz	1 fl oz = 29.575 mL
microliter (µL) = 0.000 001 L		
Mass		
kilogram (kg) = 1,000 g	1 kg = 2.205 lb	1 lb = 0.454 kg
gram (g) = 1,000 mg	1 g = 0.035 oz	1 oz = 28.329 g
milligram (mg) = 0.001 g		
microgram (µg) = 0.000 001 g		

Temperature Scales

Temperature can be expressed with three different scales: Fahrenheit, Celsius, and Kelvin. The SI unit for temperature is the kelvin (K). Although 0 K is much colder than 0°C, a change of 1 K is equal to a change of 1°C.

Three Temperature Scales

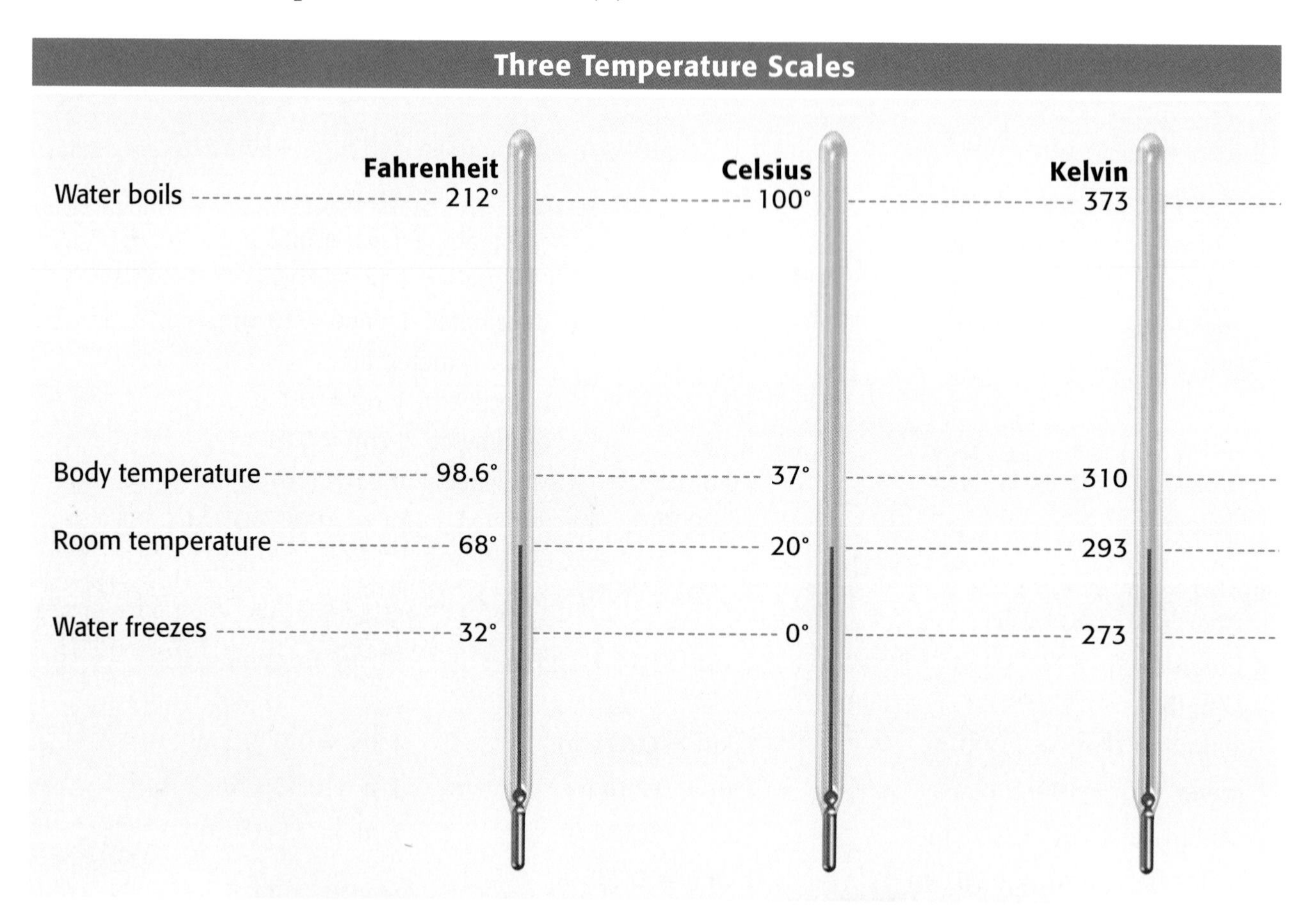

Temperature Conversions Table

To convert	Use this equation:	Example
Celsius to Fahrenheit °C → °F	$°F = \left(\frac{9}{5} \times °C\right) + 32$	Convert 45°C to °F. $°F = \left(\frac{9}{5} \times 45°C\right) + 32 = 113°F$
Fahrenheit to Celsius °F → °C	$°C = \frac{5}{9} \times (°F - 32)$	Convert 68°F to °C. $°C = \frac{5}{9} \times (68°F - 32) = 20°C$
Celsius to Kelvin °C → K	$K = °C + 273$	Convert 45°C to K. $K = 45°C + 273 = 318\ K$
Kelvin to Celsius K → °C	$°C = K - 273$	Convert 32 K to °C. $°C = 32\ K - 273 = -241°C$

Measuring Skills

Using a Graduated Cylinder

When using a graduated cylinder to measure volume, keep the following procedures in mind:

1. Make sure the cylinder is on a flat, level surface.
2. Move your head so that your eye is level with the surface of the liquid.
3. Read the mark closest to the liquid level. On glass graduated cylinders, read the mark closest to the center of the curve.

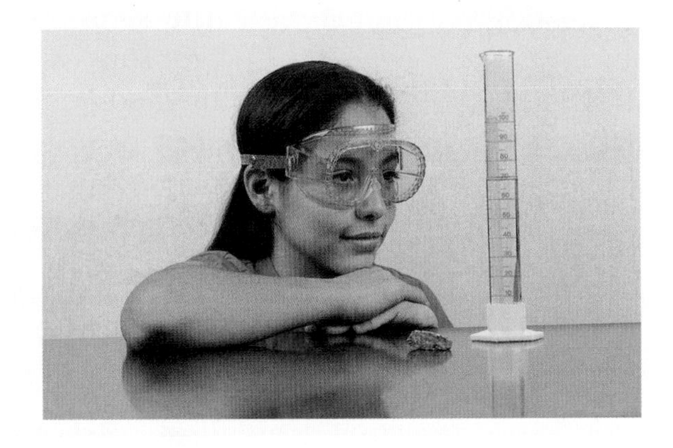

Using a Meterstick or Metric Ruler

When using a meterstick or metric ruler, keep the following procedures in mind:

1. Place the ruler firmly against the object you are measuring.
2. Align one edge of the object exactly with the zero end of the ruler.
3. Look at the other edge of the object to see which of the marks on the ruler is closest to that edge. **Note:** Each small slash between the centimeters represents a millimeter, which is one-tenth of a centimeter.

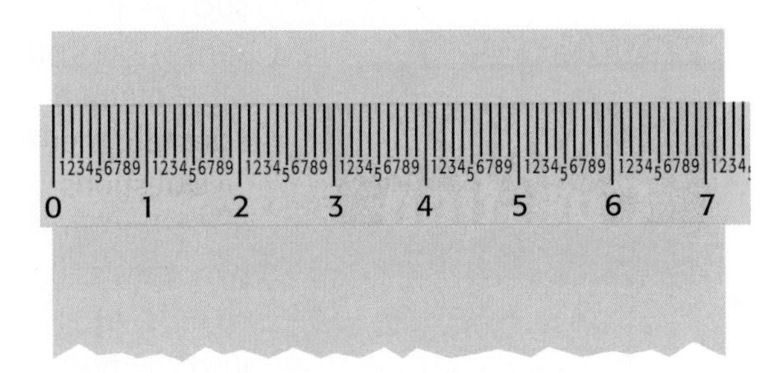

Using a Triple-Beam Balance

When using a triple-beam balance, keep the following procedures in mind:

1. Make sure the balance is on a level surface.
2. Place all of the countermasses at zero. Adjust the balancing knob until the pointer rests at zero.
3. Place the object you wish to measure on the pan. **Caution:** Do not place hot objects or chemicals directly on the balance pan.
4. Move the largest countermass along the beam to the right until it is at the last notch that does not tip the balance. Follow the same procedure with the next-largest countermass. Then move the smallest countermass until the pointer rests at zero.
5. Add the readings from the three beams together to determine the mass of the object.
6. When determining the mass of crystals or powders, use a piece of filter paper. First mass the paper. Then add the crystals or powder to the paper and remass. The actual mass of the crystals or powder is the total mass minus the mass of the paper. When finding the mass of liquids, first mass the empty container. Then mass the liquid and container together. The mass of the liquid is the total mass minus the mass of the container.

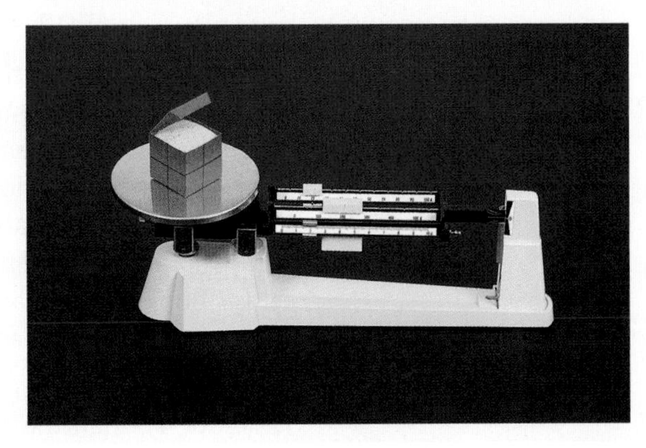

Scientific Method

The steps that scientists use to answer questions and solve problems is often called the **scientific method.** The scientific method is not a rigid procedure. Scientists may use all of the steps or just some of the steps of the scientific method. They may even repeat some of the steps. The goal of a scientific method is to come up with reliable answers and solutions.

Six Steps of a Scientific Method

1 **Ask a Question** Good questions come from careful **observations.** You make observations by using your senses to gather information. Sometimes you may use instruments, such as microscopes and telescopes, to extend the range of your senses. As you observe the natural world, you will discover that you have many more questions than answers. These questions drive the scientific method.

Questions beginning with *what, why, how,* and *when* are very important in focusing an investigation, and they often lead to a hypothesis. (You will learn what a hypothesis is in the next step.) Here is an example of a question that could lead to further investigation.

Question: How does acid rain affect plant growth?

2 **Form a Hypothesis** After you come up with a question, you need to turn the question into a **hypothesis.** A hypothesis is a clear statement of what you expect the answer to your question to be. Your hypothesis will represent your best "educated guess" based on your observations and what you already know. A good hypothesis is one that is testable. If observations and information cannot be gathered or if an experiment cannot be designed to test your hypothesis, it is untestable, and the investigation can go no further.

Here is a hypothesis that could be formed from the question, "How does acid rain affect plant growth?"

Hypothesis: Acid rain causes plants to grow more slowly.

Notice that the hypothesis provides some specifics that lead to methods of testing. The hypothesis can also lead to predictions. A **prediction** is what you think will be the outcome of your experiment or data collection. Predictions are usually stated in an "if . . . then" format. For example, **if** meat is kept at room temperature, **then** it will spoil faster than meat kept in the refrigerator. More than one prediction can be made for a single hypothesis. Here is a sample prediction for the hypothesis that acid rain causes plants to grow more slowly.

Prediction: If a plant is watered with only acid rain (which has a pH of 4), then the plant will grow at half its normal rate.

3 **Test the Hypothesis** After you have formed a hypothesis and made a prediction, you should test your hypothesis. There are different ways to do this. Perhaps the most familiar way is to conduct a **controlled experiment.** A controlled experiment tests only one factor at a time. A controlled experiment has a **control group** and one or more **experimental groups.** All the factors for the control and experimental groups are the same except one factor, which is called the **variable.** By changing only one factor (the variable), you can see the results of just that one change.

Sometimes, the nature of an investigation makes a controlled experiment impossible. For example, dinosaurs have been extinct for millions of years, and the Earth's core is surrounded by thousands of meters of rock. It would be difficult if not impossible to conduct controlled experiments on such things. Under such circumstances, a hypothesis may be tested by making detailed observations. Taking measurements is one way of making observations.

Test Your Hypothesis

APPENDIX

4 **Analyze the Results** After you have completed your experiments, made your observations, and collected your data, you must analyze all the information you have gathered. Tables and graphs are often used in this step to organize the data.

Analyze the Results

5 **Draw Conclusions** Based on the analysis of your data, you should conclude whether or not your results support your hypothesis. If your hypothesis is supported, you (or others) might want to repeat the observations or experiments to verify your results. If your hypothesis is not supported by the data, you may have to check your procedure for errors. You may even have to reject your hypothesis and make a new one. If you cannot draw a conclusion from your results, you may have to try the investigation again or carry out further observations or experiments.

Draw Conclusions

Do they support your hypothesis?

No

Yes

6 **Communicate Results** After any scientific investigation, you should report your results. By doing a written or oral report, you let others know what you have learned. They may want to repeat your investigation to see if they get the same results. Your report may even lead to another question, which in turn may lead to another investigation.

Communicate the Results

Scientific Method in Action

A scientific method is not a "straight line" of steps. It contains loops in which several steps may be repeated over and over again, while others may not be necessary. For example, sometimes scientists will find that testing one hypothesis raises new questions and new hypotheses to be tested. And sometimes, testing the hypothesis leads directly to a conclusion. Furthermore, the steps in a scientific method are not always used in the same order. Follow the steps in the diagram below, and see how many different directions a scientific method can take you.

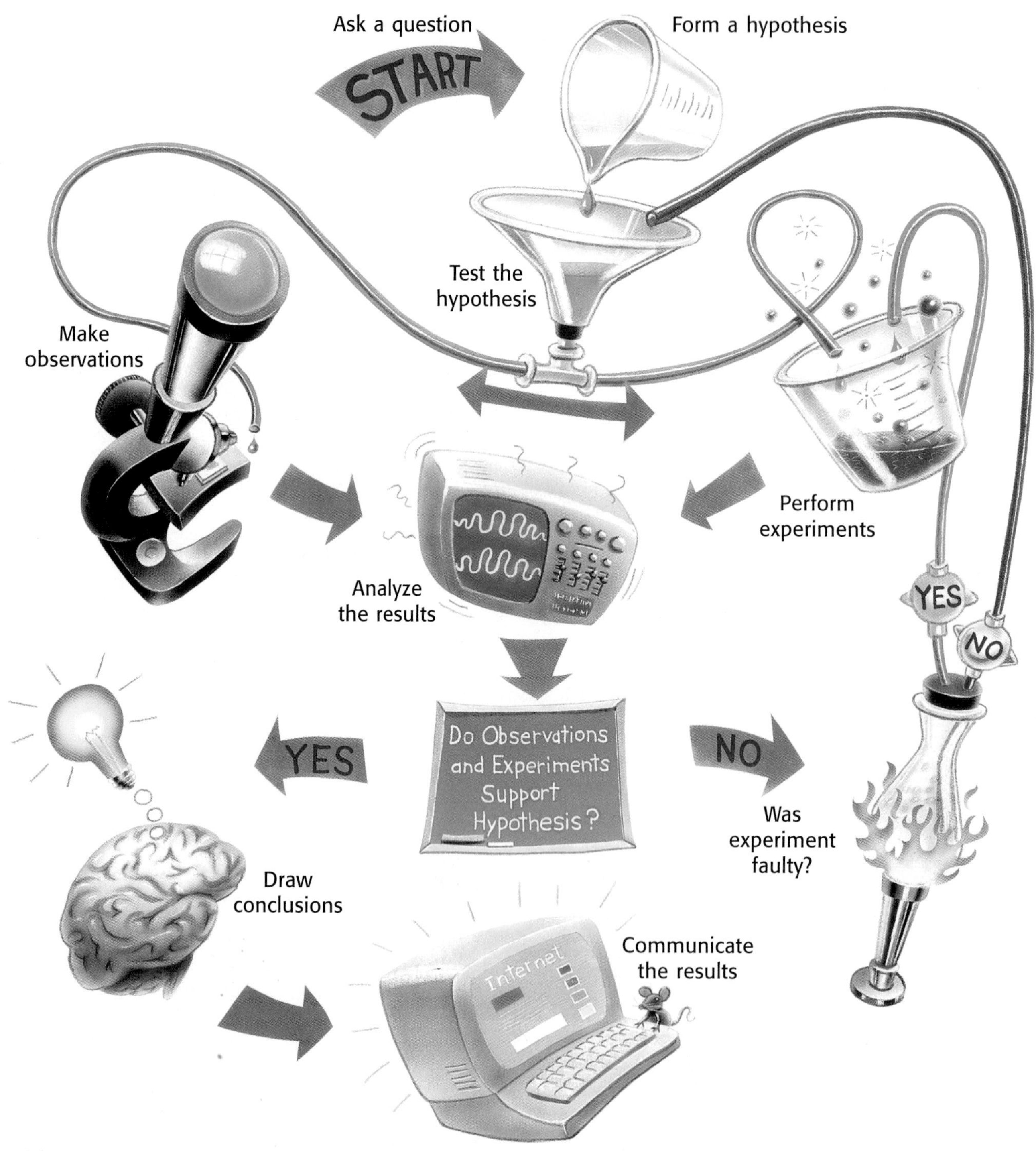

Making Charts and Graphs

Circle Graphs

A circle graph, or pie chart, shows how each group of data relates to all of the data. Each part of the circle represents a category of the data. The entire circle represents all of the data. For example, a biologist studying a hardwood forest in Wisconsin found that there were five different types of trees. The data table at right summarizes the biologist's findings.

Wisconsin Hardwood Trees	
Type of tree	**Number found**
Oak	600
Maple	750
Beech	300
Birch	1,200
Hickory	150
Total	3,000

How to Make a Circle Graph

1. In order to make a circle graph of this data, first find the percentage of each type of tree. To do this, divide the number of individual trees by the total number of trees and multiply by 100.

$$\frac{600 \text{ Oak}}{3{,}000 \text{ Trees}} \times 100 = 20\%$$

$$\frac{750 \text{ Maple}}{3{,}000 \text{ Trees}} \times 100 = 25\%$$

$$\frac{300 \text{ Beech}}{3{,}000 \text{ Trees}} \times 100 = 10\%$$

$$\frac{1{,}200 \text{ Birch}}{3{,}000 \text{ Trees}} \times 100 = 40\%$$

$$\frac{150 \text{ Hickory}}{3{,}000 \text{ Trees}} \times 100 = 5\%$$

2. Now determine the size of the pie shapes that make up the chart. Do this by multiplying each percentage by 360°. Remember that a circle contains 360°.

$20\% \times 360° = 72°$ $25\% \times 360° = 90°$

$10\% \times 360° = 36°$ $40\% \times 360° = 144°$

$5\% \times 360° = 18°$

3. Then check that the sum of the percentages is 100 and the sum of the degrees is 360.

$20\% + 25\% + 10\% + 40\% + 5\% = 100\%$

$72° + 90° + 36° + 144° + 18° = 360°$

4. Use a compass to draw a circle and mark its center.

5. Then use a protractor to draw angles of 72°, 90°, 36°, 144°, and 18° in the circle.

6. Finally, label each part of the graph and choose an appropriate title.

A Community of Wisconsin Hardwood Trees

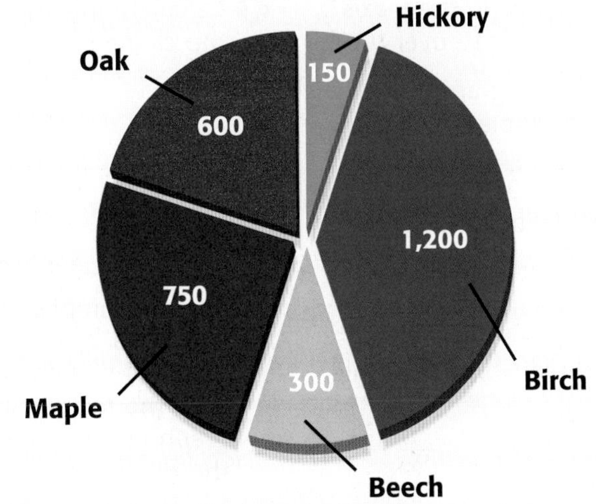

APPENDIX

Population of Appleton, 1900–2000	
Year	**Population**
1900	1,800
1920	2,500
1940	3,200
1960	3,900
1980	4,600
2000	5,300

Line Graphs

Line graphs are most often used to demonstrate continuous change. For example, Mr. Smith's science class analyzed the population records for their hometown, Appleton, between 1900 and 2000. Examine the data at left.

Because the year and the population change, they are the *variables*. The population is determined by, or dependent on, the year. Therefore, the population is called the **dependent variable,** and the year is called the **independent variable.** Each set of data is called a **data pair.** To prepare a line graph, data pairs must first be organized in a table like the one at left.

How to Make a Line Graph

1. Place the independent variable along the horizontal (x) axis. Place the dependent variable along the vertical (y) axis.
2. Label the x-axis "Year" and the y-axis "Population." Look at your largest and smallest values for the population. Determine a scale for the y-axis that will provide enough space to show these values. You must use the same scale for the entire length of the axis. Find an appropriate scale for the x-axis too.
3. Choose reasonable starting points for each axis.
4. Plot the data pairs as accurately as possible.
5. Choose a title that accurately represents the data.

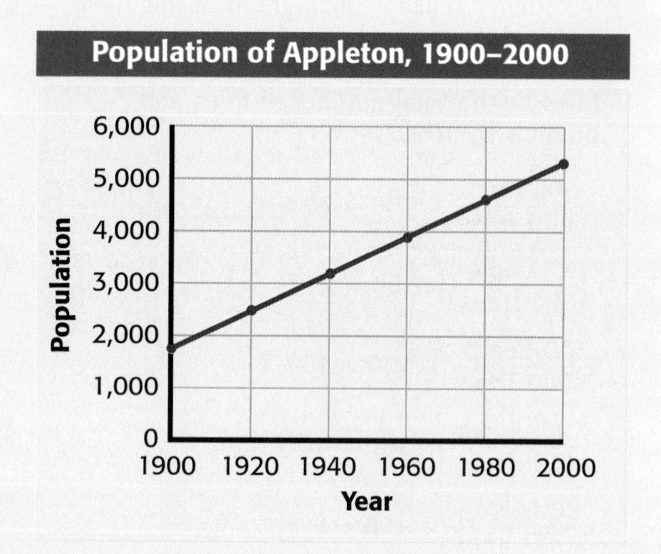

How to Determine Slope

Slope is the ratio of the change in the y-axis to the change in the x-axis, or "rise over run."

1. Choose two points on the line graph. For example, the population of Appleton in 2000 was 5,300 people. Therefore, you can define point a as (2000, 5,300). In 1900, the population was 1,800 people. Define point b as (1900, 1,800).
2. Find the change in the y-axis.
   (y at point a) − (y at point b)
   5,300 people − 1,800 people = 3,500 people
3. Find the change in the x-axis.
   (x at point a) − (x at point b)
   2000 − 1900 = 100 years
4. Calculate the slope of the graph by dividing the change in y by the change in x.

$$\text{slope} = \frac{\text{change in } y}{\text{change in } x}$$

$$\text{slope} = \frac{3{,}500 \text{ people}}{100 \text{ years}}$$

$$\text{slope} = 35 \text{ people per year}$$

In this example, the population in Appleton increased by a fixed amount each year. The graph of this data is a straight line. Therefore, the relationship is **linear.** When the graph of a set of data is not a straight line, the relationship is **nonlinear.**

Using Algebra to Determine Slope

The equation in step 4 may also be arranged to be:

$$y = kx$$

where y represents the change in the y-axis, k represents the slope, and x represents the change in the x-axis.

$$\text{slope} = \frac{\text{change in } y}{\text{change in } x}$$

$$k = \frac{y}{x}$$

$$k \times x = \frac{y \times x}{x}$$

$$kx = y$$

Bar Graphs

Bar graphs are used to demonstrate change that is not continuous. These graphs can be used to indicate trends when the data are taken over a long period of time. A meteorologist gathered the precipitation records at right for Hartford, Connecticut, for April 1–15, 1996, and used a bar graph to represent the data.

Precipitation in Hartford, Connecticut April 1–15, 1996

Date	Precipitation (cm)	Date	Precipitation (cm)
April 1	0.5	April 9	0.25
April 2	1.25	April 10	0.0
April 3	0.0	April 11	1.0
April 4	0.0	April 12	0.0
April 5	0.0	April 13	0.25
April 6	0.0	April 14	0.0
April 7	0.0	April 15	6.50
April 8	1.75		

How to Make a Bar Graph

1. Use an appropriate scale and reasonable starting point for each axis.
2. Label the axes, and plot the data.
3. Choose a title that accurately represents the data.

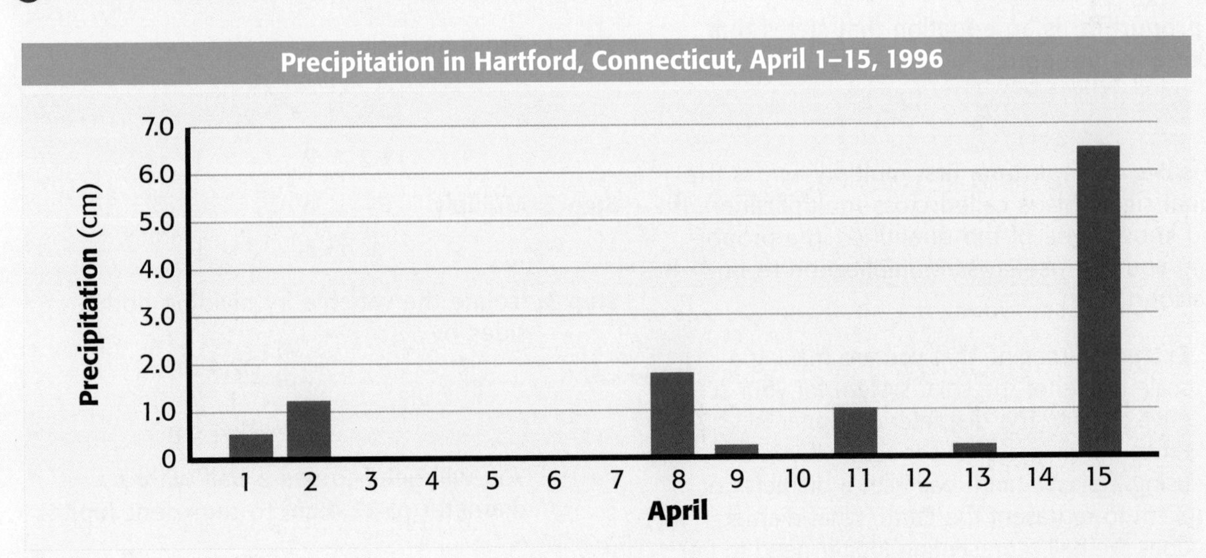

Math Refresher

Science requires an understanding of many math concepts. The following pages will help you review some important math skills.

Averages

An **average,** or **mean,** simplifies a list of numbers into a single number that *approximates* their value.

Example: Find the average of the following set of numbers: 5, 4, 7, 8.

Step 1: Find the sum.

$$5 + 4 + 7 + 8 = 24$$

Step 2: Divide the sum by the amount of numbers in your set. Because there are four numbers in this example, divide the sum by 4.

$$\frac{24}{4} = 6$$

The average, or mean, is **6.**

Ratios

A **ratio** is a comparison between numbers, and it is usually written as a fraction.

Example: Find the ratio of thermometers to students if you have 36 thermometers and 48 students in your class.

Step 1: Make the ratio.

$$\frac{36 \text{ thermometers}}{48 \text{ students}}$$

Step 2: Reduce the fraction to its simplest form.

$$\frac{36}{48} = \frac{36 \div 12}{48 \div 12} = \frac{3}{4}$$

The ratio of thermometers to students is **3 to 4,** or $\frac{3}{4}$. The ratio may also be written in the form 3:4.

Proportions

A **proportion** is an equation that states that two ratios are equal.

$$\frac{3}{1} = \frac{12}{4}$$

To solve a proportion, first multiply across the equal sign. This is called cross-multiplication. If you know three of the quantities in a proportion, you can use cross-multiplication to find the fourth.

Example: Imagine that you are making a scale model of the solar system for your science project. The diameter of Jupiter is 11.2 times the diameter of the Earth. If you are using a plastic-foam ball with a diameter of 2 cm to represent the Earth, what diameter does the ball representing Jupiter need to be?

$$\frac{11.2}{1} = \frac{x}{2 \text{ cm}}$$

Step 1: Cross-multiply.

$$\frac{11.2}{1} \times \frac{x}{2}$$

$$11.2 \times 2 = x \times 1$$

Step 2: Multiply.

$$22.4 = x \times 1$$

Step 3: Isolate the variable by dividing both sides by 1.

$$x = \frac{22.4}{1}$$

$$x = 22.4 \text{ cm}$$

You will need to use a ball with a diameter of **22.4 cm** to represent Jupiter.

Percentages

A **percentage** is a ratio of a given number to 100.

Example: What is 85 percent of 40?

Step 1: Rewrite the percentage by moving the decimal point two places to the left.

.85

Step 2: Multiply the decimal by the number you are calculating the percentage of.

$$0.85 \times 40 = 34$$

85% of 40 is **34**

Decimals

To **add** or **subtract decimals,** line up the digits vertically so that the decimal points line up. Then add or subtract the columns from right to left, carrying or borrowing numbers as necessary.

Example: Add the following numbers: 3.1415 and 2.96.

Step 1: Line up the digits vertically so that the decimal points line up.

$$\begin{array}{r} 3.1415 \\ +\ 2.96 \\ \hline \end{array}$$

Step 2: Add the columns from right to left, carrying when necessary.

$$\begin{array}{r} \scriptstyle 1\ 1 \\ 3.1415 \\ +\ 2.96 \\ \hline 6.1015 \end{array}$$

The sum is **6.1015**

Fractions

Numbers tell you how many; **fractions** tell you *how much of a whole.*

Example: Your class has 24 plants. Your teacher instructs you to put 5 in a shady spot. What fraction does this represent?

Step 1: Write a fraction with the total number of parts in the whole as the denominator.

$$\frac{?}{24}$$

Step 2: Write the number of parts of the whole being represented as the numerator.

$$\frac{5}{24}$$

$\mathbf{\frac{5}{24}}$ of the plants will be in the shade.

Reducing Fractions

It is usually best to express a fraction in simplest form. This is called *reducing* a fraction.

Example: Reduce the fraction $\frac{30}{45}$ to its simplest form.

Step 1: Find the largest whole number that will divide evenly into both the numerator and denominator. This number is called the greatest common factor (GCF).

factors of the numerator 30: 1, 2, 3, 5, 6, 10, 15, 30

factors of the denominator 45: 1, 3, 5, 9, 15, 45

Step 2: Divide both the numerator and the denominator by the GCF, which in this case is 15.

$$\frac{30}{45} = \frac{30 \div 15}{45 \div 15} = \frac{2}{3}$$

$\frac{30}{45}$ reduced to its simplest form is $\mathbf{\frac{2}{3}}$.

APPENDIX

Adding and Subtracting Fractions

To **add** or **subtract fractions** that have the **same denominator,** simply add or subtract the numerators.

Examples:

$$\frac{3}{5} + \frac{1}{5} = ? \text{ and } \frac{3}{4} - \frac{1}{4} = ?$$

Step 1: Add or subtract the numerators.

$$\frac{3}{5} + \frac{1}{5} = \frac{4}{} \text{ and } \frac{3}{4} - \frac{1}{4} = \frac{2}{}$$

Step 2: Write the sum or difference over the denominator.

$$\frac{3}{5} + \frac{1}{5} = \frac{4}{5} \text{ and } \frac{3}{4} - \frac{1}{4} = \frac{2}{4}$$

Step 3: If necessary, reduce the fraction to its simplest form.

$\mathbf{\frac{4}{5}}$ cannot be reduced, and $\frac{2}{4} = \mathbf{\frac{1}{2}}$

To **add** or **subtract fractions** that have **different denominators,** first find a common denominator (LCD).

Examples:

$$\frac{1}{2} + \frac{1}{6} = ? \text{ and } \frac{3}{4} - \frac{2}{3} = ?$$

Step 1: Write the equivalent fractions with a common denominator.

$$\frac{3}{6} + \frac{1}{6} = ? \text{ and } \frac{9}{12} - \frac{8}{12} = ?$$

Step 2: Add or subtract.

$$\frac{3}{6} + \frac{1}{6} = \frac{4}{6} \text{ and } \frac{9}{12} - \frac{8}{12} = \frac{1}{12}$$

Step 3: If necessary, reduce the fraction to its simplest form.

$\frac{4}{6} = \mathbf{\frac{2}{3}}$ and $\mathbf{\frac{1}{12}}$ cannot be reduced

Multiplying Fractions

To **multiply fractions,** multiply the numerators and the denominators together, and then reduce the fraction to its simplest form.

Example:

$$\frac{5}{9} \times \frac{7}{10} = ?$$

Step 1: Multiply the numerators and denominators.

$$\frac{5}{9} \times \frac{7}{10} = \frac{5 \times 7}{9 \times 10} = \frac{35}{90}$$

Step 2: Reduce.

$$\frac{35}{90} = \frac{35 \div 5}{90 \div 5} = \mathbf{\frac{7}{18}}$$

Dividing Fractions

To **divide fractions**, first rewrite the divisor (the number you divide *by*) upside down. This is called the reciprocal of the divisor. Then you can multiply and reduce if necessary.

Example:

$$\frac{5}{8} \div \frac{3}{2} = ?$$

Step 1: Rewrite the divisor as its reciprocal.

$$\frac{3}{2} \rightarrow \frac{2}{3}$$

Step 2: Multiply.

$$\frac{5}{8} \times \frac{2}{3} = \frac{5 \times 2}{8 \times 3} = \frac{10}{24}$$

Step 3: Reduce.

$$\frac{10}{24} = \frac{10 \div 2}{24 \div 2} = \mathbf{\frac{5}{12}}$$

Scientific Notation

Scientific notation is a short way of representing very large and very small numbers without writing all of the place-holding zeros.

Example: Write 653,000,000 in scientific notation.

Step 1: Write the number without the place-holding zeros.

653

Step 2: Place the decimal point after the first digit.

6.53

Step 3: Find the exponent by counting the number of places that you moved the decimal point.

6.53000000

The decimal point was moved eight places to the left. Therefore, the exponent of 10 is positive 8. Remember, if the decimal point had moved to the right, the exponent would be negative.

Step 4: Write the number in scientific notation.

$\mathbf{6.53 \times 10^8}$

Area

Area is the number of square units needed to cover the surface of an object.

Formulas:

Area of a square = side × side

Area of a rectangle = length × width

Area of a triangle = $\frac{1}{2}$ base × height

Examples: Find the areas.

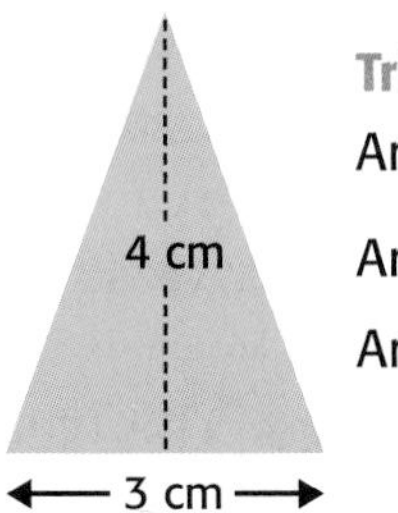

Triangle

Area = $\frac{1}{2}$ × base × height

Area = $\frac{1}{2}$ × 3 cm × 4 cm

Area = **6 cm²**

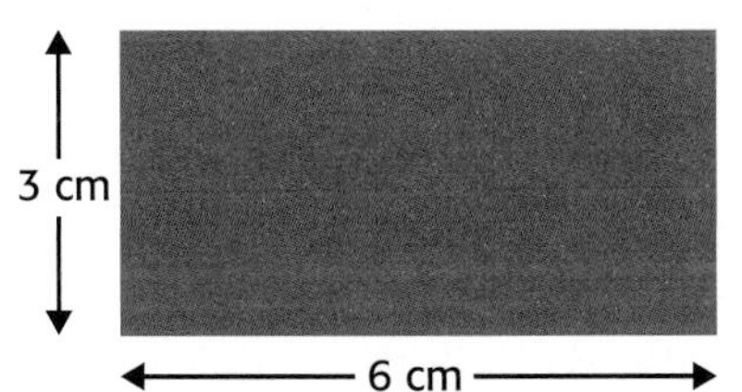

Rectangle

Area = length × width

Area = 6 cm × 3 cm

Area = $\mathbf{18\ cm^2}$

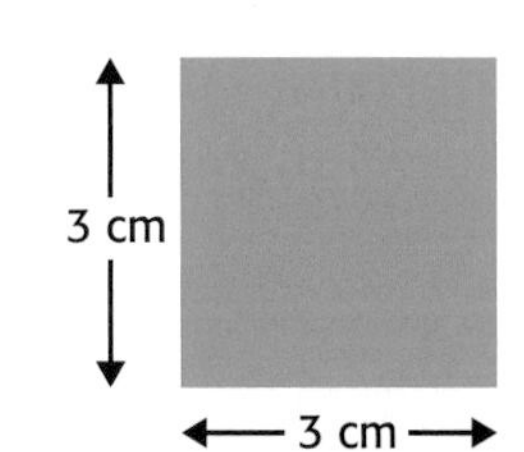

Square

Area = side × side

Area = 3 cm × 3 cm

Area = $\mathbf{9\ cm^2}$

Volume

Volume is the amount of space something occupies.

Formulas:

Volume of a cube = side × side × side

Volume of a prism = area of base × height

Examples:

Find the volume of the solids.

Cube

Volume = side × side × side

Volume = 4 cm × 4 cm × 4 cm

Volume = $\mathbf{64\ cm^3}$

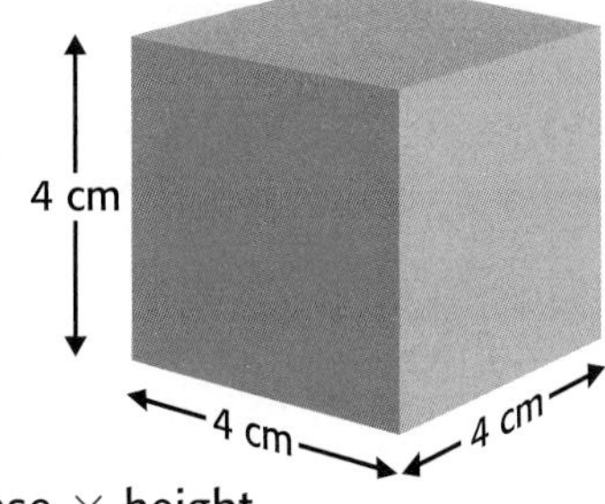

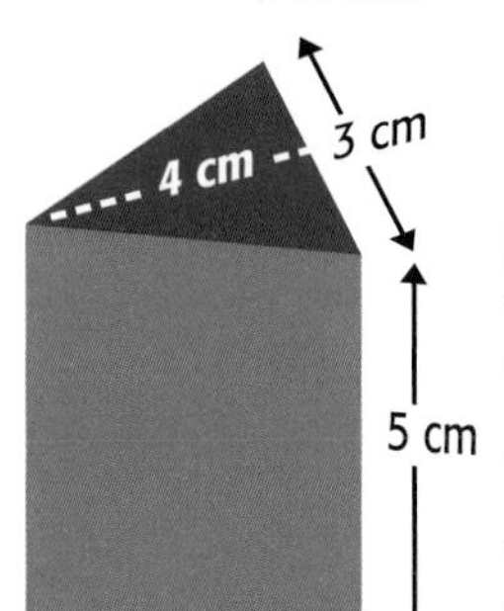

Prism

Volume = area of base × height

Volume = (area of triangle) × height

Volume = $\left(\frac{1}{2} \times 3\text{ cm} \times 4\text{ cm}\right) \times 5\text{ cm}$

Volume = $6\ cm^2 \times 5$ cm

Volume = $\mathbf{30\ cm^3}$

APPENDIX

Physical Science Refresher

Atoms and Elements

Every object in the universe is made up of particles of some kind of matter. **Matter** is anything that takes up space and has mass. All matter is made up of elements. An **element** is a substance that cannot be separated into simpler components by ordinary chemical means. This is because each element consists of only one kind of atom. An **atom** is the smallest unit of an element that has all of the properties of that element.

Atomic Structure

Atoms are made up of small particles called subatomic particles. The three major types of subatomic particles are **electrons, protons,** and **neutrons.** Electrons have a negative electrical charge, protons have a positive charge, and neutrons have no electrical charge. The protons and neutrons are packed close to one another to form the **nucleus.** The protons give the nucleus a positive charge. The electrons of an atom move in a region around the nucleus known as an **electron cloud.** The negatively charged electrons are attracted to the positively charged nucleus. An atom may have several energy levels in which electrons are located.

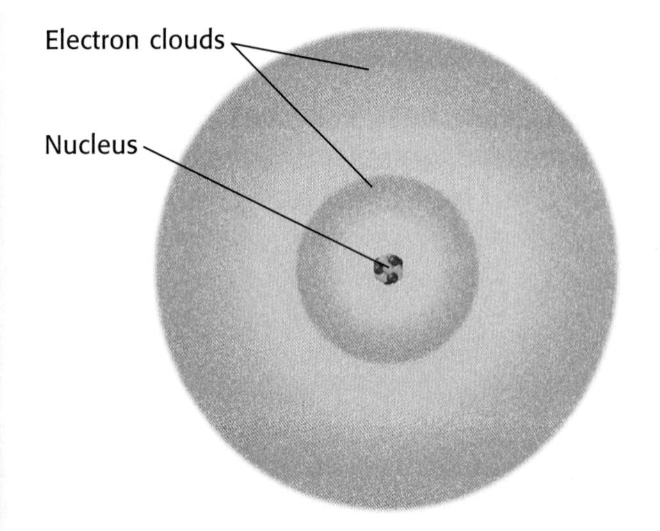

Atomic Number

To help in the identification of elements, scientists have assigned an **atomic number** to each kind of atom. The atomic number is equal to the number of protons in the atom. Atoms with the same number of protons are all the same kind of element. In an uncharged, or electrically neutral, atom there are an equal number of protons and electrons. Therefore, the atomic number also equals the number of electrons in an uncharged atom. The number of neutrons, however, can vary for a given element. Atoms of the same element that have different numbers of neutrons are called **isotopes.**

Periodic Table of the Elements

In the periodic table, the elements are arranged from left to right in order of increasing atomic number. Each element in the table is in a separate box. Each element has one more electron and one more proton than the element to its left. Each horizontal row of the table is called a **period.** Changes in chemical properties across a period correspond to changes in the elements' electron arrangements. Each vertical column of the table, known as a **group,** lists elements with similar properties. The elements in a group have similar chemical properties because they have the same number of electrons in their outer energy level. For example, the elements helium, neon, argon, krypton, xenon, and radon all have similar properties and are known as the noble gases.

Molecules and Compounds

When the atoms of two or more elements are joined chemically, the resulting substance is called a **compound.** A compound is a new substance with properties different from those of the elements that compose it. For example, water (H_2O) is a compound formed when atoms of hydrogen (H) and oxygen (O) combine. The smallest complete unit of a compound that has all of the properties of that compound is called a **molecule.** A chemical formula indicates the elements in a compound. It also indicates the relative number of atoms of each element present. The chemical formula for water is H_2O, which indicates that each water molecule consists of two atoms of hydrogen and one atom of oxygen. The subscript number is used after the symbol for an element to indicate how many atoms of that element are in a single molecule of the compound.

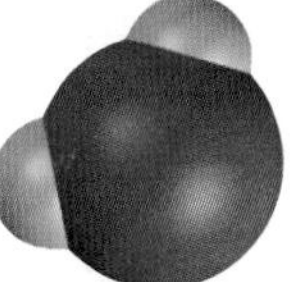

Acids, Bases, and pH

An ion is an atom or group of atoms that has an electrical charge because it has lost or gained one or more electrons. When an acid, such as hydrochloric acid (HCl), is mixed with water, it separates into ions. An **acid** is a compound that produces hydrogen ions (H^+) in water. The hydrogen ions then combine with a water molecule to form a hydronium ion (H_3O^+). A **base,** on the other hand, is a substance that produces hydroxide ions (OH^-) in water.

To determine whether a solution is acidic or basic, scientists use pH. The **pH** is a measure of the hydronium ion concentration in a solution. The pH scale ranges from 0 to 14. The middle point, pH = 7, is neutral, neither acidic nor basic. Acids have a pH less than 7; bases have a pH greater than 7. The lower the number is, the more acidic the solution. The higher the number is, the more basic the solution.

Chemical Equations

A chemical reaction occurs when a chemical change takes place. (In a chemical change, new substances with new properties are formed.) A chemical equation is a useful way of describing a chemical reaction by means of chemical formulas. The equation indicates what substances react and what the products are. For example, when carbon and oxygen combine, they can form carbon dioxide. The equation for the reaction is as follows: $C + O_2 \longrightarrow CO_2$.

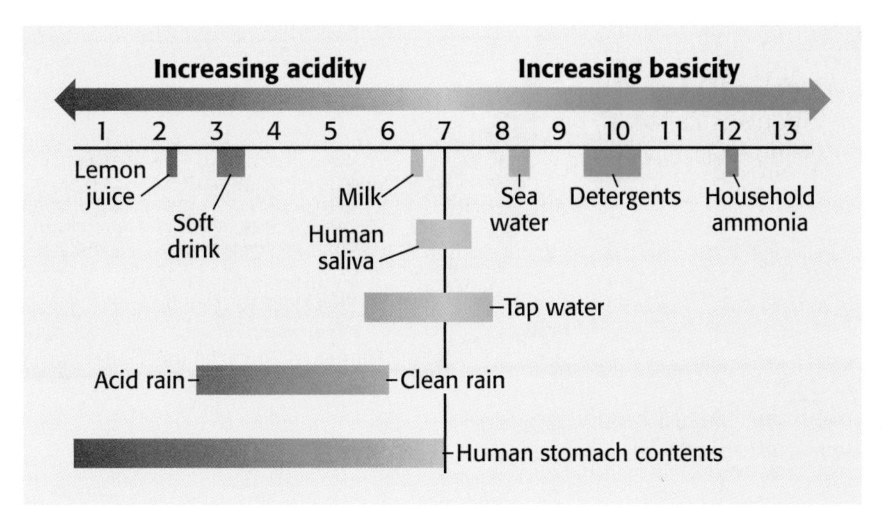

Periodic Table of the Elements

Each square on the table includes an element's name, chemical symbol, atomic number, and atomic mass.

Atomic number — 6
Chemical symbol — C
Element name — Carbon
Atomic mass — 12.0

The background color indicates the type of element. Carbon is a nonmetal.

The color of the chemical symbol indicates the physical state at room temperature. Carbon is a solid.

Background
- Metals
- Metalloids
- Nonmetals

Chemical symbol
- Solid
- Liquid
- Gas

	Group 1	Group 2	Group 3	Group 4	Group 5	Group 6	Group 7	Group 8	Group 9
Period 1	1 H Hydrogen 1.0								
Period 2	3 Li Lithium 6.9	4 Be Beryllium 9.0							
Period 3	11 Na Sodium 23.0	12 Mg Magnesium 24.3							
Period 4	19 K Potassium 39.1	20 Ca Calcium 40.1	21 Sc Scandium 45.0	22 Ti Titanium 47.9	23 V Vanadium 50.9	24 Cr Chromium 52.0	25 Mn Manganese 54.9	26 Fe Iron 55.8	27 Co Cobalt 58.9
Period 5	37 Rb Rubidium 85.5	38 Sr Strontium 87.6	39 Y Yttrium 88.9	40 Zr Zirconium 91.2	41 Nb Niobium 92.9	42 Mo Molybdenum 95.9	43 Tc Technetium (97.9)	44 Ru Ruthenium 101.1	45 Rh Rhodium 102.9
Period 6	55 Cs Cesium 132.9	56 Ba Barium 137.3	57 La Lanthanum 138.9	72 Hf Hafnium 178.5	73 Ta Tantalum 180.9	74 W Tungsten 183.8	75 Re Rhenium 186.2	76 Os Osmium 190.2	77 Ir Iridium 192.2
Period 7	87 Fr Francium (223.0)	88 Ra Radium (226.0)	89 Ac Actinium (227.0)	104 Rf Rutherfordium (261.1)	105 Db Dubnium (262.1)	106 Sg Seaborgium (263.1)	107 Bh Bohrium (262.1)	108 Hs Hassium (265)	109 Mt Meitnerium (266)

A row of elements is called a period.

A column of elements is called a group or family.

Lanthanides	58 Ce Cerium 140.1	59 Pr Praseodymium 140.9	60 Nd Neodymium 144.2	61 Pm Promethium (144.9)	62 Sm Samarium 150.4
Actinides	90 Th Thorium 232.0	91 Pa Protactinium 231.0	92 U Uranium 238.0	93 Np Neptunium (237.0)	94 Pu Plutonium 244.1

These elements are placed below the table to allow the table to be narrower.

This zigzag line reminds you where the metals, nonmetals, and metalloids are.

Group 10	Group 11	Group 12	Group 13	Group 14	Group 15	Group 16	Group 17	Group 18
								2 **He** Helium 4.0
			5 **B** Boron 10.8	6 **C** Carbon 12.0	7 **N** Nitrogen 14.0	8 **O** Oxygen 16.0	9 **F** Fluorine 19.0	10 **Ne** Neon 20.2
			13 **Al** Aluminum 27.0	14 **Si** Silicon 28.1	15 **P** Phosphorus 31.0	16 **S** Sulfur 32.1	17 **Cl** Chlorine 35.5	18 **Ar** Argon 39.9
28 **Ni** Nickel 58.7	29 **Cu** Copper 63.5	30 **Zn** Zinc 65.4	31 **Ga** Gallium 69.7	32 **Ge** Germanium 72.6	33 **As** Arsenic 74.9	34 **Se** Selenium 79.0	35 **Br** Bromine 79.9	36 **Kr** Krypton 83.8
46 **Pd** Palladium 106.4	47 **Ag** Silver 107.9	48 **Cd** Cadmium 112.4	49 **In** Indium 114.8	50 **Sn** Tin 118.7	51 **Sb** Antimony 121.8	52 **Te** Tellurium 127.6	53 **I** Iodine 126.9	54 **Xe** Xenon 131.3
78 **Pt** Platinum 195.1	79 **Au** Gold 197.0	80 **Hg** Mercury 200.6	81 **Tl** Thallium 204.4	82 **Pb** Lead 207.2	83 **Bi** Bismuth 209.0	84 **Po** Polonium (209.0)	85 **At** Astatine (210.0)	86 **Rn** Radon (222.0)
110 **Uun** Ununnilium (271)	111 **Uuu** Unununium (272)	112 **Uub** Ununbium (277)						

The names and symbols of elements 110–112 are temporary. They are based on the atomic number of the element. The official name and symbol will be approved by an international committee of scientists.

63 **Eu** Europium 152.0	64 **Gd** Gadolinium 157.3	65 **Tb** Terbium 158.9	66 **Dy** Dysprosium 162.5	67 **Ho** Holmium 164.9	68 **Er** Erbium 167.3	69 **Tm** Thulium 168.9	70 **Yb** Ytterbium 173.0	71 **Lu** Lutetium 175.0
95 **Am** Americium (243.1)	96 **Cm** Curium (247.1)	97 **Bk** Berkelium (247.1)	98 **Cf** Californium (251.1)	99 **Es** Einsteinium (252.1)	100 **Fm** Fermium (257.1)	101 **Md** Mendelevium (258.1)	102 **No** Nobelium (259.1)	103 **Lr** Lawrencium (262.1)

A number in parentheses is the mass number of the most stable isotope of that element.

Physical Science Laws and Principles

Law of Conservation of Energy

The law of conservation of energy states that energy can be neither created nor destroyed.

The total amount of energy in a closed system is always the same. Energy can be changed from one form to another, but all the different forms of energy in a system always add up to the same total amount of energy, no matter how many energy conversions occur.

Law of Universal Gravitation

The law of universal gravitation states that all objects in the universe attract each other by a force called gravity. The size of the force depends on the masses of the objects and the distance between them.

The first part of the law explains why a bowling ball is much harder to lift than a table-tennis ball. Because the bowling ball has a much larger mass than the table-tennis ball, the amount of gravity between the Earth and the bowling ball is greater than the amount of gravity between the Earth and the table-tennis ball.

The second part of the law explains why a satellite can remain in orbit around the Earth. The satellite is carefully placed at a distance great enough to prevent the Earth's gravity from immediately pulling it down but small enough to prevent it from completely escaping the Earth's gravity and wandering off into space.

Newton's Laws of Motion

Newton's first law of motion states that an object at rest remains at rest and an object in motion remains in motion at constant speed and in a straight line unless acted on by an unbalanced force.

The first part of the law explains why a football will remain on a tee until it is kicked off or until a gust of wind blows it off.

The second part of the law explains why a bike's rider will continue moving forward after the bike tire runs into a crack in the sidewalk and the bike comes to an abrupt stop until gravity and the sidewalk stop the rider.

Newton's second law of motion states that the acceleration of an object depends on the mass of the object and the amount of force applied.

The first part of the law explains why the acceleration of a 4 kg bowling ball will be greater than the acceleration of a 6 kg bowling ball if the same force is applied to both.

The second part of the law explains why the acceleration of a bowling ball will be larger if a larger force is applied to it.

The relationship of acceleration (a) to mass (m) and force (F) can be expressed mathematically by the following equation:

$$\text{acceleration} = \frac{\text{force}}{\text{mass}}, \text{ or } a = \frac{F}{m}$$

This equation is often rearranged to the form:

$$\text{force} = \text{mass} \times \text{acceleration},$$

or

$$F = m \times a$$

Newton's third law of motion states that whenever one object exerts a force on a second object, the second object exerts an equal and opposite force on the first.

This law explains that a runner is able to move forward because of the equal and opposite force the ground exerts on the runner's foot after each step.

Law of Reflection

The law of reflection states that the angle of incidence is equal to the angle of reflection. This law explains why light reflects off of a surface at the same angle it strikes the surface.

Charles's Law

Charles's law states that for a fixed amount of gas at a constant pressure, the volume of the gas increases as its temperature increases. Likewise, the volume of the gas decreases as its temperature decreases.

If a basketball that was inflated indoors is left outside on a cold winter day, the air particles inside of the ball will move more slowly. They will hit the sides of the basketball less often and with less force. The ball will get smaller as the volume of the air decreases. If a basketball that was inflated outdoors on a cold winter day is brought indoors, the air particles inside of the ball will move more rapidly. They will hit the sides of the basketball more often and with more force. The ball will get larger as the volume of the air increases.

Boyle's Law

Boyle's law states that for a fixed amount of gas at a constant temperature, the volume of a gas increases as its pressure decreases. Likewise, the volume of a gas decreases as its pressure increases.

This law explains why the pressure of the gas in a helium balloon decreases as the balloon rises from the Earth's surface.

Pascal's Principle

Pascal's principle states that a change in pressure at any point in an enclosed fluid will be transmitted equally to all parts of that fluid.

When a mechanic uses a hydraulic jack to raise an automobile off the ground, he or she increases the pressure on the fluid in the jack by pushing on the jack handle. The pressure is transmitted equally to all parts of the fluid-filled jacking system. The fluid presses the jack plate against the frame of the car, lifting the car off the ground.

Archimedes' Principle

Archimedes' principle states that the buoyant force on an object in a fluid is equal to the weight of the volume of fluid that the object displaces.

A person floating in a swimming pool displaces 20 L of water. The weight of that volume of water is about 200 N. Therefore, the buoyant force on the person is 200 N.

Bernoulli's Principle

Bernoulli's principle states that as the speed of a moving fluid increases, its pressure decreases.

Bernoulli's principle explains how a wing gives lift to an airplane or even how a Frisbee® can fly through the air. Because of the shape of the Frisbee, the air moving over the top of the Frisbee must travel farther than the air below the Frisbee in the same amount of time. In other words, the air above the Frisbee is moving faster than the air below it. This faster-moving air above the Frisbee exerts less pressure than the slower-moving air below it. The resulting increased pressure below exerts an upward force, pushing the Frisbee up.

Useful Equations

Average speed

$$\text{Average speed} = \frac{\text{total distance}}{\text{total time}}$$

Example: A bicycle messenger traveled a distance of 136 km in 8 hours. What was the messenger's average speed?

$$\frac{136 \text{ km}}{8 \text{ h}} = 17 \text{ km/h}$$

The messenger's average speed was **17 km/h.**

Average acceleration

$$\text{Average acceleration} = \frac{\text{final velocity} - \text{starting velocity}}{\text{time it takes to change velocity}}$$

Example: Calculate the average acceleration of an Olympic 100 m dash sprinter who reaches a velocity of 20 m/s south at the finish line. The race was in a straight line and lasted 10 s.

$$\frac{20 \text{ m/s} - 0 \text{ m/s}}{10 \text{ s}} = 2 \text{ m/s/s}$$

The sprinter's average acceleration is **2 m/s/s south.**

Net force

Forces in the Same Direction

When forces are in the same direction, add the forces together to determine the net force.

Example: Calculate the net force on a stalled car that is being pushed by two people. One person is pushing with a force of 13 N northwest and the other person is pushing with a force of 8 N in the same direction.

$$13 \text{ N} + 8 \text{ N} = 21 \text{ N}$$

The net force is **21 N northwest.**

Forces in Opposite Directions

When forces are in opposite directions, subtract the smaller force from the larger force to determine the net force.

Example: Calculate the net force on a rope that is being pulled on each end. One person is pulling on one end of the rope with a force of 12 N south. Another person is pulling on the opposite end of the rope with a force of 7 N north.

$$12 \text{ N} - 7 \text{ N} = 5 \text{ N}$$

The net force is **5 N south.**

Work

Work is done by exerting a force through a distance. Work has units of joules (J), which are equivalent to Newton-meters.

$$\mathbf{W = F \times d}$$

Example: Calculate the amount of work done by a man who lifts a 100 N toddler 1.5 m off the floor.

$$W = 100 \text{ N} \times 1.5 \text{ m} = 150 \text{ N•m} = 150 \text{ J}$$

The man did **150 J** of work.

Power

Power is the rate at which work is done. Power is measured in watts (W), which are equivalent to joules per second.

$$\mathbf{P = \frac{W}{t}}$$

Example: Calculate the power of a weightlifter who raises a 300 N barbell 2.1 m off the floor in 1.25 s.

$$W = 300 \text{ N} \times 2.1 \text{ m} = 630 \text{ N•m} = 630 \text{ J}$$

$$P = \frac{630 \text{ J}}{1.25 \text{ s}} = 504 \text{ J/s} = 504 \text{ W}$$

The weightlifter has **504 W** of power.

Pressure

Pressure is the force exerted over a given area. The SI unit for pressure is the pascal, which is abbreviated Pa.

$$\textbf{Pressure} = \frac{\textbf{force}}{\textbf{area}}$$

Example: Calculate the pressure of the air in a soccer ball if the air exerts a force of 10 N over an area of 0.5 m^2.

$$\text{Pressure} = \frac{10 \text{ N}}{0.5 \text{ m}^2} = 20 \text{ N/m}^2 = 20 \text{ Pa}$$

The pressure of the air inside of the soccer ball is **20 Pa.**

Density

$$\textbf{Density} = \frac{\textbf{mass}}{\textbf{volume}}$$

Example: Calculate the density of a sponge with a mass of 10 g and a volume of 40 mL.

$$\frac{10 \text{ g}}{40 \text{ mL}} = 0.25 \text{ g/mL}$$

The density of the sponge is **0.25 g/mL.**

Concentration

$$\textbf{Concentration} = \frac{\textbf{mass of solute}}{\textbf{volume of solvent}}$$

Example: Calculate the concentration of solution in which 10 g of sugar is dissolved in 125 mL of water.

$$\frac{10 \text{ g of sugar}}{125 \text{ mL of water}} = 0.08 \text{ g/mL}$$

The concentration of this solution is **0.08 g/mL.**

APPENDIX

Astronomical Data

Spring constellations

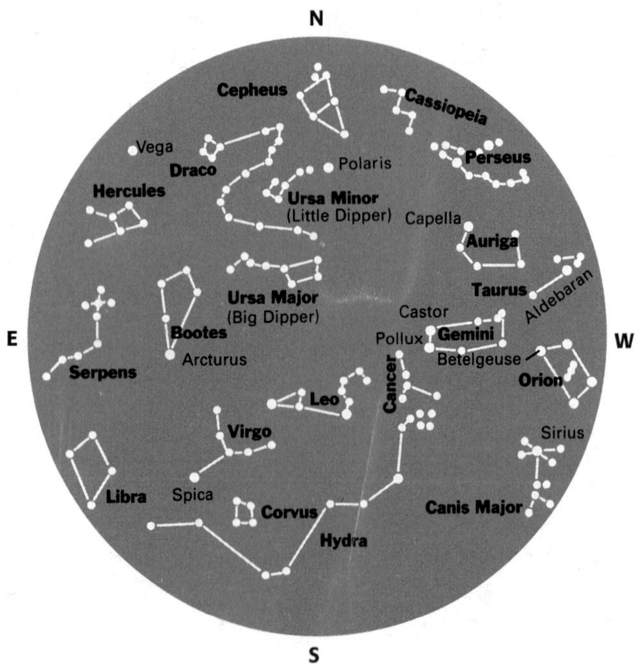

Summer constellations

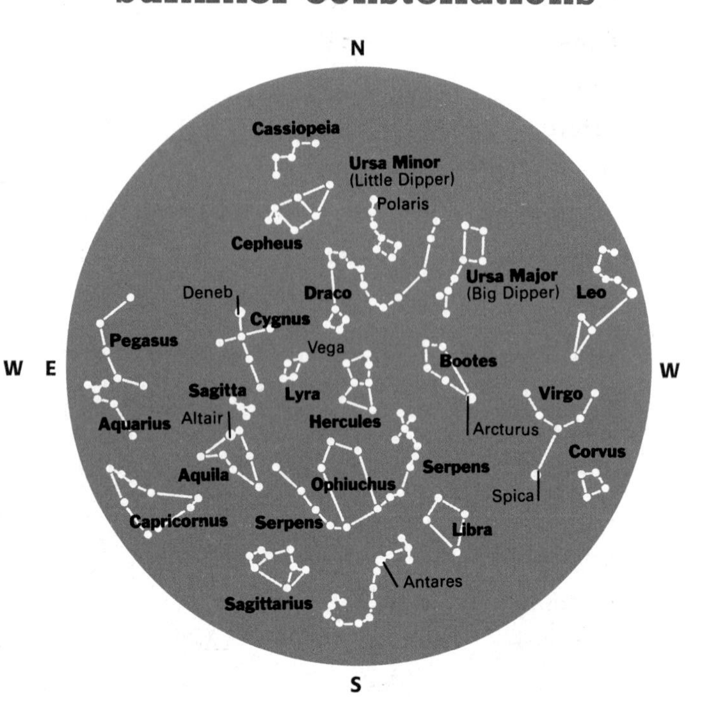

Autumn constellations

Winter constellations

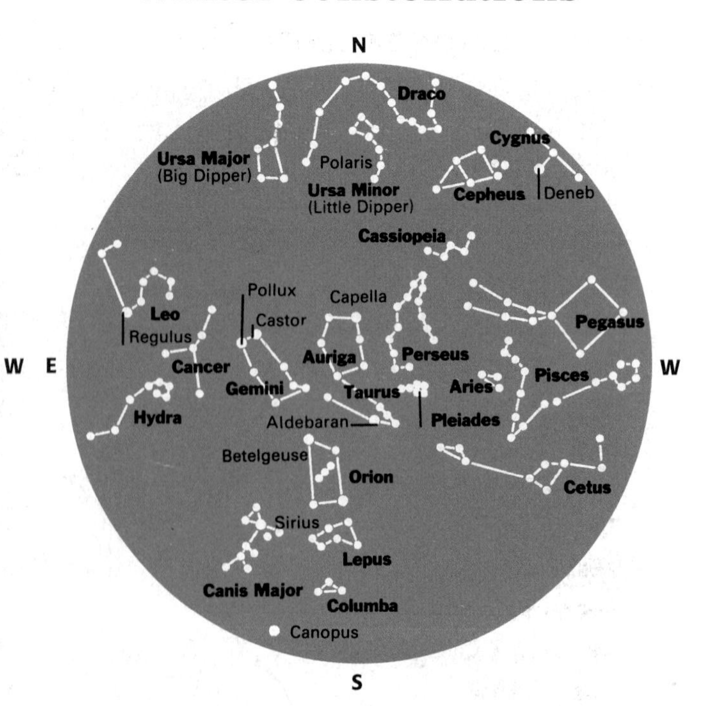

More Planetary Data

Planet	Distance from sun (AU)	Orbital period (Earth years)	Mean orbital speed (km/s)	Inclination of orbit to ecliptic (°)
Mercury	0.39	0.24	47.9	7.0
Venus	0.72	0.62	35.0	3.4
Earth	1.00	1.00	29.8	0.0
Mars	1.52	1.88	24.1	1.8
Jupiter	5.20	11.86	13.1	1.3
Saturn	9.54	29.46	9.6	2.5
Uranus	19.19	84.07	6.8	0.8
Neptune	30.06	164.82	5.4	1.8
Pluto	39.53	248.6	4.7	17.2

Planetary Data in Relation to the Earth

Planet	Diameter	Mass	Surface gravity
Mercury	0.38	0.06	0.38
Venus	0.95	0.82	0.91
Earth	1.00	1.00	1.00
Mars	0.53	0.11	0.38
Jupiter	11.2	317.8	2.53
Saturn	9.41	94.3	1.07
Uranus	4.11	14.6	0.92
Neptune	3.81	17.2	1.18
Pluto	0.17	0.003	0.09

Solar Eclipses from Present Through the Year 2019

Date	Duration of totality (min)	Location
June 21, 2001	4.9	Southern Africa
December 4, 2002	2.1	South Africa, Australia
November 23, 2003	2.0	Antarctica
April 8, 2005	0.7	South Pacific Ocean
March 29, 2006	4.1	Africa, Asia Minor, Russia
August 1, 2008	2.4	Arctic Ocean, Siberia, China
July 22, 2009	6.6	India, China, South Pacific
July 11, 2010	5.3	South Pacific Ocean, Southern South America
November 13, 2012	4.0	Northern Australia, South Pacific
November 3, 2013	1.7	Atlantic Ocean, Central Africa
March 20, 2015	4.1	North Atlantic, Arctic Ocean
March 9, 2016	4.5	Indonesia, Pacific Ocean
August 21, 2017	2.7	Pacific Ocean, United States, Atlantic Ocean
July 2, 2019	4.5	South Pacific, South America

Glossary

A

absolute magnitude the actual brightness of a star (488)

acceleration (ak SEL uhr AY shuhn) the rate at which velocity changes; an object accelerates if its speed changes, if its direction changes, or if both its speed and its direction change (112)

acid any compound that increases the number of hydrogen ions when dissolved in water and whose solution tastes sour and can change the color of certain compounds; acids turn blue litmus red, react with metals to produce hydrogen gas, and react with limestone or baking soda to produce carbon dioxide gas (401)

activation energy the minimum amount of energy needed for substances to react (386)

alkali metals the elements in Group 1 of the periodic table; they are the most reactive metals, and their atoms have one electron in their outer level (334)

alkaline-earth metals the elements in Group 2 of the periodic table; they are reactive metals but less reactive than alkali metals; their atoms have two electrons in their outer level (335)

alloys solid solutions of metals or nonmetals dissolved in metals (93)

amplitude the maximum distance a wave vibrates from its rest position (252)

annular (AN yoo luhr) **eclipse** a solar eclipse during which the outer ring of the sun can be seen around the moon (466)

aphelion (uh FEE lee uhn) the point in the orbit of a planet at which the planet is farthest from the sun (472)

apparent magnitude how bright a light appears to an observer (488)

Archimedes' (ahr kuh MEE deez) **principle** the principle that states that the buoyant force on an object in a fluid is an upward force equal to the weight of the volume of fluid that the object displaces (168)

area a measure of how much surface an object has (26)

asteroid a small, rocky body that revolves around the sun (473)

asteroid belt the region of the solar system most asteroids occupy; roughly between the orbits of Mars and Jupiter (473)

astronomical unit (AU) the average distance between the Earth and the sun, or approximately 150,000,000 km (430, 452)

atmospheric pressure the pressure caused by the weight of the atmosphere (163)

atom the smallest particle into which an element can be divided and still retain all of the properties of that element (304)

atomic mass the weighted average of the masses of all the naturally occurring isotopes of an element (316)

atomic mass unit (amu) the SI unit used to express the masses of particles in atoms (312)

atomic number the number of protons in the nucleus of an atom; the atomic number is the same for all atoms of an element (314)

average speed the overall rate at which an object moves; can be calculated by dividing total distance by total time (109)

B

balanced forces forces on an object that cause the net force to be zero; balanced forces do not cause a change in motion or acceleration (118)

base any compound that increases the number of hydroxide ions when dissolved in water and whose solution tastes bitter, feels slippery, and can change the color of certain compounds; bases turn red litmus blue (403)

battery a device that is made of several cells and that produces an electric current by converting chemical energy into electrical energy (278)

Bernoulli's (buhr NOO leez) **principle** the principle that states that as the speed of a moving fluid increases, its pressure decreases (173)

big bang theory the theory that states the universe began with a tremendous explosion (500)

biochemicals organic compounds made by living things (408)

biomass organic matter, such as plants, wood, and waste, that contains stored energy (236)

black hole an object with more than three solar masses squeezed into a ball only 10 km across whose gravity is so strong that not even light can escape (495)

block and tackle a fixed pulley and a movable pulley used together; it can have a large mechanical advantage if several pulleys are used (204)

boiling vaporization that occurs throughout a liquid (70)

boiling point the temperature at which a liquid boils and becomes a gas (70)

Boyle's law the law that states that for a fixed amount of gas at a constant temperature, the volume of a gas increases as its pressure decreases (65)

buoyant force the upward force that fluids exert on all matter; buoyant force opposes gravitational force (168)

C

carbohydrates biochemicals composed of one or more simple sugars bonded together that are used as a source of energy and for energy storage (408)

catalyst (KAT uh LIST) a substance that speeds up a reaction without being permanently changed (389)

cell a device that produces an electric current by converting chemical energy into electrical energy (278)

centripetal (sen TRIP uht uhl) **acceleration** the acceleration that occurs in circular motion; an object traveling in a circle is constantly changing directions, so acceleration occurs continuously (114)

change of state the conversion of a substance from one physical form to another (68)

characteristic property a property of a substance that is always the same whether the sample observed is large or small (48)

Charles's law the law that states that for a fixed amount of gas at a constant pressure, the volume of a gas increases as its temperature increases (66)

chemical bond a force of attraction that holds two atoms together (352)

chemical bonding the joining of atoms to form new substances (352)

chemical change a change that occurs when one or more substances are changed into entirely new substances with different properties; cannot be reversed using physical means (49)

chemical energy the energy of a compound that changes when its atoms are rearranged to form a new compound; chemical energy is a form of potential energy (218)

chemical equation a shorthand description of a chemical reaction using chemical formulas and symbols (378)

chemical formula a shorthand notation for a compound or a diatomic element using chemical symbols and numbers (376)

chemical property a property of matter that describes a substance based on its ability to change into a new substance with different properties (47)

chemical reaction the process by which one or more substances undergo change to produce one or more different substances (374)

chromosphere (KROH muh SFIR) a thin region of the sun's atmosphere between the corona and the photosphere, too thin to see unless there is a total solar eclipse (433)

circuit a complete, closed path through which electric charges flow (288)

closed system a well-defined group of objects that transfer energy between one another; energy is always conserved within a closed system (230)

coefficient (KOH uh FISH uhnt) a number placed in front of a chemical symbol or formula; used to balance a chemical equation (380)

colloid (KAWL OYD) a mixture in which the particles are dispersed throughout but are not heavy enough to settle out (97)

comet a small body of ice and rock that gives off gas and dust in the form of a tail as it passes close to the sun (471)

compound a pure substance composed of two or more elements that are chemically combined (86)

compound machines machines that are made of two or more simple machines (204)

concentration a measure of the amount of solute dissolved in a solvent (94)

condensation the change of state from a gas to a liquid (71)

condensation point the temperature at which a gas becomes a liquid (71)

GLOSSARY

conduction a method of charging an object that occurs when electrons are transferred from one object to another by direct contact (273)

conductor a material in which charges can move easily (275)

convective zone a region of the sun where hot and cooler gases circulate in convection currents, bringing the sun's energy to the surface (433)

core the center of the sun where the sun's energy is produced (433); the central, spherical part of the Earth below the mantle (439)

corona the sun's outer atmosphere, which can extend outward a distance equal to 10–12 times the diameter of the sun (433)

cosmic background radiation radiation left over from the big bang that fills all of space (501)

cosmology the study of the origin and future of the universe (500)

covalent (KOH VAY luhnt) **bond** the force of attraction between the nuclei of atoms and the electrons shared by the atoms (360)

covalent compounds compounds that are composed of elements that are covalently bonded; these compounds are composed of independent molecules, tend to have low melting and boiling points, do not usually dissolve in water, and form solutions that do not conduct an electric current when they do dissolve (399)

crust the thin, outermost layer of the Earth, or the uppermost part of the lithosphere (439)

crystal lattice (LAT is) a repeating three-dimensional pattern of ions (359)

current a continuous flow of charge caused by the motion of electrons; specifically, the rate at which charge passes a given point; expressed in amperes (281)

D

data any pieces of information acquired through experimentation (16)

decomposition reaction a reaction in which a single compound breaks down to form two or more simpler substances (383)

density the amount of matter in a given space; mass per unit volume (27, 44, 165)

diffraction the bending of waves around a barrier or through an opening (257)

dimension a measurement in one direction (38)

double-replacement reaction a reaction in which ions in two compounds switch places; one of the products is often a gas or a precipitate (384)

drag the force that opposes or restricts motion through a fluid; drag opposes thrust (176)

ductility (duhk TIL uh tee) the ability of a substance to be drawn or pulled into a wire (44)

E

eclipse an event in which the shadow of one celestial body falls on another (466)

electric discharge the loss of static electricity as charges move off an object (276)

electric force the force between charged objects (271)

electric generator a device that changes kinetic energy into electrical energy (234)

electric power the rate at which electrical energy is used to do work; expressed in watts (285)

electrical energy the energy of electric charges (219, 278)

electron clouds the regions inside an atom where electrons are likely to be found (310)

electrons the negatively charged particles found in all atoms; electrons are involved in the formation of chemical bonds (307)

element a pure substance that cannot be separated or broken down into simpler substances by physical or chemical means (82)

ellipse a closed curve in which the sum of the distances from the edge of the curve to two points inside the ellipse is always the same (430)

elliptical galaxy a spherical or elongated galaxy with a bright center and very little dust and gas (497)

endothermic the term used to describe a change in which energy is absorbed; the change can be a physical change or a chemical change (69, 386)

energy the ability to do work (214)

energy conversion a change from one form of energy into another; any form of energy can be converted into any other form of energy (222)

energy efficiency (e FISH uhn see) a comparison of the amount of energy before a conversion and the amount of useful energy after a conversion (228)

energy resource a natural resource that can be converted by humans into other forms of energy in order to do useful work (232)

evaporation (ee VAP uh RAY shuhn) vaporization that occurs at the surface of a liquid below its boiling point (70)

exothermic the term used to describe a change in which energy is released or removed; the change can be a physical change or a chemical change (69, 385)

F

fixed pulley a pulley that is attached to something that does not move; fixed pulleys change the direction of the force but do not increase the force (203)

fluid any material that can flow and that takes the shape of its container (162)

force a push or a pull; all forces have both size and direction (115)

fossil fuels nonrenewable energy resources that form in the Earth's crust over millions of years from the buried remains of once-living organisms (232)

free fall the condition an object is in when gravity is the only force acting on it (141)

freezing the change of state from a liquid to a solid (69)

freezing point the temperature at which a liquid changes into a solid (69)

frequency the number of waves produced in a given amount of time (254)

friction a force that opposes motion between two surfaces that are touching (119)

fulcrum the fixed point about which a lever pivots (198)

G

galaxy a large grouping of stars in space (496)

gas the state in which matter changes in both shape and volume (63)

gas giants the large, gaseous planets of the outer solar system (458)

geothermal energy energy resulting from the heating of the Earth's crust (236)

globular cluster a group of older stars that looks like a ball of stars (498)

gravitational potential energy energy due to an object's position above the Earth's surface (217)

gravity a force of attraction between objects that is due to their masses (39, 125)

greenhouse effect the natural heating process of a planet, such as the Earth, by which gases in the atmosphere trap thermal energy (454)

group a column of elements on the periodic table (333)

H

H-R diagram Hertzsprung-Russell diagram; a graph that shows the relationship between a star's surface temperature and its absolute magnitude (491)

halogens the elements in Group 17 of the periodic table; they are very reactive nonmetals, and their atoms have seven electrons in their outer level (340)

heterogeneous (HET uhr OH JEE nee uhs) **mixture** a combination of substances in which different components are easily observed (96)

homogeneous (HOH moh JEE nee uhs) **mixture** a combination of substances in which the appearance and properties are the same throughout (92)

hydraulic device (hie DRAW lik) a device that uses liquids to transmit pressure from one point to another (167)

hydrocarbons organic compounds that are composed of only carbon and hydrogen (412)

hydroelectricity electrical energy produced from falling water (235)

hypothesis a possible explanation or answer to a question (14)

I

ideal machine a 100 percent efficient machine (197)

inclined plane a simple machine that is a straight, slanted surface; a ramp (200)

induction a method of charging an object that occurs when charges in an uncharged object are rearranged without direct contact with a charged object (273)

inertia the tendency of all objects to resist any change in motion (42, 147)

inhibitor a substance that slows down or stops a chemical reaction (389)

input force the force applied to a machine (193)

insulator a material in which charges cannot easily move (275)

interference the result of two or more waves overlapping (258)

ionic (ie AHN ik) **bond** the force of attraction between oppositely charged ions (356)

ionic compounds compounds that contain ionic bonds; composed of ions arranged in a crystal lattice, they tend to have high melting and boiling points, are solid at room temperature, and dissolve in water to form solutions that conduct an electric current (398)

ions charged particles that form during chemical changes when one or more valence electrons transfer from one atom to another (356)

irregular galaxy a galaxy that does not fit into any other category; one with an irregular shape (497)

isotopes atoms that have the same number of protons but have different numbers of neutrons (315)

J

joule the unit used to express work and energy; equivalent to the newton-meter (N•m) (190)

K

kinetic (ki NET ik) **energy** energy of motion; kinetic energy depends on speed and mass (215)

Kuiper (KIE puhr) **belt** the region of the solar system outside the orbit of Neptune that is occupied by small, icy, cometlike bodies (472)

L

law a summary of many experimental results and observations; a law tells you how things work (19)

law of conservation of energy the law that states that energy is neither created nor destroyed (230, 386)

law of conservation of mass the law that states that mass is neither created nor destroyed in ordinary chemical and physical changes (381)

law of electric charges the law that states that like charges repel and opposite charges attract (271)

law of universal gravitation the law that states that all objects in the universe attract each other through gravitational force; the size of the force depends on the masses of the objects and the distance between them (126, 431)

lever a simple machine consisting of a bar that pivots at a fixed point, called a fulcrum; there are three classes of levers, based on where the input force, output force, and fulcrum are placed in relation to the load: first-class levers, second-class levers, and third-class levers (198)

lift an upward force on an object (such as a wing) caused by differences in pressure above and below the object; lift opposes the downward pull of gravity (174)

light energy the energy produced by the vibrations of electrically charged particles (220)

light-minute a unit of length equal to the distance light travels in space in one minute, or 18,000,000 km (452)

light-year a unit of length equal to the distance that light travels in space in one year (489)

lipids biochemicals that do not dissolve in water; their functions include storing energy and making up cell membranes; lipids include waxes, fats, and oils (409)

liquid the state in which matter takes the shape of its container but has a definite volume (62)

load a device that uses electrical energy to do work (288)

longitudinal wave a wave in which the particles of the medium vibrate back and forth along the path that the wave travels (250)

lubricant (LOO bri kuhnt) a substance applied to surfaces to reduce the friction between them (123)

lunar eclipse an event in which the shadow of the Earth falls on the moon's surface (466)

M

machine a device that helps make work easier by changing the size or direction (or both) of a force (192)

main sequence a diagonal pattern of stars on the H-R diagram (492)

malleability (MAL ee uh BIL uh tee) the ability of a substance to be pounded into thin sheets (44)

mantle the layer of the Earth between the crust and the core (439)

mass the amount of matter that something is made of; its value does not change with the object's location in the universe (26, 38, 129)

mass number the sum of the protons and neutrons in an atom (315)

matter anything that has volume and mass (7, 36)

mechanical advantage a number that tells how many times a machine multiplies force; can be calculated by dividing the output force by the input force (196)

mechanical efficiency (e FISH uhn see) a comparison expressed as a percentage of a machine's work output with the work input; can be calculated by dividing work output by work input and then multiplying by 100 (197)

mechanical energy the total energy of motion and position of an object (217)

medium a substance through which a wave can travel (247)

melting the change of state from a solid to a liquid (69)

melting point the temperature at which a substance changes from a solid to a liquid (69)

meniscus (muh NIS kuhs) the curve at a liquid's surface by which you measure the volume of the liquid (37)

metallic bond the attraction between a positively charged metal ion and the electrons in a metal (363)

metalloids elements that have properties of both metals and nonmetals; sometimes referred to as semiconductors (85)

metals elements that are shiny and are good conductors of thermal energy and electric current; most metals are malleable and ductile (85)

meteor a streak of light caused when a meteoroid or comet dust burns up in the Earth's atmosphere before it reaches the ground (474)

meteorite a meteoroid that reaches the Earth's surface without burning up completely (474)

meteoroid a very small, rocky body that revolves around the sun (474)

meter the basic unit of length in the SI system (25)

mixture a combination of two or more substances that are not chemically combined (90)

model a representation of an object or system (20, 307)

molecule (MAHL i KYOOL) a neutral group of atoms held together by covalent bonds (360)

momentum the property of a moving object that depends on the object's mass and velocity (152)

moon a natural satellite of a planet (463)

motion an object's change in position over time when compared with a reference point (108)

movable pulley a pulley attached to the object being moved; movable pulleys increase force (203)

N

nebula (NEB yuh luh) a large cloud of dust and gas in interstellar space; the location of star formation (424, 498)

negative acceleration acceleration in which velocity decreases; also called deceleration (113)

net force the force that results from combining all the forces exerted on an object (116)

neutron star a star in which all the particles have become neutrons; collapsed remains of a supernova (495)

neutrons the particles of the nucleus that have no charge (312)

newton (N) the SI unit of force (41, 115)

noble gases the unreactive elements in Group 18 of the periodic table; their atoms have eight electrons in their outer level (except for helium, which has two electrons) (340)

nonmetals elements that are dull (not shiny) and that are poor conductors of thermal energy and electric current (85)

nonrenewable resource a natural resource that cannot be replaced or that can be replaced only over thousands or millions of years (232)

nuclear (NOO klee uhr) **energy** the form of energy associated with changes in the nucleus of an atom; an alternative energy resource (221)

nuclear fission the process in which a large nucleus splits into two smaller nuclei (235)

nuclear fusion the process by which two or more nuclei with small masses join together, or fuse, to form a larger, more massive nucleus, along with the production of energy (435)

nucleic acids biochemicals that store information and help to build proteins and other nucleic acids; made up of subunits called nucleotides (411)

nucleus (NOO klee uhs) the tiny, extremely dense, positively charged region in the center of an atom; made up of protons and neutrons (309)

O

observation any use of the senses to gather information (12)

Oort (ohrt) **cloud** a spherical region of space that surrounds the solar system in which distant comets revolve around the sun (472)

open cluster a group of stars that forms when large amounts of gas and dust come together (498)

orbit the elliptical path a body takes as it travels around another body in space; the motion itself (429, 453)

organic compounds covalent compounds composed of carbon-based molecules (407)

output force the force applied by a machine (193)

P

parallax an apparent shift in the position of an object when viewed from different positions (489)

parallel circuit a circuit in which different loads are on separate branches (291)

pascal the SI unit of pressure; equal to the force of one newton exerted over an area of one square meter (162)

Pascal's principle the principle that states that a change in pressure at any point of an enclosed fluid is transmitted equally to all parts of that fluid (167)

perihelion (PER i HEE lee uhn) the point in the orbit of a planet at which the planet is closest to the sun (472)

period a horizontal row of elements on the periodic table (333)

period of revolution the time it takes for one body to make one complete orbit, or *revolution*, around another body in space (429, 453)

period of rotation the time it takes for a body to rotate once as it spins about its axis (453)

periodic having a regular, repeating pattern (326)

periodic law a law that states that the chemical and physical properties of elements are periodic functions of their atomic numbers (327)

perpetual (puhr PECH oo uhl) **motion machine** a machine that runs forever without any additional energy input; a machine whose energy output would equal its energy input; perpetual motion machines are impossible to create (231)

pH a measure of hydronium ion concentration in a solution; a pH of 7 is neutral; a pH less than 7 is acidic; a pH greater than 7 is basic (404)

phases the different appearances of the moon due to varying amounts of sunlight on the side of the moon that faces the Earth; results from the changing relative positions of the moon, Earth, and sun (465)

photocell the part of a solar panel that converts light into electrical energy (280)

photosphere the layer of the sun at which point the gases get thick enough to see; the surface of the sun (433)

physical change a change that affects one or more physical properties of a substance; most physical changes are easy to undo (48)

physical property a property of matter that can be observed or measured without changing the identity of the matter (43)

physical science the study of matter and energy (7)

planetesimal (PLAN i TES i muhl) the tiny building blocks of the planets that formed as dust particles stuck together and grew in size (426)

plasma the state of matter that does not have a definite shape or volume and whose particles have broken apart; plasma is composed of electrons and positively charged ions (67)

positive acceleration acceleration in which velocity increases (113)

potential difference energy per unit charge; specifically, the difference in energy per unit charge as a charge moves between two points in an electric circuit (same as voltage); expressed in volts (279)

potential energy energy of position or shape (216)

power the rate at which work is done (191)

pressure the amount of force exerted on a given area; the SI unit for pressure is the pascal (64, 162)

products the substances formed from a chemical reaction (378)

prograde rotation the counterclockwise spin of a planet or moon as seen from above the planet's north pole (454)

projectile (proh JEK tuhl) **motion** the curved path an object follows when thrown or propelled near the surface of Earth (143)

proteins biochemicals that are composed of amino acids; their functions include regulating chemical activities, transporting and storing materials, and providing structural support (410)

protons the positively charged particles of the nucleus; the number of protons in a nucleus is the atomic number, which determines the identity of an element (312)

pulley a simple machine consisting of a grooved wheel that holds a rope or a cable; there are two kinds of pulleys—fixed and movable (203)

pulsar a spinning neutron star that emits rapid pulses of light (495)

pure substance a substance in which there is only one type of particle; includes elements and compounds (82)

Q

quasar a "quasi-stellar" object; a starlike source of light and radio waves that is extremely far away; one of the most powerful sources of energy in the universe (499)

R

radiative zone a very dense region of the sun in which the atoms are so closely packed that light takes a long time to travel through (433)

reactants (ree AKT UHNTS) the starting materials in a chemical reaction (378)

red giant a star that expands and cools once it runs out of hydrogen fuel (493)

reference point an object that appears to stay in place in relation to an object being observed for motion (108)

reflection the bouncing back of a wave after it strikes a barrier or object (256)

refraction the bending of a wave as it passes at an angle from one medium to another (257)

renewable resource a natural resource that can be used and replaced over a relatively short time (235)

resistance the opposition to the flow of electric charge; expressed in ohms (283)

resonance what occurs when an object vibrating at or near the resonant frequency of a second object causes the second object to vibrate (260)

resultant velocity the combination of two or more velocities (111)

retrograde orbit the clockwise revolution of a satellite around a planet as seen from above the north pole of the planet (470)

retrograde rotation the clockwise spin of a planet or moon as seen from above the planet's or moon's north pole (454)

revolution the elliptical motion of a body as it orbits another body in space (429, 453)

rotation the spinning motion of a body on its axis (429, 453)

S

salt an ionic compound formed from the positive ion of a base and the negative ion of an acid (406)

satellite a natural or artificial body that revolves around a planet (463)

saturated hydrocarbon a hydrocarbon in which each carbon atom in the molecule shares a pair of electrons with each of four other atoms (412)

saturated solution a solution that contains all the solute it can hold at a given temperature (94)

scientific method a series of steps that scientists use to answer questions and solve problems (11)

screw a simple machine that is an inclined plane wrapped in a spiral (201)

series circuit a circuit in which all parts are connected in a single loop (290)

simple machines the six machines from which all other machines are constructed: a lever, an inclined plane, a wedge, a screw, a wheel and axle, and a pulley (198)

single-replacement reaction a reaction in which an element takes the place of an element in a compound; this can occur only when a more-reactive element takes the place of a less-reactive one (383)

solar eclipse an event in which the shadow of the moon falls on the Earth's surface (466)

solar nebula the nebula that formed into the solar system (425)

solar system the system composed of the sun (a star) and the planets and other bodies that travel around the sun (424)

solid the state in which matter has a definite shape and volume (61)

solubility (SAHL yoo BIL uh tee) the ability to dissolve in another substance; more specifically, the amount of solute needed to make a saturated solution using a given amount of solvent at a certain temperature (44, 94)

solute the substance that is dissolved to form a solution (92)

solution a mixture that appears to be a single substance but is composed of particles of two or more substances that are distributed evenly amongst each other (92)

solvent the substance in which a solute is dissolved to form a solution (92)

sound energy the energy caused by an object's vibrations (220)

spectrum the rainbow of colors produced when white light passes through a prism or spectrograph (484)

speed the rate at which an object moves; speed depends on the distance traveled and the time taken to travel that distance (109)

spiral galaxy a galaxy with a nuclear bulge in the center and very distinctive spiral arms (496)

standing wave a wave that forms a stationary pattern in which portions of the wave do not move and other portions move with a large amplitude (260)

states of matter the physical forms in which a substance can exist; states include solid, liquid, gas, and plasma (60)

static electricity the buildup of electric charges on an object (275)

sublimation (SUHB luh MAY shuhn) the change of state from a solid directly into a gas (72)

subscript a number written below and to the right of a chemical symbol in a formula (376)

sunspot an area on the photosphere of the sun that is cooler than surrounding areas, showing up as a dark spot (437)

supernova the death of a large star by explosion (494)

surface gravity the percentage of your Earth weight you would experience on another planet; the weight you would experience on another planet (453)

surface tension the force acting on the particles at the surface of a liquid that causes the liquid to form spherical drops (63)

surface wave a wave that occurs at or near the boundary of two media and that is a combination of transverse and longitudinal waves (251)

suspension a mixture in which particles of a material are dispersed throughout a liquid or gas but are large enough that they settle out (96)

synthesis (SIN thuh sis) **reaction** a reaction in which two or more substances combine to form a single compound (382)

T

technology the application of knowledge, tools, and materials to solve problems and accomplish tasks; technology can also refer to the objects used to accomplish tasks (11)

temperature a measure of how hot (or cold) something is; specifically, a measure of the average kinetic energy of the particles in an object (26)

terminal velocity the constant velocity at which a falling object travels when the size of the upward force of air resistance matches the size of the downward force of gravity (140)

terrestrial planets the small, dense, rocky planets of the inner solar system (453)

theory a unifying explanation for a broad range of hypotheses and observations that have been supported by testing (19, 304, 352)

thermal energy the total kinetic energy of the particles that make up an object (218)

thermocouple a device that converts thermal energy into electrical energy (280)

thrust the forward force produced by an airplane's engines; thrust opposes drag (175)

transverse wave a wave in which the particles of the wave's medium vibrate perpendicular to the direction the wave is traveling (249)

turbulence an irregular or unpredictable flow of fluids that can cause drag; lift is often reduced by turbulence (176)

U

unbalanced forces forces on an object that cause the net force to be other than zero; unbalanced forces produce a change in motion or acceleration (117)

unsaturated hydrocarbon a hydrocarbon in which not all carbon atoms have four single bonds; at least one double or triple bond is present (412)

V

valence (VAY luhns) **electrons** the electrons in the outermost energy level of an atom; these electrons are involved in forming chemical bonds (353)

vaporization the change of state from a liquid to a gas; includes boiling and evaporation (70)

velocity (vuh LAHS uh tee) the speed of an object in a particular direction (110)

viscosity (vis KAHS uh tee) a liquid's resistance to flow (63)

voltage the difference in energy per unit charge as a charge moves between two points in an electric circuit (same as potential difference); expressed in volts (282)

volume the amount of space that something occupies or the amount of space that something contains (25, 36)

W

watt the unit used to express power; equivalent to joules per second (J/s) (191)

wave a disturbance that transmits energy through matter or space (246)

wave speed the speed at which a wave travels (255)

wavelength the distance between one point on a wave and the corresponding point on an adjacent wave in a series of waves; for example, the distance between two adjacent crests or compressions (253)

wedge a simple machine that is a double inclined plane that moves; a wedge is often used for cutting (201)

weight a measure of the gravitational force exerted on an object, usually by the Earth (40, 128)

wheel and axle a simple machine consisting of two circular objects of different sizes; the wheel is the larger of the two circular objects (202)

white dwarf a small, hot star near the end of its life; the leftover center of an old star (492)

work the action that results when a force causes an object to move in the direction of the force (188)

work input the work done on a machine; the product of the input force and the distance through which it is exerted (193)

work output the work done by a machine; the product of the output force and the distance through which it is exerted (193)

Index

A **boldface** number refers to an illustration on that page.

A

B

C

D

E

G

H

I

J

K

N

Q

R

S

T

U

V

W

X

Y

Z

Credits

Abbreviations used: (t) top, (c) center, (b) bottom, (l) left, (r) right, (bkgd) background

ILLUSTRATIONS

All illustrations, unless noted below, by Holt, Rinehart and Winston.

Table of Contents Page v(bl), Kristy Sprott; vii(tl), ix(tr), Stephen Durke/Washington Artists; ix(b), Keith Locke/Suzanne Craig; x(b), Marty Roper/Planet Rep; xi(tl), Blake Thornton/Rita Marie; xi(cr), Stephen Durke/Washington Artists; xi(b), Terry Kovalcik; xii(b), Stephen Durke/Washington Artists; xiii(t), Blake Thornton/Rita Marie; xiv(b), Dan McGeehan/Koralik Associates; xiv(t), xv, Paul DiMare.

Chapter One Page 6, 9, Rainey Kirk/The Neis Group; 13, John Huxtable/Black, Inc.; 14, Will Nelson/Sweet Reps; 16, Preface, Inc.; 17, Terry Guyer; 18(b), Brian White; 20(tr), Kristy Sprott; 20(bl) Morgan Cain & Associates; 21(t), Stephen Durke/Washington Artists; 21(c), Keith Locke/Suzanne Craig; 21(b) Gary Antonetti/Ortelius Design; 22(cl), Kristy Sprott; 22(b), Blake Thornton/Rita Marie; 23(t), Stephen Durke/Washington Artists and Preface, Inc.; 24(all), Stephen Durke/Washington Artists; 26(t), Morgan Cain & Associates; 29, Blake Thornton/Rita Marie; 31(tl), David Merrell/Suzanne Craig.

Chapter Two Page 38(t), 39(b), Stephen Durke/Washington Artists; 42(lc), Gary Locke/Suzanne Craig; 43(c), Blake Thornton/Rita Marie; 50(t), 51, 53, Marty Roper/Planet Rep; 55(lc), Terry Kovalcik; 57(tc), Daniels & Daniels.

Chapter Three Page 60(t), Mark Heine; 60(b), 61(b), 62(t), 63(c), 64, 65, 66(c), Stephen Durke/Washington Artists; 66(bl), Preface, Inc.; 68(c), David Schleinkofer/Mendola Ltd.; 70(t), Marty Roper/Planet Rep; 70(b), Mark Heine; 73, David Schleinkofer/Mendola Ltd. and Preface, Inc.; 74(t), Stephen Durke/Washington Artists; 74(b), Preface, Inc.; 75, Marty Roper/Planet Rep; 77(cr), Preface, Inc.

Chapter Four Page 82(c), Marty Roper/Planet Rep; 84(b), Preface, Inc.; 88(c), Blake Thornton/Rita Marie; 95(t), Preface, Inc.

Chapter Five Page 109, Preface, Inc.; 111, Marty Roper/Planet Rep; 112, Gary Locke/Suzanne Craig; 113(t), Mike Carroll/Steve Edsey & Sons; 114(cr), Preface, Inc.; 119(t), Blake Thornton/Rita Marie; 119(b), 120, 122(b), Gary Ferster; 126, Doug Henry/American Artists; 127, Stephen Durke/Washington Artists; 128(t), Craig Attebery/Frank & Jeff Lavaty; 129, 130(b), Stephen Durke/Washington Artists; 131, Terry Guyer; 133(r), Preface, Inc.

Chapter Six Page 139(l), 140(tr), Gary Ferster; 142, Craig Attebery/Frank & Jeff Lavaty; 144(tl), Mike Carroll/Steve Edsey & Sons; 146(t), Marty Roper/Planet Rep; 148, Charles Thomas; 151(t), Gary Ferster; 154(b), Craig Attebery/Frank & Jeff Lavaty; 155(c), Marty Roper/Planet Rep; 158(c), James Pfeffer.

Chapter Seven Page 160, 161(tl), Rainey Kirk/The Neis Group; 162, 163(b), Stephen Durke/Washington Artists; 164, 165, Rainey Kirk/The Neis Group; 166(b), Christy Krames; 167, Mark Heine; 168(l), Preface, Inc.; 169, Will Nelson/Sweet Reps; 171, Preface, Inc.; 172(c), Sam Collins/Art & Science, Inc.; 174(c), Craig Attebery/Frank & Jeff Lavaty; 174(b), Will Nelson/Sweet Reps; 176(l), Marty Roper/Planet Rep; 177(c), Terry Guyer; 178, Craig Attebery/Frank & Jeff Lavaty; 181(tr), Jared Schneidman/Wilkinson Studios ; 181(b), Keith Locke/Suzanne Craig.

Chapter Eight Page 189(t), Blake Thornton/Rita Marie; 190, John White/The Neis Group; 191(b), Blake Thornton/Rita Marie; 195(l), Annie Bissett; 195(r), John White/The Neis Group; 196(c), Keith Locke/Suzanne Craig; 198(c), 199(l), Annie Bissett; 201(cl), Preface, Inc.; 203(t), Gary Ferster; 203(c,b), 204(tr), John White/The Neis Group; 206(c), Blake Thornton/Rita Marie.

Chapter Nine Page 215(b), Dave Joly; 216(b), John White/The Neis Group; 218(c), Stephen Durke/Washington Artists; 219(t), Kristy Sprott; 219(b), Stephen Durke/Washington Artists; 220(t), Gary Ferster; 224(c), Will Nelson/Sweet Reps; 225, Dan Stuckenschneider/Uhl Studios Inc.; 226(t), Blake Thornton/Rita Marie; 227(b), Dan Stuckenschneider/Uhl Studios Inc.; 229, Dan McGeehan/Koralik Associates; 230(t), Marty Roper/Planet Rep; 230(b), 232(b), Dan Stuckenschneider/Uhl Studios Inc.; 233(r), Preface, Inc.; 234(b), Patrick Gnan/Deborah Wolfe; 235(br), Michael Moore; 236(cr), Dan Stuckenschneider/Uhl Studios Inc.; 237(b), Preface, Inc.; 241(c), Dave Joly.

Chapter Ten Page 244(t), Gary Antonetti/Ortelius Design; 246(b), Will Nelson/Sweet Reps; 249(t), Preface, Inc.; 249(b), 250(c), John White/The Neis Group; 250(b), Sidney Jablonski; 251(t), Stephen Durke/Washington Artists; 251(b), Jared Schneidman/Wilkinson Studios; 252(t), Marty Roper/Planet Rep; 252(b), 253(t,b), 254(c), Sidney Jablonski; 254(c), 255(tc), Mike Carroll/Steve Edsey & Sons; 255(c), 257, Will Nelson/Sweet Reps; 259(tl, tc, tr, cl, c, cr), John White/The Neis Group; 259(br), Terry Guyer; 262(b), John White/The Neis Group; 265(r), Sidney Jablonski.

Chapter Eleven Page 270(t), Blake Thornton/Rita Marie; 270(b), Stephen Durke/Washington Artists; 271, John White/The Neis Group; 273(l), Stephen Durke/Washington Artists; 276, Dan Stuckenschneider/Uhl Studios Inc.; 278, 280(cl), Mark Heine; 281, 282, Geoff Smith/Scott Hull; 284(t), Will Nelson/Sweet Reps; 285(t), Preface, Inc.; 287(cr), Boston Graphics; 292(b), Dan McGeehan/Koralik Associates.

Chapter Twelve Page 305(c), Preface, Inc.; 306(c), Mark Heine; 307(c), Stephen Durke/Washington Artists; 308(c), Mark Heine; 308(b), Preface, Inc.; 309(t), Stephen Durke/Washington Artists; 309(br), Preface, Inc.; 310, 312, 313 (tr), Stephen Durke/Washington Artists; 313(cr), Terry Kovalcik; 313(b), 314, 315(b), 317, 318(b), Stephen Durke/Washington Artists; 319, Terry Kovalcik; 320, Mark Heine; 321(r), Stephen Durke/Washington Artists.

Chapter Thirteen Page 326, Michael Jaroszko/American Artists; 328,329, Kristy Sprott; 330(tr), 331(t,bc), Stephen Durke/Washington Artists; 333(t), Preface, Inc.; 334 (bc), 335 (bl), 336(t), 337(t), 338(tc, b), Preface, Inc.; 338(l), Gary Locke/Suzanne Craig; 339(c, b), 340(tr, b), 341(cl), Preface, Inc.; 343, Gary Locke/Suzanne Craig; 345(tr), Preface, Inc.; 345(cl), Keith Locke/Suzanne Craig; 345(cr), Annie Bissett; 347(l), Dan Stuckenschneider/Uhl Studios Inc.

Unit Five Page 348(t), Kristy Sprott.

Chapter Fourteen Page 353, Stephen Durke/Washington Artists; 354, Preface, Inc.; 355, 357, 358, Stephen Durke/Washington Artists; 359(t), Keith Locke/Suzanne Craig; 359(br), Kristy Sprott; 360(c), 361(tl), Stephen Durke/Washington Artists; 361(tr,c), Preface, Inc.; 362(tr,br), Kristy Sprott; 362(bl), Stephen Durke/Washington Artists; 363(tr), Kristy Sprott; 364(cl), 365(tc), Kristy Sprott; 365(b), Preface, Inc.; 366(t), Keith Locke/Suzanne Craig; 366(b), Stephen Durke/Washington Artists; 367, Kristy Sprott; 368(b), Stephen Durke/Washington Artists; 370(tl), Kristy Sprott.

Chapter Fifteen Page 376(t), 380, Kristy Sprott; 382(b), 383(c,b), 384(b), Blake Thornton/Rita Marie; 387(t), Preface, Inc.; 389(b), Preface, Inc.; 390, Blake Thornton/Rita Marie; 391, Kristy Sprott; 393(cr), Preface, Inc.

Chapter Sixteen Page 404(t), Dave Joly; 404(b), 407, Preface, Inc.; 408(c), 409(b), Morgan Cain & Associates; 412, 413(t), 417, Preface, Inc.

Chapter Seventeen Page 422, Stephen Durke/Washington Artists; 426, 427, Paul DiMare; 429, Sidney Jablonski; 430(t), Mark Heine; 430(b), 431(cr), Sidney Jablonski; 433, Uhl Studios Inc.; 434, 435(t), Marty Roper/Planet Rep; 435(bl, br), 436(all), Stephen Durke/Washington Artists; 437(l), Sidney Jablonski; 439(t), Stephen Durke/Washington Artists; 439(b), Uhl Studios Inc.; 440, 441, Paul DiMare; 443, Uhl Studios Inc.; 444(t), Sidney Jablonski; 444(b), 445, Uhl Studios Inc.; 448, John Huxtable/Black, Inc.

Chapter Eighteen Page 450(tl), Uhl Studios Inc.; 452(b), 453(t), 458(t), 460(b), Sidney Jablonski; 461(t), Dan McGeehan/Koralik Associates; 462(cr), Paul DiMare; 464(all), Stephen Durke/Washington Artists; 465, Sidney Jablonski; 466(t), 467(l), Paul DiMare; 472(t), Stephen Durke/Washington Artists; 472(b), Paul DiMare; 473(b), Craig Attebery/Frank & Jeff Lavaty; 476(b), Sidney Jablonski; 477, Stephen Durke/Washington Artists; 480(tc), Paul DiMare.

Chapter Nineteen Page 482(b), Craig Attebery/Frank & Jeff Lavaty; 485, 486(tl), Stephen Durke/Washington Artists; 489, Sidney Jablonski; 490(tl), Sidney Jablonski; 490(b), 492(all), 493(all), Stephen Durke/Washington Artists; 498, Paul DiMare; 499(tr), Stephen Durke/Washington Artists; 500(all), Paul DiMare; 501, 503(all), Craig Attebery/Frank & Jeff Lavaty; 504(b), Stephen Durke/Washington Artists; 507(cr), Sidney Jablonski; 509(br), Paul DiMare.

LabBook Page 532, Blake Thornton/Rita Marie; 551(r), Preface, Inc.; 555(cl), John White/The Neis Group; 559(b), Dan McGeehan/Koralik Associates; 561(b), Marty Roper/Planet Rep.; 567(t), Gary Ferster; 567(b), Dave Joly; 571(t), Preface, Inc.; 574, Stephen Durke/Washington Artists; 583(c), Preface, Inc.; 586, Marty Roper/Planet Rep.; 594, Mark Heine.

Appendix Page 602(t), Terry Guyer; 606(b), Mark Mille/Sharon Langley Artist Rep.; 614, Stephen Durke/Washington Artists; 615(c), Kristy Sprott; 615(b), Bruce Burdick; 616, 617, Kristy Sprott; 619(t), Dan Stuckenschneider/Uhl Studios Inc.

PHOTOGRAPHY

Front Cover (tl) Ed Young/Science Photo Library/Photo Researchers, Inc. (also on title page); (tr) FPG International; (bl) Henry Kaiser/Leo de Wys; (br) Firefly Productions/The Stock Market; (cr) Stephen Dalton/Photo Researchers, Inc.; owl (front cover, back cover, spine and title page) Kim Taylor/Bruce Coleman, Inc.

Table of Contents Page v(tr), Robert Daemmrich/Tony Stone Images; vi(cl), Richard Megna/Fundamental Photographs; (bl), Joseph Drivas/The Image Bank; vii(bl), Richard Megna/Fundamental Photographs; viii(tl), James Balog/Tony Stone Images; (cr), Sergio Purtell/FOCA; (bl) NASA; ix(cl), T. Mein/N&M Mischler/Tony Stone Images; x(tr), NASA; xi(tr), Stephanie Morris/HRW Photo; xiv(bl), World Perspective/Tony Stone Images; xv(tl), Dr. Christopher Burrows, ESA/STScl/NASA

Feature Borders Unless otherwise noted below, all images ©2001 PhotoDisc/HRW: "Across the Sciences" Pages 56, 135, 242, 267, 322, 370, 418, 448, all images by HRW; "Careers", 32, 159, 243, 323, 395, 509, sand bkgd and saturn, Corbis Images, DNA, Morgan Cain & Associates, scuba gear, ©1997 Radlund & Associates for Artville; "Eureka" 79, 158, 182, 211, 371, ©2001 PhotoDisc/HRW; "Eye on the Environment" 394, clouds and sea in bkgd, HRW, bkgd grass and red eyed frog, Corbis Images, hawks and pelican, Animals Animals/Earth Scenes, rat, John Grelach/Visuals Unlimited, endangered flower, Dan Suzio/Photo Researchers, Inc.; "Health Watch" 57, dumbell, Sam Dudgeon/HRW Photo, aloe vera and EKG, Victoria Smith/HRW Photo, basketball, ©1997 Radlund & Associates for Artville, shoes and Bubbles, Greg Geisler; "Scientific Debate" 449, 480, Sam Dudgeon/HRW Photo; "Science Fiction" 33, 103, 183, 481, saucers, Ian Christopher/Greg Geisler, book, HRW, bkgd, Stock Illustration Source ; "Science, Technology, and Society" 78, 102, 134, 210, 266, 298, 346, robot, Greg Geisler; "Weird Science" 299, 347, 419, 508, mite, David Burder/Tony Stone, atom balls, J/B Woolsey Associates, walking stick and turtle, EclectiCollection

Unit One Page 2(t), Corbis-Bettman; 2(cl), Photosource; 2(b), Enrico Tedeschi; 3(tl), Sam Shere/Corbis Bettmann; 3(tr), Brown Brothers/HRW Photo Library; 3(bl), Natalie Fobes/Tony Stone Images; 3(br), Noble Proctor/Science Source/Photo Researchers, Inc.

Chapter One Page 4(t), David Lawrence/The Stock Market; 4(b), Donna Coveney/MIT News; 5(t), Tom McHugh/Steinhart Aquarium/Photo Researchers, Inc.; 7(t), Jeff Hunter/The Image Bank; 7(b), Scala/Art Resource, NY; 8(b), Dr. E.R. Degginger/Color-Pic, Inc.; 9(tr), Chris Madley/Science Photo Library/Photo Researchers, Inc.; 9(bkgd), Joseph Nettis/Photo Researchers, Inc.; 9(cl), Stuart Westmorland/Photo Researchers, Inc.; 9(br), M.H. Sharp/Photo Researchers, Inc.; 9(cr), David R. Frazier Photolibrary; 9(b), Norbert Wu; 10 (tl, b), John Langford/HRW Photo; 10(tr), courtesy of the U.S. Nuclear Regulatory Commission; 10(cr), Michele Forman; 12(br), HRW photo by Stephen Maclone; 12(bl), Barry Chin/Boston Globe; 15, Donna Coveney/MIT News; 19(t), Chris Butler/Science Photo Library/Photo Researchers, Inc.; 19(b), Richard Megna/Fundamental Photographs; 20(br), Rosenfeld Images LTD/Science Photo Library/Photo Researchers, Inc.; 23(b), courtesy of FHWA/NHTSA National Crash Analysis/The George Washington University; 25(t), SuperStock; 25(bl,r), HRW photo by Peter Van Steen; 27, Image ©2001 PhotoDisc, Inc.; 28(t), Tom McHugh/Steinhart Aquarium/Photo Researchers, Inc.; 31(cr), HRW photo by Victoria Smith; 32(all), HRW photos by Art Louis.

Chapter Two Page 34(l), Hartmann/Sachs/Phototake, NY; 34(c), Ken Lucas/Visuals Unlimited; 34(r), The Granger Collection, NY; 34(tr), Image ©2001 PhotoDisc, Inc.; 36(b), NASA, Media Services Corp.; 40(all), John Morrison/Morrison Photography;

41(t), HRW photo by Michelle Bridwell; 43-44(all), John Morrison/Morrison Photography; 45, Neal Nishler/Tony Stone Images; 46(all), John Morrison/Morrison Photography; 47, Rob Boudreau/Tony Stone Images; 48(t), Brett H. Froomer/The Image Bank; 48(cl,cr), John Morrison/Morrison Photography; 49, HRW photo by Lance Schriner; 50(tl,tr), John Morrison/Morrison Photography; 50(bl), Joseph Drivas/The Image Bank; 50(br), SuperStock; 52(r), HRW photo by Michelle Bridwell; 55, HRW photo by Lance Schriner; 56, David Malin/©Anglo-Australian Observatory/Royal Observatory, Edinburgh.

Chapter Three Page 58(r), Tony Stone Images; 58(l), Peter Menzel; 59(t), James L. Amos/Corbis; 61(t), Gilbert J. Charbonneau; 63(t), Dr. Harold E. Edgerton/©The Harold E. Edgerton 1992 Trust/courtesy Palm Press, Inc.; 67(b), Pekka Parviainen/Science Photo Library/Photo Researchers, Inc.; 68, Union Pacific Museum Collection; 69(t), Richard Megna/Fundamental Photographs; 76(t), Myrleen Ferguson/PhotoEdit; 77, Charles D. Winters/Photo Researchers, Inc.; 78(l), Kennan Ward Photography; 78(r), Dr. Jean-Claude Diels/University of New Mexico; 79, Union Pacific Museum Collection.

Chapter Four Page 80(l), Fr. Browne S.J. Collection; 80(r), RMS Titanic Inc.; 82(r), David R. Frazier Photolibrary; 82(l), Jonathan Blair/Woodfin Camp & Associates; 83(b), Russ Lappa/Photo Researchers, Inc.; 83(t,c), Charles D. Winters/Photo Researchers, Inc.; 84(t,l), Walter Chandoha; 84(r), Zack Burris; 84(b), Yann Arthus-Bertrand/Corbis; 85(tc), HRW photo by Victoria Smith; 85(cl), Dr. E.R. Degginger/Color-Pic, Inc.; 85(cr), Runk/Schoenberg/Grant Heilman Photography Inc.; 85(br), Joyce Photographics/Photo Researchers, Inc.; 85(l), Russ Lappa/Photo Researchers, Inc.; 85(bc), Charles D. Winters/Photo Researchers, Inc.; 86(c), Runk/Schoenberger/Grant Heilman; 87(l), Runk/Shoenberger/Grant Heilman Photography; 87(b), Richard Megna/Fundamental Photographs; 88, Runk/Schoenberg/Grant Heilman Photography Inc.; 88, Richard Megna/Fundamental Photographs; 89(t), John Kapriellan/Photo Researchers, Inc.; 89(b), Stuart Westmoreland/Tony Stone Images; 90(l), HRW photo by Victoria Smith; 91(t), Charles D. Winters/Timeframe Photography; 91(c), Charles Winters/Photo Researchers, Inc.; 91(cr), Klaus Guldbrandsen/Science Photo Library/Photo Researchers, Inc.; 93(b), HRW photo by Richard Haynes; 93(c), Image ©2001 PhotoDisc, Inc.; 94(c), Dr. E.R. Degginger/Color-Pic, Inc.; 96(c), HRW photo by Michelle Bridwell; 97(t), HRW photo by Lance Schriner; 97(b), Dr. E.R. Degginger/Color-Pic, Inc.; 98(b), HRW photo by Victoria Smith; 98(t), David R. Frazier Photolibrary; 100(t), Yann Arthus-Bertrand/Corbis; 101(t), Richard Megna/Fundamental Photographs; 102(l), Anthony Bannister/Photo Researchers, Inc.; 102(r), Richard Steedman/The Stock Market.

Unit Two Page 104(tl), Archive Photos; 104(tr), Image ©1998 PhotoDisc, Inc.; 104(bl), Photo Researchers, Inc.; 104(br), Underwood & Underwood/Corbis-Bettman; 105(tl), The Vittoria, colored line engraving, 16th century/The Granger Collection, NY; 105(tr), Stock Montage, Inc.; 105(cl), NASA/Science Source/Photo Researchers, Inc.; 105(cr), W.A. Mozart at the age of 7: oil on canvas, 1763, by P.A. Lorenzoni/The Granger Collection, NY; 105(b), NASA.

Chapter Five Page 106(c), George Catlin, Ball of Play of the Choctaw/National Museum of American Art, Washington D.C./Art Resource, NY; 106(b), Lawrence Migdale; 108(all), SuperStock; 110(t), Tom Tietz/Tony Stone Images; 110(b), Robert Ginn/PhotoEdit; 113, Sergio Putrell/Foca; 114(t), Gene Peach/The Picture Cube; 115, 116(b), HRW photos by Michelle Bridwell; 117, HRW photo by Daniel Schaefer; 118(t), David Young-Wolff/PhotoEdit; 118(b), Arthur C. Smith/Grant Heilman Photography; 121(all), HRW photos by Michelle Bridwell; 121(insets), HRW photos by Stephanie Morris; 122, Tony Freedman/PhotoEdit; 124(cl), HRW photo by Michelle Bridwell; 125, NASA; 128, Image ©2001 PhotoDisc; 130(c), Superstock; 133, HRW photo by Mavournea Hay; 134, Hunter Hoffman; 135, Bruce Hands/Tony Stone Images.

Chapter Six Page 136(b), David R. Frazier Photography; 136(all), 137(t), NASA; 138, Richard Megna/Fundamental Photographs; 139, Doug Armand/Tony Stone Images; 140, Robert Daemmrich/Tony Stone Photography; 141(t), James Sugar/Black Star; 141(b), NASA; 143(t), James Balog/Tony Stone Images; 143(l), Michelle Bridwell/Frontera Fotos; 143(r), Image ©2001 PhotoDisc, Inc.; 144(t), Richard Megna/Fundamental Photographs; 146, Marc Asnin/SABA Press Photos, Inc.; 147(l), HRW photo by Mavournea Hay; 147(r), Michelle Bridwell/Frontera Fotos; 149(all), Image ©2001 PhotoDisc, Inc.; 150, David Madison; 151(tl), Gerard Lacz/Animals Animals/Earth Sciences; 151(tc), HRW photo by Coronado Rodney Jones; 151(tr), Image ©2001 PhotoDisc, Inc.; 151(bl), NASA; 151(br), HRW photo by Lance Schriner; 152(all), HRW photos by Michelle Bridwell; 153(t), Zigy Kaluzny/Tony Stone Images; 153(b), HRW Photo by Michelle Bridwell; 154, Robert Daemmrich/Tony Stone Photography; 156(t), James Balog/Tony Stone Images; 156(b), David Madison/Tony Stone Images; 157(l), NASA; 158(t), courtesy of Steve Okamoto; 158(b), Lee Schwabe; 159, R.N. Metheny/The Christian Science Monitor.

Chapter Seven Page 160, Michael Sexton; 164(t), I. M. House/Tony Stone Images; 164(tc), David R. Frazier Photolibrary, Inc.; 164(c), Deiter and Mary Plage/Bruce Coleman; 164(bc), Wolfgang Kaeler/Corbis; 164(b), Martin Barraud/Tony Stone Images; 165(t), SuperStock; 165(tc), Daniel A. Nord; 165(c), Ken Marschall/Madison Press Books; 165(bc), Bassot/Photo Researchers, Inc.; 165(b), Bettmann Archive; 168, HRW photo by Michelle Bridwell; 170, David R. Frazier Photolibrary; 173, Richard Megna/Fundamental Photographs; 175(t), Larry L. Miller/Photo Researchers, Inc.; 175(c), Richard Neville/Check Six; 175(bl), T. Mein/N&M Mishler/Tony Stone Images; 176(t), Ron Kimball/Ron Kimball Photography, Inc.; 176(b), HRW photo by Michelle Bridwell; 177(b), George Hall/Check Six; 179(t), Ron Kimball/Ron Kimball Photography, Inc.; 182(t), Victor Malafronte.

Unit Three Page 184(t,c), The Granger Collection, NY; 184(b), Conley Photography, Inc./American Solar Energy Society; 185(tl), Phil Degginger/Color-Pic, Inc.; 185(tr), Corbis-Bettman; 185(cl), The Granger Collection, NY; 185(cr), HRW photo by Robert Wolf; 185(b), David Madison.

Chapter Eight Page 186(b), Yoram Lehman/Peter Arnold Inc., NY; 186, 187(t), Erich Lessing/Art Resource, NY; 192(t,cl,cr,bc,br), Robert Wolf; 192(bl), Image ©2001 PhotoDisc, Inc.; 197, Robert Wolf; 198(l,r), Robert Wolf; 199(c), Robert Wolf; 200, HRW photo by Lisa Davis; 201(b,c), 202(all), Robert Wolf; 204(t), HRW photo by Russell Dian; 204(b), Robert Wolf; 205(tl), Image ©2001 PhotoDisc, Inc.; 205(tc), Robert Wolf; 205(b), Photo Researchers, Inc.; 208(t), Robert Wolf; 209(l), Helmut Gritscher/Peter Arnold, Inc.; 209(br), HRW photo by Stephanie Morris; 210(all), courtesy of IBM Corp., Almaden Research Division; 211, A.W. Stegmeyer/Upstream.

Chapter Nine Page 212, courtesy of L. Stern and J. Pinkston, US Geological Survey; 214, Al Bello/Allsport; 216, Earl Kowall/Corbis; 218(c), Paul A. Souders/Corbis; 218(r), Tony Freeman/PhotoEdit; 220(br), Ira Wyman/Sygma; 221(t), NASA; 221(b), Mark C. Burnett/Photo Researchers, Inc.; 222(all), HRW photos by Peter Van Steen; 223(t), Richard Megna/Fundamental Photographs; 228, Morton Beebe/Corbis; 231, HRW photo by Victoria Smith; 232, Ted Clutter/Photo Researchers, Inc.; 233(t), Tom Carroll/Phototake, NY; 233(cl), Mark E. Gibson; 233(br), John Kaprielian/Photo Researchers, Inc.; 234(t), Robert Wolf; 235(t), D.O.E./Science Source/Photo Researchers, Inc.; 235(cl), H.P. Merton/The Stock Market; 235(b), Tom Carroll/Phototake, NY; 236(tl), Bob Gomel/The Stock Market; 236(tr), Ed Young/Corbis; 236(c), Richard Nowitz/Phototake; 235(bl), HRW photo by Coronado Rodney Jones; 235(br), Kevin R. Morris/Corbis; 239, Ed Young/Corbis; 240(b), Ted Clutter/Photo Researchers, Inc.; 241(bl), Mike Powell/Allsport; 242, Solar Survival Architecture; 243(all), Robert Wolf.

Chapter Ten Page 244, AP/Wide World Photos; 247(t), Phil Degginger/Color-Pic, Inc.; 247(b), Emil Muench/Photo Researchers; 248, Fred Bravendam/Minden Pictures; 253, Hokusai, The Great Wave of Kanagawa, from 36 views of Mount Fugi, Private Collection Art Resource, NY; 256(t), Chris Butler/Astrostock; 256(b), Erich Schrempp/Photo Researchers, Inc.; 258(b), Corbis/Richard Hamilton Smith; 258(tl,tr), 260(t), Richard Megna/Fundamental Photographs; 261(all), AP/Wide World Photos; 262, Fred Bravendam/Minden Pictures; 264, Richard Megna/Fundamental Photographs; 265, Martin Bough/Fundamental Photographs; 266, Pete Saloutos/The Stock Market; 267, Betty K. Bruce/Animals Animals/Earth Scenes.

Chapter Eleven Page 268(all), Norbert Wu; 272(all), HRW photos by Victoria Smith; 274(b), HRW photo by Stephanie Morris; 275(b), HRW photo by Peter Van Steen; 277, Jack Thunderhead Corso; 283(t), Patricia Ceisel/Visuals Unlimited; 283(b), David R. Frazier Photolibrary; 284, Takeshi Takahara/Photo Researchers, Inc.; 285, Science Photo Library/Photo Researchers, Inc.; 286(br), HRW photo by Richard Hanes; 287, Visuals Unlimited; 288(l), Richard T. Nowitz/Photo Researchers, Inc.; 293(t), Paul Silverman/Fundamental Photographs; 296(t), HRW photo by Victoria Smith; 298(t), Corbis-Bettman; 298(b), Bill Ross/Corbis; 299(l), Daniel Osborne, University of Alaska/Detlev Ban Ravenswaay/Science Photo Library/Photo Researchers, Inc.; 299(r), STARLab, Stanford University.

Unit Four Page 300(t), The Granger Collection, NY; 300(b), Reuters/Mark Cardwell/Archive Photos; 301(t), SuperStock; 301(bl), IBM/Visuals Unlimited; 301(br), Bisson/Sygma.

Chapter Twelve Page 302(t), ©2001 by Universal City Studios, Inc., courtesy of Universal Studios Licensing, Inc. All rights reserved; 302(b), James Martin/Tony Stone Images; 304(t), Dr. Mitsuo Ohtsuki/Science Photo Library/Photo Researchers, Inc.; 304(b), William Nawrocki Stock Photography; 305, Corbis-Bettman; 308, HRW photo by Stephen Maclone; 309(l), John Zoiner; 309(r), HRW photo by Mavournea Hay; 311(t), Lawrence Berkeley National Laboratory; 311(b), John Morrison/Morrison Photography; 315, Charles D. Winter/Timeframe Photography Inc./Photo Researchers, Inc.; 316, SuperStock; 318, William Nawrocki Stock Photography; 321, Fermilab National Laboratory; 323(t), HRW photo by Stephen Maclone; 323(b), Fermilab NationalLaboratory/Corbis; 322, NASA/Ames.

Chapter Thirteen Page 324(tl), B.G. Murray Jr./Animals Animals/Earth Scenes; 324(tr), Joe McDonald/Animals Animals/Earth Scenes; 324(bl), Dr. E.R. Degginger/Color-Pic, Inc.; 324(br), Richard Megna/Fundamental Photographs; 331(tr), Richard Megna/Fundamental Photographs; 331(bl), Russ Lappa/Photo Researchers, Inc.; 331(br), Lester V. Bergman/Corbis-Bettman; 332(r), From Chemical Dictionary, ©Morikita Publishing Co., Tokyo, 1981, photo by Sam Dudgeon/HRW Photo; 333(l,c), Dr. E.R. Degginger/Color-Pic, Inc.; 333(r), Richard Megna/Fundamental Photography; 334, Charles D. Winters/Photo Researchers, Inc.; 335(l,c,r), Richard Megna/Fundamental Photographs; 336(tr), P. Petersen/Custom Medical Stock Photo; 336(tc), HRW photo by Victoria Smith; 337, David Parker/Science Photo Library/Photo Researchers, Inc.; 339(tr), Image ©1998 PhotoDisc, Inc.; 339(tl), David Parker/Science & Photo Library/Photo Researchers, Inc.; 340(t,c), Richard Megna/Fundamental Photographs; 340(b), Dr. E.R. Degginger/Color-Pic, Inc.; 341(t), Michael Dalton/Fundamental Photographs; 341(b), NASA; 342(b), Richard Megna/Fundamental Photographs; 344, Richard Megna/Fundamental Photographs; 347, Image ©2001 PhotoDisc, Inc.

Unit Five Page 348(t), Archive France/Archive Photos; 348(b), Wally McNamee/Corbis; 349(t), James Foote/Photo Researchers, Inc.; 349(l), Argonne National Laboratory/Corbis-Bettman; 349(r), Sygma; 349(b), Reuters/NASA/Archive Photos.

Chapter Fourteen Page 350(b), Photri; 350(t), Jim Sugar Photography/Corbis; 356(t), Kevin Schafer/Peter Arnold, Inc.; 356(l), HRW photo by Peter Van Steen; 359, Paul Silverman/Fundamental Photographs; 363(r), Calder Sculpture, National Museum, Washington D.C. Photo ©Ted Mahiec/The Stock Market; 368, HRW photo by Victoria Smith.

Chapter Fifteen Page 372(t), Romilly Lockyer/The Image Bank; 374(t), Rob Matheson/The Stock Market; 374(br), Dorling Kindersley LTD; 375(tl), Charles D. Winters/Timeframe Photography, Inc.; 375(tr), Richard Megna/Fundamental Photographs; 375(br), Dr. E.R. Degginger/Color-Pic, Inc.; 378(b), HRW photo by Richard Haynes; 379(r), Charles D. Winters/Photo Researchers, Inc.; 379(l), Mark C. Burnett/Photo Researchers, Inc.; 381(all), Michael Dalton/Fundamental Photographs; 382, Richard Megna/Fundamental Photographs; 383, Charles D. Winters/Photo Researchers, Inc.; 384(l), Peticolas/Megna/Fundamental Photographs; 384(r), Richard Megna/Fundamental Photographs; 385(l,c), HRW photos by Peter Van Steen; 385(r), Tom Stewart/The Stock Market; 387(b), John Deeks/Photo Researchers, Inc.; 388(all), Richard Megna/Fundamental Photographs; 389, Dorling Kindersley, LTD, courtesy of the Science Museum, London; 392(t), Richard Megna/Fundamental Photographs; 393, Richard Megna/Fundamental Photographs; 394, John Deeks/Photo Researchers, Inc.; 395(all), Bob Parker/Austin Fire Investigation.

Chapter Sixteen Page 396(b), SuperStock; 398(t), Yoav Levy/Phototake, NY; 399(all), 400, Richard Megna/Fundamental Photographs; 401(b), Charles D. Winters/Timeframe Photography Inc.; 405(t), Runk/Schoenberger/Grant Heilman Photography Inc.; 405(b), Peter Arnold Inc., NY; 406, Miro Vinton/Stock Boston/PNI; 410(br), Hans Reinhard/Bruce Coleman Inc.; 411, David M. Phillips/Visuals Unlimited; 412(br), Charles D. Winters/Timeframe Photography Inc.; 414, Runk Schoenberger/Grant Heilman Photography; 416(t), Richard Megna/Fundamental Photographs; 416(b), Charles D. Winters/Timeframe Photography Inc.; 418(b),

Sygma; 419(t), Tom McHugh/Photo Researchers, Inc.; 419(b), Dennis Kunkel, University of Hawaii.

Unit Six Page 420(t), Ruggero Vanni/Corbis Media; 420(b), NASA; 421(tl), Corbis Media; 421(tr), NASA; 421(bl), Anglo-Australian Observatory 1994; 421(br), ESA/Sygma.

Chapter Seventeen Page 424, David Malin/Anglo-Australian Observatory/Royal Observatory, Edinburgh; 428(c), Royal Observatory, Edinburgh/AATB/Science Photo Library/Photo Researchers, Inc.; 428(t,bl,br), NASA/Liaison International; 431, Michael Freeman/Bruce Coleman Inc.; 437, John Bova/Photo Researchers, Inc.; 438, Earth Imaging/Tony Stone Images; 441, SuperStock; 442(l), Breck P. Kent/Animals Animals/Earth Scenes; 442(r), John Reader/Science Photo Library/Photo Researchers, Inc.; 446(t), David Hardy/Science Photo Library/Photo Researchers, Inc.; 446(b), NASA/TSADO/Tom Stack & Associates; 447, John T. Whatmough/JTW Incorporated; 448, NASA/NSO; 449, Dean Congerngs/National Geographic Image Collection.

Chapter Eighteen Page 450(l), USGS; 450(cl), NASA/Peter Arnold, Inc.; 450(c), Paul Morrell/Tony Stone Images; 450(cr), USGS/TSADO/Tom Stack & Associates; 450(r), Reta Beebe (New Mexico State University) and NASA; 451(r,cr), NASA/JPL; 451(l), Hubble Heritage Team (AURA/STScl) and NASA; 451(cl), NASA; 452(l), Hulton Getty/Liaison Agency; 452(r), NASA/JPL; 453, Mark S Robinson; 454(t), Calvin J Hamilton; 454(b), NASA/Mark Marten/Science Source/Photo Researchers, Inc.; 455(t), Frans Lanting/Minden Pictures; 455(b), NASA; 456(t), World Perspective/Tony Stone Images; 456(c), courtesy of Lunar & Planetary Institute; 456, 457(b), NASA; 457(r), NASA/Science Photo Library/Photo Researchers, Inc.; 458, NASA/Peter Arnold, Inc.; 459(t), NASA/JPL; 459(b), 460, NASA; 461-462, NASA/JPL; 463(t), NASA; 463(b), NASA/JPL; 465(all), John Bova/Photo Researchers, Inc.; 466, Fred Espenak; 467, Jerry Lodriguss/Photo Researchers, Inc.; 468(t), Dr. Edwin Bell/NSSDC/GSFC/NASA; 468(c), NASA/Viking Orbiter; 468(bl,br), NASA/JPL; 469(t), NASA/JSC; 469(b), USGS/Science Photo Library/Photo Researchers, Inc.; 470, World Perspective/Tony Stone Images; 471(t), Bob Yen/Liaison International; 471(b), Bill & Sally Fletcher/Tom Stack & Associates; 473, NASA/Science Photo Library/Photo Researchers, Inc.; 474(bc), Breck P. Kent/Animals Animals/Earth Scenes; 474(bl), E.R. Degginger/Bruce Coleman Inc.; 474(br), Ken Nichols/Institute of Meteoritics; 474(t), Dennis Milon/Science Photo Library/Photo Researchers, Inc.; 475(t), NASA; 475(b), Jonathan Blair/Woodfin Camp & Associates; 476, NASA; 478,(bl), NASA/JPL; 478,(tr), E.R. Degginger/Bruce Coleman Inc.

Chapter Nineteen Page 482(all), NASA/Science Source/Photo Researchers, Inc.; 483(t), David Malin/Anglo-Australian Observatory; 484(l), Dr. E.R. Degginger/Color-Pic Inc.; 484(c), John Chumack/Photo Researchers, Inc.; 484(r), Dr. E.R. Degginger/Color-Pic Inc.; 486, Allan Morton/Science Photo Library/Photo Researchers, Inc.; 487, Magrath Photography/Science Photo Library/Photo Researchers, Inc.; 488, Andre Gallant/The Image Bank; 491(r), Astrophysics Library, Princeton University; 494(l,c), David Malin/Anglo-Australian Observatory; 494(r), Dr. Christopher Burrows, ESA/STSCL/NASA; 495(t), Detlev Van Ravenswaay/Science Photo Library; 495(b), David Hardy/Science Photo Library/Photo Researchers, Inc.; 496, Bill & Sally Fletcher/Tom Stack & Associates; 497(t), Brian Parker/Tom Stack & Associates; 497(cr), courtesy of Dr. Larry Marks, University of Missouri Bands; 497(cl), David Malin/Anglo-Australian Observatory; 497(b), Dennis Di Cicco/Peter Arnold, Inc.; 498(t), I. M. House/Tony Stone Images; 498(br), Bill & Sally Fletcher/Tom Stack & Associates; 498(bl), Jerry Lodriguss/Photo Researchers, Inc.; 499, X-Ray Astronomy Group, Leicester University/Science Photo Library/Photo Researchers, Inc.; 501 Property of AT&T Archives. Reprinted with permission of AT&T; 504, John Chumack/Photo Researchers, Inc.; 505, I. M. House/Tony Stone Images; 506(t), David Nunuk/Science Photo Library/Photo Reseachers, Inc.; 506(l), 508(all), NASA; 509, Open University.

LabBook "LabBook Header": "L," Corbis Images, "a," Letraset Phototone, "b" and "B," HRW, "o" and "k," images ©2001 PhotoDisc/HRW; 513(c), HRW photo by Michelle Bridwell; 513(br), Image copyright ©2001 PhotoDisc, Inc.; 514(cl), HRW photo by Victoria Smith; 514(bl), HRW photo by Stephanie Morris; 515(b), HRW photo by Peter Van Steen; 515(tl), Patti Murray/Animals Animals; 515(tr), HRW photo by Jana Birchum; Page 516, HRW photo by Victoria Smith; 521(b), NASA; 528-529, HRW photos by Victoria Smith; 530, Stuart Westmoreland/Tony Stone Images; 534, Gareth Trevor/Tony Stone Images; 539, Image ©1998 PhotoDisc; 543, NASA; 552, HRW photo by Stephanie Morris; 553, Paul Dance/Tony Stone Images; 555(l,r), HRW photo by Robert Wolf; 560, HRW photo by Victoria Smith; 562, James H. Karales/Peter Arnold Inc., NY; 565(all), Richard Megna/Fundamental Photographs; 569, HRW photo by Victoria Smith; 570, HRW photo by Victoria Smith; 579, Rob Boudreau/Tony Stone Images; 581, 584, 588, HRW photo by Victoria Smith; 592, John Sanford/Astrostock.

Appendix Page 599(t), 599(b), 603(b), HRW photo by Sam Dudgeon; 603(t), HRW photo by Peter Van Steen; 620, HRW photo by Sam Dudgeon.

Sam Dudgeon/HRW Photo Page vii(tl,tr,br); xii(tl), xvi(bl); xvii(tl,br); xviii(tr); xix(br); T6(tl,tr); T11(br); T12(bl); T15(tr); 2(cr); 5(b); 8(tl,tr); 9(bl); 12(t,c); 22(cl); 23(t); 30(all); 31(l); 37(bl); 59(b); 61(bl); 64(r); 66; 76(b); 81(tr); 85(tl,tr,c); 87(r); 90(t); 91(cl); 93(t); 94(bl); 96(t,b); 99; 100(b); 101(br); 106(tr); 107; 114(b); 116(t); 118(c); 124(bl); 132; 137(b); 144(b); 157(r); 161; 177(t); 180; 182(b); 187(b); 195(c); 198(c); 199(bl); 213; 215; 217; 220(t); 238; 240(t); 245; 255; 263; 280; 286(t,bl,bc); 288(b,c); 293(b); 295; 297(b); 303; 306-307; 327; 330; 331(c,tl); 332(l); 335(b); 336(tl,b); 339(c,b); 342(t); 360(b); 362; 363(l); 369; 371; 373; 374(bl); 380; 397; 398(b); 402(b); 410(l); 412(t); 417; 422(all); 423; 483(b); 491(l); 502(all); 510(t,cl,b); 512; 513(b); 514(t,br); 518; 520(all); 522-527; 531-533; 535-538; 540-542; 544-551; 554; 563-564; 566; 571-572; 575-578; 582; 587; 590-591; 593-595.

John Langford/HRW Photo Page T7(r); T8(bl); T12(tr); 6(l); 10(tl,b); 20(cl); 21; 22(tr); 28(b); 31(t); 35; 36(t); 37(t); 38-39; 41(b); 42; 52(l); 54(all); 64(l); 91(bl,bc,br); 95(all); 145; 188; 189; 191(all); 193; 195(t,b); 199(t); 201; 205(tr); 206-207; 208; 209(tr); 218(l); 219; 220(bl); 223(bl,bc,br); 224; 226; 233(cr,bl); 235(cr); 260(b); 269; 273; 274(tl,tr); 275(t); 279(t); 286(cl); 288(r); 289-291; 292; 294; 296; 297(t); 325; 351-352; 356(b); 360(t); 364-365; 372(b); 378(t); 379(c); 386; 387(t); 392(b); 396(t); 401; 402(t); 403(all); 412(bl); 413; 513(t); 521(t); 555(b); 556; 568; 573; 585.

Scott Van Osdol/HRW Photo Page T17(br); 61(br); 62(all); 63(b); 67(t); 69(b); 71-72; 163; 166; 175(br); 194(all); 196; 199(br); 375(bl); 408-409; 410(t); 415; 510(cr); 517; 519.